Phase Transitions and Critical Phenomena

Phase Transitions and Critical Phenomena

Volume 6

Edited by

C. Domb
Kings College
University of London, England

and

M. S. Green
Temple University
Philadelphia
Pennsylvania, U.S.A.

1976

ACADEMIC PRESS · London · New York · San Francisco

A Subsidiary of Harcourt Brace Jovanovich, Publishers

ACADEMIC PRESS INC. (LONDON) LTD.
24/28 Oval Road,
London NW1

United States Edition published by
ACADEMIC PRESS INC.
111 Fifth Avenue
New York, New York 10003

Library of Congress Catalog Card Number: 77–170760

ISBN: 0–12–220306–2

PRINTED IN GREAT BRITAIN BY

PAGE BROS (NORWICH) LTD
Mile Cross Lane
Norwich

Contributors

AMNON AHARONY, *Department of Physics, Harvard University, Cambridge, Mass. 02138, USA (now at Department of Physics, University of Tel Aviv, Israel)*

E. BRÉZIN, *Service de Physique Théorique, Centre d'Etudes Nucléaires de Saclay, BP n° 2–91190 Gif-sur-Yvette, France*

C. DI CASTRO, *Istituto di Fisica, dell'Università, Roma Gruppo G.N.S.M., Roma, Italy*

G. JONA-LASINIO, *Istituto di Fisica dell'Università, Roma Gruppo G.N.S.M., Roma, Italy*

J. C. LE GUILLOU, *Laboratoire de Physique Théorique et Hautes Energies Tour 16, Université Paris VI, 4 place Jussieu, 75230 Paris Cedex 05*

SHANG-KENG MA, *Department of Physics and Institute for Pure and Applied Physical Sciences, University of California, San Diego, La Jolla, California 92037, USA*

TH. NIEMEIJER, *Laboratorium voor Technische Natuurkunde, Technische Hogeschool, Delft, the Netherlands*

J. M. J. VAN LEEUWEN, *Laboratorium voor Technische Natuurkunde, Technische Hogeschool, Delft, the Netherlands*

D. J. WALLACE, *Department of Physics, The University, Southampton SO9 5NH, England*

F. J. WEGNER, *Institut für Theoretische Physik, Universität Heidelberg, Germany*

KENNETH G. WILSON, *Newman Laboratory, Cornell University, Ithaca, New York 14853, USA*

J. ZINN-JUSTIN, *Service de Physique Théorique, Centre d'Etudes Nucléaires de Saclay, BP n° 2–91190 Gif-sur-Yvette, France*

General Preface

The discontinuities in physical behaviour which occur when a system undergoes a phase transition have claimed the attention of scientists for many years. Particular interest has been focussed on phenomena associated with critical points such as that of liquid–vapour equilibrium, or ferromagnetism (the Curie point). It was recognized at an early stage that the discontinuities were associated with interactions between the microscopic constituents of the system, but an exact treatment by means of statistical mechanics appeared prohibitively difficult. During the latter part of the 19th and the earlier part of the 20th century, phenomenological explanations were devised which seemed to account adequately for the discontinuities and to provide insight into their detailed character.

The modern era in phase transitions started in 1944 when Onsager succeeded in carrying out an exact statistical mechanical calculation for a simple interacting atomic model, the two-dimensional Ising model. His solution showed that the previous "classical" theories were unreliable in their quantitative predictions, and provided a great stimulus to explore the true behaviour near discontinuities and critical points. The problems to be faced in extending Onsager's work were formidable, and at first progress was slow. However, there were many different aspects to be investigated and a number of alternative lines of theoretical approach to be pursued. The field began to attract new experimentalists and theoreticians of ability.

During the 1960's there was considerable progress towards a greater understanding of phase transitions and critical phenomena, and the research literature grew rapidly. Much of this literature is of ephemeral value and has been rendered obsolete by new developments. However, a body of established results has accumulated steadily, but remains scattered in the literature. After discussions with many of those engaged in active research in the field, we felt that the time was ripe for an attempt to incorporate these results into a publication which could then serve as a basic text. The aim of the publication would be to present a coherent account of all that is definitely known about phase transitions and critical phenomena, and to provide a standard reference for some time to come, particularly for graduate students.

We therefore listed a number of topics of major importance, and contacted an expert who has been particularly concerned with each topic with the request that he should contribute an article. We were gratified by the favourable response, and by the calibre of the manuscripts we received.

It soon became clear that several volumes would be needed to achieve our objective. In fairness to contributors who produced their manuscripts on time, we have tried to avoid unnecessary delay in publication, but at the same time we have endeavoured as far as possible to maintain coherence in the contents of individual volumes. Volume 1 deals with exact results and solutions, Volume 3 with series expansions, and Volume 4 with critical correlations; Volume 2 contains a number of different articles each throwing its own light on phase transitions and critical phenomena, and related topics are grouped together.

We are well aware of the many gaps which still remain even after the publication of these 4 volumes. We shall endeavour to cover them in later volumes of the publication.

C. Domb
M. S. Green

Preface to Volume 6

In the preface to Volume 5 of this series the editors promised a volume devoted entirely to the broad synthesis of our subject represented by the renormalization group approach. The present volume, Volume 6, is the promised presentation. In contrast to previous volumes of this series, which are in the main concerned with material which has developed slowly over a period of 10 to 20 years, the renormalization group approach has grown from a doubtful, speculative idea to a complete theoretical synthesis in a period of a few years. We have, therefore, felt a certain pressure to complete the present volume on a much shorter time scale than previous volumes. We wish to thank our authors who made every effort to meet a short deadline as well as the printers and editorial staff of Academic Press for their help in getting out this volume in good time.

As pointed out by Kenneth Wilson, in his introductory article, there are three main elements in the new approach: the renormalization group proper, which establishes the conceptual framework, and the ε- and $1/n$ expansions which permit concrete, albeit approximate, results to be obtained. Since Wilson's introduction presents an overview of the new approach in which each of the articles in the present volume is placed in context we will not comment extensively in this preface. We will, however, attempt to classify the articles according to the above scheme. The most complete presentation of the renormalization group proper is given in the articles of Wegner and Niemeijer and van Leeuwen; the first in a continuous field context, the second in a lattice context. The idea of renormalization and of the renormalization group first arose in quantum field theory. The articles by Brézin, Le Guillou and Zinn-Justin, and by Di Castro and Jona-Lasinio make the closest contact with the known methods of quantum field theory. The former exploits the Callen-Symanzik equation in conjunction with the ε-expansion. The latter has a more synthetic purpose, attempting to show the interconnections among many different trends including the authors' own early insight into the relation between renormalization and critical phenomena. Noteworthy are the connections to other branches of mathematics such as the connection with the central limit theorem of probability theory and the idea of fixed points and of diffeomorphisms to the invariance of critical exponents. An extensive exposition of ε-expansion and physical results obtainable by it is given in the article by Wallace, and of the $1/n$

expansion and physical insights obtainable by it in the article by Ma. One of the most beautiful aspects of the renormalization group approach is the wealth of useful physical information which can be obtained from multiple fixed points, flow diagrams and crossover effects. The article by Aharony exploits these concepts to give a thorough discussion of anisotropy, randomness and polycritical points in magnets and other solid systems.

There is one subject which is conspicuous by its absence in this volume, the application of the renormalization group approach to dynamical critical phenomena. At the time the present volume was planned this subject was not yet in a form suitable for review. This situation has perhaps changed by now, and the editors hope to include an article on this subject in a future projected volume.

Despite the efforts mentioned above the volume has taken longer to appear than we initially hoped. We apologize to our contributors and thank them for their patience.

February, 1976

C. Domb
M. S. Green

Contents

3. Field Theoretical Approach to Critical Phenomena

E. BREZIN, J. C. LE GUILLOU AND J. ZINN-JUSTIN

4. The $1/n$ Expansion

SHANG-KENG MA

5. The ε-Expansion for Exponents and the Equation of State in Isotropic Systems
D. J. Wallace

6. Dependence of Universal Critical Behaviour on Symmetry and Range of Interaction

AMNON AHARONY

7. Renormalization Theory for Ising-like Spin Systems

TH. NIEMEIJER AND J. M. J. VAN LEEUWEN

8. Renormalization Group Approach to Critical Phenomena

C. Di Castro and G. Jona-Lasinio

Contents of Volume 1 (Exact Results)

Contents of Volume 2

Contents of Volume 3 (Series Expansions for Lattice Models)

Contents of Volume 4

Contents of Volume 5a

Contents of Volume 5b

1. The Renormalization Group—Introduction

Kenneth G. Wilson

Newman Laboratory, Cornell University, Ithaca, New York 14853

The theory of critical phenomena has been transformed in the last few years due to three simultaneous developments. The efforts of many years to apply Feynman diagram methods to critical phenomena finally succeeded. Feynman diagrams are based on an expansion; the breakthrough came with the discovery that the expansion parameter to use is $\varepsilon = 4 - d$ where d is the dimension of space. The ε expansion differs from the earlier high-temperature expansion in that the expansion is valid in the critical region itself; one can compute expansions for critical exponents, the equation of state, and many more.

The variety of phenomena that can be computed from the ε expansion will be evident in the articles of Brézin *et al* (Chapter 3), Wallace (Chapter 5), and Aharony (Chapter 6). Furthermore, the starting point for the ε expansion is Landau's mean field theory, which is exact apart from logarithms in four dimensions ($\varepsilon = 0$). As d decreases, critical exponents vary smoothly away from their mean field values. For the simplest (Ising-like) case these exponents move in the direction of the 2-dimensional values obtained by Onsager, and in three dimensions agree well with high-temperature calculations.

Independently, Abe discovered another expansion for critical exponents, namely an expansion in $1/n$ where n is the number of internal degrees of freedom in the system. This expansion is less accurate than the ε expansion for realistic problems but is very useful for more formal studies of critical behaviour. (See Ma, Chapter 4.)

A more fundamental development, occurring at the same time, is the set of renormalization group ideas. These ideas are concerned with the basic physics of a critical point, namely the long-wavelength fluctuations which are the cause of critical singularities. The starting point in the renormalization group approach is to realize that the most important fluctuations at the critical point do not have a characteristic length. Instead the important fluctuations have *all* wavelengths ranging from the atomic spacing up to the correlation length; close to the critical point the correlation length is much larger than the atomic spacing. Thus near the critical point the important wavelengths cover many decades.

Physicists do not realize how crucially their thinking depends on the existence of characteristic lengths (or characteristic energies, times, etc.). In the ancient past, physicists would use dimensional analysis to identify the important length or energy scale explicitly. Now the practice is to use dimensionless units from the beginning. In consequence physicists have been slow to recognize the syndrome of many length scales without a characteristic dominating the physics of a problem. An exception is in the case of turbulence in classical physics where the role of many different length scales has been recognized for some time. (For a recent discussion of the problems of turbulence, see Kraichnan (1975).) It is only in the more "modern" subjects like elementary particle physics and statistical mechanics that the myopia has persisted.

A common symptom of a problem without a characteristic length is a logarithmically divergent integral. Let k be a momentum variable, for example, and consider the two integrals

$$\int_0^{k_F} \frac{1}{k^2 + k_F^2}\, k\mathrm{d}k \tag{1}$$

$$\int_0^{k_F} \frac{\mathrm{d}k}{k} \tag{2}$$

where k_F is a characteristic momentum. The first integral is finite; it is also dominated by momenta $k \sim k_F$. In other words a problem which leads to the integral (1) has k_F as a characteristic momentum. The integral (2) diverges. The reason it diverges is that all momenta less than k_F are equally important. Any individual order of magnitude range of momentum makes a finite contribution to the integral, for example the range $10^{-6} < k < 2 \cdot 10^{-6}$ contributes ln 2. But there are an infinite number of factor of 2 ranges of k below $k = k_F$, each trying to contribute ln 2 to the integral. Naturally the sum diverges. In this situation there is a momentum parameter (k_F) but it is not the characteristic momentum for the problem—the problem has no characteristic momentum.

Field theorists have learned how to remove divergences from their theories in some cases. They do this by "renormalization theory". This theory, as developed by Feynman, Schwinger, and others, involves essentially no physics. Instead one manipulates expansions with nominally infinite coefficients, taking ratios of quantities which individually are unobservable and substituting new expansion parameters for old, etc., until the divergences miraculously disappear. Unfortunately these methods only work for formally exact expansions; they fail completely when one tries to use non-exact approximations, such as variational methods. Thus the art of making simple but inexact estimates of physical quantities has all but disappeared from quantum field theory, the victim of the formalism of renormalization (see Feynman (1961)).

Part two of the renormalization group approach is a strategy for dealing with a problem with many different length (or energy or time or momentum or . . .) scales. The strategy is by fair means or, perhaps, foul means, to break the problem into a sequence of subproblems, with each subproblem involving only a few length scales. In other words, each subproblem does have a characteristic length. A very elegant example of this is described by Niemeijer and Van Leeuwen later in this volume. In their approach, the subproblems are a sequence of generalized Ising models. The lattice spacing for the nth model is $3^{n/2}$ times the original atomic spacing; the lattice spacing is the characteristic length for each subproblem. By solving the whole sequence of subproblems one takes into account all length scales up to the correlation length, yet without facing all these length scales at once. There are not even hints of divergences in any individual step of their calculation.

It will also be evident from Niemeijer and Van Leeuwen's article (Chapter 7) that achieving a reduction to subproblems of manageable size is a challenge and an art. There are many problems in physics awaiting the bright young physicist who will achieve this reduction (the problem of turbulence being one example).

The third part of the renormalization group apprach is a body of lore and theorems governing the outcome of renormalization group calculations. The basic feature of renormalization group calculations is that they are iterative. That is, whatever calculation one performs for one length scale, the same type of calculation will be repeated for all other length scales. If many length scales are important, there will be many iterations. The consequence is that one sees fixed points, exponential growth or decay near fixed points, crossover effects and more, all of which will be explained in Wegner's and Aharony's articles (Chapters 2 and 6). All these phenomena are the consequence of many length scales being important in the problem.

This is my view of the renormalization group approach. The process of breaking the problem down to finite subproblems is a form of renormalization; the iterations which generate the solution of the problem have a group-like character. Jona-Lasinio and di Castro will describe a more general but abstract view of the renormalization group.

Since I have no responsibility to present detailed physics in this introduction, I will close with a little history, briefly outlining how I became interested in critical phenomena and worked out some of the renormalization group ideas. My principal research has been in elementary particle physics. I first encountered the renormalization group in its field theoretic setting, as discussed by Gell-Mann and Low (1954), Stueckelberg and Petermann (1953), and Bogoliubov and Shirkov (1959). In the early 60's, a study of the fixed source theory of pions coupled to nucleons (Wilson, 1963)† taught me much about the basic physical ideas of the renormalization group; these ideas were not explained in the earlier papers. An overlapping set of ideas was being discovered simultaneously by Kadanoff (1966). Being at Cornell, I had the good fortune to learn about critical phenomena from Ben Widom and especially Michael Fisher. Further, in the summer of 1966 I worked through the solution of the two-dimensional Ising model, using the paper of Schultz, Mattis, and Lieb (1964). Onsager's solution is one of the triumphs of theoretical physics, and I had promised myself long ago that I would study it some day. This gave me enough background to realize that there must be some relevance of the renormalization group to critical phenomena. Kadanoff's work made this even more evident.

There remained the problem of making the renormalization group a practical tool for solving real problems. Kadanoff had proposed that a ferromagnet near its critical point could be described in terms of blocks of spins interacting through an effective Ising-like coupling. In Kadanoff's picture a single parameter K characterized the effective interaction. I tried for several years to learn how to calculate K, with no success. Then I learned from a more extended study of the fixed source theory (Wilson, 1970) that the effective interaction would be much more complex than Kadanoff suggested, containing next-nearest-neighbour couplings, four-spin couplings, etc., leading to an infinite number of parameters in an exact treatment. (See the article by Niemeijer and Van Leeuwen, Chapter 7, for examples of such effective interactions.) With this idea it became possible to think of many formulations of the renormalization group, although it took a while to find a procedure which gave actual numbers.

When I started studying the renormalization group, my hope was that it would be useful for the problem of strong interactions of elementary particles where (at the time I began the study) Feynman diagram methods seemed hopeless. Again, in the case of critical phenomena I assumed that diagrammatic methods had and would fail. History has passed me by. Some of you

† The static theory died as a topic of research five years earlier, but it is difficult to abandon the subject of one's Ph.D. thesis.

who are struggling with the field theoretic methods as applied to critical phenomena may wish that history had been kinder to me. To you I can only apologize for my part in bringing the diagrammatic methods into the problem. Even the strong interactions have since turned out to be rather weak, at least at short distances—this was shown by the electron-scattering experiments at the Stanford Linear Accelerator Center, and recent theories of strong interactions can be solved in part by graphical methods (Weinberg, 1974). Only in the case of the Kondo problem in solid-state physics has a genuine strong coupling problem, previously unsolved, been solved by renormalization group methods (see Wilson, 1974, 1975).

References

Bogoliubov, N. N. and Shirkov, D. V. (1959). "Introduction to the Theory of Quantized Fields", Ch. VIII. Interscience, New York.

Feynman, R. P. (1961). *In* "The Quantum Theory of Fields" Solvay Conference Proceedings, 1961 (R. Stoops, ed.), pp. 75–76. Interscience, New York.

Gell-Mann, M. and Low, F. E. (1954). *Phys. Rev.* **95**, 1300.

Kadanoff, L. (1966). *Physics* **2**, 263.

Kraichnan, R. H. (1975). *Adv. Mathematics* **16**, 305.

Schultz, T. D., Mattis, D. C. and Lieb, E. H. (1964). *Rev. Mod. Phys.* **36**, 856.

Stueckelberg, E. C. G. and Petermann, A. (1953). *Helv. Phys. Acta* **26**, 499.

Weinberg, S. (1974). *Rev. Mod. Phys.* **46**, 255

Wilson, K. G. (1963). *Phys. Rev.* **140**, B445.

Wilson, K. G. (1970). *Phys. Rev.* **D2**, 1438.

Wilson, K. G. (1974). *In* "Nobel Symposia—Medicine and Natural Sciences", Vol. 24. Academic Press, New York. More details are supplied in the Cargèse Lecture Notes 1973 (E. Brézin and J. M. Charap, eds.) Gordon and Breach.

Wilson, K. G. (1975). *Rev. Mod. Phys.* **47**, 773.

2. The Critical State, General Aspects

F. J. Wegner

Institut für Theoretische Physik, Universität Heidelberg, Germany

I. Introduction

The equilibrium properties (like energy, magnetization, density, etc.) of a piece of matter are almost everywhere in the phase diagram analytic functions of the external variables (like temperature, a magnetic field, chemical potential, etc.). There may be however exceptional points (or lines) in the phase diagram where the properties change in a drastic way. We know first order phase transitions where these properties change discontinuously. We know the singularities of the susceptibility (compressibility) and of the specific heat at critical points of continuous (second order) phase transitions. We know nonanalyticities at zero temperature (paramagnons, Kondo effect).

The renormalization group (RG) approach as developed during recent years has provided insight into the understanding of some of these phenomena. In particular we have now an elaborate picture of the behaviour close to continuous phase transitions (for review articles see Fisher, 1974; Ma, 1973; Wegner, 1973b, 1975b; Wilson and Kogut, 1974, and this volume). It transpires that the same framework includes a description of the noncritical (analytic) region (Section III.G). RG ideas can also be applied to the Kondo effect (Anderson, 1970, compare Wilson, 1973 and Zawadowski, 1973). Until now we have not gained insight from this theory into first order transitions either for systems of discrete or of continuous symmetry.

The term RG approach is used here in the sense of Wilson (1971a, b) as a theory of the transformation properties of effective Hamiltonians under contraction. Certain Hamiltonians called fixed point Hamiltonians remain invariant under this transformation. The transformation properties of Hamiltonians close to such a fixed point reflect the critical behaviour of the system. Attention is focussed to the general properties of this transformation and the implied nonanalytic behaviour of the free energy and its derivatives at critical points, and to the invariance properties due to the fact that the RG transformation is not uniquely defined. To exemplify the conclusions a RG equation with smooth momentum cut-off is used.

A few warnings are appropriate: (i) The reader will find that although the theory is described in mathematical terms, it is not rigorous. Besides we have to make several assumptions: We deal with formal series expansions without knowing anything about their convergence or divergence and the term limit will be used in Section III.A without having defined a metric. Despite this unpleasant mathematical status, we hope that the theory reflects some of the underlying structure (that is, gives a reasonable explanation of the observations in experimental systems and models) and that it may serve as a guide for further investigation. We would be very happy if some of the assumptions and formal statements could be proven right or wrong, since in either case we may expect a deeper understanding of critical phenomena.

(ii) An enormous amount of work has been done recently in ε (= 4 − space dimensionality) and $1/\hat{n}$ (order parameter dimensionality) expansions. Although the methods presented here are capable of deriving at least ε expansions (examples will be given), we do not claim that they are the most effective ones and the reader is referred to other chapters in this volume for this purpose. The emphasis of this paper is on the basic features which should hold, in general, for critical phenomena and are not restricted to the limit of small ε and large $\hat{n}$, respectively.

In Section II the renormalization group is introduced. It begins with a short introduction to critical phenomena and critical exponents (Section II.A) and reviews some aspects of the Landau theory of critical phenomena which provides an intuitive picture for the transformation from discrete to continuous variables (Section II.B). The observed scale invariance of the correlation functions suggests that at criticality the effective Hamiltonian is invariant under scale transformations. This leads to the renormalization group transformation as a transformation of the Hamiltonian under the change of the length scale (Section II.C). Section II.D is technical, providing the definitions and notations necessary for the derivation of the renormalization group equation with smooth momentum cut-off (Section II.E) which will serve as an example throughout this paper. Other RG equations are briefly reviewed in Section II.F.

Section III deals with the RG transformation linearized around a fixed point. The eigenoperators of this linear transformation grow or decay by a factor b^y where the length scale is changed by a factor b (Section III.A). These exponents y determine the critical exponents in the homogeneity laws for the free energy (Section III.C) and for the correlation functions (Sections III.D, III.E). Since the RG equation contains arbitrary functions, the fixed point and the eigenperturbations are not uniquely defined. Instead there is a class of equivalent fixed points with two types of eigenoperators: (i) scaling operators whose exponents y are invariants and thus universal and (ii) redundant operators whose exponents y depend on the choice of the RG. The redundant operators do not contribute to the critical behaviour (Section III.B). The homogeneity law for the correlation functions in momentum space is given in Section III.D. Section III.E deals with three aspects of the correlations in coordinate space: homogeneity law, the basic ideas of the operator product expansion, and conformal invariance.

The trivial (Gaussian) fixed point Hamiltonian and its eigenoperators are calculated in Section III.F. The linearized theory closes with some considerations on the noncritical fixed point, the stress tensor, and the Bose condensation (Section III.G). Arguments are given that the (quantum mechanical) Bose condensation is governed by the same critical exponents as classical systems.

The full RG equation which is nonlinear in the interaction is exploited in the remainder of this chapter. The quadratic term of the RG equation is parametrized in Section IV.A. Since the RG is nonlinear more than one class of equivalent fixed point Hamiltonians may exist (Section IV). If a scaling operator becomes marginal as function of a parameter (like dimensionality), then this is normally a bifurcation point for two (or more) fixed points. This is demonstrated by the calculation of the non-trivial fixed point (Section IV.B). The exponent η and its invariance in order ε^2 are discussed in Section IV.C. The results for one-component systems are generalized to isotropic $\hat{n}$-component systems in Section IV.D.

The general properties of homogeneous systems are considered in Section V. Sections V.A and V.B are counterparts of the nonlinear theory to Sections IV.A and IV.B for the linear theory. Provided certain inequalities for the exponents y hold, the free energy is exactly a homogeneous function of scaling fields g which are curvilinear coordinates (Section V.A). The free energy does not depend on the scaling fields g of the redundant operators (Section V.B). Different phase transitions which are described by equivalent fixed point Hamiltonians have not only the same critical exponents (Section III.B) but are also governed by a universal leading singular part of the free energy in terms of scaling fields (Section V.C). Only the scale of the scaling fields is arbitrary. The behaviour along a coexistence curve ending at a critical point is discussed in Section V.D. If certain equalities for the exponents y hold, then logarithmic corrections arise (Section V.E) and a vanishing exponent y_E of the energy operator (crudely speaking) yields a phase transition of infinite order (Section V.F).

Section VI deals with correlation functions. Wilson and Kogut have given an exact solution for the order–parameter correlation (Section VI.A). The RG equation yields a recurrence relation for general correlation functions which allows us to express the n point correlations in terms of $(n - 1)$ point correlations (Section VI.B). Thus the correlation function (for nonexceptional momenta) can be expressed in terms of the expectation values of translational invariant operators. The scaling properties of these correlations are discussed in Section VI.C. In Section VI.D scaling fields for inhomogeneous perturbations are introduced and the universality of the scaling functions for correlations is derived. Finally arguments are summarized that the renormalization group equations have to be nonlinear in the Hamiltonian since linear equations lead to ultraviolet divergencies.

To facilitate reading, several sections are divided into a descriptive part and a technical part. The phrase "In the remainder of this section..." indicates the beginning of the technical part. Part of this paper was presented as a graduate course at the University of Heidelberg in the winter of 1974/75. It covered approximately Sections II.A–F, III.A, C, D, V.A and C completely, and the Sections III.B, E, F, IV.A, B, V.B, E and VI.B in excerpts. This selection may serve as a guide for introductory reading.

In view of the extensively growing literature in this field, in several cases the references are meant as examples and are not complete. I apologize to those who have done similar work and are not cited.

II. The Renormalization Group

A. Order parameter, critical exponents

This section provides a phenomenological introduction to observations near critical points.

We describe the system by a Hamiltonian $\mathscr{H}$ which is a function of the local variables $S(r)$. $S(r)$ is chosen so that the expectation value $m = \langle S(r) \rangle$ is the order parameter, and $\langle S(r) \rangle$ is independent of r in a homogeneous system. As is well-known (see Fisher, 1967 and Kadanoff *et al.*, 1967), most phase transitions can be described in terms of such an order parameter. In many cases the Hamiltonian is invariant under certain transformations of $S(r)$ like either of the three

$$\begin{array}{ll} \text{reversal} & S(r) \to -S(r) \\ \text{change of phase} & S(r) \to e^{i\varphi} S(r) \\ \text{rotation} & S(r) \to US(r) \end{array} \tag{2.1}$$

(U is a rotation matrix). This holds for example for the Ising model, superfluid helium, and the Heisenberg model, resp. Then we expect $m = \langle S(r) \rangle = 0$. Actually this applies (in most cases) above the critical temperature T_c whereas below T_c the system is in the ordered phase $m = \langle S(r) \rangle \neq 0$. The amount of m is uniquely defined, but the sign, phase and direction of m, resp., is not defined. It depends on the history (preparation in an external field which removes the symmetry (2.1) or accidental fluctuations) of the system. One says: the symmetry of the system is broken.

On approaching the critical point from below T_c the order parameter vanishes continuously. This behaviour can often be described by a power law

$$m = A_m |T - T_c|^{\beta} \quad \text{for} \quad T < T_c, \tag{2.2}$$

where T is the temperature. In the case of a liquid-vapour transition, there is no obvious symmetry which is broken. One finds however that the difference between the density and its critical value shows a similar behaviour along the coexistence curve below T_c. The situation is similar for mixtures close to the critical point where the miscibility gap vanishes.

We review a few more results observed in most model calculations and many experiments. The response of the order parameter to an external symmetry breaking field

$$-h \int \mathrm{d}^d r\, S(r) \tag{2.3}$$

is the susceptibility

$$\chi = \frac{\partial m}{\partial h} = (kT)^{-1} \int \mathrm{d}^d r\, G(r, T) \tag{2.4}$$

with the correlation function

$$G(r, T) = \langle S(0)\, S(r)\rangle - \langle S(0)\rangle \langle S(r)\rangle \tag{2.5}$$

(if the operator (2.3) does not commute with $\mathscr{H}$ then the eqn (2.5) has to be modified. We restrict ourselves to classical variables $S(r)$). One observes the power laws

$$\chi(T) = A_\chi |T - T_c|^{-\gamma}, \tag{2.6}$$

$$G(r, T_c) = c r^{-d+2-\eta} \text{ for large } r, \tag{2.7}$$

where η denotes the deviation from the Ornstein–Zernike behaviour. At $T \neq T_c$ the correlation function (2.5) decays in many cases (crudely) like

$$G(r, T) \sim \exp(-r/\xi(T)) \tag{2.8}$$

for large distances. One observes the power law

$$\xi(T) = A_\xi |T - T_c|^{-\nu}. \tag{2.9}$$

The specific heat behaves, close to T_c, in most cases like

$$c(T) = A_c |T - T_c|^{-\alpha} + B \quad \text{for} \quad \alpha \neq 0 \tag{2.10a}$$

or

$$c(T) = -A \ln|T - T_c| + B \quad \text{for} \quad \alpha = 0. \tag{2.10b}$$

The exponents α, β, γ, ν and η are called critical exponents. The exponents α, γ, ν are equal above and below T_c for all model calculations (this does not apply for all experimental data). The coefficients A in (2.6), (2.9), (2.10*a*) and the coefficients B in (2.10*b*) are in general different above and below T_c. For theoretical reasons we expect the coefficients B in (2.10*a*) and A in eqn (2.10*b*), respectively, to be equal above and below T_c.

A digression on phase transitions without local order parameter may be allowed. We distinguish two types of transitions

(i) Mermin and Wagner (1966) showed that the magnetization of two-dimensional systems with rotational symmetry vanishes at any finite temperature. The transverse fluctuations are so strong that they destroy the magnetization. On the other hand there is evidence for a phase transition in these systems since the susceptibility apparently diverges at some temperature T_c from above (Stanley and Kaplan 1966) and approximations indicate an infinite susceptibility at low temperature (Wegner 1967). Kosterlitz and Thouless (1973) explained this phase transition in terms of dislocations which are above T_c unbound, whereas below T_c they form bound pairs. They introduced a Burgers vector for closed paths and found for large paths that the mean square of the Burgers vector grows proportional to the enclosed area above T_c and proportional to the circumference below T_c. Therefore the system shows qualitatively a different behaviour in both temperature regimes. Kosterlitz (1974) derived RG equations according to which the system undergoes a phase transition of infinite order (we come back to these equations in Section V.E. See however the calculation by Lublin (1975)).

(ii) For certain gauge invariant Ising models (Wegner 1971), it can be shown that a phase transition exists, that is the specific heat shows a nonanalytic behaviour (provided one accepts a nonanalytic behaviour of the specific heat at the phase transition of usual Ising models). Due to this gauge invariance there exists no local order parameter. However, there are more complicated correlation functions (like expectation values of products of spins along a closed extended line) which exhibit qualitatively different behaviour above and below T_c. I am not aware of any renormalization group calculations for these systems. However according to the duality transformation, one knows that for a certain class of systems the gauge invariant operators have correlations which decay like those of the Ising model with two-spin interactions. Therefore one expects that some aspects of the scale invariance apply to these systems also. We note that for our general theory the concept of an order parameter is not necessary and can be eliminated. The duality transformation itself is not yet understood within the renormalization group, therefore the hierarchy of operators obtainable as yet for the self-dual two-dimensional Ising model (Kramers and Wannier 1941) is much smaller than that for Kadanoff's operator algebra (Kadanoff, 1969) which incorporates duality.

In the following, the quantum mechanical nature of the microscopic origin of many phase transitions will be neglected (superfluid He, superconductors, spin and exchange interaction in magnets, etc.). Since critical phenomena become apparent on a macroscopic scale, it is assumed that the commutators can be neglected and $S(r)$ can be handled like a classical variable. We will comment on this in Section III.G. In many cases $S(r)$ will be considered to be a continuous variable even if it is discrete in the original model since the discrete eigenvalues of the total sum $\Sigma S(r)$ become very dense on a macroscopic scale. This is discussed on the basis of Landau's model in Section II.B.

In the examples we will restrict ourselves to systems with short range interactions. It should be mentioned however that the renormalization group has been applied to systems with a $r^{-d-\sigma}$ interaction, with dipolar interactions, and with effective interaction due to the coupling to lattice distortions (for references see Aharony, 1975). For the sake of simplicity we will, in general, use the language of magnetism. Thus we have mainly in mind the critical behaviour in terms of a ferromagnet consisting of classical spins $S(r)$ on a rigid lattice governed by an exchange interaction without dipolar interaction.

B. From discrete to continuous models

In this section we introduce and discuss a Hamiltonian of the Landau type, eqn (2.27).

The variable $S(r)$ is continuous in some systems. This applies for the displacements of the ions in a ferroelectric, for example. In such a case the interaction is anharmonic (we have to eliminate first the degrees of freedom which do not become critical). In the case of magnetic isolators, we deal with spins which allow only for certain quantized values. The spins of the Ising model, for example, adopt the values $S(r) = \pm 1$. One may either handle these discrete variables explicitly (compare Section II.F) or one may apply a transformation (Baker, 1962; Mühlschlegel and Zittartz, 1963; Siegert, 1963) whose relevance for the RG has been pointed out by Hubbard (1972).

A different and physically more intuitive way to attack the problem comes close to the Landau model (1937). Let us start from the Hamiltonian (2.11) for the Ising model

$$H = -\tfrac{1}{2}\sum K(r - r')\, S(r)\, S(r') - \hat{h}\sum S(r) \tag{2.11}$$

where

$$H = \mathscr{H}/(kT). \tag{2.12}$$

One introduces the Fourier components

$$S_q = v \sum S(r)\, e^{-iqr} \tag{2.13}$$

where v is the volume per lattice site r. Then one obtains

$$H = -\frac{1}{2N} \sum K_q S_q S_{-q} - \hat{h} S_0 / v \tag{2.14}$$

with

$$K_q = v^{-2} \sum_r K(r)\, e^{iqr}. \tag{2.15}$$

For short range interactions one expands

$$K_q = K_0 - K_2 q^2 + \ldots \tag{2.16}$$

with

$$K_2 = \frac{1}{6v^2} \sum_r K(r)\, r^2 \tag{2.17}$$

(cubic symmetry is assumed for simplicity). For the moment let us forget all terms save the homogeneous component S_0. Then one obtains the interaction

$$H_{mf} = -\frac{1}{2N} K_0 S_0^2 - \hat{h} S_0 / v \tag{2.18}$$

It is easy to calculate the critical behaviour within this approximation. Note that the free energy has to be minimized as a function of S_0. The free energy $\mathscr{F}$ is given by

$$\beta \mathscr{F} = \beta E - \mathscr{S}/k \tag{2.19}$$

with $\beta = 1/(kT)$ where E and $\mathscr{S}$ are the energy and the entropy as a function of S_0, resp. Introducing the mean magnetization m per spin one obtains from (2.18)

$$\beta E = -\frac{Nv^2}{2} K_0 m^2 - N \hat{h} m \tag{2.20}$$

and the entropy

$$\mathscr{S} = k \ln W \tag{2.21}$$

with

$$W = \frac{N!}{N_\uparrow! N_\downarrow!} \approx \left(\frac{N}{N_\uparrow}\right)^{N_\uparrow} \left(\frac{N}{N_\downarrow}\right)^{N_\downarrow} \tag{2.22}$$

where W is the number of configurations for a given number $N_\uparrow$ and $N_\downarrow$ of spins pointing up and down, resp.

$$N_\uparrow = N\frac{1+m}{2}, \qquad N_\downarrow = N\frac{1-m}{2}. \tag{2.23}$$

Thus one obtains

$$\begin{aligned}\mathscr{S} &= -kN\left[\left(\frac{1+m}{2}\right)\ln\left(\frac{1+m}{2}\right)+\left(\frac{1-m}{2}\right)\ln\left(\frac{1-m}{2}\right)\right]\\ &= -kN\left(-\ln 2+\frac{m^2}{2}+\frac{m^4}{12}+\dots\right)\end{aligned} \tag{2.24}$$

which yields (apart from an additive constant)

$$\frac{\beta\mathscr{F}}{N} = \tfrac{1}{2}(1-K_0\,v^2)m^2+\frac{1}{12}m^4+\dots-\hat{h}m. \tag{2.25}$$

Magnetization is obtained by minimizing the free energy. Note that K_0 depends on the temperature, $K_0 \sim 1/(kT)$. The restriction $S(r) = \pm 1$ leads via the entropy to terms proportional to m^2, m^4 etc., where m is now a practically continuous variable since we may neglect the fact that only discrete values are allowed according to (2.23). Therefore both anharmonic forces and discrete variables lead to higher order terms in m.

In the molecular field approximation (2.25), all fluctuations are neglected, that is all Fourier components with $q \neq 0$ are suppressed. Combining (2.16) and (2.25) one obtains

$$\beta\mathscr{F} = \frac{1}{v}\int \mathrm{d}^d r\left[\tfrac{1}{2}(1-K_0 v^2)m^2(r)+\tfrac{1}{2}K_2 v^2[\nabla m(r)]^2+\frac{1}{12}m^4(r)-\hat{h}m(r)\right]. \tag{2.26}$$

Now two terrible things have happened:

(i) We assumed on the one hand a long-range interaction (2.18) to determine the entropy, on the other hand we introduced the short range interaction (2.14). One may compromise by saying one chooses cells of the sample which are sufficiently large (but finite) so that one may introduce continuous variables $m(r)$ to describe the mean magnetization of the cells. Then it seems plausible to use the Landau form (2.26) provided the coefficients $(1-K_0 v^2)/v$, $K_2 v$, $1/(3v)$, $\hat{h}/v$ are replaced by $a(T)$, $c(T)$, $b(T)$, h, respectively. One expects these coefficients to be smooth functions of the temperature since the cells are chosen of finite size.

(ii) Now the free energy is no longer a function of one variable but of a large number of variables. Although this number is small in comparison to N, it is still of order N. In such a situation it is no longer correct to minimize the free energy. In particular, close to the critical point the fluctuations in these variables may be so large as to lead to a qualitatively different behaviour from that given by the minimum of (2.26). Therefore

$$H = \int d^d r[\tfrac{1}{2}a(T)m^2(r) + \tfrac{1}{2}c(T)(\nabla m(r))^2 + \tfrac{1}{4}b(T)m^4(r) - hm(r)] \tag{2.27}$$

should be called an effective (free) Hamiltonian for which the integration over the cell spins remains to be done. The concept of thinning out the number of variables which we used to gain the expression (2.27) will be a further guide in the RG procedure.

C. Scale invariance and basic properties of the renormalization group

In this section we formulate the basic RG ideas.

We consider spins $S(r)$ on a lattice. At the critical point the correlation function decays according to (2.7)

$$\langle S(0)S(r)\rangle_{\text{crit}} = \frac{c}{r^{d-2+\eta}} \tag{2.28}$$

for large distances. This correlation function is scale invariant in the following sense. Let us consider the same ferromagnet under a different length scale. To accomplish this we divide the sample into cubic cells of length b units of lattice spacings in each direction. Then the magnetization of a cell

$$s = \sum_{\text{cell}} S(r) \tag{2.29}$$

obeys asymptotically

$$\langle s(0)S(r)\rangle = \frac{cb^{2d}}{r^{d-2+\eta}} \tag{2.30}$$

since each cell contains b^d spins. Now change the length scale by a factor b which yields a spacing of the cells equal to the original lattice spacing

$$r = bR. \tag{2.31a}$$

Moreover change the scale for the magnetization by a factor $b^{(d+2-\eta)/2}$

$$s(r) = b^{(d+2-\eta)/2}S_1(R) \tag{2.31b}$$

Then one obtains the asymptotic behaviour of the new spin variables

$$\langle S_1(0)S_1(R)\rangle_{\text{crit}} = \frac{c}{R^{d-2+\eta}}. \tag{2.32}$$

Therefore the correlation function is invariant under the change of the scale (2.31). This scale invariance of the correlation function suggests that the effective interaction at criticality is invariant with respect to the change of the length scale. We call the procedure which changes the scale of the Hamiltonian (effective interaction) renormalization group (RG) procedure and the corresponding transformation is called RG transformation.

We outline some requirements and properties of the RG transformation and start with a few definitions. Since the partition function Z depends on $\beta\mathscr{H}$ we use as before eqn (2.12)

$$H = \beta\mathscr{H}$$

so that the partition function Z reads

$$Z = \text{trace}\exp(-H) \tag{2.33}$$

We introduce the density for the free energy $\mathscr{F}$

$$F = \frac{\beta\mathscr{F}}{V} = -\frac{1}{V}\ln\text{trace}\exp(-H) \tag{2.34}$$

where V is the volume of the system. For simplicity's sake we call H and F Hamiltonian and free energy density, respectively.

The RG transformation consists of:

(i) A change of the length scale by a factor $b = \mathrm{e}^{l}$ in all linear dimensions (we leave the partition function (2.33) invariant). Since the volume shrinks by a factor e^{-dl} we obtain

$$F_0 = \mathrm{e}^{-dl}F_l\,, \tag{2.35}$$

(ii) a transformation and/or elimination of the spin variables S which leaves the free energy invariant. This is the thinning out process mentioned at the end of Section II.B. The transformation shall not generate long-range interactions. The new Hamiltonian H_l has to be comparable with the original Hamiltonian H_0 (same Hilbert or function space). This demands an extension of the system to the original volume for finite systems (we use the fact that the free energy density F does not depend on the volume in the thermodynamic limit). The RG transformation transforms H_0 into H_l

$$H_l = \mathscr{R}_l(H_0) \tag{2.36}$$

$$F(H_0) = \mathrm{e}^{-dl}F(H_l). \tag{2.37}$$

D. Definitions and notations

This section contains a collection of definitions for the partition function as a functional integral, for the representation of the Hamiltonian and for functional derivatives. They will be necessary and useful to derive the RG equation in the next section.

The partition function Z for a system with continuous variables $S(r)$ at lattice points r can be written

$$Z = \prod_r \left(\int \mathrm{d} \sqrt{\left(\frac{v}{2\pi}\right)} S(r) \right) \exp(-H) \tag{2.38}$$

where v is the volume per lattice site. The factor $v^{1/2}(2\pi)^{-1/2}$ is introduced for convenience. Introducing the Fourier components (2.13)

$$S_q = v \sum_r S(r)\, \mathrm{e}^{-iqr}$$

and

$$S(r) = \frac{1}{V} \sum_q S_q\, \mathrm{e}^{iqr} \tag{2.39}$$

one obtains the partition function

$$Z = \int [\mathrm{d}S] \exp(-H) \tag{2.40}$$

with

$$\int [\mathrm{d}S] = \prod_q \left(\int \frac{\mathrm{d}S_q}{\sqrt{2\pi V}} \right) \tag{2.41}$$

The transformation (2.13) is not unitary. The Jacobian

$$\left| \frac{\partial \{S(r)\}}{\partial \{S_q\}} \right| = \left(\frac{1}{v\sqrt{N}} \right)^N = \left(\frac{1}{\sqrt{vV}} \right)^N \tag{2.42}$$

produces the factors $(2\pi V)^{-1/2}$ in (2.41) (N is the total number of sites r). All factors are chosen for convenience. For $q \neq 0$ the components S_q and S_{-q} are conjugate (for real $S(r)$)

$$S_q = S^*_{-q} \tag{2.43}$$

Since

$$\left| \frac{\partial(S_q, S_{-q})}{\partial(\mathrm{Re}\, S_q, \mathrm{Im}\, S_q)} \right| = 2 \tag{2.44}$$

one has

$$\int \mathrm{d}S_q\, \mathrm{d}S_{-q} = 2 \int \mathrm{d}(\mathrm{Re}\, S_q)\, \mathrm{d}(\mathrm{Im}\, S_q) \tag{2.45}$$

which supplements (2.41).

Next we consider the representation for the Hamiltonian. In (2.13) one may easily perform the limit to variables $S(r)$ depending on the continuum of points r

$$S_q = \int d^d r S(r) \, e^{-iqr}. \tag{2.46}$$

We may now express the Hamiltonian (2.27)

$$H = \int d^d r \left(\frac{a}{2} S^2(r) + \frac{c}{2} (\nabla S(r))^2 + \frac{b}{4} S^4(r) - hS(r) \right) \tag{2.47}$$

in terms of the Fourier components (2.46) using (2.39) and obtain

$$H = \frac{1}{2V} \sum_q (a + cq^2) S_q S_{-q} + \frac{b}{4V^3} \sum S_{q_1} S_{q_2} S_{q_3} S_{-q_1-q_2-q_3} - hS_0. \tag{2.48}$$

The spacing of the q-vectors is

$$(\Delta q)^d = \frac{(2\pi)^d}{V}. \tag{2.49}$$

Since one has the correspondence

$$\sum_q (\Delta q)^d f(q) \triangleq \int d^d q \, f(q) \tag{2.50}$$

one obtains

$$\frac{1}{V} \sum f(q) \triangleq \int_q f(q) \tag{2.51}$$

with the definition

$$\int_q f(q) = \frac{1}{(2\pi)^d} \int d^d q \, f(q). \tag{2.52}$$

Relation (2.51) becomes an equality in the thermodynamic limit. From this one finds the correspondence of the Kronecker–Delta to the δ-distribution

$$V \delta_{q,0} \triangleq (2\pi)^d \delta^d(q). \tag{2.53}$$

Therefore the Hamiltonian (2.48) becomes

$$H = \frac{1}{2} \int_q (a + cq^2) \, S_q S_{-q} + \frac{b}{4} \int_{q_1 q_2 q_3} S_{q_1} S_{q_2} S_{q_3} S_{-q_1-q_2-q_3} - hS_0. \tag{2.54}$$

There is no explicit volume dependence in (2.54). In general we will write

$$H = Vu_0 + u_1 S_0 + \frac{1}{2}\int_q u_2(q)\, S_q S_{-q} + \frac{1}{3!}\int_{q_1 q_2} u_3(q_1, q_2)\, S_{q_1} S_{q_2} S_{-q_1-q_2}$$

$$+ \frac{1}{4!}\int u_4(q_1, q_2, q_3)\, S_{q_1} S_{q_2} S_{q_3} S_{-q_1-q_2-q_3} + \ldots \quad (2.55)$$

for translational invariant interactions. We require that the u_n's are invariant under permutations of their arguments q including $q_n = -q_1 - q_2 - \ldots - q_{n-1}$ which is often not explicitly written.

$$u_2(q) = u_2(-q) \quad (2.56)$$

$$u_3(q_1, q_2) = u_3(q_2, q_1) = u_3(q_1, -q_1 - q_2) \text{ etc.} \quad (2.57)$$

For a finite volume the Hamiltonian (2.55) reads

$$H = Vu_0 + u_1 S_0 + \frac{1}{2V}\sum u_2(q)\, S_q S_{-q}$$

$$+ \frac{1}{3!\,V^2}\sum u_3(q_1, q_2)\, S_{q_1} S_{q_2} S_{-q_1-q_2} + \ldots \quad (2.58)$$

The factors (2π) were chosen so that for

$$H = Vu_0 + u_1 S_0 + \frac{1}{2}\int_q u_2(q)\, S_q S_{-q} \quad (2.59)$$

one obtains the density of the free energy (definition (2.34))

$$F = u_0 - \frac{u_1^2}{2u_2(0)} + \frac{1}{2}\int_q \ln u_2(q). \quad (2.60)$$

We supplement our definitions by introducing the functional derivative. One defines

$$V\frac{\partial}{\partial S_q} = \frac{(2\pi)^d}{(\Delta q)^d}\frac{\partial}{\partial S_q} \triangleq \frac{\delta}{\partial S_q}. \quad (2.61)$$

Therefore one obtains from the Hamiltonian (2.55)

$$\frac{\delta H}{\delta S_q} = Vu_1 \delta_{q,0} + u_2(q)\, S_{-q} + \frac{1}{2!}\int_q u_3(q_1, q)\, S_{q_1} S_{-q-q_1}$$

$$+ \frac{1}{3!}\int_{q_1 q_2} u_4(q_1, q_2, q)\, S_{q_1} S_{q_2} S_{-q-q_1-q_2} + \ldots \quad (2.62)$$

and

$$\frac{\delta^2 H}{\delta S_q \delta S_{-q}} = V u_2(q) + u_3(q, -q) S_0 + \frac{1}{2!} \int_{q_1} u_4(q_1, q, -q) S_{q_1} S_{-q_1} + \dots \quad (2.63)$$

where use is made of the symmetry relations (2.56) and (2.57).

Similarly one defines

$$\frac{1}{v} \frac{\partial}{\partial S(r)} \triangleq \frac{\delta}{\delta S(r)}. \quad (2.64)$$

Then one obtains

$$\frac{\delta}{\delta S(r)} \triangleq \frac{1}{v} \frac{\partial}{\partial S(r)} = \frac{1}{v} \sum_q \frac{\partial S_q}{\partial S(r)} \frac{\partial}{\partial S_q} = \sum_q \mathrm{e}^{-iqr} \frac{\partial}{\partial S_q} \triangleq \int_q \mathrm{e}^{-iqr} \frac{\delta}{\delta S_q} \quad (2.65)$$

and

$$\frac{\delta}{\delta S_q} = \int \mathrm{d}^d r \, \mathrm{e}^{iqr} \frac{\delta}{\delta S(r)}. \quad (2.66)$$

Sometimes local perturbations will be considered. For these perturbations the parametrization

$$\hat{\mathcal{O}} = v_0 + \int_q v_1(q) S_q + \frac{1}{2!} \int_{q_1 q_2} v_2(q_1, q_2) S_{q_1} S_{q_2} + \dots . \quad (2.67)$$

is appropriate.

E. Renormalization group equation with smooth momentum cut-off

In this section we derive a RG equation with smooth momentum cut-off (Wegner, 1974). It is a generalization of a RG equation given by Wilson (Wilson, 1970; Wilson and Kogut, 1974). This RG equation serves as an example in many sections. To derive the equation we use the definitions of Section II.D and perform explicitly the steps of Section II.C. The transformation of the Hamiltonian consists of two contributions: a dilatation transformation and a transformation of the variables.

First let us consider the *dilatation transformation*: Under this transformation the length scale changes by a factor b. If the original quantities are denoted by V, q, S etc. and the quantities after the dilatation by V', q', S' etc, then one has

$$V = b^d V', \quad (2.68a)$$

$$q' = bq. \quad (2.68b)$$

Finally one chooses

$$\frac{1}{\sqrt{V}} S_q = \frac{1}{\sqrt{V'}} S'_{q'} \quad (2.68c)$$

so that the partition function (2.40) remains unchanged since

$$\int [\mathrm{d}S] = \int [\mathrm{d}S']. \tag{2.69}$$

We consider the transformation (2.68) for infinitesimal dilatation (infinitesimal l)

$$b = 1 + l \tag{2.70}$$

$$V = (1 + ld)\, V' \tag{2.71a}$$

$$q' = (1 + l)\, q \tag{2.71b}$$

$$S_q = \left(1 + l\frac{d}{2}\right) S'_{q'} = \left(1 + l\frac{d}{2}\right) S'_q + lq\nabla S'_q. \tag{2.71c}$$

One substitutes equation (2.71c) into the Hamiltonian (2.58) using the variables V, q, S'_q which yields

$$H = Vu_0 + \left(1 + l\frac{d}{2}\right) S'_0 + \frac{1}{2V}(1 + ld) \sum_q u_2(q)(S'_q + lq\nabla S'_q) \times (S'_{-q} - lq\nabla S'_{-q}) + \dots \tag{2.72}$$

Since the summation runs over q (not q'), one obtains with equation (2.51)

$$H = V'(1 + ld)\, u_0 + \left(1 + l\frac{d}{2}\right) u_1 S'_0 + \frac{1}{2}(1 + ld) \int_q u_2(q)(S'_q + lq\nabla S'_q) \times (S'_{-q} - lq\nabla S'_{-q}) + \dots \tag{2.73}$$

If the volume and the variables of the new Hamiltonian (2.73) are denoted by V and S, then one finds that the Hamiltonian (2.55) has changed by $l\mathscr{G}_{\mathrm{dil}}H$ with the generator of the dilatation

$$\mathscr{G}_{\mathrm{dil}}H = dV\frac{\partial H}{\partial V} + \int_q \left(\frac{d}{2} S_q \frac{\delta H}{\delta S_q} + q\nabla S_q \frac{\delta H}{\delta S_q}\right). \tag{2.74}$$

According to eqn (2.55), only the constant term contributes to

$$\frac{\partial H}{\partial V} = u_0. \tag{2.75}$$

On the other hand one may substitute all eqns (2.71) into the Hamiltonian (2.58) using the variables V', q', $S'_{q'}$. Then one obtains

$$H = V'(1 + ld)\, u_0 + \left(1 + l\frac{d}{2}\right) u_1 S'_0 + \frac{1}{2V'} \sum_{q'} u_2(q' - lq') S'_{q'} S'_{-q'} + \dots \tag{2.76}$$

which in the representation (2.55) yields

$$H = V'(1 + ld)\, u_0 + \left(1 + l\frac{d}{2}\right) u_1 S'_0 + \frac{1}{2}\int_{q'} u_2(q' - lq')\, S'_{q'} S'_{-q'} + \ldots \qquad (2.77)$$

If one denotes the variables in the new Hamiltonian (2.77) by V, S, q, then one obtains for u_n

$$\left.\frac{\partial u_n}{\partial l}\right|_{\text{dil}} = \left(d - \frac{d}{2}n\right) u_n - \sum_i q_i \nabla_i u_n(\ldots q_i \ldots). \qquad (2.78)$$

Similarly one obtains for the local perturbation (2.67)

$$\left.\frac{\partial v_n}{\partial l}\right|_{\text{dil}} = -\frac{d}{2} n v_n - \sum_i q_i \nabla_i v_n(\ldots q_i \ldots). \qquad (2.79)$$

The factors in front of u_n and v_n differ by d. This comes from the extra integral

$$\int_{q'_n} = (1 + ld)\int_{q_n}. \qquad (2.80)$$

Wilson has used an incomplete elimination procedure to obtain a RG equation without sharp momentum cut-off. This incomplete elimination process can also be considered as a transformation of the variables. Instead of an elimination of the Fourier components for large wave vectors q, the amplitudes u_n of the interaction are diminished. Then one obtains a Hamiltonian with the following properties: (i) The Hamiltonian contains all Fourier components S_q (whatever the magnitude of q might be), however (ii) the amplitudes $u_n(q_1, q_2, \ldots)$ decay rapidly for $q_i \gg q_0$ where q_0 is some cut-off momentum, except for (iii) the amplitude $u_2(q)$ which approaches unity rapidly for $q \gg q_0$. Therefore the interactions of the Fourier components for $q \gg q_0$ are extremely small and only the term $u_2(q)S_q S_{-q}$ with $u_2(q) \to 1$ survives. Therefore the Fourier components $q \gg q_0$ are practically free and since $u_2(q)$ approaches unity, they do not contribute to the free energy, equation (2.60). Therefore they are practically eliminated.

Next the *transformation of the variables* is considered. Let us replace S_q by

$$S_q = S'_q + l\Psi_q\{S'\} \qquad (2.81)$$

where Ψ_q is a function which may depend on all Fourier components S'. Then one has

$$H\{S\} = H\{S'\} + l\int_q \Psi_q\{S'\}\frac{\delta H\{S'\}}{\delta S'_q}. \qquad (2.82)$$

The transformation must leave the partition function Z invariant. Equation (2.81) yields

$$\int [\mathrm{d}S] = \int [\mathrm{d}S'] \frac{\partial\{S\}}{\partial\{S'\}} = \int [\mathrm{d}S'] \left(1 + l \int_q \frac{\delta \Psi_q\{S'\}}{\delta S'_q}\right) \tag{2.83}$$

Therefore one obtains

$$\begin{aligned} Z &= \int [\mathrm{d}S] \exp(-H\{S\}) \\ &= \int [\mathrm{d}S'] \exp(-H\{S'\} - l\mathscr{G}_{\mathrm{tra}}\{\Psi\}\, H\{S'\}) \end{aligned} \tag{2.84}$$

with the generator $\mathscr{G}_{\mathrm{tra}}$ for the transformation of the variables

$$\mathscr{G}_{\mathrm{tra}}\{\Psi\}\, H\{S\} = \int_q \left(\Psi_q \frac{\delta H}{\delta S_q} - \frac{\delta \Psi_q}{\delta S_q}\right). \tag{2.85}$$

In eqn (2.85) S' is replaced by S. The generator $\mathscr{G}$ of the RG consists of the contributions (2.74) and (2.85)

$$\mathscr{G}\{\Psi\} = \mathscr{G}_{\mathrm{dil}}\, H + \mathscr{G}_{\mathrm{tra}}\{\Psi\}\, H \tag{2.86}$$

and the RG equation reads

$$\frac{\mathrm{d}H}{\mathrm{d}l} = \mathscr{G}\{\Psi\}\, H. \tag{2.87}$$

One cannot expect that every Ψ_q will eliminate the Fourier components for large wave-vectors. However, one is free to vary Ψ_q within regions. Wilson's choice (which leads to an elimination) is

$$\Psi_q = h(q)\left(S_q - \frac{\delta H}{\delta S_{-q}}\right) \tag{2.88}$$

which leads to

$$\begin{aligned} \mathscr{G}H = \int_q \left(q\nabla S_q + \frac{d}{2} S_q\right) \frac{\delta H}{\delta S_q} + \int_q h(q) \left[\left(S_q - \frac{\delta H}{\delta S_{-q}}\right) \frac{\delta H}{\delta S_q} \right. \\ \left. + \frac{\delta^2 H}{\delta S_q \delta S_{-q}} - 1\right] + dV \frac{\partial H}{\partial V} \end{aligned} \tag{2.89}$$

where he uses mainly

$$h(q) = c + 2q^2. \tag{2.90}$$

Wilson and Kogut (1974) give the following argument that eqn (2.88) with a sufficiently large $h(q)$ leads practically to an elimination of the degrees of freedom S. They consider the differential equation

$$\frac{\partial \rho(S, l)}{\partial l} = h \frac{\partial}{\partial S}\left(\frac{\partial}{\partial S} + S\right) \rho(S, l). \tag{2.91}$$

Since the right hand side is a derivation with respect to S, one finds that the integral

$$Z = \int_{-\infty}^{+\infty} \rho(S, l)\, \mathrm{d}S \tag{2.92}$$

is independent of l. If one defines $H_l(S)$ by

$$\rho(S, l) = \exp(-H_l(S)) \tag{2.93}$$

then the partition function of the "Hamiltonian $H_l(S)$" is independent of l (a generalization to many variables S is straightforward). Equation (2.91) can be solved by means of a Green's function $\hat{G}$

$$\rho(S', l) = \int \hat{G}(S', S, l)\, \rho(S, 0)\, \mathrm{d}S \tag{2.94}$$

with

$$\hat{G}(S', S, l) = \frac{1}{\sqrt{2\pi(1 - \mathrm{e}^{-2hl})}} \exp\left(-\frac{(S' - S\,\mathrm{e}^{-hl})^2}{2(1 - \mathrm{e}^{-2hl})}\right). \tag{2.95}$$

For $l = 0$ one has

$$\hat{G}(S', S, 0) = \delta(S' - S) \tag{2.96}$$

and for $l \to \infty$ and positive h one obtains

$$\hat{G}(S', S, \infty) = \frac{1}{\sqrt{2\pi}} \exp(-\tfrac{1}{2} S'^2) \tag{2.97}$$

which yields

$$\rho(S', \infty) = \frac{1}{\sqrt{2\pi}} \exp\left(-\frac{1}{2} S'^2\right) \int_{-\infty}^{+\infty} \mathrm{d}S\, \rho(S, 0)$$

$$= \frac{Z}{\sqrt{2\pi}} \exp\left(-\frac{1}{2} S'^2\right) \tag{2.98}$$

Therefore the original interaction $H_0(S)$ is reduced to an interaction $\frac{1}{2}S'^2$ + constant. Generalization from $H(S)$ to $H\{S_q\}$ yields eqn (2.85) with (2.81) and a reduced interaction $\frac{1}{2}\int_q S'_q S'_{-q}$ + const. According to (2.60) the constant yields the free energy. Actually the dilatation generator (2.74) acts on $H\{S_q\}$, also. The function ρ is extremely spread out as a function of S_0 at the critical point. This leads to the behaviour of $H\{S'_q\}$ for large l as described after (2.80).

In general one has to expect that Ψ_q depends on H as for example in eqn (2.88). Therefore one obtains RG equations nonlinear in H. We will see in Section VI.D that these nonlinearities are quite essential for useful calculations.

F. Other renormalization group transformations

The original field-theoretic problem of renormalization dealt with the behaviour at fixed wave-vectors q for a system with a momentum cut-off q_0 tending towards infinity. In the critical phenomena of solid state physics, there is most often a natural cut-off (like the Brillouin zone) and the interest is focused on the limit $q \to 0$. Both problems are similar as they are concerned with the limit $q/q_0 \to 0$ and—as it turns out—the behaviour for small q does not depend on the cut-off momentum q_0. As mentioned in the introduction, we do not consider here the field theoretic methods and related equations like the Callan–Symanzik equation (Callan, 1970; Symanzik, 1970) or the RG equation in the original sense. They are treated in other chapters of this volume. Jona-Lasinio (1973) introduced generalized RG transformations by requiring a Legendre transform of the free energy to be an invariant of the transformation.

We review however some other RG procedures which transform Hamiltonians under dilatations. The idea of such a RG transformation was formulated by Kadanoff (1966). He used it to derive the homogeneity properties and relations between critical exponents. However Kadanoff did not explain how to obtain such a transformation. Anderson (1970) was the first to derive explicitly transformation equations for the Kondo problem by thinning out degrees of freedom. An explicit RG equation for critical phenomena was first given by Wilson. The original derivation of this recurrence relation can be found in Wilson (1971b); a diagrammatic derivation by Polyakov is published in the review by Wilson and Kogut (1974). This approximate formula has been used to derive ε expansions (Wilson and Fisher, 1972; Fisher and Pfeuty, 1972; Wegner 1972b; Wilson 1972; Chang, Tuthill and Stanley, 1974) and to perform numerical calculations for $d = 3$ (Wilson, 1971b; Grover, Kadanoff and Wegner, 1972; Grover, 1972; Swift and Grover, 1974). Golner (1973) has given a modified version which allows for $\eta \neq 0$.

A further possibility is to perform a RG transformation with sharp momentum cut-off at $q_0 = 1$. There one has to integrate over the Fourier components in a small shell with $1 - l \leqslant q < 1$. This can be done explicitly (Wegner and Houghton, 1973) and one obtains the generator for infinitesimal l

$$\mathscr{G}_{\mathrm{eli}} H = \frac{V}{2(2\pi)^d} \int \mathrm{d}\Omega \left[\ln \left(P \frac{\delta^2 H}{\delta S_e \, \delta S_{-e}} \right) - \left(P \frac{\delta H}{\delta S_e} \right) \left(P \frac{\delta H}{\delta S_{-e}} \right) \Big/ \left(P \frac{\delta^2 H}{\delta S_e \, \delta S_{-e}} \right) \right] \tag{2.99}$$

where the integral runs over the surface in momentum space and P indicates that only components with $q < 1$ are retained. One has to add the generator (2.74) of the dilatation and to perform an additional scale transformation for S

$$\mathscr{G}_{\mathrm{tra}} H = c \int_q \left(S_q \frac{\delta H}{\delta S_q} - 1 \right). \tag{2.100}$$

The total transformation reads

$$\mathscr{G} H = \mathscr{G}_{\mathrm{eli}} H + \mathscr{G}_{\mathrm{tra}} H + \mathscr{G}_{\mathrm{dil}} H. \tag{2.101}$$

A disadvantage of this equation is the occurrence of long-range interactions (in coordinate space) due to the sharp cut-off. Nicoll, Chang and Stanley (1974b) have derived and applied an approximation of eqn (2.101).

Besides this differential type of RG transformation with sharp momentum cut-off, one may also consider a large change of the momentum cut-off, an approach used for example by Wilson and Fisher (1972), Wilson and Kogut (1974), Bell and Wilson (1974), Aharony and Fisher (1973), Aharony (1973a–d). For a justification of this method see Bruce, Droz, and Aharony (1974).

A third group of RG equations treats Ising models for two-dimensional systems. (Niemeijer and van Leeuwen, 1973, 1974; van Leeuwen, 1975; Nauenberg and Nienhuis, 1974; Bell and Wilson, 1975; Knops, 1975; Kadanoff and Houghton, 1975). Kadanoff (1975) treated the three-dimensional Ising model, Lublin (1975) the two-dimensional XY-model. For the Ising model one starts out from a Hamiltonian H and defines a new Hamiltonian H'

$$\exp(-H'\{S'\}) = \sum_{\{S\}} \hat{G}\{S', S\} \exp(-H\{S\}) \tag{2.102}$$

where $\hat{G}$ has the property

$$\sum_{\{S'\}} \hat{G}\{S', S\} = 1. \tag{2.103}$$

Note that this $\hat{G}$ corresponds to $\hat{G}$, equation (2.94). Then

$$Z(H') = Z(H) \tag{2.104}$$

is automatically fulfilled.

At first glance two possible choices for the new variables seem to be natural (for more details see Kadanoff and Houghton, 1975):

(i) *decimation:* The new lattice is part of the old lattice and the new spins are chosen to equal the old ones, $S'(r) = S(r)$. This does not work, however, for the following reason: if it would work then at the fixed point

$$\langle S(0)\, S(r)\rangle = \langle S'(0)\, S'(r)\rangle = \langle S(0)\, S(br)\rangle, \tag{2.105}$$

where we use the fact that the old and the new system are equal apart from the factor b of the length scale. Thus the correlation should not depend on the distance, which is not quite feasible.

(ii) *block transformation:* One defines the new spins in terms of the surrounding old spins. Niemeijer and van Leeuwen (1973), for example, choose S' to equal the sign of the sum of the three surrounding spins on a triangular lattice. These block transformations yield good numerical results.

III. Linearized Theory

Sections III.C, III.D, and III.E contain the basic ideas of critical phenomena: Scaling exponents and homogeneity laws for the free energy and the correlation functions. They are obtained from the RG linearized around the fixed point Hamiltonian (Section III.A). Sections III.B, III.F and III.G supplement with invariance properties and examples.

A. Fixed point, linearized renormalization group equations

The scale invariance of the correlation function $\langle S(0)\, S(r)\rangle$ at the critical point has led us in Section II.C to the assumption that the effective interaction at the critical point is invariant under the change of the length scale, more precisely under the RG procedure as derived in Section II.E, for example. Therefore we might expect that at the critical point the Hamiltonian maps into itself under the RG transformation. We call such a Hamiltonian a fixed point Hamiltonian. It obeys

$$\mathcal{G} H^* = 0 \tag{3.1}$$

Obviously if $H_0 = H^*$, then any $H_l = H^*$ since according to (2.87)

$$\frac{\mathrm{d}H}{\mathrm{d}l} = \mathcal{G} H.$$

Therefore the Hamiltonian H^* is invariant under change of the length scale. However we expect that (3.1) imposes a strong condition on H^* and we will find that the fixed points H^* are (for given $\mathcal{G}\{\Psi\}$) normally isolated points or low dimensional manifolds in the space of the Hamiltonians H. The dimensionality of this space is infinite even if we allow only for short range interactions. Indeed if H is such a Hamiltonian then, for example,

$$H + \frac{1}{2} C_n \int_q q^{2n} \mathrm{e}^{-q^2} S_q S_{-q} \tag{3.2}$$

is a short range interaction for any positive integer n, too. Since only very few variables like temperature and magnetic field are experimentally accessible, one can hardly require $H_0 = H^*$ for the critical point. As will be seen it is sufficient that at criticality H_l approaches a fixed point for large l. Therefore a Hamiltonian H_0 will be called critical if

$$\lim_{l \to \infty} H_l = \lim_{l \to \infty} \mathrm{e}^{\mathscr{G} l} H_0 = H^* \tag{3.3}$$

converges to a fixed point. This means the larger the length scale one applies, the closer is the effective Hamiltonian H_l to the fixed point Hamiltonian H^*.

Let us follow H under the renormalization group procedure. To do this one separates H into $H^* + \Delta H$, thus obtaining

$$H^* + \Delta H_l = \mathrm{e}^{\mathscr{G} l}(H^* + \Delta H_0) \tag{3.4}$$

Further one separates $\mathscr{G}$ in a linear and a quadratic part in ΔH

$$\mathscr{G}(H^* + \Delta H) = \mathscr{L}\Delta H + \mathscr{Q}\Delta H \tag{3.5}$$

From (2.89) one obtains

$$\mathscr{L}\Delta H = \int_q \left(q\nabla S_q + \frac{d}{2} S_q \right) \frac{\delta \Delta H}{\delta S_q} + dV \frac{\partial \Delta H}{\partial V} + \int_q h(q) \left(S_q - 2 \frac{\delta H^*}{\delta S_{-q}} + \frac{\delta}{\delta S_{-q}} \right) \frac{\delta \Delta H}{\delta S_q} \tag{3.6}$$

$$\mathscr{Q}\Delta H = - \int_q h(q) \frac{\delta \Delta H}{\delta S_q} \frac{\delta \Delta H}{\delta S_{-q}} \tag{3.7}$$

One expects that close to H^* the quadratic contribution (3.7) can be neglected. (For a more general Ψ there may be contributions of higher orders in ΔH. They will not alter our discussion). In Section III the linearized RG equation

$$\frac{\mathrm{d}(H^* + \Delta H)}{\mathrm{d}l} \approx \mathscr{L}\Delta H \tag{3.8}$$

is discussed. The nonlinear contributions will be discussed in Sections IV–VI.

Since the Hamiltonian H_0 may be far away from H^*, one first applies the full RG procedure which yields H_p. One chooses p so that starting from the critical Hamiltonian H_0 one obtains an H_p which is sufficiently close to H^* so that the linearized form (3.8) for $l > p$ may be used (Fig. 1). Since p is a finite number, the transformation $\exp(p\mathscr{G})$ is an analytic mapping which implies that a small change in temperature or magnetic field in H_0 corresponds

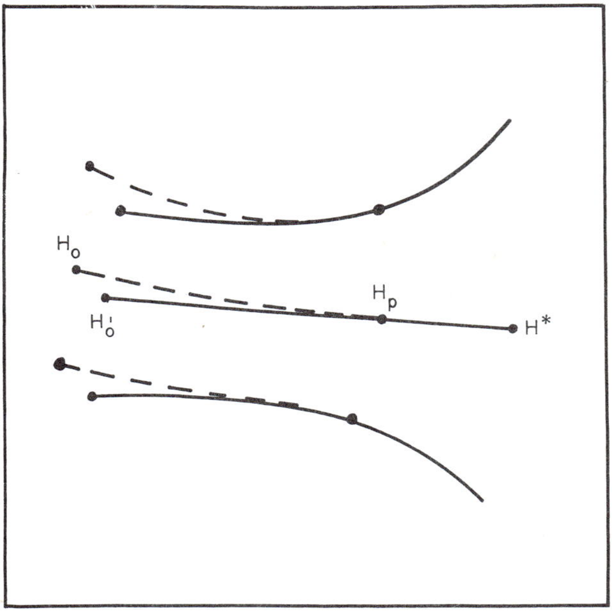

FIG. 1. Trajectories for the full RG and the linearized RG.

to a linear change in H_p, although the change in H_p may be larger than that in H_0. Now we transform back with the linearized form

$$\Delta H_0' = \exp(-p\mathscr{L})\,\Delta H_p \tag{3.9}$$

thus obtaining a Hamiltonian H_0' for which (3.8) is a good approximation. In the following we will not distinguish between H_0 and H_0'. As long as the non-linear terms cannot be neglected, one may replace H_l by H_l'.

The eigenvalue equation

$$\mathscr{L}\mathscr{O}_i^* = y_i \mathscr{O}_i^* \tag{3.10}$$

defines eigenoperators (or eigenperturbations) $\mathscr{O}_i^*$ and scaling exponents y_i.

We require that the $\mathcal{O}_i^*$ are translationally invariant. We assume that the eigenoperators form a complete set of operators so that any Hamiltonian H_0 can be expanded

$$H_0 = H^* + \sum \mu_i \mathcal{O}_i^*. \tag{3.11}$$

($\mathcal{L}$ may be expressed in terms of a matrix $L, \mathcal{L}\mathcal{O}_i^{*\prime} = \Sigma_j L_{ij} \mathcal{O}_j^{*\prime}$ where the $\mathcal{O}_j^{*\prime}$ form a complete set of operators. A matrix of finite order can always be transformed by a similarity transformation into the form $L_{ij} = 0$ for $i < j$, $L_{ii} = y_i$, $L_{ij} = 0$ for $i > j$ and $y_i \neq y_j$. Finite nondiagonal terms L_{ij} with $i > j$ and $y_i = y_j$ cannot be eliminated by a similarity transformation. In such a case we do not obtain a complete set of operators obeying (3.10). This leads to logarithmic corrections, Section V.E).

In eqn (3.11) the Hamiltonian is parametrized by the fields μ_i (in field theory one calls the μ_i sources). One obtains from eqns (3.8) and (3.10) in linear order in μ

$$\frac{\mathrm{d}\mu_i(l)}{\mathrm{d}l} = y_i \mu_i(l). \tag{3.12}$$

This yields

$$H_l = H^* + \sum \mu_i(l)\, \mathcal{O}_i^* = H^* + \sum \mu_i \,\mathrm{e}^{y_i l} \mathcal{O}_i^*. \tag{3.13}$$

Let us now consider the perturbations $\mathcal{O}^*$ and their eigenvalues y. In most cases the eigenvalues y are real. There are exceptions (we give an example in Section III.B) but we are not aware of one which is relevant for the physics. Therefore for simplicity's sake we will normally assume that the eigenvalues are real. If $y_i > 0$ then a perturbation $\mu_i \mathcal{O}_i^*$ grows exponentially with l. If the Hamiltonian H_0 contains such a perturbation, then H_l does not approach the fixed point for $l \to \infty$ and the system is not at criticality. Such an operator $\mathcal{O}_i^*$ is called *relevant*. At criticality the fields μ_i of the relevant operators have to vanish. If on the other hand a perturbation $\mathcal{O}_i^*$ with $y_i < 0$ contributes to H_0, then this perturbation decays exponentially so that H_l finally approaches H^*. These perturbations are called *irrelevant* operators. We summarize

$y_i > 0$	$\mathcal{O}_i^*$ is relevant
$y_i < 0$	$\mathcal{O}_i^*$ is irrelevant
criticality: $\mu_i = 0$	for all relevant operators

The exponent y_i of an operator $\mathcal{O}_i^*$ may vanish. Then $\mathcal{O}_i^*$ is called *marginal* operator. Within the linear approximation $H^* + \mu_i \mathcal{O}_i^*$ is a fixed point for any μ_i. (One expects such a situation for the eight-vertex model (Baxter, 1971, 1972)). In general however non-linear terms in the RG will destroy this property. It depends on these higher order terms whether μ_i of the marginal operator may be arbitrary or has to vanish or has to have a certain sign at criticality.

The constant term $u_0 V$ in the Hamiltonian (2.55) plays a special role. From eqn (3.5) one sees that $\mu_0(l)$ in the perturbation grows with an exponent d

$$\mu_0(l) = \mu_0\, e^{dl}. \tag{3.14}$$

This is consistent with

$$V(l) = e^{-dl}\, V(0) \tag{3.15}$$

leaving $\mu_0(l)\, V(l)$ invariant Therefore one has formally $y_0 = d$. But the addition of the term $\mu_0 V$ to the Hamiltonian does not change its critical behaviour. Therefore $\mu_0 = 0$ is not required for the critical point. This term is the origin of the regular part of the free energy. Because of this particular property, we call the volume a *special operator* and do not count it as a relevant operator. We note that Wilson and other authors often treat this constant contribution separately.

The type of the critical behaviour depends on the number of symmetry conserving relevant operators. (Symmetry conserving means that the symmetry of the Hamiltonian is conserved, it does not exclude a spontaneous symmetry breaking of the system.) Let us expand

$$\mathscr{H} = \sum \mu_i^0 \mathcal{O}_i^* \tag{3.16}$$

$$H^* = \sum \mu_i^* \mathcal{O}_i^* \tag{3.17}$$

then one obtains from eqns (2.12) and (3.11)

$$\mu_i = \beta\mu_i^0 - \mu_i^*. \tag{3.18}$$

For a normal critical point one has one relevant symmetry conserving operator (apart from the volume) $\mathcal{O}_E^*$ which determines the critical temperature

$$\mu_E \equiv \tau = \beta\mu_E^0 - \mu_E^* = \mu_E(\beta - \beta_c) \tag{3.19}$$

with

$$\beta_c = \mu_E^*/\mu_E^0. \tag{3.20}$$

Crudely speaking $\mathcal{O}_E^*$ is proportional to the Hamiltonian minus its expectation value at the critical point.

At a tricritical point (compare Riedel and Wegner, 1972) one has two relevant symmetry conserving operators (apart from the volume) and consequently two conditions for criticality. (In antiferromagnets these conditions determine the critical temperature and the critical homogeneous magnetic field for the tricritical point, for example.)

We remember that the RG equation (2.87) is not uniquely defined. It depends on the choice of the functional Ψ_q. This has a number of consequences which will modify the conclusions we have derived in this section and which will be discussed in the next section.

B. Redundant operators

1. *Introduction*

To obtain a fixed point one has to find a solution (H^*, Ψ^*) of

$$\mathscr{G}(\Psi^*)\, H^* = \mathscr{G}_{\text{dil}}\, H^* + \mathscr{G}_{\text{tra}}(\Psi^*)\, H^* = 0. \tag{3.21}$$

The Hamiltonian H^* and the function Ψ^* are not uniquely defined from this equation. We will see that there is a whole set of equivalent solutions to any fixed point H^*. Moreover even for given Ψ^* and H^* the eigenoperators $\mathcal{O}_i^*$ are not uniquely defined since Ψ itself depends on H and one is free to choose this dependence. As a consequence the eigenoperators $\mathcal{O}_i^*$ will fall into two classes:

(i) scaling operators whose scaling exponents are independent of the fixed point within the class of equivalent fixed points,

(ii) operators which can be expressed in the form $\mathscr{G}_{\text{tra}}(\Phi)\, H^*$. We call these operators redundant operators. The exponents y of the redundant operators depend on the choice of $\Psi(H)$. These exponents are spurious since the free energy does not depend on the fields μ_i of these operators (in linear order) according to our construction in section II.E.

Then the following picture emerges: The eigenoperators $\mathcal{O}_i^*$ (3.10) can be classified into

- (i) scaling operators
  - (a) special operating (volume)
  - (b) relevant operators
  - (c) marginal operators
  - (d) irrelevant operators
- (ii) redundant operators.

At the critical point the fields μ_i of the relevant (and maybe marginal) operators have to vanish. Special and scaling operators have invariant and physically meaningful exponents y_i. These exponents are universal for a class of equivalent fixed point Hamiltonians. Jona-Lasinio (1973) considers a more general transformation than (2.87). He distinguishes two categories of equivalent transformations. For one category the critical exponents are invariants, for the other category this is not guaranteed. He does not introduce the concept of redundant operations.

Let us first consider the operators S_0 and $\delta H^*/\delta S_0$ as examples for the invariance properties. (Part of the discussion is only meaningful for fixed points with $\delta H^*/\partial S_0 \neq 0$). The operator $\delta H^*/\delta S_0$ is a redundant operator since one has

$$\frac{\delta H^*}{\delta S_0} = \mathscr{G}_{\text{tra}}(\Phi)\, H^* \tag{3.22}$$

with

$$\Phi_q = (2\pi)^d\, \delta^d(q) \tag{3.23}$$

One obtains from the linearized generator (3.6) of the RG eqn (2.89)

$$\begin{aligned}
\mathscr{L}\frac{\delta H^*}{\delta S_0} &= \int_q \left(q\nabla S_q + \frac{d}{2} S_q\right)\frac{\delta^2 H^*}{\delta S_0\, \delta S_q} \\
&\quad + \int_q h(q)\left(S_q - 2\frac{\delta H^*}{\delta S_{-q}} + \frac{\delta}{\delta S_{-q}}\right)\frac{\delta^2 H^*}{\delta S_0\, \delta S_q} \\
&= \frac{\delta}{\delta S_0}\mathscr{G}(\Psi^*)\, H^* + \left(\frac{d}{2} - h(0)\right)\frac{\delta H^*}{\delta S_0}.
\end{aligned} \tag{3.24}$$

Therefore the redundant operator $\delta H^*/\delta S_0$ shows formally up with the eigenvalue $y = (d/2) - h(0)$. Hubbard and Schofield (1972) pointed out that this operator does not contribute to the critical behaviour. It can be eliminated by the substitution $S_0 \to S_0 - \mu$, since

$$H^*(S_0 + \mu) = H^*(S_0) + \mu\frac{\delta H^*}{\delta S_0} + O(\mu^2). \tag{3.25}$$

As will be seen later (Sections III.F and IV.C), in many cases the value $h(0)$ is not at our disposal. Therefore one cannot conclude from (3.24) that y can be chosen arbitrarily. We may however introduce the generator $\mathscr{G}'$ with arbitrary constant c

$$\mathscr{G}'H = \mathscr{G}H - cu_1(0)\frac{\delta H}{\delta S_0} \tag{3.26}$$

and $\mathscr{G}$ given by (2.89). It yields for fixed point Hamiltonians even in S

$$\mathscr{G}'H^* = 0 \tag{3.27}$$

from $\mathscr{G}H^* = 0$ and

$$\mathscr{L}'\frac{\delta H^*}{\delta S_0} = \left(\frac{d}{2} - h(0) - cu_2^*(0)\right)\frac{\delta H^*}{\delta S_0}. \tag{3.28}$$

Thus for even fixed points with $u_2^*(0) \neq 0$ the operator $\delta H^*/\delta S_0$ can adopt any eigenvalue

$$y = \frac{d}{2} - h(0) - cu_2^*(0) \tag{3.29}$$

as a function of c. Bell and Wilson (1974) gave another example of an eigenoperator whose exponent y depends on the choice of the RG.

The operator S_0 of the magnetization yields

$$\mathscr{L}'S_0 = \left(\frac{d}{2} + h(0)\right) S_0 - (2h(0) + c)\frac{\delta H^*}{\delta S_0}. \tag{3.30}$$

Thus one obtains the eigenoperator

$$\mathscr{O}_1^* = S_0 - \frac{2h(0) - c}{2h(0) + cu_2^*(0)} \frac{\delta H^*}{\delta S_0} \tag{3.31}$$

with eigenvalue

$$y_h = y_1 = \frac{d}{2} + h(0). \tag{3.32}$$

This eigenvalue does not depend on c, but the eigenoperator is not uniquely determined. It differs for various c's by terms proportional to the redundant operator $\delta H^*/\delta S_0$.

2. *Derivation of the invariance properties*

In the remainder of this section the invariance properties of the RG are shown explicitly (see Wegner 1974). From (3.21) one finds that the Hamiltonian generated by the dilatation transformation

$$H^* + \delta\mathscr{G}_{\text{dil}} H^*$$

equals the Hamiltonian

$$H^* + \delta\mathscr{G}_{\text{tra}}(-\Psi^*) H^*$$

generated by the transformation $\mathscr{G}_{\text{tra}}$. Hamiltonians generated from H by the generators $\mathscr{G}_{\text{tra}}(\Phi)$ are called equivalent to H. Thus a fixed point Hamiltonian transforms under the dilatation transformation into an equivalent Hamiltonian.

There is however an important difference between $\mathscr{G}_{\text{dil}} H^*$ and $\mathscr{G}_{\text{tra}}(-\Psi^*)H^*$: The application of $\exp(l\mathscr{G}_{\text{dil}})$ transforms the system of linear dimensions L into one of linear dimensions Le^{-l} whereas the generator $\mathscr{G}_{\text{tra}}(-\Psi^*)$ shall be dilatation free. This means: The application of $\exp(l\mathscr{G}_{\text{tra}}(-\Psi^*))$ shall not change the size of the system under consideration. Therefore the total transformation $\exp(l(\mathcal{G}_{\text{dil}} + \mathcal{G}_{\text{tra}}(\Psi^*))$ leaves the local interaction invariant but changes the volume by a factor e^{-dl}.

The discussion of the invariance properties of eigenoperators and eigenvalues y is divided in three sections: (i) H^* and Ψ^* fixed, (ii) H^* only fixed, and (iii) H^* replaced by an equivalent fixed point.

(i) *H^* and Ψ^* fixed.* The eigenvalue problem for fixed H^* and Ψ^* is considered. Since in general Ψ is a functional of H, one has

$$\Psi = \Psi\{\mu\} = \Psi^* + \sum \mu_i \Psi_i + O(\mu^2) \tag{3.33}$$

and one obtains the eigenvalue equation

$$\mathcal{G}(\Psi^* + \sum \mu_i \Psi_i)(H^* + \sum \mu_i \mathcal{O}_i^*) = \sum y_i \mu_i \mathcal{O}_i^* + O(\mu^2). \tag{3.34}$$

Some simple algebra shows that a perturbation $\mu \mathcal{G}_{\text{tra}}(\Phi) H^*$ to H^* generates

$$\mathcal{G}(\Psi^* + \mu\hat{\Psi})(H^* + \mu\mathcal{G}_{\text{tra}}(\Phi) H^*)$$
$$= \mu\mathcal{G}_{\text{tra}}(K(\Psi^* + \mathcal{D}, \Phi) + \hat{\Psi}) H^* + O(\mu^2) \tag{3.35}$$

with

$$K_q(\Psi, \Phi) = \int_{q'} \left(\Psi_{q'} \frac{\delta \Phi_q}{\delta S_{q'}} - \Phi_{q'} \frac{\delta \Psi_q}{\delta S_{q'}} \right) \tag{3.36}$$

and

$$\mathcal{D}_q = q \nabla S_q + \frac{d}{2} S_q. \tag{3.37}$$

The function K is called a commutator since

$$\int_{q'} \Psi_{q'} \frac{\delta}{\delta S_{q'}} \left(\int_q \Phi_q \frac{\delta}{\delta S_q} \right) - \int_{q'} \Phi_{q'} \frac{\delta}{\delta S_{q'}} \left(\int_q \Psi_q \frac{\delta}{\delta S_q} \right) = \int_q K_q(\Psi, \Phi) \frac{\delta}{\delta S_q}. \tag{3.38}$$

From (3.35) one sees that a redundant operator (that is $\mathcal{G}_{\text{tra}}(\Phi) H^*$) transforms again into a redundant operator (l infinitesimal)

$$e^{l\mathcal{G}}(H^* + \mu\mathcal{G}_{\text{tra}}(\Phi) H^*) = H^* + \mu\mathcal{G}_{\text{tra}}(\Phi + lK(\Psi^* + \mathcal{D}, \Phi) + l\hat{\Psi}) H^*$$
$$+ O(l^2, \mu^2). \tag{3.39}$$

If $\mathcal{G}_{\text{tra}}(\Phi)$ and $\mathcal{G}_{\text{tra}}(\Psi^*)$ are dilatation free, then $\mathcal{G}_{\text{tra}}(K(\Psi^*, \Phi))$ is dilatation free. This does not hold for $\mathcal{G}_{\text{tra}}(K(\mathcal{D}, \Phi))$. Since the whole system is contracted by a factor e^{-l} in length scale, we expect that the perturbations are contracted, too. Indeed the first term of the right hand side of

$$\mathscr{G}_{\text{tra}}(K(\Psi^* + \mathscr{D}, \Phi))\, H^* = \int_q q\nabla S_q \frac{\delta}{\delta S_q} \mathscr{G}_{\text{tra}}(\Phi)\, H^*$$

$$+ \int_q \Psi_q^* \frac{\delta}{\delta S_q} \mathscr{G}_{\text{tra}}(\Phi)\, H^* \tag{3.40}$$

describes a contraction, the second a local transformation. It is obvious from (3.39) that any perturbation $\mathscr{G}_{\text{tra}}(\Phi)\, H^*$ can be an eigenoperator with any eigenvalue y by choosing

$$\hat{\Psi} = y\Phi - K(\Psi^* + \mathscr{D}, \Phi). \tag{3.41}$$

The scaling exponents y of the perturbations $\mathscr{G}_{\text{tra}}\, H^*$ depend therefore on the choice of the renormalization group. There are no inherent (universal) quantities. Since the free energy is invariant, to order μ, under the change in the Hamiltonian due to these perturbations (the corresponding result for higher-order perturbations is given in Section V.B), we call these operators redundant operators.

We return to the general eigenvalue problem for the operator $\mathscr{O}'$

$$\mathscr{G}(\Psi\{\mu\})\,(H^* + \mu\mathscr{O}') = y\mu\mathscr{O}' + O(\mu^2). \tag{3.42}$$

Suppose that the operators $\mathscr{O}_i^*$ in (3.34) form a complete set of linearly independent nonredundant eigenoperators for a certain choice of $\Psi\{\mu\}$. They are linearly independent of the redundant generators, so that the operator

$$\mathscr{O}' = \sum \alpha_i \mathscr{O}_i^* + \mathscr{G}_{\text{tra}}(\Phi)\, H^* \tag{3.43}$$

vanishes only if $\alpha_i = 0$ and $\mathscr{G}_{\text{tra}}(\Phi)\, H^* = 0$. Then solve (3.42) with

$$\Psi' = \Psi + \mu\hat{\Psi}\{\alpha\} \tag{3.44}$$

One finds that

$$\mathscr{G}(\Psi')\,(H^* + \mu\mathscr{O}') = \mu[\sum y_i \alpha_i \mathscr{O}_i^* + \mathscr{G}_{\text{tra}}(K(\Psi^* + \mathscr{D}, \Phi) + \bar{\Psi})\, H^*] \tag{3.45}$$

Since $\mathscr{O}_i^*$ and $\mathscr{G}_{\text{tra}}\, H^*$ are linearly independent, it follows that

$$(y_i - y)\,\alpha_i = 0 \tag{3.46}$$

and

$$\bar{\Psi} = y\Phi - K(\Psi^* + \mathscr{D}, \Phi). \tag{3.47}$$

Any operator $\mathscr{O}'$, eqn (3.43) with at least one nonvanshing α_i is called a scaling operator. The scaling eigenoperators are determined uniquely (apart from degeneracies) except for a possible additional redundant operator. The eigenvalues y_i do not depend on the choice of Ψ_i.

(ii) H^* *fixed.* Until now the eigenvalue equation for fixed H^* and Ψ^* has been discussed. It may be however that for a certain fixed point H^* we may choose various Ψ^*'s. Suppose that besides (3.21) also

$$\mathcal{G}(\Psi^* + \Theta)H^* = 0 \tag{3.48}$$

holds. Then the Hamiltonian H^* is an invariant of the generator

$$\mathcal{G}_{\text{tra}}(\Theta)\, H^* = 0. \tag{3.49}$$

With

$$\int_q \Theta_q \frac{\delta \mathcal{O}_i^*}{\delta S_q} = \sum \Theta_{ij} \mathcal{O}_j^* \tag{3.50}$$

one obtains the invariance property

$$F\{\mu_i\} = F\left\{\mu_i + \omega \sum_j \Theta_{ij}\mu_j\right\} \tag{3.51}$$

for infinitesimal ω. The symmetry (3.49) has the further consequence that the exponents y are no longer invariants. We give an example. Let us consider a two-component system (components S_x and S_y). Then the transformation

$$\mathcal{G}_{\text{tra}}(\Theta)\, H = \int_q \left(S_{qx} \frac{\delta}{\delta S_{qy}} - S_{qy} \frac{\delta}{\delta S_{qx}} \right) H \tag{3.52}$$

describes a rotation of the spins. It yields zero if applied on a rotational invariant fixed-point Hamiltonian

$$\mathcal{G}(\Psi^* + \omega\Theta)\, H^* = 0 \tag{3.53}$$

where

$$\Psi_{q\alpha} = h(q)\left(S_{q\alpha} - \frac{\delta H}{\delta S_{-q\alpha}} \right), \qquad \alpha = x, y \tag{3.54}$$

If $\mathcal{L}(\Psi + \omega\Theta)$ is applied on the components of a vector like

$$\mathcal{O}_{1\pm}^* = \mathcal{O}_{1x}^* \pm i\mathcal{O}_{1y}^* \tag{3.55}$$

and $\mathcal{O}_{1\alpha}^*$ is defined by (3.31) ($c = 0$) where S_0 is replaced by $S_{0\alpha}$, then one obtains

$$\mathcal{L}(\Psi + \omega\Theta)\, \mathcal{O}_{1\pm}^* = \left(\frac{d}{2} + h(0) \pm i\omega \right) \mathcal{O}_{1\pm}^* \tag{3.56}$$

where ω is at our choice. The arbitrariness of the exponent

$$y_{1\pm} = \frac{d}{2} + h(0) \pm i\omega \tag{3.57}$$

reflects only the rotational invariance of the system. The free energy of the Hamiltonian

$$\mu^* + \mu_+ \mathcal{O}^*_{h+} + \mu_- \mathcal{O}^*_{h-} \tag{3.58}$$

depends only on the scalar $\mu_+\mu_-$. Therefore the derivatives $\partial^2 F/\partial\mu_+^2$ and $\partial^2 F/\partial\mu_-^2$ vanish at $\mu_+ = \mu_- = 0$ (as long as this limit is uniquely defined) whereas the critical exponent of $\partial^2 F/\partial\mu_+\partial\mu_- \neq 0$ is invariant since

$$\begin{aligned} \mu_+(l)\,\mu_-(l) &= \exp\left[(y_{1+} + y_{1-})\,l\right] \mu_+(0)\,\mu_-(0) \\ &= \exp\left[(d + 2h(0))\,l\right] \mu_+(0)\,\mu_-(0) \end{aligned} \tag{3.59}$$

does not depend on ω. The operator

$$\int_q \Theta_q \frac{\delta}{\delta S_q} \tag{3.60}$$

as defined in (3.52) commutes with $\mathscr{G}\{\Psi\}$. Therefore all eigenoperators of $\mathscr{L}(\Psi^* + \omega\Theta)$ can be chosen as eigenoperators of

$$\int \Theta_q \frac{\delta \mathcal{O}_i^*}{\delta S_q} = i m_i \mathcal{O}_i^*, \qquad m_i \text{ integer}. \tag{3.61}$$

Therefore one obtains

$$\mathscr{L}(\Psi + \omega\Theta)\,\mathcal{O}_i^* = (y_{io} + i m_i \omega)\,\mathcal{O}_i^*. \tag{3.62}$$

The free energy of the Hamiltonian (8.11) depends only on the products $\mu_{i_1} \ldots \mu_{i_n}$ with $\Sigma m_i = 0$. We are not aware of a general discussion of these invariance properties.

(iii) *H^* replaced by an equivalent fixed point.* Consider the eigenvalue problem for an equivalent fixed point $H^* + l\mathscr{G}_{\text{tra}}(\Phi)\,H^*$ (l infinitesimal). One easily verifies

$$\begin{aligned} &\mathscr{G}(\Psi - lK(\Psi + \mathscr{D}, \Phi))\,(H + l\mathscr{G}_{\text{tra}}(\Phi)\,H) \\ &\qquad = \mathscr{G}(\Psi)\,H + l \int_q \Phi_q \frac{\delta}{\delta S_q} \mathscr{G}(\Psi)\,H + O(l^2). \end{aligned} \tag{3.63}$$

Equation (3.35) is a special form of this equation. One applies (3.63) to the eigenvalue problem (3.34) and obtains

$$\mathscr{G}(\Psi^{*\prime} + \textstyle\sum \mu_i \Psi_i')\,(H^{*\prime} + \sum \mu_i \mathcal{O}_i') = \sum y_i \mu_i \mathcal{O}_i' + O(\mu^2, l^2) \tag{3.64}$$

with

$$H^{*\prime} = H^* + l\mathscr{G}_{\text{tra}}(\Phi)\,H^* \tag{3.65a}$$

$$\mathcal{O}'_i = \mathcal{O}^*_i + l \int_q \Phi_q \frac{\delta \mathcal{O}^*_i}{\delta S_q} \tag{3.65b}$$

$$\Psi^{*\prime} = \Psi^* - lK(\Psi^* + \mathscr{D}, \Phi) \tag{3.65c}$$

$$\Psi'_i = \Psi_i - lK(\Psi_i, \Phi) \tag{3.65d}$$

Therefore eqn (3.34) is invariant under the transformation (3.65). In particular the scaling exponents y_i of the scaling operators are invariants of this transformation. For a variation of Ψ for the equivalent fixed point, we may apply our discussion for case (i) and (ii).

C. Scaling of the free energy

In this section the homogeneity relation for the free energy is derived.

Since the volume shrinks by a factor e^{-dl} during the RG procedure and since the free energy is kept fixed, one obtains, (2.35)

$$F_0 = e^{-dl} F_l$$

for the density of the free energy. On the other hand the Hamiltonian $H_0 = H\{\mu_i\}$ transforms into $H_l = H\{\mu_i e^{y_i l}\}$ (Section III.A). Therefore one obtains the homogeneity law

$$F\{\mu_i\} = e^{-dl} F\{\mu_i e^{y_i l}\}. \tag{3.66}$$

The extraction of the constant term $\mu_0 V$ from the Hamiltonian yields the singular part $F_{\text{sing}} = F(\mu_0 = 0)$

$$F\{\mu_i\} = \mu_0 + e^{-dl} F_{\text{sing}}\{\mu_i e^{y_i l}\}. \tag{3.67}$$

We denote the field μ_E of the relevant symmetry conserving operator $\mathcal{O}^*_E$ by τ, eqn (3.19). It describes (roughly) the temperature difference from the critical point. The parameter l in (3.67) can be chosen arbitrarily. Requiring that

$$|\tau| e^{y_E l} = 1 \tag{3.68}$$

one obtains the homogeneity law

$$F\{\mu_i\} = \mu_0 + |\tau|^{2-\alpha} F_{\text{sing}}\{\mu_i |\tau|^{-\Delta_i}\} \tag{3.69}$$

where

$$2 - \alpha = \frac{d}{y_E} \tag{3.70}$$

and

$$\Delta_i = y_i / y_E. \tag{3.71}$$

On the right-hand side of (3.69) one has the free energy of the Hamiltonian H_l with $\tau_l = \pm 1$. One may object that for such a large value of τ_l one is far from the critical point and that therefore the linearized theory does not apply. The condition (3.68) can however be replaced by

$$|\tau| e^{y_E l} = 10^{-3} \tag{3.72}$$

which yields

$$F\{\mu_i\} = \mu_0 + |10^3 \tau|^{2-\alpha} F(\mu_0 = 0, |10^3 \tau|^{-\Delta_i} \mu_i\}). \tag{3.73}$$

Therefore in any case F can be expressed as $|\tau|^{2-\alpha}$ multiplied by a function which depends only on the reduced fields $\mu_i |\tau|^{-\Delta_i}$.

Consider now a simplified situation and assume that the Hamiltonian H_0 can be expressed as

$$H_0 = H^* + \tau \mathcal{O}_E^* + \mu_h \mathcal{O}_h^* \tag{3.74}$$

with no further operator present (we have already subtracted the regular part $\mu_0 V$). This simplification corresponds to the parametrization of the Hamiltonian in Kadanoff's cell model (Kadanoff, 1966). Crudely speaking $\mathcal{O}_h^*$ is the operator of the magnetization, and τ and μ_h are proportional to $T - T_c$ and the magnetic field, resp.

$$\tau = c_2 (T - T_c), \tag{3.75a}$$

$$\mu_h = c_1 h. \tag{3.75b}$$

From the linearized RG equation one obtains

$$H_l = H^* + \tau \exp(y_E l)\, \mathcal{O}_E^* + \mu_h \exp(y_h l)\, \mathcal{O}_h^* \tag{3.76}$$

and according to (3.69)

$$F_{\text{sing}}(\tau, \mu_h) = |\tau|^{2-\alpha} F_{\text{sing}}(\pm 1, \mu_h |\tau|^{-\Delta}) = |\tau|^{2-\alpha} f_\pm \left(\frac{\mu_h}{|\tau|^\Delta} \right) \tag{3.77}$$

where

$$\Delta = y_h / y_E. \tag{3.78}$$

This is the homogeneity law as predicted by Widom (1965) and derived by Kadanoff (1966). It yields for $h = 0$ the specific heat

$$c_{\text{sing}} = -kT \frac{\partial^2}{\partial T^2} (T F_{\text{sing}}) = A_{c\pm} |T - T_c|^{-\alpha} \cdot (1 + O(\tau)) \tag{3.79}$$

with

$$A_{c\pm} = -k T_c^2 c_2^{2-\alpha} (2 - \alpha)(1 - \alpha) F(\pm 1, 0). \tag{3.80}$$

Note that the amplitudes may differ above and below T_c.

The nonanalytic part of the magnetization is obtained from

$$m_{\text{sing}} = -kT\frac{\partial F_{\text{sing}}}{\partial h} = -kT_c c_1 |\tau|^{\beta} f'_{\pm}\left(\frac{c_1 h}{|\tau|^{\Delta}}\right) \tag{3.81}$$

with

$$\beta = 2 - \alpha - \Delta \tag{3.82}$$

and

$$f'_{\pm}(x) = \frac{\partial}{\partial x} f_{\pm}(x). \tag{3.83}$$

This homogeneity law (3.81) has been checked against experimental measurements on several magnetic substances: for a discussion and references see for example the article by Gaunt and Domb (1970).

The singular part of the susceptibility χ yields

$$\chi_{\text{sing}} = \frac{\partial m_{\text{sing}}}{\partial h} = -kT_c c_1^2 |\tau|^{-\gamma} \; f''_{\pm}\left(\frac{c_1 h}{|\tau|^{\Delta}}\right) \tag{3.84}$$

with

$$\gamma = 2\Delta - 2 + \alpha. \tag{3.85}$$

The results (3.79), (3.81), (3.84) are in agreement with the observations (2.10a), (2.2) and (2.6) resp. where

$$A_m = -kT_c c_1 c_2^{\beta} f'_{-}(0) \tag{3.86}$$

and

$$A_{\chi\pm} = -kT_c c_1^2 c_2^{-\gamma} f''_{\pm}(0) \tag{3.87}$$

The amplitudes A_χ may be different above and below T_c.

In general the Hamiltonian H_0 cannot be written in the simple form (3.74) since it will contain contributions of many operators $\mathcal{O}_i^*$. For simplicity's sake we will allow for one more operator $\mathcal{O}_i^*$ since the inclusion of more operators $\mathcal{O}_i^*$ is straightforward (apart from questions of convergence which we are not able to discuss). Then one obtains from

$$H_0 = H^* + \tau\mathcal{O}_E^* + \mu_h \mathcal{O}_h^* + \mu_i \mathcal{O}_i^* \tag{3.88}$$

within the linear approximation the Hamiltonian

$$H_l = H^* + \tau \exp(y_E l)\, \mathcal{O}_E^* + \mu_h \exp(y_h l)\mathcal{O}_h^* + \mu_i \exp(y_i l)\mathcal{O}_i^* \tag{3.89}$$

and the free energy

$$F_{\text{sing}}(\tau, \mu_h, \mu_i) = |\tau|^{2-\alpha} F_{\text{sing}}\left(\pm 1, \frac{\mu_h}{|\tau|^{\Delta}}, \frac{\mu_i}{|\tau|^{\Delta_i}}\right) \tag{3.90}$$

with Δ_i given by (10.6). One is interested in the behaviour for small τ. Since $y_E > 0$ (in almost all cases) one obtains

$$\lim_{\tau \to 0} \frac{\mu_i}{|\tau|^{\Delta_i}} = \begin{cases} 0 & \text{for } y_i < 0 \text{ or } \mu_i = 0 \\ \infty & \text{for } y_i > 0 \text{ and } \mu_i \neq 0. \end{cases} \tag{3.91}$$

If $\mathcal{O}_i^*$ is a relevant operator, then the expression (3.91) diverges except for $\mu_i = 0$. Therefore the field μ_i has to be included in this case in the homogeneity law as in (3.90). If $\mathcal{O}_i^*$ is an irrelevant operator, the term $\mu_i|\tau|^{-\Delta_i}$ vanishes for small τ. If the free energy can be expanded in powers of $\mu_i(l)$ at $\tau(l) = \pm 1$, then one obtains

$$F_{\text{sing}} = |\tau|^{2-\alpha} f_{\pm}\left(\frac{\mu_h}{|\tau|^{\Delta}}\right) + \mu_i|\tau|^{2-\alpha-\Delta_i} f_{i\pm}\left(\frac{\mu_h}{|\tau|^{\Delta}}\right) + \ldots \tag{3.92}$$

where

$$f_{i\pm}(\mu_h) = \frac{\partial}{\partial \mu_i} F_{\text{sing}}(\tau_l = \pm 1, \mu_h, \mu_i)\bigg|_{\mu_i = 0} \tag{3.93}$$

and the second term in (3.92) provides a correction to the leading scaling behaviour (3.77). This expansion is plausible above T_c since one expects an analytic behaviour of F in this region. This is in agreement with experimental observations in superfluid He (Ahlers 1973), on EuO (Kornblit and Ahlers, 1975), and with numerical calculations by Saul, Wortis and Jasnow (1975) for magnetic susceptibilities.

Below T_c the situation may be more complicated since we expect several phases in equilibrium. The phase boundary is characterized by a nonanalytic behaviour of F. Therefore if (τ, μ_h, μ_i) is a point at which two phases are in equilibrium, then according to (3.90) the point $(\tau' = c\tau, \mu_h' = c^{\Delta}\mu_h, \mu_i' = c^{\Delta_i}\mu_i)$ lies on the phase boundary, too. In general the phase boundary obeys the scaling law

$$\mu_h = |\tau|^{\Delta} w(\mu_i|\tau|^{-\Delta_i}). \tag{3.94}$$

If one expands F within the one-phase domains, then (3.92) holds, too. It is not known whether such an expansion is possible at the phase-boundary.

The expansion (3.92) implies the existence of the derivative (3.93). If this derivative does not exist, then the expansion (3.92) is not possible. We give an example. As will be seen in Section III.F systems with $d > 4$ have a fixed point (crudely)

$$H^* = \int d^d r (\nabla \hat{S}(r))^2 \tag{3.95}$$

with symmetry conserving operators

$$\mathcal{O}_E^* = \int \mathrm{d}^d r (\hat{S}^2(r) - k_1) \tag{3.96}$$

and

$$\mathcal{O}_4^* = \int \mathrm{d}^d r (\hat{S}^4(r) - k_2 \hat{S}^2(r) + k_3) \tag{3.97}$$

$\mathcal{O}_E^*$ is relevant ($y_E = 2$) and $\mathcal{O}_4^*$ is irrelevant ($y_4 = 4 - d$). Obviously $H^* + \mu_E \mathcal{O}_E^*$ has a well defined free energy for $\mu_E \geqslant 0$. For $\mu_E < 0$ this Hamiltonian has no lower bound and one has to add a term $\mu_4 \mathcal{O}_4^*$ with $\mu_4 > 0$ to obtain a finite free energy. Therefore $F(\mu_E, \mu_4)$ cannot be expanded for $\mu_E < 0$ in powers of μ_4. As a consequence the exponent β cannot be obtained from (3.70), (3.78), (3.82) which would yield $\beta = (d - y_h)/y_E = (d - 2)/4$. In this case Fisher's idea (Fisher 1973) of the anomalous dimension of the vacuum applies which yields the molecular field exponent $\beta = \frac{1}{2}$.

In general one obtains from (3.69) the expectation values

$$\frac{1}{V} \langle \mathcal{O}_i^* \rangle = \frac{\partial F}{\partial \mu_i} = |\tau|^{\beta_i} F_{i\pm}\{\mu|\tau|^{-\Delta}\} \tag{3.98}$$

with

$$\beta_i = 2 - \alpha - \Delta_i \tag{3.99}$$

and

$$\chi_{ij} = -\frac{\partial^2 F}{\partial \mu_i \, \partial \mu_j} = -|\tau|^{-\gamma_{ij}} F_{ij\pm}\{\mu|\tau|^{-\Delta}\} \tag{3.100}$$

with

$$\gamma_{ij} = \Delta_i + \Delta_j - 2 + \alpha \tag{3.101}$$

where F_i and F_{ij} are first and second derivatives of F with respect to its ith (and jth) argument.

D. Correlation functions in momentum space

In this section the homogeneity properties of the wave-vector dependent correlation functions are derived.

Until now only translationally invariant perturbations $\mathcal{O}_i^*$ to the fixed point Hamiltonian have been considered. Within the linear approximation one obtains the scaling behaviour starting from the eigenoperators, equation (3.10).

$$\mathcal{L}\mathcal{O}_i^* = y_i \mathcal{O}_i^*.$$

In this section we start from the local perturbations $\tilde{\mathcal{O}}_i$ (an operator which decays within some distance $1/q_0$ from the origin) which obeys

$$\mathscr{L}\tilde{\mathcal{O}}_i = -x^i\tilde{\mathcal{O}}_i. \tag{3.102}$$

A parametrization for such an operator is given in (2.67). Explicit examples are calculated in Section III.F. The operator $\tilde{\mathcal{O}}_i$ is a function of the Fourier components S_q

$$\tilde{\mathcal{O}}_i = \tilde{\mathcal{O}}_i\{S_q\} \tag{3.103}$$

One defines the operators $\tilde{\mathcal{O}}_i(r)$ by replacing S_q by $e^{iqr} S_q$ (compare Wilson and Kogut, 1974; Wegner, 1975a).

$$\tilde{\mathcal{O}}_i(r) = \tilde{\mathcal{O}}_i\{e^{iqr} S_q\} \tag{3.104}$$

Let us calculate $\mathscr{L}\tilde{\mathcal{O}}_i(r)$. Defining

$$S'_q = e^{iqr} S_q \tag{3.106}$$

one obtains

$$\frac{\delta}{\delta S_q} = e^{iqr}\frac{\delta}{\delta S'_q} \tag{3.106}$$

and

$$q\nabla S_q = e^{-iqr}(q\nabla S'_q - iqr\, S'_q). \tag{3.107}$$

The translational invariance of H^* yields

$$H^*\{S'_q\} = H^*\{S_q\}. \tag{3.108}$$

Using (3.105) to (3.108) one obtains from (3.6)

$$\mathscr{L} = \mathscr{L}' - ir\int_q q\, S'_q \frac{\delta}{\delta S'_q} \tag{3.109}$$

where $\mathscr{L}'$ is obtained from $\mathscr{L}$ by replacing S by S'. From (3.104) one finds

$$i\int_q qS'_q \frac{\delta}{\delta S'_q}\tilde{\mathcal{O}}_i\{e^{iqr} S_q\} = \nabla\tilde{\mathcal{O}}_i(r) \tag{3.110}$$

and finally with (3.102)

$$\mathscr{L}\tilde{\mathcal{O}}_i(r) = -x_i\tilde{\mathcal{O}}_i(r) - r\nabla\tilde{\mathcal{O}}_i(r). \tag{3.111}$$

Therefore one has

$$\exp(\mathscr{L}l)\,\tilde{\mathcal{O}}_i(r) = \exp(-x_i l)\,\tilde{\mathcal{O}}_i(r\, e^{-l}). \tag{3.112}$$

Under the RG procedure the perturbation $\tilde{\mathcal{O}}_i$ is multiplied by a factor $\exp(-x_i l)$ but aslo shifted to re^{-l} due to the change of the length scale.

It follows from (3.112) that the wave-vector dependent operator

$$\mathcal{O}_i^*(q) = \int \mathrm{d}^d r\, \tilde{\mathcal{O}}_i(r)\, \mathrm{e}^{-iqr} \tag{3.113}$$

obeys the transformation law

$$\exp(\mathcal{L}l)\, \mathcal{O}_i^*(q) = \exp[(d - x_i)l]\, \mathcal{O}_i^*(q\,\mathrm{e}^l). \tag{3.114}$$

Comparison with (3.10) for $q = 0$ yields ($\mathcal{O}_i^* \equiv \mathcal{O}_i^*(0)$)

$$y_i = d - x_i \tag{3.115}$$

provided $\mathcal{O}_i^*(0)$ does not vanish.

From (3.111) one obtains

$$\mathcal{L}\nabla\tilde{\mathcal{O}}_i = -(x_i + 1)\nabla\tilde{\mathcal{O}}_i. \tag{3.116}$$

In the following only operators $\mathcal{O}_i^*(q)$ with $\mathcal{O}_i^*(0) \neq 0$ will be considered since the Fourier transform of $\nabla\tilde{\mathcal{O}}_i(r)$ can be expressed by

$$\int \mathrm{d}^d r\, \nabla\tilde{\mathcal{O}}_i(r)\, \mathrm{e}^{-iqr} = iq\mathcal{O}_i^*(q). \tag{3.117}$$

Starting out from the Hamiltonian

$$H_0 = H^* + \sum_i \mu_i \mathcal{O}_i^* + \lambda_j \mathcal{O}_j^*(q) + \lambda_k \mathcal{O}_k^*(-q) \tag{3.118}$$

one obtains

$$H_l = H^* + \sum \mu_i \exp(y_i l)\, \mathcal{O}_i^* + \lambda_j \cdot \exp(y_j l)\, \mathcal{O}_j^*(q\,\mathrm{e}^l) + \lambda_k \exp(y_k l)\, \mathcal{O}_k^*(-q\,\mathrm{e}^l). \tag{3.119}$$

The correlation function

$$VG_{jk}(q, \ldots \mu_i \ldots) = \langle \mathcal{O}_j^*(q)\, \mathcal{O}_k^*(-q)\rangle_0 - \langle \mathcal{O}_j^*(q)\rangle_0 \langle \mathcal{O}_k^*(-q)\rangle_0 \tag{3.120}$$

(the subscript refers to the Hamiltonian H_0) is obtained from the free energy

$$G_{jk} = -\frac{\partial^2 F}{\partial\lambda_j\, \partial\lambda_k}. \tag{3.121}$$

Note that we introduced a factor V in the definition of G so that G does not depend on the volume. F in (3.121) is now the free energy averaged over the whole volume. Equations (2.37), (3.119) and (3.121) yield

$$G_{jk}(q, \ldots \mu_i \ldots) = \exp[(y_j + y_k - \mathrm{d})\, l]\, G_{jk}(q\,\mathrm{e}^l, \ldots \mu_i \exp(y_i l) \ldots). \tag{3.122}$$

Obviously the corresponding relation holds for the n-point cumulant

$$G_{j_1 \dots j_n}(q_1, \dots q_n, \dots \mu_i \dots) = \\ = \exp\left[(y_{j_1} + \dots + y_{j_n} - d)\, l\right] G_{j_1 \dots j_n}(q_1\, e^l, \dots q_n\, e^l, \dots \mu_i \exp(y_i l) \dots) \tag{3.123}$$

which is defined by

$$\lambda_i G_i - \frac{1}{2!} \sum \lambda_i \lambda_j G_{ij} + \dots = F(\lambda) - F(0). \tag{3.124}$$

In particular one obtains for the correlation function of the order parameter as a function of the temperature difference τ

$$G(q, \tau) = \langle \mathcal{O}_h^*(q)\, \mathcal{O}_h^*(-q) \rangle - \langle \mathcal{O}_h^*(q) \rangle \langle \mathcal{O}_h^*(-q) \rangle \\ = \exp\left[(2y_h - d)\, l\right] G(q\, e^l, \tau \exp(y_E l)). \tag{3.125}$$

(In Section VI.D it is shown that redundant contributions to $\mathcal{O}_h^*$ do not contribute to the leading scaling behaviour.)

Let us discuss equation (3.125) for the correlation function:

(i) At the critical point ($\tau = 0$) one obtains with $|q| e^l = 1$

$$G(q, 0) = |q|^{d - 2y_h} G(1, 0) = c \cdot |q|^{-2+\eta} \tag{3.126}$$

where

$$2 - \eta = 2y_h - d \tag{3.127}$$

and according to (3.32)

$$2 - \eta = 2h(0). \tag{3.128}$$

(ii) For $\tau \neq 0$ one chooses $|\tau| \exp(y_E l) = 1$ and obtains

$$G(q, \tau) = |\tau|^{-\gamma} G(q|\tau|^{-1/y_E}, \pm 1) = |\tau|^{-\gamma} g_0(q\xi). \tag{3.129}$$

Away from the critical point the correlation functions decay normally with an exponential (2.8). Therefore the Fourier transform is a function of $q\xi$ where ξ is the correlation length. From (3.129) it follows that

$$\xi = A\, |T - T_c|^{-\nu} \tag{3.130}$$

with

$$\nu = 1/y_E \tag{3.131}$$

in agreement with (2.9). The critical exponents α, β, γ, η and ν depend only on y_h and y_E (compare Section III.C) which implies the well-known scaling exponent relations

$$\alpha + 2\beta + \gamma = 2, \tag{3.132}$$

$$d\nu = 2 - \alpha, \tag{3.133}$$

$$\gamma = \nu(2 - \eta). \tag{3.134}$$

According to the theory presented here the exponents α, γ and υ, respectively are equal above and below T_c.

A generalization of these results to other correlations is obvious. The two-point function $\langle \mathcal{O}_i^*(q)\, \mathcal{O}_i^*(-q) \rangle$ diverges at the critical point for $q \to 0$ only if $y_i > d/2$ (we do not discuss the case $y_i = d/2$ here). In particular the correlation $\langle \mathcal{O}_E^*(q)\, \mathcal{O}_E^*(-q) \rangle$ does not converge for $y_E < d/2$. In this case a regular part is dominant at the critical point. This will be discussed in context with the nonlinear terms of the RG.

E. Correlation functions in coordinate space

Here we consider shortly (1) the correlations of local operators $\tilde{\mathcal{O}}(r)$, (2) the product expansion for the operators $\tilde{\mathcal{O}}(r)$, and (3) the conformal covariance. The discussion of this section is restricted to the correlations for the fixed point Hamiltonian. We refer the reader to the paper by Fisher (1973).

1. *Correlations of local operators*

These correlations can simply be found from (3.112). At the critical point (more precisely for the fixed point Hamiltonian), one obtains from

$$H_0 = H^* + \sum_i \lambda_i \tilde{\mathcal{O}}_i(r_i) \tag{3.135}$$

the Hamiltonian

$$H_l = H^* + \sum_i \lambda_i \exp(-x_i l)\, \tilde{\mathcal{O}}_i\,(r_i\, \mathrm{e}^{-l}). \tag{3.136}$$

The cumulant

$$G_{i_1 \ldots i_n}(r_1, \ldots r_n) = \langle \tilde{\mathcal{O}}_{i_1}(r_1) \ldots \tilde{\mathcal{O}}_{i_n}(r_n) \rangle_c \tag{3.137}$$

is now defined by

$$\sum \lambda_i G_i(r_i) - \frac{1}{2!} \sum \lambda_i \lambda_j G_{ij}(r_i, r_j) + \ldots = VF(\lambda) - VF(0) \tag{3.138}$$

for which one obtains the homogeneity law

$$G_{i_1 \ldots i_n}(r_1, \ldots r_n) = \exp\left[-(x_{i_1} + \ldots + x_{i_n})\, l\right] \times G_{i_1 \ldots i_n}(r_1\, \mathrm{e}^{-l}, \ldots r_n\, \mathrm{e}^{-l}). \tag{3.139}$$

Apart from the constant all operators $\tilde{\mathcal{O}}_i$

$$\langle \tilde{\mathcal{O}}_i \rangle = \frac{1}{V} \langle \mathcal{O}_i \rangle \propto |\tau|^{(d-y_i)/y_E} \tag{3.140}$$

becomes zero at the fixed point since in all known cases $y_i < d$. Therefore one obtains

$$G_{ij}(r_1, r_2) = \langle \tilde{\mathcal{O}}_i(r_1) \tilde{\mathcal{O}}_j(r_2) \rangle \tag{3.141}$$

$$G_{ijk}(r_1, r_2, r_3) = \langle \tilde{\mathcal{O}}_i(r_1) \tilde{\mathcal{O}}_j(r_2) \tilde{\mathcal{O}}_k(r_3) \rangle. \tag{3.142}$$

One finds in particular from (3.139)

$$G_{ij}(0, r) = c_{ij} r^{-x_i - x_j} \tag{3.143}$$

where c_{ij} may depend on the direction of r. For the order parameter one has $x_i + x_j = 2x_h = d - 2 + \eta$ according to (3.127).

2. *Operator product expansion*

If the operators $\tilde{\mathcal{O}}_i(r)$ form a complete set of operators for fixed r, then it should be possible to expand any product (Kadanoff 1969, Polyakov 1969, Wilson 1969) in these operators. We follow a method widely due to Nachtmann (private communication).

$$\tilde{\mathcal{O}}_j(r_1) \tilde{\mathcal{O}}_k(r_2) = \sum_i U_{ijk}(r_1 - r_2) \tilde{\mathcal{O}}_i\left(\frac{r_1 + r_2}{2}\right). \tag{3.144}$$

To determine the coefficients U one applies the generator $\mathscr{L}$, eqn (3.6) on both sides of the eqn (3.144).

$$(\mathscr{L}\tilde{\mathcal{O}}_j(r_1)) \tilde{\mathcal{O}}_k(r_2) + \tilde{\mathcal{O}}_j(r_2) \mathscr{L}\tilde{\mathcal{O}}_k(r_2) + 2 \int_q h(q) \frac{\delta \tilde{\mathcal{O}}_j(r_1)}{\delta S_q} \frac{\delta \tilde{\mathcal{O}}_k(r_2)}{\delta S_{-q}}$$

$$= \sum_i U_{ijk}(r_1 - r_2) \mathscr{L}\tilde{\mathcal{O}}_i\left(\frac{r_1 + r_2}{2}\right) \tag{3.145}$$

The integral term in (3.145) can be expressed as

$$\int_q h(q) \frac{\delta \tilde{\mathcal{O}}_j(r_1)}{\delta S_q} \frac{\delta \tilde{\mathcal{O}}_k(r_2)}{\delta S_{-q}} = \int \mathrm{d}^d r\, \mathrm{d}^d r'\, \tilde{h}(r - r') \frac{\delta \tilde{\mathcal{O}}_j(r_1)}{\delta S(r)} \frac{\delta \tilde{\mathcal{O}}_k(r_2)}{\delta S(r')} \tag{3.146}$$

where the functional derivative $\delta/\delta S(r)$ has been defined in (2.65) and

$$h(q) = \int \mathrm{d}^d r\, \mathrm{e}^{-iqr}\, \tilde{h}(r) \tag{3.147}$$

which for $h(q) = c + 2q^2$, eqn (2.90) yields

$$\tilde{h}(r) = \left(c - 2\frac{\partial^2}{\partial r^2}\right)\delta^d(r). \tag{3.148}$$

Since the operators $\tilde{\mathcal{O}}_j(r_1)$ and $\tilde{\mathcal{O}}_k(r_2)$ are localized around r_1 and r_2 the functional derivatives on the right hand side of eqn (3.146) vanish unless r is close to r_1 and r' is close to r_2. Moreover $\tilde{h}(r)$ can be chosen of short range as in (3.148). Thus the expression (3.146) decays rapidly if $|r_1 - r_2| \gg 1/q_0$. Therefore this term can be neglected for large distances $|r_1 - r_2|$ which yields with (3.112)

$$\begin{aligned}\exp(\mathscr{L}l)(\tilde{\mathcal{O}}_j(r_1)\tilde{\mathcal{O}}_k(r_2)) &= \exp(-(x_j + x_k)l)\,\tilde{\mathcal{O}}_j(r_1\,\mathrm{e}^{-l})\,\tilde{\mathcal{O}}_k(r_2\,\mathrm{e}^{-l})\\ &= \sum_i U_{ijk}(r_1 - r_2)\exp(-x_i l)\,\tilde{\mathcal{O}}_i\left(\frac{r_1 + r_2}{2}\mathrm{e}^{-l}\right)\end{aligned} \tag{3.149}$$

from which it follows that $U(r)$ is a homogeneous function of r

$$U_{ijk}(r) \doteq \frac{u_{ijk}(r/|r|)}{|r|^{x_j + x_k - x_i}} \text{ for } |r| \gg 1/q_0. \tag{3.150}$$

Consequently the two-point correlation function can be expressed as

$$\langle\tilde{\mathcal{O}}_j(0)\,\tilde{\mathcal{O}}_k(r)\rangle = \sum_i \frac{u_{ijk}(r/|r|)}{|r|^{x_j + x_k - x_i}}\,\frac{1}{V}\langle\mathcal{O}_i^*\rangle \tag{3.151}$$

since the translational invariance implies

$$\langle\tilde{\mathcal{O}}_i(r)\rangle = \frac{1}{V}\langle\mathcal{O}_i^*\rangle. \tag{3.152}$$

Equation (3.151) indicates a complicated temperature behaviour since $\langle\mathcal{O}_i^*\rangle \sim |\tau|^{2-\alpha-\Delta_i}$.

3. *Conformal covariance*

The conformal transformation performs an isotropic change of the length scale which varies locally and allows for local rotations. The group of this transformation can be composed of

$$r' = r + c \quad \text{(displacement)} \tag{3.153a}$$

$$r' = \lambda r \quad \text{(enlargement)} \tag{3.153b}$$

$$r' = Ur \quad \text{(rotation, } U \text{ orthogonal)} \tag{3.153c}$$

$$r' = \frac{r + ar^2}{1 + 2ar + a^2r^2} \quad \text{(special conformal transformation)} \tag{3.153d}$$

(For $d = 1$ and $d = 2$ the group of the conformal transformations is richer). For infinitesimal a the special transformation can be written

$$r' = r + u(r) = r + ar^2 - 2(ra)\,r. \tag{3.154}$$

Since

$$\frac{\partial r'_i}{\partial r_j} = \delta_{ij}(1 - 2(\mathrm{ar})) + 2(\mathrm{a}_i\mathrm{r}_j - \mathrm{a}_j\mathrm{r}_i) \tag{3.155}$$

a volume element at point r is stretched in the linear dimensions by $1 - \lambda(r)$ with

$$\lambda(r) = 2(ar) \tag{3.156}$$

and rotated by the orthogonal matrix

$$u_{ij} = \delta_{ij} + 2(a_i r_j - a_j r_i). \tag{3.157}$$

The effect of this transformation on critical phenomena was first described by Polyakov (1970) and Migdal (1971). More recent work within this RG approach has been done by Wolsky, Green and Gunton (1973) and Schäfer (1975).

The generator $\mathscr{G}$, eqn (2.89), describes a homogeneous dilatation transformation. It can be rewritten

$$\lambda\mathscr{G}H = \int \mathrm{d}^d r\left(-\lambda r\nabla S(r) - \lambda\frac{d}{2}S(r)\right)\frac{\delta H}{\delta S(r)} + \int \mathrm{d}^d r\,\mathrm{d}^d r'\,\lambda h(r - r')$$
$$\left[\left(S(r') - \frac{\delta H}{\delta S(r')}\right)\frac{\delta H}{\delta S(r)} + \frac{\delta^2 H}{\delta S(r)\,\delta S(r')} - \delta^d(r - r')\right] \tag{3.158}$$

This equation describes the change of the Hamiltonian under the homogeneous contraction

$$r' = r + u(r) \tag{3.159}$$

$$u(r) = -\lambda r \tag{3.160}$$

We replace this homogeneous contraction by that of the conformal group

$$a\mathscr{G}_{\mathrm{con}}H = \int \mathrm{d}^d r\left(u(r)\,\nabla S(r) - \frac{d}{2}\lambda(r)\,S(r)\right)\frac{\delta H}{\delta S(r)} + \int \mathrm{d}^d r\,\mathrm{d}^d r'\,\lambda(r)\,h(r - r')$$
$$\left[\left(S(r') - \frac{\delta H}{\delta S(r')}\right)\frac{\delta H}{\delta S(r)} + \frac{\delta^2 H}{\delta S(r)\,\delta S(r')} - \delta^d(r - r')\right] \tag{3.161}$$

If H^* and the operators $\tilde{\mathcal{O}}_i(r)$ were local operators acting only at a point, then

$$a\mathcal{G}_{\text{con}}(H^* + \sum \rho_i \tilde{\mathcal{O}}_i(r)) = \sum \rho_i(-x_i\lambda(r_i) + u(r_i)\nabla)\tilde{\mathcal{O}}_i(r) \qquad (3.162)$$

Since the free energy is conserved, the correlation functions obey

$$G_{ij}(r_1, r_2) = (1 - x_1\lambda(r_1) - x_j\lambda(r_2)) \times G_{ij}(r_1 + u(r_1), r_2 + u(r_2)) \qquad (3.163)$$

and

$$G_{ijk}(r_1, r_2, r_3) = (1 - x_i\lambda(r_1) - x_j\lambda(r_2) - x_k\lambda(r_3)) \times G_{ijk}(r_1 + u(r_1), r_2 + u(r_2), r_3 + u(r_3)). \qquad (3.164)$$

Actually H^* and $\tilde{\mathcal{O}}_i(r)$ are not strictly local operators. Therefore the operators are deformed under the transformation. For a discussion see Schäfer (1975), In particular $\mathcal{G}_{\text{con}}$ must not generate a perturbation with $y = 1$ from H^* since otherwise the conformal covariance may be violated.

With the choice

$$r_1 = \frac{r}{2}, \qquad r_2 = -\frac{r}{2} \qquad (3.165)$$

eqn (3.163) yields

$$G_{ij}\left(\frac{r}{2}, -\frac{r}{2}\right) = (1 + (x_j - x_i)(ar))\, G_{ij}\left(\frac{r}{2} + u, -\frac{r}{2} + u\right) \qquad (3.166)$$

with

$$u(r_1) = u(r_2) = \frac{1}{4}ar^2 - \frac{1}{2}(ar)\, r. \qquad (3.167)$$

Since the correlation functions are translationally invariant, the correlation functions on both sides of eqn (3.166) are equal which implies

$$G_{ij}(r) = 0 \quad \text{for} \quad x_i \neq x_j. \qquad (3.168)$$

Taking internal deformations of the operators into account, Schäfer shows that $G_{ij}(r) = 0$ follows if x_i and x_j do not differ by an integer.

The three point function (3.164) depends because of translational and rotational invariance only on the distances $|r_{12}|$, $|r_{13}|$, $|r_{23}|$ (we assume the operators $\mathcal{O}$ to be rotational invariant). The distance $|r_{12}|$ transforms into

$$|r'_{12}| = |r_{12}|(1 - ar_1 - ar_2) \qquad (3.169)$$

With the choice $ar_1 = ar_2 = -ar_3$ which can always be arranged by a proper choice of the origin and of a, eqn (3.164) yields

$$G_{ijk}(|r_{12}|, |r_{13}|, |r_{23}|) = (1 + 2(x_i + x_j - x_k)(ar_3)) \times G_{ijk}(|r_{12}|(1 + 2(ar_3)), |r_{13}|, |r_{23}|) \qquad (3.170)$$

Thus for fixed $|r_{13}|$ and $|r_{23}|$ the correlation function is proportional to $|r_{12}|^{-x_i-x_j+x_k}$. Similar arguments hold for $|r_{23}|$ and $|r_{13}|$, thus

$$G_{ijk}(|r_{12}|, |r_{13}|, |r_{23}|) = C_{ijk} |r_{12}|^{-x_i-x_j+x_k} |r_{13}|^{-x_i+x_j-x_k} |r_{23}|^{x_i-x_j-x_k} \quad (3.171)$$

Note that for $|r_{23}| \ll |r_{12}|$ one has $|r_{13}| \approx |r_{12}|$ and

$$G_{ijk} = C_{ijk} |r_{12}|^{-2x_i} |r_{23}|^{-x_j-x_k+x_i} \quad (3.172)$$

which can also be concluded from the product expansion (3.144), (3.150) and (3.168).

F. Trivial (Gaussian) fixed point

It is rather easy to calculate fixed points of the form

$$H^* = Vu_0^* + \frac{1}{2}\int_q u_2^*(q)\, S_q S_{-q} \quad (3.173)$$

and their eigenoperators $\tilde{\mathcal{O}}$ from the RG rquation (2.89). It turns out that $h(0)$ has to be an integer. The fixed points for $h(0) \leqslant 0$ describe a noncritical behaviour. They will be discussed in Section III.G. In this section primarily the case $h(0) > 0$ will be discussed. The fixed point for $h(0) = 1$ describes a critical behaviour. It is called trivial or Gaussian fixed point. The fixed points for larger values of $h(0)$ have an infinite number of relevant eigenoperators for $d \leqslant 4$. For small q one finds for $h(0) \geqslant 0$

$$u_2^*(q) \sim q^{2h(0)} \quad (3.174)$$

since $u_2^*(q)$ obeys

$$2h(q)\,(1 - u_2^*(q))\, u_2^*(q) - q\nabla u_2^*(q) = 0. \quad (3.175)$$

The local eigenoperators can be written

$$\tilde{\mathcal{O}}_{nr} = \exp(-\mathcal{Y})\frac{1}{n!}\int_{q_1 \ldots q_n} P_r(q_1, \ldots q_n) \prod_{i=1}^{n} (\varphi(q_i)\, S_{q_i}) \quad (3.176)$$

where

$$\mathcal{Y} = \frac{1}{2}\int_q \frac{1}{u_2^*(q)} \frac{\delta^2}{\delta S_q \delta S_{-q}} \quad (3.177)$$

$$\varphi(q) = [u_2^*(q)\,(1 - u_2^*(q))/Cq^{2h(0)}]^{1/2} \quad (3.178)$$

and P_r is a homogeneous polynomial of order r

$$P_r(\lambda q_1, \ldots \lambda q_n) = \lambda^r P(q_q, \ldots q_n). \quad (3.179)$$

The exponent x of $\tilde{\mathcal{O}}_{nr}$ is

$$x = \left(\frac{d}{2} - h(0)\right) n + r. \tag{3.180}$$

The translational invariant eigenoperators $\mathcal{O}^*$ can be easily obtained from (3.176) by insertion of the factor $(2\pi)^d \delta^d(q_1 + \ldots + q_n)$. Part of this calculation has been given by Wilson and Kogut (1974).

From (2.78) and (2.89) one obtains for a Hamiltonian

$$H = Vu_0 + \frac{1}{2}\int_q u_2(q)\, S_q S_{-q} \tag{3.181}$$

the derivatives

$$\frac{du_0}{dl} = du_0 + \int_q h(q)\,(u_2(q) - 1) \tag{3.182a}$$

$$\frac{\partial u_2(q)}{\partial l} = 2h(q)\,(1 - u_2(q))\, u_2(q) - q\nabla u_2(q) \tag{3.182b}$$

The fixed point obeys $\partial u_2^*/\partial l = 0$ which yields (3.175) with the solution

$$u_2^*(q) = \frac{1}{1 + \exp[-\int^q h(p)\, dp^2/p^2]} \tag{3.183}$$

Separating $h(0)$ from $h(q)$ one obtains

$$u_2^*(q) = \frac{Cq^{2h(0)}}{Cq^{2h(0)} + \exp[-\int_0^q (h(p) - h(0))/p^2\, dp^2]} \tag{3.184}$$

where C is an arbitrary positive constant. For short-range interactions $u_2^*(q)$ has to be analytic. Therefore $h(0)$ is an integer. The usual critical fixed point is described by $h(0) = 1$. This corresponds to $u_2^*(q) \sim q^2$ as it appears in the Landau model.

The local perturbations $\tilde{\mathcal{O}}$ are obtained from (3.102)

$$\mathcal{L}\tilde{\mathcal{O}} = -x\tilde{\mathcal{O}}$$

Substituting H^* into (3.6) yields

$$\int_q \Gamma_q \frac{\delta\tilde{\mathcal{O}}}{\delta S_q} + \int_q h(q) \frac{\delta^2\tilde{\mathcal{O}}}{\delta S_q\, \delta S_{-q}} = -x\tilde{\mathcal{O}} \tag{3.185}$$

with

$$\Gamma_q = \left(q\nabla S_q + \frac{d}{2} S_q\right) + h(q)\,(1 - 2u_2^*(q))\, S_q. \tag{3.186}$$

After some straightforward calculation one finds the commutator

$$\left[\mathscr{Y}, \int_q \Gamma_q \frac{\delta}{\delta S_q}\right] = -\int_q h(q) \frac{\delta^2}{\delta S_q \, \delta S_{-q}} \tag{3.187}$$

where one makes use of (3.175). From (3.187) one deduces

$$\exp \mathscr{Y} \left(\int_q \Gamma_q \frac{\delta \tilde{\mathcal{O}}}{\delta S_q} + \int_q h(q) \frac{\delta^2 \tilde{\mathcal{O}}}{\delta S_q \, \delta S_{-q}} \right) = \int_q \Gamma_q \frac{\delta}{\delta S_q} \exp(\mathscr{Y}) \, \tilde{\mathcal{O}}. \tag{3.188}$$

Substitution of (3.188) into (3.185) yields the eigenvalue problem

$$\int_q \Gamma_q \frac{\delta}{\delta S_q} \exp(\mathscr{Y}) \, \tilde{\mathcal{O}} = -x \exp(\mathscr{Y}) \, \tilde{\mathcal{O}}. \tag{3.199}$$

This equation can be solved easily for $\exp(\mathscr{Y})\tilde{\mathcal{O}}$. The solutions are homogeneous polynomials in S

$$\exp(\mathscr{Y}) \, \tilde{\mathcal{O}} = \frac{1}{n!} \int_{q_1 \ldots q_n} v_n(q_1 \ldots q_n) \, S_{q_1} \ldots S_{q_n} \tag{3.190}$$

since the operator $\int \Gamma \delta / \delta S$ conserves the degree in S. Moreover the equation for v_n separates in the variables

$$\sum_{i=1}^{n} \left[-\frac{d}{2} - q_i \nabla_i + h(q_i)\,(1 - 2u_2^*(q_i)) \right] v_n = -x v_n. \tag{3.191}$$

Using (3.175) one shows that the general solution of (3.191) reads

$$v_n(q_1, \ldots q_n) = P_r \; (q_1, \ldots q_n) \sum_i \varphi(q_i) \tag{3.192}$$

with x given by (3.180) which confirms (3.176)–(3.180). The function $\varphi(q)$ is chosen so that it approaches $\varphi(0) = 1$ for $h(0) > 0$. The complete set of eigenoperators $\tilde{\mathcal{O}}$ has been calculated.

The function $u_2^*(q)$ tends to unity for large q since the integral in (2.60) has to converge. To avoid ultraviolet divergencies for the operators $\tilde{\mathcal{O}}$ one requires that $\varphi(q)$ decays for large q faster than any power of q. Then it follows from (3.178) that $1 - u_2^*(q)$ approaches zero faster than any power of q. This can be easily obtained since Wilson's choice of $h(q) = c + 2q^2$, eqn (1.90) does it. Equation (3.184) yields (for $c = 1$)

$$u_2^*(q) = \frac{Cq^2}{Cq^2 + \exp(-q^2)} \tag{3.193}$$

The operator $\mathscr{Y}$, eqn (3.177), has a simple meaning in terms of diagrams.

The factor $1/u_2^*(q)$ is the bare Green's function. Application of $\mathscr{Y}$ on the four-point interaction yields

$$\mathscr{Y}\frac{1}{4!}\int_q v_4(q_1, q_2, q_3, -q_1 - q_2 - q_3)\, S_{q_1}S_{q_2}S_{q_3}S_{-q_1-q_2-q_3}$$

$$= \frac{1}{2!}\int_q \Sigma'(q)\, S_q S_{-q} \tag{3.194}$$

where

$$\Sigma'(q) = \frac{1}{2}\int_{q'} \frac{1}{u_2^*(q')} v_4(q, -q, q', -q') \tag{3.195}$$

is the first order contribution in v_4 to the self energy Σ. Therefore

$$\exp(-\mathscr{Y})\frac{1}{4!}\int_q v_4 S^4 \tag{3.196}$$

does not contribute to the self-energy in first order. In general $\mathscr{O}_n^*$ does not contribute to the self-energy in the first order. Even more general $\mathscr{O}_n^*$ does not contribute to the irreducible m-vertex function in first order provided $n \neq m$.

The operators with $r = 0$, that is $P_0(q_1, \ldots q_n) = 1$ can be explicitly given. With

$$I_0 = \frac{1}{2}\int_q \varphi^2(q)/u_2^*(q) \tag{3.197}$$

equation (3.176) yields (3.198)

$$\tilde{\mathscr{O}}_{n0} = \frac{1}{n!}\tilde{\mathscr{O}}_1^n - \frac{I_0}{(n-2)!}\tilde{\mathscr{O}}_1^{n-2} + \ldots$$

$$= \frac{I_0^{n/2}}{n!} H_n\left(\tilde{\mathscr{O}}_1/(2I_0^{1/2})\right)$$

where

$$\tilde{\mathscr{O}}_1 = \int_q \varphi(q)\, S_q \tag{3.199}$$

and H_n is the Hermite polynomial of degree n. Hermite polynomials for the eigenoperators occur also within approximations (Wegner, 1972b; Nicoll, Chang and Stanley, 1974b).

The scaling exponents y are obtained from (3.115) and (3.180)

$$y = d - \left(\frac{d}{2} - h(0)\right)n - r \tag{3.200}$$

The operators $\tilde{\mathcal{O}}_{n0}$, eqn (3.198) are scaling operators for $h(0) > 0$. This can be seen as follows: From (3.173) one obtains

$$\mathcal{G}_{\text{tra}}(\Phi)\, H^* = \int_q \left(\Phi_q u_2^*(q)\, S_{-q} - \frac{\delta \Phi_q}{\delta S_q} \right) \tag{3.201}$$

By means of the operator $\mathcal{Y}$ the term $\delta\Phi/\delta S$ can be transformed away

$$\mathcal{G}_{\text{tra}}(\Phi)\, H^* = \exp(-\mathcal{Y}) \int_q u_2^*(q)\, S_{-q} \exp(\mathcal{Y})\, \Phi_q. \tag{3.202}$$

If $\tilde{\mathcal{O}}_{n0}$ were a redundant operator, that is, if the relation

$$\tilde{\mathcal{O}}_{n0} = \mathcal{G}_{\text{tra}}(\Phi)\, H^* \tag{3.203}$$

would hold then this would imply

$$\frac{1}{n!} \int_{q_1 \ldots q_n} \prod_i (\varphi(q_i)\, S_{q_i}) = \int_q u_2^*(q)\, S_{-q} \exp(\mathcal{Y})\, \Phi_q \tag{3.204}$$

Apparently $\exp(\mathcal{Y})\Phi$ has to be of order $(n-1)$ in S

$$\exp(\mathcal{Y})\, \Phi_q = \frac{1}{(n-1)!} \int_{q_2 \ldots q_n} w(q; q_2 \ldots q_n)\, S_{q_2} \ldots S_{q_n} \tag{3.205}$$

Then the equation

$$\varphi(q_1) \ldots \varphi(q_n) = u_2^*(q_1)\, w(q_1; q_2 \ldots q_n) + u_2^*(q_2)\, w(q_2; q_1, q_3 \ldots q_n) + \ldots \tag{3.206}$$

had to hold which for $q_1 = q_2 = \ldots = 0$ leads to the contradiction

$$1 = 0 \tag{3.207}$$

since $u_2{}^*(0) = 0$ provided $h(0) > 0$. (We allow only for functions which are analytic since otherwise $\mathcal{G}_{\text{tra}}(\Phi)$ will generate long-range interactions for general H). Therefore the operators $\tilde{\mathcal{O}}_{n0}$ are scaling operators. For these operators we have

$$y = d - \left(\frac{d}{2} - h(0) \right) n. \tag{3.208}$$

If $d \leqslant 2h(0)$ then there is an infinite number of relevant operators and one can hardly approach the critical point. Therefore

$$d > 2h(0) \tag{3.209}$$

has to be required. Operators even in S have even n. These operators conserve the symmetry. Since $n = 0$ describes the special operator the number of symmetry conserving relevant operators for $h(0) = 1$ is

$$\begin{aligned} &1 &&\text{for } 4 \leqslant d \\ &2 &&\text{for } 3 \leqslant d \leqslant 4 \\ &\sigma - 1 &&\text{for } d_\sigma \leqslant d < d_{\sigma-1} \end{aligned} \tag{3.210}$$

with

$$d_\sigma = \frac{2\sigma}{\sigma - 1}. \tag{3.211}$$

Actually we considered only $r = 0$. For $n = 0$ and 1 only $r = 0$ yields a translational invariant eigenoperator. All operators with $r = 1$ can be expressed as gradients. Thus they do not yield translational invariant eigenoperators. Therefore one is left with $n \geqslant 2$ and $r \geqslant 2$. The largest exponent y within this class of operators is

$$y = 0 \text{ for } n = 2, r = 2. \tag{3.212}$$

All other exponents y are negative. The non-vanishing translationally invariant operators for $n = 2$, $r = 2$ are ($h(0) = 1$)

$$\mathcal{O}^*_{2,2,\alpha\beta} = \frac{1}{2}\int_q q_\alpha q_\beta \varphi^2(q)\, S_q S_{-q} + \text{const.} \tag{3.213}$$

The operator

$$\mathcal{O}^*_{2,2} = \sum_\alpha \mathcal{O}^*_{2,2,\alpha\alpha} = \frac{1}{2}\int_q q^2\varphi^2(q)\, S_q S_{-q} + \text{const.} \tag{3.214}$$

is redundant since

$$\mathcal{O}^*_{2,2} = \int_q \left(\Phi_q \frac{\delta H^*}{\delta S_q} - \frac{\delta \Phi_q}{\delta S_q}\right) \tag{3.215}$$

with

$$\Phi_q = \frac{1}{2}(1 - u_2^*(q))\, S_q \tag{3.216}$$

The operators $\Sigma C_{\alpha\beta}\mathcal{O}^*_{22\alpha\beta}$ apart from $\mathcal{O}^*_{22}$ are scaling operators. The proof can be given analogously to that for $\tilde{\mathcal{O}}_n$. There is a second way to see that the operators $\mathcal{O}^*_{2,2\alpha\beta}$ have eigenvalues $y = 0$. The fixed point, eqn (3.184), is not uniquely defined since the constant C is arbitrary. Therefore the perturbation

$$\frac{\partial H^*}{\partial C} = \frac{1}{C}\mathcal{O}^*_{22} \tag{3.217}$$

is eigenoperator with $y = 0$. Moreover we may replace Cq^2 by $\Sigma C_{\alpha\beta} q_\alpha q_\beta$. Then

$$\left.\frac{\partial H^*}{\partial C_{\alpha\beta}}\right|_{C_{\alpha\beta} = C\delta_{\alpha\beta}} = \frac{1}{C}\,\mathcal{O}^*_{2,2,\alpha\beta} \tag{3.218}$$

which again implies that $\mathcal{O}^*_{2,2,d\beta}$ has vanishing exponent y.

G. Comments

Finally we comment on three topics: on the Bose condensation, on the non-critical fixed point and on the stress tensor.

1. *Comment on Bose condensation*

It is often assumed that quantum corrections are negligible for critical exponents since the condensation takes place on a macroscopic level where the quantization can be neglected. This can be supported for the Bose condensation by the following argument. One may eliminate certain Fourier components a_q^+, a_q of the creation and annihilation operators from $\exp(-H)$ since any Fourier component commutes with all other Fourier components. The procedure may be similar to the RG equation with sharp momentum cut-off equation (2.101). Of course the generator of the elimination will be more complicated since we deal with non-commuting operators. Again we hope to approach a fixed point. For this purpose the constant c in (2.100) has to be adjusted properly. If we add a source term ha_q^+ to H_0 then from

$$H_0' = H_0 + ha_q^+ \tag{3.219}$$

it follows that

$$H_l' = H_l + h\exp\left[\left(\frac{d}{2} + c\right)l\right] a^+_{qe^l} \text{ for } q\,e^l < 1. \tag{3.220}$$

Thus we have

$$y_h = \frac{d}{2} + c \tag{3.221}$$

and eqn (3.127) yields

$$2 - \eta = 2c \tag{3.222}$$

Therefore the Hamiltonians H_l is expressed in terms of operators

$$a_q^{(l)+} = \exp\left[-\left(\frac{d}{2} + c\right)l\right] a^+_{qe^{-l}}. \tag{3.223}$$

Since one starts out from the commutator

$$[a_k, a_q^+] = (2\pi)^d \, \delta^d(q - k) \tag{3.224}$$

one obtains

$$[a_k^{(l)}, a_q^{(l)\,+}] = (2\pi)^d \exp(-\,2cl)\, \delta^d(q - k). \tag{3.225}$$

Therefore besides the parameters u of the Hamiltonian in a form similar to (2.55) where S has to be replaced by a^+ and a, one needs a number t for the description of the system

$$[a_k, a_q^+] = (2\pi)^d \, t\delta^d(q - k) \tag{3.226}$$

which according to (3.225) decays with the exponent

$$y_t = \eta - 2. \tag{3.227}$$

Therefore we have *one* extra parameter for the quantization which turns out to be irrelevant and the effective Hamiltonian for the Bose gas may very well approach the critical fixed point of a classical system under the RG. Because of its smallness one can hardly expect to observe the quantum corrections due to y_t. This agrees with the calculations of η by Abe and Hikami (1974) for a Bose system.

2. *The noncritical fixed point*

In Section III.F the trivial fixed point for $h(0) = 1$ has been discussed. Here it will be shown that the fixed points for $h(0) \leqslant 0$ have the property that any operator can be expressed as the sum of a constant (special operator) and a redundant operator. There is no scaling operator to this fixed point apart from the volume. This implies that any infinitesimal perturbation added to the fixed point Hamiltonian can be absorbed into the fixed point Hamiltonian by means of a transformation of the variables leaving only an additional constant term. The free energy consists only of the regular contribution of (3.69).

$$F = \mu_0 \tag{3.228}$$

The scaling part is absent since F does not depend on fields μ of the redundant operators. Such a regular behaviour is expected for a system away from criticality. (Taking into account the nonlinear terms of the RG equation, then (3.228) has to be replaced by $F = g_0$, compare (5.17).) Equation (3.184) can be written

$$u_2^*(q) = \frac{C}{C + q^{-2h(0)} \exp\left(-\int_0^q (h(p) - h(0))/p^2 \, dp^2\right)} \tag{3.229}$$

and requiring

$$\varphi(0) = 1 \tag{3.230}$$

one obtains

$$\varphi(q) = [u_2^*(q)(1 - u_2^*(q))\, Cq^{2h(0)}]^{1/2}, \qquad h(0) < 0 \tag{3.231}$$

$$\varphi(q) = (C + 1)\,[u_2^*(q)(1 - u_2^*(q))/C]^{1/2} \qquad h(0) = 0 \tag{3.232}$$

instead of (3.178) and

$$x = \left(\frac{d}{2} + h(0)\right) n + r \tag{3.233}$$

instead of (3.180).

In the limit $C \to \infty$ the functions $u_2^*(q)$ and $\varphi(q)$ converge uniformly to the limit

$$u_2^*(q) = 1 \tag{3.234}$$

$$\varphi(q) = \exp\left(-\frac{1}{2}\int_0^q (h(p) - h(0))/p^2\, \mathrm{d}p^2\right) \tag{3.235}$$

(The functions $u_2^*(q)$ and $\varphi(q)$ for $h(0) > 0$, eqns (3.184) and (3.178), do not have this property). It is not really necessary to perform this limit but it yields a simple form for the fixed point

$$H^* = \frac{1}{2}\int_q S_q S_{-q} = \frac{1}{2}\int S^2(r)\, \mathrm{d}^d r \tag{3.236}$$

which describes a system of noninteracting spins.

It remains to be shown that any operator is the sum of a constant and a redundant operator

$$\mathcal{O} = c + \mathcal{G}_{\mathrm{tra}}(\Phi)\, H^* \tag{3.237}$$

We proceed as in Section III.F and use (3.202). Then

$$\exp(\mathcal{Y})\, \mathcal{O} = c + \int_q u_2^*(q)\, S_{-q} \exp(\mathcal{Y})\, \Phi_q \tag{3.238}$$

has to be shown. Expanding

$$\exp(\mathcal{Y})\, \mathcal{O} = \sum \frac{1}{n!}\int_{q_1 \ldots q_n} p_n(q_1, \ldots q_n)\, S_{q_1} \ldots S_{q_n} \tag{3.239}$$

and using the parametrization (3.205) one obtains with eqn (3.234)

$$p_n(q_1, \ldots q_n) = c\delta_{n0} + (w_{n-1}(q_1; q_2, \ldots) + w_{n-1}(q_2; q_1, \ldots) + \ldots) \tag{3.240}$$

a solution

$$w_{n-1}(q_1; q_2 \ldots q_n) = \frac{1}{n} p_n(q_1, \ldots q_n) \tag{3.241}$$

$$c = p_0 \tag{3.242}$$

which proves that apart from a constant any operator of this fixed point is redundant.

3. *The stress tensor*

A homogeneous transformation of the sample

$$r'_i = (\delta_{ij} - A_{ij})\, r_j \tag{3.243}$$

$$q'_i = (\delta_{ij} + A_{ij})\, q_j \tag{3.244}$$

$$V' = (1 - \Sigma A_{ii})\, V \tag{3.245}$$

(A infinitesimal) has three effects:

(i) it changes the volume of the system. Since the isotropic dilatation has already been considered, the restriction of constant volume, that is

$$\Sigma A_{ii} = 0 \tag{3.246}$$

will be made in this section.

(ii) it rotates the sample. This rotation is given by

$$r'_i = (\delta_{ij} - \tfrac{1}{2}(A_{ij} - A_{ji}))\, r_j, \tag{3.247}$$

(iii) it distorts the sample anisotropically. This distortion is described by

$$r'_i = (\delta_{ij} - \tfrac{1}{2}(A_{ij} + A_{ji}))\, r_j. \tag{3.248}$$

In terms of the new variables

$$S'_{q'} = S'_{(1+A)q} = S_q \tag{3.249}$$

the new system is described by the Hamiltonian

$$H'\{S'\} = H\{S\{S'\}\} \tag{3.250}$$

which is obtained from H by replacing S_q by $S'_{q'}$. Then obviously

$$\frac{\delta H'}{\delta S'_{(1+A)q}} = \frac{\delta H}{\delta S_q} \tag{3.251}$$

holds. Further Ψ_q is substituted by

$$\Psi'_{q'}\{S'\} = \Psi_q\{S\{S'\}\} \tag{3.252}$$

from which one obtains

$$\frac{\delta \Psi'_{q'}}{\delta S'_{q'}} = \frac{\delta \Psi_q}{\delta S_q}. \tag{3.253}$$

Due to the transformation at constant volume one has

$$\int_{q'} = \int_q \tag{3.254}$$

so that

$$\mathscr{G}_{\text{tra}}\{\psi'\}H' = \mathscr{G}_{\text{tra}}\{\psi\}H \tag{3.255}$$

holds. Moreover one has

$$\mathscr{G}_{\text{dil}}H'\{S'\} = \mathscr{G}_{\text{dil}}H\{S\} \tag{3.256}$$

since the dilatation transformation $\exp(l\mathscr{G}_{\text{dil}})$ multiplies any S by $\exp(\frac{1}{2}dl)$ and any q by e^l which commutes with the transformation (3.249). Thus from $\mathscr{G}\{\Psi^*\}H^* = 0$ it follows that

$$\mathscr{G}\{\Psi'^*\}H'^* = 0 \tag{3.257}$$

This after all is not surprising since the critical behaviour of a system should not depend on the chosen coordinates in r-space, (3.243). Therefore we may say that H and H' are equivalent. (This is an extension of the equivalence defined in Section III.B).

Since the transformation from H^* to H'^* is not local, one cannot conclude that $H'^* - H^*$ is a redundant operator (in the sense of Section III.B). Application of $\mathscr{G}\{\psi\}$ on H'^* yields

$$\begin{aligned}\mathscr{G}\{\Psi\}H'^* &= \mathscr{G}_{\text{dil}}H'^* + \mathscr{G}_{\text{tra}}\{\Psi\}H'^* \\ &= \mathscr{G}_{\text{dil}}H'^* + \mathscr{G}_{\text{tra}}\{\Psi'^*\}H'^* + \mathscr{G}_{\text{tra}}\{\Psi - \Psi'^*\}H'^* \\ &= \mathscr{G}_{\text{tra}}\{\Psi - \Psi'^*\}H^* + O(A^2).\end{aligned} \tag{3.258}$$

Since $\Psi - \Psi'^* = \Psi - \Psi^* + \Psi^* - \Psi'^*$ is of order A the Hamiltonian H'^* could be replaced by H^* in the last line of (3.258). (Note that $\Psi = \Psi\{H'^*\}$). Since the application of $\mathscr{G}$ on H'^* yields a redundant operator (3.259) the perturbation

$$H' - H^* = \sum A_{ij} \int_q q_i \nabla_j S_q \frac{\partial H}{\delta S_q} \tag{3.260}$$

is either redundant or a marginal scaling operator apart from a redundant operator. This agrees with the discussion by Kawasaki (1966) and is one of the ingredients of Kadanoff's operator algebra for the two-dimensional Ising model (Kadanoff, 1969). The operator

$$T_{ij} = \tfrac{1}{2} \int_q \left(q_i \nabla_j S_q + q_j \nabla_i S_q - \frac{2}{d} \delta_{ij} q \nabla S_q \right) \frac{\delta H}{\delta S_q} \tag{3.261}$$

is called a stress tensor.

As a translationally invariant operator acting on the whole volume, the stress tensor can be transformed away using the homogeneous transformation as defined above. So far the stress tensor behaves similarly as a redundant operator and the exponent $y = 0$ has no physical meaning. The corresponding local operators however cannot be generated by such a transformation. So far the exponent $y = 0$ is meaningful and enters in the exponents of correlations with the (local or wave-vector dependent) stress tensor. The stress tensor of the trivial fixed point is

$$T_{ij} = -\tfrac{1}{2} \int_q \left(q_i \frac{\partial u^*}{\partial q_j} + q_j \frac{\partial u^*}{\partial q_i} - \frac{2}{d} \delta_{ij} q \nabla u^* \right) S_q S_{-q} \tag{3.262}$$

which can be written with (3.175) and (3.213), (3.214)

$$T_{ij} = -2Ch(0) \left(\mathcal{O}^*_{2,2,ij} - \frac{1}{d} \delta_{ij} \mathcal{O}^*_{2,2} \right) - C \int_q (h(q) - h(0))\left(q_i q_j - \frac{1}{d} \delta_{ij} q^2 \right) \varphi^2(q) S_q S_{-q} \tag{3.263}$$

The first term on the right hand side of (3.263) is a scaling operator with $y = 0$ as explained after (3.216), the second term is a redundant operator with

$$\Phi_q = -C \frac{h(q) - h(0)}{u_2^*(q)} \left(q_i q_j - \frac{1}{d} \delta_{ij} q^2 \right) \varphi^2(q) \tag{3.264}$$

since $h(q) - h(0) = O(q^2)$.

IV. Nonlinear Theory: The Nontrivial Fixed Point

A. The nonlinear term

In the previous sections the behaviour of the free energy and of the correlation functions close to criticality has been discussed on the basis of the linearized RG equations. In the following sections the quadratic term (3.7)

$$\mathcal{Q}\Delta H = -\int_q h(q) \frac{\delta \Delta H}{\delta S_q} \frac{\delta \Delta H}{\delta S_{-q}}$$

in eqn (3.5)

$$\mathcal{G}(H^* + \Delta H) = \mathcal{L}\Delta H + \mathcal{Q}\Delta H$$

will be taken into account. The terms "linearized theory" and "nonlinear theory" have to be distinguished from the terms "linear RG" and "nonlinear RG" which indicate whether

$$H'_0 = H_0 + \lambda S_q \tag{4.1}$$

transforms exactly into

$$H'_l = H_l + \lambda \, e^{y_h l} S_{qe^l} \tag{4.2}$$

or not (compare Bell and Wilson, 1974). Equation (2.101) is a linear RG equation for $qe^l < 1$ (compare (3.219) and (3.220)), equation (2.89) is a nonlinear RG equation. We will use the term "RG nonlinear in H" equivalent to "nonlinear theory". The quadratic term is parametrized by substituting

$$\Delta H = \sum \mu_i \mathcal{O}_i^* \tag{4.3}$$

in (3.7) and expanding

$$\mathcal{Q}(\Sigma_i \, \mu_i \mathcal{O}_i^*) = \tfrac{1}{2} \Sigma \, a'_{njk} \mu_j \mu_k \mathcal{O}_n^* \tag{4.4}$$

with

$$\sum a'_{njk} \mathcal{O}_n^* = -2 \int_q h(q) \frac{\delta \mathcal{O}_j^*}{\delta S_q} \frac{\delta \mathcal{O}_k^*}{\delta S_{-q}}. \tag{4.5}$$

This yields the basic nonlinear equation for the RG transformation

$$\frac{d\mu_i}{dl} = y_i \mu_i + \tfrac{1}{2} \sum a'_{ijk} \mu_j \mu_k. \tag{4.6}$$

On the basis of this RG equation we will consider:

(i) The bifurcation of fixed points at the value of a parameter (like the dimensionality) where an operator becomes marginal (IV). The ε-expansion for nontrivial fixed points serves as an example.

(ii) Exact scaling of the free energy in terms of scaling fields g, and the breakdown of scaling under certain conditions (logarithms, phase transitions of infinite order) (Section V).

(iii) Correlation functions. To discuss them the quadratic term $\mathcal{Q}$ will be considered for inhomogeneous perturbations (Section VI).

Sections IV, V, and VI are to a large extent independent of each other. Section V is based on (4.6). In Section V.D. the coefficients (4.8) will be used. Section VI makes use of Sections V.A and to some extent of Section V.B. Section VI.A can be read immediately after Section III.D.

Branching off of the nontrivial fixed points is considered in section IV on the basis of the RG equation with smooth momentum cut-off. Shukla and Green (1974, 1975) calculated the nontrivial fixed point in powers of ε up to order ε^2, critical exponents to order ε and η to order ε^2. For the calculation of η to order ε^2 see also Golner and Riedel (1975a), Rudnick (1975). The trivial fixed point and its eigenoperators as derived in section III.F serves as a starting point.

Golner and Riedel (1975b) derived scaling field equations of the type (4.6) and used them to calculate critical exponents for $d = 3$.

In the remainder of this section some of the coefficients a' for the trivial fixed point are determined which will be used in sections IV.B, C, and V.D.

Evaluation of some Coefficients a′ for the Trivial Fixed Point

The coefficients $a'_{n0,\, j0,\, k0}$ and $a'_{2,\,2,\,j0,\,k0}$, of the operators, eqn (3.214)

$$\mathcal{O}^*_{2,\,2} = exp\,(-\mathscr{Y})\tfrac{1}{2}\int_q q^2\varphi^2(q)S_qS_{-q}$$

and (compare (3.198))

$$\mathcal{O}^*_{n0} = \exp\,(-\mathscr{Y})\frac{(2\pi)^d}{n!}\int_{q_1\ldots q_n}\prod_{i=1}^{n}(\varphi(q_i)S_{q_i})\delta^d(q_1 + \ldots + q_n) \tag{4.7}$$

are evaluated for the trivial fixed point.

Provided that any of the three numbers $j - 1$, $k - 1$, and n does not exceed the sum of the two other numbers and provided the sum of these three numbers is even then one obtains

$$a'_{n0,\, j0,\, k0} = -\frac{2n!}{m!(j - 1 - m)!(k - 1 - m)!}\,\rho_m(0) \tag{4.8}$$

with

$$2m = j + k - n - 2 \tag{4.9}$$

$$\rho_m(q) = \int_{q'}\rho_{m-1}(q - q')\frac{\varphi^2(q')}{u_2^*(q')} \tag{4.10}$$

$$\rho_0(q) = h(q)\varphi^2(q). \tag{4.11}$$

Otherwise $a'_{n0,\, j0,\, k0}$ vanishes. The coefficients a'_{22} are obtained to be

$$a'_{2,\,2,\,j0,\,j0} = -\frac{4}{(j-2)!}\frac{\partial}{\partial q^2}\rho_{j-2}(q)\big|_{q=0}. \tag{4.12}$$

Golner and Riedel (1975b) calculated the expansion coefficients a' using a projection operator formula derived from the integrated form of the RG equation (2.89). Here we use that according to (3.176) the operators $\exp\,(\mathscr{Y})\mathcal{O}^*_{n,r}$ have a particular simple form. They contain only terms of order S^n and can be written

$$\exp(\mathscr{Y})\mathcal{O}^*_{nr} = \frac{(2\pi)^d}{n!}\int_{q_1\ldots q_n} P_r(q_1,\ldots q_m)\prod_{i=1}^{n}(\varphi(q_i)S_{q_i})\delta^d(q_1+\ldots+q_n). \tag{4.13}$$

Thus one calculates preferably $\exp(\mathscr{Y})\mathcal{O}$ for an expansion of $\mathcal{O}$ into the eigenoperators. The coefficients a' can be determined from

$$\sum_{nr} a'_{nr,\,jr',\,kr''}\exp(\mathscr{Y})\mathcal{O}^*_{nr} = -2\exp(\mathscr{Y})\int_q h(q)\frac{\delta\mathcal{O}^*_{jr'}}{\delta S_q}\frac{\delta\mathcal{O}^*_{kr''}}{\delta S_{-q}}. \tag{4.14}$$

$\exp(\mathscr{Y})$ acts on both $\mathcal{O}^*_{jr'}$ and $\mathcal{O}^*_{kr''}$. To separate these actions one expresses for the time being $\mathcal{O}^*_{jr'}$ as a function of S and $\mathcal{O}^*_{kr''}$ as a function of T. At the end one remembers $T = S$. Then $\mathscr{Y}$ is written

$$\mathscr{Y} = \mathscr{Y}_S + \mathscr{Y}_T + \mathscr{Y}_{ST} \tag{4.15}$$

with

$$\mathscr{Y}_S = \tfrac{1}{2}\int_q \frac{1}{u_2^*(q)}\frac{\delta^2}{\delta S_q \delta S_{-q}}, \tag{4.16}$$

similarly for $\mathscr{Y}_T$, and

$$\mathscr{Y}_{ST} = \int_q \frac{1}{u_2^*(q)}\frac{\delta^2}{\delta S_q \delta T_{-q}} \tag{4.17}$$

Equation (4.14) reads

$$\sum_{nr} a'_{nr,\,jr',\,kr''}\exp(\mathscr{Y})\mathcal{O}^*_{nr} = -2\exp(\mathscr{Y}_{ST})$$

$$\int_q h(q)\frac{\delta}{\delta S_q}(\exp(\mathscr{Y}_S)\mathcal{O}^*_{jr'}(S))\frac{\delta}{\delta T_{-q}}(\exp(\mathscr{Y}_T)\mathcal{O}^*_{kr''}(T))\Big|_{T=S} \tag{4.18}$$

The integral in (4.18) is of order S^{j+k-2} (after putting $T = S$). On the other hand $\exp(\mathscr{Y})\mathcal{O}^*_{nr}$ is of order S^n. Therefore the contributions of order S^n are obtained by applying $\mathscr{Y}_{ST}$ m times with

$$n = j + k - 2 - 2m, \tag{4.19}$$

since the operator $\mathscr{Y}$ lowers the exponent of S by 2. Thus a' vanishes if $j+k-n$ is odd. It vanishes if $m < 0$, that is, if $n > j + k - 2$. Moreover m must not be larger than j and k since otherwise $\delta^{m+1}\exp(\mathscr{Y})\mathcal{O}^*_{j,k}/\delta S^{m+1} = 0$. Therefore any of the three numbers $j - 1$, $k - 1$, and m must not exceed the sum of the two other numbers (triangular condition) and the sum of al' three has to be even. Then

$$\sum_r a'_{nr,\, jr',\, kr''} \exp(\mathscr{Y})\mathcal{O}^*_{nr} = -\frac{2}{m!}(\mathscr{Y}_{ST})^m \int_q h(q)$$

$$\frac{\delta}{\delta S_q}(\exp(\mathscr{Y}_S)\mathcal{O}^*_{jr'})\frac{\delta}{\delta T_{-q}}(\exp(\mathscr{Y}_T)\mathcal{O}^*_{kr''})\bigg|_{T=S} \tag{4.20a}$$

otherwise

$$a'_{nr,\, jr',\, kr''} = 0 \tag{4.20b}$$

For

$$\exp(\mathscr{Y})\mathcal{O}^*_{jr'} = \frac{(2\pi)^d}{j!}\int_{q_1 \ldots q_j} u_{jr'}(q_1, \ldots q_{j-1}) \prod S_q \, \delta^d(q_1 + \ldots + q_j) \tag{4.21}$$

and similarly for $\mathcal{O}^*_{kr''}$ one obtains

$$\int_q h(q)\frac{\delta}{\delta S_q}(\exp(\mathscr{Y}_S)\mathcal{O}^*_{jr'})\frac{\delta}{\delta T_{-q}}(\exp(\mathscr{Y}_T)\mathcal{O}^*_{kr''})$$

$$= \frac{(2\pi)^d}{(j-1)!(k-1)!}\int_{q_1 \ldots q_{j+k-2}} u_{jr'}(q_1, \ldots q_{j-1})u_{kr''}(q_j, \ldots q_{j+k-2}) \tag{4.22}$$

$$\times h(q_1 + \ldots + q_{j-1}) \prod_1^{j-1} S_q \prod_j^{j+k-2} T_q \delta^d(q_1 + \ldots + q_{j+k-2})$$

In particular for $\mathcal{O}^*_{j0}$ and $\mathcal{O}^*_{k0}$ this yields the expression

$$\frac{(2\pi)^d}{(j-1)!(k-1)!}\int_{q_1 \ldots q_{j+k-2}} \rho_0(q_1 + \ldots + q_{j-1}) \prod_{i=1}^{j+k-2} \varphi(q_i)$$

$$\prod_1^{j-1} S_q \prod_j^{j+k-2} T_q \delta^d(q_1 + \ldots + q_{j+k-2}) \tag{4.23}$$

where ρ_0 is given by (4.11). The application of $(\mathscr{Y}_{ST})^m$ yields

$$\sum_r a'_{nr,\, j0,\, k0} \exp(\mathscr{Y})\mathcal{O}^*_{nr} = -\frac{2(2\pi)^d}{m!(j-m-1)!(k-m-1)!}$$

$$\int_{q_1 \ldots q_n} \rho_m(q_1 + \ldots + q_{j-m-1}) \prod_{i=1} (\varphi(q_i)S_{q_i})\delta^d(q_1 + \ldots + q_n) \tag{4.24}$$

where ρ_m is given by (4.10). From (4.24) one concludes (4.8) since

$$u_{nr}(q_1 = 0, \ldots q_n = 0) = \delta_{r0}. \tag{4.25}$$

Finally the coefficient $a'_{2,2,j0,j0}$ will be determined. Equation (4.19) yields $m = j - 2$ and (4.24) reduces to

$$\sum_r a'_{2,r,j0,j0}\,\mathcal{O}^*_{2,r} = -\frac{2}{(j-2)!}\int_q u(q)S_qS_{-q} \tag{4.26}$$

with

$$u(q) = \rho_{j-2}(q)\varphi^2(q). \tag{4.27}$$

Since the operators $\exp(\mathscr{Y})\mathcal{O}^*_{2,r}$ are of the type

$$\tfrac{1}{2}\int_q P_r(q)\varphi^2(q)S_qS_{-q} \tag{4.28}$$

the factor $-(4/(j-2)!)\rho_{j-2}(q)$ has to be expanded in powers of q. The coefficient in front of q^2 yields a'_{22} eqn (4.12).

B. The nontrivial fixed point in order ε

The trivial fixed point has at the dimensions d_σ eqn (3.211), a marginal operator with the following symmetry of the Hamiltonian: rotational invariance in space and invariance against reversal of the spins, $S_q \to -S_q$. This is the operator $\mathcal{O}^{*(0)}_{2\sigma,0}$, eqn (4.7), where the superscript 0 indicates that the operator refers to the trivial fixed point. The exponent $y^{(0)}_{2\sigma,0}$ of $\mathcal{O}^{*(0)}_{2\sigma,0}$ is

$$y^{(0)}_{2\sigma,0} = (\sigma - 1)\varepsilon \tag{4.29}$$

with

$$d = d_\sigma - \varepsilon. \tag{4.30}$$

Due to this marginal operator a second fixed point exists which can be expanded in powers of ε. The basic idea is the following: Consider (4.6)

$$\frac{\mathrm{d}\mu_i}{\mathrm{d}l} = y^{(0)}_i\mu_i + \tfrac{1}{2}\sum a'_{ijk}\mu_j\mu_k.$$

The exponent $y^{(0)}_i$ as well as the coefficients a' depend on the dimensionality. If now *one* exponent $y^{(0)}_u$ becomes zero at a certain dimensionality but differs from zero at other dimensionalities and

$$a'_{uuu} \neq 0 \tag{4.31}$$

then one obtains a second fixed point for which

$$\mu^*_u = O(y^{(0)}_u) \tag{4.32}$$

and

$$\mu^*_i = O(y^{(0)2}_u) \qquad i \neq u. \tag{4.33}$$

Indeed substituting the estimations (4.32) and (4.33) into the fixed point equation

$$\frac{d\mu_i^*}{dl} = y_i^{(0)}\mu_i^* + \tfrac{1}{2}\sum a'_{ijk}\mu_j^*\mu_u^* = 0 \tag{4.34}$$

and keeping all terms up to order ε^2 one obtains

$$y_u^{(0)}\mu_u^* + \tfrac{1}{2}a'_{uuu}\mu_u^{*2} = 0 \tag{4.35}$$

$$y_i^{(0)}\mu_i^* + \tfrac{1}{2}a'_{iuu}\mu_u^{*2} = 0 \tag{4.36}$$

from which one obtains the nontrivial solution

$$\mu_u^* = -\frac{2y_u^{(0)}}{a'_{uuu}} + O(y_u^{(0)2}) \tag{4.37}$$

$$\mu_i^* = -\frac{a_{iuu}}{2y_i^{(0)}}(\mu_u^*)^2 + O(y_u^{(0)3}) \tag{4.38}$$

which proves (4.32) and (4.33). Linearization around the new fixed point yields for

$$\delta\mu_i = \mu_i - \mu_i^* \tag{4.39}$$

to order ε

$$\frac{\partial\delta\mu_i}{\partial l} = \sum Y_{ij}\delta\mu_j \tag{4.40}$$

with

$$Y_{ij} = y_i^{(0)}\delta_{ij} + a'_{iju}\mu_u^*. \tag{4.41}$$

The eigenvalues of the matrix Y are the exponents of the eigenoperators for the new fixed point. To obtain the eigenvalues in first order in $y_u^{(0)}$ it is sufficient to diagonalize the matrices consisting of the matrix elements of operators which are degenerate at $y_u^{(0)} = 0$. As far as these matrices are triangular matrices, one obtains

$$y_i = y_i^{(0)} + a'_{iiu}\mu_u^* = y_i^{(0)} + y_u^{(0)}p_i \tag{4.42}$$

where

$$p_i = -\frac{2a'_{iiu}}{a'_{uuu}}. \tag{4.43}$$

In particular one obtains

$$y_u = -y_u^{(0)} \tag{4.44}$$

Therefore a *relevant* operator $\mathcal{O}_u^{(0)*}$ corresponds to an *irrelevant* operator $\mathcal{O}_u^*$ for the new fixed point and vice versa.

We return to the calculation of the nontrivial fixed points from the trivial fixed point. The new fixed point reads in order ε

$$H^* = H^{*(0)} - 2(\sigma - 1)\varepsilon \frac{1}{a'_{2\sigma,\,0,\,2\sigma,\,0,\,2\sigma,\,0}} \mathcal{O}^{*(0)}_{2\sigma\;0}\,. \tag{4.45}$$

One obtains the exponent

$$y_{n,\,0} = d - \frac{n}{2}(d - 2) + (\sigma - 1)\varepsilon p_n + \mathcal{O}(\varepsilon^2) \tag{4.46}$$

for the operator $\mathcal{O}^*_{n,\,0}$ with

$$p_n = -\frac{2a'_{n,\,0,\,n,\,0,\,2\sigma,\,0}}{a'_{2\sigma,\,0,\,2\sigma,\,0,\,2\sigma,\,0}} = -2\frac{n!\sigma!}{(n - \sigma)!(2\sigma)!} \tag{4.47}$$

from (4.8) which yields

$$a'_{n,\,0,\,n,\,0,\,2\sigma,\,0} = -\frac{2n!}{(n - \sigma)!\sigma!(\sigma - 1)!}\rho_{\sigma-1}(0) \tag{4.48}$$

the exponents p_n do not depend on the choice of the function $u_2^{*(0)}(q)$ since the integral $\rho_{\sigma-1}(0)$ cancels in the final expression and only a combinatorial factor is left.

Equation (4.46) holds provided the matrix Y for degenerate eigenvalues is triangular. According to (3.180) and (3.211) at $d = d_\sigma$ the operators

$$\mathcal{O}^*_{n-k(\sigma-1),\,k} \tag{4.49}$$

are degenerate for $k = 1, 2, 3, \ldots$. Actually to show that $y_{n,\,0}$, eqn (4.46) is correct it is sufficient to show that

$$a'_{n,\,0,\,n-k(\sigma-1),\,k,\,2\sigma,\,0} = 0 \qquad \text{for} \qquad k > 0 \tag{4.50}$$

From eqn (4.20a) it is obvious that a', equation (4.50), vanishes for odd k. Due to the triangular condition after (4.19) a' vanishes for $k > 2$. For $k = 2$, eqn (4.20a) has to be applied with $m = 2$. One finds that

$$\sum_r a'_{nr} \exp(\mathcal{Y})\mathcal{O}^*_{nr} = \int_{q_1 \ldots q_{n-1}} \tilde{v}(q_1, \ldots q_{n-1}) S^n_q \tag{4.51}$$

with $\tilde{v} = \mathcal{O}(q^2)$. Therefore (4.50) is fulfilled and (4.46) holds.

Until now we did not take into account that the operators $\mathcal{O}^{*(0)}_{2,\,2,\,\alpha\beta}$ have the exponent $y^{(0)}_{2,\,2} = 0$. The triangular condition after (4.19) guarantees that

$$a'_{2,\,2,\,2,\,2,\,2,\,2\sigma,\,0} = 0 \tag{4.52}$$

for $\sigma > 2$. For $\sigma = 2$ one deduces easily from (4.20a) that (4.52) is fulfilled. Therefore (4.34) is fulfilled with $\mu^*_{i,i} = 0$.

The exponents (4.46) were calculated: for $\sigma = 2$ approximately from Wilson's Recurrence Relation (Wilson, 1971b) by Wilson and Fisher (1972) for some n and by Wegner (1972b) for general n, exactly for some n by Wilson and Fisher (1972), from the RG equation with smooth momentum cut-off by Shukla and Green (1974) for some n, for $\sigma = 3$ by Stephen and McCauley (1973), for $\sigma = 3$ and $\sigma = 4$ for some n by Chang, Tuthill, and Stanley (1974), from an approximate Recursion Relation for general σ and n by Nicoll, Chang and Stanley (1974b). It is remarkable that all the mentioned approximate RG equations yield correct exponents y in first order in ε.

C. Exponent η

In the preceding section the nontrivial fixed point has been calculated to lowest order in ε. One may continue (at least in principle) to calculate the fixed point to higher order by iterating (4.34)

$$\mu^*_i = -\frac{1}{2y_i^{(0)}} \sum a'_{i\,jk} \mu^*_j \mu^*_k \tag{4.53}$$

provided $y_i^{(0)} \neq 0$. However the exponent $y_{2,2}^{(0)}$ of the operator $\mathcal{O}^{*(0)}_{2,2}$ vanishes for any ε. Since $a'_{2,\,2,\,2\sigma,\,0,\,2\sigma,\,0}$ does not vanish the equation

$$y_{2,2}^{(0)} \mu^*_{2,\,2} + \tfrac{1}{2} \sum a'_{2,\,2,\,j,\,k} \mu^*_j \mu^*_{u_k} = 0 \tag{4.54}$$

cannot be fulfilled in order ε^2. On the other hand one should not expect that eqn (4.34) can be solved in any order in ε since $h(0)$ is related to the critical exponent η (eqn (3.128))

$$h(0) = 1 - \tfrac{1}{2}\eta$$

Therefore one should allow for

$$h(q) = h_0(q) - \tfrac{1}{2}\eta \tag{4.55}$$

where $h_0(q)$ is the cut-off function for the trivial fixed point. Then the generator of the RG obeys

$$\mathcal{G}_\eta H - \mathcal{G}_0 H = -\tfrac{1}{2}\eta \int_q \left[\left(S_q - \frac{\delta H}{\delta S_{-q}} \right) \frac{\delta H}{\delta S_q} + \frac{\delta^2 H}{\delta S_q \delta S_{-q}} - 1 \right] \tag{4.56}$$

In lowest order one substitutes $H^{*(0)}$ on the right-hand side of (4.56) which yields

$$\mathcal{G}_\eta H - \mathcal{G}_0 H = -\eta C \mathcal{O}^{*(0)}_{2,2} + \eta O(\mu) \tag{4.57}$$

Substitution in (4.54) yields

$$-\eta C + \tfrac{1}{2} a'_{2,2,2\sigma,0,2\sigma,0} (\mu_u^*)^2 + O(\mu_u^{*3}, \eta\mu_u^*) = 0 \tag{4.58}$$

from which one obtains with (4.37) and (4.29)

$$\eta = \frac{2(\sigma - 1)^2 a'_{2,2,2\sigma,0,2\sigma,0}}{C a'^2_{2\sigma,0,2\sigma,0,2\sigma,0}} \varepsilon^2 + O(\varepsilon^3) \tag{4.59}$$

In the remainder of this section eqn (4.59) will be evaluated: From (4.8) and (4.12) one obtains

$$\eta = -\frac{2(\sigma - 1)^2 (\sigma!)^4 (\sigma - 1)!^2}{(2\sigma - 2)!(2\sigma)!^2 C \rho^2_{\sigma-1}(0)} \frac{\partial}{\partial q^2} \rho_{2\sigma-2}(q)\bigg|_{q=0} \varepsilon^2 + \dots \tag{4.60}$$

where $\rho_m(q)$ is defined (eqns (4.10) and (4.11))

$$\rho_m(q) = \frac{(2\pi)^d}{C^m} \int_{q_1 \dots q_{m+1}} \tilde{\rho}(q_1) \dots \tilde{\rho}(q_m) h(q_{m+1}) \varphi^2(q_{m+1}) \times \delta^d(q - q_1 - \dots - q_{m+1}) \tag{4.61}$$

with (eqn (3.178))

$$\tilde{\rho}(q) = \frac{C\varphi^2(q)}{u_2^*(q)} = \frac{1 - u_2^*(q)}{q^2} \tag{4.62}$$

On the other hand $h\varphi^2$ can be expressed due to (3.175)

$$Ch(q)\varphi^2(q) = -\tfrac{1}{2} q \nabla \tilde{\rho}(q) - \tilde{\rho}(q), \tag{4.63}$$

thus

$$\rho_m(q) = \frac{(2\pi)^d}{C^{m+1}} \int_{q_1 \dots q_{m+1}} \tilde{\rho}(q_1) \cdot \dots \tilde{\rho}(q_m) \times (-\tfrac{1}{2} q_{m+1} \nabla \tilde{\rho}(q_{m+1}) - \tilde{\rho}(q_{m+1}))\, \delta^d(q - q_1 - \dots q_{m+1}). \tag{4.64}$$

This multiple convolution integral can be simplified by a transformation into real space. With

$$\hat{\rho}(r) = \int_q e^{iqr} \tilde{\rho}(q) \tag{4.65}$$

one obtains

$$\rho_m(q) = \frac{1}{C^{m+1}} \int d^d r\, (\hat{\rho}(r))^m \left(\left(\frac{d}{2} - 1 \right) \hat{\rho}(r) + \tfrac{1}{2} r \nabla \hat{\rho}(r) \right) e^{iqr} \tag{4.66}$$

which yields for $d = d_\sigma$, $m = \sigma - 1$, $q = 0$ the integral of a complete derivative

$$\rho_{\sigma-1}(0) = \frac{\Omega}{2\sigma C^\sigma} \int_0^\infty \mathrm{d}r \frac{\partial}{\partial r} (r^{d-2}\,\hat{\rho}(r))^\sigma \qquad (4.67)$$

$$= \frac{\Omega}{2\sigma C^\sigma} \lim_{r\to\infty} (r^{d-2}\,\hat{\rho}(r))^\sigma$$

where Ω is the surface of the unit sphere in d_σ dimensions. The derivative

$$\frac{\partial}{\partial q^2} \rho_{2\sigma-2}(q)\bigg|_{q=0} = \frac{1}{2d} \frac{\partial^2}{\partial q^2} \rho_{2\sigma-2}(q)\bigg|_{q=0}$$

$$= -\frac{1}{2dC^{2\sigma-1}} \int \mathrm{d}^d r\, r^2 (\hat{\rho}(r))^{2\sigma-2} \left(\left(\frac{d}{2} - 1\right)\hat{\rho}(r) + \frac{1}{2} r \nabla \hat{\rho}(r)\right) \qquad (4.68)$$

is the integral of a complete derivative, too

$$\frac{\partial}{\partial q^2} \rho_{2\sigma-2}(q)\bigg|_{q=0} = -\frac{\Omega}{4d(2\sigma-1)\,C^{2\sigma-1}} \int_0^\infty \mathrm{d}r \frac{\partial}{\partial r} (r^{d-2}\,\hat{\rho}(r))^{2\sigma-1}$$

$$= -\frac{\Omega}{4d(2\sigma-1)\,C^{2\sigma-1}} \lim_{r\to\infty} (r^{d-2}\,\hat{\rho}(r))^{2\sigma-1} \qquad (4.69)$$

Insertion of (4.67) and (4.69) in (4.60) yields

$$\eta = \frac{2(\sigma-1)^2\,\sigma!^6\,\varepsilon^2}{(2\sigma-1)!\,(2\sigma)!^2\,d\Omega} \lim_{r\to\infty} (r^{d-2}\,\hat{\rho}(r))^{-1} \qquad (4.70)$$

The calculation of this limit is a problem of electrostatics in d dimensions. From

$$\Delta\hat{\rho}(r) = -\int_q q^2\, \mathrm{e}^{iqr}\, \hat{\rho}(q) = -\int_q (1 - u_2^*(q))\, \mathrm{e}^{iqr} = -p(r) \qquad (4.71)$$

it becomes apparent that the "charge" $p(r)$ is a regular function in r rapidly decaying for large r. To obtain the "electrostatic potential" $\hat{\rho}(r)$ one calculates the "electric field"

$$E(r) = -\nabla\hat{\rho}(r) \qquad (4.72)$$

and obtains for a sphere of radius r around the origin

$$\int \mathrm{div}\, E(r')\, \mathrm{d}^d r' = r^{d-1}\,\Omega E(r) = \int p(r')\, \mathrm{d}^d r'. \qquad (4.73)$$

For large r the "field" $E(r)$ approaches

$$E(r) = \frac{1}{r^{d-1}\,\Omega}\int p(r')\,\mathrm{d}^d r' = \frac{1}{r^{d-1}\,\Omega} \tag{4.74}$$

where the integral is the "total charge" $1 - u_2^*(0) = 1$. Integration of eqn (4.72) yields

$$\lim_{r\to\infty}(r^{d-2}\,\hat{\rho}(r)) = \frac{1}{\Omega(d-2)} \tag{4.75}$$

from which we obtain finally

$$\eta = \frac{\sigma!^6}{(2\sigma)!^3}[2(\sigma-1)\,\varepsilon]^2 + O(\varepsilon^3). \tag{4.76}$$

For $\sigma = 2$ one obtains the well-known result

$$\eta = \frac{1}{54}\varepsilon^2 + O(\varepsilon^3) \tag{4.76}$$

derived at first by Wilson and Fisher (1972). From the RG equation with smooth momentum cut-off it has been derived for arbitrary cut-off function $h(q)$ by Shukla and Green (1975) and Rudnick (1975) and numerically for several functions $h(q)$ by Golner and Riedel (1975a). That η does not depend on the cut-off function is a consequence of the invariance properties of the RG (Section III.B and Wegner (1974)). For $\sigma = 3$ one obtains

$$\eta = \frac{1}{500}\varepsilon^2 + O(\varepsilon^3) \tag{4.78}$$

as obtained by Stephen and McCauley (1973).

Three remarks are in order:

(i) *Higher order calculations.* Although the calculation of the other exponents in order ε^2 and of η in order ε^3 along these lines should be possible, it is probably easier to calculate H^* parametrized by the $u_n^*(q_1, \ldots q_n)$ in order ε^2 first (as has been done by Shukla and Green (1974)) and to linearize around this fixed point.

(ii) *Odd-dominated fixed points.* The operator $\mathcal{O}^*_{2\sigma-1,0}$ becomes marginal at dimensionality $d_{\sigma-1/2}$. One may expand around these dimensions (at least formally) nontrivial fixed points. In this case however a'_{iiu} vanishes so that for all eigenoperators (including y_h) $y_i - y_i^{(0)} = O(\mu_u^{*2})$ and $\mu_u^* = O(\varepsilon^{1/2})$. Calculations for these odd-dominated fixed points have been performed exactly for $\sigma = 2$ by Mack (1973) and approximately for general σ by Nicoll, Chang, and Stanley (1974b).

(iii) *Remark on* $h(0)$ *and* $\mathcal{O}^*_{2,2}$. The parameter $h(0)$ defines the exponent y_h uniquely. Therefore the RG equation can only be solved for $h(0) = 1 - \frac{1}{2}\eta$. In the calculation presented above $h(0)$ is defined uniquely due to the appearance of the redundant operator $\mathcal{O}^*_{2,2}$ with $y = 0$. One shows: If the fixed point equation

$$\mathcal{G}\{\Psi(a, H^*)\}\, H^* = 0 \tag{4.79}$$

can be solved only for discrete values a, then there exists a redundant operator which for this RG equation has exponent $y = 0$. (Compare Bell and Wilson (1974)). The proof runs as follows: From (4.79) one obtains

$$\begin{aligned}\mathcal{G}\{\Psi\,(a + \delta, H^* + \delta \textstyle\sum \mu_i \mathcal{O}^*_i)\}\,(H^* + \delta \textstyle\sum \mu_i \mathcal{O}^*_i)\\ = \delta \mathcal{G}_{\text{tra}}\{\Psi'\,(a, H^*)\}\, H^* + \delta \textstyle\sum y_i \mu_i \mathcal{O}^*_i + \mathcal{O}(\delta^2)\\ = \delta \sum_i (\alpha_i + y_i \mu_i)\, \mathcal{O}^*_i + \mathcal{O}(\delta^2)\end{aligned} \tag{4.80}$$

where $\Psi'(a) = \partial\Psi(a)/\partial a$ and the α_i are the expansion coefficients of $\mathcal{G}_{\text{tra}}\{\Psi'\}H^*$ in the eigenoperators $\mathcal{O}^*_i$ of $\mathcal{G}\{\Psi\}$.

The left-hand side of (4.80) does not vanish for infinitesimal δ since (4.79) can only be solved for discrete values a. Therefore at least one of the coefficients $\alpha_i + y_i\mu_i$ does not vanish for whatever μ_i. This implies $\alpha_i \neq 0$, $y_i = 0$. Since $\mathcal{G}_{\text{tra}}\{\Psi'\}\, H^*$ is a redundant operator, the coefficients α vanish for all scaling operators. Therefore one of the redundant operators, in our case $\mathcal{O}^*_{22}$, has exponent $y_i = 0$.

D. Isotropic $\hat{n}$-component model

The calculations of the previous sections can be generalized to isotropic $\hat{n}$-component models. Denoting the components by S_α, the RG equatoin (2.89) can be written

$$\begin{aligned}\mathcal{G}H = \sum_\alpha \int_q \left(q\nabla S_{q\alpha} + \frac{d}{2} S_{q\alpha}\right) \frac{\delta H}{\delta S_{q\alpha}}\\ \times \sum_\alpha \int_q h(q) \left[\left(S_{q\alpha} - \frac{\delta H}{\delta S_{-q\alpha}}\right) \frac{\delta H}{\delta S_{q\alpha}} + \frac{\delta^2 H}{\delta S_{q\alpha}\, \delta S_{-q\alpha}} - 1\right] + dV \frac{\partial H}{\partial V}\end{aligned} \tag{4.81}$$

The trivial fixed point Hamiltonian reads

$$H^* = V u^*_0 + \tfrac{1}{2} \int_q u^*_2(q)\, S_q S_{-q} \tag{4.82}$$

where $u_2^*(q)$ obeys (3.175). The eigenperturbations can be written

$$\tilde{\mathcal{O}} = \exp(-\mathcal{Y}) \int_{q_1 \ldots q_n} \varphi(q_1) \ldots \varphi(q_n) P_r(q_1, \ldots q_n) \times \sum_\alpha Y_{\alpha_1, \ldots \alpha_n} S_{q_1 \alpha_1} \cdot \ldots S_{q_n \alpha_n} \tag{4.83}$$

where now

$$\mathcal{Y} = \frac{1}{2} \int_q \frac{1}{u_2^*(q)} \sum_\alpha \frac{\delta^2}{\delta S_{q\alpha} \delta S_{-q\alpha}} \tag{4.84}$$

and P_r and φ are given by (3.178) and (3.179). It is useful to classify the eigenoperators according to their behaviour under rotation. For this purpose one chooses

$$Y^{(\hat{l}, m)}_{\alpha_1, \ldots \alpha_{\hat{l}+2m}} = \mathrm{sym}\,(Y^{(\hat{l})}_{\alpha_1, \ldots \alpha_{\hat{l}}} \delta_{\alpha_{\hat{l}+1}, \alpha_{\hat{l}+2}} \cdot \ldots \delta_{\alpha_{\hat{l}+2m-1}, \alpha_{\hat{l}+2m}}) \tag{4.85}$$

with

$$\sum_\alpha Y^{(\hat{l})}_{\alpha_1, \ldots \alpha_{\hat{l}-2}, \alpha, \alpha} = 0 \tag{4.86}$$

where $Y^{(\hat{l})}$ is symmetric in the indices $\alpha_1, \ldots, \alpha_{\hat{l}}$ and sym denotes a summation over all $(\hat{l} + 2m)!$ permutations of $\alpha_1, \ldots \alpha_{\hat{l}+2m}$ and division by $(\hat{l} + 2m)!$ Apparently $\hat{l} = 0$ describes a scalar, $\hat{l} = 1$ a vector, $\hat{l} = 2$ a symmetric tensor of rank 2, etc. We do not consider antisymmetric tensors here since we are mainly interested in the operators with $r = 0$, that is $P_r = 1$. We denote these operators by $\tilde{\mathcal{O}}_{m\hat{l}}$. At $d = d_\sigma$ all operators with $2\sigma = \hat{l} + 2m$ become marginal. The isotropic fixed point contains only scalar contributions

$$H^* = H^{*(0)} + \mu^*_{\sigma 0} \mathcal{O}^*_{\sigma 0} + O(\varepsilon^2). \tag{4.87}$$

To calculate the exponents $y_{m\hat{l}}$ one has to evaluate $a'_{m\hat{l}, m\hat{l}, \sigma 0}$ (4.42). In analogy to (4.20a) one obtains

$$\sum_r a'_{m\hat{l}r, m\hat{l}, \sigma 0} \mathcal{O}^*_{m\hat{l}r} = -\frac{2}{(\sigma - 1!)} (\mathcal{Y}_{ST})^{\sigma - 1}$$

$$\sum_\alpha \int_q h(q) \frac{\delta}{\delta S_{q\alpha}} (\exp(\mathcal{Y}_S) \mathcal{O}^*_{m\hat{l}}) \frac{\delta}{\delta T_{-q\alpha}} (\exp(\mathcal{Y}_T) \mathcal{O}^*_{\sigma 0})$$

$$= -\frac{2(2\pi)^d}{(\sigma - 1)!} \int_{q_1 \ldots q_{2m+\hat{l}}} \delta^d(q_1 + \ldots + q_{2m+\hat{l}}) \rho_{\sigma-1}(q_1 + \ldots + q_{2m+\hat{l}-\sigma})$$

$$\prod_i \varphi(q_i) \cdot U(S_{q_1}, \ldots S_{q_{2m+\hat{l}}}) \tag{4.88}$$

where U is symmetric in the first $2m + \hat{l} - \sigma$ and the last σ variables, respectively and where for

$$S_i = \begin{cases} S & i \leqslant \hat{l} + 2m - \sigma \\ T & i > \hat{l} + 2m - \sigma \end{cases} \tag{4.89}$$

one obtains

$$U = \left(\sum_\alpha \frac{\partial^2}{\partial S_\alpha \, \partial T_\alpha} \right)^\sigma \hat{Y}^m{}_l(S)\,(T^2)^\sigma \tag{4.90}$$

with

$$\hat{Y}_{m\hat{l}}(S) = \sum_\alpha Y^{\hat{l}}_{\alpha_1 \ldots \alpha_l} S^\alpha{}_1 \cdot \ldots S_{\alpha_{\hat{l}}} (S^2)^m \tag{4.91}$$

One shows by induction in k that

$$\left(\sum_\alpha \frac{\partial^2}{\partial S_\alpha \, \partial T_\alpha} \right)^k \hat{Y}(S)\,(T^2)^\sigma = \sum_r 2^{k-2r} \frac{\sigma!\,k!}{(k-2r)!\,r!\,(\sigma - k + r)!} \Delta^r_S (T\nabla_S)^{k-2r} \times (T^2)^{\sigma - k + r}\; \hat{Y}(S). \tag{4.92}$$

With

$$\Delta\, \hat{Y}_{m\hat{l}} = 4m\left(\frac{\hat{n}}{2} + \hat{l} + m - 1 \right) \hat{Y}_{m-1\hat{l}} \tag{4.93}$$

and

$$(T\nabla_S)^{k-2r}\, \Delta^r Y_{m\hat{l}} \bigg|_{T=S} = \frac{(2m + \hat{l} - 2r)!}{(2m + \hat{l} - k)!} \Delta^r Y_{m\hat{l}} \tag{4.94}$$

one obtains finally

$$U\big|_{T=S} = 2^\sigma \sigma! \zeta_{m\hat{l}}\, Y_{m\hat{l}} \tag{4.95}$$

with

$$\zeta_{m\hat{l}} = \sum_r \frac{\sigma!\,m!\,((\hat{n}/2) + \hat{l} + m - 1)!\,(\hat{l} + 2m - 2r)!}{(\sigma - 2r)!\,r!^2 (m-r)!\,((\hat{n}/2) + \hat{l} + m - r - 1)!\,(\hat{l} + 2m - \sigma)!}. \tag{4.96}$$

This yields

$$a'_{m\hat{l},\, m\hat{l},\, \sigma 0} = -2^{\sigma+1} \sigma \rho_{\sigma-1}(0)\, \zeta_{m\hat{l}} \tag{4.97}$$

and

$$y_{m\hat{l}} = d - (d-2)\left(m + \frac{\hat{l}}{2} \right) - 2(\sigma - 1) \frac{\zeta_{m\hat{l}}}{\zeta_{\sigma 0}} \varepsilon + O(\varepsilon^2) \tag{4.98}$$

For $\sigma = 2$ this reduces to

$$y_{m\hat{l}} = 4 - \hat{l} - 2m + \varepsilon(\tfrac{1}{2}\hat{l} + m - \zeta_{m\hat{l}}/(\hat{n} + 8)) = O(\varepsilon^2) \quad (4.99)$$

$$\zeta_{m\hat{l}} = (\hat{l} + 2m)(\hat{l} + 2m - 1) + m(\hat{n} + 2m + 2\hat{l} - 2) \quad (4.10)$$

which has been derived for some values m, $\hat{l}$ by Fisher and Pfeuty (1972) and for general m, $\hat{l}$ by Wegner (1972b) from Wilson's approximate recurrence relation (Wilson, 1971b), for some values m, $\hat{l}$ by Wilson (1972) and $m = 0$, general $\hat{l}$ by Houghton and Wegner (1974) from exact RG techniques.

For $\sigma = 3$ one obtains

$$y_{m\hat{l}} = 3 - \frac{\hat{l}}{2} - m + \varepsilon(\tfrac{1}{2}\hat{l} + m - \zeta_{m\hat{l}}/(3(3\hat{n} + 22))) + O(\varepsilon^2), \quad (4.101)$$

$$\zeta_{m\hat{l}} = (\hat{l} + 2m - 2)[(\hat{l} + 2m)(\hat{l} + 2m + 1) + 3m(\hat{n} + 2m + 2\hat{l} - 2)] \quad (4.102)$$

Some of these exponents were derived by Stephen and McCauley (1973).

A similar combinatorial analysis yields η in order ε^2

$$\eta = 2\varepsilon^2(\sigma - 1)^2 \frac{\sigma!((\hat{n}/2) + \sigma - 1)!}{(\hat{n}/2)!\zeta_{\sigma 0}^2} + O(\varepsilon^3) \quad (4.103)$$

which reduces to

$$\eta = \frac{(\hat{n} + 2)}{2(\hat{n} + 8)^2}\varepsilon^2 + O(\varepsilon^3), \quad \sigma = 2 \quad (4.104)$$

$$\eta = \frac{(\hat{n} + 2)(\hat{n} + 4)}{12(3\hat{n} + 22)^2}\varepsilon^2 + O(\varepsilon^3), \quad \sigma = 3 \quad (4.105)$$

(Wilson, 1972; Stephen and McCauley, 1973). One notes that η approaches zero for $\hat{n} \to \infty$ and even σ but it does not vanish for odd σ in the limit $\hat{n} \to \infty$.

V. Nonlinear Theory: Homogeneous Systems

A. Scaling fields

Apart from certain exceptions which will be discussed in Sections V.E and V.F, the nonlinearities can be absorbed in scaling fields g_i (Wegner, 1972a)

$$\exp(\mathscr{G}l)\, H\{g_i\} = H\{g_i \exp(y_i l)\} \quad (5.1)$$

which leads to the exact homogeneity relation

$$F\{g_i\} = \exp(-dl)\, F\{g_i \exp(y_i l)\} \quad (5.2)$$

instead of the approximate one (3.66). In differential form these equations read

$$\mathscr{G}H = DH \quad (5.3)$$

and

$$\mathrm{d}F = DF \tag{5.4}$$

with the differential operator

$$D = \sum_i y_i g_i \frac{\partial}{\partial g_i} \tag{5.5}$$

One expects that in linear order $\mu_i = g_i$ and expands

$$\mu_i = g_i + \frac{1}{2!}\sum b_{ijk} g_j g_k + \frac{1}{3!}\sum b_{ijkm} g_j g_k g_m + \ldots \tag{5.6}$$

The coefficients b are obtained from (4.6)

$$\frac{\mathrm{d}\mu_i}{\mathrm{d}l} = y_i \mu_i + \tfrac{1}{2}\sum a'_{ijk}\, \mu_j \mu_k.$$

One substitutes eqn (5.6) in (4.6) using

$$\frac{\mathrm{d}g_i}{\mathrm{d}l} = y_i g_i \tag{5.7}$$

and obtains

$$\begin{aligned} y_i g_i + \tfrac{1}{2}\sum (y_j + y_k)\, b_{ijk} g_j g_k + \ldots \\ = y_i g_i + \tfrac{1}{2}\sum (a'_{ijk} + y_i b_{ijk})\, g_j g_k + \ldots \end{aligned} \tag{5.8}$$

Comparison of the coefficients in front of $g_j g_k$ yields the equations for the coefficients b

$$(y_j + y_k - y_i)\, b_{ijk} = a'_{ijk} \tag{5.9}$$

and similarly

$$(y_j + y_k + y_m - y_i)\, b_{ijkm} = \sum_s a'_{ijs} b_{skm} + 2 \text{ permutations} \tag{5.10}$$

etc. These equations can be solved provided $y_i \neq y_j + y_k$ and $y_i \neq y_j + y_k + y_m$, respectively. Similarly the terms of n-th order in g can be calculated if y_i differs from any sum of n exponents y. Thus one obtains a formal expansion in powers of g. We do not know whether sums like that on the right hand side of (5.10) exist.

Let us review the situation: Experimentally a small number of parameters $\tilde{\mu}_i^{\text{expt}}$ are available. The fields μ_i of the linearized theory depend on these parameters

$$\mu_i = \mu_i\{\tilde{\mu}^{\text{expt}}\} \tag{5.11}$$

in a simple way since the fields μ_i are linear in $1/T$ (temperature) and $\hbar/T$ where $\hbar$ may be a magnetic field or a chemical potential or any other action on the system which linearly enters the Hamiltonian. Inversion of the series (5.6) yields an expansion for g

$$g_i = \mu_i - \frac{1}{2}\sum b_{ijk}\mu_j\mu_k + \dots. \tag{5.12}$$

Equations (5.11) and (5.12) allow g_i to be expressed in terms of the experimental parameters $\tilde{\mu}^{\text{expt}}$

$$g_i = g_i\{\tilde{\mu}^{\text{expt}}\} \tag{5.13}$$

Criticality is now determined by the condition that all relevant scaling fields vanish

$$g_i^{\text{rel}}\{\tilde{\mu}^{\text{expt}}\} = 0. \tag{5.14}$$

This yields a finite set of equations for a finite number of experimentally available parameters. We assume that the fields g_i can be expanded in a Taylor series of the fields $\tilde{\mu}^{\text{expt}}$. They are curvilinear coordinates in the space of the fields $\tilde{\mu}^{\text{expt}}$.

One of the eigenoperators is the volume $\mathcal{O}_0^* = V$. From (4.5) it is obvious that

$$a'_{ij0} = 0. \tag{5.15}$$

Therefore $g_i - \mu_i$ does not depend on μ_o which yields with (3.67)

$$\left.\frac{\partial F}{\partial \mu_0}\right|_{\mu_i = \text{const}} = 1 = \left.\frac{\partial F}{\partial g_0}\right|_{g_i = \text{const}} \tag{5.16}$$

from which it follows that

$$F = g_0 + F_{\text{sing}}(g_0 = 0, \dots g_j \dots) \tag{5.17}$$

The function $g_0\{\mu\}$ is the regular part of the free energy. In analogy to (3.68) the choice

$$|g_E| \exp(y_E l) = 1 \tag{5.18}$$

yields the homogeneity law

$$F = g_0 + |g_E|^{2-\alpha} F_{\text{sing}}\left(g_0 = 0, g_E = \pm 1, \dots \frac{g_i}{|g_E|^{\Delta_i}} \dots\right) \tag{5.19}$$

The following example may serve as an illustration for the scaling fields: Suppose it is sufficient to consider only the two operators $\mathcal{O}_0^*$ and $\mathcal{O}_E^*$. Then the eqn (4.6) reduces to

$$\frac{\partial \mu_0}{\partial l} = d\mu_0 + q_0 \mu_E^2 \tag{5.20}$$

$$\frac{\partial \mu_E}{\partial l} = y_E \mu_E + a_E \mu_E^2 \tag{5.21}$$

with the abbreviation

$$a_i = \tfrac{1}{2} a'_{iEE}. \tag{5.22}$$

μ_E depends only on g_E. It can be determined from

$$\frac{\partial \mu_E}{\partial l} = y_E g_E \frac{\partial \mu_E}{\partial g_E} \tag{5.23}$$

which with (5.21) yields

$$\mu_E = \frac{g_E}{1 - (a_E/y_E)\, g_E} \tag{5.24}$$

The difference $\mu_0 - g_0$ depends only on g_E

$$\mu_0 = g_0 + s(g_E). \tag{5.25}$$

The equation for s is obtained by substituting (5.20) and (5.25) in

$$\frac{\partial \mu_0}{\partial l} = d g_0 \frac{\partial \mu_0}{\partial g_0} + y_E g_E \frac{\partial \mu_0}{\partial g_E} \tag{5.26}$$

which yields

$$y_E g_E \frac{\partial s}{\partial g_E} = ds + a_0 \frac{g_E^2}{(1 - (a_E/y_E)\, g_E)^2}. \tag{5.27}$$

Expanding s in powers of g_E one obtains

$$g_0 = \mu_0 + a_0 \sum_{n=2}^{\infty} \frac{n-1}{d - n y_E} \left(\frac{a_E}{y_E}\right)^{n-2} g_E^n \tag{5.28}$$

provided d/y_E is not integer.

The derivatives of F with respect to g_j are homogeneous functions

$$\frac{\partial F}{\partial g_j} = |g_E|^{2-\alpha-\Delta_j} F_j\left(g_0 = 0, g_E = \pm 1, \ldots \frac{g_i}{|g_E|^{\Delta_i}} \cdots\right) \tag{5.29}$$

On the other hand one has

$$\frac{\partial F}{\partial g_j} = \frac{1}{V}\left\langle \frac{\partial H}{\partial g_j} \right\rangle = \frac{1}{V} \langle \mathcal{O}_j \rangle \tag{5.30}$$

with the notation

$$\mathcal{O}_j = \partial H/\partial g_j. \tag{5.31}$$

Therefore the expectation value of $\mathcal{O}_j$ is given by the homogeneous function (5.29). It has to be noted however that $\mathcal{O}_j$ depends on the fields g

$$\mathcal{O}_j = \sum_i \frac{\partial \mu_i}{\partial g_j} \mathcal{O}_i^* = \mathcal{O}_j^* + \sum_{ik} b_{ijk} g_k \mathcal{O}_i^* + \ldots \tag{5.32}$$

Therefore the operators $\mathcal{O}_j^*$ do not obey an exact scaling law since correction terms proportional to g and higher powers of g occur.

The expectation value of the operator

$$\mathcal{O}^* = \sum_i \lambda_i \mathcal{O}_i^* \tag{5.33}$$

can be reduced to the expectation values of the operators $\mathcal{O}_j$ by inverting (5.32)

$$\mathcal{O}_i^* = \sum_j \frac{\partial g_j}{\partial \mu_i} \mathcal{O}_j = \mathcal{O}_i - \sum_{jk} b_{ijk} g_k \mathcal{O}_j + \ldots \tag{5.34}$$

to the expectation value of

$$\mathcal{O}^* = \sum_j \kappa_j \mathcal{O}_j \tag{5.35}$$

where

$$\kappa_j = \sum_i p_{ji}\{g\}\,\lambda_i \tag{5.36}$$

and

$$p_{ji}\{g\} = \partial g_j/\partial \mu_i. \tag{5.37}$$

κ_j itself may be expanded in the experimental variables. If for example one measures $\langle \mathcal{O}^* \rangle$ as a function of temperature then

$$\langle \mathcal{O}^* \rangle = \sum \kappa_j^{(0)} \langle \mathcal{O}_j \rangle + \tau \sum \kappa_j^{(1)} \langle \mathcal{O}_j \rangle + \ldots \tag{5.38}$$

with

$$\kappa_j = \kappa_j^{(0)} + \tau \kappa_j^{(1)} + \ldots \tag{5.39}$$

For $\langle \mathcal{O}_j \rangle$ itself one applies (5.30) which in leading order yields a contribution proportional $|\tau|^{\beta j}$, equation (3.99).

The irrelevant scaling fields g yield corrections similarly as discussed in connection with (3.92) which now reads

$$F = g_0 + f(g_E, g_h) + \sum_i{}' g_i f_i(g_E, g_h) + \tfrac{1}{2}\sum{}' g_i g_j f_{ij}(g_E, g_h) + \dots$$

$$= g_0 + |g_E|^{2-\alpha} f\left(\pm 1, \frac{g_h}{|g_E|^{\Delta}}\right)$$

$$+ \sum_i{}' g_i |g_E|^{2-\alpha-\Delta_i} f_i\left(\pm 1, \frac{g_h}{|g_E|^{\Delta}}\right) + \dots \tag{5.40}$$

where the prime indicates summation over indices of the irrelevant operators. In Section V.B it will be shown that F does not depend on the scaling fields of the redundant operators. Equation (5.40) applies only if a Taylor expansion in powers of the irrelevant operators is possible. The second derivative

$$\frac{\partial^2 F}{\partial g_j \, \partial g_k} = |g_E|^{-\gamma_{jk}} F_{jk}\left(g_0 = 0, g_E = \pm 1, \dots \frac{g_i}{|g_E|^{\Delta_i}} \dots\right) \tag{5.41}$$

equals

$$-\frac{1}{v}(\langle \mathcal{O}_j \mathcal{O}_k \rangle - \langle \mathcal{O}_j \rangle \langle \mathcal{O}_k \rangle - \langle \mathcal{O}_{jk} \rangle) \tag{5.42}$$

where

$$\mathcal{O}_{jk} = \frac{\partial^2 H}{\partial g_j \, \partial g_k} = \sum_i b_{ijk} \mathcal{O}_i + \dots \tag{5.43}$$

Thus the generalized susceptibility $\chi_{jk} = \langle \mathcal{O}_j \mathcal{O}_k \rangle - \langle \mathcal{O}_j \rangle \langle \mathcal{O}_k \rangle$ differs from the second derivative of F with respect to the scaling fields by the expectation value of an operator $\mathcal{O}_{jk}$ (Wegner, 1972; Ma, 1974).

B. Invariance properties

In Section III.B the operators are classified into two classes, scaling operators and redundant operators. The critical exponents of the redundant operators are spurious since they depend on the arbitrary choice of the renormalization group. Within the linear approximation the free energy does not depend on the fields μ of the redundant operators. Therefore the critical exponents of the redundant operators are meaningless. In Section V.A the nonlinear contributions were included in the homogeneity law for the free energy which is a homogeneous function of the scaling fields g. Accordingly one conjectures that the free energy depends only on the scaling fields of the scaling operators but not on those of the redundant operators. Indeed under certain presumptions listed below one shows that the operator $\mathcal{O}_\kappa\{g\}$ can be expressed as

$$\mathcal{O}_\kappa\{g\} = \mathcal{G}_{\text{tra}}(\Phi_\kappa\{g\})\, H \tag{5.44}$$

for the scaling fields g_κ of all redundant operators. Equation (5.44) implies that F does not depend on g_κ since

$$\frac{\partial F}{\partial g_\kappa} = \langle \mathcal{O}_\kappa\{g\}\rangle = -\frac{\partial}{\partial \delta} \ln \text{trace} \exp\,(-H - \delta\mathcal{G}_{\text{tra}}(\Phi_\kappa)\, H) = 0 \tag{5.45}$$

The transformation $H \to H + \delta\mathcal{G}_{\text{tra}}(\Phi_\kappa)\, H$ has been constructed in Section II.E as to leave the partition function invariant (in order δ).

The presumptions read: It is presupposed that

(i) the equations

$$K(\mathcal{D} + \Psi^*, \chi) - z\chi = \Theta \tag{5.46a}$$

$$\mathcal{G}_{\text{tra}}(\chi)\, H^* = 0 \tag{5.46b}$$

can be solved for given Θ which obeys

$$\mathcal{G}_{\text{tra}}(\Theta) H^* = 0 \tag{5.47}$$

and for any $z \neq z_i$ (for a set of numbers z_i),

(ii) the inequalities

$$y_i \neq y_{j_1} + \ldots + y_{j_n} \quad \text{for } n > 1 \tag{5.48}$$

and

$$z \neq y_{j_1} + \ldots + y_{j_n} \quad \text{for } n \geqslant 1 \tag{5.94}$$

hold.

The functions Θ, eqn (5.47) describe the equivalence transformations $\mathcal{G}_{\text{tra}}(\Theta)$ which leave the fixed point Hamiltonian invariant. If all function Θ are linear combinations of eigenfunctions Θ_i

$$K(\mathcal{D} + \Psi^*, \Theta_i) = z_i\Theta_i \tag{5.50}$$

then eqn (5.46) can be solved for $z \neq z_i$. It follows from (3.35) and (5.47) that

$$\mathcal{G}_{\text{tra}}\,(K\,(\mathcal{D} + \Psi^*, \Theta))\, H^* = 0. \tag{5.51}$$

Therefore the commutator in (5.50) transforms within the class of functions Θ defined by (5.47).

If for at least one scaling operator $\mathcal{O}_i^*$ the operator

$$\mathcal{O}'_j = \int_q \Theta_{iq}\, \delta\mathcal{O}_i^*/\delta S_q \tag{5.52}$$

is not redundant then $\mathcal{O}'$ is a scaling eigenoperator (apart from redundant contributions) with

$$y'_j = z_i + y_j \tag{5.53}$$

since one easily verifies using (5.47) and (5.50)

$$\int_q (\mathscr{D}_q + \Psi^*_q) \frac{\delta}{\delta S_q} \int_p \Theta_{ip} \frac{\delta \mathcal{O}^*_j}{\delta S_p} = (z_i + y_j) \int_q \Theta_{iq} \frac{\delta \mathcal{O}^*_j}{\delta S_q}$$
$$+ \text{ redundant operators.} \tag{5.54}$$

In this case (eqn (5.53)) the inequality (5.49) reduces to the inequality (5.48). Inequality (5.48) guarantees the existence of the scaling fields. If Θ_i transforms several scaling operators into scaling operators, then the exponents y obey

$$y_{j'} - y_j = y_{k'} - y_k. \tag{5.55}$$

This raises the question after the existence of the eigenfunctions Θ_i, eqn (5.50), and in case they exist after their properties.

In Section III.G it has been shown that the stress tensor is a marginal operator unless it is a redundant operator. In view of this result, the condition (5.48) cannot be fulfilled. However if we restrict ourselves to a fixed point and to operators which are isotropic in space, then the stress tensor does not show up. Here only the properties of the homogeneous system are considered. Therefore a term $\Sigma c_{ij} T_{ij}$ may be added to the generator of the RG where c_{ij} depends on H. Then the stress tensor behaves like a redundant operator and its eigenvalues can be chosen arbitrarily.

In the remainder of this section a comment on the proof is given. The proof is performed by induction. It also serves as a prescription how a RG equation and scaling fields can be constructed which fulfil (5.44) in case the presumptions (5.46), (5.47), (5.49) do not hold. From (3.41) and the presumptions listed above one shows that

$$\Psi_\kappa\{g\} = \partial\Psi/\partial g_\kappa \tag{5.56}$$

obeys

$$\Psi_\kappa = (y_\kappa + \mathscr{D})\,\Phi_\kappa - K(\Psi + \mathscr{D}, \Phi_\kappa) \tag{5.57}$$

in order g^0. One shows that provided (5.44) and (5.57) hold in order g^n, they also hold in order g^{n+1} (Wegner, 1974). Equation (4.15) of this reference is unproven. We omit it by introduction of the presumptions (5.46) to (5.49) and reformulation of the essential step. The main difficulty is to conclude (5.57) from

$$\mathscr{G}_{\text{tra}}(\Psi_\kappa)\,H^* = \mathscr{G}_{\text{tra}}((y_\kappa + \mathscr{D})\,\Phi_\kappa - K(\psi + \mathscr{D}, \Phi_\kappa))\,H^*. \tag{5.58}$$

Equation (5.58) yields (5.57) apart from an additional function Θ, eqn (5.47), in order g^{n+1} (since (5.57) already holds in order g^n). On the other hand Φ_κ, (5.44) is not uniquely defined. Adding χ, eqn (5.46), to Φ_κ leaves $\mathcal{O}_\kappa$, eqn (5.44), invariant and yields the eqn (5.57).

C. Universality

Under certain conditions it has been shown in Sections V.A and V.B that scaling fields g_i can be constructed so that the scaling law (5.2)

$$F\{g_i \exp(y_i l)\} = \exp(\mathrm{d}l) F\{g_i\}$$

holds and that the free energy does not depend on the fields g_κ of the redundant operators. Provided the free energy can be expanded in powers of the scaling fields g_i of the irrelevant operators, eqn (5.40), the leading scaling function $f(g_E, g_h)$ does not depend on the choice of the RG equation and thus is a universal function. Only the scale of the scaling fields g_E and g_h is not uniquely defined. Therefore for two different choices $\mathcal{G}$ and $\hat{\mathcal{G}}$ for the generator of the RG and consistent choices of the scaling fields g one has

$$\hat{f}(\hat{g}_E, \hat{g}_h) = f(\hat{\lambda}_E g_E, \hat{\lambda}_h g_h). \tag{5.59}$$

This property (in terms of fields μ) has been conjectured by various authors (see Betts, Guttmann and Joyce, 1971). Similar relations can be obtained for the scaling functions $f_i, f_{ij}, \ldots$ where the scales of the fields g_i may vary from RG to RG

$$\hat{f}_i(\hat{g}_E, \hat{g}_h) = \hat{\lambda}_i f_i(\hat{\lambda}_E g_E, \hat{\lambda}_h g_h) \tag{5.60}$$

$$\hat{f}_{ij}(\hat{g}_E, \hat{g}_2) = \hat{\lambda}_i \hat{\lambda}_j f_{ij}(\hat{\lambda}_E g_E, \hat{\lambda}_h g_h) \tag{5.61}$$

Therefore different phase transitions which are described by equivalent fixed point Hamiltonians have the same (physical) exponents and are governed by the same scaling functions f. The property that a large number of quantities and in particular those describing the nonanalyticities agree for various systems belonging to the same "critical class" is called universality. It makes model calculations useful and comparison of data on various systems meaningful.

In Section VI similar properties will be derived for the correlation functions. There explicit equations for the calculation of correlation functions from the expectation values of homogeneous operators will be given. Similarly one expects that the free energy and its derivatives can be calculated. Within this RG approach the theory for the calculation of the free energy has not been worked out at the same level of sophistication as for the correlation functions. (For calculations from field theoretic methods see Brezin, Wallace and Wilson (1972, 1973), Avdeeva and Migdal (1972), Aharony (1974)).

However within approximate schemes the free energy or its derivatives have been evaluated. Nauenberg and Nienhuis (1974) calculated the free energy of the two-dimensional Ising model on the basis of an approximate RG equation. If the RG equation has been chosen appropriately, then in the limit $l \to \infty$ the constant $\exp(-dl)\,\mu_0(l)$ approaches the free energy (for an example see Section V.F). This idea of integrating the RG equation far into the noncritical region has been used by Riedel and Wegner (1974) to discuss the crossover from one critical behaviour to another. Recently Nicoll, Chang and Stanley (1974a) extended this idea to include two critical fixed points and the noncritical fixed point (which for their RG equation is the infinite Gaussian fixed point).

In the remainder of this section eqn (5.99) is derived. We introduce a class of RG equations characterized by a parameter v so that

$$\mathscr{G} = \mathscr{G}(0), \qquad \hat{\mathscr{G}} = \mathscr{G}(1) \tag{5.62}$$

and the derivative $d\mathscr{G}(v)/dv$ exists. Then the free energy is expressed in terms of the scaling fields $g_i(v)$

$$F = g_0(v) + f(g_E(v), g_h(v), v) + \sum{}' g_i(v) f_i(g_E(v), g_h(v), v) + O(g_{\text{irr}}^2). \tag{5.63}$$

(The prime indicates summation over indices i of irrelevant operators).

Due to the homogeneity law one has

$$\begin{aligned}\exp(dl)F &= \exp(dl)g_0 + f(g_E \exp(y_E l), g_h \exp(y_h l), v) \\ &\quad + \sum{}' g_i \exp(y_i l)\, f_i(g_E \exp(y_E l), g_h \exp(y_h l), v) + O(g_{\text{irr}}^2).\end{aligned} \tag{5.64}$$

For a given Hamiltonian H the free energy is fixed, however the scaling fields depend on the choice of the RG and the choice of the scales. Thus one has $dF/dv = 0$ which yields

$$\begin{aligned}&\exp(dl)g_0' + f' + \exp(y_E l)g_E' f^E + \exp(y_h l)g_h' f^h \\ &\qquad + \sum{}' g_i' \exp(y_i l)\, f_i + O(g_{\text{irr}}) = 0\end{aligned} \tag{5.65}$$

with the abbreviations

$$g_i' = \partial g_i/dv \tag{5.66a}$$

$$f' = \partial f/\partial v \tag{5.66b}$$

$$f^E = \partial f(x_E, x_h, v)/\partial x_E \tag{5.66c}$$

and similarly for f^h. The arguments of f and the derivatives are $g_E(v)\exp(y_E l)$, $g_h(v)\exp(y_h l)$, v.

Since the fields g can be formally expanded in the μ's and the μ's in the g's, one obtains a formal expansion of g_i' in powers of the fields g. Thus

$$g_i'\big|_{\substack{g_{\mathrm{irr}}=0\\ g_{\mathrm{red}}=0}} = \sum t_{ijk} g_E^j g_h^k. \tag{5.67}$$

Considering (5.65) for $g_{\mathrm{irr}} = g_{\mathrm{red}} = 0$, substituting (5.67) and replacing $g_i \exp(y_i l)$ by g_i one obtains

$$f'(g_E, g_h) + \sum g_E^j g_h^k \exp[(-jy_E - ky_h)l]$$
$$(t_{0jk} \exp(dl) + t_{Ejk} \exp(y_E l) f^E + t_{hjk} \exp(y_h l) f^h + {\textstyle\sum_i'} t_{ijk} \exp(y_i l) f_i) = 0 \tag{5.68}$$

This equation can be written

$$\sum \exp(xl) U(g_E, g_h, x) = 0 \tag{5.69}$$

with $x \leqslant d$. Since it holds for any l it implies

$$U(g_E, g, x) = 0 \tag{5.70}$$

for any x. Let us evaluate $U(g_E, g_h, 0) = 0$. If

$$d = jy_E + ky_h \tag{5.71a}$$

has no solution (j, k) for $j, k = 0, 1, 2, \ldots$ and if the equations

$$y_E = jy_E + ky_h \tag{5.71b}$$

and

$$y_h = jy_E + ky_h \tag{5.71c}$$

have the only solutions $(j,k) = (1,0)$ and $(j,k) = (0,1)$ respectively for $j, k = 0, 1, 2, \ldots$ then $U(g_E, g_h, 0) = 0$ reads

$$\frac{\partial f}{\partial v} + \sigma_E(v) g_E \frac{\partial f}{\partial g_E} + \sigma_h(v) g_h \frac{\partial f}{\partial g_h} = 0 \tag{5.72a}$$

with

$$\sigma_E(v) = t_{E10}(v), \qquad \sigma_h(v) = t_{h01}(v). \tag{5.72b}$$

If the eqns (5.71) have further solutions then the scaling fields are not uniquely defined and the proof does not apply. Equation (5.72) yields

$$f(g_E, g_h, v) = f(\hat{\lambda}_E g_E, \hat{\lambda}_h g_h, 0) \tag{5.73}$$

with

$$\hat{\lambda}_E = \exp\left(-\int_0^v \sigma_E(v')\mathrm{d}v'\right) \tag{5.74}$$

and similarly for $\hat{\lambda}_h$ which proves (5.59) since

$$f(g_E, g_h, 1) = \hat{f}(g_E, g_h) \tag{5.75a}$$

$$f(g_E, g_h, 0) = f(g_E, g_h). \tag{5.75b}$$

Generalization to more fields for relevant operators is obvious. In case the eigenvalues of the operators $\mathcal{O}_1^*$ and $\mathcal{O}_2^*$ are degenerate one obtains

$$\hat{f}(\hat{g}_1, \hat{g}_2) = f(\hat{\lambda}_{11}g_1 + \hat{\lambda}_{12}g_2, \hat{\lambda}_{21}g_1 + \hat{\lambda}_{22}g_2). \tag{5.76}$$

In such a case symmetry requirements on the scaling fields can often remove some of the arbitrariness in the choice of the scale factors.

D. Coexistence curve

At many phase transitions (like magnetic phase transitions), the Hamiltonian without symmetry breaking field is invariant under a change of sign $S \rightarrow -S$. Including the (experimental) symmetry breaking field $\tilde{\mu}_h$ one has

$$H(\tilde{\mu}_h, S) = H(-\tilde{\mu}_h, -S) \tag{5.77}$$

In these cases the phase transition below T_c obeys $\tilde{\mu}_h = 0$ (unless $\tilde{\mu}_h(T)$ is not uniquely determined). This symmetry simplifies the discussion of the critical behaviour along the coexistence curve considerably.

For the liquid–vapour critical points and for the critical points of alloys and binary mixtures (at the point where the miscibility gap vanishes), this symmetry property does not apply. Here the critical behaviour at the coexistence curve is reviewed (Wegner, 1973) with particular emphasis on the case where the Hamiltonian H does not obey (5.77) but the fixed point Hamiltonian H^* is even in S

$$H^*(S) = H^*(-S). \tag{5.78}$$

Since most fixed points obey this symmetry (for exceptions see the S^3 model by Mack (1973) and odd dominated Ising models by Nicoll, Chang and Stanley (1974b)), the question arises whether the critical behaviour of Ising models and liquid–vapour systems are governed by the same fixed point Hamiltonian and therefore the same critical exponents. An analysis of the experimental data by Levelt-Sengers (1974) including the correction terms found the exponents of these two types of systems in disagreement. We use the language for the liquid–vapour transition where the density difference $\rho - \rho_c$ is the order parameter.

The coexistence curve is characterized by a non-analyticity of F_{sing}, equation (5.19). In a system with two relevant operators $\mathcal{O}_E^*$ and $\mathcal{O}_h^*$ it obeys

$$g_h = |g_E|^{\Delta} c(\ldots, g_i |g_E|^{-\Delta_i}, \ldots) \tag{5.79}$$

If one assumes that g_h can be expanded away from criticality ($|g_E| \neq 0$) then

$$g_h = c_h|g_E|^{\Delta} + \sum_i c_i g_i |g_E|^{\Delta - \Delta_i} + \tfrac{1}{2}\sum_{ij\ldots} c_{ij} g_i g_i |g_E|^{\Delta - \Delta_i - \Delta_j} + \ldots \tag{5.80}$$

Instead of $g_h/|g_E|^{\Delta}$ we use the variable

$$q = \frac{g_h}{|g_E|^{\Delta}} - c(\ldots, g_i|g_E|^{-\Delta_i}, \ldots) \tag{5.81}$$

to express the free energy

$$F = g_0 + |g_E|^{2-\alpha}\tilde{F}^{\pm}_{\text{sing}}\left(q, \ldots, \frac{g_i}{|g_E|^{\Delta_i}}, \ldots\right). \tag{5.82}$$

Then both phases are separated by the transition line $q = 0$. We choose $q > 0$ for the liquid and $q < 0$ for the gas. We assume that the free energy can be expanded in both phases provided that we are not at criticality ($g_E \neq 0$). Then one obtains

$$F = g_0 + |g_E|^{2-\alpha}\tilde{f}(q) + \sum g_i|g_E|^{2-\alpha-\Delta_i}\tilde{f}_i(q) + \tfrac{1}{2}\sum g_i g_i |g_E|^{2-\alpha-\Delta_i-\Delta_j}\tilde{f}_{ij}(q) + \ldots \tag{5.83}$$

With the notation

$$\Delta_{\mathrm{I}} = \Delta_{i_1} + \ldots + \Delta_{i_n} \tag{5.84}$$

$$g_{\mathrm{I}} = g_{i_1} \cdot \ldots \cdot g_{i_n}$$

eqs (5.81) and (5.83) can be written

$$q = g_h|g_E|^{-\Delta} - \sum c_I g_I |g_E|^{-\Delta_I} \tag{5.85}$$

$$F = g_0 + \sum g_{\mathrm{I}}|g_E|^{2-\alpha-\Delta}\tilde{f}_{\mathrm{I}}(q). \tag{5.86}$$

To obtain the density ρ the free energy F is differentiated with respect to the chemical potential (factors $\beta = 1/(kT)$ are incorporated in the potentials)

$$\rho = \partial F/\partial \mu. \tag{5.87}$$

Assuming that the fields g can be expanded (in the space of two experimentally available fields $\tilde{\mu}^{\text{expt}}$) in powers of τ and g_h one finds that any function p expandable in g_h and τ is described along the coexistence curve by an expansion

$$P = P_{\text{reg}}(\tau) + \sum_{\substack{n=0\\ \mathrm{I}}}^{\infty} E_{\mathrm{I}n}(\tau)|\tau|^{\Delta + n(\Delta-1) - \Delta_{\mathrm{I}}} \tag{5.88}$$

whereas any expectation value ρ of a homogeneous operator obeys

$$\rho = \rho_{\text{reg}}(\tau) + \sum_{\substack{n=0\\ \mathrm{I}}}^{\infty} D_{\mathrm{I}n}(\tau)|\tau|^{\Delta + n(\Delta-1) - \Delta_n}\sum_{\substack{n=0\\ \mathrm{I}}} C_{\mathrm{I}n}(\tau)|\tau|^{\beta + n(\Delta-1) - \Delta_{\mathrm{I}}} \tag{5.89}$$

The functions C, D, and E have formal Taylor expansions in τ. If the fixed point Hamiltonian (not the starting Hamiltonian) which describes the critical behaviour has no special symmetry, then one has to expect that all the terms in (5.88) and (5.89) appear unless certain coefficients vanish by chance. If however the fixed point Hamiltonian is even in some order parameter (not necessarily $\rho - \rho_c$), as it is for the Ising model, then the eigenperturbations $\mathcal{O}_i^*$ can be classified as even and odd (correspondingly one introduces $\sigma_i = \pm 1$), so that

$$F\{g_i\} = F\{\sigma_i q_i\} \tag{5.90}$$

which leads to the following selection rules (for $\sigma_h = -1$, $\sigma_E = +1$) and the symmetry relations

$$C_{\mathrm{I}} = 0 \qquad \text{for} \qquad q = +1, \tag{5.91}$$

$$\tilde{f}_{\mathrm{I}}(q) = \sigma_{\mathrm{I}}\tilde{f}_{\mathrm{I}}(-q), \tag{5.92}$$

$$C_{\mathrm{I}n}\,(\text{vapour}) = (-)^n \sigma_{\mathrm{I}} C_{\mathrm{I}n}(\text{fluid}), \tag{5.93}$$

$$D_{\mathrm{I}n} = E_{\mathrm{I}n} = 0 \qquad \text{for} \qquad \sigma_{\mathrm{I}} = (-)^n. \tag{5.94}$$

Therefore one obtains the lowest order contributions to the nonanalytic term of P, eqn (5.88), of order

$$|\tau|^{\Delta - \Delta_i}, \quad \sigma_i = -1, \tag{5.95a}$$

and

$$|\tau|^{2\Delta - 1 - \Delta_j}, \qquad \sigma_j = +1. \tag{5.95b}$$

From eqn (5.89) one obtains the lowest order contributions of $\rho_L - \rho_V$

$$|\tau|^{\beta}, |\tau|^{\beta - \Delta_i}, \qquad \sigma_i = +1 \tag{5.96a}$$

and

$$|\tau|^{1 - \alpha - \Delta_j}, \qquad \sigma_j = -1, \tag{5.96b}$$

whereas for $\rho_L + \rho_V$ one obtains besides analytic contributions the terms

$$|\tau|^{1-\alpha}, \qquad |\tau|^{1 - \alpha - \Delta_i}, \qquad \sigma_i = +1 \tag{5.97a}$$

and

$$|\tau|^{\beta - \Delta_j}, \qquad \sigma_j = -1. \tag{5.97b}$$

The term $|\tau|^{\beta}$ in (5.96a) is the well-known leading term for the critical behaviour of the order parameter. The term $|\tau|^{1-\alpha}$, eqn (5.97a), for the rectilinear diameter is known for several models (Hemmer and Stell, 1970; Widom and Rowlinson, 1970; Mermin, 1971a, b) and from a parametric equation of state by Green, Cooper and Levelt-Sengers (1971). Mermin and Rehr (1971) showed under rather weak assumptions that $\partial(\rho_L + \rho_V)/\partial\tau$ diverges at least like $|\tau|^{-\alpha}$.

E. Logarithmic anomalies

The specific heat of the two two-dimensional Ising model shows a logarithmic singularity (Onsager 1944) as described in equation (2.10b). More generally singularities

$$c_{\text{sing}}(T) = -A(T - T_c)^n \ln|T - T_c| \tag{5.98}$$

were found for the eight-vertex model for even n (Baxter 1971, 1972). The spherical model and the ideal Bose gas display logarithmic factors for the specific heat, and the susceptibility and the compressibility, respectively, in four dimensions

$$c_{\text{sing}}(T) = \frac{A}{\ln|T - T_c|} \tag{5.99a}$$

$$\chi(T) = A_\chi \frac{\ln|T - T_c|}{|T - T_c|} \tag{5.99b}$$

(Joyce, 1966; Gunton and Buckingham, 1968).

During recent years more logarithmic anomalies

$$|T - T_c|^\lambda (\ln|T - T_c|)^p \tag{5.100}$$

near the critical point were found which will be discussed below. These logarithmic singularities arise when the concept of the scaling fields breaks down since the equation

$$y_i = y_j + y_k + \ldots \tag{5.101}$$

is fulfilled. We distinguish two types of logarithmic singularities. (1) If (5.101) is fulfilled only a finite number of times, then logarithmic singularities (5.100) appear where p is a positive integer. (2) If a marginal operator has to be taken into account, then p may becomes an arbitrary number.

1. *Logarithms* $(\ln|\tau|)^p$, *p positive integer*

In Section V.A scaling fields g were constructed provided $y_i \neq y_j + y_k$, $y_i \neq y_j + y_k + y_m$, etc. If however $y_i = y_j + y_k$ and the inhomogeneity a'_{ijk} does not vanish, then we can repair this by letting b_{ijk} depend on l. In general we will allow that the coefficients b are polynomials in l of finite order whenever it is necessary. Then (5.9) and (5.10) are replaced by (Wegner, 1972a)

$$\left(y_j + y_k - y_i + \frac{\partial}{\partial l}\right) b_{ijk}(l) = a'_{ijk} \tag{5.102}$$

$$(y_j + y_k + y_m - y_i)\, b_{ijkm}(l) = \sum_s a'_{ijs}\, b_{skm}(l) + 2 \text{ permutations} \tag{5.103}$$

For $y_i = y_j + y_k$ (5.102) has the solution

$$b_{ijk} = (l + l_0)\, a'_{ijk} \tag{5.104}$$

The coefficient b_{ijkm} will depend on l if either the equation $y_i = y_j + y_k + y_m$ holds or the inhomogeneity depends on l. The Hamiltonian depends now on the scaling fields g and $l + l_0$. Therefore

$$H_0 = H(\{g_i\}, l_0) \tag{5.105}$$

transforms into

$$H_l = H(\{g_i\, e^{y_i l}\}, l_0 + l). \tag{5.106}$$

Let us choose, for example,

$$l_0 = \frac{1}{y_E} \ln g_E \tag{5.107}$$

then

$$l + l_0 = \frac{1}{y_E} \ln g_E(l). \tag{5.108}$$

Therefore $l + l_0$ can be replaced by $y_E^{-1} \ln g_E(l)$ which yields a representation for H depending on the scaling fields g only. The appearance of $\ln g_E$ produces the logarithmic anomalies.

Let us consider two examples. We assume that it is sufficient to consider only two operators, $\mathcal{O}_0^*$ and $\mathcal{O}_E^*$. In the first example which is a caricature of the behaviour of the two-dimensional Ising model, we assume

$$\frac{d\mu_0}{dl} = 2\mu_0 + a_0 \mu_E^2 \tag{5.109}$$

$$\frac{d\mu_E}{dl} = \mu_E \tag{5.110}$$

All higher order terms in μ_E are neglected. One obtains

$$\mu_E(l) = e^l\, \mu_E \tag{5.111}$$

$$\mu_0(l) = e^{2l}\, \mu_0 + l a_0\, e^{2l}\, \mu_E^2 \tag{5.112}$$

which yields

$$F(\mu_0 = 0, \mu_E) = e^{-2l}\, F(l a_0\, e^{2l} \mu_E^2, e^l \mu_E) = l a_0 \mu_E^2 + e^{-2l}\, F(0, e^l\, \mu_E). \tag{5.113}$$

With

$$|\mu_E|e^l = 1 \tag{5.114}$$

one obtains

$$F(0, \mu_E) = -a_0\mu_E^2 \ln|\mu_E| + \mu_E^2 F(0, \pm 1). \tag{5.115}$$

The amplitude of $\mu_E^2 \ln|\mu_E|$ is the same above and below T_c, whereas the amplitude $F(0, \pm 1)$ of μ_E^2 may in general be different above and below T_c. Logarithms of this type were already explained by Widom (1965) and Kadanoff (1966).

The second more general example is that of Section V.A, eqns (5.20) and (5.21). If $d/y_E = 2 - \alpha = m$ is an integer, then the expansion (5.28) becomes

$$\mu_0 = g_0 + a_0 \frac{(m-1)}{y_E}\left(\frac{a_E}{y_E}\right)^{m-2} g_E^m \ln|g_E| - a_0 \sum_{n=2}^{\infty}{}' \frac{n-1}{d - ny_E}\left(\frac{a_E}{y_E}\right)^{n-2} g_E^n \tag{5.116}$$

where the prime indicates that the term with $n = m$ is omitted. For given g_E and $\mu_0 = 0$ the free energy amounts to

$$F = g_0 + |g_E|^m F_\pm \tag{5.117}$$

where $F_\pm$ is the free energy at $g_0 = 0$, $g_E = \pm 1$. According to (5.116) the logarithmic term

$$-a_0 \frac{m-1}{y_E}\left(\frac{a_E}{y_E}\right)^{m-2} g_E^m \ln|g_E| \tag{5.118}$$

contributes to the free energy. Now g_0 can no longer be called regular part of the free energy.

In Section III.A it is assumed that a complete set of eigenoperators exists. It may however happen that the matrix L mentioned after (3.11) is not diagonal. Then one obtains logarithmic corrections, too. If for example

$$\frac{d\mu_1}{dl} = y\mu_1 \tag{5.119}$$

$$\frac{d\mu_2}{dl} = y\mu_2 + a\mu_1 \tag{5.120}$$

then one obtains

$$\mu_1(l) = e^{yl}\mu_1(0) \tag{5.120}$$

$$\mu_2(l) = e^{yl}(\mu_2(0) + al\mu_1(0)) \tag{5.122}$$

or equivalently

$$\mu_1 = g_1 \tag{5.123}$$

$$\mu_2 = g_2 + alg_1. \tag{5.124}$$

2. *Logarithms* $(\ln|\tau|)^p$, *p arbitrary*

Logarithmic corrections show up if the condition (5.101) is satisfied. As long as we are interested only in the contributions from relevant operators and if no marginal field has to be taken into account, then there are only a finite number of sets $(y_i, y_j, y_k, \ldots)$ which fulfil condition (5.101). These sets determine the logarithmic corrections.

If, however, a field μ_u with $y_u = 0$ has to be considered, then (5.101) is satisfied for an infinite number of sets. This may lead to a different anomalous behaviour. The most singular contributions are obtained from the following truncated equations (A more complete discussion has been given by Wegner and Riedel (1973)).

$$\frac{d\mu_u}{dl} = \tfrac{1}{2} a'_{uuu} \mu_u^2 \tag{5.125}$$

$$\frac{d\mu_i}{dl} = y_i\mu_i + a'_{iui} \mu_i\mu_u \tag{5.126}$$

which yield

$$\mu_2 = s(l + l_0)^{-1} \tag{5.127}$$

$$s = -2/a'_{uuu} \tag{5.128}$$

$$\mu_i(l) = \mu_i(0) \exp(y_i l) \left(\frac{l + l_0}{l_0}\right)^{p_i} \tag{5.129}$$

$$p_i = sa'_{iui} = -2a'_{iui}/a'_{uuu}. \tag{5.130}$$

Equation (5.129) can be written

$$\mu_i(\mathrm{l}) = \left(\frac{l + l_0}{l_0}\right)^{p_i} g_i(l). \tag{5.131}$$

Therefore μ_i contains in general factors of fractional powers of $\ln|g_E|$ as corrections to the fields g_i. The explicit discussion (Wegner and Riedel, 1973) shows that the fields μ_i can be expressed as polynomials of $g_i(l + l_0)^{p_i}$ in leading order provided no set of relevant exponents satisfies (5.101). If (5.101) is fulfilled for a set of relevant exponents, then further factors $(l + l_0)$ appear, as described in the first part of this section.

The operators $\mathcal{O}_{n,0}$ of the trivial fixed point obey

$$y_n = d - \left(\frac{d}{2} - 1\right) n \tag{5.132}$$

(compare (3.208). At the dimensionalities

$$d_\sigma = \frac{2\sigma}{\sigma - 1},$$

equation (3.211) (σ integer or half-integer) the operator $\mathcal{O}^*_{2\sigma,0}$ becomes marginal. If σ is integer, then the above mentioned logarithmic corrections appear. Of particular interest are the cases

$$\sigma = 2 \qquad d = 4 \tag{5.133}$$

(normal critical point) which was first discussed by Larkin and Khmelnitskii (1969). This critical point and the tricritical point

$$\sigma = 3, \qquad d = 3 \tag{5.134}$$

were discussed by Wegner and Riedel (1973). The exponents p_n of the logarithmic factors agree with the coefficients p_n of (4.47).

As long as $\mu_2 > 0$ for $\sigma = 2$ and $\mu_4 > 0$ for $\sigma = 3$ the Hamiltonian has a lower bound even for $\mu_{2\sigma} = 0$ (fields of the marginal operators). If $\mu_{2\sigma-2} < 0$ then $\mu_{2\sigma} > 0$ is necessary to obtain a finite free energy. This can produce further logarithmic factors for the spontaneous magnetization for example (Wegner and Riedel, 1973). A complete discussion for this phenomenon has not yet been given. The problem is connected to that of the anomalous dimension of the vacuum (Fisher 1973) mentioned at the end of Section III.C.

If a'_{uuu} vanishes then one obtains other types of singularities. It may happen that μ_u decays like $(l + l_0)^{-1/2}$ which implies

$$\mu_i(l) = e^{y_i l} \mu_i(0) \exp\left(a_i\left((l + l_0)^{1/2} - l_0^{1/2}\right)\right) \tag{5.135}$$

(Jegerlehner, 1975). There is also the possibility that a whole line of fixed points emerges along which the critical exponents may vary continuously as observed in the eight-vertex model.

F. The limit case $y_E = 0$: Phase transitions of infinite order

Until now the exponent y_E was assumed to be positive. There exist however systems with vanishing y_E, that is, in these systems there is no symmetry conserving relevant operator $\mathcal{O}^*_E$. The F-model is such a limit case of the eight-vertex model since the critical exponent α approaches $-\infty$ in this limit. Anderson and Yuval (1971) treated the Kondo-problem and a one-dimensional Ising model the long-range interactions in terms of RG equations. Similar equations are applied to the critical behaviour of the XY-model by Kosterlitz (1974). These RG equations exhibit a line of fixed points (which implies a marginal operator). At the critical point which is here of interest, a second operator becomes marginal.

It is characteristic for these phase transitions of infinite order that any derivative of the free energy with respect to the temperature exists at T_c. The expansions of the free energy at the critical temperature are asymptotic series. Three models will be described: The first model contains only one marginal operator (Wegner 1972a). It yields a free energy

$$F = c_{\pm} \exp\left(-\frac{1}{p\tau^k}\right) + c' \sum (n-1)!\,(p\tau^k)^n \tag{5.136}$$

with $k = 1, 2$. It cannot simulate the F-model which shows an asymptotic series $\Sigma(2n-1)!\,(p^2\tau)^n$ (Lieb, 1967). This behaviour is obtained from the second model which is the one proposed by Anderson and Yuval. It yields

$$F = c_{\pm} \exp\left(-\frac{1}{p\sqrt{\tau}}\right) + c' \sum (2n-1)!\,(p^2\tau)^n. \tag{5.137}$$

However this model is not yet a reasonable description for the F-model since it cannot be considered as the limit of models with continuously varying y_E (as for the eight-vertex model). This is achieved by a third model which we call parabolic model since the trajectories form parabola. The trajectories of the second model are hyperbola. Therefore we will call it hyperbolic model.

1. *One marginal operator*

The first model is described in terms of fields μ_0, μ_E and μ_h governed by

$$\frac{d\mu_0}{dl} = d\mu_0 + a_0\mu_E^2 \tag{5.138}$$

$$\frac{d\mu_E}{dl} = a_E\mu_E^{k+1} \quad k = 1, 2 \tag{5.139}$$

$$\frac{d\mu_h}{dl} = y_h\mu_h \tag{5.140}$$

Differentiating

$$F(\mu_0(l), \mu_E(l)) = e^{dl}\, F(\mu_0, \mu_E) \tag{5.141}$$

with respect to l yields

$$\frac{\partial F}{\partial \mu_0}\frac{d\mu_0}{dl} + \frac{\partial F}{\partial \mu_E}\frac{d\mu_E}{dl} = dF \tag{5.142}$$

Substituting

$$F = \mu_0 + f(\mu_E) \tag{5.143}$$

and eqns (5.138) and (5.139) in (5.142) yields

$$df(\mu_E) = a_E \mu_E^{k+1} \frac{\mathrm{d}f(\mu_E)}{\mathrm{d}\mu_E} + a_0 \mu_E^2 \tag{5.144}$$

with the solution

$$f(\mu_E) = c_\pm \exp\left(-\frac{1}{p\mu_E^k}\right) + c' \sum_n (n-1)!\,(p\mu_E^k)^n \tag{5.145}$$

for $k = 1, 2$ where

$$p = \frac{ka_E}{d}, c' = \begin{cases} \dfrac{a_0 d}{2a_E^2} & k = 1 \\ \dfrac{a_0}{2a_E} & k = 2 \end{cases} \tag{5.146}$$

and the sum (5.145) starts with $n = 2$ for $k = 1$ and $n = 1$ for $k = 2$, resp.

The leading behaviour of the susceptibility is obtained from (5.140) which yields

$$\chi(0) = \exp\left[(2y_h - d)\, l\right] \chi(l). \tag{5.147}$$

From (5.139) one obtains

$$l = \frac{1}{ka_E \mu_E^k} - \frac{1}{ka_E \mu_E(l)^k} \tag{5.148}$$

Choosing $\mu_E(l) = \text{const}$ yields the susceptibility

$$\chi = c''_\pm \exp\left(a/(p\tau^k)\right) \tag{5.149}$$

with

$$a = (2y_h - d)/d. \tag{5.150}$$

2. *Hyperbolic model*

It is defined by the RG equations

$$\frac{\mathrm{d}\mu_0}{\mathrm{d}l} = d\mu_0 + a\mu_2^2 \tag{5.151}$$

$$\frac{\mathrm{d}\mu_1}{\mathrm{d}l} = -\mu_2^2 \tag{5.152}$$

$$\frac{\mathrm{d}\mu_2}{\mathrm{d}l} = -\mu_1\mu_2 \tag{5.153}$$

and we may add (5.140). There is a line of fixed points $\mu_2 = 0$. The operator $\mathcal{O}_2^*$ is relevant for $\mu_1 > 0$ and irrelevant for $\mu_1 < 0$. Unless $\mu_1 > |\mu_2|$ or $\mu_2 = 0$ no fixed point is approached. Accordingly $\mu_1 = |\mu_2|$ is a critical line.

The difference $\mu_2^2 - \mu_1^2$ is invariant under the RG transformation and vanishes at the critical line. We define

$$\tau = \mu_2^2 - \mu_1^2 \tag{5.154}$$

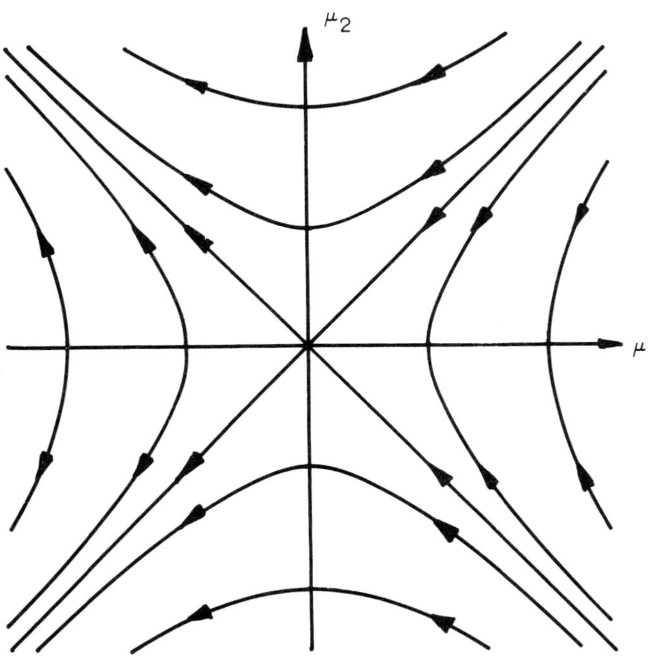

FIG. 2. Trajectories of the hyperbolic model.

The trajectories (lines of constant τ) are hyperbola and plotted in Fig. 2. Then (5.152) and (5.153) are solved by

$$\mu_1 = \sqrt{\tau} + \tan\varphi \tag{5.155}$$

$$\mu_2 = \frac{\sqrt{\tau}}{\cos\varphi} \tag{5.156}$$

$$\varphi(l) = \varphi(0) - l\sqrt{\tau} \tag{5.157}$$

Suppose the critical point is approached at

$$\mu_{1c} = \mu_{2c} = \frac{1}{c} \tag{5.158}$$

and (for simplicity) along the line

$$\varphi(0) = \frac{\pi}{2} - c\sqrt{\tau} \tag{1.159}$$

that is

$$\mu_1 = \frac{1}{c} - \frac{1}{6}c\tau + O(\tau^2) \tag{5.160}$$

$$\mu_2 = \frac{1}{c} + \frac{1}{3}c\tau + O(\tau^2) \tag{5.161}$$

The free energy is obtained from

$$F(\mu_0 = 0, \mu_1, \mu_2) = \mathrm{e}^{-dl}\,\mu_0(l) + \mathrm{e}^{-dl}\,F(0, \mu_1(l), \mu_2(l)) \tag{5.162}$$

which may be integrated out to the noncritical region

$$\varphi(l) = -\frac{\pi}{2} + \sqrt{\tau} \tag{5.163}$$

that is

$$\mu_1(l) = -1 + \tfrac{1}{6}\tau + O(\tau^2) \tag{5.164}$$

$$\mu_2(l) = 1 + \tfrac{1}{3}\tau + O(\tau^2) \tag{5.165}$$

To evaluate $\mathrm{e}^{-dl}\mu_0(l)$ in (5.162)

$$\frac{\mathrm{d}}{\mathrm{d}l}\left(\mathrm{e}^{-dl}\,\mu_0(l)\right) = a\,\mathrm{e}^{-dl}\,\mu_2^2(l) = \frac{a\tau e^{-dl}}{\cos^2\varphi} \tag{5.166}$$

we use the representation

$$\frac{1}{\cos^2\varphi} = \sum_k \frac{1}{(\varphi - (\pi/2) + k\pi)^2} \tag{5.167}$$

The term $k = 0$ yields a constant, the terms $k = \pm 1$ the leading asymptotic behaviour

$$\mathrm{e}^{-dl}\,\mu_0(l) = \text{const.} + \frac{2ad}{1 + cd}\sum (2n - 1)!\left(\frac{\tau}{\pi^2 d^2}\right)^n \tag{5.168}$$

Adding the terms which vanish proportional e^{-dl} one obtains (5.137) with

$$p = (\pi d)^{-1}. \tag{5.169}$$

From (5.147), (5.157), (5.613) one obtains

$$\chi = c'' \exp\left(a/(p\sqrt{\tau})\right) \tag{5.170}$$

with a given by (5.150).

This model cannot serve as an explanation for the F-model since all critical points along the line (5.158) show the same behaviour. The situation is different if criticality is approached along the line

$$\mu_1 = 0, \qquad \mu_2 = \tau. \tag{5.171}$$

Then however $\sqrt{\tau}$ has to be replaced by τ in (5.155) and (5.156). An analogous calculation yields

$$F = c_+ \exp\left(-\frac{1}{p\tau}\right) + c' \sum_n (2n-1)!\,(p\tau)^{2n} \tag{5.172}$$

with

$$p = 2/(\pi d). \tag{5.173}$$

However (5.172) does not yield the asymptotic series for the F-model.

3. *Parabolic model*

It is defined by the RG equations

$$\frac{d\mu_0}{dl} = d\mu_0 + \tfrac{1}{2}a_{11}\mu_1^2 + a_{12}\mu_1\mu_2 + \tfrac{1}{2}a_{22}\mu_2^2, \tag{5.174}$$

$$\frac{d\mu_1}{dl} = -\mu_2, \tag{5.175}$$

$$\frac{d\mu_2}{dl} = -2\mu_1\mu_2. \tag{5.176}$$

Again their is a line of fixed points for $\mu_2 = 0$, and $\mathcal{O}_2^*$ is relevant for $\mu_1 < 0$ and irrelevant for $\mu_1 > 0$. The trajectories are parabola obeying

$$\tau = \mu_2 - \mu_1^2 \tag{5.177}$$

(Fig. 3). Approaching the critical point of the F-model means aproaching the origin

$$\mu_1 = 0, \qquad \mu_2 = \tau \tag{5.178}$$

(The more general case $\mu_1 = \tau$, $\mu_2 = c\tau$ can be easily discussed). One obtains

$$\mu_1(l) = \sqrt{\tau}\tan\varphi \tag{5.179}$$

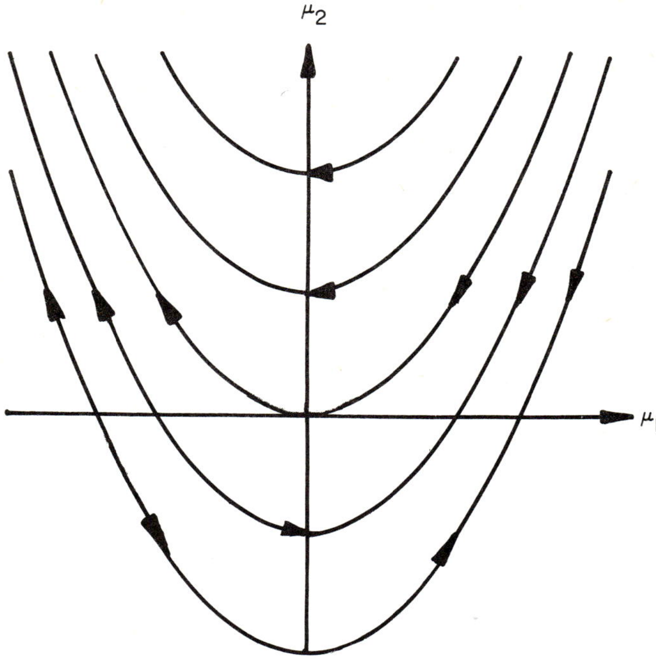

FIG. 3. Trajectories of the parabolic model.

$$\mu_2(l) = \frac{\tau}{\cos^2\varphi} \tag{5.180}$$

with

$$\varphi(l) = -l\sqrt{\tau}. \tag{5.181}$$

Integrating up to

$$\varphi(l) = \frac{\pi}{2} - \sqrt{\tau} \tag{5.182}$$

that is

$$\mu_1 = -1 + \frac{\tau}{3} + \mathcal{O}(\tau^2) \tag{5.183}$$

$$\mu_2 = 1 + \frac{\tau}{3} + \mathcal{O}(\tau^2) \tag{5.184}$$

yields similarly to the calculation for the hyperbolic model an asymptotic behaviour

$$F = c_+ \exp\left(-\frac{1}{p\sqrt{\tau}}\right) + c' \sum_{n=1}^{\infty} (2n-1)!(p^2\tau)^n \tag{5.185}$$

with

$$p = 2/(\pi d) \tag{5.186}$$

$$c' = a_{11}d + a_{12}d^2 + \tfrac{1}{6}a_{22}d^3 \tag{5.187}$$

and

$$\chi = c'' \exp(-a/(p\sqrt{\tau})) \tag{5.188}$$

with a given by (5.150).

VI. Nonlinear Theory: Correlations

A. Order parameter correlations

Wilson and Kogut (1974) have derived an exact relation between the correlations of the order parameter for Hamiltonians H_0 and H_l related by (2.87) and (2.89). In particular for the two-point cumulant

$$VG(q, -q) = \langle S_q S_{-q}\rangle - \langle S_q\rangle\langle S_{-q}\rangle \tag{6.1}$$

they obtain

$$w^2(q)(G_0(q, -q) - 1) = e^{-dl}w^2(qe^l)(G_l(ge^l, -qe^l) - 1) \tag{6.2}$$

with

$$w(q) = q^{\frac{1}{2}(d+2-\eta)} \exp\left(\int_0^q \frac{h(k) - h(0)}{k}\, dk\right) \tag{6.3}$$

where the index l of G_l indicates that the expectation value is taken for H_l. Close to T_c the correlation G is large in comparison to 1, and for small q the exponential in equation (5.3) is close to unity which yields with $\tau_l = \tau \exp(y_E l)$ the scaling law (3.125). On the other hand for large q-values the factor $w^2(q)$ becomes very large so that $G - 1$ approaches zero. Therefore

$$\lim_{q\to\infty} G(q) = 1 \tag{6.4}$$

which agrees with the observation that large Fourier components enter the Hamiltonian only via $u_2^*(q)S_q S_{-q}$ which yields with $\lim_{q\to\infty} u_2^*(q) = 1$, eqn (6.4).

To derive (6.2) one starts from the Hamiltonian

$$H'_0 = H_0 + \sum \lambda_i w(q_i) S_{q_i} + Vc(0). \tag{6.5}$$

The function Ψ'_k in the generator $\mathcal{G}(\Psi')$ is chosen to be

$$\Psi'_k = h(k)\left(S_k - \frac{\delta H}{\delta S_{-k}} + (2\pi)^d \sum_i \lambda_i w(q_i)\delta^d(k + q_i)\right). \tag{6.6}$$

With this choice of Ψ' all mixed terms $(\delta H/\delta S_k)(\delta S_q/\delta S_{-k})$ cancel and one obtains

$$\begin{aligned}\mathcal{G}(\Psi')H' = d(\Psi)H &+ \sum \lambda_i w(q_i)(q_i \nabla S_{q_i} + (d/2)S_{q_i} + h(q_i)S_{q_i}) \\ &+ V \sum \lambda_i \lambda_j h(q_i) w(q_i) w(q_j) \delta_{q_i + q_j, 0} + dVc(0).\end{aligned} \tag{6.7}$$

This generator produces an interaction

$$H'_l = H_l + \sum \lambda_i w(q_i e^l) S_{q_i e^l} + Vc(l) \tag{6.8}$$

where w and c obey

$$\frac{dw(qe^l)}{dl} = \left(\frac{d}{2} + h(qe^l)\right) w(qe^l) \tag{6.9}$$

$$\frac{\partial c(l)}{\partial l} = dc(l) + \sum \lambda_i \lambda_j h(qe^l) w^2(qe^l) \delta_{q_i + q_j, 0}. \tag{6.10}$$

These equations are solved by (6.3) and

$$c(l) = \tfrac{1}{2} \sum \lambda_i \lambda_j w^2(q_i e^l) \delta_{q_i + q_j, 0} \tag{6.11}$$

where (3.128) is used.

As in Section III.D the correlation functions are determined from the free energy by differentiation with respect to λ. This yields

$$\begin{aligned}-\frac{\partial^2 F_0}{\partial \lambda_i \partial \lambda_j} &= w(q_i) w(q_j)(G_0(q_i, q_j) - \delta_{q_i + q_j, 0}) = -e^{-dl} \frac{\partial^2 F_l}{\partial \lambda_i \partial \lambda_j} \\ &= e^{-dl} w(q_i e^l) w(q_j e^l)(G_l(q_i e^l, q_i e^l) - \delta_{q_i + q_j, 0})\end{aligned} \tag{6.12}$$

which reduces for $q_i = -q_j = q$ to (6.2). Note that c depends on λ via (6.11). For the n-point cumulant, $n > 2$, one obtains

$$w(q_1) \ldots w(q_2) G_0(q_1, \ldots q_2) = e^{-dl} w(q_1 e^l) \ldots w(q_2 e^l) G_l(q_1 e^l, \ldots q_2 e^l) \tag{6.13}$$

which for small wave-vectors yields the homogeneity law

$$G_0(q_1, \ldots, q_n) = \exp\left[-d + \tfrac{1}{2} n(d + 2 - \eta))l\right] G_l(q_1 e^l, \ldots q_n e^l) \tag{6.14}$$

Since $w(0)$ vanishes the perturbation by a homogeneous magnetic field has to be considered separately. One obtains from

$$H'_0 = H_0 + \lambda S_0 + \tfrac{1}{2} V \lambda^2 \tag{6.15}$$

the Hamiltonian

$$H'_l = H_l + \lambda \exp[\tfrac{1}{2}(d + 2 - \eta)l]S_0 + \tfrac{1}{2}V\lambda^2 \exp[(d + 2 - \eta)l] \tag{6.16}$$

Therefore the spontaneous magnetization obeys

$$\frac{1}{V}\langle S_0\rangle_l = \exp[\tfrac{1}{2}(d - 2 + \mu)l]\,.\frac{1}{V}\langle S_0\rangle_0. \tag{6.17}$$

B. Recursion equation for correlation functions

The correlation functions show besides the scaling behaviour (Sections III.D and VI.A) a second characteristic behaviour: at T_c they are nonanalytic even for finite wave-lengths as a function of the thermodynamic variables. This is not unexpected in view of (3.151) obtained from the operator product expansion. First we sketch the calculation of these nonanalytic contributions. (Wegner, 1975b).

Within the linear approximation of Section III.D one obtains the correlation function

$$\langle \mathcal{O}_j^*(q)\mathcal{O}_k^*(-q)\rangle = \frac{c}{q^{y_j + y_k - d}} \tag{6.18}$$

at T_c. According to the calculations in Section III.F, the operators become extremely small for large q. As an example consider

$$\mathcal{O}_1^*(q) = \varphi(q)S_q \tag{6.19}$$

where $\varphi(q)$ decays faster then any power in q for large q with the choice $h(q) = c + 2q^2$. Therefore one cannot expect that (6.18) holds for large q. The reason is that the RG equation is nonlinear in H and the perturbations $\lambda_j\mathcal{O}_j^*(q) + \lambda_k\mathcal{O}_k^*(-q)$ will generate contributions $\mathcal{O}_0^*, \mathcal{O}_E^*$ etc. Similarly homogeneous perturbations and nonhomogeneous perturbations generate contributions nonlinear in μ and λ. To discuss these effects we give a derivation under the following simplifying assumptions:

(i) The linear approximation is good for $q \lesssim q_0$.

(ii) In a narrow region around q_0 the nonlinear contributions dominate.

(iii) For $q \gtrsim q_0$ the operators $\mathcal{O}^*(q)$ are negligible.

From

$$H_0 = H^* + \tau\mathcal{O}_E^* + \lambda_1\mathcal{O}_h^*(q) + \lambda_2\mathcal{O}_h^*(-q) \tag{6.20}$$

one obtains according to (i) with $e^l = q_0/q$

$$H_{l-} = H^* + \tau\left(\frac{q_0}{q}\right)^{y_E}\mathcal{O}_E^* + \lambda_1\left(\frac{q_0}{q}\right)^{y_h}\mathcal{O}_h^*(q_{0-}) + \lambda_2\left(\frac{q_0}{q}\right)^{y_h}\mathcal{O}_h^*(-q_{0-}) \tag{6.21}$$

where q_{0-} is slightly smaller than q_0. Increasing l slightly one obtains due to (ii) nonlinear contributions

$$H_{l+} = H^* + \tau\left(\frac{q_0}{q}\right)^{y_E} \mathcal{O}_E^* - \lambda_1\lambda_2\left(\frac{q_0}{q}\right)^{2y_h}\left(AV + B\mathcal{O}_E^* + C\tau\left(\frac{q_0}{q}\right)^{y_E} V + \ldots\right) \tag{6.22}$$

The perturbations $\mathcal{O}_h(\pm q_{0+})$ are already neglected in accordance with (iii). The Hamiltonian H_{l+} is translational invariant and application of the inverse RG transformation yields

$$\tilde{H}_0 = H^* + \tau\mathcal{O}_E^* - \lambda_1\lambda_2\left(\frac{q_0}{q}\right)^{2y_h}\left(\left(\frac{q_0}{q}\right)^{-d} AV + \left(\frac{q_0}{q}\right)^{-y_E} B\mathcal{O}_E^* + C\left(\frac{q_0}{q}\right)^{y_E - d} \tau V + \ldots\right) \tag{6.23}$$

Since the free energy is invariant under the total of these transformations, one obtains

$$\frac{1}{V}\langle\mathcal{O}_h^*(q)\mathcal{O}_h^*(-q)\rangle_c = -\frac{\partial^2 F(\tilde{H}_0)}{\partial\lambda_1\partial\lambda_2}$$

$$= \left(\frac{q_0}{q}\right)^{2y_h - d}\left(A + \frac{B}{V}\left(\frac{q_0}{q}\right)^{d-y_E}\langle\mathcal{O}_E^*\rangle + C\tau\left(\frac{q_0}{q}\right)^{y_E} + \ldots\right) \tag{6.24}$$

Therefore besides the temperature independent term proportional q^{d-2y_h} terms proportional $|\tau|^{1-\alpha}q^{y_E-2y_h}$ and $\tau q^{d-y_E-2y_h}$ appear. This result was suggested by Fisher and Langer (1968) and has also been derived by means of the Callan–Symanzik equation (Brezin, Le Guillou and Zinn-Justin (1974), compare Symanzik (1971)). Starting from the more general Hamiltonian

$$H_0 = H^* + \sum \mu_i\mathcal{O}_i^* + \lambda_j\mathcal{O}_j^*(q) + \lambda_k\mathcal{O}_k^*(-q) \tag{6.25}$$

one obtains terms to $\langle\mathcal{O}_j^*(q)\mathcal{O}_k^*(-q)\rangle$ proportional to

$$\mu_m\mu_n \cdot \ldots q^{d-y_j-y_k-y_m-y_n-\ldots+y_i}\langle\mathcal{O}_i^*\rangle. \tag{6.26}$$

All these terms obey the scaling relation (3.123). The occurrence of the expectation values $\langle\mathcal{O}_i^*\rangle$ causes the nonanalytic behaviour at T_c.

Actually the assumptions (i) and (ii) are not fulfilled. However consideration of the nonlinear contributions allows for similar conclusions at small q (provided certain inequalities hold for the exponents y). One proceeds in three steps (Wegner 1975a):

(i) To account for the contributions proportional μ^2 to $\mathrm{d}\mu/\mathrm{d}l$ one uses the scaling fields g (Section V.A).

(ii) To account for the contributions proportional $\mu\lambda$ to $d\lambda/dl$ one introduces appropriate operators $\mathcal{O}_j(q)$. These operators depend on the fields g. They coincide at the fixed point with $\mathcal{O}_j^*(q)$

$$\mathcal{O}_j(\{g = 0\}, q) = \mathcal{O}_j^*(q) \tag{6.27}$$

and at $q = 0$ with $\mathcal{O}_j$, eqn (5.31)

$$\mathcal{O}_j(\{g\}, q = 0) = \mathcal{O}_j\{g\}. \tag{6.28}$$

Then

$$H_0 = H\{g\} + \sum \varkappa_j \mathcal{O}_j(\{g\}, q_j) \tag{6.29}$$

transforms into

$$H_l = H\{g\,e^{yl}\} + \sum_j \varkappa_j \mathcal{O}_j(\{g\,e^{yl}\}, q_j\,e^l) + \mathcal{O}(\varkappa^2). \tag{6.30}$$

If no terms of $\mathcal{O}(\varkappa^2)$ were present in (6.30), then the n-point cumulant for the operators $\mathcal{O}(q)$ were an exact scaling function obeying

$$(D + \sum (g_j \nabla_j + y_j) - d) G_{j_1 \dots j_n}(\{g\}, q_1, \dots q_n) = 0 \tag{6.31}$$

where D is defined in (5.5).

(iii) The terms proportional $\varkappa^2$ to $d\varkappa/dl$ generate contributions proportional $\varkappa_j \varkappa_k \mathcal{O}_i(q_j + q_k)$ etc. in (6.30). They produce an inhomogeneity to (6.31)

$$\begin{aligned}(D + \sum_j (q_j \nabla_j + y_j) - d) G_{j_1 \dots j_n}(\{g\}, q_1, \dots q_n) \\ = \sum A_{i j_1 j_2}(q_1, q_2) G_{i j_3 \dots j_n}(\{g\}, q_1 + q_2, q_3, \dots q_n) \\ + \text{permutations.}\end{aligned} \tag{6.32}$$

This recursion equation serves as a starting point for the calculation of the correlation functions in Section VI.C. Note that the inhomogeneity depends on the $(n - 1)$-point function which allows recurrence to the one-point function. In particular for $n = 2$ and $q_1 = -q_2 = q$ one obtains

$$(D + q\nabla + y_j + y_k - d) G_{jk}(\{g\}, q) = \sum A_{ijk}(q, -q)\, 1/V \langle \mathcal{O}_i \rangle_g \tag{6.33}$$

which allows to express G in terms of the expectation values $\langle \mathcal{O}_i \rangle$ by integration up to large q.

In the remainder of this section we introduce explicitly the operators $\mathcal{O}_j(q)$ and derive (6.32).

From the nonlinear part $\mathcal{O}$, ((3.7), of the RG equation, one defines in analogy to (4.5) the coefficients a'

$$-2\int_q h(q)\frac{\delta\mathcal{O}_j^*(q_j)}{\delta S_q}\frac{\delta\mathcal{O}_k^*(q_k)}{\delta S_{-q}} = \sum_i a'_{ijk}(q_j, q_k)\mathcal{O}_i^*(q_j + q_k)$$
$$+ \delta_{q_j+q_k,0}a'_{0jk}(q_j, q_k)V \tag{6.34}$$

where i runs over the indices of all eigenoperators but $\mathcal{O}_0^* = V$. Then the fields λ of

$$H_l = H^* + \sum \mu_i(l)\mathcal{O}_i^* + \sum\lambda_i(q_i\,e^l, l)\mathcal{O}_i^*(q_i e^l) \tag{6.35}$$

obey

$$\frac{d\lambda_i(q)}{dl} = y_i\lambda_i(q) + \sum a'_{ijk}(q, 0)\lambda_j(q)\mu_k + \tfrac{1}{2}\sum a'_{ijk}(q_j, q_k)\lambda_j(q_j)\lambda_k(q_k)\delta_{q,q_j+q_k} \tag{6.36}$$

where

$$\frac{d}{dl} = \frac{\partial}{\partial l} + q\nabla. \tag{6.37}$$

The fields μ obey (4.6) with

$$a'_{ijk} = a'_{ijk}(0, 0). \tag{6.38}$$

To obtain (6.30) from (6.29) one requires

$$\frac{d\kappa_i(q)}{dl} = y_i\kappa_i(q) + \tfrac{1}{2}\sum A_{ijk}(\{g\}, q_j, q_k)\,\kappa_j(q_j)\,\kappa_k(q_k)\,\delta_{q,q_j+q_k} \tag{6.39}$$

and expands

$$\mathcal{O}_j(\{g\}, q) = \sum_i t_{ij}(\{g\}, q)\,\mathcal{O}_i^*(q). \tag{6.40}$$

Insertion of (6.40) in (6.36) and requiring (6.39) yields for the coefficients of $\kappa_j(q)$

$$(D + q\nabla + y_j - y_i)\,t_{ij} = \sum_{ks} a'_{iks}(q, 0)\,\mu_s\{g\}\,t_{kj} \tag{6.41}$$

and for the coefficients of $\kappa_j(q_1)\kappa_k(q_2)$

$$\sum_{j'k'} a'_{ij'k'}(q_1, q_2)\,t_{j'j}(q_1)t_{k'k}(q_2) = \sum_{i'} t_{ii'}(q_1 + q_2)\,A_{i'jk}(q_1, q_2) \tag{6.42}$$

where

$$\frac{d}{dl} = D + q\nabla \tag{6.43}$$

is substituted in (6.41) since t depends on g and q. From (6.27) one obtains

$$t_{ij}(q) = \delta_{ij} + O(g). \tag{6.44}$$

Iteration of (6.41) yields

$$t_{ij} = \delta_{ij} - \sum_s q^{y_i - y_j - y_s} g_s \left(\int_q^\infty \mathrm{d}p \, p^{y_s + y_j - y_i - 1} \, a'_{ijs}(p, 0) \right.$$

$$\left. + \text{ const.} \right) + O(g^2) \qquad (6.45)$$

where the constant is chosen so that t_{ij} is regular for $q = 0$. This can be done provided $y_i - y_j - y_s + 1$ is not a positive integer in which later case a logarithmic term would appear. We disregard logarithmic terms in our discussion.

Similarly one obtains for the inverse matrix p

$$\kappa_j(q) = \sum_i p_{ji}(q) \, \lambda_i(q) \qquad (6.46)$$

the differential equation

$$(D + q\nabla + y_i - y_j) \, p_{ji} + \sum_{sk} p_{jk} \, a'_{kis}(q, 0) \, \mu_s\{g\} = 0. \qquad (6.47)$$

Multiplication of (6.35) by p yields

$$A_{ijk}(q_1, q_2) = \sum_{i'j'k'} p_{ii'}(q_1 + q_2) \, a'_{i'j'k'}(q_1, q_2) \, t_{j'j}(q_1) \, t_{k'k}(q_2). \qquad (6.48)$$

It is expected that $A_{ijk}(q_1, q_2)$ is a smooth function of the wave vectors q and of the fields g since these properties are expected for a', p, and t.

According to (6.30) and (6.39) one obtains for infinitesimal $l = \delta$

$$H_\delta = H\{g \exp(y\delta)\} + \sum_j \kappa_j \exp(y_j \delta) \, \mathcal{O}_j \, (g \exp(y\delta), q_j \, \mathrm{e}^\delta)$$

$$+ \, \delta \sum_i A_{ij_1j_2}(q_{j_1}, q_{j_2}) \, \kappa_{j_1} \kappa_{j_2} \mathcal{O}_i(q_{j_1} + q_{j_2}) + \text{permutations}. \qquad (6.49)$$

Since $F(H_\delta) = \exp(d\delta) \, F(H_0)$ and the cumulant

$$G_{j_1 \ldots j_n}(q_1, \ldots q_n) = (-)^{n+1} \frac{\partial^n F}{\partial \kappa_{j_1} \cdot \ldots \partial \kappa_{j_n}} \qquad (6.50)$$

one obtains (6.32).

C. Correlation functions for finite wave-lengths near T_c

The solution of the recursion eqn (6.32) yields the correlation functions for non-exceptional momenta in terms of the expectation values $\langle \mathcal{O}_i \rangle$

$$\langle \mathcal{O}_j(q) \, \mathcal{O}_k(-q) \rangle = \sum_i M_{ijk}(q, -q) \langle \mathcal{O}_i \rangle \qquad (6.51)$$

$$\langle \mathcal{O}_j(q_1) \, \mathcal{O}_k(q_2) \, \mathcal{O}_m(q_3) \rangle = \sum_i M_{ijkm}(q_1, q_2, q_3) \langle O_i \rangle \qquad (6.52)$$

for $q_1 + q_2 + q_3 = 0$ etc. (The momenta $q_1, q_2, \ldots q_n$ are called non-exceptional if no subset of the wave-vectors q obeys $\Sigma q = 0$). The expansion coefficients M can be expressed in terms of regular functions R (provided that no logarithmic singularities appear) and singular functions S

$$M_{ijk}(q, -q) = R_{ijk}(q, -q) + S_{ijk}(q, -q) \tag{6.53}$$

$$\begin{aligned} M_{ijkm}(q_1, q_2, q_3) &= R_{ijkm}(q_1, q_2, q_3) + S_{ijkm}(q_1, q_2, q_3) \\ &\quad + \sum_s S_{ijs}(q_1)\, R_{skm}(q_2, q_3) + 2 \text{ permutations} \end{aligned} \tag{6.54}$$

where the functions S are homogeneous functions

$$\begin{aligned} S_{i, j_1, \ldots j_n}(q_1, \ldots q_n, \{g\}) &= \exp\left[(\Sigma y_j - y_i)\, l\right] \\ &\quad \times S_{ij_1, \ldots j_n}(q_1\, \mathrm{e}^l, \ldots q_n\, \mathrm{e}^l, \{g\, \mathrm{e}^{yl}\}). \end{aligned} \tag{6.55}$$

The two-point correlation function is calculated from (6.33)

$$\begin{aligned} G_{jk}(q, -q) = q^{-y_j - y_k + d} \int_q^\infty \mathrm{d}p\, p^{y_j + y_k - d - 1} \\ A_{ijk}\left(p, -p, \left\{g\left(\frac{p}{q}\right)^y\right\}\right) \langle \mathcal{O}_i \rangle_{g(p/q)^y}. \end{aligned} \tag{6.56}$$

The exact scaling relation

$$\langle \mathcal{O}_i \rangle_{g(p/q)^y} = \left(\frac{p}{q}\right)^{d - y_i} \langle \mathcal{O}_i \rangle_g \tag{6.57}$$

yields

$$G_{ij}(q, -q, \{g\}) = \sum_i M_{ijk}(q, -q, \{g\}) \langle \mathcal{O}_i \rangle_g \tag{6.58}$$

with

$$\begin{aligned} M_{ijk}(q, -q, \{g\}) = q^{y_i - y_j - y_k} \int_q^\infty \mathrm{d}p\, p^{y_j + y_k - y_i - 1} \\ \times A_{ijk}\left(p, -p, \left\{g\left(\frac{p}{q}\right)^y\right\}\right) \end{aligned} \tag{6.59}$$

The integration is performed up to infinite p since it is assumed that the correlation function vanishes rapidly in the limit $q \to \infty$.

To discuss the behaviour of M for small q one expands

$$A_{ij_1j_2}(p, -p, \{g\}) = \sum c_{ij_1j_2K}(p, -p)\, g_K \tag{6.60}$$

with

$$g_K = g_{k_1} \cdot \ldots \cdot g_{k_n}. \tag{6.61}$$

Note that for the empty set $K = \varnothing$ one has

$$c_{ij_1j_2}(p, -p) = a'_{ij_1j_2}(p, -p). \tag{6.62}$$

This yields the expansion

$$M_{ij_1j_2} = \sum q^{y_i - y_J - y_K} g_K \int_q^\infty \mathrm{d}p\, p^{y_J + y_K - y_i - 1}\, c_{ij_1ji_K}(p, -p) \tag{6.63}$$

with

$$y_J = y_{j_1} + y_{j_2}, y_K = y_{k_1} + \ldots + y_{k_n}. \tag{6.64}$$

Therefore one has to evaluate integrals

$$I(q) = q^z \int_q^\infty \mathrm{d}p\, p^{-z-1}\, c(p). \tag{6.65}$$

The functions $c(p)$ are regular at small p and we split off

$$c(p) = c_0 + c_1 p + \ldots c_v p^v + c^R(p) \tag{6.66}$$

where

$$c^R(p) = O(p^{v+1}) \tag{6.67}$$

with

$$z - 1 < v < z \tag{6.68}$$

We exclude the discussion for integer z which yields logarithmic terms. Then one obtains

$$I(q) = -\sum_{m=0}^{v} \frac{c_m q^m}{z - m} + q^z \int_q^\infty \mathrm{d}p\, p^{-z-1}\, c^R(p) \tag{6.69}$$

with the nonanalytic contribution

$$q^z \int_0^\infty \mathrm{d}p\, p^{-z-1}\, c^R(p) \tag{6.70}$$

so that $I(q)$ consists of the regular contribution $I_{\mathrm{reg}}(q)$ and the nonanalytic term (6.70)

$$I(q) = I_{\mathrm{reg}}(q) + q^z \int_0^\infty \mathrm{d}p\, p^{-z-1}\, c^R(p). \tag{6.71}$$

Substitution of this result into (6.63) yields (6.53) where R comes from the regular part $I_{\text{reg}}(q)$ and S contains the nonanalytic terms

$$S_{ij_1j_2}(q) = \sum_K q^{y_i - y_J - y_K} g_K \int_0^\infty dp\, p^{-y_i + y_J + y_K - 1} c^R_{ij_1j_2K}(p) \tag{6.72}$$

from which (6.55) for $n = 2$ follows. The singular part

$$G^s_{jk}(q, -q) = \sum S_{ijk}(q, -q) \langle \mathcal{O}_i \rangle \tag{6.73}$$

obeys scaling.

On approaching the critical point, the scaling fields of the relevant operators tend to zero $g_i^{\text{rel}} \to 0$ whereas the scaling fields of the irrelevant operators approach some constant. Therefore one is interested in the scaling behaviour for the transformation

$$q \to b^{-1} q, \qquad g_i^{\text{rel}} \to b^{-y_i} g_i^{\text{rel}}, \qquad g^{\text{irr}} \to g^{\text{irr}}. \tag{6.74}$$

Since y is positive for relevant and negative for irrelevant operators, one finds in the critical limit $b \to \infty$ that the terms g/q^y are constant for relevant scaling fields and vanish for irrelevant scaling fields. This yields the leading scaling behaviour

$$G^{\text{scal}}_{jk}(q, -q) = \sum_i |q|^{y_i - y_j - y_k} S_{ijk}\left(\frac{q}{|q|}, \frac{-q}{|q|}, \left\{\frac{g_{\text{rel}}}{|q|^y}\right\}\right) \langle \mathcal{O}_i \rangle_{\text{scal}} \tag{6.75}$$

where $\langle \mathcal{O}_i \rangle_{\text{scal}}$ denotes the leading scaling behaviour of $\langle \mathcal{O}_i \rangle$. All operators $\mathcal{O}_i$ contribute in (6.75) indicating a complicated behaviour of the leading scaling function.

If $y_j + y_k > d$, then the scaling part (6.75) of the correlation function dominates the regular contribution

$$G^{\text{reg}}_{jk}(q, -q, \{g\}) = \sum_i R_{ijk}(q, -q) \langle \mathcal{O}_i \rangle \tag{6.76}$$

If however $y_j + y_k < d$, then the regular contribution may yield the leading term. If the critical exponent α of the specific heat is negative ($2 - \alpha = d/y_E$), then the energy-energy correlation does not become singular since its scaling part behaves like $q^{d - 2y_E} = q^{-\alpha y_E}$. In this case the regular part yields a finite limit for the correlation whereas the singular part vanishes for $b \to \infty$, (6.74).

The expansion coefficients M of the three-point correlation function (6.52), obey ($q_1 + q_2 + q_3 = 0$)

$$Q_3 M_{ijkm}(q_1, q_2, q_3) = \sum_s A_{sjk}(q_1, q_2)\, M_{ims}(q_3, -q_3) + 2 \text{ permutations} \tag{6.77}$$

(see eqn (6.32)) with the differential operator

$$Q_3 = D + q_1\nabla_1 + q_2\nabla_2 + q_3\nabla_3 + y_j + y_k + y_m - y_i. \tag{6.78}$$

To solve (6.77) one first introduces

$$M_{ijk}(q_1, q_2) = \int_1^\infty \mathrm{d}\lambda\, \lambda^{y_j + y_k - y_i - 1} A_{ijk}(q_1\lambda, q_2\lambda, \{g\lambda^y\}). \tag{6.79}$$

This equation reduces for $q_1 = -q_2 = q$ to the definition (6.59) of $M_{ijk}(q, -q)$ with $p = \lambda q$. From the definition (6.79) it follows that

$$Q_2 M_{ijk}(q_1, q_2) = A_{ijk}(q_1, q_2) \tag{6.80}$$

where

$$Q_2 = D + q_1\nabla_1 + q_2\nabla_2 + y_j + y_k - y_i. \tag{6.81}$$

Separation of M into a contribution R regular at $q_1 = q_2 = 0$ and a homogeneous contribution S yields

$$M_{ijk}(q_1, q_2) = R_{ijk}(q_1, q_2) + S_{ijk}(q_1, q_2) \tag{6.82}$$

$$Q_2 S_{ijk}(q_1, q_2) = 0 \tag{6.83}$$

$$Q_2 R_{ijk}(q_1 . q_2) = A_{ijk}(q_1, q_2). \tag{6.84}$$

Substitution of the ansatz (6.54) into (6.77) yields with (6.83) and (6.84)

$$Q_3(R_{ijkm} + S_{ijkm}) = \sum_s A_{sjk}(q_1, q_2) R_{ims}(q_3, -q_3) + 2 \text{ permutations}. \tag{6.85}$$

Again one requires R to be regular at $q_1 = q_2 = q_3 = 0$ and S to be a homogeneous function

$$Q_3 R_{ijkm} = \sum_s A_{sjk} R_{ims} + 2 \text{ permutations} \tag{6.86}$$

$$Q_3 S_{ijkm} = 0 \tag{6.87}$$

where S is determined for non-exceptional momenta by the boundary condition

$$\lim_{\lambda \to \infty} (\lambda^{y_j + y_k + y_m - y_i} M_{ijkm}(\lambda q_1, \lambda q_2, \lambda q_3, \{g\lambda^y\}) = 0 \tag{6.88}$$

In a similar way one can proceed to calculate higher order correlations.

D. Scaling fields for inhomogeneous perturbations, universality of scaling functions

In this section scaling functions w for the inhomogeneous perturbations are introduced so that

$$H_0 = H(\{g_i(0)\}, \{w_i(q, 0)\}) \tag{6.89}$$

transforms into

$$H_l = H(\{g_i(l)\}, \{w_i(q, l)\}) \tag{6.90}$$

with

$$g_i(l) = \exp(y_i l)\, g_i(0) \tag{6.91}$$

$$w_i(q, l) = \exp(y_i l)\, w_i(q\, \mathrm{e}^{-l}, 0) \tag{6.92}$$

Four statements can be derived (For (i) and (ii) see Wegner (1975a)):

(i) Requiring that the scaling fields are regular at $q = 0$ one obtains

$$w_i(q) = \kappa_i(q) - \frac{1}{2!} \sum R_{ijk}(q_1, q_2)\, \kappa_j(q_1)\, \kappa_k(q_2)\, \delta_{q, q_1+q_2}$$
$$+ \frac{1}{3!} \sum R_{ijkm}(q_1, q_2, q_3)\, \kappa_j(q_1)\, \kappa_k(q_2)\, \kappa_m(q_3)\, \delta_{q, q_1+q_2+q_3} - \cdots \tag{6.93}$$

The summation runs over all wave-vectors.

(ii) The free energy of the Hamiltonian (6.29) equals the free energy of of $H\{\tilde{g}\}$ with

$$\tilde{g}_i = g_i + w_i(o) - \frac{1}{2!} \sum{}' S_{ijk}(q, -q)\, w_j(q)\, w_k(-q)$$
$$+ \frac{1}{3!} \sum{}' S_{ijkm}(q_1, q_2, q_3)\, w_j(q_1)\, w_k(q_2)\, w_m(q_3)\, \delta_{q_1+q_2+q_3, 0} - \cdots \tag{6.94}$$

The summation runs over non-exceptional wave-vectors.

(iii) Under similar provisions as for the statement that F does not depend on the scaling fields g of redundant operators, it is argued that F does not depend on the scaling fields $w_\lambda(q)$ of the redundant operators $O_\lambda(q)$.

$$\partial F / \partial w_\lambda(q) = 0. \tag{6.95}$$

In particular one concludes from this statement that the scaling part (6.73) of the correlation function of an operator with a redundant operator vanishes.

(iv) The scaling function G^{scal}, equation (6.75) is universal

$$\hat{G}^{\mathrm{scal}}_{j_1 j_2}(\hat{g}_E, \hat{g}_h, q) = \hat{\lambda}_{j_1} \hat{\lambda}_{j_2}\, G^{\mathrm{scal}}_{j_1 j_2}(\hat{\lambda}_E g_E, \hat{\lambda}_h g_h, q) \tag{6.96}$$

provided the Hamiltonian and the operators are rotationally invariant in coordinate space.

Finally we close with a summary why RG equations have to be nonlinear.

1. *Scaling fields*

According to (6.92) the scaling fields obey

$$\frac{\mathrm{d}w_i}{\mathrm{d}l} = \frac{\partial w_i}{\partial l} + q\nabla w_i = y_i w_i. \tag{6.97}$$

In view of (6.39) one expects

$$w_i(q) = \kappa_i(q) + O(\kappa^2).$$

Substitution of the expansion

$$w_i(q) = \kappa_i(q) - \frac{1}{2!} \sum P_{ijk}(q_1, q_2)\, \kappa_j(q_1)\, \kappa_k(q_2)\, \delta_{q, q_1+q_2} + \dots \tag{6.98}$$

in (6.97) yields with (6.39)

$$\frac{1}{2!} \sum_{jk} (A_{ijk}(q_1, q_2) - Q_2 P_{ijk}(q_1, q_2))\, \kappa_j(q_1)\, \kappa_k(q_2)\, \delta_{q, q_1+q_2} + \dots = 0 \tag{6.99}$$

Thus $P_{ijk}(q_1, q_2)$ obeys

$$Q_2 P_{ijk}(q_1, q_2) = A_{ijk}(q_1, q_2). \tag{6.100}$$

From the comparison with (6.84) and the requirement that $P_{ijk}(q_1, q_2)$ is regular at $q_1 = q_2 = 0$, it follows that

$$P_{ijk}(q_1, q_2) = R_{ijk}(q_1, q_2). \tag{6.101}$$

Similarly one shows

$$P_{ijkm}(q_1, q_2, q_3) = R_{ijkm}(q_1, q_2, q_3) \tag{6.102}$$

etc.

2. *Free energy as a function of the scaling fields*

From (6.89) to (6.92) and (2.37) one obtains the homogeneity relation

$$\frac{\partial^2 F\{g\}}{\partial w_j(q)\, \partial w_k(-q)} = \exp\left[(y_j + y_k - d)\, l\right] \frac{\partial^2 F\{g\, \mathrm{e}^{yl}\}}{\partial w_j(q\, \mathrm{e}^l)\, \partial w_k(-q\, \mathrm{e}^l)}. \tag{6.103}$$

We find that

$$\begin{aligned}
\frac{\partial^2 F}{\partial w_j(q)\, \partial w_k(-q)} &= \frac{\partial^2 F}{\partial \kappa_j(q)\, \partial \kappa_k(-q)} + \sum_i \frac{\partial F}{\partial \kappa_i(0)} \frac{\partial^2 \kappa_i}{\partial w_j(q)\, \partial w_k(-q)} \\
&= -\frac{1}{V} \langle \mathcal{O}_j(q)\, \mathcal{O}_k(-q) \rangle_c + \sum R_{ijk}(q, -q) \frac{1}{V} \langle \mathcal{O}_i \rangle \\
&= -\sum_i S_{ijk}(q, -q) \frac{1}{V} \langle \mathcal{O}_i \rangle
\end{aligned} \tag{6.104}$$

which indeed is a homogeneous function. Ma (1974) who used these scaling functions has pointed out that $\mathcal{O}_j(q)\mathcal{O}_k(-q)$ does not show the exact scaling behaviour but $\mathcal{O}_j\mathcal{O}_k + \mathcal{O}_{jk}$ does it with

$$\mathcal{O}_{jk}(q, -q) = -\frac{\partial^2 H}{\partial w_j(q)\,\partial w_k(-q)}. \tag{6.105}$$

One finds

$$\mathcal{O}_{jk}(q, -q) = -\sum_k R_{ijk}(q, -q)\,\mathcal{O}_i \tag{6.106}$$

in agreement with Ma.

Similarly one shows that

$$\frac{\partial^3 F}{\partial w_j(q_1)\partial w_k(q_2)\,\partial w_m(q_3)} = \sum_i S_{ijkm}(q_1, q_2, q_3)\frac{1}{V}\langle\mathcal{O}_i\rangle \tag{6.107}$$

etc. Since $\langle\mathcal{O}_i\rangle/V = \partial F/\partial g_i$ the free energy $F\{g, w\}$ can be expressed

$$F\{g, w\} = F\{\tilde{g}, 0\} \tag{6.108}$$

with (6.94). This confirms and refines a conjecture by Fisher (1962) (compare Riedel and Wegner, 1969) according to which inhomogeneous perturbations to a system change the critical parameters but leave the structure of the free energy basically unchanged. The condition for criticality is $\tilde{g}_i = 0$ for all relevant operators. Thus at criticality one has

$$g_i^{\text{rel}} = \tfrac{1}{2}\sum M_{ijk}(q, -q)\,\kappa_j(q)\,\kappa_k(-q) + \dots. \tag{6.109}$$

3. *Redundant operators*

In Section V.B it is argued that the free energy does not depend on the fields q_λ of the redundant operators. Similar arguments can be given for the fields $w_\lambda(q)$ (6.95). Since $\mathcal{O}_\lambda$ depends now on q, the operator D has to be replaced by $D + q\nabla$. The assumptions (i) and (ii) read now: It is presupposed that

(i) the equations

$$K(\mathcal{D} + \Psi^*, \chi(q)) - (z + q\nabla)\,\chi(q) = \Theta(q) \tag{6.110a}$$

$$\mathcal{G}_{\text{tra}}(\chi(q))\,H^* = 0 \tag{6.110b}$$

can be solved for any function $\Theta(q)$ analytic in q which obeys

$$\mathcal{G}_{\text{tra}}(\Theta(q))\,H^* = 0 \tag{6.111}$$

and for any $z \neq z_i$ (for a set of numbers z_i),

(ii) the inequalities

$$y_i - y_{j_1} - \ldots - y_{j_n} \neq 0, 1, 2, \ldots \tag{6.112}$$

and

$$z \neq y_{j_1} + \ldots + y_{j_n} \tag{6.113}$$

hold.

If the inequalities (6.112) hold, then the RG equations can be determined so that (6.95) holds without the condition (i) and (6.113).

From (6.95) it follows that the correlation of an operator with a redundant operator obeys

$$G_{j\lambda}(q, -q) = -\frac{\partial^2 F}{\partial \kappa_j(q)\, \partial \kappa_\lambda(-q)} = -\frac{\partial^2 F}{\partial w_j(q)\, \partial w_\lambda(-q)}$$

$$-\sum_i \frac{\partial w_i(0)}{\partial \kappa_j(q)\, \partial \kappa_\lambda(-q)} \frac{\partial F}{\partial w_i(0)} = \sum_i R_{ij\lambda}(q, -q) \frac{1}{V} \langle \mathcal{O}_i \rangle \tag{6.114}$$

Therefore the singular part G^S_{jk}, eqn (6.73), vanishes, and since $R(q, -q)$ is regular in q the correlation function $\langle \mathcal{O}_j(r_1)\, \mathcal{O}_\lambda(r_2) \rangle$ is short-range.

There remains an obstacle in the proof of (6.95). Since the stress tensor is marginal, $y = 0$ (Section III.G) unless it is redundant the inequality (6.112) is violated for $y_i = y_{j_1}, y_{j_2} = 0$. The condition (6.112) is introduced to ensure the existence of regular solutions t_{ij} and p_{ij} at $q = 0$. If we restrict ourselves to a Hamiltonian $H\{g\}$ and perturbations $\mathcal{O}_j(q) = \int \mathrm{d}^d r\, \tilde{\mathcal{O}}_j(r)\, \mathrm{e}^{-iqr}$, similarly $\mathcal{O}_k(-q)$ where $H\{g\}$ and $\tilde{\mathcal{O}}_j(0)$, $\tilde{\mathcal{O}}_k(0)$ are isotropic in coordinate space, then one can show: $a'_{ijk}(q, 0)$ and consequently $t_{ij}(q)$, $p_{ij}(q)$ and $A_{ijk}(q, 0)$ vanish unless $\tilde{\mathcal{O}}_i(0)$ is isotropic. Since the stress-tensor is no scalar, it does not enter the calculation in this isotropic case and the conclusions remain valid.

4. *Universality of the singular part of the scaling function*

In Section V.C arguments have been given that the scaling functions f, f_i, f_{ij}, etc. of the free energy are universal up to arbitrary scales of the fields g. Similarly one shows that the functions $S_{ij_1j_2}$ (6.73), are universal. For different generators of the RG $\mathcal{G}$ and $\hat{\mathcal{G}}$ one obtains

$$\hat{S}_{ij_1j_2}(\hat{g}_E, \hat{g}_h, q) = \hat{\lambda}_i^{-1} \hat{\lambda}_{j_1} \hat{\lambda}_{j_2} S_{ij_1j_2}(\hat{\lambda}_E g_E, \hat{\lambda}_h g_h, q) \tag{6.115}$$

From this equation one deduces with (5.59) and (5.60) that the leading scaling function (6.75) obeys (6.96). Note that no new scale variable for q appears. One deduces for $g_h = 0$ that the correlation length ξ obeys

$$\xi = \frac{c}{\hat{g}_E^\nu} = \frac{c}{(\hat{\lambda}_E g_E)^\nu} \tag{6.116}$$

Since on the other hand the singular part of the density of the free energy obeys (for $g_{\text{irr}} = 0$, $g_h = 0$)

$$F_{\text{sing}} = c'(\hat{g}_E)^{2-\alpha} = c'(\hat{\lambda}_E g_E)^{2-\alpha} \tag{6.117}$$

one obtains that $\xi^d F_{\text{sing}}$ is a universal constant in the limit $g_E \to 0$, which has been conjectured by Stauffer, Ferer and Wortis (1972) and verified in order ε^2 by Aharony (1974).

The universality of the singular part of the correlations (and of the free energy) has been suggested and discussed by several authors. Numerical calculations and ε-expansions for these functions are available. In view of the large number of publications on this subject, we refer only to a few references: Brezin, Le Guillou, and Zinn–Justin (this volume), Ferer and Wortis (1972), Ferer, Moore and Wortis (1973), and Fisher and Tarko (1975).

In the remainder of this section we outline the main steps of the proof. It runs similar to that in Section V.C. One takes the derivative

$$\exp(dl)\frac{\mathrm{d}F}{\mathrm{d}v} = \tilde{g}'_0 \exp(dl) + (\tilde{g}'_E - \sigma_E \tilde{g}_E)\exp(y_E l)\, f^E$$

$$+ (\tilde{g}'_h - \sigma^h \tilde{g}_h)\exp(y_h l)\, f^h + \sum_i{}' (\tilde{g}'_i - \sigma_i \tilde{g}_i)\exp(y_i l)\, f_i + \ldots = 0 \tag{6.118}$$

where one uses (5.72a)

$$f' = -\sigma_E \tilde{g}_E f^E - \sigma_h \tilde{g}_h f^h \tag{6.119}$$

and similarly

$$f'_i = -\sigma_i f_i - \sigma_E \tilde{g}_E f^E_i - \sigma_h \tilde{g}_h f^h_i \tag{6.120}$$

From (6.94) one obtains

$$\tilde{g}'_i = g'_i + w'_i(0) - \tfrac{1}{2}\sum s_{ij_1j_2K}\, g_K\, w_{j_1}(q)\, w_{j_2}(-q)$$

$$\times \left(\frac{w'_{j_1}(q)}{w_{j_1}(q)} + \frac{w'_{j_2}(-q)}{w_{j_2}(-q)} + \frac{g'_K}{g_K} + \frac{s'}{s}\right) \tag{6.121}$$

Substituting of expansions for g'_i and $w'_i(q)$ similar to (5.67), setting $g_{\text{irr}} = 0$, substituting

$$g_i = g_i(l)\exp(-y_i l) \tag{6.122}$$

$$w_j(q) = w_j(q\,\mathrm{e}^l, l)\exp(-y_j l) \tag{6.123}$$

yields an equation of the form

$$\sum U(q\,\mathrm{e}^l, x)\,\mathrm{e}^{xl} = 0 \tag{6.124}$$

for the expansion coefficient of $w_{j_1}(q\,e^l)\,w_{j_2}(-q\,e^l)\,g_K$. From this equation it follows that $U(g\,e^l, x) = 0$. Under the presumption (6.112) one obtains

$$U(q\,e^l, 0) = -\frac{1}{2}\sum((\sigma_{j_1} + \sigma_{j_2} + \sum\sigma_k - \sigma_i)\,s_{ij_1j_2K} + s'_{ij_1j_2K})\,f_i = 0 \qquad (6.125)$$

where it is used that

$$w'_j(q) = \sigma_j(q)\,w_j(q) + \ldots \qquad (6.126)$$

with $\sigma_j(0) = \sigma_j$. Provided the exponents y_i are not degenerate, (6.125) yields (6.115).

5. *Why a* RG *equation nonlinear in H?*

We have used RG equations which are nonlinear in H, that is, Ψ depends on H. What happens if Ψ is chosen independent of H? First of all the scaling laws for the free energy (3.66) and for the correlation functions (3.123), (3.139) in terms of the fields μ and λ would hold exactly which seems desirable. On the other hand we see no possibility for logarithmic factors. Moreover the operators $\mathcal{O}^*_j(q)$ would not become negligible for large q which indicates slowly decaying or even increasing contributions of the Fourier components S_q to $\tilde{\mathcal{O}}_j$ for large q. This can be easily checked for the trivial fixed point. Let us choose

$$\Psi_q = h(q)\left(S_q - \lambda\frac{\delta H}{\delta S_{-q}} - (1-\lambda)\frac{\delta H^*}{\delta S_{-q}}\right) \qquad (6.127)$$

which for $\lambda = 1$ reduces to (2.89) and for $\lambda = 0$ is independent of H. Then (3.176), (3.177), (3.179) and (3.180) apply with

$$\varphi(q) = [u^*_2(q)\,(1 - u^*_2(q))^\lambda/Cq^2]^{1/2} \qquad (6.128)$$

Thus $\varphi(q)$ decays faster than any power in q for positive λ, it grows faster than any power in q for negative λ and it decays proportional q^{-1} for vanishing λ. The slow decay of $\varphi(q)$ for $\lambda = 0$ causes ultraviolet divergencies. For example

$$\tilde{\mathcal{O}}_{2,0} = \exp(-\mathcal{Y})\int_q \varphi^2(q)\,S_q S_{-q} \qquad (6.129)$$

(eqn (3.176)) and the free energy of

$$H = H^* + \mu\int_q \varphi^2(q)\,S_q S_{-q} \qquad (6.130)$$

(compare (2.60)) do not exist because of ultraviolet divergencies. After all the renormalization group was originally invented to overcome these divergencies. The renormalization group equation (in Wilson's sense) is essentially non-linear in the interaction.

References

Abe, R. and Hikami, S. (1974). *Phys. Letters* **47A**, 341.
Aharony, A. (1973a–d). *Phys. Rev.* **B8**, 3342, 3349, 3358, 3363.
Aharony, A. (1974). *Phys. Rev.* **B9**, 2107.
Aharony, A. (1975), this volume.
Aharony, A. and Fisher, M. E. (1973). *Phys. Rev.* **B8**, 3323.
Ahlers, G. (1973). *Phys. Rev.* **A8**, 530.
Anderson, P. W. (1970). *J. Physics* **C3**, 2436.
Anderson, P. W. and Yuval, G. (1971). *J. Phys.* **C4**, 607.
Avdeeva, G. M. and Migdal, A. A. (1972). *Zh. ETF Pis. Red.* **16**, 253; (1972). *JEPT Letters* **16**, 178.
Baker, G. A. (1962). *Phys. Rev.* **126**, 2071.
Baxter, R. J. (1971). *Phys. Rev. Ltt.* **26**, 832.
Baxter, R. J. (1972). *Annals of Physics N.Y.* **70**, 193.
Bell, T. L. and Wilson, K. G. (1974). *Phys. Rev.* **B10**, 3935.
Bell, T. L. and Wilson, K. G. (1975) *Phys. Rev.* **B11**,3431.
Betts, D. D., Guttmann, A. J. and Joyce, G. S. (1971). *J. Phys.* **C4**, 1994.
Brezin, E., Le Guillou, J. C. and Zinn–Justin, J. (1974). *Phys. Rev. Letts* **32**, 473.
Brezin, E., Le Guillou, J. C. and Zinn–Justin, J. (1975). this volume.
Brezin, E., Wallace, D. J. and Wilson, K. G. (1972). *Phys. Rev. Letts* **29**, 591.
Brezin, E., Wallace, D. J. and Wilson, K. G. (1973). *Phys. Rev.* **B7**, 232.
Bruce, A. D., Droz, M. and Aharony, A. (1974). *J. Physics* **C7**, 3673.
Callan, C. G. (1970). *Phys. Rev.* **D2**, 1541.
Chang, T. S., Tuthill, G. F. and Stanley, H. E. (1974). *Phys. Rev.* **B9**, 4882.
Ferer, M., Moore, M. and Wortis, M. (1973). *Phys. Rev.* **B8**, 5205.
Ferer, M., Stauffer, D. and Wortis, M. (1972). AIP Conference Proceedings USA (18th International Conference on Magnetism) **10**, 836.
Ferer, M. and Wortis, M. (1972). *Phys. Rev.* **B6**, 3426.
Fisher, M. E. (1962). *Phil. Mag.* **7**, 1731.
Fisher, M. E. (1967). *Progr. Phys.* **30**, 615.
Fisher, M. E. (1973). *In* Lundqvist, B. and Lundqvist, S. (Eds), "Collective Properties of Physical Systems", Academic Press, New York, p. 16.
Fisher, M. E. (1974). *Rev. Mod. Phys.* **46**, 597.
Fisher, M. E. and Langer, J. S. (1968). *Phys. Rev. Letts.* **20**, 665.
Fisher, M. E. and Pfeuty, P. (1972). *Phys. Rev.* **B6**, 1889.
Fisher, M. E. and Tarko, H. B. (1975). *Phys. Rev.* **B11**, 1217.
Gaunt, D. S. and Domb, C. (1970). *J. Physics* **C3**, 1442.
Golner, G. R. (1973). *Phys. Rev.* **B8**, 339.
Golner, G. R. and Riedel, E. K. (1975a). *Phys. Rev. Letts.* **34**, 271.
Golner, G. R. and Riedel, E. K. (1975b). *Phys. Rev. Letts* **34**, 856.
Green, M. S., Cooper, M. J. and Levelt-Sengers, J. M. H. (1971). *Phys. Rev. Letts.* **27**, 492.
Grover, M. K. (1972). *Phys. Rev.* **B6**, 3546.
Grover, M. K., Kadanoff, L. P. and Wegner, F. J. (1972). *Phys. Rev.* **B6**, 311.
Gunton, J. D. and Buckingham, M. J. (1968). *Phys. Rev.* **166**, 152.
Hemmer, P. C. and Stell, G. (1970). *Phys. Rev. Letts.* **24**, 1284.
Houghton, A. and Wegner, F. J. (1974). *Phys. Rev.* **A10**, 435.
Hubbard, J. (1972). *Phys. Letters.* **39A**, 365.
Hubbard, J. and Schofield, P. (1972). *Phys. Letters.* **40A**, 245.

Jegerlehner, F. (1975). "Lecture Notes in Physics", Springer-Verlag, Berlin, **37**, 114.
Jona-Lasinio, G. (1973) in Lundqvist, B. and Lundqvist, S. (Eds), "Collective Properties of Physical Systems", Academic Press, New York, p. 38.
Joyce, G. S. (1966). *Phys. Rev.* **146**, 349.
Kadanoff, L. P. (1966). *Physics* **2**, 263.
Kadanoff, L. P. (1969). *Phys. Rev. Letts.* **23**, 1430.
Kadanoff, L. P. (1975). *Phys. Rev. Letts.* **34**, 1005.
Kadanoff, L. P., Goetze, W. *et al.* (1967). *Rev. Mod. Phys.* **39**, 395.
Kadanoff, L. P. and Houghton, A. (1975). *Phys. Rev.* **B11**, 377.
Kawasaki, K. (1966). *Phys. Rev.* **150**, 291.
Knops, H. J. F. (1975). Preprint.
Kornblit, A. and Ahlers, G. (1975). *Phys. Rev.* **B11**, 2678.
Kosterlitz, J. M. (1974). *J. Physics* **C7**, 1046.
Kosterlitz, J. M. and Thouless, D. J. (1973). *J. Physics* **C6**, 1181.
Kramers, H. A. and Wannier, G. H. (1941). *Phys. Rev.* **60**, 252.
Landau, L. D. (1937). *Phys. Z. Sowjetunion* **11**, 26.
Larkin, A. I. and Khmel'nitskii, D. E. (1969). *Zh. Eksp. Teor. Fiz.* **56**, 2087, (1969) *JETP* **29**, 1123.
Levelt-Sengers, J. M. H. (1974). *Physica* **73**, 73.
Lieb, E. H. (1967). *Phys. Rev. Letts.* **18**, 1046.
Lublin, D. M. (1975). *Phys. Rev. Letts.* **34**, 568.
Ma, S. (1973). *Rev. Mod. Phys.* **45**, 589.
Ma, S. (1974). *Phys. Rev.* **A10**, 1818.
Mack, G. (1973). *In* Rühl, W. and Vancura, A. (Eds), "Lecture Notes in Physics", Springer-Verlag Berlin, **17**, 300.
Mermin, N. D. (1971a, b) *Phys. Rev. Letts.* **26**, 169, 957.
Mermin, N. D. and Rehr, J. J. (1971). *Phys. Rev. Letts.* **26**, 1155.
Mermin, N. D. and Wagner, H. (1966). *Phys. Rev. Letts.* **22**, 1133.
Migdal, A. A. (1971). *Physics Letters* **37B**, 386.
Mühlschlegel, B. and Zittartz, H. (1963). *Z. Physik* **175**, 553.
Nauenberg, M. and Nienhuis, B. (1974). *Phys. Rev. Letts.* **33**, 1598.
Nicoll, J. F., Chang, T. S. and Stanley, H. E. (1974a). *Phys. Rev. Letts.* **32**, 1446; **E33**, 1524.
Nicoll, J. F., Chang, T. S. and Stanley, H. E. (1974b). *Phys. Rev. Letts.* **33**, 540; **E33**, 1525.
Niemeijer, Th. and van Leeuwen, J. M. J. (1973). *Phys. Rev. Letts.* **31**, 1411.
Niemeijer, Th. and van Leeuwen, J. M. J. (1974). *Physica* **71**, 17.
Onsager, L. (1944). *Phys. Rev.* **65**, 117.
Polyakov, A. M. (1969). *Zh. Eksper. i Teor. Fiz.* **57**, 271; (1970) *JETP* **30**, 151.
Polyakov, A. M. (1970). *Zh. ETF Pis. Red.* **12**, 538; (1970) *JETP Letters* **12**, 381.
Riedel, E. and Wegner, F. (1969). *Physics Letters* **29A**, 77.
Riedel, E. K. and Wegner, F. J. (1972). *Phys. Rev. Letts.* **29**, 349.
Riedel, E. K. and Wegner, F. J. (1974). *Phys. Rev.* **B9**, 294.
Rudnick, J. (1975). *Phys. Rev. Letts.* **34**, 438.
Saul, D. M., Wortis, M. and Jasnow, D. (1975). *Phys. Rev.* **B11**, 2571.
Schäfer, L. (1975). Preprint.
Shukla, P. and Green, M. S. (1974). *Phys. Rev. Letts.* **33**, 1263.
Shukla, P. and Green, M. S. (1975). *Phys. Rev. Letts.* **34**, 436.
Siegert, A. J. F. (1963). "Brandeis Lectures", Benjamin, 3, 159.
Stanley, H. E. and Kaplan, T. A. (1966). *Phys. Rev. Letts.* **17**, 913.

Stauffer, D., Ferer, M. and Wortis, M. (1972). *Phys. Rev.* **B6**, 3426.
Stephen, M. J. and McCauley, Jr. J. L. (1973). *Physics Letters* **44A**, 89.
Swift, J. and Grover, M. K. (1974). *Phys. Rev.* **A9**, 2579.
Symanzik, K. (1970). *Commun. Math. Phys.* **18**, 227.
Symanzik, K. (1971). *Commun. Math. Phys.* **23**, 49.
van Leeuwen, J. M. J. (1975). *Phys. Rev. Letts.* **34**, 1056.
Wegner, F. (1967). *Z. Physik* **206**, 465.
Wegner, F. J. (1971). *J. Math. Phys.* **12**, 2259.
Wegner, F. J. (1972a). *Phys. Rev.* **B5**, 4529.
Wegner, F. J. (1972b). *Phys. Rev.* **B6**, 1891.
Wegner, F. J. (1973a). *In* Gunton, J. D. and Green, M. S. (Eds), "Proceedings of the Conference on the Renormalization Group in Critical Phenomena and Quantum Field Theory", Temple University, Philadelphia, Pa., USA.
Wegner, F. J. (1973b). "Critical Phenomena and the Renormalization Group", VIII Finnish Summer School in Theoretical Solid State Physics, University of Helsinki.
Wegner, F. J. (1974). *J. Physics* **C7**, 2098.
Wegner, F. J. (1975a). *J. Physics* **A8**, 710.
Wegner, F. J. (1975b). "Lecture Notes in Physics", Springer-Verlag Berlin, **37**, 171.
Wegner, F. J. and Houghton, A. (1973). *Phys. Rev.* **A8**, 401.
Wegner, F. J. and Riedel, E. K. (1973). *Phys. Rev.* **B7**, 248.
Widom, B. (1965). *J. Chem. Phys.* **43**, 3898.
Widom, B. and Rowlinson, J. S. (1970). *J. Chem. Phys.* **52**, 1670.
Wilson, K. G. (1969). *Phys. Rev.* **179**, 1499.
Wilson, K. G. (1970). Irvine Conference (unpublished).
Wilson, K. G. (1971a, b). *Phys. Rev.* **B4**, 3174, 3184.
Wilson, K. G. (1972). *Phys. Rev.* **28**, 548.
Wilson, K. G. (1973). *In* Lundqvist, B. and Lundqvist, S. (Eds), "Collective Properties of Physical Systems", Academic Press, New York, p. 68.
Wilson, K. G. and Fisher, M. E. (1972). *Phys. Rev. Letts.* **28**, 240.
Wilson, K. G. and Kogut, J. B. (1974). *Phys. Reports* **12C**, 75.
Wolsky, A. M., Green, M. S. and Gunton, J. D. (1973). *Phys. Rev. Letts.* **31**, 1193.
Zawadowski, A. (1973). In Lundqvist, B. and Lundqvist, S. (Eds), "Collective Properties of Physical Systems", Academic Press, New York, p. 68.

Note added in proof

Some work noticed by the author after completion of this article requires brief mention. Migdal (1975) gives an approximate recursion equation for the gauge-invariant models mentioned in II.A. Nelson (1975) calculates cross-over scaling functions by expressing the free energy as a line integral along a renormalization group trajectory (V.C.). Tuthill *et al.* (1975) calculated the critical exponent η for arbitrary σ and n to order ε^2 using diagram techniques obtaining the final results of Section IV.C and D.

References

Migdal, A. A. (1975). *Zh.ETF Pis. Red.* (in the press).
Nelson, D. R. (1975). *Phys. Rev.* **B11**, 3504.
Tuthill, G. F., Nicoll, J. F., and Stanley, H. E. (1975). *Phys. Rev.* **B11**, 4579.

3. Field Theoretical Approach to Critical Phenomena

E. Brézin, J. C. Le Guillou* and J. Zinn-Justin

Service de Physique Théorique,
Centre d'Etudes Nucléaires de Saclay,
BP n° 2—91190 Gif-sur-Yvette, France

* Present address: Laboratoire de Physique Théorique et Hautes Energies Tour 16, Université Paris VI, 4 place Jussieu, 75230 Paris Cedex 05.

I. Introduction

The theory of critical phenomena has advanced considerably during the last few years by the introduction of Wilson's renormalization group ideas (1971). This method, which establishes the existence of flow equations governing the parameter of the interaction, has been a very powerful tool in the understanding of scaling laws and their universal character. Furthermore the critical behaviour, characterized by exponents, an equation of state, etc., has become for the first time calculable with reasonable accuracy by analytical methods. A thorough and detailed review of these ideas, of their relations with Kadanoff's (1966) original picture, and of their consequences has been written by Wilson and Kogut (1974), so we are not going to repeat its content here. Instead, we shall show how techniques initially developed for the renormalization of relativistic quantum field theories turn out to be of central interest for the understanding of critical behaviour.

The reason for the appearance of field theory in this context may be schematized in the following manner: near the critical point, the system is characterized by two lengths, a lattice spacing a, and a correlation length ξ. The existence of this elementary length a is at the origin of a cut-off Λ in momentum space of order $1/a$. Critical behaviour is expected only when (i) the correlation length ξ is much larger than a, (ii) one explores distances large compared to a in correlation functions. In these conditions one expects, and we shall argue that it is indeed the case, that the discrete theory may be replaced by a continuous one described in terms of fields. For this theory we are in a regime which, in momentum space, is described by a momentum cut-off Λ large compared to $m \equiv \xi^{-1}$ (which enters into the basic propagator), and large compared to the typical momenta p. The existence of this limit in well-specified cases is the central result of renormalization theory and many consequences follow directly just from this property, as for instance the proof of the strong scaling hypothesis for the correlation functions. Furthermore, the existence of a renormalized theory independent of Λ in the large Λ-limit will be seen as the basic reason for the understanding of the flow equations for the interaction with their fixed points from which follow scaling laws and universality. Clearly it is of central importance for such problems to realize that the theory is equivalent to another, which may be described by one length parameter fewer. We shall stress that the renormalization group equations may be understood by this same fact; we shall develop them in two equivalent ways:

(i) as in field theory, we derive differential equations for the renormalized quantities, by expressing that they are Λ-independent;

(ii) or, staying within the usual unrenormalized quantities (which in this problem are the physical ones), we express that when Λ varies the equivalent renormalized theory remains fixed. This second method is closer in spirit to Wilson's procedure.

We shall discuss in some detail the roles of the ε-expansion (introduced by Wilson and Fisher (1972) and developed by Wilson (1972)) and of the $1/n$ expansion (Abe, 1972, 1973a; Wilson, 1973; Ma, 1972, 1973a, b; Ferrell and Scalapino, 1973); from now on ε will denote the difference $4 - d$ (d being the dimension of space), and n the number of components of the order parameter. The first role is to provide simple fixed points of the flow equations, which enables one for instance to calculate exponents by perturbative methods. A more basic role is to give meaning to the critical theory; for it is easy to convince oneself that the perturbation theory for a Landau–Ginzburg–Wilson interaction, which describes the fluctuations of a continuous parameter, is meaningless at the critical temperature for any fixed dimension smaller than four. It will be argued that this difficulty disappears within the ε and $1/n$ expansions. Though the formalism that we are using may sometimes look more abstract we believe that it has several advantages: (i) the renormalization group equations have a much simpler form; all the "irrelevant" operators in the sense of Wilson and Kogut (1974) and Wegner (1972a, b; 1973) may be ignored in the critical domain; their contribution leads to corrections to scaling laws which are studied in Section VIII; (ii) one can derive all the static scaling laws, both for correlation functions and thermodynamic scaling, to all orders in ε; therefore if a discrepancy persists in one of the scaling laws—for instance in the δ, η relation—it cannot be attributed to some higher order in ε; (iii) calculation of the diagrams in renormalized perturbation theory, in which no cut-off is needed, seems to us easier; (iv) one can use the results of renormalization theory on composite operators and short-distance expansions of a product of operators to construct a more detailed analysis of the critical behaviour.

This Chapter is arranged as follows:

In this first Section we give a brief summary of the definition of critical exponents, scaling laws, universality, etc. without discussion.

Section II is a review of perturbation theory. Throughout this article we shall make use of the saddle-point expansion of Feynman path integrals, which generates the loop expansion of field theory. We shall also frequently introduce a Legendre transformation to express all quantities in terms of the expectation value of the order parameter rather than of its source.

Section III gives brief account of the results of renormalization theory. No proof has been included, but we have tried instead to give examples of how the difficulties appear and how they are solved. The nature of some problems encountered later makes it necessary to discuss the renormalization of composite operators and the equations of motion which relate them.

Section IV contains a discussion of mean field theory. A perturbation theory is constructed which at lowest order—the tree approximation—corresponds to mean field theory. It is argued that, order by order, this expansion does not modify the nature of the singularities but merely their coefficients above four dimensions. Conversely this expansion is meaningless below four dimensions; the successive terms are more and more divergent and the strongest singularity is generated by a four-spin local interaction.

Section V discusses the simplest case of renormalization group equations at the critical temperature, and their consequences on the corresponding correlation functions.

Section VI deals with the same problem above T_c, and we use a version of the renormalization group equations which shows explicitly the role of the deviations of the temperature from T_c. It contains also a derivation of the relations $\gamma = \nu(2 - \eta)$, $\alpha = 2 - \nu d$.

Section VII is devoted to scaling laws above and below T_c, the existence of a scaling equation of state, with exponents β and δ related to γ and η, the derivation of the relations $\nu = \nu'$, $\alpha = \alpha'$, $\gamma = \gamma'$ (when defined).

Section VIII deals with correction to scaling laws when one moves away from the critical region. These deviations occur from several origins, either because the renormalized interaction is not at a fixed point, or because one has to consider the effect of "irrelevant" operators. It is shown that the leading corrections are all governed by one new exponent. Logarithmic deviations from power behaviour are considered for the ordinary critical point in four dimensions, or for the tricritical point in three dimensions.

Section IX contains a summary of results of calculations with the ε-expansion. It gives the equations for the fixed point, the critical exponents, the equation of state, the correlation functions and some universal ratios of critical amplitudes.

Section X is devoted to the large n limit and the $1/n$ expansion. Rather than give many details about the actual calculations, we have preferred to discuss why the renormalization group equations are valid in this limit.

Section XI is a discussion of a general n-vector with an arbitrary non-rotationally invariant coupling. The stability of the various fixed points is discussed.

Section XII contains a few applications of the Wilson–Kadanoff short distance expansion of a product of operators to the characterization of the next-to-leading singular behaviour above or below T_c.

A. Characterization of the static critical behaviour

For a review, see for instance Fisher (1967); Kadanoff *et al.* (1967); and Stanley (1971).

Throughout this Chapter we shall use for convenience the language of ferromagnetic transitions. It is a continuous transition which by lowering the temperature gives rise below the critical temperature T_c to a spontaneous magnetization M; the relation between M and $t \equiv T - T_c$ is singular near $t = 0$, and characterized by an exponent β as

$$M \propto (-t)^{\beta}. \tag{1.1}$$

The specific heat in zero field is singular according to

$$C \propto \begin{cases} A_+(t)^{-\alpha} + B_+ & (t > 0) \quad (1.2) \\ A_-(-t)^{-\alpha'} + B_- & (t < 0). \quad (1.2a) \end{cases}$$

For $\alpha < 0$ a cusp is meant by (1.2); for $\alpha = 0$ a logarithmic singularity.

The magnetic susceptibility in zero field is divergent as

$$\chi \propto \begin{cases} C_+(t)^{-\gamma} & (t > 0) \quad (1.3) \\ C_-(-t)^{-\gamma'} & (t < 0) \quad (1.3a) \end{cases}$$

However, (1.3a) makes sense only for the $n = 1$ Ising-like systems which have a scalar order parameter. For $n > 1$, the susceptibility blows up when the applied magnetic field vanishes.

At the critical temperature itself the relation between the applied magnetic field H and the magnetization is

$$H \propto M^{\delta} \qquad (t = 0). \tag{1.4}$$

The spin–spin correlation function has an exponential fall-off characterized by the correlation length ξ

$$\langle S(x)\, S(0)\rangle_{C.} \underset{x \to \infty}{\simeq} \exp - \frac{x}{\xi},$$

TABLE I. Values of critical exponents.

	Mean field theory	Ising model $d = 2$	Ising model $d = 3$*	Heisenberg model $d = 3$*	Spherical model $\varepsilon > 0$
α, α'	0	0, 0	0·12	−0·06	$-\frac{\varepsilon}{2-\varepsilon}$
β	$\frac{1}{2}$	$\frac{1}{8}$	0·31	0·38	$\frac{1}{2}$
γ, γ'	1	$\frac{7}{4}$	1·25	1·38; not defined	$\frac{2}{2-\varepsilon}$
ν, ν'	$\frac{1}{2}$	1	0·64	0·7	$\frac{1}{2-\varepsilon}$
δ	3	15	5·0		$\frac{6-\varepsilon}{2-\varepsilon}$
η	0	$\frac{1}{4}$	0·04	0·05	0

* Approximate results from numerical extrapolations of high-temperature and low-temperature series. The uncertainties are higher for $n = 3$ than $n = 1$ and for the primed indices.

in which the subscript $C.$ in (1.5) means that below T_c one subtracts from the correlation function the square of the magnetization. Near T_c, the correlation length diverges as

$$\xi \propto \begin{cases} f_+(t)^{-\nu} & t > 0 \qquad (1.6) \\ f_-(-t)^{-\nu'} & t < 0. \qquad (1.6a) \end{cases}$$

In the intermediate range, in which x is much smaller than ξ, but still much larger than the lattice constant, one observes the correlation scaling

$$\langle S(x)\, S(0)\rangle_{C.\ a \ll x \ll \xi} \sim (x)^{-(d-2+\eta)}, \tag{1.7}$$

or equivalently

$$\int \mathrm{d}^d x \exp(ik \cdot x) \langle S(x)\, S(0)\rangle_{C.\ m=\xi^{-1} \ll k \ll a^{-1}=\Lambda} \sim (k)^{-(2-\eta)}. \tag{1.8}$$

The nine exponents β, α, α', γ, γ', δ, ν, ν', η are listed in Table I for a few cases.

B. Correlation and thermodynamic scaling

Several observations go under the name of scaling laws. First of all the power behaviour of the singularities of IA is by itself a scaling property. For instance (1.1) means that there exists a rescaling of $T - T_c$ and of the magnetization which leaves this equation invariant. At many stages in the following we shall be concerned with the derivation of such properties. In particular (1.8) will be of central interest.

Secondly, Widom (1965), Domb and Hunter (1965), Kadanoff *et al.* (1967), and others have made a series of phenomenological observations. The nine critical exponents previously defined satisfy seven relations and thus two of them are sufficient to determine all the others; these relations are

$$\alpha = \alpha', \qquad \nu = \nu', \qquad \gamma = \gamma' \tag{1.9}$$

$$\alpha = 2 - \nu d \tag{1.10}$$

$$\delta = \frac{d + 2 - \eta}{d - 2 + \eta} \tag{1.11}$$

$$\beta = \frac{\gamma}{\delta - 1} \qquad \alpha + 2\beta + \gamma = 2 \tag{1.12}$$

$$\gamma = \nu(2 - \eta). \tag{1.13}$$

They are exactly verified for the two-dimensional Ising model and the spherical model for any value of $d \leqslant 4$. In three dimensions the accuracy is not equally good for all these relations, and in particular up to now the relation (1.11) seems to lie beyond the range allowed by the numerical uncertainties.

Thirdly, the magnetic equation of state, i.e. the relation between applied field, temperature and magnetization

$$H = H(M, T) \tag{1.14}$$

takes the Widom (1965) scaling form

$$\frac{H}{M^{\delta}} = f\left(\frac{t}{M^{1/\beta}}\right), \tag{1.15}$$

in which the function $f(x)$ is regular around $x = 0$. Such a relation, again true for the exactly soluble cases, is well supported by experiment and by numerical extrapolations.

Finally, the critical behaviour characterized in particular by the nine exponents and the equation of state (1.15) is remarkably "universal". By this, we mean that one observes throughout a large class of systems the same critical behaviour, provided the dimension of space and the number of components of the order parameter is fixed, though the nature of the interaction, the lattice structure, etc., vary widely. This extraordinarily puzzling observation has been the main reason for the recent wide interest in the understanding of critical phenomena. We believe that the renormalization group method has provided a major breakthrough towards the explanation of these facts.

II. Perturbation Theory

In this Section we shall review the perturbation theory, starting from Feynman's path integral formulation for the partition function Z, in presence of an external magnetic field H (Feynman and Hibbs, 1965; Schwinger, 1951, 1958; Abers and Lee, 1973). In field theoretic language, $Z(H)$ is the generating functional of the complete euclidean Green function, the Schwinger functions, $G^{(N)}$:

$$Z(H) = \sum_{0}^{\infty} \frac{1}{N!} \int \mathrm{d}x_1 \ldots \mathrm{d}x_N \, H(x_1) \ldots H(x_N) \, G^{(N)}(x_1, \ldots, x_N). \tag{2.1}$$

It is given by the functional integral

$$Z(H) = K^{-1} \int [\mathrm{d}S] \exp\left(-\int \mathrm{d}x[\mathscr{H}(S) - S(x)\, H(x)]\right), \tag{2.2}$$

in which the Hamiltonian density $\mathscr{H}$ is a function of the field variables $S(x)$. The constant K is chosen such that Z is normalized to

$$Z(0) = 1. \tag{2.3}$$

A. Perturbation expansion of the partition function

In order to construct an expansion of $Z(H)$, we separate the Hamiltonian into two pieces:

(i) a quadratic part

$$\int \mathrm{d}x\, \mathscr{H}_0(S) \equiv \tfrac{1}{2}\int \mathrm{d}x\, \mathrm{d}x'\, S(x)\, A(x-x')\, S(x'); \tag{2.4}$$

(ii) an "interaction"

$$\int \mathrm{d}x\, V[S(x)].$$

The basic remark is that $S(x)$ and $H(x)$ are conjugate variables in the sense that

$$\frac{\delta}{\delta H(x)}\int [\mathrm{d}S] \exp\left(-\int \mathrm{d}x\, [\mathscr{H}_0(S) - H(x)\, S(x)]\right)$$
$$= \int [\mathrm{d}S]\, S(x) \exp\left(-\int \mathrm{d}x\, [\mathscr{H}_0(S) - H(x)\, S(x)]\right). \tag{2.5}$$

Repeated use of this formula leads to the equation

$$Z(H) = K^{-1} \exp\left[-\int \mathrm{d}x\, V\left(\frac{\delta}{\delta H(x)}\right)\right]\int [\mathrm{d}S] \exp\left(-\int \mathrm{d}x\, [\mathscr{H}_0(S) - HS]\right). \tag{2.6}$$

The remaining functional integral in the r.h.s. of (2.6) is Gaussian, and our definition of such integrals is an extension to a continuous number of variables of the formula, which holds for a discrete number of variables:

$$\int_{-\infty}^{+\infty} \mathrm{d}S_1 \ldots \mathrm{d}S_n \exp -(\tfrac{1}{2}S_i A_{ij} S_j - H_i S_i) = (2\pi)^{n/2}(\det A)^{-1/2}$$
$$\times \exp(\tfrac{1}{2}H_i \Delta_{ij} H_j) \tag{2.7}$$

in which Δ is the inverse of A in the matrix sense $\Delta_{ik}A_{kj} = \delta_{ij}$.

Application of (2.7) to (2.6) gives

$$Z(H) = K'^{-1} \exp\left(-\int \mathrm{d}x\, V\left(\frac{\delta}{\delta H(x)}\right)\right)$$
$$\times \exp\left(\tfrac{1}{2}\int \mathrm{d}x\, \mathrm{d}x'\, H(x)\, \Delta(x-x') H(x')\right), \tag{2.8}$$

in which K' incorporates the factors present in (2.7). Through the normalization condition (2.3) the infinite factors, which appear when one takes the limit of (2.7) for an infinite numbers of variables, cancel. The free propagator $\Delta(x - x')$ is the inverse of A in the operator sense:

$$\int \mathrm{d}x''\, A(x - x'')\, \Delta(x'' - x') = \delta(x - x'). \tag{2.9}$$

In fact, (2.8) is the precise definition that we shall give to the path integral (2.2).

Now, in order to recover the usual perturbation theory for the Schwinger functions, one has simply to note that the expression

$$\frac{\delta}{\delta H(x_1)} \cdots \frac{\delta}{\delta H(x_{2n})} \exp\left(\tfrac{1}{2} \int \mathrm{d}x\, \mathrm{d}x'\, H(x)\, \Delta(x - x')\, H(x')\right)\Bigg|_{H=0} \tag{2.10}$$

is given by Wick's theorem: it is a sum of products of all possible pair contractions. The elementary contraction is the propagator $\Delta(x_i - x_j)$ since

$$\frac{\delta}{\delta H(x_1)} \frac{\delta}{\delta H(x_2)} \exp\left(\tfrac{1}{2} \int \mathrm{d}x\, \mathrm{d}x'\, H(x)\, \Delta(x - x')\, H(x')\right)\Bigg|_{H=0} = \Delta(x_1 - x_2). \tag{2.11}$$

Thus the perturbation expansion of $Z(H)$ is generated in the following way: one first expands (2.8) in powers of V, then V itself in powers of its argument, and one is led to a sum of terms of the type (2.10); Wick's theorem is then applied. The result is conveniently described in terms of diagrams (Feynman diagrams). It is easy to check that with the condition (2.3) there are no vacuum-to-vacuum parts in the diagrams of the $G^{(N)}$.

B. Generating functionals for the connected and the one-particle irreducible correlation functions

The diagrams constructed in the previous section are not in general connected. It is obviously sufficient to know the connected functions $G_{C.}^{(N)}$ from which the $G^{(N)}$ are expressed by simple algebraic relations. The generating functional $W(H)$ of these connected functions

$$W(H) = \sum_{1}^{\infty} \frac{1}{N!} \int \mathrm{d}x_1 \ldots \mathrm{d}x_N\, H(x_1) \ldots H(x_N)\, G_C^{(N)}(x_1, \ldots, x_N) \tag{2.12}$$

is proportional to the free energy. It is well known that W is related to Z by

$$W(H) = \ln Z(H). \tag{2.13}$$

Explicitly, relation (2.13) means

$$G_{C.}^{(1)}(x) = G^{(1)}(x)$$

$$G_{C.}^{(2)}(x_1, x_2) = G^{(2)}(x_1, x_2) - G^{(1)}(x_1) G^{(1)}(x_2)$$

$$G_{C.}^{(3)}(x_1, x_2, x_3) = G^{(3)}(x_1, x_2, x_3) - G^{(2)}(x_1, x_2)G^{(1)}(x_3) \\ - G^{(2)}(x_1, x_3)G^{(1)}(x_2) - G^{(2)}(x_2, x_3)G^{(1)}(x_1) \\ + 2G^{(1)}(x_1)G^{(1)}(x_2)G^{(1)}(x_3) \tag{2.14}$$

etc.

In the following it will be useful to study separately the N-point vertex functions $\Gamma^{(N)}(x_1 \ldots x_N)$ which consist only of one-particle irreducible diagrams (1-P.I.), i.e. the diagrams which do not fall into two pieces by cutting one internal line. The construction of the generating functional of the $\Gamma^{(N)}$ may be described algebraically through a Legendre transformation. Consider the "magnetization"

$$M(x) = \frac{\delta W(H)}{\delta H(x)} \tag{2.15}$$

which is the expectation value of $S(x)$ since

$$M(x) = \frac{\int [\mathrm{d}S]\, S(x) \exp\left(-\int \mathrm{d}x\, [\mathscr{H}(S) - SH]\right)}{\int [\mathrm{d}S] \exp\left(-\int \mathrm{d}x\, [\mathscr{H}(S) - SH]\right)}, \tag{2.16}$$

and the functional $\Gamma(M)$ defined as

$$\Gamma(M) = \sum_2^\infty \frac{1}{N!} \int \mathrm{d}x_1 \ldots \mathrm{d}x_N\, [M(x_1) - G^{(1)}(x_1)] \ldots [M(x_N) - G^{(1)}(x_N)] \\ \times \Gamma^{(N)}(x_1, \ldots, x_N) \tag{2.17}$$

The relation between Γ and W is then:

$$\Gamma(M) + W(H) = \int \mathrm{d}x\, H(x)\, M(x), \tag{2.18}$$

in which H is expressed in terms of M by solving (2.15). The Legendre transformation (2.18) is reciprocal, and in particular

$$H(x) = \frac{\delta \Gamma(M)}{\delta M(x)}. \tag{2.19}$$

This is not the appropriate place to derive the 1-P.I. character of $\Gamma(M)$ (De Dominicis, 1963; De Dominicis and Martin, 1964; Jona-Lasinio, 1964); we shall simply give here the lowest orders.

Since all the systems that we shall consider are translationally invariant (in the absence of an external field) it is simpler to deal with the Fourier transforms of the N-point functions in d-dimensional space:

$$\delta^{(d)}(p_1 + \ldots + p_N)G^{(N)}(p_1, \ldots, p_N) = (2\pi)^{-d(N-1)} \int dx_1 \ldots dx_N \times \exp\left(i \sum_1^N x_j p_j\right) G^{(N)}(x_1, \ldots, x_N) \quad (2.20)$$

and similar definitions for the $G_{C.}^{(N)}$ and $\Gamma^{(N)}$.

Explicit use of (2.15) and (2.18) gives

$$\Gamma^{(2)}(p, -p) = \frac{1}{G_{C.}^{(2)}(p, -p)}. \quad (2.21)$$

Note that with our conventions, the 1-P.I. two-point function $\Gamma^{(2)}$ is the inverse of the full propagator. For $N = 3$, one finds:

$$\Gamma^{(3)}(p_1, p_2, p_3) = -\frac{G_{C.}^{(3)}(p_1, p_2, p_3)}{G_{C.}^{(2)}(p_1, -p_1)G_{C.}^{(2)}(p_2, -p_2)G_{C.}^{(2)}(p_3, -p_3)}$$

which is indeed the amputated three-point function. The equation for $\Gamma^{(4)}$ is quite long and is more simply described by a diagram (Fig. 1).

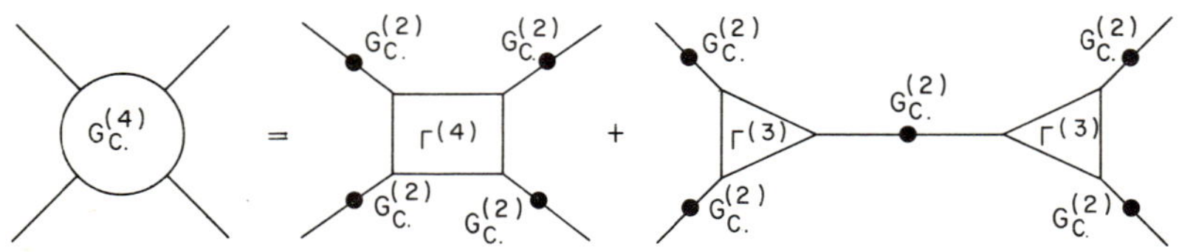

FIG. 1. Relation between connected and 1-P.I. four-point functions.

C. Loop expansion

This topic has been discussed by De Witt (1967), Lee and Zinn-Justin (1972), and Coleman and Weinberg (1973), etc.

Let us assume that we multiply arbitrarily the interaction $\mathscr{H}(S) - S(x)H(x)$ by a constant l (which plays here a role similar to the inverse of Planck's constant in front of the action integral in Quantum Field Theory).

We now expand the functional (2.2) for l large and then characterize this expansion in terms of diagrams. The standard method to obtain this expansion is the steep descent procedure; the first step is to determine the saddle-point $S_0(H)$ defined as the solution of the equations

$$\left.\frac{\delta \mathscr{H}}{\delta S(x)} - H(x)\right|_{S=S_0(H)} = 0 \quad (2.22)$$

In fact $S_0(H)$ is nothing but the magnetization in the mean field approximation (see Section IV), and if there are several solutions to (2.22), one must choose that which gives the lowest potential energy.

We now perform the translation

$$S(x) = S_0 + \phi(x) \tag{2.23}$$

and expand the Hamiltonian in powers of ϕ

$$\int \mathrm{d}x\,[\mathscr{H}(S) - S(x)H(x)]$$

$$= \int \mathrm{d}x\,[\mathscr{H}(S_0) - S_0 H] + \frac{1}{2!}\int \mathscr{H}^{(2)}(x_1, x_2)\,\phi(x_1)\,\phi(x_2) + \ldots + \frac{1}{N!}\int \mathscr{H}^{(N)}(x_1, \ldots, x_N)\phi(x_1)\ldots\phi(x_N) + \ldots. \tag{2.24}$$

in which $\mathscr{H}^{(N)}$ stands for

$$\mathscr{H}^{(N)}(x_1, \ldots, x_N) \equiv \left.\frac{\partial^{(N)}\mathscr{H}}{\partial S(x_1)\ldots\partial S(x_N)}\right|_{S=S_0(H)}. \tag{2.25}$$

The $\mathscr{H}^{(N)}$ depend on the magnetic field H through $S_0(H)$.

We keep in the exponential of (2.2) the first two terms of (2.24) which are constant and quadratic in ϕ, and expand all the higher powers. Their contribution is given again by a Gaussian average and thus one applies Wick's theorem with a "propagator" $l\Delta$ given by

$$\int \mathrm{d}z\, l\Delta(x, z; H)\,\mathscr{H}^{(2)}(z, y) = \delta(x - y). \tag{2.26}$$

At lowest orders one finds

$$Z(H) = K^{-1}\exp\left\{-l\int \mathrm{d}x\,[\mathscr{H}[S_0(H)] - H\,S_0(H)] - \tfrac{1}{2}\operatorname{tr}\ln \mathscr{H}^{(2)}\right\}$$
$$\times \left\{1 + (1/l)\left[-\tfrac{1}{8}\int \mathrm{d}x_1\ldots\mathrm{d}x_4\,\mathscr{H}^{(4)}(x_1, \ldots, x_4)\Delta(x_1, x_2; H)\Delta(x_3, x_4; H)\right.\right.$$
$$+ \int \mathrm{d}x_1\ldots\mathrm{d}y_3\,\mathscr{H}^{(3)}(x_1, x_2, x_3)\,\mathscr{H}^{(3)}(y_1, y_2, y_3)\,(\tfrac{1}{8}\Delta(x_1, x_2; H)$$
$$\left.\times\,\Delta(y_1, y_2; H)\,\Delta(x_3, y_3; H) + \tfrac{1}{12}\Delta(x_1, y_1; H)\,\Delta(x_2, y_2; H)\,\Delta(x_3, y_3; H)\,)\right]$$
$$\left. + O(l^{-2})\right\}. \tag{2.27}$$

In expression (2.27) the following operator formula has been used:

$$\det \mathscr{H}^{(2)}(x, y) = \exp \operatorname{tr} \ln \mathscr{H}^{(2)}. \tag{2.28}$$

This immediately gives the expansion of $W(H) = l^{-1} \ln Z(H)$, and it remains to perform the Legendre transformation (2.18). From the definition (2.15) of $M(x)$ one finds

$$S_0(x) = M(x) + (1/2l) \int \mathscr{H}^{(3)}(y, y_1, y_2; M)\, D(x, y; M)\, D(y_1, y_2; M) \times \mathrm{d}y\, \mathrm{d}y_1\, \mathrm{d}y_2 + O(l^{-2}) \tag{2.29}$$

in which the propagator D is given as

$$\int \mathrm{d}y'\, D(x, y'; M) \left[\frac{\partial^2 \mathscr{H}}{\partial S(y')\partial S(y)}\right]\Bigg|_{S=M} = \delta(x - y).$$

and is now a function of the magnetization.

Similarly the magnetic field is related to the magnetization by†

$$H(x) = \frac{\partial \mathscr{H}}{\partial S}\bigg|_{S=M(x)} + \frac{1}{2l} \int \mathscr{H}^{(3)}(x, y_1, y_2; M)\, D(y_1, y_2; M) + O(l^{-2}) \tag{2.30}$$

We are now in position to perform the Legendre transformation by using the expansions (2.27, 29, 30). The effect is, as expected, to cancel the reducible part of (2.27), and one obtains

$$\begin{aligned}\Gamma(M) = {} & \int \mathrm{d}x\, \mathscr{H}(M) + (1/2l) \operatorname{tr} \ln \mathscr{H}^{(2)}(M) \\ & + 1/l^2 \Bigg[\tfrac{1}{8} \int \mathscr{H}^{(4)}(x_1, x_2, x_3, x_4; M)\, D(x_1, x_2; M)\, D(x_3, x_4; M) \\ & - \tfrac{1}{12} \int \mathscr{H}^{(3)}(x_1, x_2, x_3; M)\, \mathscr{H}^{(3)}(y_1, y_2, y_3; M)\, D(x_1, y_1; M)\, D(x_2, y_2; M) \\ & \times D(x_3, y_3; M) \Bigg] \\ & - \{\text{same expression evaluated at } M = 0\} + O(1/l^3),\end{aligned} \tag{2.31}$$

in which

$$\mathscr{H}^{(N)}(x_1, \ldots, x_N; M) \equiv \frac{\partial^{(N)} \mathscr{H}}{\partial S(x_1) \ldots \partial S(x_N)}\bigg|_{S=M}.$$

† From now on, integration over repeated variables is meant.

In Fig. 2 we give a diagrammatic representation of these equations.

It remains to show (Nambu, 1966; Brown and Boulware, 1968) that the coefficients of the expansion of $\Gamma(M)$ in powers of l^{-1} are the generating functionals of all the 1-P.I. Feynman diagrams with a prescribed number of loops (i.e. the number of independent integration momenta after use of the conservation law at each vertex) but an arbitrary number of external legs.

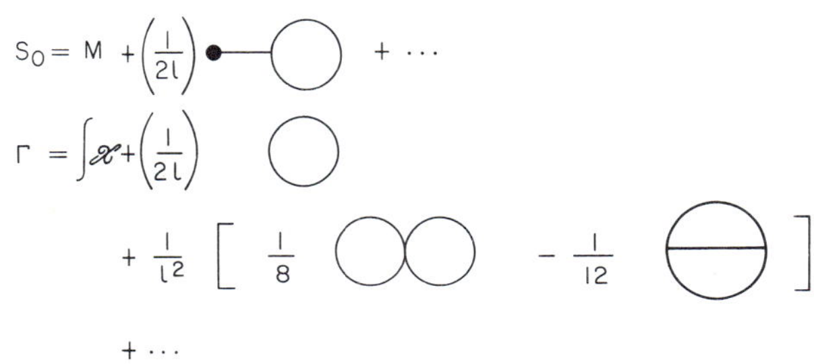

FIG. 2. Diagrammatic representation of equations (2.29) and (2.31).

Indeed, consider a diagram consisting of I internal lines and n_i vertices of type i. The number of loops L is given by counting the number of (independent) conservation laws

$$L = I - \left(\sum_i n_i - 1\right). \tag{2.32}$$

In this diagram, each internal line brings a propagator $l^{-1} D(x_1, x_2)$, and each vertex one power of l, and a factor $1/l$ from the definition of W, thus combining these powers a global factor $(l^{-1})^L$ as stated above.

D. Dimensional continuation of Feynman diagrams

Reference should first be made to the work in this field of Wilson and Fisher (1972), 't Hooft and Veltman (1972), and Bollini and Giambiagi (1972).

It will be necessary in the following to make explicit the dependence of Feynman diagrams in the space-dimension d, in order to define a continuation to non-integer values of d.

This will be done as follows. First, each propagator of the diagram is written in terms of its Laplace transform

$$\Delta(k^2) = \int_0^\infty \mathrm{d}a\, \rho(a) \exp(-ak^2) \tag{2.33}$$

as for instance for the simple propagator

$$(k^2 + m^2)^{-1} = \int_0^\infty \mathrm{d}a \exp(-am^2) \exp(-ak^2). \tag{2.34}$$

Then an arbitrary diagram is expressed as

$$\int \prod_{i=1}^{I} \mathrm{d}a_i\, \rho_i(a_i) \prod_{l=1}^{L} \mathrm{d}^d k_l \exp - [M_{ll'}(a_i)(k_l \cdot k_{l'}) + 2v_l(a_i)\cdot k_l + c(a_i)] \tag{2.35}$$

in which the $L \times L$ matrix $M_{ll'}$ is only function of the a_i's, whereas the vectors v_l are linear combinations of the external momenta and the constant c is quadratic in these momenta.

In order to perform the integral of the Gaussian form (2.35) in the k_l's, we take the following definition of a d-dimensional integral

(i) $$\int \mathrm{d}^d k\, F(k + q) = \int \mathrm{d}^d k\, F(k)$$

(ii) $$\int \mathrm{d}^d k\, F(\lambda k) = |\lambda|^{-d} \int \mathrm{d}^d k\, F(k)$$

(iii) $$\int \mathrm{d}^d k\, \mathrm{e}^{-k^2} = \left[\int_{-\infty}^{+\infty} \mathrm{d}x\, \mathrm{e}^{-x^2}\right]^d = \pi^{d/2}.$$

With these three properties, the k-integrations in (2.35) can be performed and they lead to

$$\int \prod_1^L \mathrm{d}^d k_l \exp - (M_{ll'} k_l \cdot k_{l'} + 2v_l \cdot k_l) = \left(\frac{\pi^L}{\det M}\right)^{d/2} \exp (M^{-1})_{ll'} v_l \cdot v_{l'}. \tag{2.36}$$

Example: let us compute with these prescriptions the integral

$$\mathcal{T} = \int \frac{\mathrm{d}^d k}{k^2 (k + q)^2}.$$

The transformation (2.33, 34) gives

$$\mathcal{T} = \int_0^\infty \mathrm{d}a_1\, \mathrm{d}a_2\, \mathrm{d}^d k \exp - ((a_1 + a_2)k^2 + 2a_2 k \cdot q + a_2 q^2)$$

and from (2.36)

$$\mathcal{T} = \int_0^\infty \mathrm{d}a_1\, \mathrm{d}a_2 \left(\frac{\pi}{a_1 + a_2}\right)^{d/2} \exp - \left(\frac{a_1 a_2}{a_1 + a_2} q^2\right).$$

With the change of variables $a_1 + a_2 = t$, $a_1 = \lambda t$, the variables decouple and the result is

$$\mathcal{T} = \pi^{d/2}(q^2)^{(d/2)-2} \frac{\Gamma(2 - \frac{1}{2}d)\,\Gamma^2(\frac{1}{2}d - 1)}{\Gamma(d - 2)}.$$

It has a simple pole at $d = 4$, which corresponds to the logarithmic divergence of $\mathcal{T}$ in four dimensions.

III. Renormalization Scheme

This Section gives some results of renormalization theory but their derivation goes beyond the scope of this article; the reader interested in the proofs is referred to standard textbooks such as those of Bogoliubov and Shirkov (1959) and Bjorken and Drell (1965), and to the lectures of Callan (1975).

A. The problem of divergences in local field theories

From now on we shall assume that the Hamiltonian density $\mathcal{H}(x)$ is a local polynomial in the field $S(x)$ and its derivatives. Let us first assume that the quadratic form $\mathcal{H}_0$ consists simply of

$$\mathcal{H}_0(x) = \tfrac{1}{2}\{[\nabla S(x)]^2 + m_0^2 S^2(x)\}, \tag{3.1}$$

i.e. an associated propagator which in momentum space is

$$\Delta(p) = [p^2 + m_0^2]^{-1}. \tag{3.2}$$

Again x and p and d-dimensional vectors, and we assume here that the interaction is rotationally invariant in d dimensions. An arbitrary monomial in the interaction V consists of v_i fields $S(x)$ on which act k_i derivatives

Example:	$S^4(x)$	$v = 4$	$k = 0$
	$(\nabla S)^2$	$v = 2$	$k = 2.$

Let us consider an arbitrary diagram with E external legs, n_i vertices of type i, I internal lines. The number L of loops of this diagram is

$$L = I - \left(\sum_i n_i - 1\right). \tag{3.3}$$

The obvious topological relation expressing that an internal line is common to two vertices, whereas an external one goes only through one vertex, gives

$$E + 2I = \sum_i n_i v_i. \tag{3.4}$$

We are now in position to study the properties of convergence—or divergence—of a given diagram γ in momentum space. First we note that it is sufficient to characterize the 1-P.I. diagrams, from which all the others can be obtained without new integration. As long as m_0 does not vanish, which we explicitly assume, the only possible source of divergence of a diagram comes from the large momentum region. Let us define an index $\delta(\gamma)$ attached to the diagram γ characteristic of its overall behaviour when all momenta go to infinity at the same rate. All we have to do is to count the powers of momentum appearing in numerator (coming from the derivatives in the interaction) and in denominator (from the propagators) plus the number of integration variables. The result is

$$\delta(\gamma) = \sum_i k_i n_i - 2I + dL. \tag{3.5}$$

It is convenient to eliminate L and I, using (3.3–4) in order to express $\delta(\gamma)$ in terms of the properties of the vertices; this gives

$$\delta(\gamma) = d - [(d/2) - 1]E + \sum_i n_i \delta_i, \tag{3.6}$$

where δ_i characterizes the interaction vertex i

$$\delta_i = k_i + v_i[(d/2) - 1] - d. \tag{3.7}$$

A diagram for which $\delta(\gamma) \geqslant 0$, is divergent. When $\delta(\gamma) < 0$, the diagram is "superficially" convergent, i.e. no divergence is associated with the global integration, but it is possible to have sub-integrations on some subset of the momenta which are divergent. These divergences are associated with sub-diagrams of γ.

On formula (3.6) one sees that, if an interaction vertex has a positive δ_i, the degree of divergence $\delta(\gamma)$ will increase with the order of perturbation theory. This kind of interaction leads to a non-renormalizable theory, i.e. a theory which cannot be made finite by the procedure indicated in the following. We shall therefore assume for the moment that all our vertices are such that $\delta_i \leqslant 0$. If all vertices are such that $\delta_i < 0$ (super-renormalizable theory), the degree of divergence decreases with the order of the interaction and thus there is only a finite number of superficially divergent diagrams. If $\delta_i = 0$, $\delta(\gamma)$ is independent of n_i; the situation looks the same to all orders (renormalizable theory). In both cases notice that

$$\delta(\gamma) \leqslant d - [(d/2) - 1]E$$

and if the dimension is larger than 2, the divergence will be present for only a finite number of external legs.

Let us now examine what theories are allowed in view of the restriction $\delta_i \leqslant 0$

(i) $d = 4$ implies $k_i + v_i \leqslant 4$ leaving the following possibilities S, S^2, S^3, S^4, $(\nabla S)^2$.

(ii) $d = 3$ $k_i + v_i/2 \leqslant 3$ leaves the same possibility as above plus S^5, S^6.

(iii) $d = 2$ $k_i \leqslant 2$, but no additional restriction; S^n with n arbitrary is allowed, as well as $(\nabla S)^2 S^n$.

B. Regularization

We have seen that, if one just applies perturbation theory as described before, divergences are present. Nevertheless in order to give a meaning to this theory, there is a well-known procedure in Quantum Field Theory which consists in cutting off the integrals in some way (regularization), and then the infinities which would appear when the cut-off goes to infinity are absorbed in a change of parameters, and of normalization. We choose here a particular procedure of regularization which consists in modifying the free-propagator $(p^2 + m_0^2)^{-1}$, for large values of p:

$$\Delta_\Lambda(p) = \int_0^\infty \mathrm{d}x\,[1 - \rho(\Lambda^2 x)] \exp[-x(p^2 + m_0^2)] \tag{3.8}$$

with a smooth function $\rho(z)$ on which the only requirements are that it vanishes for large z and is close to 1 for z small. Quantitatively it is necessary to have $|1 - \rho(z)| < Az^k$ $(z \to 0)$ where k depends on the structure of the theory, and fixes the behaviour of $\Delta_\Lambda(p)$ for p larger than Λ. In order to see which minimum value of k is required, one has to go back to the power-counting argument of the previous section. The regularized propagator modifies the coefficient of I in (3.5). For the S^4 coupling in $d = 4$ it is necessary to take $k > 1$. We are now in position to study what to do when Λ goes to infinity.

C. Renormalization

To illustrate the mechanism of renormalization, we shall deal with a specific theory of an S^4 interaction in four dimensions

$$\mathscr{H}(x) = \tfrac{1}{2}[(\nabla S)^2 + m_0^2 S^2]_\Lambda + \frac{g_0}{4!} S^4, \tag{3.9}$$

in which the index Λ is there to remind us that the propagator is replaced by its regularized version (3.8).

1. *Superficial behaviour of the diagrams*

The power-counting formula (3.6) applied to this theory is simply $\delta = 4 - E$. It indicates that the divergences (when Λ goes back to infinity) are:

—quadratic for the two-point function ($E = 2$)

—logarithmic for the four-point function.

When $E > 4$ the diagrams are superficially convergent.

2. *One-loop diagrams*

Diagrams involving only one loop of integration are simple since they do not have subdiagrams, and thus power-counting corresponds to their true nature of convergence or divergence.

(a) *Two-point function.* At the one-loop approximation, there is only one diagram (Fig. 3)

$$\Gamma^{(2)}(p) = p^2 + m_0^2 + \tfrac{1}{2}g_0 \int \frac{d^4k}{(2\pi)^4} \Delta_\Lambda(k). \tag{3.10}$$

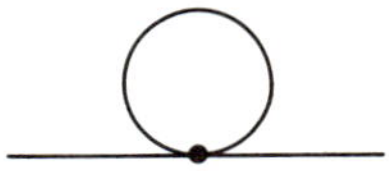

FIG. 3. One-loop contribution to the two-point function.

Let us call m_1^2 the value at zero-momentum of $\Gamma^{(2)}$:

$$\Gamma^{(2)}(0) = m_1^2 \tag{3.11}$$

i.e.

$$m_0^2 = m_1^2 - \tfrac{1}{2}g_0 \int \frac{d^4k}{(2\pi)^4} \Delta_\Lambda(k). \tag{3.11a}$$

It is clear that if one chooses to express $\Gamma^{(2)}$ in terms of the parameter m_1 instead of m_0, all the divergence is absorbed in the relation (3.11a), and m_0^2 is infinite with Λ for fixed m_1. In terms of m_1, $\Gamma^{(2)}$ reads

$$\Gamma^{(2)}(p) = p^2 + m_1^2. \tag{3.12}$$

(b) *Four-point function.* At the same order there are three diagrams (Fig. 4)

$$\Gamma^{(4)}(p_1, p_2, p_3, p_4) = g_0 - \frac{g_0^2}{2}[I_\Lambda(p_1 + p_2) + I_\Lambda(p_1 + p_3) + I_\Lambda(p_1 + p_4)] \tag{3.13}$$

$$I_\Lambda(p) = \int \frac{d^4k}{(2\pi)^4} \Delta_\Lambda(k)\Delta_\Lambda(p + k) \tag{3.14}$$

For large Λ the coefficient of $\ln \Lambda$ in $I_\Lambda(p)$ is independent of p. Therefore

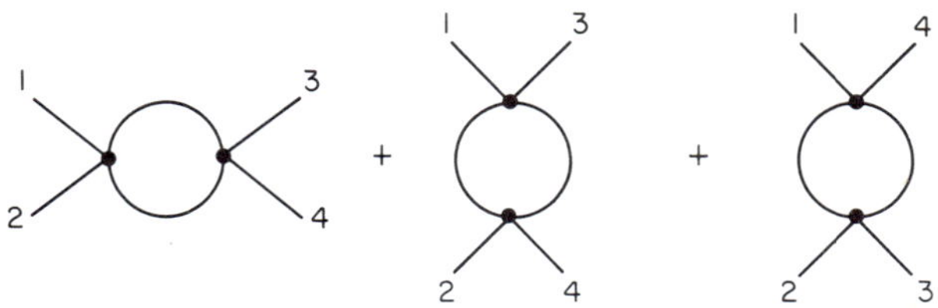

FIG. 4. One-loop contributions to the four-point function.

$I_\Lambda(p) - I_\Lambda(0)$ has a finite limit when Λ goes to infinity. Let us then define a parameter g_1 as:

$$g_1 \equiv \Gamma^{(4)}(0, 0, 0, 0) \tag{3.15}$$

i.e.

$$g_0 = g_1 + \frac{3g_1^2}{2} I_\Lambda(0) + 0(g_1^3). \tag{3.15a}$$

Up to this order the logarithmic divergence of $\Gamma^{(4)}$ is absorbed in (3.15a) and $\Gamma^{(4)}$ is finite in terms of the parameter g_1. Furthermore at this order one can replace m_0 by m_1 and

$$\Gamma^{(4)}(p_i; m_1, g_1) = g_1 - \frac{g_1^2}{2} \lim_{\Lambda \to \infty} [I_\Lambda(p_1 + p_2; m_1) - I_\Lambda(0; m_1) + 2 \text{ terms}] \tag{3.16}$$

(c) *N-point function.* No divergence appears when N is larger than 4. Furthermore since the 1-P.I. diagrams of lowest order involve at least one loop, one can replace m_0 by m_1 and g_0 by g_1 since the difference would come at the two-loop order.

3. *Two-loop diagrams*

When there are more loops a new feature appears. As in the previous case, we shall make a change of parameters, but a new renormalization will also be necessary. To illustrate this point, let us examine the two-point function (Fig. 5)

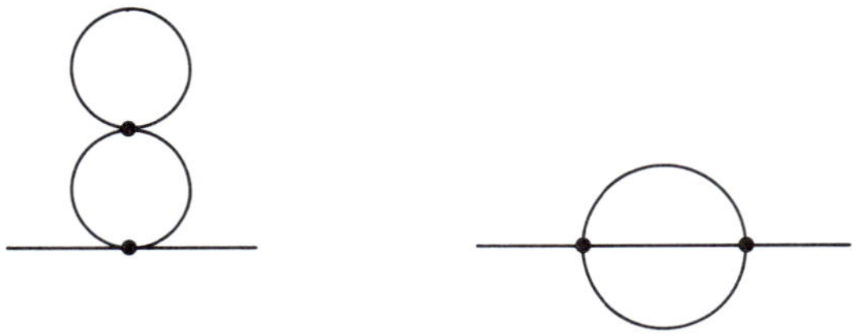

FIG. 5. Two-loop contributions to the two-point function.

$$\Gamma^{(2)}(p) = p^2 + m_0^2 + \frac{g_0}{2}\int \Delta_\Lambda(k)\frac{d^4k}{(2\pi)^4} - \frac{g_0^2}{4}\left[\int \Delta_\Lambda(k)\frac{d^4k}{(2\pi)^4}\right]^2$$
$$- \frac{g_0^2}{6}\int \frac{d^4p_1\, d^4p_2}{(2\pi)^8}\Delta_\Lambda(p + p_1 + p_2)\,\Delta_\Lambda(p_1)\,\Delta_\Lambda(p_2). \qquad (3.17)$$

The conditions (3.11), (3.15) allow us to replace (3.17) by

$$\Gamma^{(2)}(p) = p^2 + m_1^2 - \frac{g_1^2}{6}\int \frac{d^4p_1\, d^4p_2}{(2\pi)^8}\{\Delta_\Lambda(p + p_1 + p_2)\,\Delta_\Lambda(p_1)\,\Delta_\Lambda(p_2)$$
$$- \Delta_\Lambda(p_1 + p_2)\,\Delta_\Lambda(p_1)\,\Delta_\Lambda(p_2)\} \qquad (3.18)$$

(in which we now use the mass m_1 in Δ_Λ).

The last integral in (3.17) was quadratically divergent; the subtraction yielded by the change from (m_0, g_0) to (m_1, g_1) has cancelled the Λ^2-terms, but there is still a logarithmic divergence present which we are now going to eliminate. It is sufficient to note that the integral in (3.18) behaves as $p^2 \ln \Lambda$ for large Λ, and thus absorb this term in a normalization of $\Gamma^{(2)}(p)$. Let us introduce:

$$\Gamma_R^{(2)}(p) = \Gamma^{(2)}(p)\left[\frac{\partial}{\partial p^2}\Gamma^{(2)}(p)\bigg|_{p^2=0}\right]^{-1} \qquad (3.19)$$

and thus by definition

$$\frac{\partial}{\partial p^2}\Gamma_R^{(2)}(p)\bigg|_{p^2=0} = 1. \qquad (3.19a)$$

At this order we have then

$$\Gamma_R^{(2)}(p) = p^2 + m_1^2 - \frac{g_1^2}{6}\int \frac{d^4p_1\, d^4p_2}{(2\pi)^8}\Delta_\Lambda(p_1)\,\Delta_\Lambda(p_2)$$
$$\times\Bigg[\Delta_\Lambda(p + p_1 + p_2) - \Delta_\Lambda(p_1 + p_2) - (p^2 + m_1^2)$$
$$\times\left(\frac{\partial}{\partial p^2}\Delta_\Lambda(p + p_1 + p_2)\right)_{p^2=0}\Bigg]. \qquad (3.20)$$

This change (3.19) of normalization of $\Gamma^{(2)}$ may be realized by a change in the normalization of the field

$$S(x) = Z^{1/2}S_R(x) \qquad (3.21)$$

in which the field strength renormalization Z is

$$Z = \left[\frac{\partial}{\partial p^2}\Gamma^{(2)}(p)\bigg|_{p^2=0}\right]^{-1}. \qquad (3.22)$$

In this change all the vertex functions are modified:

$$\Gamma_R^{(N)}(p_1, \ldots, p_N) = Z^{N/2}\Gamma^{(N)}(p_1, \ldots, p_N) \tag{3.23}$$

and the final finite functions are these $\Gamma_R^{(N)}$. It is therefore necessary to replace the normalizations (3.11) and (3.15) by conditions on $\Gamma_R^{(2)}$ and $\Gamma_R^{(4)}$:

$$\Gamma_R^{(2)}(0) = m^2 \tag{3.24}$$

$$\Gamma_R^{(4)}(0, 0, 0, 0) = g. \tag{3.25}$$

If one now expresses the $\Gamma_R^{(N)}$ in terms of m and g instead of the intermediate parameters m_1 and g_1, they become finite. For instance, the expression (3.20), which still contains a logarithmically divergent part, is replaced by the finite equation:

$$\Gamma_R^{(2)}(p) = p^2 + m^2 - \frac{g^2}{6}\int \frac{d^4p_1\, d^4p_2}{(2\pi)^8}\Delta(p_1; m)\,\Delta(p_2; m)$$

$$\times \left[\Delta(p + p_1 + p_2; m) - \Delta(p_1 + p_2; m) - p^2\left(\frac{\partial}{\partial p^2}\Delta(p + p_1 + p_2; m)\right)_{p^2=0}\right]$$

At this order, diagrams with more than four external legs have divergences, but they correspond to subdiagrams of the two- and four-point functions. They are automatically finite when the two- and four-point functions have been renormalized.

4. *The Hamiltonian and its counter-terms*

The foregoing analysis suggests therefore that the expansion should be made for the Schwinger functions of the field S_R in terms of renormalized parameters m and g. The parameters in terms of which $\mathscr{H}$ is written are the bare ones: $S(x)$, m_0, g_0, and they will be expanded in powers of g. This is done by the following procedure. We re-write the Hamiltonian (3.9) as

$$\mathscr{H}(x) = \tfrac{1}{2}((\nabla S_R)^2 + m^2 S_R^2)_\Lambda + \mathscr{H}_{\text{int}}(x), \tag{3.26}$$

$$\mathscr{H}_{\text{int}}(x) = \frac{g}{4!}(S_R^2)^2 + \tfrac{1}{2}(Zm_0^2 - m^2)S_R^2 + \tfrac{1}{2}(Z - 1)(\nabla S_R)^2 + \frac{1}{4!}g(Z_{(4)} - 1)(S_R^2)^2, \tag{3.27}$$

where the last three terms in the interaction (the counter terms) are power series in g

$$Z - 1 = \sum_2^\infty A_n g^n \tag{3.28a}$$

$$Zm_0^2 - m^2 = \Lambda^2 \sum_1^\infty B_n g^n \tag{3.28b}$$

$$g(Z_{(4)} - 1) \equiv g_0 Z^2 - g = \sum_2^\infty C_n g^n. \tag{3.28c}$$

The A_n, B_n, C_n are dimensionless functions of the ratio m/Λ. These counter terms are determined order by order in powers of g by expressing the three renormalization conditions:

$$\Gamma_R^{(2)}(0) = m^2 \tag{3.29a}$$

$$\frac{\partial}{\partial p^2}\Gamma_R^{(2)}(p^2)\bigg|_0 = 1 \tag{3.29b}$$

$$\Gamma_R^{(4)}(0, 0, 0, 0) = g. \tag{3.29c}$$

The central result of renormalization theory is that the $\Gamma_R^{(N)}(p_1, \ldots, p_N; m, g; \Lambda)$, calculated with the Hamiltonians (3.26), (3.27) in which the counter terms have been determined by the conditions (3.29), have finite limits when Λ goes to infinity. Furthermore this limit is independent of the particular choice of a cut-off function $\rho(x)$ for the propagator in (3.8).

For a general renormalizable Hamiltonian—in which all the δ_i of (3.7) are non-positive—the same theorem holds, i.e. the theory is finite in terms of renormalized mass and coupling constants for the renormalized fields (provided all possible vertices of degree at most equal to the bare ones and compatible with the symmetry of the problem have been added to the Hamiltonian).

D. Composite operators

A useful background to the material in this and in the following subsection (E) will be found in lectures by Zimmermann (1971) and by Lowenstein (1970, 1972).

Up to now we have described the renormalization scheme of correlation functions $\langle S(x_1) \ldots S(x_N)\rangle$ in which the points x_i are distinct. In the following it will also be necessary to study what happens when sets of two or more points coincide, for instance when we deal with the energy density correlation function. In that case, even if we deal with the renormalized functions, the limit $x_i \to x_j$ can be singular and additional renormalizations have to be performed in order to obtain a finite result.

1. *Operator* S^2

In order to define a renormalized S^2 operator, let us first consider the bare 1-P.I. function

$$\delta\left(q + \sum_1^N p_i\right)\Gamma^{(1,N)}(q; p_1, \ldots, p_N) = \int \mathrm{d}y\, \mathrm{d}x_1 \ldots \mathrm{d}x_N \exp i\left[yq + \sum_1^N p_i x_i\right]$$

$$\times \langle \tfrac{1}{2}S^2(y)S(x_1) \ldots S(x_N)\rangle_{1\text{-P.I.}}. \tag{3.30}$$

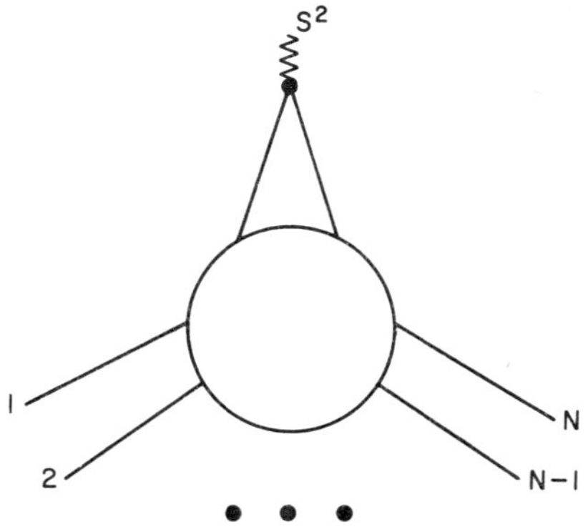

FIG. 6. S^2-insertion with N external lines.

As before the problem will be characterized by the superficial degree of divergence of the corresponding diagrams, which is trivially deduced from (3.6) by noting that an S^2 insertion with N external lines is obtained by closing two lines of an $(N + 2)$-point correlation function (Fig. 6). The result is simply

$$\delta(\gamma) = d - N\frac{d-2}{2} - 2 + \Sigma\, n_i\delta_i, \tag{3.31}$$

and for an S^4 interaction in four dimensions the primitive divergences are

—an infinite constant $\langle S^2 \rangle$ which plays no role in the following
—a logarithmic divergence associated to $N = 2$, i.e.

$$\langle S^2(y) S(x_1)\, S(x_2) \rangle_{1\text{-P.I.}}.$$

It should be clear that this divergence is new and that we cannot get rid of it just by the previous renormalizations. To convince ourselves of this, let us examine the first contributions to

$$\langle \tfrac{1}{2}(S_R)^2(q) S_R(p_1) S_R(p_2) \rangle_{1\text{-P.I.}} = 1 - \frac{g}{2} I_\Lambda(q, m) + O(g^2). \tag{3.32}$$

The integral I_Λ diverges logarithmically in spite of the fact that at this order the field strength and coupling constant renormalization are not present. It is, however, possible to define a finite $(S^2)_R$ operator proportional to $(S_R)^2$, but with a divergent ratio $Z_{(2)}$

$$(S^2)_R = Z_{(2)}(S_R)^2 = \frac{Z_{(2)}}{Z} S^2. \tag{3.33}$$

This S^2-operator renormalization constant is determined by requiring the condition

$$\langle \tfrac{1}{2}(S^2)_R(q)\, S_R(p_1)\, S_R(p_2)\rangle_{1\text{-P.I.}}\big|_{q=p_1=p_2=0} = 1. \tag{3.34}$$

Let us show on the first orders that the condition (3.34) provides the subtraction needed to cancel the infinities in (3.32): given the expansion

$$Z_{(2)} = 1 + \sum_1^{\infty} g^n D_n, \tag{3.35}$$

at first order in g, we find from (3.32), (3.33)

$$\langle \tfrac{1}{2}(S^2)_R(q)\, S_R(p_1)\, S_R(p_2)\rangle_{1\text{-P.I.}} = 1 - \frac{g}{2} I_\Lambda(q; m) + gD_1 + O(g^2) \tag{3.36}$$

with the normalization

$$1 = 1 - \frac{g}{2} I_\Lambda(O; m) + gD_1 + O(g^2)$$

from which follows

$$\langle \tfrac{1}{2}(S^2)_R(q)\, S_R(p_1)\, S_R(p_2)\rangle_{1\text{-P.I.}} = 1 - \frac{g}{2}[I_\Lambda(q) - I_\Lambda(O)] + O(g^2) \tag{3.37}$$

which has a finite limit when Λ goes to infinity. It is possible to show that this persists to all orders.

For diagrams with more than two external S-legs, only subdiagrams of the previous type are divergent and one can show that once they have been renormalized according to the above prescriptions, these functions are finite.

We are now in position to discuss the general situation of a correlation function for L insertions of an S^2-operator in an N-point function:

$$\Gamma^{(L,N)}(q_1, \ldots, q_L; p_1, \ldots, p_N).$$

The power-counting formula is similarly

$$\delta(\gamma) = d - N\frac{d-2}{2} - 2L + \Sigma\, n_i \delta_i, \tag{3.38}$$

and for an S^4 interaction only one new divergence appears corresponding to $N = 0$, $L = 2$: $\langle S^2(x)\, S^2(0)\rangle$ is logarithmically divergent. This function which has no external S line corresponds to diagrams which can never appear as diverging subdiagrams. Consequently this divergence can be removed by a simple subtraction on $\Gamma^{(2,0)}$ and has no effect on the other $\Gamma^{(L,N)}$ (Coleman and Jackiw, 1971; Symanzik, 1973a). We shall then define $\Gamma_R^{(2,0)}(p, -p)$ so that

$$\Gamma_R^{(2,0)}(p, -p)\big|_{p=0} = 0. \tag{3.39}$$

Consequently the relation between bare and renormalized functions is

$$\Gamma_R^{(L,N)}(q_1, \ldots, q_L; p_1, \ldots, p_N; m, g) = \left(\frac{Z_{(2)}}{Z}\right)^L Z^{N/2}\Gamma^{(L,N)}(q_1, \ldots, p_N; m_0, g_0, \Lambda) + \delta_{L2}\delta_{N0}A(m_0, g_0, \Lambda) \tag{3.40}$$

and we summarize the whole set of the renormalization conditions:

$$\Gamma_R^{(0,2)}(p, -p; m, g)\big|_{p=0} = m^2 \tag{3.41a}$$

$$\frac{\partial}{\partial p^2}\Gamma_R^{(0,2)}\big|_{p=0} = 1 \tag{3.41b}$$

$$\Gamma_R^{(0,4)}(p_1, p_2, p_3, -p_1 - p_2 - p_3; m, g)\big|_{p_i=0} = g \tag{3.41c}$$

$$\Gamma_R^{(1,2)}(-p_1 - p_2; p_1, p_2; m, g)\big|_{p_i=0} = 1 \tag{3.41d}$$

$$\Gamma_R^{(2,0)}(p, -p; m, g)\big|_{p=0} = 0. \tag{3.41e}$$

Another convenient way of defining the S^2-insertions is to introduce a source $t(x)$ for this composite field in the functional integral associated to the energy density

$$\mathscr{H}(x) - H(x)\,S(x) + \tfrac{1}{2}t(x)\,S^2(x).$$

From this, one obtains a generating functional $Z(H, t)$. The 1-P.I. functional $\Gamma(M, t)$, which generates the $\Gamma^{(L,N)}$ defined above, is the Legendre transform of $\ln Z$ with respect to the H-source. Another way of stating the previous results is to realize that the corresponding renormalized Hamiltonian is obtained by adding all the possible counter-terms of dimension $\leqslant 4$ in both $S(x)$ and $t(x)$. From the expression of the energy density we see that $t(x)$ has dimension 2. Thus local counter-terms involving $t(x)$ are of the form

$$\tfrac{1}{2}(Z_{(2)} - 1)\,t(x)(S_R)^2(x) + at(x) + bt^2(x).$$

The first term is the additional renormalization of the S^2-insertions; the second and third provide the counter-terms necessary to remove the infinities from $\langle S^2 \rangle$ and $\Gamma^{(2,0)}$. This method allows one to deal in a simple way with all the insertions of operators whose addition to the Hamiltonian gives a renormalizable theory.

2. *Composite operators of higher canonical dimension*

The limit of correlation functions when more than two points coincide is still singular and we have to follow a similar procedure if we want to define operators like $(S^4(x))_R$, $(\partial^\mu S\partial_\mu S)_R$, etc. . . The method consists again in considering the operator as an insertion in an N-point function. For the corresponding 1-P.I. diagrams the power-counting formula is for $\Gamma_A^{(N)}$

$$\delta(\gamma) = d - N\frac{d-2}{2} + \Sigma\, n_i\delta_i + \delta_A, \tag{3.42}$$

in which δ_A is the degree of the operator $A(x)$; if $A(x)$ consists of p fields on which act q derivatives

$$\delta_A = \frac{p}{2}(d-2) + q - d. \tag{3.43}$$

This degree δ_A is the canonical dimension of A minus d in mass units.

Examples:

$$A = S^2(x) \qquad \delta_A = -2$$
$$A = \partial^\mu S(x)\partial_\mu S(x) \qquad \delta_A = 0$$
$$A = S^4(x) \qquad \delta_A = d - 4.$$

In four dimensions and for an S^4-interaction, formulae (3.42), (3.43) give

$$\delta(\gamma) = 4 - N + (p + q - 4). \tag{3.44}$$

For all values of N leading to a positive value of $\delta(\gamma)$ new superficial divergences appear. All functions $\Gamma_A^{(N)}$ for which $N \leqslant p + q$ have to be subtracted, the number of subtractions being related to the corresponding value of $\delta(\gamma)$. In order to illustrate these features, let us choose the example of $A = S^4(x)$. $\Gamma_{S^4}^{(2)}$ is quadratically divergent and $\Gamma_{S^4}^{(4)}$ logarithmically divergent.

As in the case of the logarithmic divergence of $\Gamma_{S^2}^{(2)}$, a multiplicative renormalization of $S^4(x)$ is sufficient to remove the divergence of $\Gamma_{S^4}^{(4)}$ but the two-point function $\Gamma_{S^4}^{(2)}$ is still divergent. The structure of the superficial divergence is

$$\Gamma_{S^4}^{(2)}(q; p_1, p_2) = C_1(\Lambda) + C_2(\Lambda)(p_1^2 + p_2^2) + C_3(\Lambda)p_1 \cdot p_2$$
$$+ \text{ finite contributions} \tag{3.45}$$

in which $C_1(\Lambda)$ behaves as Λ^2 for large Λ, and C_2 and C_3 are logarithmic in Λ. The C_1-term may be eliminated by subtraction from $S^4(x)$ of the operator $C_1S^2(x)/2$; C_2 and C_3 are compensated by subtraction of a linear combination involving the operators $S(x)\nabla^2S(x)$ and $(\nabla S)^2$ respectively. The operator $(S^4)_R$ is thus a combination of S^4, $S\nabla^2S$, $(\nabla S)^2$ and S^2.

The same argument would hold for the other operators of canonical dimension four: $S\nabla^2 S$, $(\nabla S)^2$. We then arrive at the conclusion that

$$(\Gamma^{(N)}_{A_i})_R = Z^{N/2}\left(\sum_{j=1}^{3} Z_{ij}\Gamma^{(N)}_{A_j} + Y_i\Gamma^{(N)}_{S^2}\right), \tag{3.46}$$

in which the renormalization constants Z_{ij} are determined by the 12 conditions:

$$\begin{aligned}
&(\Gamma^{(4)}_{S^4})_R(p_i = 0) = 4!\\
&(\Gamma^{(4)}_{S\nabla^2 S})_R(p_i = 0) = 0\\
&(\Gamma^{(4)}_{(\nabla S)^2})_R(p_i = 0) = 0\\
&(\Gamma^{(2)}_{S^4})_R(p_i)_{p_i\to 0} = \mathcal{O}(p_i^4)\\
&(\Gamma^{(2)}_{S\nabla^2 S})_R(q; p_1, p_2)_{p_i\to 0} = p_1^2 + p_2^2 + \mathcal{O}(p_i^4)\\
&(\Gamma^{(2)}_{(\nabla S)^2})_R(q; p_1, p_2)_{p_i\to 0} = 2p_1\cdot p_2 + \mathcal{O}(p_i^4).
\end{aligned} \tag{3.47}$$

This example shows that an operator of given dimension is not simply multiplicatively renormalizable in general. The renormalized operator is a linear combination of all the operators of same or lower canonical dimension, except those which are decoupled from symmetry arguments. For example, we have not considered here in the renormalization of $S^4(x)$ the operator S^3, since the interaction preserves the $S \to -S$ symmetry, or the operator $\partial_\mu S \partial_\nu S$, which would break the $O(4)$-symmetry of the interaction.

Remark. In spite of the fact that we have defined renormalized operators, which inserted in correlation functions give finite result, their product even at distinct points is not necessarily finite. Again power-counting would show this phenomenon, but let us simply notice that even in a non-interacting theory the function $\langle S^4(x_1)S^2(x_2)S(y_1)S(y_2)\rangle_{1-\text{P.I.}}$ is infinite.

E. Equations of motion

The operators introduced above do not lead to independent correlation functions. There are linear relations among them which are consequences of the equations of motion that we are now going to establish.

The starting point is the algebraic identity:

$$\int [\mathrm{d}S(x)]\left\{\frac{\delta\mathscr{H}}{\delta S(x)} - H(x)\right\}\exp\left(-\int \mathrm{d}x\,[\mathscr{H}(S) - H(x)S(x)]\right) = 0 \tag{3.48}$$

obtained by noticing that the integrand is a total derivative. This means that the functional $Z(H)$ (2.2) satisfies the equation

$$\left\{\left.\frac{\delta\mathscr{H}}{\delta S(x)}\right|_{S=\delta/\delta H(x)} - H(x)\right\} Z(H) = 0. \tag{3.49}$$

For the S^4-theory, (3.49) means

$$\left[(-\nabla^2 + m_0^2)_\Lambda \frac{\delta}{\delta H(x)} + \frac{g_0}{3!}\left(\frac{\delta}{\delta H(x)}\right)^3\right] Z(H) = H(x)Z(H). \qquad (3.50)$$

This equation is the basis of various relations among composite operators.

1. *Renormalization of* $S^3(x)$

The simplest relation concerns the operator $S^3(x)$. First we shall examine the connected diagrams in which one operator S^3 is inserted, and let us define the corresponding generating functional

$$W_{S^3(y)}(H) = \sum_1^\infty \frac{1}{N!}\int H(x_1)\ldots H(x_N)\langle S^3(y)S(x_1)\ldots S(x_N)\rangle_C. \qquad (3.51)$$

For the connected functionals $W(H)$, $W_{S^3}(H)$, (3.50) gives

$$(-\nabla^2 + m_0^2)_\Lambda \frac{\delta W(H)}{\delta H(x)} + \frac{g_0}{3!} W_{S^3(x)}(H) = H(x). \qquad (3.52)$$

The renormalized $S^3(x)$ operator will be defined by its 1-P.I. insertions, whose generating functional $\Gamma_{S^3(x)}(M)$ satisfies the equation obtained by a Legendre transformation on (3.52):

$$(-\nabla^2 + m_0^2)_\Lambda M(x) + \frac{g_0}{3!}\Gamma_{S^3(x)}(M) = \frac{\delta\Gamma(M)}{\delta M(x)}. \qquad (3.53)$$

More explicitly this equation means

$$\delta_{N1}(q^2 + m_0^2)_\Lambda + \frac{g_0}{3!}\Gamma_{S^3}^{(N)}(q; p_1, \ldots, p_N) = \Gamma^{(N+1)}(q, p_1, \ldots, p_N). \qquad (3.54)$$

This shows that the insertion of S^3 corresponds essentially to the insertion of $S(x)$ itself. If we express $\Gamma^{(N+1)}$ in terms of the corresponding renormalized function

$$\Gamma_R^{(N+1)}(q; p_1, \ldots, p_N) = g_0 \frac{Z^{(N+1)/2}}{3!}\Gamma_{S^3}^{(N)}(q; p_1, \ldots, p_N), \qquad N > 1 \qquad (3.55)$$

and define $(S^3)_R = Z_{(3)}(S_R)^3$ by

$$\frac{1}{3!}(\Gamma_{S^3}^{(3)})_R\big|_{p_i=0} = 1 \qquad (3.56)$$

we obtain simply

$$\Gamma_R^{(N+1)}(q; p_1, \ldots, p_N) = \frac{g}{3!}(\Gamma_{S^3}^{(N)})_R(q; p_1, \ldots, p_N), \qquad N > 1, \qquad (3.57)$$

with $$Z_{(3)} = Z_{(4)}.$$

For $N = 1$ additional subtractions are needed in order to define a finite function $\langle (S^3)_R(x) S_R(x) \rangle$; the conditions are

$$(\Gamma^{(1)}_{S^3})_R\,(p, -p)\big|_{p=0} = 0 \tag{3.58a}$$

$$\frac{\partial}{\partial p^2}(\Gamma^{(1)}_{S^3})_R\,(p, -p)\big|_{p=0} = 0. \tag{3.58b}$$

Finally this yields the equations

$$\Gamma^{(N+1)}_R(q; p_i) - \frac{g}{3!}(\Gamma^{(N)}_{S^3})_R\,(q; p_i) = \delta_{N1}(q^2 + m^2), \tag{3.59}$$

which summarize how $S^3(x)$ is related to $S(x)$.

2. Other consequences of the equation of motion

The general technique applied to derive relations among composite operators consists in acting on (3.50) with a polynomial in $\delta/\delta H$. Assume that we simply act with $\delta/\delta H(y)$ on this equation; it gives

$$\left\{\frac{\delta}{\delta H(y)}(-\nabla_x^2 + m_0^2)_\Lambda \frac{\delta}{\delta H(x)} + \frac{g_0}{3!}\frac{\delta}{\delta H(y)}\frac{\delta^3}{\delta H^3(x)}\right\} Z = \delta(y - x)Z + H(x) \times \frac{\delta Z}{\delta H(y)}. \tag{3.60}$$

On the generating functionals of connected diagrams, this corresponds to, when y approaches x,

$$W_{S(x)(-\nabla^2 + m_0^2)_\Lambda S(x)} + \frac{g_0}{3!} W_{S^4(x)} = H(x)\frac{\delta W}{\delta H(x)}, \tag{3.61}$$

in which W_A is the generating functional of the connected Green functions with one insertion of the operator A. On the corresponding 1-P.I. functionals, it gives:

$$\Gamma_{S(x)(-\nabla^2 + m_0^2)_\Lambda S(x)} + \frac{g_0}{3!}\Gamma_{S^4(x)} = M(x)\frac{\delta\Gamma}{\delta M(x)}. \tag{3.62}$$

We now expand to the bare operators in terms of the renormalized ones. One has to be careful with the operator $S(x)\,(-\nabla^2 + m_0^2)_\Lambda S(x)$ which changes the coefficients of (3.62). In fact the regularization, which modifies the propagator, is equivalent to the introduction of operators quadratic in S with more than two derivatives. These operators appear thus in the equation of motion; as will be seen in Section VIII.C, at leading order in Λ^{-1} they can be replaced by linear combinations of S^2, S^4, $S\nabla^2 S$, $(\nabla S)^2$. Thus the renormalized version of this equation becomes

$$a_1(\Gamma^{(N)}_{S\nabla^2 S})_R + a_2(\Gamma^{(N)}_{m^2 S^2})_R + a_3(\Gamma^{(N)}_{S^4})_R + a_4(\Gamma^{(N)}_{(\nabla S)^2})_R = \sum_{i=1}^{N} \Gamma^{(N)}_R (p_1, \ldots, p_i + q, \ldots, p_N) \tag{3.63}$$

in which the operators are inserted in the $\Gamma^{(N)}$ at momentum q. The four numbers a_i are functions only of the coupling constant. This equation has implications when one deals with the short distance expansion of a product of fields which generates such composite operators. Details of this problem may be found in papers by Brézin *et al.* (1974b) and Mitter (1973).

F. Massless theory

Throughout the previous sections we have found it convenient to renormalize all the correlation functions at zero momentum. In a massless theory this is impossible because the correlation functions are singular at this point. One is therefore forced to introduce an arbitrary mass parameter μ which will characterize the renormalization points. For example for the $S^4(x)$ theory we can choose the following renormalization conditions:

$$\begin{aligned} \Gamma^{(2)}_R(p; \mu, g)\big|_{p^2=0} &= 0 \\ \frac{\partial}{\partial p^2}\Gamma^{(2)}_R(p; \mu, g)\big|_{p^2=\mu^2} &= 1 \\ \Gamma^{(4)}_R(p_i; \mu, g)\big|_{\mathrm{S.P.}(\mu)} &= g, \end{aligned} \tag{3.64}$$

where the S.P.(μ) stands for the point:

$$p_i \cdot p_j = \tfrac{1}{3}(4\delta_{ij} - 1)\,\mu^2.$$

The relation between renormalized and bare correlation functions then reads:

$$\Gamma^{(N)}_R(p_i; \mu, g) = Z^{N/2}\left(\frac{\mu}{\Lambda}, g_0\right)\Gamma^{(N)}(p_i; \Lambda, g_0). \tag{3.65}$$

We have omitted in $\Gamma^{(N)}$ the argument m_0^2 which is now fixed at its critical value m_{0c}^2, function of Λ and g_0. For the same reason the insertions of composite operators have to be renormalized at momenta of order μ. In particular the insertion of $\int S^2(x)\,dx$ is infrared-divergent (see VI.C). Later we shall also have to define insertions of operators of non-negative degree at zero momentum $\int A(x)\,dx$. In order to avoid the infrared divergencies which would normally occur, one has then to add to this operator a term proportional to $\int S^2(x)\,dx$ such that the new operator $\tilde{A}(x)$ satisfies:

$$\tilde{A}(x) = A(x) + aS^2(x) \tag{3.66}$$

$$\Gamma^{(2)}_{\tilde{A}(x)}(q; p_1, p_2)\big|_{q^2 = p_1^2 = p_2^2 = 0} = 0 \tag{3.67}$$

An operator of this form has non-infrared-divergent insertions at zero momentum. An example of such an operator is provided by the derivative with respect to the coupling constant of the Hamiltonians of the massless theory.

IV. Mean Field Theory and the Role of Dimension Four

The simplest models which exhibit a critical behaviour are classical spin systems on a d-dimensional lattice with an attractive and translationally invariant interaction. In presence of an external magnetic field H_i, the Hamiltonian reads

$$-\beta\mathscr{H} = \sum_{i \neq j} V_{ij} S_i S_j + \sum_i H_i S_i \tag{4.1}$$

Beyond one dimension, the only system of this type which has been solved is the Ising–Onsager model (Onsager, 1944): a two-dimensional lattice with nearest-neighbour interactions, spin one-half ($S_i = \pm\frac{1}{2}$), and no external field. In order to obtain a rough idea of what happens for a more general case, we have then to make some approximations and we shall use the simple molecular field approximation (or mean field theory) of P. Weiss (1907). For reviews of this theory see, for example, Fisher (1967), Kadanoff *et al.* (1967), and Stanley (1971).

A. Mean field theory

There are several ways of deriving the Weiss theory, and we shall describe here the simple variational method of Peierls (1958), which consists in finding the best approximate one-body Hamiltonian. It uses the convexity inequality

$$\langle \exp Q \rangle \geqslant \exp\langle Q \rangle \tag{4.2}$$

with the mean value defined by the one-body density matrix associated to

$$-\beta\mathscr{H}_0 = \sum_i X_i S_i \tag{4.3}$$

i.e.

$$\langle Q \rangle = \frac{\Sigma_{Si}\, Q(S_i) \exp(-\beta\mathscr{H}_0)}{\Sigma_{Si} \exp(-\beta\mathscr{H}_0)}. \tag{4.4}$$

Applying the inequality (4.2) to the operator $Q = -\beta(\mathscr{H} - \mathscr{H}_0)$ one obtains

$$\ln \sum_{S_i} \exp(-\beta\mathscr{H}) \geqslant \ln \sum_{S_i} \exp(-\beta\mathscr{H}_0) - \langle \beta\mathscr{H} - \beta\mathscr{H}_0 \rangle. \tag{4.5}$$

The maximization of the r.h.s. of (4.5) with respect to the parameters X_i which define $\mathscr{H}_0$ leads to the best one-body approximation of the free energy $-(1/\beta)W, W = \ln \Sigma_{S_i} \exp(-\beta\mathscr{H})$.

The spin distribution at a given lattice site is characterized by an even measure $\rho(S)$, for the Ising model $\delta(S^2 - 1)$, and it is convenient to define its transform $A(X)$

$$A(X) \equiv \ln \int \mathrm{d}S\, \rho(S)\, \mathrm{e}^{SX} \tag{4.6}$$

(= ln cosh X for the Ising model).

The Weiss variational free-energy is then

$$W = \sum_i A(X_i) + (H_i - X_i)\, A'(X_i) + \sum_{i \neq j} V_{ij}\, A'(X_i)\, A'(X_j) \tag{4.7}$$

and the X_i are given by the stationary conditions

$$O = (H_i - X_i) + 2 \sum_j V_{ij}\, A'(X_j). \tag{4.8}$$

The magnetization M_i is related to the applied field by

$$M_i = \frac{\partial W}{\partial H_i} = A'(X_i). \tag{4.9}$$

It is more convenient to discuss the problem in terms of the magnetization M_i and of the corresponding thermodynamic potential $\Gamma(M)$ obtained by the Legendre transformation

$$\Gamma(M) + W(H) = \sum_i H_i M_i \tag{4.10}$$

in which the H_i are the solutions of (4.9). This leads to

$$\Gamma(M) = -\sum_{i \neq j} V_{ij} M_i M_j + \sum_i B(M_i) \tag{4.11}$$

in which

$$B(M) = MX(M) - A(X); \tag{4.12}$$

in particular for the Ising model this gives

$$\Gamma(M) = -\sum_{i \neq j} V_{ij} M_i M_j + \sum_i \left[\left(\frac{1 + M_i}{2}\right) \ln(1 + (M_i) + \right. \\ \left. + \left(\frac{1 - M_i}{2}\right) \ln(1 - M_i) \right] \tag{4.13}$$

The magnetic field is now

$$H_i = \frac{\partial \Gamma}{\partial M_i} = -2\sum_j V_{ij} M_j + \frac{\partial B}{\partial M_i}. \tag{4.14}$$

The minimum of $\Gamma(M)$ gives the magnetization, i.e. the spin average, in the absence of an external field. In this case, translation invariance implies that the magnetization is uniform and

$$\frac{1}{N}\Gamma(M) = -VM^2 + B(M) \tag{4.15}$$

in which N is the number of lattice sites, and the positive constant V is

$$V = \sum_j V_{ij}. \tag{4.16}$$

This V is proportional, as V_{ij} itself, to the inverse temperature β. From its definition (4.6), one sees that $A(X)$ is a convex function of X and thus $B(M)$ defined in (4.12) is a convex function of M since $B''(M) = 1/A''(X)$.

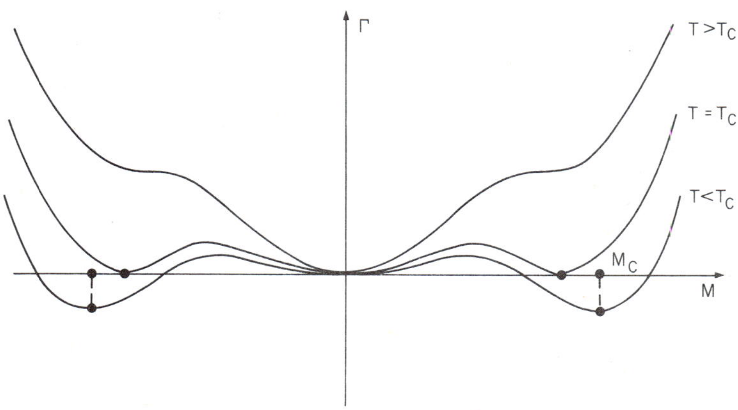

FIG. 7. Free-energy in the first-order transition case.

At high temperatures, i.e. small V, $-VM^2$ is negligible with respect to $B(M)$ and thus $\Gamma(M)$ has only one minimum at $M = 0$. When the temperature decreases in general an absolute minimum will appear away from the origin (Fig. 7). In this case the magnetization jumps from zero to some finite value and one has *a first-order transition*, i.e. a transition where the order parameter has a discontinuity in the temperature. This will occur for temperatures for which V satisfies the condition $\Gamma''(0) = -2V + B''(0) > 0$ since for smaller temperature the origin becomes a maximum.

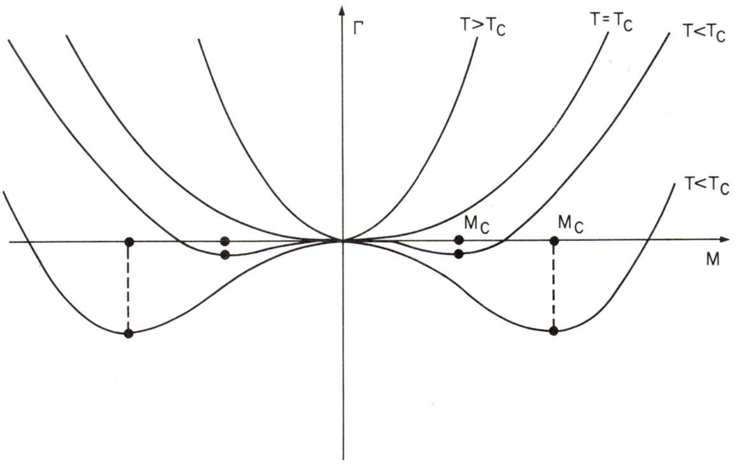

FIG. 8. Free-energy in the second-order transition case.

In the special case where the transition occurs exactly at

$$V_c = \tfrac{1}{2}B''(0) \tag{4.17}$$

and when the coefficient of M^4 in the expansion of $B(M)$ is positive, the minimum moves continuously away from the origin when V increases and one has *a second-order phase transition* (Fig. 8). In the following we shall only be concerned with such continuous transitions. Their characteristic is that for T close to T_c the order parameter is small and therefore only the first terms of the Taylor expansion of $\Gamma(M)$ are relevant. Thus it is only for this kind of transitions that one expects a universal behaviour.

Let us now expand Γ as

$$\Gamma(M_i) = -\sum_{i,j} V_{ij}M_iM_j + \sum_i \left(\frac{a}{2!}M_i^2 + \frac{b}{4!}M_i^4\right), \tag{4.18}$$

with a and $b > 0$. The Ising model corresponds to the particular values $a = 1, b = 2$.

The critical temperature is given by

$$V - \frac{a}{2} = 0 \tag{4.19}$$

and the spontaneous magnetization is

$$M = [12(V - (a/2))/b]^{1/2} \propto (T_c - T)^\beta \tag{4.20}$$

with the critical exponent

$$\beta = \tfrac{1}{2}. \tag{4.21}$$

The inverse of the magnetic susceptibility $\chi = \partial M/\partial H$ is obtained as the second derivative of Γ with respect to M. In zero-field this gives

$$\chi_+^{-1} = 2\left(\frac{a}{2} - V\right) \qquad T > T_c$$

$$\chi_-^{-1} = 4\left(V - \frac{a}{2}\right) \qquad T < T_c$$

and therefore they diverge as

$$\chi_+ = C_+(T - T_c)^{-\gamma}, \qquad \gamma = 1 \tag{4.22a}$$

$$\chi_- = C_-(T - T_c)^{-\gamma'}, \qquad \gamma' = 1 \tag{4.22b}$$

and

$$C_+/C_- = 2. \tag{4.23}$$

At $T = T_c$ the field temperatue relation is

$$H = \frac{b}{6} M^3 \propto M^\delta, \tag{4.24}$$

with the exponent

$$\delta = 3. \tag{4.25}$$

The equation of state, which is the relation between an applied constant field, the magnetization and the temperature, is here

$$H = 2\left(V - \frac{a}{2}\right) M + \frac{b}{6} M^3. \tag{4.26}$$

If the normalizations of fields and temperatures are adjusted so that

$$\frac{H}{M^\delta} = 1 \quad \text{at} \quad T = T_c \tag{4.27a}$$

$$x \equiv \frac{(T - T_c)}{M^{1/\beta}} = -1 \quad \text{for} \quad H = 0, \tag{4.27b}$$

Eq. (4.26) has the scaling form proposed by Widom (1965):

$$\frac{H}{M^\delta} = f(x) \tag{4.28}$$

with

$$f(x) = 1 + x. \tag{4.29}$$

From the functional (4.18) one obtains the spin–spin connected correlation function

$$G_{ij}^{-1} = \frac{\delta^2 \Gamma}{\delta M_i \, \delta M_j}$$

whose Fourier transform is

$$G(q) = \sum_j G_{ij} \exp(i\mathbf{q}\cdot\mathbf{r}_{ij}) = \left[a + \frac{b}{2} M^2 - 2\tilde{V}(q)\right]^{-1} \tag{4.30}$$

in which

$$\tilde{V}(q) = \sum_j V_{ij} \exp(i\mathbf{q}\cdot\mathbf{r}_{ij}) \tag{4.31}$$

and the components of the momentum are by definition in the range $|q_\alpha| \leqslant \pi$. If the interaction is short-ranged, $\tilde{V}(q)$ is regular around $q = 0$, and if we assume that V_{ij} preserves the cubic symmetry, one has

$$\tilde{V}(q) = V[1 - \rho^2 \mathbf{q}^2] + O(q^4), \tag{4.32}$$

in which V is defined by (4.16) and ρ^2 is some constant (1/2d for the Ising model). The correlation function (4.30) is thus of the type described by Ornstein and Zernike (1914).

Let us now define the correlation length ξ as the square-root of the second moment of G_{ij} or equivalently by the small momentum behaviour of its Fourier transform:

$$G(q) = G(0)\,[1 - \xi^2 \mathbf{q}^2] + O(q^4). \tag{4.33}$$

This length ξ characterizes, up to a purely numerical factor, the exponential fall-off of the correlations.

From (4.30–4.33) one finds

$$\xi^{-2} = \frac{1}{2V\rho^2}\left[a + \tfrac{1}{2} b M^2 - 2V\right]$$

and in zero-field this gives

$$\xi_+^{-2} = \frac{1}{V\rho^2}\left(\frac{a}{2} - V\right) \quad \text{for} \quad T > T_c$$

$$\xi_-^{-2} = \frac{2}{V\rho^2}\left(V - \frac{a}{2}\right) \quad \text{for} \quad T < T_c.$$

Equivalently these relations may be written

$$\xi_+ = f_+(T - T_c)^{-\nu} \tag{4.34a}$$

$$\xi_- = f_-(T_c - T)^{-\nu'} \tag{4.34b}$$

with the classical values of the exponents

$$\nu = \nu' = \tfrac{1}{2} \tag{4.35}$$

and the ratio of amplitudes

$$f_+/f_- = \sqrt{2}. \tag{4.36}$$

At the critical temperature itself $G(q)$ is proportional to $1/q^2$ and thus

$$\langle S_i S_j \rangle_{T_c} \underset{r_{ij}\to\infty}{\sim} \frac{1}{|r_{ij}|^{d-2+\eta}} \tag{4.37}$$

with the critical index

$$\eta = 0. \tag{4.38}$$

When the order parameter is an n-component vector M^α ($\alpha = 1, \ldots, n$), and if the thermodynamical potential $\Gamma(M)$ is rotationally invariant in this n-dimensional space, most of the previous results remain unchanged. Nevertheless the correlation functions, in non-zero magnetization, have a somewhat different structure. Let us consider a free energy which, instead of (4.18) reads

$$\Gamma(M_i^\alpha) = -\sum_{i,j} V_{ij}\mathbf{M}_i\cdot\mathbf{M}_j + \sum_i \left(\frac{a}{2!}\mathbf{M}_i^2 + \frac{b}{4!}(\mathbf{M}_i^2)^2\right) \tag{4.39}$$

from which one finds the magnetic field

$$H_i^\alpha = -2V_{ij}M_j^\alpha + aM_i^\alpha + \frac{b}{6}\mathbf{M}_i^2 M_i^\alpha. \tag{4.40}$$

For a constant field the magnetization is along the field and

$$\frac{H}{M} = -2V + a + \frac{b}{6}M^2. \tag{4.41}$$

Let us now compute the correlation function $G_{ij}^{\alpha\beta}$

$$[G^{-1}]_{ij}^{\alpha\beta} = \left.\frac{\partial^2\Gamma}{\partial M_i^\alpha\,\partial M_j^\beta}\right|_{M_i^\alpha = M^\alpha}$$

$$= (-2V_{ij} + a\delta_{ij})\,\delta_{\alpha\beta} + \frac{b}{6}\delta_{ij}\,[\mathbf{M}^2\,\delta_{\alpha\beta} + 2M_\alpha M_\beta]. \tag{4.42}$$

We see from this expression that the magnetization introduces a distinguished direction in space and that the correlations of spin-operators parallel or transverse to the field are different. One can write this two-point function as a sum of a longitudinal part and of a transverse part:

$$[G^{-1}]_{ij}^{\alpha\beta} = u_\alpha u_\beta (G_L)_{ij}^{-1} + (\delta_{\alpha\beta} - u_\alpha u_\beta)(G_T)_{ij}^{-1}, \tag{4.43}$$

in which u_α is the unit vector along the magnetization M_α, with $\Gamma_L^{(2)}$ and $\Gamma_T^{(2)}$ given by

$$\Gamma_L^{(2)} = [G_L^{-1}]_{ij} = -2V_{ij} + a\delta_{ij} + \frac{b}{2}\delta_{ij}M^2 \tag{4.44}$$

$$\Gamma_T^{(2)} = [G_T^{-1}]_{ij} = -2V_{ij} + a\delta_{ij} + \frac{b}{6}\delta_{ij}M^2. \tag{4.45}$$

The longitudinal part coincides, as expected, with the result (4.30) for $n = 1$. In order to analyse the transverse part, let us write its Fourier transform

$$G_T^{-1}(q) = -2\tilde{V}(q) + a + \frac{b}{6}M^2. \tag{4.46}$$

Using the relation (4.41) for H, this equation may be written

$$G_T^{-1}(q) = \frac{H}{M} - 2[\tilde{V}(q) - \tilde{V}(0)]. \tag{4.47}$$

Thus below T_c, in the absence of an applied field, since H/M vanishes, $G_T^{-1}(q)$ behaves as q^2 for low momentum. This corresponds to a "massless" Nambu–Goldstone excitation (Nambu (1960); Goldstone (1961)), i.e. to an infinite transverse correlation length. Another way of expressing this fact is to say that an infinitesimal amount of energy is required to create these excitations, the spin waves corresponding to the degeneracy of the minimum of the free energy. These excitations tend to restore the symmetry broken by the choice of a particular solution.

The previous results have been derived by using the variational principle on specific models, but following Landau (1937) one can abstract the general assumptions which would lead to these results:

(a) the functional Γ may be expanded in M around $M = 0$;

(b) the coefficients of this expansion are regular in T near T_c, the temperature for which the coefficient of M^2 vanishes;

(c) if one expands $\Gamma(M)$ in powers of $\tilde{M}(q)$, the Fourier transform of M_i, as

$$\Gamma(M) = \sum_N \frac{1}{N!}\int \mathrm{d}q_1 \ldots \mathrm{d}q_N\, \tilde{M}(q_1)\ldots\tilde{M}(q_N)\,\Gamma^{(N)}(q_1,\ldots,q_N)$$

the $\Gamma^{(N)}$ are regular for small q's and $\Gamma^{(4)}(0,0,0,0) > 0$.

B. Beyond mean field theory

The previous calculations have been done with the use of mean field theory, and we have now to compute corrections to this approximation. In fact, up to now the fluctuations of the magnetization have been ignored. They are particularly large near T_c, since at T_c for second-order transitions the minimum of Γ is very flat. We shall thus construct a perturbation expansion whose first term is the mean field theory.

The method has been described by Baker (1962), Siegert (1963), and Hubbard (1972), and the reader is also referred to Brout (1974). It relies on the algebraic identity

$$\int \prod_i (\mathrm{d}X_i) \exp\left(-\frac{1}{4} X_i V_{ij}^{-1} X_j + S_i X_i \right) = \text{const. } \exp(S_i V_{ij} S_j) \tag{4.48}$$

Thus, for the partition function of the Hamiltonian (4.1), one can write

$$Z(H) = \sum_{S_i} \int \prod_i (\mathrm{d}X_i) \exp\left(-\frac{1}{4}(X_i - H_i) V_{ij}^{-1}(X_j - H_j) + X_i S_i \right). \tag{4.49}$$

This equation makes sense only for a positive V_{ij} matrix. But even for a simple nearest-neighbour interaction the matrix V is not positive, and the following method has to be regarded as an algebraic formal way to derive the perturbation expansion.

Performing the sum on the S_i's we obtain

$$Z(H_i) = \int \prod_i (\mathrm{d}X_i) \exp\left(-\tfrac{1}{4}(X_i - H_i) V_{ij}^{-1}(X_j - H_j) + \sum_i A(X_i) \right). \tag{4.50}$$

The integral (4.50) will be calculated using the steepest-descent method. At lowest order it will yield again mean field theory. In fact the saddle-point is given by

$$\tfrac{1}{2} V_{ij}^{-1}(\bar{X}_j - H_j) = A'(\bar{X}_i) \tag{4.51}$$

and the first approximation to $W = \ln Z$ is

$$W(H_i) = -\tfrac{1}{4}(\bar{X}_i - H_i) V_{ij}^{-1}(\bar{X}_j - H_j) + \sum_i A(\bar{X}_i). \tag{4.52}$$

The magnetization is

$$M_i = A'(\bar{X}_i) \tag{4.53}$$

and the corresponding Γ functional does coincide with the mean field expression (4.11).

Systematic corrections to the approximate mean field theory are now obtained by expanding the integrand (4.50) around the saddle-point as set out in section II. It is natural at this stage to introduce a parameter in which this procedure becomes a power series. Such a parameter has been introduced by Wallace (private communication); at each lattice site i one introduces l independent spin variables $S_i^{(1)} \ldots S_i^{(l)}$, distributed according to the same measure $\rho(s)$ as above, and we call S_i their sum:

$$S_i = \sum_{k=1}^{l} S_i^{(k)}. \tag{4.54}$$

The interaction is replaced by

$$-\beta\mathscr{H} = \frac{1}{l} \sum_{i \neq j} V_{ij} S_i S_j + \sum_i H_i S_i \tag{4.55}$$

and instead of (4.50) we obtain by the same procedure

$$Z(H_i) = \int \prod_i (\mathrm{d}X_i) \exp\left(-\frac{l}{4}(X_i - H_i) V_{ij}^{-1}(X_j - H_j) + l \sum_i A(X_i) \right) \tag{4.56}$$

and for convenience we now define W as $(1/l) \ln Z$. The method then yields an expansion in powers of $1/l$.

The first correction in $1/l$ is given by the square-root of the determinant of the quadratic form around the saddle-point as shown in Section II.C. After Legendre transformation this gives

$$\Gamma(M_i) = -V_{ij} M_i M_j + \sum_i B(M_i) + \frac{1}{2l}\left(\operatorname{tr} \ln \left[\delta_{ij} - 2V_{ij} \frac{1}{B''(M_i)} \right] - \text{value at } M_i = 0 \right) + O\left(\frac{1}{l^2}\right) \tag{4.57}$$

where as usual the logarithm in tr ln is taken in the operator sense.

We are now in position to study the first correction to mean field theory. Rather than treating all the quantities, one after the other, let us illustrate the main features by an example.

The magnetic susceptibility in zero field above T_c is obtained as

$$\chi^{-1} = \left.\frac{\partial^2 \Gamma}{\partial M^2}\right|_{M=0}, \tag{4.58}$$

and this formula involves only the first coefficients of the Taylor expansion of $B(M)$ at the origin. If $B(M)$ is written as

$$B(M) = \frac{a}{2!} M^2 + \frac{b}{4!} M^4 + \frac{c}{6!} M^6 + O(M^8), \tag{4.59}$$

then

$$\chi^{-1} = -2V + a + \frac{1}{l}\frac{b}{a^2}\left[\frac{1}{1-(2/a)V}V\right]_{ii} + O\left(\frac{1}{l^2}\right). \tag{4.60}$$

The vanishing of χ^{-1} defines T_c as the solution of

$$0 = -2V_c + a + \frac{1}{l}\frac{b}{a^2}\left[\frac{1}{1-(2/a)V_c}V_c\right]_{ii}. \tag{4.61}$$

Subtraction of (4.61) from (4.60) gives (in terms of the Fourier transform (4.31) of the interaction):

$$\chi^{-1} = -2(V - V_c) + \frac{(V - V_c)}{lV_c}\frac{b}{a^2}\int\frac{d^dq}{(2\pi)^d}$$
$$\times \frac{\tilde{V}_c(q)}{(1-(2/a)\tilde{V}(q))(1-(2/a)\tilde{V}_c(q))} \tag{4.62}$$

We now have to study the behaviour of the integral in the r.h.s. of (4.62) when V approaches V_c. The only possible singularities come from the vanishing of the denominator, which occurs only for $q^2 = 0$ since for small q^2

$$\tilde{V}_c(q) = \frac{a}{2}[1 - \rho^2 q^2] + O\left(q^4, \frac{1}{l}\right). \tag{4.63}$$

We see that, as long as the dimension of space d is greater than four, the integral (4.62) has a limit for T near T_c, and therefore χ^{-1} still vanishes as $(T - T_c)$, as in mean field theory, the only modification being in the amplitude.

However for $2 < d \leqslant 4$, the integral (4.62) diverges as $(T - T_c)^{(d/2-2)}$, and therefore for an arbitrary large l, there still exist small enough values of $(T - T_c)$ for which the $1/l$ term is dominant. Consequently mean field theory is certainly not valid below four dimensions.

We have studied the $1/l$ contribution to χ^{-1}; the divergence at $q = 0$ when d is smaller than four makes perturbation theory clearly meaningless. An inspection of higher-order contributions in $1/l$ indicates that the same features persist:

(a) no "infra-red" divergence above four dimensions,

(b) singular powers below $d = 4$.

Since the first statement implies that the classical mean field exponents are not modified above $d = 4$, we shall try in the next section to give a hint of the proof of this property. More rigorous arguments may be found in subsequent Sections.

C. Validity of Landau theory above four dimensions

The perturbation expansion in powers of $1/l$, constructed in the previous section, has exactly the same algebraic structure as the loopwise expansion introduced in Section II. Our propagator is now proportional to $[1 - (2/a)\tilde{V}(q)]^{-1}$. The expansion generates diagrams with four, six, eight . . . lines leaving each vertex; and the vertices are momentum-dependent, but regular around momentum zero. In order to cancel the multiple poles which express the displacement of T_c order by order, it is necessary to invert the relation between T and χ^{-1}—the equation (4.62) at the one-loop order—and to express all the propagators in terms of χ^{-1}. For q small, the propagator now behaves as $[q^2 + \chi^{-1}]^{-1}$, and the problem is to examine the divergence which may appear when χ^{-1} vanishes.

For a given diagram, this problem reduces to the counting of the total power of momenta around $q = 0$ generated by the propagators and vertices, analogous to the power-counting of ultra-violet divergences of Section III. If we study the momentum dependence at the critical temperature of a diagram for the pair correlation function for instance, this gives a degree of the integrand

$$\delta(\gamma) = 2 + \sum_i n_i \left[k_i + v_i \left(\frac{d}{2} - 1 \right) - d \right], \tag{4.64}$$

in which v_i is the number of lines at the ith vertex and k_i the number of powers of momenta at this vertex. When $\delta(\gamma)$ is positive, the diagram is infra-red convergent, and we are going to verify on (4.64) that indeed $\delta(\gamma) > 0$ provided d is larger than four. In fact the minimum of the r.h.s. of (4.64) occurs when:

$k_i = 0$, i.e. a constant interaction (in momentum space). Any momentum dependence of the vertex would obviously make the diagram more convergent.

$v_i = 4$, i.e. a four-leg vertex;

and

$$\delta_{\min}(\gamma) = 2 + n(d - 4) \tag{4.65}$$

for a diagram of order n.

The coefficient of n is indeed positive for $d > 4$, and thus, order by order, the theory remains of the Landau type, classical exponents with modified amplitudes.

In contrast, when $d < 4$, infra-red divergences do appear, and one can suspect that mean field theory is no longer valid. Furthermore, this shows that, order by order, the most divergent diagrams are due to a constant four-point vertex. These diagrams have the same divergences as those of the Euclidean correlation functions of a local relativistic field theory in d dimensions with an interaction $\int S^4(X)\, d^dX$ and a vanishing mass. In addition the lattice structure induces a bounded domain of variation of the momenta which fixes a momentum cut-off of order one in the Feynman diagrams.

Therefore we are going to study now in detail these particular cut-off field theories because, as we shall see, there exists among them a particular one which is exactly scale-invariant. Then we shall show that, as expected, a different theory would only differ from that one by corrections which are negligible in the critical region.

We are thus led to study, as first shown by K. Wilson (1971), the $S^4(X)$ theory below four dimensions.

D. Remark concerning tricritical points

So far we have assumed that the coefficient b of the M^4 term in the expansion (4.18) of the free energy was strictly positive at T_c. There are situations, as noted by Landau (1937), in which an additional physical parameter allows one to make this b equal zero or some negative value. This occurs for instance in He^3–He^4 mixtures, as shown by Graf *et al.* (1967), and in metamagnetic transitions, of which Birgeneau *et al.* (1972) give examples; and the model described by Blume (1966) and Capel (1966) provides an example of a spin system having this property.

If we are in a domain of the physical parameters in which $T - T_c$ and the coefficient of M^4 are both small, the magnetization will also be small. Thus the critical behaviour will be sensitive to the first terms only of the expansion of $\Gamma(M)$, but now the coefficient of M^6 has to be retained and we assume that it is positive (in general one has to expand up to the first term whose coefficient is not small). Landau's mean field theory follows from the same considerations and predicts a tricritical point at the separation between a line of critical points and a line of first-order transitions. Corrections to mean field theory can be studied by the same methods as before. Using power-counting arguments one can again look for the most singular terms near the tricritical point. It is given by (4.64), except that by definition of the tricritical point there is no effective 4-point interaction. Thus the most singular part is generated then by the local S^6-interaction, and the same dimensional considerations will now indicate that above *three* dimensions mean field theory predicts the correct singular behaviour, whereas it is not valid below three (Riedel and Wegner, 1972).

E. Perturbation theory for an arbitrary spin Hamiltonian

Up to now we have considered only the case of two-body interactions; let us briefly show how one can derive perturbation theory for a general-spin Hamiltonian, Horwitz and Callen (1961), and Englert (1963), have published relevant work. Already for a two-spin interaction, the previous method fails when V_{ii} is non-zero because even with a one-body problem $\langle S^2 \rangle \neq \langle S \rangle^2$. Therefore the correct one-body trial density matrix must include terms in S_i^2

in that case. In order to construct perturbation theory for a general Hamiltonian, one first considers the successive powers S_i, S_i^2, etc ... as independent variables

$$\beta\mathscr{H} = \mathscr{H}(S_i, S_i^2, \ldots) \tag{4.66}$$

One then replaces in $\mathscr{H}$ the variables S_i^k by $\sigma_i^{(k)}$ and the partition function is

$$Z = \int \prod_{i,k} \mathrm{d}\sigma_i^{(k)}\, \mathrm{d}\alpha_i^{(k)} \exp\left[-\mathscr{H}(\sigma_i^{(1)}, \sigma_i^{(2)}, \ldots) + \sum_{i,k} \sigma_i^{(k)} \alpha_i^{(k)} + \sum_i F(\alpha_i^{(k)})\right]$$

with

$$\exp F(\alpha_i^{(k)}) = \sum_{\{S\}} \exp - \sum_k S^k \alpha_i^{(k)}. \tag{4.67}$$

The $\alpha_i^{(k)}$ variables have been introduced to constrain the $\sigma_i^{(k)}$ to have the value S_i^k. The integral on $\sigma_i^{(k)}$ and $\alpha_i^{(k)}$ is then calculated by the steepest-descent method. The main term associated with the value of the exponent at the saddle-point coincides with the variational principle with the freedom of an arbitrary one-body potential. The next terms generate corrections as above. In the particular case where $\mathscr{H}$ is quadratic in S_i, the integral on the σ variables can be performed and one recovers the formalism discussed in the preceeding sections.

V. Critical Phenomena near Dimension Four

The conclusions of the previous Section may be summarized as follows:

(a) above four dimensions it seems that one may trust perturbation theory and correspondingly the results of Landau theory for the critical behaviour;

(b) below four dimensions the successive orders of perturbation theory are more and more divergent and the results are no longer of the mean-field type;

(c) in the vicinity of a critical point can one replace the actual Hamiltonian by an effective local interaction which consists simply of the first relevant terms of the Landau–Ginzburg–Wilson functional

$$\beta\mathscr{H} = \int \mathscr{H}(x)\, \mathrm{d}^d x \tag{5.1}$$

$$\mathscr{H}(x) = \tfrac{1}{2}(c(\nabla S)^2 + aS^2) + \frac{b}{4!}(S^2)^2 \tag{5.2}$$

with finite parameters a, b and c regular near T_c in the temperature.

The corresponding Feynman integrals meant (see Section II) are those with a cut-off of order 1 in momentum space generated by the lattice structure.

In this Section, following the ideas of Wilson (1971), we shall use renormalization group equations (Stueckelberg and Petermann, 1953; Gell-Mann and Low, 1954; Bogoliubov and Shirkov 1959; Callan, 1970; Symanzik, 1970a, 1971) in order to study this field theory near dimension four. Many workers have contributed such studies: for example, Di Castro (1972), Mack (1972), Brézin *et al.* (1973a, b), Mitter (1973), Parisi (1973), Schroer (1973), Zinn-Justin (1973), Di Castro *et al.* (1974), and Jegerlehner and Schroer (1973).

A. Renormalization group equations at T = T$_c$

We have seen in the discussion of mean field theory that the critical domain consists of momenta much smaller than one, i.e. distances large compared to the lattice spacing, temperatues near T_c, and magnetization small compared to one. In this critical domain it is more appropriate to rescale the length and the spin magnitude in order to give them finite rather than infinitesimal values. Let us change them according to:

$$x \to \Lambda x \tag{5.3a}$$

$$S(x) \to \zeta S(x) \tag{5.3b}$$

so that the coefficient of $\frac{1}{2}(\nabla S)^2$ becomes unity. This corresponds to a propagator which is just $1/q^2$ at T_c in the mean field approximation.

The Hamiltonian thus becomes

$$\mathscr{H}(x) = \tfrac{1}{2}[(\nabla S)^2 + r_0 S^2] + \frac{u_0}{4!} S^4 \tag{5.4}$$

with $\zeta = c^{-1/2}\,\Lambda^{1-d/2}$

$$r_0 = \frac{a}{c}\Lambda^2 \tag{5.5a}$$

$$u_0 = \frac{b}{c^2}\Lambda^{4-d}. \tag{5.5b}$$

The theory may be trivially extended to an n-component order parameter $S_\alpha(x)$ ($\alpha = 1, \ldots, n$) with an O(n)-symmetric interaction:

$$\mathscr{H}(x) = \tfrac{1}{2}\sum_\alpha (\nabla S_\alpha \nabla S_\alpha + r_0 S_\alpha^2) + \frac{u_0}{4!}\left(\sum_\alpha S_\alpha^2\right)^2.$$

This index α will be omitted in the following, except if necessary.

The critical domain is more conveniently defined now by fixing the scale of the critical parameters, $T - T_c$, the magnetization and the momenta, and taking instead the large Λ limit, i.e.:

$$r_0 - r_{0c} \ll \Lambda^2$$
$$\text{momenta} \ll \Lambda \tag{5.6}$$
$$\text{magnetization} \ll \zeta^{-1} \sim \Lambda^{d/2-1}.$$

We are now, as in Section III, studying the limit of a theory with a large momentum cut-off. The additional important constraint that we have here is the equation (5.5b), which shows that the bare coupling constant u_0 has to be taken proportional to Λ^{ε}. It is thus convenient to define a dimensionless coupling constant g_0 as

$$g_0 = u_0\Lambda^{-\varepsilon}. \tag{5.7}$$

The idea of the method of Wilson and Fisher (1972) for dimension smaller than four is to use the fact that one knows the result for $d \geqslant 4$. Then if there is some continuity of the physical quantities one can try to make an expansion in powers of $\varepsilon \equiv 4 - d$. Thus the successive singularities, which for ε fixed are more and more divergent powers when the order of perturbation increases, become, once expanded in ε, powers of logarithms. Consequently it is possible to construct a theory at $T = T_c$ as a double expansion in the coupling constant g_0 and in ε, whereas it would not exist as a power series of g_0 for fixed ε.

The large Λ limit has been discussed in Section III and the results are given by the theory of renormalization. Let us first consider the correlation functions at T_c. Since this corresponds to an infinite correlation length, in field-theoretic language we are dealing with a theory with zero renormalized mass. In order to describe the renormalized theory we have to introduce a mass parameter μ. The renormalized correlation functions are defined through the conditions

$$\Gamma_R^{(2)}(p; g, \mu)\big|_{p^2=0} = 0 \tag{5.8a}$$

$$\frac{\partial}{\partial p^2}\Gamma_R^{(2)}(p; g, \mu)\big|_{p^2=\mu^2} = 1 \tag{5.8b}$$

$$\Gamma_R^{(4)}(p_i; g, \mu)\big|_{\text{S.P.}} = \mu^{\varepsilon} g, \tag{5.8c}$$

in which S.P. stands for the point $p_i \cdot p_j = (\mu^2/3)(4\delta_{ij} - 1)$.
The bare functions are related to the renormalized ones by

$$\Gamma^{(N)}(p_i; g_0, \Lambda) = Z^{-N/2}\left(g_0, \frac{\Lambda}{\mu}\right)\Gamma_R^{(N)}(p_i; g, \mu, \Lambda). \tag{5.9}$$

Renormalization theory tells us that $\Gamma_R^{(N)}$ has a limit when Λ goes to infinity in four dimensions, and below. Therefore, at each fixed finite order of perturbation one has

$$\Lambda \frac{d}{d\Lambda}\bigg|_{g,\mu} \Gamma_R^{(N)} = \mathcal{O}\left(\frac{(\ln \Lambda)^p}{\Lambda^2}\right) \tag{5.10}$$

in which p depends on the order. The importance of this bound (5.10) is that it is uniform in $d \leqslant 4$ and remains true when d approaches four, and thus holds order by order in the double expansion in g and ε.

From (5.10) one gets easily a condition on the bare function themselves:

$$\Lambda \frac{d}{d\Lambda}\bigg|_{g,\mu} \left[Z^{N/2}\left(g_0, \frac{\Lambda}{\mu}\right)\Gamma^{(N)}(p_i; g_0, \Lambda)\right] \simeq 0. \tag{5.11}$$

The conditions (5.8b) and (5.8c) may be used to express g_0 as a function of g and Λ/μ; thus when we vary Λ with fixed g and μ, g_0 varies. Let us call†:

$$W\left(g_0, \varepsilon; \frac{\Lambda}{\mu}\right) = \Lambda \frac{d}{d\Lambda}\bigg|_{g,\mu} g_0. \tag{5.12}$$

It is also necessary to introduce

$$\eta\left(g_0, \varepsilon; \frac{\Lambda}{\mu}\right) = -\Lambda \frac{d}{d\Lambda}\bigg|_{g,\mu} \ln Z\left(g_0, \frac{\Lambda}{\mu}\right), \tag{5.13}$$

and it is then elementary to write (5.11) in the form

$$\left[\Lambda \frac{\partial}{\partial\Lambda} + W\left(g_0, \varepsilon; \frac{\Lambda}{\mu}\right)\frac{\partial}{\partial g_0} - \frac{N}{2}\eta\left(g_0, \varepsilon; \frac{\Lambda}{\mu}\right)\right]\Gamma^{(N)}(p_i; g_0, \Lambda) = 0. \tag{5.14}$$

Since the bare $\Gamma^{(N)}$ functions are independent of the arbitrary parameter μ, the two functions W and η, which may be calculated from (5.14) itself in terms of the $\Gamma^{(N)}$, are independent of Λ/μ. Nevertheless, notice that these limits $W(g_0, \varepsilon)$ and $\eta(g_0, \varepsilon)$ functionally depend on the precise way in which the propagator has been cut-off in momentum space. Thus if in the double expansion in g_0 and ε of the $\Gamma^{(N)}$ one neglects all the terms of the form $(\ln \Lambda)^p/\Lambda^2$, assuming that they do not build up by resummation finite contributions, we end up with the renormalization group equation in differential form (Zinn-Justin, 1973):

$$\left[\Lambda \frac{\partial}{\partial\Lambda} + W(g_0,\varepsilon)\frac{\partial}{\partial g_0} - \frac{N}{2}\eta(g_0, \varepsilon)\right]\Gamma^{(N)}(p_i; g_0, \Lambda) = 0. \tag{5.15}$$

The functions W and η are calculable from their definitions (5.12), (5.13), in powers of g_0 and ε. The problem is to calculate the relevant Feynman

† Notation W, instead of the standard β-function, avoids confusion with critical exponents.

diagrams and the results of a few orders are given in Section IX. Let us only notice that at lowest orders in g_0 and ε, the relation between g and g_0 is simply

$$\mu^\varepsilon g = \Lambda^\varepsilon \left(g_0 - \frac{3g_0^2}{16\pi^2} \ln \frac{\Lambda}{\mu} \right) + \mathcal{O}(g_0^3, g_0^2\varepsilon).$$

Hence from the definition (5.12)

$$W(g_0, \varepsilon) = -\varepsilon g_0 + \frac{3g_0^2}{16\pi^2} + \mathcal{O}(g_0^3, \varepsilon g_0^2). \tag{5.16}$$

B. Consequences of the renormalization group equations

The linear partial differential equation (5.15) can be easily integrated. One first eliminates the η-term by a change of function

$$\Gamma^{(N)}(g_0, \Lambda) = \Phi(g_0, \Lambda) \exp \frac{N}{2} \int^{g_0} \frac{\eta(g')\, dg'}{W(g')},$$

where $\Phi(g_0, \Lambda)$, the solution of

$$\left[\Lambda \frac{\partial}{\partial \Lambda} + W \frac{\partial}{\partial g_0} \right] \Phi(g_0, \Lambda) = 0,$$

depends on an arbitrary function of one single variable

$$\Phi = \Phi\left(\ln \Lambda - \int^{g_0} \frac{dg'}{W(g')} \right).$$

Alternatively, one may express the dependence of Φ on a single variable by the equivalent equation

$$\Gamma^{(N)}(p_i; g_0, \lambda\Lambda) = \Gamma^{(N)}(p_i; g_0(\lambda), \Lambda) \exp \frac{N}{2} \int_1^\lambda \frac{d\sigma}{\sigma} \eta[g_0(\sigma)] \tag{5.17}$$

in which one has defined an effective coupling constant $g_0(\lambda)$, function of the dilatation parameter λ (and also of g_0 itself) by the "flow" equation

$$\frac{dg_0(\lambda)}{d \ln \lambda} = -W[g_0(\lambda)]$$

$$g_0(1) = g_0. \tag{5.18}$$

The content of (5.17) is that to vary the cut-off at fixed g_0 is equivalent—apart from a multiplicative factor—to fixing the cut-off and letting the coupling constant vary. The equation (5.18) shows that the rate of change of the effective $g_0(\lambda)$ is governed by the function W. These equations are in this theory the expression of Wilson's ideas concerning the flow of interaction induced by a renormalization group transformation.

In particular, from (5.17) one sees that the large cut-off limit is reduced to study the behaviour of $g_0(\lambda)$ for large λ. Assume that one is given a very small value of g_0, then $g_0(\lambda)$ starts by moving away from the origin since W has a negative slope at $g_0 = 0$ (Fig. 9). If W happens to have a zero g_0^* at which its

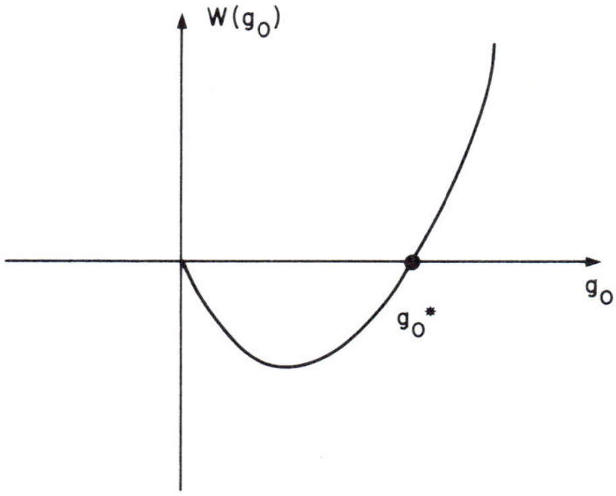

FIG. 9. Shape of $W(g_0)$ near and below four dimensions.

derivative is positive, when λ increases $g_0(\lambda)$ will ultimately reach that fixed point. The linearization of (5.18) around this fixed point

$$W(g_0^*) = 0$$

$$\omega \equiv W'(g_0^*) > 0 \tag{5.19}$$

gives the large λ behaviour of $g_0(\lambda)$:

$$g_0(\lambda) - g_0^* \propto \lambda^{-\omega}. \tag{5.20}$$

In fact the function W has indeed such a zero (Fig. 9) since on equation (5.16) one verifies that near and below four dimensions

$$g_0^* = \frac{16\pi^2}{3}\varepsilon + O(\varepsilon^2) \tag{5.21a}$$

$$\omega = \varepsilon + O(\varepsilon^2). \tag{5.21b}$$

Furthermore, if $\Gamma^{(N)}(g_0^*)$ and $\eta(g_0^*)$ are finite, one has

$$\Gamma^{(N)}(p_i; g_0, \lambda\Lambda) \underset{\lambda\to\infty}{\to} \Gamma^{(N)}(p_i; g_0^*, \Lambda)\lambda^{N\eta/2}, \tag{5.22}$$

in which

$$\eta \equiv \eta(g_0^*). \tag{5.23}$$

In addition simple dimensional considerations give the relation

$$\Gamma^{(N)}\left(\frac{p_i}{\lambda}; g_0, \Lambda\right) = \lambda^{-d+\frac{1}{2}N(d-2)}\,\Gamma^{(N)}(p_i; g_0, \lambda\Lambda), \tag{5.24}$$

and finally with (5.22)

$$\Gamma^{(N)}\left(\frac{p_i}{\lambda}; g_0, \Lambda\right) \underset{\lambda\to\infty}{\sim} \lambda^{\frac{1}{2}N(d-2+\eta)-d}\,\Gamma^{(N)}(p_i; g_0^*, \Lambda). \tag{5.25}$$

This equation shows that the correlation functions at T_c have a power behaviour for small momenta. In the previous Section, we argued that the only relevant part of the interaction was a local four-point coupling. We discover now, in (5.25), an even more universal character of the power behaviour: it does not depend on g_0, the initial value of $g_0(\lambda)$.

If we set $N = 2$ in (5.25) in order to obtain the inverse propagator, we find that

$$G(p) \underset{(p/\Lambda)\ll 1}{\sim} \frac{1}{p^{2-\eta}}; \tag{5.26}$$

and thus the quantity defined by (5.23) coincides with the critical exponent η.

In perturbation theory, $\eta(g_0)$ starts with a term proportional to g_0^2; hence η behaves as ε^2.

Another way of expressing the result (5.25) is to say that the field $S(x)$, which has a canonical dimension that one reads from (5.24) as

$$d_S^c = \tfrac{1}{2}(d-2), \tag{5.27}$$

acquires a dynamical or anomalous dimension (Wilson, 1970)

$$d_S = \tfrac{1}{2}(d-2+\eta). \tag{5.28}$$

It now remains to study all the scaling laws and more generally what happens in the critical domain, away from T_c. It would not be difficult to extend the previous method to that situation. However it is simpler, both theoretically and for practical calculations, to use the corresponding renormalized theory. The main reason is that the bare correlation functions still contain a negligible dependence in the cut-off function, and therefore satisfy only approximately the R.G. equation (5.15). In fact, even if $g_0 = g_0^*$ the correlation functions at T_c have not a pure scaling form, the correction terms being, order by order, of the form $(\ln \Lambda)^p/\Lambda^2$. It is nevertheless possible to construct an equivalent renormalized theory which is exactly scale-invariant, and this will be the subject of the next Section.

VI. Renormalized Theory; Derivation of Scaling Laws above the Critical Temperature

It is convenient, then, to study the renormalized theory in which the large Λ limit has been performed and which is thus directly adapted to the critical region. Let us recall that in order to go from the bare to the renormalized theory two modifications are required:

(a) a change of normalization of the field;

(b) elimination of the bare parameters r_0, g_0 in favour of renormalized quantities, which are values of the new correlation functions at given momenta. The existence of such a renormalized theory proves in this framework the strong scaling hypothesis, as will be shown in the next subsection.

A. Renormalization at zero momentum

On this topic see Brézin *et al.* (1973a) and Mitter (1973).

If T is not equal to T_c, the correlation length $\xi \equiv m^{-1}$ is large but not infinite, and one can choose it as one of the renormalized parameters. This theory may be described by the following conditions

$$\Gamma^{(2)}(p; m, g) \underset{p \to 0}{=} m^2 + p^2 + O(p^4) \tag{6.1}$$

$$\Gamma^{(4)}(p_i = 0; m, g) = m^{\varepsilon} g \tag{6.2}$$

(in this Section $\Gamma^{(N)}$ stands for the renormalized correlation functions). These renormalized functions are obtained from the bare, i.e. physical in this problem, correlation functions by the relations

$$\Gamma^{(N)}(p_i; m, g) = \lim_{\Lambda \to \infty} Z^{N/2}\left(\frac{m}{\Lambda}, g\right) \Gamma^{(N)}_{\text{bare}}(p_i; r_0^{1/2}, g_0, \Lambda). \tag{6.3}$$

Using the facts that:

(*a*) the $\Gamma^{(N)}$ are independent of Λ

(b) g is dimensionless

and dimensional analysis, one obtains the relation

$$\Gamma^{(N)}(p_i; m, g) = m^{d-(N/2)(d-2)}\,\Gamma^{(N)}(p_i/m; 1, g). \tag{6.4}$$

Together with (6.3), this equation constitutes the proof of the strong scaling property of this theory: the bare functions depend on p_i, m and Λ, and in (6.3) the dependence in Λ is factored out, the resulting function depending only on p_i and m.

Since the correlations length $1/m$ is large but not infinite, one expects to see scaling laws only for distances small compared to m^{-1}—but of course large compared to the lattice spacing Λ^{-1}. Therefore we are interested in the limit in which the ratios p_i/m are large. Thus we are going to examine what happens when T varies, i.e. for the bare theory r_0 varies, at g_0 and Λ fixed. In the corresponding renormalized theory this induces a change both of m and of g. The corresponding differential equation has been first derived by Callan (1970) and Symanzik (1970a, 1971). The method consists in applying on both sides of (6.3) the differential operator $m(\mathrm{d}/\mathrm{d}m)|_{g_0,\Lambda\,\mathrm{fixed}}$. If one characterizes the corresponding variation of g and of Z by the functions

$$W\left(g, \frac{m}{\Lambda}\right) = m \frac{\mathrm{d}}{\mathrm{d}m} g \Big|_{g_0, \Lambda} \tag{6.5}$$

$$\eta\left(g, \frac{m}{\Lambda}\right) = m \frac{m}{\mathrm{d}m} \ln Z \Big|_{g_0, \Lambda} \tag{6.6}$$

the equation reads:

$$\left[m \frac{\partial}{\partial m} + W\left(g, \frac{m}{\Lambda}\right) \frac{\partial}{\partial g} - \frac{N}{2} \eta\left(g, \frac{m}{\Lambda}\right)\right] \Gamma^{(N)}(m, g)$$
$$= m \frac{\partial r_0}{\partial m} Z^{N/2} \frac{\partial}{\partial r_0} \Gamma^{(N)}_{\mathrm{bare}}(r_0, g_0, \Lambda). \tag{6.7}$$

In the right hand side appears the derivative with respect to r_0 of a bare correlation function: $(\partial/\partial r_0)\,\Gamma^{(N)}_{\mathrm{bare}}$.

If we consider the derivative with respect to r_0 of the generating functional of the bare correlation functions $Z(H)$ given by (2.2), we obtain

$$\frac{\partial}{\partial r_0} Z(H) = -\frac{1}{2K} \int [\mathrm{d}S_B] \left(\int S_B^2(y)\,\mathrm{d}y\right) \exp\left(-\int \mathrm{d}x\,[\mathscr{H}(S_B) - S_B H]\right) \tag{6.8}$$

which is thus the generating functional of the correlation functions in which one S_B^2 operator has been inserted at zero momentum. Correspondingly $(\partial/\partial r_0)\Gamma^{(N)}_{\mathrm{bare}}$ is the one-particle irreducible function $\Gamma^{(1,N)}_{\mathrm{bare}}(0; p_1, \ldots, p_N)$ with one such insertion. The renormalization of such functions has been described in Section III; in addition to the previous renormalization, a new one is needed for the S^2 insertion. The relation between bare and renormalized functions has been given in Section III, (3.40):

$$\Gamma^{(1,N)}(q; p_1, \ldots, p_N; m, g) = \frac{Z_{(2)}}{Z} Z^{N/2} \Gamma^{(1,N)}_{\mathrm{bare}}(q; p_i; r_0, g_0, \Lambda) \tag{6.9}$$

in which the new renormalization constant $Z_{(2)}$ is defined by the condition

$$\Gamma^{(1,2)}(0; 0, 0; m, g) = 1. \tag{6.10}$$

Therefore the r.h.s. of (6.7) is proportional to $\Gamma^{(1,N)}(0; p_i; m, g)$ with a coefficient independent of N. In order to find the factor in front of $\Gamma^{(1,N)}$, one sets all momenta equal to zero for $N = 2$ and uses the renormalization conditions (6.1), (6.10). The equation (6.7) is now

$$\left[m\frac{\partial}{\partial m} + W\left(g, \frac{m}{\Lambda}\right)\frac{\partial}{\partial g} - \frac{N}{2}\eta\left(g, \frac{m}{\Lambda}\right)\right]\Gamma^{(N)}(m, g)$$
$$= m^2\left(2 - \eta\left(g, \frac{m}{\Lambda}\right)\right)\Gamma^{(1,N)}. \qquad (6.11)$$

Furthermore since the $\Gamma^{(N)}$ and $\Gamma^{(1,N)}$ are independent of m/Λ, and since the functions W and η may be calculated from (6.11) itself in terms of these finite functions, it follows that W and η are independent of m/Λ: this leads to the final form of the Callan–Symanzik equation

$$\left[m\frac{\partial}{\partial m} + W(g)\frac{\partial}{\partial g} - \frac{N}{2}\eta(g)\right]\Gamma^{(N)}(p_1, \ldots, p_N; m, g)$$
$$= m^2(2 - \eta(g))\,\Gamma^{(1,N)}(0; p_1, \ldots, p_N; m, g). \qquad (6.12)$$

In this equation we have kept the same names for the coefficients, though W and η differ from the functions used in (5.15).

From this equation we can now obtain information on the large p_i/m limit when all momenta are scaled by a common factor λ. Weinberg's theorem (1960) tells us that the diagrams of $\Gamma^{(1,N)}$ are of the order λ^{-2} (up to powers of ln λ) order by order in g *and* ε, compared to those of $\Gamma^{(N)}$. This statement is equivalent to that used above in Section V about the existence of a massless theory. As usual, we shall assume that $\Gamma^{(1,N)}$ is still negligible after resummation on g and ε. In this asymptotic region, we obtain an homogeneous equation:

$$\left[m\frac{\partial}{\partial m} + W(g)\frac{\partial}{\partial g} - \frac{N}{2}\eta(g)\right]\Gamma^{(N)}_{\text{as}} = 0. \qquad (6.13)$$

Again the asymptotic regime is governed by the zeros of $W(g)$. Near the origin, W has two zeros:

one with a negative slope:

$$g = 0 \qquad W'(0) = -\varepsilon \qquad (6.14)$$

another with a positive slope

$$g = g^* \qquad W'(g^*) = \omega > 0. \qquad (6.15)$$

In the large momentum limit one would normally expect the origin to be the relevant zero, except if g takes exactly the value g^*. If this is so one finds immediately from (6.13) and dimensional analysis that

$$\Gamma^{(N)}_{\text{as}}(\lambda p_i; m, g^*) = \lambda^{d-(N/2)(d-2+\eta)}\,\Gamma^{(N)}_{\text{as}}(p_i; m, g^*). \tag{6.16}$$

Let us show that the coupling constant g actually is equal to g^*. We have to remember that in the physics of this problem, the bare coupling constant u_0 is of the form $g_0\Lambda^\varepsilon$ with fixed g_0 (5.7). Let us write the relation between g_0 and g as

$$g_0 = f\left(g; \frac{m}{\Lambda}\right), \tag{6.17}$$

and let us take the derivative of this equation with respect to m with fixed g_0 and Λ. From the definition (6.5) of $W(g)$ we obtain

$$\left[m\frac{\partial f}{\partial m} + W(g)\frac{\partial f}{\partial g}\right] = 0 \tag{6.18}$$

and integration of this equation in a similar way to what was done for (5.15) leads to the relation

$$g_0\left(g; \frac{m}{\Lambda}\right) = g_0\left(g(\lambda); \lambda\frac{m}{\Lambda}\right). \tag{6.19}$$

As before, in this derivation of (6.19), the main information used is that $W(g)$ does not depend on m/Λ. From this equation one sees that in order to increase the cut-off, for *fixed* g_0, one has to decrease λ and simultaneously take as renormalized coupling constant $g(\lambda)$. Therefore $g(\lambda)$ tends toward the infrared stable zero g^* of $W(g)$

$$g(\lambda) - g^* \underset{\lambda\to 0}{\sim} \lambda^\omega \tag{6.20}$$

Thus the renormalized theory which describes the critical region is *the* theory in which $g = g^*$. Let us insist on the fact that g is fixed at the infrared stable fixed point g^*, because the theory is described in terms of renormalized parameters associated with the large length scale m^{-1}, whereas the bare coupling constant raised to the power $-1/\varepsilon$ is a finite length.

B. Renormalization group equations for the critical theory

The previous renormalization is convenient because it is written in terms of the smallest number of dimensional parameters. Unfortunately, it is not convenient to study the theory exactly at T_c, since all momenta are measured in the scale m, which vanishes at T_c. Furthermore the normalization of the correlation functions becomes singular. Since we are going to be interested in what follows in scaling laws above and below T_c and in continuous variations of $T - T_c$, as for instance in the equation of state, we shall follow a somewhat different procedure. First we shall describe the theory at $T = T_c$ and then we shall deal with the neighbourhood of T_c in the next subsection.

In order to avoid the problem of singular normalizations at T_c, we first have to introduce an arbitrary non-zero mass parameter μ. The renormalized massless theory is then described by the conditions

$$\Gamma^{(2)}(p; g, \mu)\big|_{p^2=0} = 0 \tag{6.21a}$$

$$\frac{\partial}{\partial p^2}\Gamma^{(2)}(p; g, \mu)\big|_{p^2=\mu^2} = 1 \tag{6.21b}$$

$$\Gamma^{(4)}(p_1, \ldots, p_4; g, \mu)\big|_{\text{S.P.}} = \mu^{\varepsilon} g, \tag{6.21c}$$

in which the symmetry point S.P. is defined as

$$p_i \cdot p_j = \tfrac{1}{3}\mu^2(4\delta_{ij} - 1). \tag{6.22}$$

The bare theory is related to this one by (5.9)

$$\Gamma^{(N)}_{\text{bare}}(p_i; g_0, \Lambda) = Z^{-N/2}\left(g, \frac{\Lambda}{\mu}\right)\Gamma^{(N)}(p_i; g, \mu) \tag{6.23}$$

in which we are expressing the field strength renormalization Z in terms of g rather than g_0. From (6.23) one sees that, the ratio Λ/μ being large but fixed, the behaviour of the bare theory for $|p_i|/\Lambda$ small corresponds to the small $|p_i|/\mu$ limit of the renormalized one.

Here the bare and renormalized theories depend on the same number of parameters, but the renormalization group equations are exact for the renormalized theory, once the limit of a large Λ/μ ratio has been taken. These R.G. equations are the expression in differential form of the independence with respect to μ of the bare theory:

$$\mu \frac{\mathrm{d}}{\mathrm{d}\mu}\bigg|_{g_0, \Lambda} \Gamma^{(N)}_{\text{bare}} = 0.$$

The same procedure as in Section V.A yields the equation

$$\left[\mu\frac{\partial}{\partial\mu} + W(g)\frac{\partial}{\partial g} - \frac{N}{2}\eta(g)\right]\Gamma^{(N)}(p_i; g, \mu) = 0, \tag{6.24}$$

in which again we have kept the same names for the new coefficients of this R.G. equation which are here:

$$W(g) = \mu\frac{\mathrm{d}}{\mathrm{d}\mu}\bigg|_{g_0, \Lambda} g \tag{6.25}$$

$$\eta(g) = \mu\frac{\mathrm{d}}{\mathrm{d}\mu}\bigg|_{g_0, \Lambda} \ln Z\left(g, \frac{\Lambda}{\mu}\right). \tag{6.26}$$

This equation (6.24) is analogous to (5.15) on the bare theory and yields identical conclusions concerning the scaling behaviour of correlation functions at T_c. Notice that, since the renormalized theory is also independent of the particular cut-off function of the bare theory, it follows that the exponent η calculated as $\eta(g^*)$ in Section V and here, but from different functions $\eta(g)$ and different g^*, is the same.

It now remains to derive the R.G. equations away from T_c. The first step consists in generalizing the R.G. equation (6.24), to correlation functions at T_c involving products of S and S^2 operators. The renormalization of the S^2 operators has been discussed in Section III. Its multiplicative character reflects on the equation

$$\Gamma^{(L,N)}(q_1, \ldots, q_L; p_1, \ldots, p_N; g, \mu)$$
$$= Z^{N/2}\left(\frac{Z_{(2)}}{Z}\right)^L \Gamma^{(L,N)}_{\text{bare}}(q_i; p_j; g_0, \Lambda) + \delta_{N0}\delta_{L2}\mu^{-\varepsilon}A\left(g_0, \frac{\Lambda}{\mu}\right) \tag{6.27}$$

in which now the extra-renormalizations $Z_{(2)}/Z$ and A are fixed by the conditions

$$\Gamma^{(1,2)}(q; p_1, p_2; g, \mu)\big|_{p_1^2 = p_2^2 = \mu^2,\ p_1 \cdot p_2 = -\frac{1}{3}\mu^2} = 1 \tag{6.28}$$

$$\Gamma^{(2,0)}(q, -q; g, \mu)\big|_{q^2 = \frac{4}{3}\mu^2} = 0. \tag{6.29}$$

The corresponding R.G. equation is

$$\left[\mu\frac{\partial}{\partial\mu} + W(g)\frac{\partial}{\partial g} - \frac{N}{2}\eta(g) - L\left(\frac{1}{\nu(g)} - 2\right)\right]\Gamma^{(L,N)} = \delta_{N0}\delta_{L2}\mu^{-\varepsilon}B(g), \tag{6.30}$$

in which, for reasons related to usual conventions concerning critical exponents, one has defined

$$\frac{1}{\nu(g)} - 2 = \mu\frac{\partial}{\partial\mu}\bigg|_{g_0, \Lambda} \ln\left(\frac{Z_{(2)}}{Z}\right), \tag{6.31}$$

and

$$B(g) = -\left[\varepsilon + 2\left(\frac{1}{\nu(g)} - 2\right) - \mu\frac{\partial}{\partial\mu}\right]A\left(g_0, \frac{\Lambda}{\mu}\right). \tag{6.32}$$

The solution of (6.30) is obtained in the same way as before:

$$\Gamma^{(L,N)}\left(\frac{p_i}{\lambda};\frac{q_j}{\lambda}; g, \mu\right) \underset{\lambda\to\infty}{\sim} \lambda^{-(d-(N/2)(d-2+\eta)-(L/\nu))}, \tag{6.33}$$

in which

$$\nu \equiv \nu(g^*), \tag{6.34}$$

except for the energy–energy correlation function, in which the r.h.s. of (6.30) generates an additional power zero of λ:

$$\Gamma^{(2,0)}(q/\lambda; g, \mu) \underset{\lambda\to\infty}{\sim} C_1\lambda^{-(d-2/\nu)} + C_2. \tag{6.35}$$

C. Expansion around the critical theory

Material relevant to this sub-section will be found in the work of Zinn-Justin, Di Castro *et al.*, already cited, and in a paper by Weinberg (1973).

Now that we have constructed the correlation functions at T_c with an arbitrary number L of S^2 insertions we are in position to study the theory away from T_c. In fact let us write the Hamiltonian density in terms of the renormalized field $S(x)$:

$$\mathscr{H}(x) = \frac{Z}{2}(\nabla S)^2 + \frac{1}{4!}g_0 Z^2(S^2)^2 + \tfrac{1}{2}r_{0c}ZS^2 + \tfrac{1}{2}tZ_{(2)}S^2 \tag{6.36}$$

in which the deviation t from the critical temperature is defined by

$$r_0 - r_{0c} = t\frac{Z_{(2)}}{Z}. \tag{6.37}$$

Formally, the finite temperature correlation functions $\Gamma^{(N)}(t)$ may be obtained as an expansion in powers of t from the $\Gamma^{(L,N)}$ $(t = 0)$ defined in the previous section. However, this expansion generates insertions of S^2 at zero momentum which are infinite in the massless theory. Thus one has to be more cautious and first let the parameter t of the Hamiltonian (6.36) be space-dependent. The expansion in powers of t is then

$$\Gamma^{(N)}(p_i; t, g, \mu) = \sum_{L=0}^{\infty}\frac{1}{L!}\int dq_1 \ldots dq_L\, \tilde{t}(q_1)\ldots\tilde{t}(q_L)\,\Gamma^{(L,N)}(q_j; p_i; t = 0, g, \mu). \tag{6.38}$$

Since (6.30) holds independently of all the momenta, one can derive from it and (6.38) the equation

$$\left[\mu\frac{\partial}{\partial\mu} + W(g)\frac{\partial}{\partial g} - \frac{N}{2}\eta(g) - \left(\frac{1}{\nu(g)} - 2\right)\int \mathrm{d}q\, \tilde{t}(q)\frac{\delta}{\delta\tilde{t}(q)}\right]\Gamma^{(N)}(p_i; t, g, \mu) = 0.$$

Before going to the limit of a uniform t, one has to perform partial resummations on the massless theory in order to obtain a massive propagator. The loopwise expansion with the Hamiltonian (6.36) would have this effect. No more infra-red divergences appear in the limit of a constant t and the final R.G. equation is

$$\left[\mu\frac{\partial}{\partial\mu} + W(g)\frac{\partial}{\partial g} - \frac{N}{2}\eta(g) - \left(\frac{1}{\nu(g)} - 2\right)t\frac{\partial}{\partial t}\right]\Gamma^{(N)}(p_i; t, g, \mu) = 0. \quad (6.40)$$

This procedure in fact constructs a massive theory as in subsection A but renormalized in a different manner. Their correlation functions differ by finite renormalizations.

D. Scaling laws above T_c

Here the reader is referred to the papers by Brézin *et al.* (1973b).

From (6.40) one can obtain all the scaling laws for the correlation functions above T_c. We first have to integrate (6.40), and the general method for such linear equations is to define functions of a scale parameter λ by

$$\mu(\lambda) = \mu\lambda \quad (6.41a)$$

$$\lambda\frac{\mathrm{d}}{\mathrm{d}\lambda}g(\lambda) = W(g(\lambda)); \qquad g(1) = g \quad (6.41b)$$

$$\frac{\lambda}{t(\lambda)}\frac{\mathrm{d}t(\lambda)}{\mathrm{d}\lambda} = -\left(\frac{1}{\nu(g(\lambda))} - 2\right); \qquad t(1) = t. \quad (6.41c)$$

The equation (6.40) is then expressed as

$$\left[\lambda\frac{\mathrm{d}}{\mathrm{d}\lambda} - \frac{N}{2}\eta(g(\lambda))\right]\Gamma^{(N)}(p_i; t(\lambda), g(\lambda), \mu(\lambda)) = 0, \quad (6.42)$$

whose solution gives immediately

$$\Gamma^{(N)}(p_i; t, g, \mu) = \left[\exp - \frac{N}{2}\int_1^{\lambda} \eta(g(x))\frac{\mathrm{d}x}{x}\right]\Gamma^{(N)}(p_i; t(\lambda), g(\lambda), \lambda\mu). \quad (6.43)$$

We can now give to λ the value which is a solution of the equation

$$\frac{t(\lambda)}{\lambda^2\mu^2} = 1, \tag{6.44}$$

which means in terms of t and μ:

$$\ln\frac{t}{\mu^2} = \int_g^{g(\lambda)} \frac{\mathrm{d}g'}{W(g')\nu(g')}. \tag{6.45}$$

Equation (6.43) can then be written in the form:

$$\Gamma^{(N)}(p_i; t, g, \mu) = \tilde{Z}^{N/2}\left(\frac{m}{\mu}, g(\lambda)\right)\tilde{\Gamma}^{(N)}(p_i; m, g(\lambda)) \tag{6.46}$$

with the definitions:

$$m = \lambda\mu = \mu\exp\int_g^{g(\lambda)} \frac{\mathrm{d}g'}{W(g')} \tag{6.47a}$$

$$\ln\tilde{Z}\left(\frac{m}{\mu}, g(\lambda)\right) = -\int_g^{g(\lambda)} \frac{\mathrm{d}g'\eta(g')}{W(g')}. \tag{6.47b}$$

Equation (6.46) shows simply the fact that the correlation functions that we have constructed in this subsection can be obtained from correlation functions $\tilde{\Gamma}^{(N)}$ analogous to those defined in Section VI.A (i.e. directly expressed in terms of the correlation length m^{-1}) by a finite field renormalization $\tilde{Z}$ and a change of parameters.

We shall now use the freedom that we have in the choice of μ to derive the scaling laws. *A priori* the critical domain was defined by $t^{1/2} \ll \Lambda$ and $|p_i| \ll \Lambda$. It is possible to choose for μ a size much smaller than Λ, but much larger than $t^{1/2}$ and the $|p_i|$. Then, in the renormalized theory, μ will appear as the natural large scale, which replaces the cut-off, in terms of which the critical domain is defined. The advantage of the replacement of Λ by μ is that the R.G. equations (6.40) are exact and that all non-universal dependence in the cut-off function has been eliminated. This transformation is therefore convenient for practical calculations and also for the systematic study of corrections to scaling laws.

Now in the large μ limit, equations (6.45) and (6.47a) show that $g(\lambda)$ tends, as discussed before, towards g^*. The ratio t/μ^2 behaves then like:

$$\ln\frac{t}{\mu^2} \sim \frac{1}{\nu}\int_g^{g(\lambda)} \frac{\mathrm{d}g'}{W(g')}, \tag{6.48}$$

where we have set $\nu(g^*) = \nu$.

Then m and $\tilde{Z}$ behave like:

$$\frac{m}{\mu} \sim \left(\frac{t}{\mu^2}\right)^{\nu} \tag{6.49a}$$

$$\tilde{Z} \sim \left(\frac{t}{\mu^2}\right)^{-\eta\nu}. \tag{6.49b}$$

In $\tilde{\Gamma}^{(N)}$, $g(\lambda)$ can be replaced by g^* provided the ratios p_i/m remain fixed. Using dimensional considerations we can write:

$$\tilde{\Gamma}^{(N)}(p_i, m, g^*) = m^{d-\frac{1}{2}N(d-2)}\tilde{F}^{(N)}\left(\frac{p_i}{m}\right). \tag{6.50}$$

In the critical domain $\Gamma^{(N)}(p_i; t, g, \mu)$ is then of the form:

$$\Gamma^{(N)}(p_i; t, g, \mu) = \mu^{d-(N/2)(d-2)}\left(\frac{t}{\mu^2}\right)^{\nu[d-\frac{1}{2}N(d-2+\eta)]}\tilde{F}^{(N)}\left[\frac{p_i}{\mu}\left(\frac{t}{\mu^2}\right)^{-\nu}\right] \tag{6.51}$$

$$\frac{t}{\mu^2} \ll 1, \qquad \frac{|p_i|}{\mu} \ll 1.$$

Equation (6.51) gives the general scaling form of the correlation functions in the critical domain.

Equation (6.49a) shows that the correlation length $\xi = m^{-1}$ diverges as $t^{-\nu}$.

At zero momentum the $\Gamma^{(N)}$ have finite limit for $t \neq 0$. This limit behaves as $t^{\nu[d-(N/2)(d-2+\eta)]}$. In particular for $N = 2$ we obtain the magnetic susceptibility χ^{-1} since

$$\chi^{-1} = \Gamma^{(2)}(p = 0; t, g, \mu)$$

and thus χ diverges as $t^{-\gamma}$ with the scaling law:

$$\gamma = \nu(2 - \eta). \tag{6.52}$$

We know that the $\Gamma^{(N)}$ have finite limits when t goes to zero. This implies that in the range $\mu \gg |p_i| \gg t^{\nu}$ the power of t in front of the r.h.s. of (6.51) must be cancelled. Thus:

$$\Gamma^{(N)}(\lambda p_i; t, g, \mu) \sim \lambda^{d-\frac{1}{2}N(d-2+\eta)} \tag{6.53}$$

$$\mu \gg \lambda|p_i| \gg t^{\nu}$$

as shown before in (6.16).

E. Correlation functions involving S^2 insertions

The same methods allow us to study a general correlation function of L S^2-fields and N S-fields. For the 1-P.I. functions $\Gamma^{(L,N)}$ one starts from the

R.G. equation (6.30) at T_c, and by the same procedure of treating the temperature as generating S^2-insertions one obtains:

$$\left[\mu\frac{\partial}{\partial\mu} + W(g)\frac{\partial}{\partial g} - \frac{N}{2}\eta(g) - \left(\frac{1}{\nu(g)} - 2\right)\left(L + t\frac{\partial}{\partial t}\right)\right] \times \Gamma^{(L,N)}(q_i; p_j; t, g, \mu) = 0 \qquad (6.54)$$

except for $N = 0$ and $L \leqslant 2$ which will be discussed below. Integration of (6.54) is done by methods identical with those used above, and yields

$$\Gamma^{(L,N)}(q_i; p_j; t, g, \mu) = \mu^{d - \frac{1}{2}N(d-2) - 2L}\left(\frac{t}{\mu^2}\right)^{-L + \nu[d - \frac{1}{2}N(d-2+\eta)]} \times F^{(L,N)}\left(\frac{q_i}{\mu}\left(\frac{t}{\mu^2}\right)^{-\nu}; \frac{p_j}{\mu}\left(\frac{t}{\mu^2}\right)^{-\nu}\right) \qquad (6.55)$$

$$\frac{t}{\mu^2} \ll 1, \qquad \frac{|q_i|}{\mu} \ll 1, \qquad \frac{|p_j|}{\mu} \ll 1$$

Again from this equation we obtain:

(a) At zero-momentum an additional divergence t^{-L} associated to the LS^2–insertions which is expected since

$$\Gamma^{(L,N)}(0; p_i; t, g, \mu) = \left(\frac{\partial}{\partial t}\right)^L \Gamma^{(N)}(p_i; t, g, \mu).$$

(b) At $t = 0$ the scaling derived in (6.33) for the massless theory.

For the energy–energy correlation function $\Gamma^{(2,0)}(q; -q; \mu, g, t)$ the additional term in the r.h.s. of the R.G. equation (6.30) gives

$$\left[\mu\frac{\partial}{\partial\mu} + W(g)\frac{\partial}{\partial g} - \left(\frac{1}{\nu(g)} - 2\right)\left(2 + t\frac{\partial}{\partial t}\right)\right]\Gamma^{(2,0)} = \mu^{-\varepsilon}B(g) \qquad (6.56)$$

which implies

$$\Gamma^{(2,0)}(q; t, g, \mu) = \mu^{-\varepsilon}\left(\frac{t}{\mu^2}\right)^{-(2-\nu d)} F\left(\frac{q}{\mu}\left(\frac{t}{\mu^2}\right)^{-\nu}\right) + \mu^{-\varepsilon}C_1(g) \qquad (6.57)$$

$$\frac{t}{\mu^2} \ll 1, \qquad \frac{|q|}{\mu} \ll 1.$$

The usual scaling laws predict at zero momentum a singular dependence in $t^{-\alpha}$ of this function, and (6.57) proves the scaling law

$$\alpha = 2 - \nu d. \qquad (6.58)$$

In the massless theory we had not defined $\Gamma^{(1,0)}$, which is some arbitrary constant; $\Gamma^{(0,0)}$ is by definition the free-energy functional $\Gamma(M)$ for $M = 0$, which is zero at T_c since the partition function has been normalized to one in zero field. The variation with t of these two quantities may be obtained from the remark that at zero momentum

$$\Gamma^{(L,0)}(t) = \left(\frac{\partial}{\partial t}\right)^L \Gamma^{(0,0)}(t). \tag{6.59}$$

Therefore, by integration of (6.57), we obtain

$$\Gamma^{(0,0)} \equiv \Gamma(M = 0, t, g, \mu) = \frac{F(0)\,\mu^d}{(2-\alpha)(1-\alpha)}\left(\frac{t}{\mu^2}\right)^{2-\alpha} + \tfrac{1}{2}\mu^d C_1(g)\left(\frac{t}{\mu^2}\right)^2 + C_2\mu^d\frac{t}{\mu^2}\left(\frac{t}{\mu^2} \ll 1\right) \tag{6.60}$$

which exhibits the singular part of the free-energy in zero-field above T_c.

VII. Scaling Laws below the Critical Temperature

In this Section we shall show that the scaling laws related to the equation of state (Widom, 1965; Domb and Hunter, 1965), first verified in the renormalization group calculations of Brézin *et al.* (1972, 1973c) and Avdeeva and Migdal (1972), the equality of exponents above and below T_c, etc., are satisfied, as described by Brézin *et al.* (1973b).

The existence of a spontaneously broken symmetry (see Lee, 1969, 1972; Symanzik, 1970b) below the critical temperature introduces a few modifications to the theory developed in the previous Section. In addition, differences will appear between the breaking of a discrete symmetry, as for instance $S \to -S$ for an Ising-like order parameter, and the breaking of a continuous symmetry for which the massless Nambu-Goldstone (Nambu 1960, Goldstone 1961) modes are present. This has already been discussed at the mean field level in Section IV.

A. Equation of state

We are first going to show that in the critical domain there is an equation of state in the Widom (1965) scaling form

$$\frac{H}{M^\delta} = f\left(\frac{t}{M^{1/\beta}}\right), \tag{7.1}$$

in which the function $f(x)$ is regular around zero. Thus the same function describes the two regions above and below T_c. In addition the two new exponents δ and β will be related to η and ν that we have introduced previously.

If the order parameter has more than one component, (7.1) stands for the magnitude of the magnetization which lies along the applied field.

We are first going to derive the R.G. equation satisfied by the magnetic field $H(M, t)$. We know that H is the derivative with respect to M of the generating functional $\Gamma(M)$ of the 1-P.I. functions at zero momentum

$$\Gamma(M, t, g, \mu) = \sum_{N=0}^{\infty} \frac{M^N}{N!} \Gamma^{(N)} (p_i = 0; t, g, \mu) \tag{7.2}$$

and thus

$$H(M, t, g, \mu) = \sum_{1}^{\infty} \frac{M^N}{N!} \Gamma^{(N+1)} (p_i = 0; t, g, \mu). \tag{7.3}$$

From the R.G. equation (6.40) satisfied by the $\Gamma^{(N)}$ we obtain for H

$$\left[\mu \frac{\partial}{\partial \mu} + W(g) \frac{\partial}{\partial g} - \tfrac{1}{2}\eta(g)\left(1 + M \frac{\partial}{\partial M}\right) - \left(\frac{1}{\nu(g)} - 2\right) t \frac{\partial}{\partial t}\right] H(M, t, g, \mu) = 0 \tag{7.4}$$

(see Brézin *et al.*, 1973b; Zinn-Justin, 1973; and Coleman and Weinberg, 1973). Integration of (7.4) by the same method gives, for a dilatation λ,

$$H(M, t, g, \mu) = H(M(\lambda), t(\lambda), g(\lambda), \lambda\mu) \exp - \tfrac{1}{2} \int_{g}^{g(\lambda)} \frac{dg'\, \eta(g')}{W(g')} \tag{7.5}$$

in which, in addition to the functions $g(\lambda)$ and $t(\lambda)$ defined by (6.41), we have used $M(\lambda)$:

$$\frac{\lambda}{M(\lambda)} \frac{dM(\lambda)}{d\lambda} = - \tfrac{1}{2}\eta[g(\lambda)], \qquad M(1) = M. \tag{7.6}$$

Use of dimensional analysis gives the relation

$$H(M, t, g, \mu) = \mu^{3 - \frac{1}{2}\varepsilon} H\left(\frac{M}{\mu^{1 - \frac{1}{2}\varepsilon}}, \frac{t}{\mu^2}, g, 1\right). \tag{7.7}$$

Applying this to the r.h.s. of (7.5) we obtain

$$H(M, t, g, \mu) = \left[\exp - \tfrac{1}{2} \int_{g}^{g(\lambda)} \frac{dg'\, \eta(g')}{W(g')}\right] (\lambda\mu)^{3 - \frac{1}{2}\varepsilon} \times H\left(\frac{M(\lambda)}{(\lambda\mu)^{1 - \varepsilon/2}}, \frac{t(\lambda)}{(\lambda\mu)^2}, g(\lambda), 1\right). \tag{7.8}$$

Up to now λ was an arbitrary parameter; if we choose the particular value of λ given by the solution of the equation

$$\frac{M(\lambda)}{(\lambda\mu)^{1-\frac{1}{2}\varepsilon}} = 1, \tag{7.9a}$$

or more explicitly

$$\mathrm{Ln}\frac{M}{\mu^{1-\frac{1}{2}\varepsilon}} = \frac{1}{2}\int_{g}^{g(\lambda)} \frac{\mathrm{d} - 2 + \eta(g')}{W(g')}\,\mathrm{d}g', \tag{7.9b}$$

we obtain

$$H(M, t, g, \mu) = \left[\exp - \frac{1}{2}\int_{g}^{g(\lambda)} \frac{\mathrm{d}g'\,\eta(g')}{W(g')}\right](\lambda\mu)^{3-\frac{1}{2}\varepsilon} H\left(1, \frac{t(\lambda)}{\lambda^2\mu^2}, g(\lambda), 1\right), \tag{7.10}$$

in which λ is to be taken from (7.9). In the critical domain in which the magnetization is small (in the scale $\mu^{1-\frac{1}{2}\varepsilon}$), equation (7.9) is satisfied provided $g(\lambda)$ approaches g^*, which implies that λ goes to zero as

$$\lambda \sim \left(\frac{\mathrm{M}}{\mu^{1-\frac{1}{2}\varepsilon}}\right)^{2/(d-2+\eta)}. \tag{7.11}$$

In the function H of the r.h.s. of (7.10) the limit $g(\lambda) \to g^*$ can be taken provided the ratio $t(\lambda)/(\lambda^2\mu^2)$ remains fixed, which, taking into account the explicit function $t(\lambda)$, means that t/μ^2 is small.

We finally obtain, for small temperatures and magnetizations, the scaling result

$$H(M, t, g, \mu) = \mu^{3-\frac{1}{2}\varepsilon}\left(\frac{M}{\mu^{1-\frac{1}{2}\varepsilon}}\right)^{(d+2-\eta)/(d-2+\eta)} \times f\left(\frac{t}{\mu^2}\left(\frac{M}{\mu^{1-\frac{1}{2}\varepsilon}}\right)^{-2/[\nu(d-2+\eta)]}\right), \tag{7.12}$$

which is indeed of the form

$$\frac{H}{M^{\delta}} = f\left(\frac{t}{M^{1/\beta}}\right),$$

with

$$\delta = \frac{d+2-\eta}{d-2+\eta} \tag{7.13}$$

$$\beta = \frac{\nu}{2}(d-2+\eta). \tag{7.14}$$

Properties of the function $f(x)$

(i) Griffiths analyticity. The coefficients of the expansion of H in powers of M, given in (7.3), are finite for t positive different from zero. This property implies, as was first shown by Grifiths (1967) that the function $f(x)$ of (7.1) has an expansion for x large and positive of the form

$$f(x) = \sum_{p=1}^{\infty} a_p x^{\gamma - 2(p-1)\beta}. \tag{7.15}$$

(ii) Regularity around $x = 0$. We are going to show that $f(x)$ may be expanded for x small in powers of x. The argument goes as follows: instead of calculating H through a power series in M, one uses the loopwise expansion (2.31) of $\Gamma(M)$. The effect of this modified expansion is to take as inverse propagator

$$\left.\frac{\delta^2 \mathscr{H}}{\delta S(x)\, \delta S(y)}\right|_{S=M}$$

which reads $(p^2 + t + aM^2)$ in momentum space. In the corresponding diagrams the propagator is no longer singular at $t = 0$ for non-zero magnetization, and $\Gamma(M)$ as well as all its derivatives is finite at $t = 0$. This implies the existence of all the derivatives of $f(x)$ at $x = 0$.

B. Correlation functions below T_c for Ising-like systems

The correlation functions for systems with more than one component of the order parameter will be considered in a separate subsection. The reason is that correlation functions carry group indices corresponding to the nature of the external legs and, below T_c, there is a distinguished direction fixed by the magnetization. Thus there are differences between longitudinal and transverse correlations.

No such problem arises for $n = 1$. We have first to extend the R.G. equation for the correlation functions in arbitrary magnetization. These 1-P.I. functions are obtained by functional differentiation (cf. (2.17)) of the free-energy $\Gamma(M(x))$. Setting $M(x)$ to the constant value M, determined by H and t

$$\Gamma^{(N)}(x_1, \ldots, x_N; t, M, g, \mu) = \frac{\delta^N}{\delta M(x_1) \ldots \delta M(x_N)} \Gamma(M(x), t, g, \mu)|_{M(x)=M}. \tag{7.16}$$

Expanding Γ in powers of M, (7.16) gives

$$\Gamma^{(N)}(p_1, \ldots, p_N; t; M, g, \mu) = \sum_{N'=0}^{\infty} \frac{M^{N'}}{N'!} \Gamma^{(N+N')}(p_1, \ldots, p_N; 0 \ldots 0; t, 0, g, \mu). \tag{7.17}$$

Since we know the R.G. equations of the $\Gamma^{(N+N')}$ in zero magnetization (6.40) we obtain an equation for $\Gamma^{(N)}(M)$:

$$\left[\mu\frac{\partial}{\partial\mu} + W(g)\frac{\partial}{\partial g} - \tfrac{1}{2}\eta(g)\left(N + M\frac{\partial}{\partial M}\right) - \left(\frac{1}{\nu(g)} - 2\right)t\frac{\partial}{\partial t}\right] \times \Gamma^{(N)}(p_i; t, M, g, \mu) = 0 \qquad (7.18)$$

The solution of this equation should be familiar by now to the reader. In the critical domain it yields the scaling relation

$$\Gamma^{(N)}(p_i; t, M, g, 1) = M^{[2d/(d-2+\eta)]-N} F^{(N)}(p_i M^{-\nu/\beta}, t M^{-1/\beta}), \qquad (7.19)$$

provided $|p_i|$, $|t|$, $M \ll 1$ (in formula (7.19) we have set the normalizing parameter μ equal to one).

For reasons identical to those given for the equation of state, the $\Gamma^{(N)}(t, M)$ may be expanded in M for t strictly positive and in t for M non-zero. In particular from (7.19) one obtains the behaviour in zero-field below T_c of the correlation function.

From the equation of state we know that $t/M^{1/\beta} = x_0$ where x_0 is the (negative) solution of $f(x) = 0$. Therefore the correlation functions behave as

$$\Gamma^{(N)}(p_i; t, M, g)\big|_{H=0, t<0} = (-t)^{\nu[d - \frac{1}{2}N(d-2+\eta)]} F^{(N)}[p_i(-t)^{-\nu}]. \qquad (7.20)$$

Thus the correlation functions have exactly the same structure as above T_c with a correlation length

$$\xi \propto (-t)^{-\nu'} \quad \text{with} \quad \nu' = \nu. \qquad (7.21)$$

The magnetic susceptibility is obtained by setting $N = 2$, $p_i = 0$ in (7.20). Thus

$$\chi \propto (-t)^{-\gamma'} \quad \text{with} \quad \gamma' = \gamma. \qquad (7.22)$$

The same method applies to $\Gamma^{(L,N)}$ functions in non-zero magnetization. The results are similar. In particular for the energy–energy correlation function $N = 0$ $L = 2$, it gives a singular behaviour in zero field below T_c

$$\Gamma^{(2,0)}(q = 0, t, M)\big|_{H=0, t<0} \propto (-t)^{-\alpha'} \qquad (7.23)$$

with $\propto$

$$\alpha' = \alpha.$$

C. Longitudinal and transverse susceptibilities below T_c for an n-component system

We have already seen in our discussion of mean field theory that the main new features are related to the existence of massless modes in the $(n - 1)$ transverse directions. We are first going to show that this phenomenon persists to all orders in the interactions. This comes as a consequence of the Ward Takahashi identity (see Lee, 1972, and Symanzik, 1970b), which, for the case of $O(n)$-symmetry, reduces to the trivial statement that $W(H)$ and $\Gamma(M)$ are rotationally invariant. The consequences are

$$H_\alpha = \frac{\partial \Gamma}{\partial M} \frac{M_\alpha}{M}, \tag{7.24}$$

i.e., M_α is parallel to H_α with magnitudes related by

$$H = \frac{\partial \Gamma}{\partial M} \tag{7.25}$$

and by differentiation of (7.24) the longitudinal and transverse inverse propagators at zero-momentum

$$\Gamma_L^{(2)}(p = 0) = \frac{\partial^2 \Gamma}{\partial M^2} \tag{7.26}$$

$$\Gamma_T^{(2)}(p = 0) = \frac{1}{M} \frac{\partial \Gamma}{\partial M} = \frac{H}{M}. \tag{7.27}$$

From (7.27) one sees that, in zero field below T_c, $\Gamma_T^{(2)}(p = 0)$ vanishes, which shows that the transverse modes are massless below T_c and that the transverse susceptibility is

$$\chi_T = \frac{M}{H}. \tag{7.28}$$

This induces a divergence of the zero-field longitudinal susceptibility.

VIII. Corrections to Scaling Laws

Up to now in deriving scaling laws in the critical domain, we have considered only the leading terms in the large cut-off limit. For instance, within the S^4-coupling theory, we have neglected deviations occurring when one moves away from the fixed point. Furthermore from the very beginning we have eliminated the part of the interaction which was not of the local S^4 type. We are now going to examine successively the corrections generated when these terms are taken into account. These have been discussed by Wegner (1972a, b; 1973) and by Brézin *et al.* (1973b).

A. Deviations from scaling in the S^4-theory below four dimensions

From the general R.G. equation (6.40) and (7.18) one sees that, at the fixed point g^*, the theory is exactly scale-invariant for any value of the scale parameter μ. For $g \neq g^*$ and for μ large but not infinite, deviations from the pure scaling theory $g = g^*$ appear. In order to characterize these deviations, one could use the general solution (6.43) of this R.G. equation, but it is more convenient to integrate it in a slightly different manner. Let us define a temperature and magnetization with a g-dependent normalization by:†

$$M(g) = M \exp \tfrac{1}{2} \int_{g^*}^{g} \frac{dg'}{W(g')} [\eta(g') - \eta] \tag{8.1}$$

$$t(g) = t \exp \int_{g^*}^{g} \frac{dg'}{W(g')} \left[\frac{1}{\nu(g')} - \frac{1}{\nu} \right], \tag{8.2}$$

and define an auxiliary function C by the relation

$$\Gamma^{(N)}(p_i; t, M, g, \mu) = \left[\exp + \frac{N}{2} \int_{g^*}^{g} dg' \frac{[\eta(g') - \eta]}{W(g')} \right] \Gamma^{(N)}(p_i; t(g), M(g), g^*, \mu) \times C^{(N)}(p_i; t(g), M(g), g, \mu). \tag{8.3}$$

The function $C^{(N)}$ satisfies the R.G. equation

$$\left[\mu \frac{\partial}{\partial \mu} + W(g) \frac{\partial}{\partial g} - \tfrac{1}{2} \eta M \frac{\partial}{\partial M} - \left(\frac{1}{\nu} - 2 \right) t \frac{\partial}{\partial t} \right] C^{(N)} = 0 \tag{8.4}$$

with the boundary condition

$$C^{(N)}(p_i; t, M, g^*, \mu) = 1. \tag{8.5}$$

The equation (8.5), which, apart from $W(g)$, has constant coefficients, may be used to calculate $C^{(N)}$ to all orders in $g - g^*$. If, for simplicity, we keep only the first order the solution is

$$C^{(N)}(p_i; t, M, g, 1) = 1 + (g - g^*) t^{\omega\nu} D^{(N)}(p_i t^{-\nu}; t M^{-1/\beta}), \tag{8.6}$$

in which

$$\omega = W'(g^*). \tag{8.7}$$

At higher order it generates successive powers of $t^{\omega\nu}$. This $C^{(N)}$ function has a limit in the various cases $t = 0$ or $p_i = 0$ or $M = 0$, so that (8.6) describes the deviation from scaling arising whenever p_i or t or M is not too small. The first correction is thus entirely characterized by this exponent ω, the powers of p_i, t and M being respectively ω, $\omega\nu$ and $\nu\omega/\beta$.

Since as noted above (e.g. (5.16)), the function $W(g)$ has the expansion

$$W(g) = -\varepsilon g + a g^2 + O(g^3)$$

† On moving away from g^*, one generates corrections to scaling and also changes of normalizations of the fields. Eqns. (8.1–8.3) are constructed to separate these irrelevant changes from the genuine corrections.

the exponent ω defined by (8.7) has the ε-expansion

$$\omega = \varepsilon + O(\varepsilon^2), \tag{8.8}$$

as first obtained by F. Wegner (1972a, b; 1973). Higher orders calculated by Brézin *et al.* (1973d) are given in Section IX.

B. Logarithmic corrections in four dimensions

These corrections have been discussed by Larkin and Khmel'nitskii (1969), Wegner and Riedel (1973), and Brézin *et al.* (1973b).

In four dimensions the fixed point g^* is at the origin, and when the cut-off goes to infinity one is left with a trivial gaussian (i.e. free-field) model. Therefore the leading term of the free energy functional $\Gamma(M)$ vanishes when $d = 4$ and its structure is now dominated by terms which appeared as corrections below four dimensions.

The R.G. equation (7.10) gives the following relation for the magnetic field in terms of temperature and magnetization:

$$H(M, t, g, \mu) = \left[\exp - \tfrac{1}{2}\int_g^{g(\lambda)} \frac{dg'\,\eta(g')}{W(g')}\right](\lambda\mu)^3\, H\left(1, \frac{t(\lambda)}{\lambda^2\mu^2}, g(\lambda), 1\right), \tag{8.9}$$

in which

$$\frac{dg(\lambda)}{d\ln\lambda} = W(g(\lambda)) \tag{8.10}$$

$$\ln\frac{t(\lambda)}{\lambda^2\mu^2} = \ln\frac{t}{\mu^2} - \int_g^{g(\lambda)} \frac{dg'}{W(g')}\frac{1}{\nu(g')} \tag{8.11}$$

and λ is fixed by the relation $M(\lambda)/\lambda\,\mu = 1$, i.e.

$$\lambda = \frac{M}{\mu}\exp - \tfrac{1}{2}\int_g^{g(\lambda)} dg'\,\frac{\eta(g')}{W(g')}. \tag{8.12}$$

From the expansions (given in Section IX) in four dimensions we obtain:

$$W(g) = \frac{n+8}{6}g^2 + O(g^3) \tag{8.13a}$$

$$\eta(g) = \frac{n+2}{72}g^2 + O(g^3). \tag{8.13b}$$

$$\frac{1}{\nu(g)} = 2 - \frac{n+2}{6}g + O(g^3). \tag{8.13c}$$

We see from (8.12) that, in the critical domain, $M/\mu \ll 1$ implies again that λ is small. Indeed for λ small from (8.10)

$$g(\lambda) = \frac{6}{n+8}\frac{1}{|\ln \lambda|} + \text{constant} \tag{8.14}$$

and the integral in (8.12) has a finite limit when λ tends to zero. In this limit from (8.11) one finds

$$\frac{t(\lambda)}{\lambda^2\mu^2} = \text{const.}\ \frac{t}{M^2}\left|\ln\frac{M}{\mu}\right|^{-[(n+2)/(n+8)]} \tag{8.15}$$

and thus, in the critical domain t, $M \ll 1$

$$H(M,t,g,1) \sim M^3 H\left(1, \frac{t}{M^2}\left|\ln M\right|^{-[(n+2)/(n+8)]}, \frac{6}{n+8}\frac{1}{|\ln M|}, 1\right). \tag{8.16}$$

Since the coupling constant $g(\lambda)$ is small, H may be calculated by perturbation theory. One has:

$$\Gamma = \frac{t}{2}M^2 + \frac{g}{4!}M^4 + \ldots$$

and thus

$$H = tM + \frac{g}{6}M^3 + \ldots .$$

Hence, from (8.16) we obtain

$$H(M,t,g,1) \propto M^3\left\{\frac{t}{M^2}\left|\ln M\right|^{-[(n+2)/(n+8)]} + \frac{1}{n+8}\frac{1}{|\ln M|}\right\} \times \left(1 + \frac{\text{const.}}{|\ln M|}\right). \tag{8.17}$$

Thus in four dimensions we find logarithmic deviations from the mean field scaling law $H/M^3 = f(t/M^2)$. Formula (8.17) holds for values of M of order $|t|^{1/2}$. If M is much smaller than $|t|^{1/2}$, one must integrate again the R.G. equation with $M = 0$. The effect is, as expected to replace $\ln M$ by $\frac{1}{2}\ln|t|$ in formula (8.17). Explicit consequences of (8.17), first derived within the parquet approximation by Larkin and Khmel'nitskii (1969), are:

(a) at T_c

$$H \propto \frac{M^3}{|\ln M|} \tag{8.18}$$

(b) a spontaneous magnetization

$$M \propto (-t)^{1/2} [\ln(-t)]^{3/(n+8)} \tag{8.19}$$

(c) a susceptibility in zero field above or below T_c

$$\chi^{-1} \propto |t| |ln|t||^{-[(n+2)/(n+8)]}. \tag{8.20}$$

Corrections to formulae (8.18–8.20) involve terms proportional to $1/\ln|t|$ or $1/|\ln M|$ (more precisely $\ln|\ln|/\ln$), and the approach to the asymptotic behaviour in the critical domain is therefore extremely slow.

For the energy–energy correlation function $\Gamma^{(2,0)}(q; t, M, g, \mu)$ the R.G. equation is inhomogeneous (see (6.30)). In zero field above T_c it reads

$$\left[\mu\frac{\partial}{\partial\mu} + W(g)\frac{\partial}{\partial g} - \left(\frac{1}{\nu(g)} - 2\right)\left(2 + t\frac{\partial}{\partial t}\right)\right]\Gamma^{(2,0)} = B(g) \tag{8.21}$$

in which

$$B(g) = n/2 + O(g). \tag{8.22}$$

Integration of (8.21) yields

$$\Gamma^{(2,0)}(q; t, g, \mu) = \Gamma^{(2,0)}(q; t(\lambda), g(\lambda), \lambda\mu) \exp - 2\int_g^{g(\lambda)} \frac{dg'}{W(g')}\left(\frac{1}{\nu(g')} - 2\right)$$

$$- \int_g^{g(\lambda)} \frac{dg'}{W(g')} B(g') \exp - 2\int_g^{g'} \frac{dg''}{W(g'')}\left(\frac{1}{\nu(g'')} - 2\right) \tag{8.23}$$

and for the specific heat in the critical domain this equation gives

$$\int dx \, \langle S^2(x) \, S^2(0) \rangle \sim |\ln t|^{(4-n)/(n+8)} \left(1 + \frac{\text{const.}}{|\ln t|}\right). \tag{8.24}$$

Let us comment briefly on the derivation of these results. They have been derived in the framework of renormalized perturbation theory in which the cut-off has been sent to infinity. If the renormalized coupling constant g is calculated in terms of the bare g_0, one finds in this limit that the allowed range of g consists of the interval $[0, g^*(\varepsilon)]$, which shrinks to zero as d approaches four. In fact, if one keeps a large but finite cut-off, the analogue of (6.19) shows that, in four dimensions, g can vary in a range of order $1/\ln \Lambda$, which is sufficient to given sense to the above derivation. Actually, one could have given an alternative derivation, which avoids this complication, using the R.G. equations of the cut-off bare theory (which are, in fact, identical to those of the μ-renormalized theory).

C. Deviations from scaling induced by irrelevant even operators

Up to now we have studied the theory with a local S^4 interaction, since we argued in Section IV that it generates the most singular part in the critical domain. We shall now try to justify more precisely this point of view and evaluate the corrections associated with interactions of higher dimension. It is more appropriate to discuss this problem on the bare cut-off theory along the lines of what was done in Section V. The main tool will be the renormalization of composite operators presented in Section III.

Let us assume that one adds to the S^4 Hamiltonian density (5.4) a small interaction term $u_i \mathcal{O}_i(x)$ in which $\mathcal{O}_i$ is some local even operator like $S^6(x)$, $S^2(x)(\nabla S(x))^2$, etc.... Let us consider the effect of this new interaction at first order in u_i. The first correction of the correlation functions is then of the form $\Gamma^{(N)}_{\mathcal{O}_i}$ which is the N-point function with one insertion of $\mathcal{O}_i$ at zero momentum. If δ_i is the degree of divergence of the operator $\mathcal{O}_i$ defined by (3.43), the dimension of the constant u_i is $-\delta_i$ and thus

$$u_i = \Lambda^{-\delta_i} g_i, \tag{8.25}$$

in which g_i is dimensionless. It is also convenient to use the notation

$$A_i = \Lambda^{-\delta_i} \mathcal{O}_i. \tag{8.26}$$

For composite operators of high dimension, the coupling will thus contain high negative powers of the cut-off, but since the insertion of these operators leads to more and more divergences, the combined effect is not quite obvious.

In order to study these bare $\Gamma^{(N)}_{A_i}$ correlation functions, we shall first establish the R.G. equations that they satisfy. We have seen in section III that the bare A_i operator is coupled to all the renormalized operators which in four dimensions are of degree smaller or equal to δ_i. In particular, it contains a term proportional to the renormalized S^2-operator which has the effect of shifting the critical temperature. It is thus possible to substract from the bare A_i operator a term proportional to S^2

$$\tilde{A}_i = A_i - C_i S^2,$$

in which C_i is chosen in order to cancel the S^2-part of the expansion of A_i on renormalized operators. The insertion of $\tilde{A}_i$ at zero momentum in correlation functions is then free of the infra-red problem mentioned in Section VI.C. Let us consider for simplicity only the critical theory at $T = T_c$, the renormalization being then defined through an external mass parameter μ and similarly to (8.26) we define A_i^R by including a factor $\mu^{-\delta_i}$.

The relation between bare and renormalized correlation functions (see Section III.D) reads

$$\Gamma^{(N)R}_{A_i}(p_i; g, \mu) = Z^{N/2}\left(g_0, \frac{\Lambda}{\mu}\right) Z_{ij}\left(g_0, \frac{\Lambda}{\mu}\right) \Gamma^{(N)}_{\tilde{A}_j}(p; g_0, \Lambda). \tag{8.27}$$

Power-counting of the corresponding diagrams shows that

$$Z_{ij} = \left(\frac{\Lambda}{\mu}\right)^{\delta_i} \zeta_{ij} \tag{8.28}$$

in which δ_i is the degree of the operator A_i, and ζ_{ij} contains only logarithmic divergences in four dimensions. The renormalization group equation

$$\Lambda \frac{\partial}{\partial \Lambda}\bigg|_{g,\mu} \Gamma^{(N)R}_{A_i} = 0 \tag{8.29}$$

gives for the bare functions in the large Λ limit

$$\left\{\left[\Lambda \frac{\partial}{\partial \Lambda} + W(g_0)\frac{\partial}{\partial g_0} - \frac{N}{2}\eta(g_0)\right]\delta_{ij} - \gamma_{ij}(g_0)\right\}\Gamma^{(N)}_{\tilde{A}_j}(p; g_0, \Lambda) = 0 \tag{8.30}$$

in which

$$\gamma_{ij}(g_0) = -(Z^{-1})_{ik}\Lambda \frac{\partial}{\partial \Lambda} Z_{kj}. \tag{8.31}$$

The specific form (8.28) gives

$$\gamma_{ij} = -(\zeta^{-1})_{ik}\Lambda \frac{\partial}{\partial \Lambda}\zeta_{kj} - (\zeta^{-1})_{ik}\delta_k \zeta_{kj}. \tag{8.32}$$

Integration of the R.G. equation (8.30) will yield for $\Gamma^{(N)}_{\tilde{A}_i}(p)$ additional powers of (p/Λ) compared to $\Gamma^{(N)}(p)$ without insertions:

$$\Gamma^{(N)}_{\tilde{A}_i}(\lambda p; g_0, \Lambda) \underset{\lambda \to 0}{\sim} \lambda^{d - \frac{1}{2}N(d-2+\eta)} \sum_{\alpha} C_{i\alpha}\lambda^{-\eta_\alpha} \tag{8.33}$$

in which η_α are the eigenvalues of the matrix $\gamma_{ij}(g_0^*)$. These eigenvalues may be evaluated by noticing that the first matrix in the r.h.s. of (8.32) does not contribute at lowest order in the coupling constant, and thus gives contributions small near d = 4, whereas the second matrix has eigenvalues obviously equal to $-\delta_k$. Furthermore the matrix γ_{ij} has a particular structure due to the coupling of an operator A_i with operators of degree at most equal to δ_i. This implies that if we consider a sequence of operators of increasing degree we shall again find the eigenvalues η_α already obtained for operators of lower degree and only add new ones close to the degree of the new operator. The lowest possible values of the η_α correspond, since S^2 has been eliminated, to the insertion of $\tilde{S}^4$ and ΔS^2 whose degree vanishes with ε. These insertions, marginal in four dimensions, lead only to a change of g_0 and of the normalization of the field, and these effects have been considered before and lead to corrections to scaling proportional to $(p/\Lambda)^\omega$.

Let us now consider operators of positive degree like S^6, $S\nabla^4 S$, etc.... The matrix γ_{ij} will acquire new eigenvalues, and correspondingly new eigenoperators. These additional eigenvalues are close for small ε to a positive integer, and thus they generate corrections much smaller than those which have been considered in Subsection A. The new eigenoperators are therefore irrelevant for the critical behaviour. The effect of the insertion of a composite operator of positive degree may thus also be reproduced by a change of the coefficient of S^4 and $(\nabla S)^2$ in the Hamiltonian. This justifies the choice of an S^4-type interaction which has been made in order to study the critical point.

Let us finally notice that the regularization of the theory introduces such operators of high dimension. If we vary the cut-off function the correlation functions are modified by powers of p/Λ corresponding to the eigenvalues associated with these higher-degree operators.

D. Renormalization group equations for the tricritical behaviour

In Section IV we presented arguments to indicate that the tricritical phenomena are correctly described by mean field theory above three dimensions and that at the tricritical point the main singular behaviour is due to an S^6 local interaction. As for the critical behaviour, we shall study the renormalization group equations for this theory for a dimension $d \leqslant 3$. In $d = 3 - \varepsilon$ dimensions, one has to use a similar ε-expansion, and in three dimensions one expects logarithmic deviations from mean field theory, as shown by Wegner and Riedel (1973). The most interesting case concerns $d = 3$ on which we shall concentrate now.

In order to derive the R.G. equations of this theory, we shall perform an expansion around the tricritical theory as in Section VI. At the tricritical point, the renormalized Hamiltonian density is:

$$\mathscr{H}_c(x) = \tfrac{1}{2}(\nabla S)^2 + \frac{v}{6!}(S^2)^3 + \tfrac{1}{2}(Z-1)(\nabla S)^2 + \frac{1}{6!}v(Z-1)(S^2)^3 + Z^2\frac{u_{0c}}{4!}(S^2)^2 + \frac{Z}{2}m_{0c}^2 S^2 \tag{8.34}$$

The field strength and vertex renormalizations, as well as the critical values of the parameters m_0^2 and u_0—which are non-singular functions of the temperature and of the non-ordering field—are determined by the renormalization conditions

$$\Gamma^{(2)}(p = 0) = 0 \tag{8.35a}$$

$$\frac{\partial}{\partial p^2}\Gamma^{(2)}\big|_{p^2=\mu^2} = 1 \tag{8.35b}$$

$$\Gamma^{(4)}(p_i = 0) = 0 \tag{8.35c}$$

$$\Gamma^{(6)}(p_i)|_{\text{S.P.}} = v \tag{8.35d}$$

(in which S.P. is the symmetry point $p_i \cdot p_j = \frac{1}{5}\mu^2(6\delta_{ij} - 1)$).

For this tricritical theory the renormalization group equations follow immediately from the multiplicative renormalizability of the field $S(x)$:

$$\left[\mu\frac{\partial}{\partial\mu} + W(v)\frac{\partial}{\partial v} - \frac{N}{2}\eta(v)\right]\Gamma^{(N)}(p_i; v, \mu) = 0 \tag{8.36}$$

and the functions W and η may be calculated in perturbation theory and start at order v^2 with positive coefficients.

In order to describe the vicinity of the tricritical point, we must add to the previous Hamiltonian $\mathscr{H}_c$ sources coupled to S^2 and S^4. It is in fact natural to consider instead of S^4 a linear combination $\tilde{S}^4$ of S^4 and S^2 which, inserted once, leaves the theory massless:

$$\Gamma^{(2)}_{\tilde{S}^4}(q = 0, p = 0) = 0. \tag{8.37}$$

The four-point function with one $\tilde{S}^4$ insertion is defined by the condition

$$\Gamma^{(4)}_{\tilde{S}^4}(q; p_1 \cdots p_4)|_{\text{S.P.}} = 1 \tag{8.38}$$

(S.P. is the same as above for the p_i's).

All N-point functions with one S^4-insertion are then finite. However, since we need to treat the S^4 operator to all orders, a new problem arises because the insertion of two S^4 operators at two different points has still a logarithmic divergence removed by the renormalization condition

$$\Gamma^{(2)}_{S^4, S^4}(q, -q; 0, 0)|_{q^2 = \mu^2} = 0, \tag{8.39}$$

which can be obtained by adding to the product $S^4(x)\, S^4(y)$ of two S^4 renormalized operators a term proportional to $S^2(x)\,\delta(x - y)$. The insertion of an arbitrary number of S^4 operators is now well defined.

The insertion of S^2 is defined by the single renormalization condition

$$\Gamma^{(2)}_{S^2}(q; p_1, p_2)|_{\text{S.P}} = 1, \tag{8.40}$$

which makes finite all the S^2 insertions.

The complete renormalized Hamiltonian reads then

$$\mathscr{H}(x) = \mathscr{H}_c(x) + \frac{Z_{(4)}u}{4!}[(S^2)^2 - \alpha S^2] + Z_{(2)}\frac{t}{2}S^2 + \zeta\frac{u^2}{2}S^2 + P(u, t), \tag{8.41}$$

in which P is a polynomial of degree three in u and one in t. These considerations lead after rather lengthy calculations to the R.G. equation

$$\left[\mu\frac{\partial}{\partial\mu} + W(v)\frac{\partial}{\partial v} - \frac{N}{2}\eta(v) - \eta_4(v)\,u\frac{\partial}{\partial u} - \left(\eta_2(v)\,t + \hat{\eta}_2(v)\,u^2\right)\frac{\partial}{\partial t}\right]$$
$$\times\ \Gamma^{(N)}(p_i; u, t, v, \mu) = 0 \qquad (8.42)$$

in which the coefficients are finite functions of v. This equation is the basis of the derivation of the scaling laws in the tricritical region. From the form of $W(v)$ one sees that the origin is an attractive fixed point in three dimensions and thus, as first pointed out by Wegner and Riedel (1973), logarithmic deviations from mean field theory are present.

IX. The ε-Expansion: Calculations and Results

We shall present in this Section the calculations of various quantities, such as the critical exponents and the equation of state, as power series in ε. All these calculations are performed for the theory renormalized as indicated in Section VI. As a first step we shall have to compute various renormalization constants. We could compute them with a cut-off in the diagrams, as power series in g and ε. But, as we are interested in cut-off independent quantities like $W(g)$, $\eta(g)$, etc . . ., it is more convenient first to compute the diagrams at ε fixed but small enough, taking the infinite cut-off limit after mass renormalization. One has then to be careful and to expand in ε only the quantities finite in four dimensions and not for example the renormalizations constants.

In these conditions the Hamiltonian will read:

$$\mathscr{H}(x) = \tfrac{1}{2}\,Z(g)\,[\partial_\mu \mathbf{S}(x)]^2 + \tfrac{1}{2}\,\delta m^2(\Lambda)\,\mathbf{S}^2(x) + \frac{1}{4!}\,g\,Z_{(4)}(g)\,[\mathbf{S}^2(x)]^2$$
$$+ \tfrac{1}{2}\,t\,Z_{(2)}(g)\,\mathbf{S}^2(x). \qquad (9.1)$$

The mass parameter μ which was used previously to give the scale of the renormalization points has been set to one.

For convenience we have introduced a vertex renormalization constant $Z_{(4)}(g)$ related to the bare coupling constant g_0 by

$$gZ_{(4)}(g) = g_0 Z^2(g) \qquad (9.2)$$

All the renormalization constants needed for the calculations of the critical exponents can be computed in the massless theory ($t = 0$), on which apply also the renormalization conditions:

$$\Gamma^{(2)}(p, -p; g)\big|_{p^2=0} = 0 \qquad (9.3a)$$

$$\frac{\partial}{\partial p^2}\,\Gamma^{(2)}(p, -p; g)\big|_{p^2=e^l} = 1 \qquad (9.3b)$$

$$\Gamma^{(4)}(p_i; g)\big|_{(\text{S.P.})} = g \tag{9.3c}$$

$$\Gamma^{(1,2)}(q; p_1, p_2; g)\big|_{q^2 = 1,\, p_1^2 = p_2^2 = \frac{3}{4}} = 1 \tag{9.3d}$$

where the symmetry point (S.P.) is defined by:

$$p_i \cdot p_j = \tfrac{1}{4}(4\delta_{ij} - 1).$$

We have chosen to introduce an arbitrary dimensionless parameter l in order to show how the various functions depend on the renormalization scheme.

A. Calculation of the critical exponents

The first two terms of the ε-expansion of the critical exponents γ and η have been given by Wilson (1972). The next order has been calculated by Brézin *et al.* (1973e).

The various renormalization constants needed for the calculation of the functions $W(g)$, $\eta(g)$, $\nu(g)$ are given in terms of the values of the diagrams given in Figs. 10 and 11.

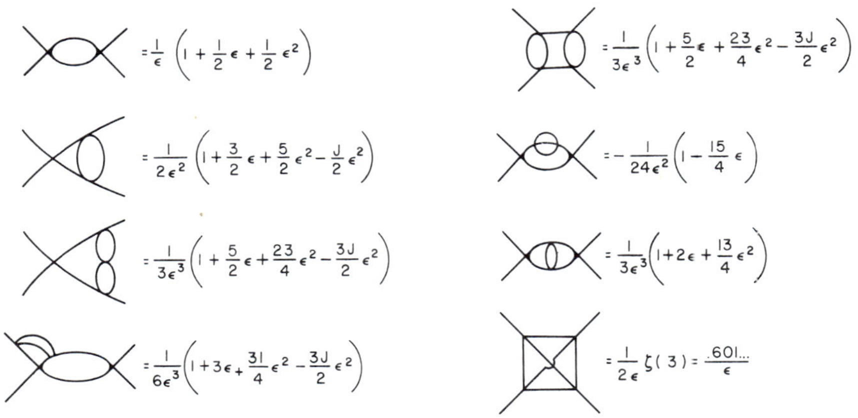

FIG. 10. Contributions to $Z_{(4)}$ and $Z_{(2)}$ expanded in powers of ε (J is given by (9.9)).

A diagram with L loops is still to be multiplied by S^L, where $S = 2\pi^{d/2} \times (2\pi)^{-d}/\Gamma(d/2)$, but we shall absorb this normalization factor into a redefinition of the coupling constant $g \to g/S$. Each diagram has been expanded up to the relevant order in ε.

Each graph of Fig. 11 stands for the derivative of the corresponding Feynman diagram, taken at the point $p^2 = e^l$.

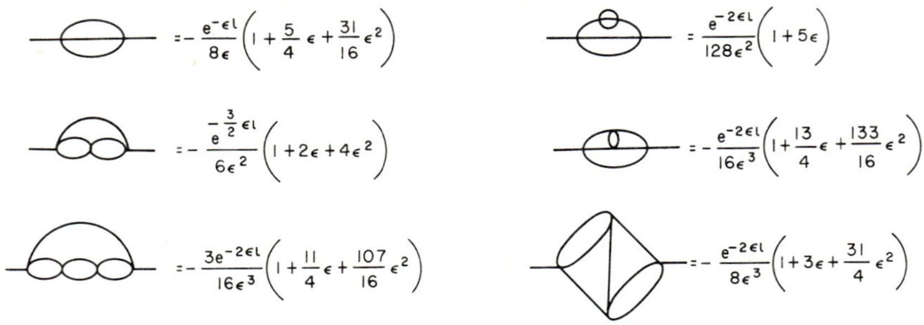

FIG. 11. Contributions to the field strength renormalization Z expanded in powers of ε.

Then the renormalization constants $Z_{(4)}$, Z and $Z_{(2)}$ of the vertex, wave function and $\mathbf{S}^2(x)$ insertion, respectively, are expanded in powers of g:

$$Z_{(4)}^{-1} = 1 = A_4 g + B_4 g^2 + C_4 g^3 + O(g^4) \tag{9.4}$$

$$Z \quad = 1 + B_1 g^2 + C_1 g^3 + D_1 g^4 + O(g^5) \tag{9.5}$$

$$Z_{(2)}^{-1} = 1 + A_2 g + B_2 g^2 + C_2 g^3 + O(g^4), \tag{9.6}$$

where the coefficients A_i, B_i, C_i, D_1 are given in Fig. 12 as functions of the diagrams.

Consequently the function $W(g)$ which is given by:

$$W(g) = -\varepsilon\left[\frac{\partial \ln g_0}{\partial g}\right]^{(-1)} \tag{9.7}$$

may be determined up to order $g^p\varepsilon^q$ with $p + q \leqslant 4$ (see Brézin *et al.*, 1974c) as

$$\begin{aligned} W(g) = {} & -\varepsilon g + \frac{(n+8)}{6}\left(1 + \tfrac{1}{2}\varepsilon + \tfrac{1}{2}\varepsilon^2\right)g^2 - \tfrac{1}{36}\{9n + 42 + \varepsilon[\tfrac{5}{4}(19n + 86) \\ & - 2(5n+22)J + (n+2)l]\}\, g^3 + \tfrac{1}{432}\{23n^2 + 414n + 1264 - (n+8) \\ & \times [2(5n+22)J - (n+2)l] + 24(5n+22)\zeta(3)\}\, g^4 \\ & + O(g^5, \varepsilon g^4, \varepsilon^2 g^3, \varepsilon^3 g^2), \end{aligned} \tag{9.8}$$

where $\zeta(3) = 1{\cdot}20206$ and J is given by the integral

$$J = \int_0^1 \frac{\mathrm{d}x}{x}\int_0^1 \mathrm{d}y \ln\left[1 - x + \tfrac{3}{4}\frac{x}{y}(1 - x + xy)\right] = 0{\cdot}7494\ldots \tag{9.9}$$

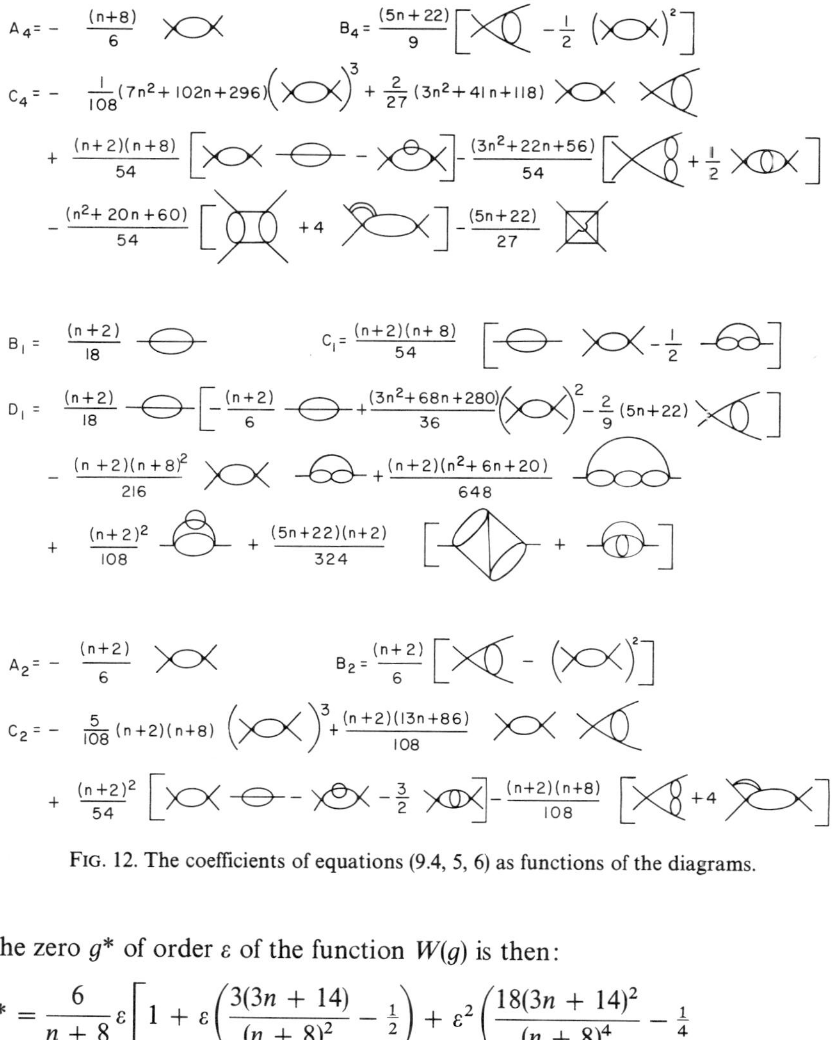

FIG. 12. The coefficients of equations (9.4, 5, 6) as functions of the diagrams.

The zero g^* of order ε of the function $W(g)$ is then:

$$g^* = \frac{6}{n+8}\varepsilon\left[1 + \varepsilon\left(\frac{3(3n+14)}{(n+8)^2} - \frac{1}{2}\right) + \varepsilon^2\left(\frac{18(3n+14)^2}{(n+8)^4} - \frac{1}{4}\right.\right.$$
$$\left.\left. - \frac{(5n^2+322n+1104)}{4(n+8)^3} - \frac{12(5n+22)}{(n+8)^3}\zeta(3) - \frac{[2(5n+22)J - (n+2)I]}{2(n+8)^2}\right)\right]$$
$$+ O(\varepsilon^4). \qquad (9.10)$$

The critical exponents η and γ are given in terms of the functions

$$\eta(g) = W(g)\frac{\mathrm{d}}{\mathrm{d}g}\ln Z(g) \qquad (9.11)$$

and

$$\eta_2(g) = W(g)\frac{\mathrm{d}}{\mathrm{d}g}\ln Z_{(2)}(g) \tag{9.12}$$

by the relations:

$$\eta = \eta(g^*) \quad \text{and} \quad \frac{1}{\nu} - 2 = \eta_2(g^*) - \eta$$

which yield γ by the scaling relation $\gamma = \nu(2 - \eta)$.

As shown by Brézin *et al.* (1974c), we obtain at the needed order:

$$\eta(g) = \frac{(n+2)}{72}g^2\Big(1 + (\tfrac{5}{4} - l)\varepsilon + \tfrac{1}{2}(l^2 - \tfrac{5}{2}l + \tfrac{31}{8})\varepsilon^2 - \frac{(n+8)}{12}$$
$$\times\, g[1 - 2l + \varepsilon(\tfrac{5}{2}l^2 - 5l + \tfrac{15}{4})] + \frac{g^2}{288}[3n^2(2l^2 - 2l + 1)$$
$$+\, n(96l^2 - 176l + 33) + 384l^2 - 736l + 162 + 16(5n + 22)J]\Big) \tag{9.14}$$

$$\eta_2(g) = -\frac{(n+2)}{6}g\Big(1 + \tfrac{1}{2}\varepsilon + \tfrac{1}{2}\varepsilon^2 - \tfrac{1}{2}g(1 + \tfrac{5}{2}\varepsilon - J\varepsilon) + \frac{g^2}{36}$$
$$\times\, [n(1 - \tfrac{1}{2}l) + 32 - l + 2J(n-1)]\Big) \tag{9.15}$$

Then we find:

$$\eta = \frac{\varepsilon^2(n+2)}{2(n+8)^2}\left[1 + \varepsilon\left(\frac{6(3n+14)}{(n+8)^2} - \frac{1}{4}\right) + \varepsilon^2\left(\frac{1}{(n+8)^4}(-\tfrac{5}{16}n^4 - \tfrac{115}{8}n^3\right.\right.$$
$$\left.\left.+\, \tfrac{281}{4}n^2 + 1120n + 2884) - \frac{24(5n+22)}{(n+8)^3}\zeta(3)\right)\right] + O(\varepsilon^5) \tag{9.16}$$

$$\eta_2 = -\varepsilon\frac{(n+2)}{(n+8)}\left[1 + \frac{6(n+3)}{(n+8)^2}\varepsilon + \varepsilon^2\left(-\frac{12(5n+22)}{(n+8)^3}\zeta(3)\right.\right.$$
$$\left.\left.-\, \frac{(n^2 + 222n + 560)}{4(n+8)^3} + \frac{36(3n+14)(n+3)}{(n+8)^4}\right)\right] + O(\varepsilon^4) \tag{9.17}$$

$$\gamma = 1 + \frac{(n+2)}{2(n+8)}\varepsilon + \frac{(n+2)(n^2 + 22n + 52)}{4(n+8)^3}\varepsilon^2 + (n+2)\,\varepsilon^3$$
$$\times\left[\frac{1}{(n+8)^5}(\tfrac{1}{8}n^4 + \tfrac{11}{2}n^3 + 83n^2 + 312n + 388) - \frac{6(5n+22)}{(n+8)^4}\zeta(3)\right] + O(\varepsilon^4). \tag{9.18}$$

We have also seen in Section VIII that the first corrections to the scaling laws are governed by the exponent $\omega = W'(g^*)$ whose ε-expansion (see Brézin *et al.*, 1973d) is:

$$\omega = \varepsilon - \frac{3(3n+14)}{(n+8)^2}\varepsilon^2 + [\tfrac{33}{4}n^2 + \tfrac{461}{2}n + 740 + 24(5n+22)\zeta(3)] \times \frac{\varepsilon^3}{(n+8)^3} - 18\frac{(3n+14)^2}{(n+8)^4}\varepsilon^3 + O(\varepsilon^4). \quad (9.19)$$

In terms of the temperature, these corrections are of order $|T - T_c|^{\omega\nu}$:

$$\omega\nu = \tfrac{1}{2}\varepsilon - \frac{(-n^2+8n+68)}{4(n+8)^2}\varepsilon^2 + \frac{(n+2)(n+3)(n+20)}{8(n+8)^3}\varepsilon^3 - \frac{9(3n+14)^2}{(n+8)^4} \times \varepsilon^3 + \frac{1}{2(n+8)^3}[\tfrac{15}{4}n^2 + \tfrac{401}{2}n + 698 + 24(5n+22)\zeta(3)]\,\varepsilon^3 + O(\varepsilon^4). \quad (9.20)$$

Notice that in the expression of the critical exponents the renormalization-dependent quantities J and l have cancelled as expected. The numerical values of η and γ are given in Table II. One may remark that the agreement with the high-temperature series is excellent at order ε^2. At order ε^3 the agreement is no longer so good. B. G. Nickel (1975) has suggested that strongly irreducible diagrams of higher order reduce the too negative contribution of the tetrahedron diagram. For ω and $\omega\nu$ the convergence of the ε-expansion is very poor. For $n = 1, 2, 3$, Padé approximants suggest $\omega \approx 0{\cdot}8$ and $\omega\nu \approx 0{\cdot}45$.

It is instructive to look for the solution of the equation $W(g) = 0$ in four dimensions. If we choose for simplicity $n = 1$, $W(g)$ becomes

$$W(g) = \tfrac{3}{2}g^2 - \tfrac{17}{12}g^3 + \tfrac{1}{16}[63 + 24\zeta(3) + l - 18J]g^4 + O(g^5). \quad (9.21)$$

We note again that the coefficient of g^4 depends on the subtraction scheme through J and l. Here the renormalization condition (9.3c) fixes J to an explicit value but l is still an independent parameter. Thus the coefficient

TABLE II. Values of η and γ obtained by the ε-expansion.

	At order ε	At order ε^2	At order ε^3	High-temperature series*
η ($n = 1$)	0·019	0·037	0·029	0·041 ± 0·010
γ ($n = 1$)	1·167	1·244	1·195	1·250 ± 0·003
η ($n = 3$)	0·021	0·039	0·032	0·043 ± 0·014
γ ($n = 3$)	1·227	1·346	1·325	1·375 ± 0·020

* Results quoted from Stanley (1971) and Ritchie and Fisher (1972).

of g^4 has a totally arbitrary sign and magnitude. Therefore the problem of a perturbative determination of a non-trivial solution of $W(g) = 0$ is somewhat ambiguous.

B. The equation of state

Using the loopwise expansion derived in Section II, (2.31), as well as the values of g^* and of the critical exponents given above, it is possible to compute the equation of state. This equation was first derived by Brézin *et al.* (1972, 1973c) at order ε^2 by a different method. We shall reproduce here only the calculation up to order ε.

In the one loop approximation $\Gamma(M, t)$ is given by the expression:

$$\Gamma(M, t) = \mathscr{H}(M) + \tfrac{1}{2}\,\mathrm{tr}_{(i,\,j;\,x,\,y)} \ln \frac{\delta^2 \mathscr{H}}{\delta S_i(x)\delta S_j(y)}. \tag{9.22}$$

Using the definition of $\mathscr{H}$ in (9.1) we obtain:

$$\begin{aligned}\Gamma(M, t) = {} & \tfrac{1}{2}tM^2 + g^* \frac{M^4}{4!} - \frac{3n}{2(4-n)}\frac{t^2}{g^*} + \tfrac{1}{2}\,\mathrm{tr}_{(ij)} \int \mathrm{d}^d p \left\{ \ln \left[(p^2 + t)\delta_{ij} \right.\right. \\ & \left. + \frac{g^*}{6}(\delta_{ij}M^2 + 2M_iM_j) \right] - \ln p^2 - \frac{1}{p^2}\left[(t + \tfrac{1}{6}g^*M^2)\delta_{ij} + \tfrac{1}{3}g^*M_iM_j\right] \\ & \left. + \tfrac{1}{2} \frac{1}{p^2(p+q)^2}\bigg|_{q^2=1} \left[(t + \tfrac{1}{6}g^*M^2)^2\delta_{ij} + \tfrac{2}{3}g^*(t + \tfrac{1}{3}g^*M^2)M_iM_j\right] \right\},\end{aligned} \tag{9.23}$$

$\Gamma(M, t)$ is defined up to a polynomial of second degree in t independent of M and therefore irrelevant for the equation of state. We have adjusted this polynomial in such a way that $\Gamma(M, t)$ satisfy exactly the R.G. equations, by introducing the term

$$-\frac{3n}{2(4-n)}\frac{t^2}{g^*}.$$

A short calculation yields:

$$\begin{aligned} g^*\Gamma(M, t) = {} & \tfrac{1}{2}ty + \frac{y^2}{24} - \frac{3n}{2(4-n)}t^2 + \frac{g^*}{16}(n-1)\left(t + \frac{y}{6}\right)^2 \\ & \times \left[1 + 2\ln\left(t + \frac{y}{6}\right)\right] + \frac{g^*}{16}\left(t + \frac{y}{2}\right)^2\left[1 + 2\ln\left(t + \frac{y}{2}\right)\right] \end{aligned} \tag{9.24}$$

where we have defined

$$y = g^*M^2.$$

In terms of the scaling variable x

$$x = ty^{-1/2\beta},$$

we obtain for $H = \partial\Gamma/\partial M$

$$H = My^{1+\varepsilon/2}\left\{(x + \tfrac{1}{6}) + \frac{3\varepsilon}{n+8}\left[\frac{(n-1)}{6}(x + \tfrac{1}{6})(\ln(x + \tfrac{1}{6}) + 1) + \tfrac{1}{2}(x + \tfrac{1}{2})(\ln(x + \tfrac{1}{2}) + 1)\right]\right\}. \quad (9.25)$$

If we now choose the following normalization conditions:

$$H = M^{\delta} \quad \text{at} \quad x = 0$$

and (9.26)

$$H = 0 \quad \text{at} \quad x = -1$$

the equation takes the form

$$H = M^{\delta} \quad f(x = t/M^{1/\beta}), \quad (9.27)$$

where $f(x)$ is given from (9.25) up to order ε.
We shall actually give the equation of state up to order ε^2

$$f(x) = 1 + x + \varepsilon f_1(x) + \varepsilon^2 f_2(x) + O(\varepsilon^3) \quad (9.28)$$

with:

$$f_1(x) = \frac{1}{2(n+8)}[(n-1)(x+1)\ln(x+1) + 3(x+3)\ln(x+3) - 9(x+1)\ln 3 + 6x\ln 2] \quad (9.29)$$

$$\begin{aligned} f_2(x) = \left[\frac{1}{2(n+8)}\right]^2 \Big\{ & [n - 1 + 6\ln 2 - 9\ln 3 + (n-1)\ln(x+1)] \\ & \times [3(x+3)\ln(x+3) + (n-1)(x+1)\ln(x+1) + 6x\ln 2 \\ & - 9(x+1)\ln 3] + \tfrac{1}{2}(10-n)(x+1)[\ln^2(x+3) - \ln^2 3] \\ & + 36[\ln^2(x+3) - (x+1)\ln^2 3 + x\ln^2 2] - 54\ln 2[\ln(x+3) \\ & + x\ln 2 - (x+1)\ln 3] + 3(n-1)(\ln\tfrac{27}{4})(x+1)\ln(x+1) \\ & + \frac{212 + 17n - 4n^2}{n+8}[(x+3)\ln(x+3) + 2x\ln 2 - 3(x+1)\ln 3] \\ & + (n-1)(x+1)\ln(x+1)\ln(x+3) - \tfrac{1}{2}n(n-1)(x+1)\ln^2(x+1) \end{aligned}$$

$$+ \frac{n-1}{n+8}(19n + 92)(x + 1)\ln(x + 1) - 2(n - 1)\{(x + 6)I_1(\rho) - 6(x + 1)I_1(\tfrac{3}{4})\} - 6(n - 1)\{I_2(\rho) - (x + 1)I_2(\tfrac{3}{4})\} + 4(n - 1)\{I_3(\rho) - (x + 1)I_3(\tfrac{3}{4})\}\Big\}, \quad (9.30)$$

where

$$\rho = \frac{(x + 3)}{4(x + 1)} \quad (9.31)$$

and

$$I_1(\rho) = \int_0^\rho \frac{du \ln u}{u(1 - u)}\left[\left(1 - \frac{u}{\rho}\right)^{1/2} - 1\right] - \int_\rho^\infty \frac{du \ln u}{u(1 - u)} \quad (9.32a)$$

$$I_2(\rho) = \rho \frac{\partial}{\partial \rho} I_1(\rho) \quad (9.32b)$$

$$I_3(\rho) = I_1(\rho) + 2I_2(\rho). \quad (9.32c)$$

The comparison with the numerical results of Gaunt and Domb (1970) for the Ising model and of Milosevic and Stanley (1972) for the Heisenberg model in three dimensions is quite good. The successive addition of the ε-term f_1 and ε^2-term f_2 do improve significantly the mean-field result. For large values of x the comparison is less satisfactory and the reason of the discrepancy may be easily understood. We know (see Section VII.A) that for large x the function $f(x)$ satisfies the analyticity condition laid down by Griffiths (1967):

$$f(x) = \sum_{p=1}^{\infty} a_p x^{\gamma - 2(p-1)\beta}. \quad (9.33)$$

In fact the function $f(x)$ given by (9.28) does have this property provided the exponents γ and β are also ε-expanded. Following Brézin (1974), we can even determine the ε-expansion of the coefficients a_p up to order ε^2 by matching (9.28) and (9.33). If we expand these coefficients as

$$a_p = a_p^{(0)} + \varepsilon a_p^{(1)} + \varepsilon^2 a_p^{(2)} + O(\varepsilon^3), \quad (9.34)$$

they are determined by the equations

$$a_1^{(0)} = a_2^{(0)} = 1, \qquad a_p^{(0)} = 0, \qquad p \geqslant 3, \quad (9.35)$$

$$\sum_1^\infty a_p^{(1)} x^{2-p} = \frac{1}{2(n+8)}\left[3(x + 3)\ln\left(1 + \frac{3}{x}\right) + (n - 1)(x + 1)\ln\left(1 + \frac{1}{x}\right) + 6x\ln 2 - 9(x + 1)\ln 3\right] \quad (9.36)$$

$$\sum_1^\infty a_p^{(2)} x^{2-p} = \left[\frac{1}{2(n+8)}\right]^2 \left\{\left[n - 1 + 3\ln\tfrac{4}{27} + (n-1)\ln\left(1+\frac{1}{x}\right)\right]\right.$$
$$\times\left[3(n+3)\ln\left(1+\frac{3}{x}\right) + (n-1)(x+1)\ln\left(1+\frac{1}{x}\right) + 6x\ln 2\right.$$
$$\left. - 9(x+1)\ln 3\right] + \tfrac{1}{2}(10-n)(x+1)\left[\ln^2\left(1+\frac{3}{x}\right) - \ln^2 3\right]$$
$$+ 36\left[\ln^2\left(1+\frac{3}{x}\right) - (x+1)\ln^2 3 + x\ln^2 2\right] - 54\ln 2\left[\ln\left(1+\frac{1}{x}\right)\right.$$
$$\left. + x\ln 2 - (x+1)\ln 3\right] - 3(n-1)\left[(\ln\tfrac{4}{27})(x+1)\ln\left(1+\frac{1}{x}\right)\right]$$
$$+ \frac{212 + 17n - 4n^2}{n+8}\left[(x+3)\ln\left(1+\frac{3}{x}\right) + 2x\ln 2 - 3(x+1)\ln 3\right]$$
$$+ (n-1)(x+1)\ln\left(1+\frac{1}{x}\right)\ln\left(1+\frac{3}{x}\right) - \frac{n(n-1)}{2}(x+1)\ln^2\left(1+\frac{1}{x}\right)$$
$$+ \frac{n-1}{n+8}(19n+92)(x+1)\ln\left(1+\frac{1}{x}\right) - 2(n-1)\left[(x+6)I_1(\rho)\right.$$
$$\left. - 6(x+1)I_1(\tfrac{3}{4})\right] - 6(n-1)\left[I_2(\rho) - (x+1)I_2(\tfrac{3}{4})\right] + 4(n-1)$$
$$\left.\times\left[I_3(\rho) - (x+1)I_3(\tfrac{3}{4})\right]\right\}. \tag{9.37}$$

Expansion of the right-hand sides of (9.36) and (9.37) in inverse powers of x gives the $a_p^{(1)}$ and $a_p^{(2)}$. For instance, for Ising-like systems we get

$$a_1 = 1 + \frac{\varepsilon}{6}\ln\tfrac{4}{27} + \varepsilon^2\left[\tfrac{25}{324}\ln\tfrac{4}{27} + \tfrac{1}{72}\ln^2\tfrac{4}{27}\right] + O(\varepsilon^3)$$
$$= 1 - 0{\cdot}318\,\varepsilon - 0{\cdot}097\,\varepsilon^2 + O(\varepsilon^3). \tag{9.38}$$

In three dimensions this gives the estimate

$$a_1 = 0{\cdot}58 \tag{9.39}$$

which is in remarkable agreement with the result deduced from the numerical analysis of Gaunt and Domb (1970), which, once comparable normalizations are chosen, reads

$$a_1 = 0{\cdot}57.$$

One can go one step further and plug back in the Griffiths expansion the values of the a_p obtained from equations (9.34–37).

This gives a modified equation of state valid for large x, in which f takes the form

$$f(x) = \tilde{f}_0 + \varepsilon\tilde{f}_1 + \varepsilon^2\tilde{f}_2 + O(\varepsilon^3) \tag{9.40}$$

which, retaining for simplicity Ising-like systems, reads:

$$\tilde{f}_0(x) = x^{\gamma} + x^{\gamma-2\beta} \tag{9.41}$$

$$\tilde{f}_1(x) = \tfrac{1}{6}x^{\gamma-2\beta}\left[(x^{2\beta}+3)\ln(1+3x^{-2\beta}) + x^{2\beta}\ln 4 - (1+x^{2\beta})\ln 27\right] \tag{9.42}$$

$$\begin{aligned}\tilde{f}_2(x) = x^{\gamma-2\beta}\{&[\tfrac{25}{324}+\tfrac{1}{36}\ln\tfrac{4}{27}][(x^{2\beta}+3)\ln(1+3x^{-2\beta}) + x^{2\beta}\ln 4\\ &-(1+x^{2\beta})\ln 27] + \tfrac{1}{9}[\ln^2(1+3x^{-2\beta}) - (1+x^{2\beta})\ln^2 3 + x^{2\beta}\ln^2 2]\\ &+\tfrac{1}{72}(1+x^{2\beta})[\ln^2(1+3x^{-2\beta}) - \ln^2 3] - \tfrac{1}{6}\ln 2[\ln(1+3x^{-2\beta})\\ &+ x^{2\beta}\ln 2 - (1+x^{2\beta})\ln 3]\}.\end{aligned} \tag{9.43}$$

The advantage of equations (9.40–43) is that one can use in them the experimental value of the critical exponents.

C. The critical spin–spin correlation function for arbitrary magnetization and temperature

One may calculate the correlation function in arbitrary magnetization† as the second derivative of the 1-P.I. functional $\Gamma(M)$:

$$\Gamma^{(2)}_{ij}(x-y;M) = \left.\frac{\delta^2\Gamma(M)}{\delta M_i(x)\delta M_j(y)}\right|_{M_i(x)=M_i}. \tag{9.44}$$

The functional $\Gamma(M)$ is calculated in the loop expansion of Section II.C. analogous to the one given above for constant M. At one-loop order it consists of the first two terms of (2.31):

$$\begin{aligned}\Gamma(M) = \int d^dx\left\{\tfrac{1}{2}(\nabla M)^2 + Z_{(2)}t\frac{M^2}{2} + Z_{(4)}g\frac{(M^2)^2}{4!} + \frac{\delta m^2}{2}M^2\right\}\\ + \tfrac{1}{2}\operatorname{tr}\ln\left[(t-\nabla^2)\delta_{ij} + \frac{g}{6}(M^2\delta_{ij} + 2M_iM_j)\right].\end{aligned} \tag{9.45}$$

The second functional derivative, with the $Z_{(2)}$, $Z_{(4)}$ taken from (9.6; 4) at order g, and δm^2 such that at the critical temperature $\Gamma^{(2)}$ ($p = 0$, $M = 0$) vanishes, is

† The spin–spin correlation function for $T > T_c$ and $H = 0$ has been calculated at order ε^2 by Fisher and Aharony (1974).

$$\Gamma_{ij}^{(2)} = (p^2 + t)\delta_{ij} + \frac{g}{6}(\delta_{ij}M^2 + 2M_iM_j) + \frac{g}{12}\Big\{[(n+1)\delta_{ij} - 2\hat{M}_i\hat{M}_j] \times \left(t + \frac{gM^2}{6}\right)\left(1 + \ln\left[t + \frac{gM^2}{6}\right]\right) + (\delta_{ij} + 2\hat{M}_i\hat{M}_j)\left(t + \frac{gM^2}{2}\right) \times \left(1 + \ln\left[t + \frac{gM^2}{2}\right]\right)\Big\} + \frac{g^2}{18}\{(M^2\delta_{ij} - M_iM_j)J_1 + \tfrac{1}{2}(n-1)M_iM_jJ_2 + \tfrac{9}{2}M_iM_jJ_3\} + \ldots \qquad (9.46)$$

in which $\hat{M}_i$ is the unit vector along the magnetization, and the J's are given by elementary integrals:

$$J_1 = 2 + \int_0^1 dx \ln\left[p^2x(1-x) + x\left(t + \frac{gM^2}{2}\right) + (1-x)\left(t + \frac{gM^2}{6}\right)\right]$$

$$J_2 = 2 + \int_0^1 dx \ln\left[p^2x(1-x) + \left(t + \frac{gM^2}{6}\right)\right]$$

$$J_3 = 2 + \int_0^1 dx \ln\left[p^2x(1-x) + \left(t + \frac{gM^2}{2}\right)\right]. \qquad (9.47)$$

Out of these formulae one can obtain directly several pieces of information: for instance, at zero momentum one gets the magnetic susceptibility and hence the ratio of the amplitudes above and below T_c in zero field; in Ising-like systems we can also calculate the true range of the correlations by looking for the zeros of $\Gamma^{(2)}$. The result in zero field reads:

$$\xi_\pm = f_\pm(\pm t)^{-\nu} \quad \text{with} \quad f_+/f_- = 2^\nu\left[1 + \frac{\varepsilon}{4}\left(\frac{\pi}{\sqrt{3}} - 1\right)\right] + O(\varepsilon^2). \qquad (9.48)$$

D. Universal ratios derived from the equation of state

From the equation of state, it is possible to calculate some universal ratios of amplitudes near T_c (see Brézin *et al.*, 1974e).

1. *Specific heat*

It has been shown in previous Sections that the specific heat C_H, which is given by the energy–energy correlation function at zero momentum, behaves like:

$$C_H = A^\pm |t|^{-\alpha}, \qquad t \to \pm 0. \qquad (9.49)$$

The ratio A^+/A^- can be calculated in terms of the derivative of the free energy with respect to the temperature, in zero field, $\partial\Gamma/\partial t|_{H=0}$. It is therefore related to the equation of state $H = \partial\Gamma/\partial M$, and to the universal function $f(x)$ defined in section 2:

$$\frac{A^+}{A^-} = \frac{\int_0^\infty \mathrm{d}x\, x^{\alpha-2}[f'(0) - f'(x)]}{\int_{-1}^0 \mathrm{d}x\,(-x)^{\alpha-2}[f'(x) - f'(0)] - f'(0)/(1-\alpha)} \tag{9.50}$$

This formula has been written for $\alpha > 0$. It is straightforward to modify it for $\alpha < 0$ by integration by part.

Using formulae (9.28–30), which give $f(x)$ at order ε^2, one obtains

$$\frac{A^+}{A^-} = 2^\alpha(1 + \varepsilon)\frac{n}{4} + O(\varepsilon^2). \tag{9.51}$$

The numerical application of this result is given in Table III.

TABLE III. Comparison of the ε-expansion for the ratios A^+/A^- and C^+/C^- with experiment and series.

		ε-expansion	Mean field	Experiments and series		
A^+/A^-	$n = 1$	0·55	0·25	0·53[a]	0·58[e]	0·62–0·7[b]
	$n = 2$	0·99	0·5	1·05[a]		
	$n = 3$	1·36	0·75	1·36[a]		1·11[c]
C^+/C^-		4·8	2	5·03 ± 0·05[d]		

[a] Ahlers (1973).
[b] Fisher and Tarko (1975).
[c] Lederman *et al.* (1974).
[d] Tarko and Fisher (1975).
[e] Salamon and Lederman (1974).

2. *Magnetic susceptibility*

From the equation of state, it is straightforward to derive the magnetic susceptibility in zero field. Below T_c it is finite only for systems without Goldstone modes. We shall therefore restrict ourselves to Ising-like systems. Defining $C^\pm$ as:

$$\chi = C^\pm |t|^{-\gamma} \qquad t \to \pm 0, \tag{9.52}$$

one obtains

$$\frac{C^+}{C^-} = 2^{\gamma-1}\frac{\gamma}{\beta} + O(\varepsilon^3). \tag{9.53}$$

Remark that in the formulae of this section we expressed the ε-expansion in terms of the already calculated critical exponents when it seemed "natural" to do so. This form could perhaps suggest an approximation scheme which goes beyond the strict ε-expansion.

The numerical result is given in Table III.

E. Other universal ratios of amplitudes in Ising-like systems

Some other ratio of amplitudes can be related to the spin–spin correlation function (see Brézin *et al.*, 1974e).

1. *Correlation length*

If we define the correlation length squared ξ_1^2 as the second (normalized) moment of the correlation function, we obtain at order ε^2:

$$\xi_1 = f_1^{\pm}|t|^{-\nu} \tag{9.54}$$

$$f_1^+/f_1^- = 2^{\nu}[1 + \tfrac{5}{24}\varepsilon + \tfrac{1}{432}\varepsilon^2(\tfrac{295}{24} + 2I)] + O(\varepsilon^3) \tag{9.55}$$

in which the integral I is given by:

$$I = \int_0^1 \mathrm{d}x \frac{\ln[x(1-x)]}{1 - x(1-x)} = -2{\cdot}3439. \tag{9.56}$$

The numerical result can be found in Table IV.

It is also possible to calculate more generally the correlation length $\xi_1(M, t)$ for an arbitrary magnetization. It has the scaling form:

$$\xi_1^2(M, t) = M^{-2\nu/\beta} g\left(\frac{t}{M^{1/\beta}}\right). \tag{9.57}$$

TABLE IV. ε-Expansion of universal ratios of amplitudes in Ising-like systems compared with the results derived from high-temperature series or with experiment.

	ε-Expansion	High-temperature series[a, b] or experiment[c]	Mean field
f_1^+/f_1^-	1·91	1·96 ± 0·03[a] 2·05 ± 0·22[c]	1·41
Q_1	0·89	0·897 ± 0·007[b]	1
Q_2	1·13	1·21 ± 0·03[b]	1
Q_3	0·96	0·896 ± 0·005[b]	1

[a] Tarko and Fisher (1973).
[b] Tarko and Fisher (1975).
[c] Lin and Schmidt (1974).

With the same normalization as above for the equation of state, we find for $g(x)$

$$g(x) = (f_1^+)^2 6^{-2\nu} z \{1 - \tfrac{1}{36}\varepsilon[z(5 + 6\ln 3) - 6(1 + z)\ln z]$$
$$+ \varepsilon^2 \left[\frac{1 + 2z^2}{72} \ln^2 z + (\tfrac{1}{18} z(z - \tfrac{1}{2})(1 - \ln 3) - \tfrac{1}{216}(16z^2 - \tfrac{47}{3}z - \tfrac{56}{3})) \ln z\right.$$
$$+ \tfrac{1}{216}(\tfrac{101}{6} + \tfrac{2}{3}I + 6\ln^2 3 + 4\ln 3 - 10) z^2 - \tfrac{1}{216}(6\ln^2 3 + \tfrac{44}{3}\ln 3$$
$$\left. + \tfrac{137}{9} + \tfrac{8}{3}I) z\right]\} + O(\varepsilon^3) \qquad (9.58)$$

in which

$$z \equiv \frac{2}{1 + \frac{1}{3}x}.$$

2. *Additional results*

As shown by Fisher and Tarko (1975) it is possible to define additional universal ratios as follows: on the critical isotherm, the correlation length behaves as

$$\xi_1 \approx f_1^c / H^{\nu/\beta\delta}, \qquad (9.59)$$

the magnetic susceptibility as

$$\chi \approx C^c / H^{1-1/\delta}; \qquad (9.60)$$

the spontaneous magnetization vanishes as

$$M \approx B(-t)^\beta \qquad (9.61)$$

and the spin–spin correlation function in momentum space behaves at T_c as

$$G_c^{-1}(k) \underset{ka \ll 1}{\approx} D(ka)^{\eta - 2}. \qquad (9.62)$$

Fisher and Tarko showed that the three ratios

$$Q_1 = C^c \delta / (B^{\delta - 1} C^+)^{1/\delta} \qquad (9.63)$$
$$Q_2 = (f_1^c / f_1^+)^{2-\eta} C^+ / C^- \qquad (9.64)$$
$$Q_3 = D (f_1^+)^{2-\eta} / C^+ \qquad (9.65)$$

are universal.

Their ε-expansions are:

$$Q_1 = 3\left(\frac{2^{1-2\beta}}{27}\right)^{(\delta-1)/2\delta} + O(\varepsilon^3) \qquad (9.66)$$

$$Q_2 = 1 + \tfrac{1}{18}\varepsilon + \tfrac{1}{54}\varepsilon^2 (\tfrac{23}{9} - \tfrac{2}{3}I) + O(\varepsilon^3) \qquad (9.67)$$

$$Q_3 = 1 + \tfrac{1}{216}\varepsilon^2 \left(\tfrac{4}{3}I - 5\right) + O(\varepsilon^3). \tag{9.68}$$

In three dimensions the results are given in Table IV.

3. *Comment about the nature of the ε-expansion*

If one considers the whole set of results obtained by the ε-expansion for the exponents, equation of state, ratios of amplitudes, one sees that in all these cases the corrections improve the mean-field result. Furthermore they give the right order of magnitude of the difference between the actual result and mean field theory. This seems to us a very important confirmation of the validity of Wilson's ideas. Furthermore it shows that the ε-expansion gives reasonable estimates of the physical quantities. Of course, its actual precision is limited, and considering the results given in Table II, one sees that the ε-expansion gives at best asymptotic series and it does not make sense to keep more than three consecutive orders in ε (unless one finds a good resummation procedure).

X. Renormalization Group for the *n*-Vector Model in the Large *n*-Limit

Up to now we have studied Wilson's theory within the framework of the ε-expansion for fixed values of *n*, the number of components of the order parameter. In 1968 Stanley was able to show that the large *n* limit of a classical *n*-vector model coincides in the critical region with the spherical model solved by Berlin and Kac in 1952 for arbitrary values of *d*. In 1972 Wilson derived an analogous result for an $(\mathbf{S}^2)^2$ field theory with a coupling constant of order $1/n$. Corrections in powers of n^{-1} may be systematically calculated both in the approach of Stanley and in that of Wilson (1973) for an arbitrary dimension and lead as expected to the same critical behaviour.

We could here develop the theory from the lattice formulation of Section IV; this would coincide with the method developed by Abe (1973). But it is simpler to study directly the continuous field theoretic model, where to go to the large *n*-limit we must evaluate the asymptotic behaviour of the weights of the Feynman diagrams as functions of *n*. For *n* large this weight behaves like a power of *n*, the power being equal to the maximum number of disjoint closed loops which may be drawn on the diagram. In this limit the dominant diagrams correspond to the series of "bubbles" drawn in Fig. 13, which forms a geometric series.

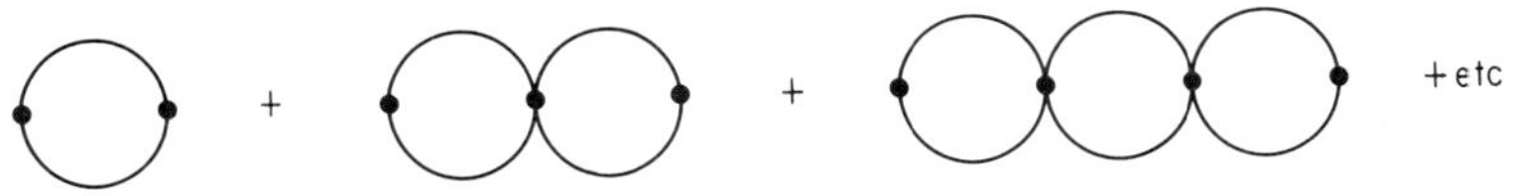

FIG. 13. Dominant diagrams in the large *n*-limit.

A. Algebraic method to generate the 1/n expansion

We could analyse the diagrams that correspond to successive powers in n^{-1}, but an algebraic construction of this expansion can easily be obtained. The idea is to introduce a variable $\alpha(x)$ conjugate of $\mathbf{S}^2(x)$ by the algebraic relation

$$\int \mathrm{d}\alpha \exp\left(\frac{n\alpha^2}{2} - \frac{\alpha}{2}\mathbf{S}^2 \sqrt{\frac{nu_0}{3}}\right) \propto \exp - \frac{u_0(\mathbf{S}^2)^2}{4!}. \tag{10.1}$$

This leads for the functional $Z(H)$ to the integral representation

$$Z(H) = \int [\mathrm{d}\mathbf{S}][\mathrm{d}\alpha] \exp - \int \mathrm{d}x \left[\tfrac{1}{2}(\nabla \mathbf{S})^2 + \tfrac{1}{2}m_0^2\mathbf{S}^2 + \tfrac{1}{2}\sqrt{(nu_0/3)}\,\alpha(x)\,\mathbf{S}^2(x) - \mathbf{H}(x)\mathbf{S}(x) - (n/2)\,\alpha^2(x)\right]. \tag{10.2}$$

The integral on the $\mathbf{S}$ variables may now be performed and we are left with

$$Z(H) = \int [\mathrm{d}\alpha] \exp\left\{\int \mathrm{d}x \left[(\tfrac{1}{2}\mathbf{H}(x)\,\Delta\mathbf{H}(x)) + (n/2)\,\alpha^2(x)\right] + (n/2)\operatorname{tr}\ln\Delta\right\} \tag{10.3}$$

in which Δ is the operator

$$\Delta^{-1} = -\partial^2 + m_0^2 + \sqrt{(nu_0/3)}\,\alpha(x). \tag{10.4}$$

On the integral representation (10.3) it is clear that the steepest-descent method will generate an expansion in powers of n^{-1} provided the operator Δ has a finite limit for large n which implies u_0 of order $1/n$, and that the magnetic field is of order $n^{1/2}$. Let us consider for instance the correlation functions in zero field above T_c; the saddle point is given by $\alpha(x) = \bar{\alpha}$ in which $\bar{\alpha}$ is the solution of

$$\bar{\alpha} - \tfrac{1}{2}\sqrt{(nu_0/3)} \int \frac{\mathrm{d}^d p}{(2\pi)^d} \frac{1}{p^2 + m_0^2 + \bar{\alpha}\sqrt{(nu_0/3)}} = 0. \tag{10.5}$$

The method is now to expand in powers of $\alpha(x) - \bar{\alpha}$ in order to evaluate the integral (10.3) keeping only in the exponential the quadratic part

$$\frac{n}{2}(\alpha(x) - \bar{\alpha})^2 + \frac{n}{4}\frac{nu_0}{3}(\alpha(x) - \bar{\alpha})\left[(\Delta(\bar{\alpha}))_{xy}\right](\alpha(y) - \bar{\alpha}), \tag{10.6}$$

which in momentum space gives an α-propagator

$$D_\alpha(p) = -\frac{1}{n}\left[1 + \frac{nu_0}{6}\int \frac{\mathrm{d}^d q}{(2\pi)^d} \frac{1}{q^2 + m_0^2 + \bar{\alpha}\sqrt{(nu_0/3)}} \times \frac{1}{(p-q)^2 + m_0^2 + \bar{\alpha}\sqrt{(nu_0/3)}}\right]^{-1}. \tag{10.7}$$

B. Existence of the critical theory

The expansion of the term tr ln Δ in (10.3) generates an infinite sum of interaction terms of the α-field. These terms correspond to one closed loop of the original S-field in presence of an external potential $\alpha(x)$; the term quadratic in H in (10.3) corresponds to an open line of S-field propagating also in the external potential $\alpha(x)$ (see Fig. 14). Individual diagrams of this theory are therefore

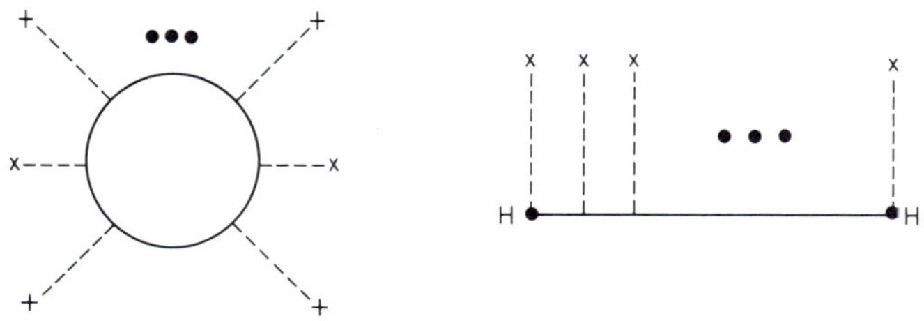

FIG. 14. Interaction terms of the external potential $\alpha(x)$ with the S-field.

the same as those of a theory of two coupled fields $\alpha(x)$ and $S(x)$ with an $\alpha(x) \times S^2(x)$ coupling; the α-propagator is given by (10·7) and the $S(x)$ field has the propagator

$$\Delta(\bar{\alpha}) = \left[-\partial^2 + m_0^2 + \sqrt{\left(\frac{nu_0}{3}\right)}\bar{\alpha}\right]^{-1}.$$

At the critical temperature the S-field becomes massless and the α-field has a propagator $D_\alpha(p)$ which behaves for small p as p^ε. We have seen that when ε and n are finite, a massless theory is infra-red-divergent in perturbation theory. Let us examine this infra-red problem for the $1/n$ expansion. Power-counting of a diagram γ of order p contributing for instance to the propagator of the S-field gives

$$\delta(\gamma) = 2 + p\delta, \tag{10.8}$$

in which δ is the dimension of the vertex αS^2. Thus

$$\delta = 2d_S + d_\alpha - d \tag{10.9}$$

in which d_S and d_α are the canonical dimensions of the two fields

$$d_S = \tfrac{1}{2}(d - 2) \tag{10.10}$$

and, from the D_α propagator

$$d_\alpha = \frac{d + \varepsilon}{2} = 2. \tag{10.11}$$

Therefore to all orders

$$\delta(\gamma) = 2$$

as for an ordinary massless theory in four dimensions and therefore at ε fixed the massless theory exists in power series of $1/n$. Consequently the conditions for the existence of renormalization group equations are satisfied.

C. Fixed point in the large n-limit

The diagrams corresponding to the $1/n$-expansion generate, beyond the spherical model limit, series of logarithms which, after exponentiation, yield the scaling laws of the theory. We have already discussed for the ε-expansion the justification of this property. It relies on the existence of the R.G. equations, that we have justified in the previous section, and on the existence of a fixed point, i.e. an infra-red zero of the function W. Let us compute W in the large n-limit. At leading order there is no field strength renormalization and the renormalized coupling constant is easily computed to be

$$g\mu^{\varepsilon} = \frac{g_0 \Lambda^{\varepsilon}}{1 + (ng_0/6)\Lambda^{\varepsilon}\left[\mu^{-\varepsilon} a(\varepsilon) - (\Lambda^{-\varepsilon}/\varepsilon)\, b(\varepsilon)\right]} \tag{10.12}$$

$$a(\varepsilon) \equiv \int \frac{d^d p}{(2\pi)^d} \frac{1}{p^2(p-q)^2}\bigg|_{q^2=1}, \quad b(\varepsilon) = \frac{2\pi^{d/2}}{\Gamma(d/2)\,(2\pi)^d}. \tag{10.13}$$

This yields for the function $W(g_0)$ (equation (5.12))

$$W(g_0) = -\,g_0\left[\varepsilon - \frac{ng_0}{6}\, b(\varepsilon)\right] + O\left(\frac{1}{n^2}\right) \tag{10.14}$$

which vanishes for

$$g_0^* = \frac{6\varepsilon}{b(\varepsilon)\, n} + O\left(\frac{1}{n^2}\right) \tag{10.15}$$

with a positive slope

$$\omega = \varepsilon + O\left(\frac{1}{n}\right). \tag{10.16}$$

The existence of a scale-invariant massless theory for large n and ε fixed shows explicitly that it is possible to construct a massless theory in a fixed dimension d smaller than four, which does not, however, exist order-by-order in the coupling constant.

D. Results of the calculations in the $1/n$ expansion

The reader is referred to the literature for all details concerning the actual calculations in the $1/n$ expansion. (Abe, 1972, 1973a; Brézin and Wallace, 1973; Ferell and Scalapino, 1973; Ma, 1972, 1973a, b; Suzuki, 1973; Wilson, 1973). Let us just mention a few of the results. The critical exponents are

$$\eta = \frac{1}{n}\varepsilon^2 \frac{\sin \frac{1}{2}\pi\varepsilon}{\pi\varepsilon/2} \frac{\Gamma(1-\varepsilon)}{\Gamma^2(1-\frac{1}{2}\varepsilon)} \frac{(1-\varepsilon)}{(2-\frac{1}{2}\varepsilon)(1-\frac{1}{2}\varepsilon)} + O\left(\frac{1}{n^2}\right)$$

$$\gamma = \frac{1}{1-\frac{1}{2}\varepsilon} - \frac{3\varepsilon}{n} \frac{\sin \frac{1}{2}\pi\varepsilon}{\pi\varepsilon/2} \frac{\Gamma(1-\varepsilon)}{\Gamma^2(1-\frac{1}{2}\varepsilon)} \frac{(1-\varepsilon)}{(1-\frac{1}{2}\varepsilon)^2} + O\left(\frac{1}{n^2}\right).$$

Abe (1973b) has calculated η up to order $1/n^2$ in three dimensions:

$$\eta = \frac{8}{3\pi^2 n} - \left(\frac{8}{3}\right)^3 \frac{1}{\pi^4 n^2} + O\left(\frac{1}{n^3}\right). \tag{10.17}$$

Several other results are known up to order $1/n$, in particular the equation of state (Brézin and Wallace, 1973), the spin–spin correlation function in zero field (Fisher and Aharony, 1973; Aharony, 1974; Theumann, 1974), and the exponent ω (Ma, 1974; Bervillier *et al.*, 1974), which governs corrections to scaling:

$$\omega = \varepsilon - \frac{9\varepsilon^2}{n} \frac{\sin \frac{1}{2}\pi\varepsilon}{\pi\varepsilon/2} \frac{\Gamma(1-\varepsilon)}{\Gamma^2(1-\frac{1}{2}\varepsilon)} \frac{(1-\varepsilon)(1-\frac{1}{3}\varepsilon)^2}{(1-\frac{1}{2}\varepsilon)(1-\frac{1}{4}\varepsilon)} + O\left(\frac{1}{n^2}\right), \tag{10.18}$$

and thus

$$\omega\nu = \frac{\varepsilon}{2-\varepsilon} - \frac{6\varepsilon^2}{n} \frac{\sin \frac{1}{2}\pi\varepsilon}{\pi\varepsilon/2} \frac{\Gamma(1-\varepsilon)}{\Gamma^2(1-\frac{1}{2}\varepsilon)} \frac{(1-\varepsilon)(1-\frac{1}{3}\varepsilon)}{(1-\frac{1}{2}\varepsilon)^2} + O\left(\frac{1}{n^2}\right).$$

Numerical applications of these formulae in three dimensions do not give very good results for the low-n systems. In particular these formulae should be applied only if the fixed point remains stable, i.e. $\omega > 0$, and since, when $d = 3$

$$\omega = 1 - \frac{64}{3\pi^2}\frac{1}{\text{n}} + O\left(\frac{1}{n^2}\right),$$

one cannot use these results as such for $n = 1$ or 2.

XI. Discussion of the General n-Vector Model

In the previous study the interaction was always invariant by rotation in the n-dimensional space spanned by the order parameter. There are physical systems for which the Hamiltonian has not this $O(n)$-symmetry. A first category, that we will not consider, consists of systems presenting several correlation lengths. In such situations when the temperature varies, in general only one length at a time becomes infinite and the behaviour of the system is not sensitive to the presence of non-critical components, which can thus be ignored.

The systems that will be considered here are thus characterized by one length. This implies that the quadratic part of Wilson's Hamiltonian is as before $\frac{1}{2}\Sigma_{i=1}^{n}((\nabla S_i)^2 + m_0^2 S_i^2)$, and that the quartic terms are invariant under a symmetry group which possesses only the single quadratic invariant $\Sigma_i S_i^2$. Examples of such systems are provided by the existence of a cubic anisotropy, which has been studied by Wallace (1973) and Aharony (1973a, b) whose Hamiltonian

$$\mathscr{H}_{\text{int}}(x) = \frac{1}{4!}\left[u_1\left(\sum_{i=1}^{n} S_i^2\right)^2 + u_2 \sum_{i=1}^{n} S_i^4\right],$$

has the discrete symmetries $S_i \to S_j$ and $S_i \to -S_i$. An example of an interaction whose symmetry is a continuous subgroup of $O(n)$ seems to be given by the p-wave superconducting transitions of He_3 (see Anderson and Morel, 1961; Balian and Werthamer, 1963). There, the order parameter is a complex 3×3 matrix M (de Gennes, 1973). The interaction has the symmetry $O(3) \times O(3) \times U(1)$ which has only one quadratic invariant, namely $\text{Tr}\, MM^+$, and five quartic invariants: $(\text{tr}\, MM^+)^2$, $\text{tr}\, MM^+MM^+$, $\text{tr}\, M\tilde{M}M^*M^+$, $\text{tr}\, MM^+M^*\tilde{M}$, $\text{tr}\, M\tilde{M}\, \text{tr}\, M^*M^+$.

We shall study here, in the framework of the ε-expansion, the most general interaction compatible with the existence of only one length scale. Several I.R. fixed points, i.e. relevant for the critical domain, are in general present. We shall discuss the stability of these fixed points with respect to small changes in the interaction.

The main results, reported by Brézin *et al.* (1974f), are the following:

(i) At lowest non-trivial order in ε, the only stable fixed point for $n < 4$ is the $O(n)$-symmetric one. This is a striking result since it means that, whatever interaction one takes, the symmetry is dynamically generated near the critical point.

(ii) In this domain, where $n < 4 + \mathcal{O}(\varepsilon)$, only two exponents characterize the first deviations to scaling when the lattice spacing increases.

(iii) Conversely, if $n > 4 + \mathcal{O}(\varepsilon)$, the symmetric fixed point is not stable, unless the interaction is $O(n)$-symmetric by itself.

(iv) If, in addition to the symmetric fixed point, a second non-trivial one is present, there is always a third fixed point.

In the linear space of interactions spanned by these three fixed points, one and only one is stable.

The critical exponent η, i.e. the anomalous dimension of the field, can be calculated for each fixed point. The largest of the three values of η always corresponds to the stable fixed point.

At order ε, it is also possible to perform a global study of the domain of attraction of a fixed point. It turns out that this domain does not cover all the

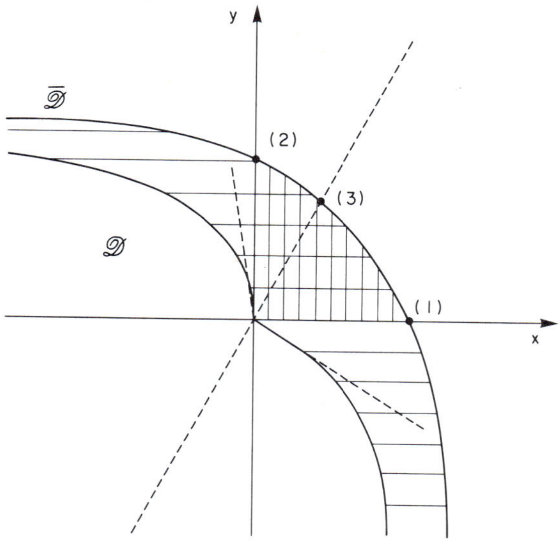

FIG. 15. Plane of the renormalized coupling constants. The domain $\mathscr{D}$ of ultra-violet attraction of the origin is below its boundary $\bar{\mathscr{D}}$. The fixed points (1), (2) and (3) are on $\bar{\mathscr{D}}$; (3) is the stable one. The domain $\mathscr{D}^+$, domain of positivity of the bare interaction, is depicted by the horizontal hatchings. The vertical ones show $\mathscr{D}^*$, domain of attraction of the infrared fixed point (3).

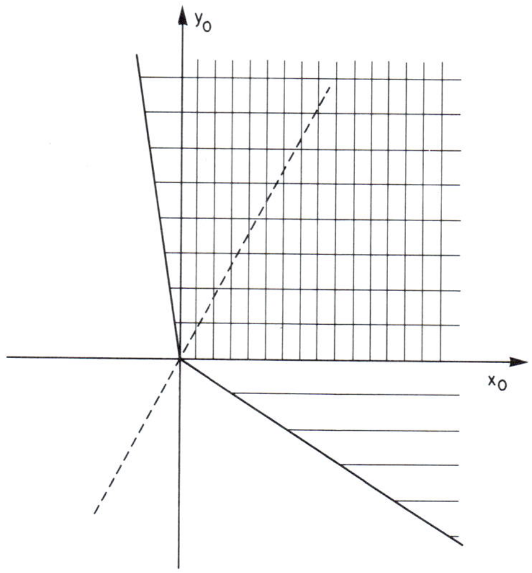

FIG. 16. Plane of the bare coupling constants. $\mathscr{D}^+$ and $\mathscr{D}^*$ are depicted with the conventions used in Fig. 15.

range of coupling constants allowed by the boundedness from below of the free energy, yielding another restriction to universality (see Figs. 15, 16).

A. The Hamiltonian density

The model is described by the renormalized Hamiltonian for n massless fields $S_i(x)$ $i = 1, \ldots, n$ with a local interaction in $d = 4 - \varepsilon$ dimension

$$\mathscr{H}(x) = \tfrac{1}{2}\partial_\mu S_i \partial^\mu S_i + (\mu^\varepsilon/4!)\{g_{ijkl}S_iS_jS_kS_l + (\zeta_{ijkl} - g_{ijkl})S_iS_jS_kS_l\} + \tfrac{1}{2}\partial_\mu S_i(Z-1)_{ij}\partial^\mu S_j + \tfrac{1}{2}S_i(\delta m^2)_{ij}S_j. \tag{11.1}$$

The dimensionless interaction g_{ijkl} is totally symmetrical in the indices and summation over repeated indices will always be meant.

The counter-terms ζ_{ijkl}, $(Z)_{ij}$ and $(\delta m^2)_{ij}$ are fixed by the renormalization conditions given on the inverse propagators $\Gamma_{ij}^{(2)}(p)$

$$\Gamma_{ij}^{(2)}(0) = 0 \tag{11.2a}$$

$$(\partial/\partial p^2)\Gamma_{ij}^{(2)}(p)\big|_{p^2=\mu^2} = \delta_{ij} \tag{11,2b}$$

and on the 4-point proper vertex function (one-particle irreducible)

$$\Gamma_{ijkl}^{(4)}\big|_{(\text{S.P.})} = \mu^\varepsilon g_{ijkl}, \tag{11.2c}$$

where (S.P.) stands for the symmetry point

$$p_i \cdot p_j = \mu^2(\delta_{ij} - \tfrac{1}{4})$$

and μ is an arbitrary renormalization mass.

We have assumed that all the fields are massless, i.e. they are critical all together; in general there is no reason to expect, when the bare masses are different, that if one parameter (the temperature) is fixed, the renormalized masses should vanish. Therefore the physically interesting case is the one in which the bare masses are the same. This situation will occur only if there is a symmetry group of the interaction which has only one quadratic invariant, namely $\Sigma_{i=1}^n (S_i)^2$.

At lowest order, the condition on the interaction to preserve this symmetry is the "trace condition"

$$g_{iijk} = G_{\delta jk}. \tag{11.3}$$

However, it will not be necessary to impose this restriction in the first part of our calculations.

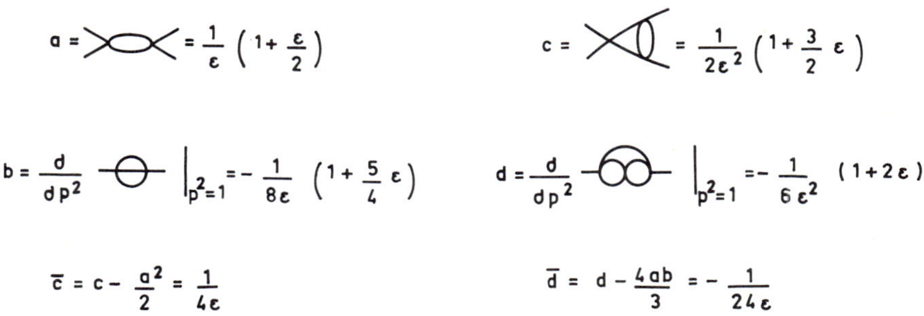

FIG. 17. Diagrams introduced in equations (11.4 and 5). The normalization of the diagrams is the same as in Section IX.

B. Renormalization group equations

The method that we are using to calculate the fixed points and the corresponding critical exponents is the same as in the preceding sections.

The relevant diagrams (denoted a, b, c, d, $\bar{c}$ and $\bar{d}$) are listed in Fig. 17, in terms of which the wave function and vertex renormalization constants are:

$$(Z)_{ij} = \delta_{ij} + (b/6)g_{ipqr}g_{jpqr} + [(ab/2) - (d/4)]g_{ipqr}g_{qrst}g_{stpj} + \mathcal{O}(g^4) \qquad (11.4)$$

$$\begin{aligned}\zeta_{ijkl} = g_{ijkl} &+ (a/2)[g_{ijpq}g_{pqkl} + 2\text{ perm. }(ijkl)] \\ &+ (a^2/4)[g_{ijpq}g_{pqrs}g_{rskl} + 2\text{ perm. }(ijkl)] \\ &+ \tfrac{1}{2}(a^2 - c)[g_{ipqr}g_{sjpr}g_{klqs} + 5\text{ perm. }(ijkl)] + \mathcal{O}(g^4).\end{aligned} \qquad (11.5)$$

The bare coupling constants $\Lambda^{\varepsilon}g^0_{ijkl}$ are expressed in terms of these quantities by

$$\Lambda^{\varepsilon}g^0_{ijkl} = \mu^{\varepsilon}\rho_{ijkl},$$

where the ρ are dimensionless and

$$\begin{aligned}\rho_{ijkl} &= (Z^{-1/2})_{ii'}(Z^{-1/2})_{jj'}(Z^{-1/2})_{kk'}(Z^{-1/2})_{ll'}\zeta_{i'j'k'l'} \\ &= \zeta_{ijkl} - (b/12)\,[g_{i'jkl}g_{i'pqr}g_{ipqr} + 3\text{ perm. }(ijkl)] + \mathcal{O}(g^4)\end{aligned} \qquad (11.6)$$

in terms of which the Callan–Symanzik functions W_{ijkl} are determined by

$$W_{i'j'k'l'}\frac{\partial \rho_{ijkl}}{\partial g_{i'j'k'l'}} = -\varepsilon\rho_{ijkl}, \tag{11.7}$$

This gives the result

$$\begin{aligned} W_{ijkl} = & -\varepsilon g_{ijkl} + \varepsilon(a/2)[g_{ijpq}g_{pqkl} + 2\text{ perm.}] \\ & - \varepsilon\bar{c}[g_{ipqr}g_{jprs}g_{klqs} + 5\text{perm.}] \\ & - \varepsilon(b/6)[g_{i'jkl}g_{i'pqr}g_{ipqr} + 3\text{ perm.}] + \mathcal{O}(g^4). \end{aligned} \tag{11.8}$$

The anomalous dimension matrix η is now given as

$$(\eta)_{ij} = 2W_{i'j'k'l'}\left(\frac{\partial Z^{1/2}}{\partial g_{i'j'k'l'}}Z^{-1/2}\right)_{ij} \tag{11.9}$$

and thus

$$(\eta)_{ij} = -\varepsilon(b/3)g_{ipqr}g_{jpqr} + (3\varepsilon/4)\bar{d}g_{ipqr}g_{qrst}g_{stpj} + \mathcal{O}(g^4). \tag{11.10}$$

Similarly, the insertions of S^2-fields in Green's functions are renormalized by the constants $(Z_{(2)})_{ij,kl}$ which makes finite the insertion of a S_iS_j pair in a (kl) two-point function. One has:

$$(Z_{(2)}^{-1})_{ij,kl} = \tfrac{1}{2}(\delta_{ik}\delta_{jl} + \delta_{il}\delta_{jk}) - (a/2)g_{ijkl} + \tfrac{1}{4}(c-a^2)[g_{ikpq}g_{jlpq}+g_{ilpq}g_{jkpq}]. \tag{11.11}$$

It is convenient to define

$$\bar{Z}_{ij,kl} = (Z_{(2)})_{ij,mn}(Z^{-1/2})_{mk}(Z^{-1/2})_{nl}, \tag{11.12}$$

in terms of which the anomalous dimension matrix $(1/\nu(g) - 2)$ of the S^2 fields is

$$[(1/\nu) - 2]_{ij,kl} = W_{i'j'k'l'}(g)\frac{\partial \bar{Z}_{ij,mn}}{\partial g_{i'j'k'l'}}(\bar{Z}^{-1})_{mn,kl}. \tag{11.13}$$

At these orders one finds

$$\begin{aligned} [(1/\nu) - 2]_{ij,kl} = & -\varepsilon(a/2)g_{ijkl} + (\varepsilon/2)\bar{c}[g_{ikmn}g_{jlmn} + g_{ilmn}g_{jkmn}] \\ & + \varepsilon(b/12)[\delta_{ik}g_{jmnp}g_{lmnp} + 3\text{ perm.}(i\leftrightarrow j, k\leftrightarrow l)]. \end{aligned} \tag{11.14}$$

With the help of these functions β, η and $1/\nu$, the renormalization group equations are for the one-particle irreducible correlation functions

$$\Gamma^{(L,N)}((q_{\tau'}, j_{\tau'}, k_{\tau'}); (p_\tau, i_\tau); \mu, g_{ijkl})$$

$$\tau' = 1,\ldots, L \qquad \tau = 1,\ldots, N$$

with $N - S$ fields and $L - S^2$ insertions:

$$\left(\mu \frac{\partial}{\partial \mu} + W_{i'j'k'l'} \frac{\partial}{\partial g_{i'j'k'l'}}\right) \Gamma^{(L,N)}(j_1 k_1, \ldots, j_L k_L; i_1, \ldots, i_N)$$

$$- \tfrac{1}{2} \sum_{n=1}^{N} (\eta)_{i_n i'_n} \Gamma^{(L,N)}(j_1 k_1, \ldots, j_L k_L; i_1, \ldots, i'_n, \ldots, i_N) \tag{11.15}$$

$$- \sum_{l=1}^{L} [(1/\nu) - 2]_{j_l k_l; j'_l k'_l} \Gamma^{(L,N)}(j_1 k_1, \ldots, j'_l k'_l, \ldots, j_L k_L; i_1, \ldots, i_N) = 0.$$

Integration of this equation gives

$$\Gamma^{(L,N)}((\lambda q_{\tau'}, j_{\tau'} k_{\tau'}); (\lambda p_\tau, i_\tau); \mu, g_{ijkl}) = \prod_{n=1}^{N} (D_1)_{i_n i'_n} \prod_{l=1}^{L} (D_2)_{j_l k_l; j'_l k'_l} \lambda^{d - \frac{1}{2}N(d-2) - 2L}$$

$$\times \Gamma^{(L,N)}((q_{\tau'}, j'_{\tau'} k'_{\tau'}); (p_\tau i'_\tau); \mu, g_{ijkl}(\lambda)) \tag{11.16}$$

where $g_{ijkl}(\lambda)$ is the solution of

$$\frac{d g_{ijkl}(\lambda)}{d \ln \lambda} = W_{ijkl}(g(\lambda)), \tag{11.17}$$

with $g_{ijkl}(1) = g_{ijkl}$, and where the matrices D_1 and D_2 are

$$[D_1(g, \lambda)]_{ij} = (Z^{1/2}(g)\, Z^{-1/2}[g(\lambda)])_{ij} \tag{11.18}$$

$$[D_2(g, \lambda)]_{ij;kl} = (\bar{Z}(g)\, \bar{Z}^{-1}[g(\lambda)])_{ij;kl}. \tag{11.19}$$

We are interested in the critical region where μ is much larger than the momenta, and we therefore look to the solution of (11.17) when $\lambda \to 0$.

In this limit a stable fixed point g^*_{ijkl} is therefore solution of $W_{ijkl}(g^*) = 0$ with eigenvalues of positive real part for the matrix B defined by:

$$B_{ijkl;i'j'k'l'} = \left.\frac{\partial W_{ijkl}}{\partial g_{i'j'k'l'}}\right|_{g=g^*}. \tag{11.20}$$

The definition of $\eta(g)$ and $(1/\nu)(g)$ implies the matrix equations

$$\lambda \frac{\partial D_1}{\partial \lambda} + \tfrac{1}{2}\eta(\lambda)\, D_1[g(\lambda), \lambda] = 0$$

$$\lambda \frac{\partial D_2}{\partial \lambda} + \left(\frac{1}{\nu(\lambda)} - 2\right) D_2[g(\lambda), \lambda] = 0. \tag{11.21}$$

When λ goes to zero, we assume as usual that the matrices η and $1/v$ reach finite limits η^* and $1/v^*$. Integration of (11.21) for λ small gives

$$D_1 \simeq \lambda^{-\eta^*/2}\,\Delta_1$$
$$D_2 \simeq \lambda^{-[(1/v^*)-2]}\,\Delta_2, \tag{11.22}$$

where Δ_1 and Δ_2 are constant matrices.

The equations (11.16) and (11.22) show, therefore, that the existence of a fixed point g^* implies scale invariance with anomalous dimensions related to the eigenvalues of η^* and $[(1/v^*) - 2]$.

A general remark due to Wallace and Zia (1974)† can be made in the framework of the ε-expansion concerning the flow equation (11.17): it is easy to verify on (11.8) that the function $W(g)$ is at order ε^2 the gradient of a potential $\Phi(g)$ given by:

$$\Phi(g) = -(\varepsilon/2)\,g_{ijkl}\,g_{ijkl} + \varepsilon(a/2)\,[g_{ijkl}\,g_{klpq}\,g_{pqij}] - (3/2)\,\varepsilon\bar{c}\,[g_{ijkl}\,g_{ijpq}\,g_{qlrs}\,g_{kprs}] - \varepsilon(b/6)\,[g_{ijkl}\,g_{ijkp}\,g_{pqrs}\,g_{lqrs}].$$

As a consequence the trajectories $g_{ijkl}(\lambda)$ are orthogonal to the equipotentials $\Phi(g) = cte$. This implies that limiting cycles and more general recurrent behaviour are excluded. The fixed points are minima of the function $\Phi(g)$ and the B matrix will always have real eigenvalues.

C. Critical exponents

1. *General expressions for η and v*

We shall now specialize for the theories satisfying the trace condition (11.3).

Let us first show that the matrix η^* is diagonal:

All the quantities are expanded in ε, and we are considering only the fixed points where the g^* are of order ε. From the equations

$$W_{ijkk}(g^*) = 0 \tag{11.23}$$

$$W_{ipqr}(g^*)\,g^*_{jpqr} = 0 \tag{11.24}$$

we obtain

$$g^*_{ipqr}\,g^*_{jpqr} = \delta_{ij}\left(\frac{G^*}{a} - \frac{G^{*2}}{2}\right)\left(1 + \frac{2b}{3a}G^* + \frac{\bar{c}}{a}\left(\frac{8}{3a} + 2G^*\right)\right) \tag{11.25}$$

and then:

$$(\eta^*)_{ij} = \eta\delta_{ij} = \varepsilon\delta_{ij}\left(\frac{\bar{d}}{2a} - \frac{b}{3}\right)\left(\frac{G^*}{a} - \frac{G^{*2}}{2}\right)\left(1 + \frac{2b}{3a}G^* + \frac{\bar{c}}{a}\left(\frac{8}{3a} + 2G^*\right)\right) + \mathcal{O}(\varepsilon^4). \tag{11.26}$$

† It has also been shown by these authors (Southampton University preprint THEP 73/4–5) that this property of gradient flow is still manifest at order ε^3.

Let us note that η^* not only is a multiple of the unit matrix, but also depends only on the value of the trace G^* of the fixed point interaction.

Considering now the dimensions of the composite operators $S_j S_k$, let us note that derivatives with respect to the temperature involve $\mathbf{S}^2$ insertions. Therefore a relation between temperature and correlation length will only exist if the $(1/\nu)^*_{ij;kl}$ matrix has the vector δ_{kl} as eigenvector.† Indeed, the same equations (11.23, 24) show that

$$(1/\nu)^*_{ij;kl}\,\delta_{kl} = (1/\nu)\,\delta_{ij}, \tag{11.27}$$

with
$$[(1/\nu) - 2] = -\,\varepsilon\,(a/2)\,G^* + \varepsilon[\bar{c} + (b/3)]\left(\frac{G^*}{a} - \frac{G^{*2}}{2}\right). \tag{11.28}$$

Note The two exponents η and ν are only functions of the trace G^*. Therefore if we eliminate G^* between their two expressions, we obtain a relation valid *for all the fixed points*, which takes the form

$$\eta = \frac{\varepsilon^2}{12}\left(1 + \frac{5\varepsilon}{6}\right)x(1 - x) + \mathcal{O}(\varepsilon^4), \tag{11.29}$$

where we have set

$$[(1/\nu) - 2] = -\varepsilon x. \tag{11.30}$$

Let us stress that this relation (11.29) is valid for Ising as well as for Heisenberg systems, with or without anisotropy. It has been obtained in the $O(n)$-symmetric case by Toulouse (1973). Its comparison with the results for η and ν obtained from high-temperature series in three dimensions is within the error bars.

2. *Bounds on G^*, η and ν*

Let us first show that G^* is bounded. Equation (11.25) summed on $i = j$ implies, using the values listed in Fig. 17:

$$g^*_{ijkl}\,g^*_{ijkl} = n\left(\frac{G^*}{a} - \frac{G^{*2}}{2}\right)\left(1 + \frac{2\varepsilon}{3} + \frac{5G^*}{12}\right) + \mathcal{O}(\varepsilon^4). \tag{11.31}$$

Let us define

$$S_{ijkl} = (\delta_{ij}\delta_{kl} + \delta_{ik}\delta_{jl} + \delta_{il}\delta_{jk})/(n + 2). \tag{11.32}$$

Now, if we write

$$g^*_{ijkl} = G^* S_{ijkl} + g'^*_{ijkl}, \tag{11.33}$$

† Other eigenvalues correspond to cross-over exponents describing the decoupling of some of the fields.

the g'^*_{ijkl} is traceless and

$$g^*_{ijkl} g^*_{ijkl} = (3n/(n+2))\, G^{*2} + g'^*_{ijkl} g'^*_{ijkl}. \tag{11.34}$$

Equations (11.31) and (11.34) yield the inequality

$$\left(\frac{G^*}{a} - \frac{G^{*2}}{2}\right)\left(1 + \frac{2\varepsilon}{3} + \frac{5G^*}{12}\right) - \frac{3}{n+2} G^{*2} \geqslant 0. \tag{11.35}$$

The left-hand side is at lowest order a parabola, consequently

$$O \leqslant G^* \leqslant G^*_s \leqslant 2\varepsilon, \tag{11.36}$$

where G^*_s is the $O(n)$-symmetric solution, given at this order by

$$G^*_s = \frac{2(n+2)}{(n+8)} \varepsilon. \tag{11.37}$$

Next, η, obtained from (11.26) as

$$\eta = \tfrac{1}{24}\left(\frac{G^*}{a} - \frac{G^{*2}}{2}\right)\left(1 + \tfrac{17}{12}\varepsilon + \frac{5G^*}{12}\right) + \mathcal{O}(\varepsilon^4), \tag{11.38}$$

appears by inequality (11.35) to be positive.

Its expression has a maximum for

$$G^*_{\max} = \varepsilon - \tfrac{7}{24}\varepsilon^2 + \mathcal{O}(\varepsilon^3) \tag{11.39}$$

which corresponds to a value $n_{\max}$ of n for the symmetric solution

$$G^*_s(n_{\max}) = G^*_{\max}, \tag{11.40}$$

with

$$n_{\max} = 4 - 4\varepsilon + \mathcal{O}(\varepsilon^2). \tag{11.41}$$

Therefore one has an upper bound on η:

$$\eta \leqslant \eta_{\max} = \frac{\varepsilon^2}{48}\left(1 + \frac{5\varepsilon}{6}\right) + \mathcal{O}(\varepsilon^4). \tag{11.42}$$

For $n \leqslant n_{\max}$, since G^* is always smaller than G^*_s and η is an increasing function of G^* for $O \leqslant G^* \leqslant G^*_{\max}$, we obtain the following bounds:

$$n \leqslant 4 - 4\varepsilon \qquad \eta \leqslant \eta_{\rm sym}(n) = \frac{(n+2)}{2(n+8)^2}\varepsilon^2\left[1 + \varepsilon\left(\frac{6(3n+14)}{(n+8)^2} - \tfrac{1}{4}\right)\right]$$

$$n > 4 - 4\varepsilon \qquad \eta \leqslant \frac{\varepsilon^2}{48}\left(1 + \frac{5\varepsilon}{6}\right). \tag{11.43}$$

A similar result holds for $[(1/\nu) - 2]$, which is essentially proportional to $(-G^*)$, so that

$$[(1/\nu_s) - 2] = -\varepsilon\frac{(n+2)}{(n+8)}\left(1 + \frac{\varepsilon}{2}\frac{(13n+44)}{(n+8)^2}\right) \leqslant [(1/\nu) - 2] \leqslant 0. \qquad (11.44)$$

D. Stability conditions

An infra-red-stable fixed point corresponds to a matrix B, defined by (11.20):

$$B_{ijkl;\, i'j'k'l'} = \left.\frac{\partial W_{ijkl}}{\partial g_{i'j'k'l'}}\right|_{g=g^*},$$

with eigenvalues of positive real part.

The trivial fixed point at the origin $g^*_{ijkl} = 0$ is always unstable.

1. *Stability of the symmetric solution*

Let us now explore the stability of the $O(n)$-symmetric fixed point

$$g^*_{ijkl} = G^*_s(n)\, S_{ijkl}, \qquad (11.45)$$

where, taking into account one more order in ε than in (11.37),

$$G^*_s(n) = \frac{2(n+2)}{(n+8)}\varepsilon\left[1 + \varepsilon\left(\frac{3(3n+14)}{(n+8)^2} - \tfrac{1}{2}\right)\right] + \mathcal{O}(\varepsilon^3). \qquad (11.46)$$

In order to linearize the functions W_{ijkl} around the fixed point, let us write

$$g_{ijkl} = (G^*_s(n) + g')\, S_{ijkl} + g'_{ijkl}, \qquad (11.47)$$

where g'_{ijkl} is traceless. Then one gets

$$W_{ijkl} \simeq \omega g' S_{ijkl} + \omega_{\text{anis.}}\, g'_{ijkl} \qquad (11.48)$$

where

$$\omega = \varepsilon - \frac{3(3n+14)}{(n+8)^2}\varepsilon^2 \qquad (11.49a)$$

$$\omega_{\text{anis.}} = \varepsilon\frac{(4-n)}{(n+8)} - \varepsilon^2\frac{(5n^2+14n+152)}{(n+8)^3}. \qquad (11.49b)$$

The linearized W matrix has thus two eigenvalues which are ω and $\omega_{\text{anis.}}$, as noted by Brézin *et al.* (1974f). The value of $\omega_{\text{anis.}}$ had been obtained in the case of the cubic anisotropy by Aharony (1973a, b), Aharony and Fisher (1973), Ketley and Wallace (1973), Wallace (1973). Therefore the symmetric solution is stable (with respect to any kind of quartic interaction) if and only if

$$\omega > 0 \quad \text{and} \quad \omega_{\text{anis.}} > 0. \qquad (11.50)$$

This gives the condition

$$n \leqslant 4 - 2\varepsilon + \mathcal{O}(\varepsilon^2). \tag{11.51}$$

Conversely, we shall show at first order in ε that for n smaller than 4 no other fixed point is stable.

Let g^*_{ijkl} be an arbitrary fixed point and B the corresponding matrix (11.20) of the derivatives of W. The subspace spanned by the two vectors g^*_{ijkl} and S_{ijkl} is invariant under the matrix B which in this subspace reduces to

$$B = \begin{bmatrix} \varepsilon & 0 \\ 6 & G^* - \varepsilon \end{bmatrix}, \tag{11.52}$$

where $G^*\delta_{kl} = g^*_{iikl}$.

The stability of the fixed point implies $G^* > \varepsilon$. Since we know that $G^* < G^*_s(n)$, this non-symmetric fixed point cannot be stable if n is such that $G^*_s(n) < \varepsilon$. This means that no other stable fixed point appears if $n < 4 + \mathcal{O}(\varepsilon)$.

Note

Actually one can study the stability of the symmetric fixed point in a more general situation with no restriction on the trace of g_{ijkl}: let us study the vicinity of g^{*s}_{ijkl} with

$$\delta g_{ijkl} = g_{ijkl} - g^{*s}_{ijkl} = uS_{ijkl} + g'_{ijkl} + \frac{1}{(n+4)}(v_{kl}\delta_{ij} + 5 \text{ terms}), \tag{11.53}$$

where

$$\begin{aligned} \delta g_{iijj} &= nu & g'_{iikl} &= 0 \\ v_{kk} &= 0 & \delta g_{iijk} &= u\delta_{jk} + v_{jk}. \end{aligned} \tag{11.54}$$

The linearization of W_{ijkl} around this point leads to a third eigenvalue ω' in addition to ω and $\omega_{\text{anis.}}$. Using (11.8) it is possible to compute ω' in the framework of the ε-expansion:

$$\omega' = \frac{\varepsilon}{(n+8)}\left[8 - \frac{\varepsilon}{(n+8)^2}(7n^2 + 68n + 336)\right] + \mathcal{O}(\varepsilon^3). \tag{11.55}$$

The physics associated with ω' is a crossover phenomenon from the totally symmetric situation to a region where some components of the order parameter become decoupled.

2. *Consequences of the existence of more than one fixed point*

The $O(n)$-symmetric fixed point is always present. Let us assume the existence of a second one. We are going to show that this implies the existence of a third infra-red point, and we shall study globally, at order ε, the resulting situation.

Let g^*_{ijkl} be a second fixed point at order ε with a trace

$$g^*_{ijkk} = G^*\delta_{ij}, \tag{11.56}$$

and let us study the renormalization group equations for coupling constants being an arbitrary linear combination of the symmetric solution and of the second one, i.e.:

$$g_{ijkl} = xG^*_s S_{ijkl} + yg^*_{ijkl}. \tag{11.57}$$

The central remark is that, at order ε, $\lambda\, dg_{ijkl}/d\lambda$ is also a linear combination of the same two quantities. The equations are at lowest order in ε

$$\begin{aligned} \frac{1}{\varepsilon}\lambda\frac{dx}{d\lambda} &= -x + x^2 + \alpha_2 xy \\ \frac{1}{\varepsilon}\lambda\frac{dy}{d\lambda} &= -y + y^2 + \alpha_1 xy, \end{aligned} \tag{11.58}$$

where

$$\begin{aligned} \alpha_1 &= 2 - \frac{1}{\varepsilon}G^*_s = \frac{12}{(n+8)} + \mathcal{O}(\varepsilon) \\ \alpha_2 &= \frac{G^*}{\varepsilon} + \mathcal{O}(\varepsilon). \end{aligned} \tag{11.59}$$

We see that, in addition to the initial fixed points ($x = 1, y = 0$) and ($x = 0$, $y = 1$), a third one is always present:

$$x = \frac{1-\alpha_2}{1-\alpha_1\alpha_2}; \qquad y = \frac{1-\alpha_1}{1-\alpha_1\alpha_2}. \tag{11.60}$$

The corresponding g'^*_{ijkl} has a trace given by

$$g'^*_{ijkk} = \varepsilon\alpha_3\delta_{ij}. \tag{11.61}$$

with

$$\alpha_3 = \frac{2-\alpha_1-\alpha_2}{1-\alpha_1\alpha_2}, \tag{11.62}$$

or equivalently

$$2 - (\alpha_1 + \alpha_2 + \alpha_3) + \alpha_1\alpha_2\alpha_3 = 0. \tag{11.63}$$

Note that the relation between the fixed points g^* and g'^* is reciprocal.

From inequalities (11.36) we obtain

$$\alpha_i \geqslant 0 \qquad i = 1, 2, 3, \tag{11.64}$$

and

$$\alpha_1 + \alpha_2 \leqslant 2 \qquad \alpha_1 + \alpha_3 \leqslant 2. \tag{11.65}$$

In addition, from the explicit expression (11.62) we find

$$\alpha_2 + \alpha_3 \leqslant 2. \tag{11.66}$$

Thus in the following considerations which depend only on the values of the α_i's, the three fixed points play a completely symmetric role.

Stability condition

To study the stability condition in the space of coupling constants spanned by these three points, one reads the eigenvalues of the matrix B corresponding to the fixed point $i = 1$, 2 or 3 in (11.52):

The stability of the ith point is equivalent to the condition

$$\alpha_i > 1. \tag{11.67}$$

The inequalities $\alpha_i + \alpha_j \leqslant 2$ show that there is at most one stable fixed point. Furthermore, there certainly exits a stable one: for intance if α_1 and α_2 are smaller than one, it is easy to verify that α_3 given by (11.62) is larger than one.

Thus, there is one and only one stable fixed point among the three.

Anomalous dimensions

At this order, the critical exponent η is directly related to the values of the α_i by the equation

$$\eta_i = \frac{\varepsilon^2}{48}\alpha_i(2 - \alpha_i) + \mathcal{O}(\varepsilon^3). \tag{11.68}$$

If, for instance, the third fixed point is the stable one:

$$\alpha_1 < 1 \qquad \alpha_2 < 1 \qquad (2 - \alpha_3) < 1.$$

In the interval $[0, 1]$, $\eta(\alpha)$ is monotonically increasing and the largest η_i corresponds to the largest of the three numbers α_1, α_2, $(2 - \alpha_3)$. Inequalities (11.65) and (11.66) show that $(2 - \alpha_3)$ is the largest one. Therefore the stable fixed point has the largest η. This leads to the intriguing conjecture that the stable fixed point could in general correspond to the largest value of η.

Note

A last remark concerning the unstable fixed points is in order; in general these fixed points are irrelevant for the critical behaviour. However, if in special situations some thermodynamical parameters can be adjusted in such a way that the coupling constant takes *exactly* an unstable fixed-point value, the system then possesses the associated scaling behaviour. This case is very similar to that of the tricritical point; by varying a physical parameter, one can adjust the four-point renormalized coupling constant to zero, which is the corresponding fixed-point value.

E. Anisotropic corrections to the equation of state for the symmetric fixed point

We have seen that for n smaller than the value given in (11.51) only the symmetric fixed point is stable. Therefore the leading term in the equation of state has the isotropic form:

$$H_i = M_i \Phi(\mathbf{M}^2).$$

In particular the magnetization is along the magnetic field and the zero-field transverse magnetic susceptibility is infinite. If the Hamiltonian itself is not $O(n)$-invariant, we know that the complete equation of state does not have this property. The spontaneous magnetization will point for very small fields in some preferred directions, and the magnetic susceptibility will not be infinite in all transverse directions.

These effects are due to anisotropic corrections to scaling laws. We shall compute these corrections using the methods set out in Section VIII.A.

Let us write the R.G. equation for the free energy defined by:

$$\Gamma(M, t, g, \mu) = \sum_{N=2}^{\infty} \frac{1}{N!} M_{i_1} \ldots M_{i_N} \Gamma^{(N)}_{i_1 \ldots i_N} \qquad (p_i = 0, t, g, \mu) \qquad (11.69)$$

Actually the true free energy regular in the temperature differs from the above definition only by a term independent of M which is irrelevant for the equation of state.

The R.G. equation reads:

$$\left[\mu \frac{\partial}{\partial \mu} + W_{ijkl} \frac{\partial}{\partial g_{ijkl}} - \frac{\eta(g)}{2} M_i \frac{\partial}{\partial M_i} - \left(\frac{1}{\nu(g)} - 2\right) t \frac{\partial}{\partial t}\right] \Gamma(M, t, g, \mu) = 0. \qquad (11.70)$$

Let us make the change of function

$$\Gamma(M, t, g, \mu) = \Gamma(MA_1(g), tA_2(g), g^*, \mu)\, C(M, t, g, \mu), \qquad (11.71)$$

where g^* is the value of the coupling constant at the symmetric fixed point, and $A_1(g)$, $A_2(g)$ and C satisfy

$$A_1(g^*) = A_2(g^*) = C(g^*) = 1. \qquad (11.72)$$

If A_1 and A_2 are solutions of the equations:

$$\left[W_{ijkl}\frac{\partial}{\partial g_{ijkl}} - \frac{(\eta(g) - \eta)}{2}\right] A_1(g) = 0$$

$$\left[W_{ijkl}\frac{\partial}{\partial g_{ijkl}} - \left(\frac{1}{\nu(g)} - \frac{1}{\nu}\right)\right] A_2(g) = 0 \qquad (11.73)$$

then $C(M, t, g, \mu)$ satisfies the same R.G. equation as $\Gamma(M, t, g, \mu)$. Let us expand C around $g_{ijkl} = g^*_{ijkl}$:

$$C(M, t, g, \mu) = 1 + (g - g^*)_{ijkl}\frac{\partial C}{\partial g_{ijkl}}(M, t, g^*, \mu). \qquad (11.74)$$

The behaviour of the first correction of the scaling form of Γ is given by the term linear in $(g - g^*)$. Let us therefore write the R.G. equation for $\partial C/\partial g$

$$\left(\mu\frac{1}{\partial\mu} - \frac{\eta}{2}M_i\frac{\partial}{\partial M_i} - [(1/\nu) - 2]\, t\frac{\partial}{\partial t}\right)\frac{\partial C}{\partial g_{ijkl}}(M, t, g^*, \mu)$$
$$+ B_{ijkl,\, i'j'k'l'}\frac{\partial C}{\partial g_{i'j'k'l'}}(M, t, g^*, \mu) = 0 \qquad (11.75)$$

As a consequence of the symmetry of the fixed point and of the trace condition satisfied by $(g - g^*)_{ijkl}$, the derivative $\partial C/\partial g_{ijkl}$ can be decomposed as

$$\frac{\partial C}{\partial g_{ijkl}}(M, t, g^*, \mu) = C_1 S_{ijkl} + C_2\frac{M_i M_j M_k M_l}{M^4}. \qquad (11.76)$$

From (11.75) we can derive the R.G. equations satisfied by C_1 and C_2:

$$\left[\mu\frac{\partial}{\partial\mu} + \omega - \frac{\eta}{2}M_i\frac{\partial}{\partial M_i} - \left(\frac{1}{\nu} - 2\right)t\frac{\partial}{\partial t}\right] C_1 + \frac{1}{n}(\omega - \omega_{\text{anis.}})\, C_2 = 0$$

$$\left[\mu\frac{\partial}{\partial\mu} + \omega_{\text{anis.}} - \frac{\eta}{2}M_i\frac{\partial}{\partial M_i} - \left(\frac{1}{\nu} - 2\right)t\frac{\partial}{\partial t}\right] C_2 = 0. \qquad (11.77)$$

C_1 leads to isotropic corrections to the scaling form of the equation of state similar to those discussed in Section VIII, but which are now governed by two exponents ω and $\omega_{\text{anis.}}$. Anisotropy effects appear only in C_2 and are therefore characterised by the critical exponent $\omega_{\text{anis.}}$ introduced in subsection D above.

Concentrating on this last effect, we can write

$$\Gamma(M, t, g, \mu) \sim M^{\delta+1}\,\Phi_1\left(\frac{t}{M^{1/\beta}}\right)$$

$$+\; t^{\nu\omega\,(\text{anis.})}(g - g^*)_{ijkl}\,\frac{M_iM_jM_kM_l}{M^4}\,M^{\delta+1}\,\Phi_2\left(\frac{t}{M^{1/\beta}}\right). \tag{11.78}$$

We then deduce for the magnetic field

$$H_i \sim \frac{M_i}{M}\,M^{\delta} f_1\left(\frac{t}{M^{1/\beta}}\right) + t^{\nu\omega\,(\text{anis.})}(g - g^*)_{ijkl}\,\frac{M_jM_kM_l}{M^3}\,M^{\delta} f_2\left(\frac{t}{M^{1/\beta}}\right) \tag{11.79}$$

We see that for small magnetic fields below T_c, the strength of the magnetization is mainly given by the isotropic part of the equation of state, while its direction is modified. This modification occurs for fields of the order

$$H \sim t^{\beta\delta + \nu\omega\,(\text{anis.})} \tag{11.80}$$

In some direction the transverse susceptibility χ_T in zero field below T_c is not infinite. The ratio of χ_T to the longitudinal susceptibility is of order $t^{-\nu\omega\,(\text{anis.})}$.

XII. Short Distance Expansions and Critical Phenomena

We have seen in the previous Sections that scaling laws operate when one explores distances x much larger than the lattice spacing a and much smaller than the correlation length ξ. Deviations from scaling when a/x is small but non-zero have been characterized in Section VIII. In this Section we do not wish to consider these deviations further—we shall remain in the critical domain in which a is infinitesimal—but to study how simple power laws are modified for growing values of the ratio x/ξ. More precisely stated the problem is the following. In Section VI we derived the strong scaling property in zero field above T_c:

$$\langle S(x)\,S(0)\rangle = \xi^{\eta}\,\frac{D(x/\xi)}{x^{d-2+\eta}} \tag{12.1}$$

and we know that

$$\lim_{x/\xi \to 0} D(x/\xi) = D(0) \tag{12.2}$$

exists. The problem is to determine the next terms in the expansion of $D(x/\xi)$ for small values of its argument. More generally, we shall study the same question in presence of an applied magnetic field or below T_c, for which the leading behaviour is still given by (12.1–2) but the subsequent terms will turn out to be somewhat different from what they are above T_c.

A. Short-distance expansion of an operator product (SDE)

To explain the nature of the problem, let us comment on the derivation of (12.2) given in Section VI. In the massive theory the inverse correlation function $\Gamma^{(2)}(p, -p; m, g)$ satisfies the Callan–Symanzik equation

$$\left[m\frac{\partial}{\partial m} + W(g)\frac{\partial}{\partial g} - \eta(g)\right]\Gamma^{(2)}(p, -p) = m^2(2 - \eta(g))\,\Gamma^{(1,2)}(0; p, -p). \quad (12.3)$$

The r.h.s. of (12.3) is negligible order-by-order in g and ε for momenta large compared to m (i.e. precisely small x/ξ ratios), and the solution of the homogenous equation is then

$$\Gamma^{(2)}(p, -p) \underset{\substack{p/m \gg 1 \\ p/\Lambda \ll 1}}{\sim} p^{2-\eta} \quad (12.4)$$

which is equivalent after Fourier transform to the property (12.2). Thus in order to find out the corrections to the single power behaviour (12.4) we must estimate the dominant contributions of the r.h.s. of (12.3) for large p/m. Indeed $\Gamma^{(1,2)}$ satisfies a Callan–Symanzik equation, namely

$$\left[m\frac{\partial}{\partial m} + W(g)\frac{\partial}{\partial g} - \eta(g) - \left(\frac{1}{\nu(g)} - 2\right)\right]\Gamma^{(1,2)}(q; p_1, p_2) = m^2(2 - \eta(g)) \times \Gamma^{(2,2)}(0, q; p_1, p_2) \quad (12.5)$$

and asymptotically the r.h.s. of (12.5) is also negligible. But this asymptotic regime consists of all three momenta p_1, p_2 and q simultaneously large compared to m, whereas in the r.h.s. of (12.3) q is zero. In those circumstances it would be wrong to neglect the r.h.s. of (12.5). The reason is roughly as follows: when all momenta go to infinity the asymptotic behaviour of a diagram may be obtained by neglecting the mass inside the propagator. However if some of the external momenta are "exceptional", as it is the case here where an external line carries zero momentum, infra-red singularities appear in the massless theory and the behaviour is modified. Thus we cannot obtain the behaviour of $\Gamma^{(1,2)}(0; p, -p)$ by this method and a new tool is needed. The solution to this problem proposed by Symanzik (1971, 1973b) relies on a technique introduced independently by Wilson (1969) and Kadanoff (1969). It is an expansion of the product of two operators $A(x)\,B(y)$ when the two points x and y are close to each other, in terms of the local composite operators $\theta_i(x)$ of the theory with C-numbers coefficients C_i—the Wilson functions—which becomes singular in the limit $x \to y$:

$$A(x)\,B(y) = \sum_i \theta_i((x + y)/2)\,C_i(x - y). \quad (12.6)$$

The existence of this expansion in perturbation theory has been derived by W. Zimmermann (1971) and Wilson and Zimmermann (1972), when A and B are the elementary fields S of the theory, and we shall not go into the proof here. The usefulness of (12.6) may be understood with the following crude arguments: if d_A, d_B, d_{θ_i} are the canonical dimensions (in mass units) of A, B and θ_i, then C_i has the dimension $d_A + d_B - d_{\theta_i}$, which means that at short distance it will behave in free-field theory as

$$C_i(x - y) \sim (x - y)^{d_{\theta_i} - d_A - d_B}. \qquad (12.7)$$

Therefore the most singular contributions come from the operators of small dimensions, i.e. the simplest operators of the theory S, S^2. etc. . ., and by including operators of increasing degree in (12.6) one can obtain a systematic expansion. Of course this argument is much too simple at several stages:

(i) As one may suspect, the behaviour of the Wilson coefficients is governed not by the canonical dimensions but by the corresponding anomalous dimensions. This is justified in the next subsection.

(ii) Thus it is not clear that the hierarchy of operators of free-field theory is maintained by the interactions. However, within the ε-expansion the difference between anomalous and canonical dimensions is infinitesimal. For the operators S and S^2, for instance, these differences are $\eta/2$ and $(1/\nu) - 2$ respectively.

(iii) when one includes higher operators, it is clear that one must at the same time obtain corrections to the most singular behaviour of the type (12.7) for the Wilson coefficients of simpler operators. This problem is solvable but its discussion is quite technical and the interested reader is referred to the literature (Callan, 1972; Mitter, 1973; Brézin *et al.*, 1974b).

B. The renormalization group equation satisfied by the first Wilson coefficient†

We follow here the method of Coleman (1971).

Let us apply the expansion (12.6) to the product of two S fields inserted in a 1-P.I. vertex function; in momentum space equation (12.6) means

$$\Gamma^{(N)}[(p/2) + k, (p/2) - k, p_3, \ldots, p_N] = \sum_{\substack{\alpha \\ k\ \text{large}}} C_\alpha(k)\, \Gamma^{(N-2)}_{A_\alpha}(p; p_3, \ldots, p_N) \qquad (12.8)$$

provided the insertion of A_α into $\Gamma^{(N-2)}$ is not primitively divergent.

By $S \to -S$ invariance the lowest operator which contributes to the r.h.s. of (12.8) is S^2 and we will keep this single term:

$$\Gamma^{(N)}[(p/2) + k, (p/2) - k, p_3, \ldots, p_N] \sim C(k)\, \Gamma^{(1, N-2)}(p; p_3, \ldots, p_N) \qquad (12.9)$$

† An elegant proof has been given recently by Callan (1975).

the error in (12.9) being terms of relative order $1/k^2$ (up to powers of log k) order-by-order in perturbation theory.

Similarly the same *operator* equation (12.6) may be used in a correlation function with one S^2 insertion and gives

$$\Gamma^{(1,N)}[q;(p/2)+k,(p/2)-k,p_3,\ldots,p_N] \sim C(k)\,\Gamma^{(2,N-2)}(q,p;p_3,\ldots,p_N). \tag{12.10}$$

In order to derive the equation for $C(k)$ it is now sufficient to use the Callan–Symanzik equations for $\Gamma^{(N)}$ and $\Gamma^{(1,N-2)}$. One first uses (12.9) in the equation for $\Gamma^{(N)}$ and one obtains

$$\Gamma^{(1,N-2)}(p;p_3,\ldots,p_N)\left[m\frac{\partial}{\partial m}+W(g)\frac{\partial}{\partial g}\right]C(k)+C(k)\left[m\frac{\partial}{\partial m}+W(g)\frac{\partial}{\partial g}\right.$$
$$\left.-\frac{N}{2}\eta(g)\right]\Gamma^{(1,N-2)} \sim m^2[2-\eta(g)]\,\Gamma^{(1,N)}(0;p_1,\ldots,p_N). \tag{12.11}$$

Using the equation satisfied by $\Gamma^{(1,N-2)}$ and (12.10) we obtain

$$\Gamma^{(1,N-2)}(p;p_3,\ldots,p_N)\left(m\frac{\partial}{\partial m}+W(g)\frac{\partial}{\partial g}\right)C(k)+C(k)\left\{\left[\frac{1}{\nu(g)}-2-\eta(g)\right]\right.$$
$$\left.\times\,\Gamma^{(1,N-2)}(p;p_3,\ldots,p_N)+m^2(2-\eta(g))\,\Gamma^{(2,N-2)}(0,p;p_3,\ldots p_N)\right\}$$
$$\sim m^2(2-\eta(g))\,C(k)\,\Gamma^{(2,N-2)}(0,p;p_3,\ldots,p_N), \tag{12.12}$$

so that finally:

$$\left[m\frac{\partial}{\partial m}+W(g)\frac{\partial}{\partial g}-\eta(g)+\left(\frac{1}{\nu(g)}-2\right)\right]C(k) \sim 0. \tag{12.13}$$

The asymptotic behaviour of $C(k)$ for k/m large and k/Λ small is given by integration of the fixed point equation and by noting that the canonical dimension of $C(k)$ is ε; one then obtains from (12.13)

$$C(k) \sim \text{const.}\, m^{\varepsilon}(k/m)^{\varepsilon-\eta+[(1/\nu)-2]}. \tag{12.14}$$

The next terms of the asymptotic behaviour of $C(k)$ may be computed as well, but this involves higher operators whose effect will not be considered here.

C. Next-to-leading terms in zero field above T_c

For a recent discussion of this topic see Brézin *et al.* (1974a).

We are now in position to apply the above considerations to the study of the behaviour of the two-point correlation function $\Gamma^{(2)}(p,-p)$. We have pointed out in the first section that we must calculate the leading behaviour of the r.h.s. of (12.3): $\Gamma^{(1,2)}(0; p, -p)$. A direct application of the short-distance expansion (SDE) at this stage is not possible ($\Gamma^{(2,0)}$ is primitively divergent)† and instead we have to use (12.5). There the SDE for $\Gamma^{(2,2)}$ gives

$$\Gamma^{(2,2)}(0,0; p, -p) \sim C(p)\,\Gamma^{(3,0)}(0,0,0) \tag{12.15}$$

and thus, using (12.14) and the canonical dimension $(d - 6)$ of $\Gamma^{(3,0)}$

$$\Gamma^{(2;2)}(0,0; p, -p) \sim \text{const.}\ (p/m)^{\varepsilon - \eta + (1/\nu) - 2}\, m^{-2}, \tag{12.16}$$

the constant in front of the r.h.s. of (12.16) being calculable in perturbation theory. Integrating (12.5) we next obtain

$$\Gamma^{(1,2)}(0; p, -p) \sim \text{const.}\,(p/m)^{2-(1/\nu)-\eta} + \text{const.}\,(p/m)^{\varepsilon - \eta + (1/\nu) - 2}. \tag{12.17}$$

We have now the leading behaviour of the r.h.s. of (12.3), which thus gives in the studied limit:

$$\Gamma^{(2)}(p, -p) \sim \text{const.}\, m^2\,(p/m)^{2-\eta} + \text{const.}\ \mathrm{m}^2\,(\mathrm{p/m})^{2-(1/\nu)-\eta}$$
$$+ \text{const.}\, m^2\,(p/m)^{\varepsilon - \eta + (1/\nu) - 2} \tag{21.18}$$

Using the relation $2 - \alpha = \nu d$, this result may be written, replacing m by t^ν

$$\Gamma^{(2)}(p) \sim m^\eta p^{2-\eta}\left[A + Bt(p)^{-1/\nu} + Ct^{1-\alpha}(p)^{-(1-\alpha)/\nu}\right]. \tag{12.19}$$

The presence of the C-term which is singular in $T - T_c$ was first discussed by Fisher and Langer (1968), who showed that the C-term is needed to explain resistive anomalies of metals. Let us note that if, as here, several powers are present in the result, it is difficult to disentangle them within the ε-expansion without prior knowledge of the result. Once the form (12.19) is known one can determine the coefficients by Feynman diagrams analysis. Fisher and Aharony (1973) determined the first two terms in the ε-expansion of A, B and C.

D. Next-to-leading terms in a field or below T_c

Brézin *et al.* (1974d) have reported on the asymptotic behaviour of the spin–spin correlation function in a field and below T_c.

Let us first discuss the $n = 1$ *case*. The two-point function for an arbitrary magnetization $\Gamma^{(2)}(p, -p; t, M)$ may be obtained as the second functional derivative of the 1-P.I. generating functional $\Gamma(M)$:

$$\Gamma^{(2)}(x - y; M) = \left.\frac{\delta^2\,\Gamma(M)}{\delta M(x)\,\delta M(y)}\right|_{M(x) = M} \tag{12.20}$$

Using the expansion of $\Gamma(M)$, (12.20) becomes

$$\Gamma^{(2)}(p, -p; M) = \sum_0^\infty (M^N/N!)\,\Gamma^{(N+2)}(p, -p, 0, \ldots, 0). \tag{12.21}$$

† See, for instance, Callan (1975).

The $N = 0$ term in the r.h.s. of (12.21) has already been examined; the $N \geqslant 2$ terms may be evaluated directly by the SDE (12.9) as

$$\Gamma^{(N+2)}(p, -p, 0, \ldots, 0) \sim C(p)\,\Gamma^{(1,N)}(0; 0, 0, \ldots, 0) \quad N \geqslant 2. \tag{12.22}$$

The functions $\Gamma^{(1,N)}(0; p_i)$ are proportional to the derivative $(\partial/\partial t)\,\Gamma^{(N)}(p_i)$ and therefore

$$\Gamma^{(2)}(p, -p; M) \sim p^{2-\eta}\{A + Bt(p)^{-1/\nu} + Ct^{1-\alpha}p^{-(1-\alpha)/\nu}\} + C(p)\frac{\partial}{\partial t}\sum_2^\infty (M^N/N!)\,\Gamma^{(N)}(0, 0, \ldots, 0). \tag{12.23}$$

Using (12.14), which gives the behaviour of $C(p)$, we have

$$\Gamma^{(2)}(p, -p; M) \sim p^{2-\eta}\left\{A + Bt(p)^{-1/\nu} + \frac{\partial}{\partial t}\left[C\frac{t^{2-\alpha}}{2-\alpha} + \sum_2^\infty \frac{M^N}{N!}\Gamma^{(N)}(0, \ldots, 0)\right]p^{-(1-\alpha)/\nu}\right\}, \tag{12.24}$$

with suitable normalizations of momenta and temperature.

We are now going to indicate why the quantity

$$\Gamma(M, t) \equiv \frac{C\,t^{2-\alpha}}{2-\alpha} + \sum_2^\infty \frac{M^N}{N!}\Gamma^{(N)}(0, \ldots, 0; t) \tag{12.25}$$

is the free energy regular in the temperature when M is non-zero, and normalized so that it satisfies exactly the scaling law

$$\Gamma(M, t) = M^{\delta+1}\,\Phi(t/M^{1/\beta}). \tag{12.26}$$

Let us write

$$\Gamma(M, t) = \Gamma^{(0)}(t) + \sum_2^\infty (M^N/N!)\,\Gamma^{(N)}(t), \tag{12.27}$$

and expand $\Gamma^{(0)}(t)$ in powers of t (or rather of a source $t(x)$ which approaches a uniform value)

$$\Gamma^{(0)}(t) = \sum_{L=0}^\infty \Gamma^{(L,0)}(0)\,(t^L/L!). \tag{12.28}$$

As we know, from power-counting, the first three terms $L = 0, 1, 2$ of (12.28) need additive renormalizations beyond the usual ones; thus $\Gamma^{(0)}(t)$ is defined up to an arbitrary polynomial of second degree in t; its third derivative $(\partial/\partial t)^3\,\Gamma^{(0)}(t) = \Gamma^{(3,0)}(t)$ is well defined and, at the fixed point, has a pure power behaviour. It is thus possible to adjust $\Gamma^{(0)}(t)$ so that it coincides with the pure singular power coming from the integration of $\Gamma^{(3,0)}(t)$. It remains to show that the constant C in (12.25) is indeed the constant which appears in (12.19):

this comes directly from the integration of the Callan–Symanzik equations (12.3, 5).

Since the magnetic field is the derivative with respect to M of $\Gamma(M, t)$, one can relate the function Φ of (12.26) to the equation of state function $f(x)$ of (7.1), by the relation

$$\Phi(x) \propto \int_1^\infty \mathrm{d}u\, u^{\delta - 1/\beta}\left[f'(0) - f'\left(\frac{x}{u^{1/\beta}}\right)\right] + \frac{f'(0)}{\delta - (1/\beta) + 1}, \tag{12.29}$$

and the final result is

$$\Gamma^{(2)}(p, t, M) \sim p^{2-\eta}\left\{A + Bt(p)^{-1/\nu} + \Phi\left(\frac{t}{M^{1/\beta}}\right)\left(\frac{p}{M^{\nu/\beta}}\right)^{-(1-\alpha)/\nu}\right\}. \tag{12.30}$$

For the *general n case* we are interested in the asymptotic behaviour of $\Gamma_{ij}^{(2)}(p; t, M)$ (where i and j are the internal indices) and thus we need the SDE of the product of two fields $S_i(x)\, S_j(y)$. A new operator of dimension two appears in addition to $\mathbf{S}^2$, which may be taken as the traceless operator

$$\mathcal{O}_{ij} = \left(S_i S_j - \frac{1}{n}\delta_{ij}\mathbf{S}^2\right)(x). \tag{12.31}$$

This operator is multiplicatively renormalizable and satisfies the renormalization group equation

$$\left[\mu\frac{\partial}{\partial\mu} + W(g)\frac{\partial}{\partial g} - \frac{N}{2}\eta(g) - \left(\frac{\phi(g)}{\nu(g)} - 2\right)\right]\Gamma^{(N)}_{\mathcal{O}\, ij} = 0, \tag{12.32}$$

in which $\phi(g^*)$ coincides with the crossover exponent ϕ (see Riedel and Wegner, 1969; Fisher and Pfeuty, 1972; Wilson, 1972) calculated by this method by Yamazaki and Suzuki (1973) and by Yamazaki (1974):

$$\phi = 1 + \varepsilon\frac{n}{2(n+8)} + \varepsilon^2\frac{n[n^2 + 24n + 68]}{4(n+8)^3} + O(\varepsilon^3).$$

Consequently a new power is present in the asymptotic behaviour which becomes:

$$\Gamma_{ij}^{(2)}(p, t, M) \sim p^{2-\eta}\left\{\left[A + Bt(p)^{-1/\nu} + \Phi\left(\frac{t}{M^{1/\beta}}\right)\left(\frac{p}{M^{\nu/\beta}}\right)^{-(1-\alpha)/\nu}\right]\delta_{ij}\right.$$
$$\left. + [(1/n)\,\delta_{ij} - \hat{M}_i\hat{M}_j]\left(\frac{p}{M^{\nu/\beta}}\right)^{-d+\phi/\nu}\psi\left(\frac{t}{M^{1/\beta}}\right)\right\}, \tag{12.33}$$

in which $\hat{M}_i$ is the unit vector along the magnetization and $\psi(x)$ is a new universal function which may be calculated directly. To first order in ε this function is

$$\psi(x) \propto 1 + \frac{\varepsilon}{2(n+8)}[(x+3)\ln(x+3) - (x+1)\ln(x+1)]. \qquad (12.34)$$

References

Abe, R. (1972). *Prog. Theor. Phys. (Kyoto)* **48**, 1414.
Abe, R. (1973a). *Prog. Theor. Phys. (Kyoto)* **49**, 113.
Abe, R. (1973b). *Prog. Theor. Phys.* **49**, 1877.
Abers, E. S. and Lee, B. W. (1973). *Phys. Rep.* **9C**, 1.
Aharony, A. (1973a). *Phys. Rev.* **B8**, 3342, 3349, 3358, 3363, 4270.
Aharony, A. (1973b). *Phys. Rev. Letts* **31**, 1494.
Aharony, A. (1974). *Phys. Rev.* **B10**, 2834.
Aharony, A. and Fisher, M. E. (1973). *Phys. Rev.* **B8**, 3323.
Ahlers, G. (1973). *Phys. Rev.* **A8**, 530.
Anderson, P. W. and Morel, P. (1961). *Phys. Rev.* **123**, 1911.
Avdeeva, G. M. and Migdal, A. A. (1972). *Zh. Eksp. Teor. Fiz. Pis'ma Red.* **16**, 253 (*JETP Lett.* **16**, 178).
Baker, G. A. (1962). *Phys. Rev.* **126**, 2071.
Balian, R. and Werthamer, N. R. (1963). *Phys. Rev.* **131**, 1553.
Berlin, T. H. and Kac, M. (1952). *Phys. Rev.* **86**, 821.
Bervillier, C.,Girardi, G. and Brézin, E. (1974). Saclay preprint.
Birgenau, R. J., Yelon, W. B., Cohen, E. and Makovsky, J. (1972). *Phys. Rev.* **B5**, 2607.
Bjorken, J. D. and Drell, S. D. (1965). "Relativistic Quantum Fields". McGraw Hill.
Blume, M. (1966). *Phys. Rev.* **141**, 517.
Bogoliubov, N. N. and Shirkov, D. V. (1959). "Introduction to the Theory of Quantized Fields". Interscience, New York.
Bollini, C. G. and Giambiagi, J. J. (1972). *Phys. Letters* **40B**, 566.
Brézin, E. (1974). Saclay preprint DPh-T.
Brézin, E. and Wallace, D. J. (1973). *Phys. Rev.* **B7**, 1967.
Brézin, E., Wallace, D. J. and Wilson, K. G. (1972). *Phys. Rev. Letts* **29**, 591.
Brézin, E., Le Guillou, J. C. and Zinn-Justin, J. (1973a). *Phys. Rev.* **D8**, 434.
Brézin, E., Le Guillou, J. C. and Zinn-Justin, J. (1973b). *Phys. Rev.* **D8**, 2418.
Brézin, E., Wallace, D. J. and Wilson, K. G. (1973c). *Phys. Rev.* **B7**, 232.
Brézin, E., Le Guillou, J. C. and Zinn-Justin, J. (1973d). *Phys. Rev.* **B8**, 5330
Brézin, E., Le Guillou, J. C., Zinn-Justin, J. and Nickel, B. G. (1973e). *Phys. Letters* **44A**, 227.
Brézin, E., Amit, D. and Zinn-Justin, J. (1974a). *Phys. Rev. Letts* **32**, 151.
Brézin, E., De Dominicis, C. and Zinn-Justin, J. (1974b). *Lett. al Nuovo Cim.* **9**, 483; **10**, 849.
Brézin, E., Le Guillou, J. C. and Zinn-Justin, J. (1974c). *Phys. Rev.* **D9**, 1121.
Brézin, E., Le Guillou, J. C. and Zinn-Justin, J. (1974d). *Phys. Rev. Letts* **32**, 473.
Brézin, E., Le Guillou, J. C. and Zinn-Justin, J. (1974e). *Phys. Letters* **47A**, 285.
Brézin, E., Le Guillou, J. C. and Zinn-Justin, J. (1974f). *Phys. Rev.* **B10**, 892.
Brout, R. (1974). *Phys. Rep.* **10C**, 1.
Brown, L. S. and Boulware, D. (1968). *Phys. Rev.* **172**, 1628.
Callan, C. G. (1970). *Phys. Rev.* **D2**, 1541.

Callan, C. G. (1972). *Phys. Rev.* **D5**, 3202.
Callan, C. G. (1975). *In* Lectures at the 1975 les Houches Summer School (R. Balian and J. Zinn-Justin, eds.). North-Holland.
Capel, H. W. (1966). *Physica* **32**, 966.
Coleman, S. (1971). *In* Lectures at the 1971 Erice Summer School.
Coleman, S. and Jackiw, R. (1971). *Ann. Phys.* (N.Y.) **67**, 552.
Coleman, S. and Weinberg, E. (1973). *Phys. Rev.* **D7**, 1888.
De Dominicis, C. (1963). *J. Math. Phys.* **4**, 255.
De Dominicis, C. and Martin, P. C. (1964). *J. Math. Phys.* **5**, 14.
De Gennes, P. G. (1973). *Phys. Letters* **44A**, 271.
De Witt, B. S. (1967). *Phys. Rev.* **162**, 1195.
Di Castro, C. (1972). *Lett. al Nuovo Cim.* **5**, 69.
Di Castro, C., Jona-Lasinio, G. and Peliti, L. (1974). *Ann. Phys.* **87**, 327.
Domb, C. and Hunter, D. L. (1965). *Proc. Phys. Soc.* (*London*) **86**, 1147.
Englert, F. (1963). *Phys. Rev.* **129**, 567.
Ferrell, R. and Scalapino, D. J. (1973). *Phys. Rev. Letts* **29**, 413.
Feynman, R. P. and Hibbs, A. R. (1965). "Quantum Mechanics and Path Integrals" McGraw Hill, New York.
Fisher, M. E. (1967). *Rept. Prog. Phys.* **30**, 615.
Fisher, M. E. and Aharony, A. (1973). *Phys. Rev. Letts* **31**, 1238.
Fisher, M. E. and Aharony, A. (1974). *Phys. Rev.* **B10**, 2818.
Fisher, M. E. and Langer, L. (1968). *Phys. Rev. Letts* **20**, 665.
Fisher, M. E. and Pfeuty, P. (1972). *Phys. Rev.* **B6**, 1889.
Fisher, M. E. and Tarko, H. B. (1975). *Phys. Rev.* **B11**, 1131.
Gaunt, D. S. and Domb, C. (1970). *J. Phys.* **C3**, 1442.
Gell-Mann, M. and Low, F. E. (1954). *Phys. Rev.* **95**, 1300.
Goldstone, J. (1961). *Nuovo Cim.* **19**, 154.
Graf, E. H., Lee, D. M. and Reppy, J. D. (1967). *Phys. Rev. Letts* **19**, 417.
Griffiths, R. B. (1967). *Phys. Rev.* **158**, 176.
't Hooft, G. and Veltman, M. (1972). *Nucl. Phys.* **B44**, 189.
Horwitz, G. and Callen, H. B. (1961). *Phys. Rev.* **124**, 1757,
Hubbard, J. (1972). *Phys. Letters* **39A**, 365.
Jegerlehner, F. and Schroer, B. (1973). *Acta Phys. Austriaca Suppl.* XI, 389. Springer-Verlag.
Jona-Lasinio, G. (1964). *Nuovo Cim.* **34**, 1790.
Kadanoff, L. P. (1966). *Physics* (*N.Y.*) **2**, 263.
Kadanoff, L. P. (1969). *Phys. Rev. Letts* **23**, 1430.
Kadanoff, L. P., Götze, W., Hamblen, D., Hecht, R., Lewis, E. A. S., Palciauskas, V. V., Rayl, M., Swift, J., Aspnes, D. and Kane, J. (1967). *Rev. Mod. Phys.* **39**, 395.
Ketley, I. J. and Wallace, D. J. (1973). *J. Phys.* **A6**, 1667.
Landau, L. D. (1937). *Phys. Z. Sowjetunion* **11**, 26. Reprinted in "Collected Papers by Landau" (D. ter Haar, ed.) p. 193. Pergamon, New York, 1965.
Larkin, A. I. and Khmel'nitskii, D. E. (1969). *Zh. Eksp. Teor. Fiz.* **56**, 2087 (*JETP* **29**, 1123).
Lederman, F. L., Salamon, M. B. and Shacklette, L. W. (1974). *Phys. Rev.* **B9**, 2981.
Lee, B. W. (1969). *Nucl. Phys.* **B9**, 649.
Lee, B. W. (1972). "Chiral Dynamics". Document in Modern Physics, Gordon and Breach, New York.
Lee, B. W. and Zinn-Justin, J. (1972). *Phys. Rev.* **D5**, 3121, appendix B.
Lin, J. S. and Schmidt, P. W. (1974). *Phys. Rev. Letts* **33**, 1265.

Lowenstein, J. (1970). *Phys. Rev.* **D4**, 2281.
Lowenstein, J. (1972). *In* Maryland Lectures—Seminars on Renormalization Theory (unpublished).
Ma, S. K. (1972). *Phys. Rev. Letts* **29**, 1311.
Ma, S. K. (1973a). *Phys. Rev.* **A7**, 2172.
Ma, S. K. (1973b). *Rev. Mod. Phys.* **45**, 589.
Ma, S. K. (1974). *Phys. Rev.* **A10**, 1818.
Mack, G. (1972). *In* "Lecture Notes in Physics", International Summer Institute on Theoretical Physics in Kaiserslautern (W. Ruhl and A. Vancura, eds.) Vol. 17. Springer, Berlin.
Milosevic, S. and Stanley, H. E. (1972). *Phys. Rev.* **B5**, 2526.
Mitter, P. K. (1973). *Phys. Rev.* **D7**, 2927.
Nambu, Y. (1960). *Phys. Rev. Letts* **4**, 380.
Nambu, Y. (1966). *Phys. Letters* **26B**, 626.
Nickel, B. G. (unpublished report).
Onsager, L. (1944). *Phys. Rev.* **65**, 117.
Ornstein, L. S. and Zernike, F. (1914). *Proc. Sect. Sci. K. med. Akad. Wet.* **17**, 793.
Parisi, G. (1973). *In* "Field Theory Approach to Second Order Phase Transitions in Three and Two Dimensional Systems", Lectures given at the 1973 Cargèse Summer School (E. Brézin and J. M. Charap, eds.), Gordon and Breach, New York. To be published.
Peierls, R. F. (1958). *Phys. Rev.* **54**, 918.
Riedel, E. and Wegner, F. (1969). *Z. Phys.* **225**, 195.
Riedel, E. and Wegner, F. (1972). *Phys. Rev. Letts* **29**, 349.
Ritchie, D. S. and Fisher, M. E. (1972). *Phys. Rev.* **B5**, 2668.
Salamon, M. B. and Lederman, F. L. (1974). *Phys. Rev.* **B10**, 4492.
Schroer, B. (1973). *Phys. Rev.* **B8**, 4200.
Schwinger, J. (1951). *Proc. Nat. Acad. Sci.* **37**, 452; 455.
Schwinger, J. (1958). *Proc. Nat. Acad. Sci.* **44**, 956.
Siegert, A. J. F. (1963). *In* "Brandeis Lectures", Vol. 3, p. 159. W. A. Benjamin, Inc.
Stanley, H. E. (1968). *Phys. Rev.* **176**, 718.
Stanley, H. E. (1971). "Introduction to Phase Transitions and Critical Phenomena". Oxford U.P., New York.
Stueckelberg, E. C. G. and Petermann, A. (1953). *Helv. Phys. Acta* **26**, 499.
Suzuki, M. (1973). *Progr. Theor, Phys.* **49**, 424; 4106.
Symanzik, K. (1970a). *Commun. Math. Phys.* **18**, 227.
Symanzik, K. (1970b). "Renormalization of Theories with Broken Symmetry". Cargèse Lectures 1970 (D. Bessis, ed.), Gordon and Breach, New York.
Symanzik, K. (1971). *Commun. Math. Phys.* **23**, 49.
Symanzik, K. (1973a). DESY Report No. 73/6. Unpublished.
Symanzik, K. (1973b). *In* "Field Theory and Critical Phenomena". Cargèse Lectures 1973 (E. Brézin and J. M. Charap, eds.), Gordon and Breach, New York.
Tarko, H. B. and Fisher, M. E. (1973). *Phys. Rev. Letts* **31**, 926.
Tarko, H. B. and Fisher, M. E. (1975). *Phys. Rev.* **B11**, 1217.
Theumann, W. (1974). Saclay preprint.
Toulouse, G. (1973). *C.R. Acad Sci.* (*Paris*) **B276**, 193.
Wallace, D. J. (1973). *J. Phys.* **C6**, 1390.
Wallace, D. J. and Zia, R. K. P. (1974). *Phys. Letters* **48A**, 325.
Wegner, F. J. (1972a). *Phys. Rev.* **B5**, 4529.
Wegner, F. J. (1972b). *Phys. Rev.* **B6**, 1891.

Wegner, F. J. (1973). *In* "Critical phenomena and the renormalization group", Lectures of the VIII Finnish Summer School.
Wegner, F. J. and Riedel, E. (1973). *Phys. Rev.* **B7**, 248.
Weinberg, S. (1960). *Phys. Rev.* **118**, 838.
Weinberg, S. (1973). *Phys. Rev.* **D8**, 3497.
Weiss, P. (1907). *J. Phys. Radium, Paris* **6**, 667.
Widom, B. (1965). *J. Chem. Phys.* **43**, 3892; 3898.
Wilson, K. G. (1969). *Phys. Rev.* **179**, 1499.
Wilson, K. G. (1970). *Phys. Rev.* **D2**, 1473.
Wilson, K. G. (1971). *Phys. Rev.* **B4**, 3174; 3184.
Wilson, K. G. (1972). *Phys. Rev. Letts* **28**, 548.
Wilson, K. G. (1973). *Phys. Rev.* **D7**, 2911.
Wilson, K. G. and Fisher, M. E. (1972). *Phys. Rev. Letts* **28**, 240.
Wilson, K. G. and Kogut, J. B. (1974). *Phys. Rep.* **12C**, 75.
Wilson, K. G. and Zimmermann, W. (1972). *Commun. Math. Phys.* **24**, 87.
Yamazaki, Y. (1974). *Phys. Letters* **49A**, 215.
Yamazaki, Y. and Suzuki, M. (1973). *Progr. Theor. Phys.* **50**, 1097.
Zimmermann, W. (1971). *In* "Lectures on Elementary Particles and Quantum Field Theory" (S. Deser *et al.*, eds.) Vol. I. MIT Press, Cambridge, Mass.
Zinn-Justin, J. (1973). *In* "Wilson's Theory of Critical Phenomena and Renormalized Perturbation Theory", Lectures given at the 1973 Cargèse Summer School (E. Brézin and J. M. Charap, eds.). Gordon and Breach, New York. To be published.

4. The $1/n$ Expansion

Shang-keng Ma†

Department of Physics and Institute for Pure and Applied Physical Sciences, University of California, San Diego, La Jolla, California 92037 USA

† Alfred P. Sloan Foundation fellow. Work supported in part by the National Science Foundation under Grant GP 38627X.

I. Introduction

A. Simplicity at large *n*

In this article, the renormalization group (RG) applied to an n-component d-dimensional spin system near its critical point will be discussed taking $1/n$ as a small expansion parameter. The reason for taking a small $1/n$ is for mathematical simplification. The expansion in $1/n$ is complementary to the expansion in $\varepsilon = 4 - d$ and is in the same spirit.

The physical reason for the mathematical complication in studying critical phenomena is the large and strongly interacting fluctuations of spins. For a large class of models, such fluctuations become weak and unimportant for $d > 4$. For d slightly below 4, i.e., for small ε, these fluctuations are important but sufficiently small and tractable by appropriately arranged perturbation theory. This is the physical basis for the ε-expansion.

The physical basis for the $1/n$ expansion is the following. When the number of components n is large, the fluctuation of the *magnitude* of the n-component spin vectors is small. In other words, even though each component may fluctuate a lot, the sum of the squares of the n spin components is a large number and is fixed at its average value apart from small fractional fluctuations $\sim 1/\sqrt{n}$. This turns out to be true for $d > 2$, for a certain class of models. When such fractional fluctuations are neglected, the spin fluctuations become independently coupled to an average field. Again, by using appropriately arranged perturbation theory, one can calculate critical behaviours in powers of $1/n$.

Under the simplification of large n, we can illustrate fundamental concepts with explicit and short calculations. Such illustrations will be the main emphasis of this article. Some technical aspects of calculating exponents will also be discussed.

In the remaining part of this section we shall define our model and show that in the limit $n \to \infty$ it is easily solved. A brief summary of exponents calculated by $1/n$ expansion is given.

In Section II and Section III, we discuss RG and related concepts such as scaling fields and scaling dimensions. We work out everything explicitly for $n \to \infty$ and thus provide a reasonably concrete picture of the RG machinery.

In Section IV, we discuss briefly the solution with the presence of a uniform magnetization and the equation of state in the $n \to \infty$ limit.

In Section V, the basis of calculating exponents by $1/n$ perturbation theory is explained. We include a few representative calculations to illustrate the general technique. The importance of very careful interpretation of results is emphasized.

In Section VI, we comment on the convergence of the $1/n$ expansion and a few other points.

B. The model and the solution above the critical point

We shall work with an n-component classical vector field $\sigma(x) = (\sigma_1(x), \ldots, \sigma_n(x))$ in a d-dimensional cubic volume L^d. The vector $\sigma(x)$ is the spin density at x and is a superposition of a discrete set of variables. More precisely, we can break the system into cubic cells or blocks of volume b^d and label the blocks with indices c. Then we define

$$\sigma(x) = \sum_c I_c(x)\sigma_c, \tag{1.1}$$

where $I_c(x) = 1$ if $x \,\varepsilon$ (the block c) and zero otherwise, σ_c is the mean spin density in the block c. Alternatively, we can define

$$\sigma(x) = \sum_{k<\Lambda} \sigma_k \exp(ik \cdot x)L^{-d/2}, \tag{1.2}$$

where σ_k are Fourier components of the spin configuration and $\Lambda \sim 2\pi/b$, or more generally

$$\sigma(x) = \sum_\lambda \phi_\lambda(x)\sigma_\lambda, \tag{1.3}$$

where $\phi_\lambda(x)$ is an orthonormal set of functions with no spatial variations over a distance smaller than b. $I_c(x)$ and $L^{-d/2}\exp(ik\cdot x)$ are examples of $\phi_\lambda(x)$.

We shall always write a probability distribution function of spins in the form

$$P \propto \exp(-\mathscr{H}[\sigma]) \tag{1.4}$$

which is a function of the variables σ_λ. We define the notation

$$\int \delta\sigma \equiv \int \prod_\lambda \mathrm{d}\sigma_\lambda. \tag{1.5}$$

Quantities of interest are

$$\exp(-FL^d) = \int \delta\sigma \exp(-\mathscr{H}[\sigma]) \tag{1.6}$$

$$\delta_{\alpha\beta}G(k) = \langle \sigma_k^\alpha \sigma_{-k}^\beta \rangle \tag{1.7}$$

$$C(x, x') = \tfrac{1}{4}(\langle \sigma^2(x)\sigma^2(x')\rangle - \langle \sigma^2 \rangle^2)$$

$$C(k) = \int \mathrm{d}^d x \exp(-ik\cdot x)\, C(x, 0), \tag{1.8}$$

where the average $\langle \ldots \rangle$ is defined by

$$\langle A \rangle = \int \delta\sigma A \exp(-\mathscr{H}) \bigg/ \int \delta\sigma \exp(-\mathscr{H})$$

and

$$\sigma^2 = \sum_{\alpha=1}^{n} (\sigma^\alpha)^2. \tag{1.9}$$

F is the free energy per volume and $G(k)$ is the simplest correlation function. The form $\delta_{\alpha\beta}G$ has to be modified if spin isotropy is broken. We shall use the symbol r to denote the inverse susceptibility, i.e.,

$$G(0) = 1/r. \tag{1.10}$$

Consider a Ginzburg–Landau model

$$\mathscr{H} = \tfrac{1}{2}\int \mathrm{d}^d x\,((\nabla\sigma)^2 + r_0\sigma^2 + \tfrac{1}{2}u\sigma^4) \tag{1.11}$$

σ^2 is the square of the length of the vector $\sigma(x)$ defined by (1.9), and

$$\sigma^4 \equiv (\sigma^2)^2 \tag{1.12}$$

$$(\nabla\sigma)^2 \equiv \sum_{\alpha=1}^{n}\sum_{a=1}^{d}\left(\frac{\partial\sigma^\alpha}{\partial x_a}\right)^2, \tag{1.13}$$

r_0 and u are numerical parameters. Clearly, $\mathscr{H}$, and hence $P \propto \mathrm{e}^{-\mathscr{H}}$, are invariant under rotations in spin space.

If the σ^4 term in $\mathscr{H}$ were absent, P would be a Gaussian and the calculation of quantities such as (1.6)–(1.8) would be trivial. The presence of σ^4 makes the calculation very difficult.

Now how would the problem simplify when n is large? The heuristic answer mentioned earlier proceeds as follows.

In the term $\frac{1}{2}u\sigma^4$, the coefficient of a given $(\sigma^\alpha)^2$ is $u(\sigma^2 - \frac{1}{2}(\sigma^\alpha)^2)$. For large n, σ^2, which is the sum of n terms, is expected to be much larger than the $-\frac{1}{2}(\sigma^\alpha)^2$ term. In other words, the coupling of each $(\sigma^\alpha)^2$ to itself is much weaker than its coupling to the total sum σ^2 of all n components.

Being the sum of a large number of terms, σ^2 is expected to have a small fractional fluctuation. More precisely, we expect

$$\langle(\sigma^2 - \langle\sigma^2\rangle)^2\rangle/\langle\sigma^2\rangle^2 = O(1/n). \tag{1.14}$$

In this sense, $\Delta\sigma^2 \ll \langle\sigma^2\rangle$. If we consider σ^2 to be $\langle\sigma^2\rangle$, a constant, then each $(\sigma^\alpha)^2$ is coupled to the constant $\langle\sigma^2\rangle$.

Now we write $\langle\sigma^2\rangle + \Delta\sigma^2$ for σ^2 in (1.11) to obtain

$$\mathscr{H} = \tfrac{1}{2}\int \mathrm{d}^d x\,((\nabla\sigma)^2 + (r_0 + u\langle\sigma^2\rangle)\sigma^2 + \tfrac{1}{2}u(\Delta\sigma^2)^2)$$
$$+ \text{ constants independent of } \sigma. \tag{1.15}$$

Note that $\langle\sigma^2\rangle$ is a constant (not a random variable), and will not affect the average values calculated via $P \propto e^{-\mathscr{H}}$. Dropping the $(\Delta\sigma^2)^2$ term, we get a quadratic form with $r_0 + u\langle\sigma^2\rangle$ as the coefficient of σ^2:

$$\begin{aligned}\mathscr{H} \approx \mathscr{H}_0 &= \tfrac{1}{2}\int d^dx((\nabla\sigma)^2 + (r_0 + u\langle\sigma^2\rangle)\sigma^2)\\ &= \tfrac{1}{2}\sum_{\alpha,\,\beta<\Lambda} |\sigma_k^\alpha|^2(k^2 + r_0 + u\bar{n}),\end{aligned} \tag{1.16}$$

where we have used (1.2) for $\sigma(x)$ in terms of σ_k and

$$\begin{aligned}\bar{n} &\equiv \langle\sigma^2\rangle\\ &= L^{-d}\sum_{\alpha,\,k<\Lambda}\langle|\sigma_k^\alpha|^2\rangle.\end{aligned} \tag{1.17}$$

The probability $P \propto e^{-\mathscr{H}_0}$ is thus a product of Gaussians, one for each (α, k). The average of $|\sigma_k^\alpha|^2$ is just

$$G(k) = \langle|\sigma_k^\alpha|^2\rangle = (r_0 + u\bar{n} + k^2)^{-1} \tag{1.18}$$

$$G(0) \equiv 1/r = 1/(r_0 + u\bar{n}). \tag{1.19}$$

Substituting (1.18) in (1.17), we find

$$\begin{aligned}\bar{n} &= nL^{-d}\sum_{k<\Lambda}(r_0 + u\bar{n} + k^2)^{-1}\\ &= nK_d\int_0^\Lambda dk\, k^{d-1}(r_0 + u\bar{n} + k^2)^{-1},\end{aligned} \tag{1.20}$$

where K_d is $(2\pi)^{-d} \times$ (area of unit sphere in d dimension):

$$K_d = 2^{-d+1}\pi^{-d/2}/\Gamma(\tfrac{1}{2}d). \tag{1.21}$$

From (1.20), $\bar{n}$ can be solved for as a function of r_0 and u.

Let r_{0c} be the value of r_0 at which $r = r_0 + u\bar{n}$ vanishes, and let

$$\begin{aligned}\bar{n}_c &= \bar{n}(r_{0c})\\ &= nK_d\int_0^\Lambda dk\, k^{d-1}/k^2\\ &= nK_d\Lambda^{d-2}/(d-2).\end{aligned} \tag{1.22}$$

Then

$$r = r_0 - r_{0c} + u(\bar{n} - \bar{n}_c). \tag{1.23}$$

Note that (1.22) makes sense only for $d > 2$. We shall always restrict our discussion to $4 > d > 2$.

When $r_0 - r_{0c}$ is small, so is r. Subtracting (1.22) from (1.20), we get, for $r \ll \Lambda^2$,

$$\bar{n} - \bar{n}_c = -nr^{d/2-1}J/(\tfrac{1}{2}d - 1), \tag{1.24}$$

where

$$J \equiv \tfrac{1}{2}K_d\pi(\tfrac{1}{2}d - 1)\csc\pi(\tfrac{1}{2}d - 1). \tag{1.25}$$

From (1.23) and (1.24), we see that

$$r_0 - r_{0c} = r^{d/2-1}unJ\left(\frac{d}{2} - 1\right) \tag{1.26}$$

for $r \to 0$. Defining γ by $1/r \propto (r_0 - r_{0c})^{-\gamma}$, we get

$$\gamma = \left(\frac{d}{2} - 1\right)^{-1}. \tag{1.27}$$

By the definition of η, $G(k) \propto k^{-2+\eta}$ at $r_0 = r_{0c}$, (1.18) says that

$$\eta = 0. \tag{1.28}$$

We define the quantity C as

$$C = -\partial^2F/\partial r_0^2, \tag{1.29}$$

which is not identical to the specific heat but contains the part of the specific heat of interest here. From the definition (1.6) of F, we have

$$\begin{aligned} C &= -\tfrac{1}{2}\partial\langle\sigma^2\rangle/\partial r_0 \\ &= -\tfrac{1}{2}\partial\bar{n}/\partial r_0 \\ &= \frac{1}{2u}(1 - \partial r/\partial r_0). \end{aligned} \tag{1.30}$$

Recall that $\langle\sigma^2\rangle \equiv \bar{n}$ and $r = r_0 + u\bar{n}$. Using (1.26) for r in terms of $r_0 - r_{0c}$, we obtain

$$C = \frac{1}{2u}(1 - r^{2-d/2}/(unJ))$$

$$+ \text{ higher order terms in } r, \tag{1.31}$$

$$r^{2-d/2} \propto (r_0 - r_{0c})^{-\alpha},$$

$$\alpha = (d - 4)/(d - 2). \tag{1.32}$$

Thus, the specific heat does not diverge at $r_0 = r_{0c}$ but has an infinite slope for $2 < d < 4$.

The above formulas (1.18) and down are good only for $r_0 > r_{0c}$, and $d > 2$.

For $d > 4$, the $O(r)$ term is more important in (1.24) and we would get $r \propto r_0 - r_{0c}$ and $\gamma = 1$. The case $r_0 < r_{0c}$ will be discussed later.

Since in the probability distribution $e^{-\mathscr{H}_0}$ spin components are independent, we have $\langle(\Delta\sigma^2)^2\rangle = O(n)$ and

$$\langle(\Delta\sigma^2)^2\rangle/\langle\sigma^2\rangle^2 = O(1/n), \tag{1.33}$$

consistent with what we assumed before. Actually different spin components are not quite independent and $\langle(\Delta\sigma^2)^2\rangle$ calculated directly via $e^{-\mathscr{H}_0}$ would not be correct. This is because spin components are related via the self-consistency equation for $\langle\sigma^2\rangle$. However, (1.33) turns out to remain valid when $\langle(\Delta\sigma^2)^2\rangle$ is correctly calculated.

Since we shall be interested only in cases where r and $r_0 - r_{0c}$ are small, $\bar{n}$ will be always taken as of $O(n)$ since it is close to $\bar{n}_c$ (see (1.22)). This means taking σ^2 as $O(n)$, $(\sigma^\alpha)^2$ as $O(1)$. We shall take r, r_0 as $O(1)$. To keep every term in (1.23) in the same order, we shall restrict

$$u = O(1/n). \tag{1.34}$$

To sum up, the approximation $\mathscr{H}_0$ given by (1.16) is simply that replacing r_0 by $r_0 + u\langle\sigma^2\rangle$ and then calculating $\langle\sigma^2\rangle$ self-consistently. $r_0 + u\langle\sigma^2\rangle$ can be viewed as a self-consistent field coupled to σ^2. This approximation is the Hartree approximation familar in various branches of physics.

C. Historical note, summary of exponents

A considerable fraction of literature on the theory of critical phenomena has been devoted to the study of models of the type

$$\mathscr{H} = -J\sum_c\sum_\delta\sum_{\alpha=1}^{n}\sigma_c^\alpha\sigma_{c+\delta}^\alpha, \tag{1.35}$$

where $c + \delta$ are nearest neighbours of the site c, and the magnitude of the spins is restricted to a fixed constant s^2:

$$\sum_{\alpha=1}^{n}(\sigma_c^\alpha)^2 = s^2. \tag{1.36}$$

We can normalize the units so that $s^2 = 1$, or n; it does not matter. This model can be solved exactly by Fourier transform were it not for the restriction (1.36), which plays the role of the σ^4 term in the Ginzburg–Landau model. To water down this restriction, Berlin and Kac (1952) proposed and solved the *spherical model*, in which (1.36) is replaced by

$$\frac{1}{N}\sum_c\langle\sigma_c^2\rangle = s^2, \tag{1.37}$$

where N is the total number of spins in the whole system (here we take

$n = 1$). Thus, the restriction on each spin is relaxed to a restriction on the mean square length of all spins. Stanley (1968) showed that if we take n-component spins and let $n \to \infty$, the the restriction (1.36) would have the same effect as (1.37).

Early work on calculating critical properties using $1/n$ as a small parameter was done by Abe (1972), Wilson (1972, private communication), Ma (1972), Brézin and Wallace (1973), Suzuki (1972), Ferrell and Scalapino (1972), etc. Later work which followed will be mentioned later in Section VI.

At present, all critical exponents of interest have been calculated to $O(1/n)$ for $2 < d < 4$. Abe (1973c) has also obtained η to $O(1/n^2)$ for $d = 3$. The calculation to $O(1/n^2)$ turns out to be very complicated. The renormalization group has been explicitly worked out only to $O(1)$, (Ma (1973b)). The thermodynamic equation of state to $O(1/n)$ was due to Brézin and Wallace (1973), and structure of the correlation function to Aharony (1974), Abe and Hikami (1974).

Let us summarize the exponents and some other information:

$$\eta = 4(4/d - 1)S_d/n + O(n^{-2}),$$

$$\eta = 8(3\pi^2 n)^{-1} - (\tfrac{8}{3})^3(\pi^4 n^2)^{-1} + O(n^{-3}), \qquad d = 3$$

$$\gamma = (\tfrac{1}{2}d - 1)^{-1}(1 - 6S_d/n) + O(n^{-2}),$$

$$\alpha = (d-4)(d-2)^{-1}(1 + 8(1-d)(4-d)^{-1}S_d/n) + O(n^{-2}), \quad (1.38)$$

$$S_d = \frac{\sin \pi(\tfrac{1}{2}d - 1)}{\pi(\tfrac{1}{2}d - 1)\, B(\tfrac{1}{2}d - 1, \tfrac{1}{2}d - 1)} = \frac{\sin \pi(\tfrac{1}{2}d - 1)\Gamma(d-1)}{2\pi\Gamma(d/2)^2}. \quad (1.39)$$

The quantity S_d as a function of d is plotted in Fig. 1. Special cases for S_d are

$$S_{4-\varepsilon} = \varepsilon/2 - \varepsilon^2/4 + O(\varepsilon^3),$$

$$S_3 = 2/\pi^2,$$

$$S_{2+\delta} = \delta/4 + O(\delta^3). \quad (1.40)$$

In Table I, we summarize exponents y_i and scaling variables D_i associated with various scaling fields g_i. These concepts and their significance will be explained in the next two sections. The exponents y_i listed are accurate to $O(1/n)$ but D_i are accurate to $O(1)$ only. Some of the symbols in Table I are defined with respect to the probability distribution $P \propto \exp(-\mathscr{H})$ with

$$\mathscr{H} = \int d^d x H. \quad (1.41)$$

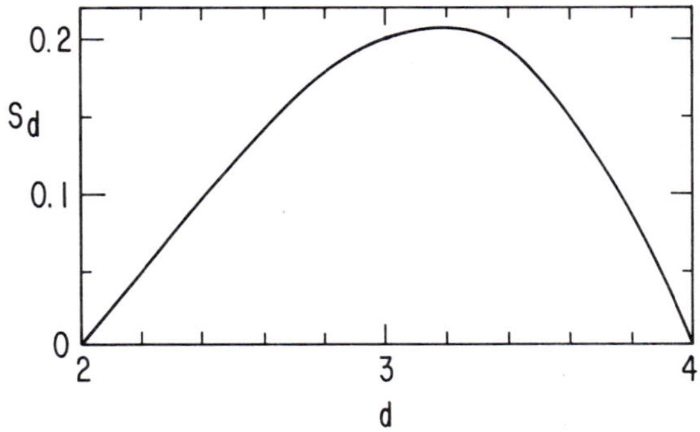

FIG. 1. S_d as a function of d. See (1.39).

TABLE I. Exponents and Scaling Variables
(See text, (1.41)–(1.46), for definitions of symbols)

Label i	Spin rotation	Coordinate rotation	Exponent y_i	Local scaling variable D_i
σ	Vector	Scalar	$\frac{1}{2}(d+2-\eta)$	σ^α
τ	Traceless tensor	Scalar	$2 - 16d^{-1} S_d/n$	$\tau_{\alpha\beta}$
1	Scalar	Scalar	$d - 2 + \frac{8}{d}(d-1)(d-2)\frac{S_d}{n}$	t
c	Cubic	Scalar	$4 - d - 16\left(1 + \frac{2}{d}\right)\frac{S_d}{n}$	$\sum_\alpha (\sigma^\alpha)^4$
$\sigma 1$	Vector	Scalar	$\frac{1}{2}(d-2+\eta)$	$\partial H/\partial\sigma^\alpha$
0	Scalar	Scalar	0	$\delta(x)$
Δ	Scalar	Scalar	0	$(\nabla\sigma)^2 - M(t)$
T	Scalar	Traceless tensor	0	T_{ab}
$\Delta\tau$	Traceless tensor	Scalar	0	$\lambda_{\alpha\beta} + t\tau_{\alpha\beta}$
$\tau 1$	Traceless tensor	Scalar	$-8(d - 3 + 4/d)\frac{S_d}{n}$	$\frac{d-4}{d(d-2)}\lambda_{\alpha\beta} + t\tau_{\alpha\beta}$
Υ	Traceless tensor	Traceless tensor	$\frac{-16(4/d - 1)S_d}{(d+2)n}$	$\Upsilon_{ab\alpha\beta}$
2	Scalar	Scalar	$d - 4 + \frac{8}{d}(4-d)(d-1)^2\frac{S_d}{n}$	$t^2 - a_2\nabla^2 t$

We define

$$\tau_{\alpha\beta} = \sigma^\alpha\sigma^\beta - (1/n)\sigma^2\delta_{\alpha\beta}, \tag{1.42}$$

$$t = \partial H/\partial\sigma^2, \tag{1.43}$$

$$T_{ab} = \sum_{\alpha=1}^{n} \frac{\partial\sigma^\alpha}{\partial x_a}\frac{\partial\sigma^\alpha}{\partial x_b} - \frac{1}{d}(\nabla\sigma)^2\delta_{ab}, \tag{1.44}$$

$$\lambda_{\alpha\beta} = \nabla\sigma^\alpha\cdot\nabla\sigma^\beta - (1/n)(\nabla\sigma)^2\delta_{\alpha\beta}, \tag{1.45}$$

$$\Upsilon_{ab\alpha\beta} = \frac{\partial\sigma^\alpha}{\partial x_a}\frac{\partial\sigma^\beta}{\partial x_b} - (1/d)\delta_{ab}\lambda_{\alpha\beta} - (1/n)\delta_{\alpha\beta}T_{ab} - (1/nd)(\nabla\sigma^2)\delta_{ab}\delta_{\alpha\beta}. \tag{1.46}$$

The quantity $M(t)$ is a series in t and a_2 is a constant. We shall have no interest in them here. [For details, see Ma (1974b).]

Some of the entries in Table I are familiar quantities. The first D_i is σ^α, the spin, which couples to $g_\sigma = h$, the magnetic field. The variable $\tau_{\alpha\beta}$ couples to $g_{\alpha\beta}$, the "quadrupole" field. The variable t can be interpreted as the local potential seen by σ^2. It couples to $g_1 \propto (r_0 - r_{0c})$. The variable $\Sigma_\alpha(\sigma^\alpha)^4$ couples to a cubic field g_c. This and other entries in Table I will be discussed later in Section III. Note that Table I is not complete. There are many other scaling fields, scaling variables and exponents not listed. The ones listed are those having the largest y_i.

II. Renormalization Group, Scaling Fields and the Fixed Point

In this section and the next we discuss the renormalization group and related concepts and show explicitly how they appear to the zeroth order in $1/n$.

Let us summarize the essentials of the RG machinery as follows:

(i) RG is a set of transformations $\{R_s; s \geqslant 1\}$ on the space of parameters $\{\mu\}$.

(ii) Scaling fields g_i are special functions of μ which have the transform properties $g_i \to g_i s^{y_i}$ under R_s. The constant y_i is the exponent of g_i.

(iii) Local scaling variables $D_i(x)$ are special random variables which have the transformation properties

$$D_i(x) \to D_i(x/s)s^{d-y_i},$$

$d - y_i$ is the *scale dimension* of D_i.

(iv) Correlation functions of scaling variables have therefore simple transformation properties which can be studied by dimensional analysis. Their critical behaviour is simply deduced from their transformation properties.

(v) The correlation function of any variables of interest can be expanded as a linear combination of correlation functions of scaling variables. Hopefully only a few terms in the combination are of interest.

The ultimate purpose of the RG machinery is to deduce critical behaviour of correlation functions of variables of physical interest.

A. Renormalization group

Consider probability distributions written in the form $P \propto \exp(-\mathscr{H}[\sigma;\mu])$, where μ denotes a set of parameters specifying $\mathscr{H}$. A renormalization group (RG) is a set of transformations $\{R_s\}$ on the space of μ, the parameter space.

R_s is defined as

$$R_s = (R_2)^l, \qquad s \equiv 2^l, \qquad l = 0, 1, 2, 3, \ldots \tag{2.1}$$

i.e., the application of the generator R_2 l times. A point μ is transformed to $R_2\mu$, which essentially involves two steps: (A) Elimination of σ_q with $\Lambda/2 < q < \Lambda$ and (B) shrink the size of the system by 2. Let us write

$$\sigma(x) \equiv \zeta(x) + \tilde{\sigma}(x) \tag{2.2}$$

where $\tilde{\sigma}(x)$ contains only Fourier components σ_q with $\Lambda/2 < q < \Lambda$. The above two steps (A) and (B) can be summarized by the formula

$$\begin{aligned}\exp\{&-\mathscr{H}[\sigma(x'); R_2\mu] - F_2^> L^d\}\\ &= \left(\int \delta\tilde{\sigma} \exp\{-\mathscr{H}[\zeta(x) + \tilde{\sigma}(x); \mu]\}\right)_{\zeta(x) = 2^{1-(d+\eta)/2}\sigma(x/2)},\\ &\qquad x' = x/2,\end{aligned} \tag{2.3}$$

or equivalently the formula

$$\exp\{-\mathscr{H}[\sigma; R_2\mu] - F_2^> L^d\} = \left(\int \delta\tilde{\sigma} \exp\{-\mathscr{H}[\sigma;\mu]\}\right)_{\sigma_k \to 2^{1-\eta/2}\sigma_{2k}} \tag{2.3'}$$

where

$$\int \delta\tilde{\sigma} \equiv \int \prod_{\alpha, \Lambda/2 < q < \Lambda} d\sigma_q^\alpha.$$

The quantity $F_2^>$ is a constant independent of σ. Note that any additive constant in $\mathscr{H}$ will be irrelevant in computing average values since it will be normalized away. However, in computing the free energy (the F defined in (1.6)), it is important. By our construction

$$R_s R_{s'} = R_{ss'}. \tag{2.4}$$

There is a great deal of flexibility in defining the generator R_2. One could also define an $R_{1+\delta}$ with very small δ to serve as the generator. The above definition is just one of many possible definitions.

A fixed point μ^* is a point in the parameter space with the invariance property

$$R_s\mu^* = \mu^* \tag{2.5}$$

when an appropriate value of η is assumed.

The operation (2.2) does not affect σ_k with $k < \Lambda/2$ except a change of name into $2^{1-\eta/2}\sigma_{2k}$. Consequently, we have

$$\langle \sigma_{k_1}\sigma_{k_2} \cdots \sigma_{k_m} \rangle_P = s^{m(1-\eta/2)} \langle \sigma_{sk_1}\sigma_{sk_2} \cdots \sigma_{sk_m} \rangle_{P'} \tag{2.6}$$

where $P' \propto \exp(-\mathscr{H}[\sigma, R_s\mu])$ and $k_1, \ldots, k_m$ are all smaller than Λ/s. We have ignored component labels on σ_k for simplicity.

B. Scaling fields

The property $R_sR_{s'} = R_{ss'}$ makes RG a semigroup. The irreducible representation of this semigroup has the form s^y, where y is a constant. We can construct a set of functions $g_i(\mu)$ of the set of parameters μ such that $g_i(\mu)$ will transform under R_s into $g_i s^{y_i}$, i.e.,

$$g_i(R_s\mu) = g_i(\mu)s^{y_i}. \tag{2.6$'$}$$

These functions form a new set of parameters which is more convenient than the old set μ. They are called *scaling fields* (Wegner (1972)). The constant y_i associated with g_i will be called the *exponent of* g_i.

We shall use the notation g for the set of scaling fields

$$g = (g_1, g_2 \ldots) \tag{2.7}$$

and

$$R_s g = (g_1 s^{y_1}, g_2 s^{y_2}, \ldots). \tag{2.8}$$

Since $\mathscr{H}$ can be regarded as a function of g as well as a function of μ, we can write $\mathscr{H}[\sigma; g]$ or simply $\mathscr{H}(g)$.

C. RG and scaling fields for $n \to \infty$; critical surface

Let us take a generalized Ginzburg–Landau form of $\mathscr{H}$:

$$\mathscr{H} = \tfrac{1}{2}\int d^d x((\nabla\sigma)^2 + U(\sigma^2; \mu)), \tag{2.9}$$

$$U(0; \mu) = 0, \tag{2.9$'$}$$

where $U(\sigma^2; \mu)$ is a polynomial in σ^2 and μ denotes the set of coefficients.

Equation (2.9′) simply excludes any additive constant from the definition of $\mathscr{H}$.

To carry out the steps (A) and (B) defining R_2, we write $\sigma = \zeta + \tilde{\sigma}$ as in (2.2). Since

$$(\nabla\sigma)^2 = (\nabla\zeta)^2 + (\nabla\tilde{\sigma})^2, \tag{2.10}$$

we have

$$\int \delta\tilde{\sigma} \exp\{-\mathscr{H}[\sigma + \zeta; \mu]\} = \exp\{-\tfrac{1}{2}\int d^d x[(\nabla\zeta)^2 - W(\zeta)]\}, \tag{2.11}$$

$$\exp\left[-\tfrac{1}{2}\int W(\zeta)\, d^d x\right] = \int \delta\tilde{\sigma} \exp\left\{-\tfrac{1}{2}\int d^d x[(\nabla\tilde{\sigma})^2 + U((\zeta + \tilde{\sigma})^2)]\right\}. \tag{2.12}$$

For large n,

$$(\zeta + \tilde{\sigma})^2 \approx \zeta^2 + \tilde{\sigma}^2 \tag{2.13}$$

because $\zeta \cdot \tilde{\sigma} = O(1)$ since only one component, that along ζ, of $\tilde{\sigma}$ contributes to it whereas $\tilde{\sigma}^2 = O(n)$. Equation (2.13) is correct to $O(n)$.

Following the arguments of Section I we conclude that the fractional fluctuation of $\tilde{\sigma}^2$ is small so that $\tilde{\sigma}^2 - \langle\tilde{\sigma}^2\rangle \ll \langle\tilde{\sigma}^2\rangle$ and

$$U(\zeta^2 + \tilde{\sigma}^2) \approx U(\zeta^2 + \tilde{n}) + t(\zeta^2 + \tilde{n})(\tilde{\sigma}^2 - \tilde{n}) \tag{2.14}$$

where

$$\tilde{n} \equiv \langle\tilde{\sigma}^2\rangle, \tag{2.15}$$

$$t(\lambda) \equiv \partial U(\lambda)/\partial\lambda. \tag{2.16}$$

The average (2.15) is taken over the probability distribution of $\tilde{\sigma}$ at a given ζ^2 and is therefore a function of ζ^2. Using the approximation (2.14), we have, with $\tfrac{1}{2}\Lambda < q < \Lambda$ understood,

$$\begin{aligned}\tfrac{1}{2}\int d^d x[(\nabla\tilde{\sigma})^2 + U(\zeta^2 + \tilde{\sigma}^2)] \approx \tilde{\mathscr{H}}_0 &= \tfrac{1}{2}\sum_{\alpha, q} |\sigma_q^\alpha|^2(q^2 + t(\zeta^2 + \tilde{n})) \\ &+ \tfrac{1}{2}\int [U(\zeta^2 + \tilde{n}) - \tilde{n}t(\zeta^2 + \tilde{n})]\, d^d x\end{aligned} \tag{2.17}$$

which is quadratic in $|\sigma_q^\alpha|^2$ and consequently $\exp(-\tilde{\mathscr{H}}_0)$ is a product of Gaussians. Thus, $\tfrac{1}{2}\int W(\zeta)\, d^d x$ of (2.12) is simply the free energy associated with $\tilde{\mathscr{H}}_0$:

$$W(\zeta) = nL^{-d}\sum_q \ln[(q^2 + t(\zeta^2 + \tilde{n}))/2\pi] + U(\zeta^2 + \tilde{n}) - \tilde{n}t(\zeta^2 + \tilde{n}). \tag{2.18}$$

The average of $\tilde{\sigma}^2$ is easily computed using $\exp(-\tilde{\mathscr{H}}_0)$:

$$\tilde{n} = \langle\tilde{\sigma}^2\rangle = nL^{-d}\sum_q (q^2 + t(\zeta^2 + \tilde{n}))^{-1}$$

$$= nK_d \int_{\Lambda/2}^{\Lambda} \mathrm{d}q\, q^{d-1}/(q^2 + t(\zeta^2 + \tilde{n})). \tag{2.19}$$

One must solve (2.19) for $\tilde{n}$ in terms of ζ^2 and substitute the result in (2.18).

Note that $W(\zeta)$ is minimized by the value of $\tilde{n}$ given by (2.19).

By the definition of $R_2\mu$ given by (2.3), we obtain from (2.11)

$$\mathscr{H}[\sigma; R_2\mu] + F_2^> L^d$$

$$= \tfrac{1}{2}\int \mathrm{d}^d x'[(\nabla\sigma(x'))^2 2^{-\eta} + s^d W(2^{1-(d+\eta)/2}\sigma(x'); \mu)] \tag{2.20}$$

$$F_2^> = \tfrac{1}{2}W(0;\mu). \tag{2.21}$$

We now write (2.20) in the form

$$\mathscr{H}[\sigma; R_2\mu] = \tfrac{1}{2}\int \mathrm{d}^d x'[(\nabla\sigma(x'))^2 + U(\sigma^2(x'); R_2\mu)]. \tag{2.22}$$

Then

$$U(\sigma^2; R_2\mu) = s^d[W(2^{1-d/2}\sigma;\mu) - W(0;\mu)]. \tag{2.23}$$

We have set $\eta = 0$ in order that the form (2.22) is possible. $R_2\mu$ can then be extracted from (2.23) after W is computed from (2.18) and (2.19).

Let us put the above results in a simpler form. Differentiating (2.23) with respect to σ^2, and using (2.18) for W, we obtain

$$t(\sigma^2; R_2\mu) \equiv \partial U(\sigma^2; R_2\mu)/\partial\sigma^2$$

$$= 2^2 t(\sigma^2 2^{2-d} + \tilde{n}; \mu). \tag{2.24}$$

Note that $\partial W/\partial\tilde{n} = 0$ at fixed ζ. Equation (2.19) supplies $\tilde{n}$. (Recall that $\zeta = 2^{2-d}\sigma^2$.) $R_2\mu$ can thus be extracted from (2.24) and (2.19).

A simple way to extract the scaling fields is as follows. Solve the equation $t = t(\sigma^2;\mu)$ for σ^2 as a function of t and μ, and write the solution as a series in t:

$$\sigma^2 = \sum_{m=1}^{\infty} b_m(\mu)t^{m-1}. \tag{2.25}$$

Also, solve $t' = t(\sigma^2; R_2\mu)$ for σ^2, and get

$$\sigma^2 = \sum_{m=1}^{\infty} b_m(R_2\mu)t'^{m-1}. \tag{2.26}$$

But, according to (2.24),

$$t'/2^2 = t(\sigma^2 2^{2-d} + \tilde{n}; \mu) \tag{2.27}$$

so we must have

$$\sigma^2 = 2^{d-2}\left(\sum_{m=1}^{\infty} b_m(\mu)\,(t'/2^2)^{m-1} - \tilde{n}\right). \tag{2.28}$$

Now we can expand (2.19) in powers of $t'/2^2$ (which is just $t(\zeta^2 + \tilde{n})$ according to (2.27)), to obtain

$$\begin{aligned}\tilde{n} &= \sum_{m=1}^{\infty} (-t'/2^2)^{m-1} nK_d \int_{\Lambda/2}^{\Lambda} dq\, q^{d-1-2m} \\ &= \sum_{m=1}^{\infty} b_m^* t'^{m-1}(2^{d-2m} - 1)2^{-d+2},\end{aligned} \tag{2.29}$$

where

$$b_m^* = -(-)^m\, nK_d\, \Lambda^{d-2m}/(d - 2m). \tag{2.30}$$

Note that $b_2^* = \bar{n}_c$ according to (1.25). Substituting (2.29) in (2.28), we obtain, for the coefficient of t'^m

$$b_m(R_2\mu) = (b_m(\mu) - b_m^*)2^{d-2m} + b_m^* \tag{2.31}$$

It is quite obvious now that we have found the scaling fields g_m and exponents y_m:

$$g_m(\mu) = b_m(\mu) - b_m^*,$$

$$g_m(R_2\mu) = g_m(\mu)2^{y_m}, \tag{2.32}$$

$$y_m = d - 2m, \qquad m = 1, 2, 3, \ldots. \tag{2.33}$$

For the space of parameters representing $\mathscr{H}$ of the form (2.9), these are all the scaling fields. There will be more scaling fields when other kinds of terms are added to (2.9). They will be discussed shortly.

Evidently, we have found a fixed point μ^* given by

$$b_m(\mu^*) = b_m^* \tag{2.34}$$

where all $g_m(\mu^*)$ vanish.

We can find $t^* \equiv t(\sigma^2; \mu^*)$ and hence $U^* \equiv U(\sigma^2; \mu^*)$ by setting $b_m = b_m^*$ in (2.25). First we sum the series to obtain

$$\sigma^2/\bar{n}_c = 1 - (d - 2)\Lambda^{2-d} \int_{\Lambda}^{\infty} dp\, p^{d-1}[(t^* + p^2)^{-1} - p^{-2}] \tag{2.35}$$

and then we have to solve for t^* as a function of σ^2. The solution cannot be expressed in a closed analytic form but a numerical plot can easily be made. Here let us list a few properties of U^*.

(i) $U^*(\sigma^2)$ is minimum at $\sigma^2 = \bar{n}_c$ for $2 < d < 4$. Near $\bar{n}_c$,

$$U^*(\sigma^2) \approx U^*(\bar{n}_c) + \frac{(\sigma^2 - \bar{n}_c)^2}{2b_2^*}, \tag{2.36}$$

$$1/b_2^* = \Lambda^{4-d}(4 - d)/(nK_d) \tag{2.37}$$

There is no other minimum and $dt^*/d\sigma^2 > 0$ always.

(ii) For $d \geqslant 4$, $U^*(\sigma^2) = t^*(\sigma^2) = 0$, i.e., a trivial fixed point.

(iii) For $d = 4 - \varepsilon$ and ε very small,

$$\begin{aligned} U^*(\sigma^2) &= u_2^*\sigma^2 + \tfrac{1}{2}u_4^*\sigma^4 + \tfrac{1}{3}u_6^*\sigma^6 + \ldots \\ u_2^* &= -(\varepsilon\Lambda^2/2) + O(\varepsilon^2), \\ u_4^* &= (8\pi^2\varepsilon/n)\Lambda^\varepsilon + O(\varepsilon^2), \\ u_{2(m+1)}^* &= \varepsilon\Lambda^2[-(8\pi^2\varepsilon/n)\Lambda^{-2+\varepsilon}]^m (m-1)^{-1} + O(\varepsilon^{m+2}), \qquad m > 1. \end{aligned} \tag{2.38}$$

For $2 < d < 4$, the exponent $y_1 = d - 2$ is positive and all the rest of y_m in (2.33) are negative.

It is instructive to evaluate the scaling fields for the Ginzburg–Landau form (1.11). We have

$$t = r_0 + u\sigma^2 \tag{2.39}$$

which means $\sigma^2 = -r_0/u + t/u$. Therefore, (2.25) and (2.32) tell us that

$$\begin{aligned} g_1 &= -r_0/u - b_1^* \\ &= -(r_0 + u\bar{n}_c)/u \\ &= -(r_0 - r_{0c})/u, \end{aligned} \tag{2.40}$$

where r_{0c} is defined above (1.22), and

$$g_2 = 1/u - b_2^*, \tag{2.41}$$

$$g_m = -b_m^*, \qquad m > 2. \tag{2.42}$$

We shall always use the notation

$$y_1 = 1/\nu. \tag{2.43}$$

The scaling fields g_m given by (2.32) are sufficient to represent all $\mathscr{H}$ of the form (2.9). When additional terms are included in $\mathscr{H}$, we need to extend the parameter space and there are more scaling fields in the extended parameter space.

The simplest extension is to include a term

$$-\int d^d x\, h \cdot \sigma(x) \tag{2.44}$$

in $\mathscr{H}$, where h is a new scaling field, which can either be a constant or a slowly varying function of x (i.e., having Fourier components of very small wavevectors). The transformation of h under R_2 is simply

$$h' = h2^{y_\sigma},$$

$$y_\sigma = \tfrac{1}{2}(d - \eta) + 1. \tag{2.45}$$

This is a result of the replacement $\sigma \to 2^{1-(d+\eta)/2}\sigma$ and $d^d x \to 2^d\, d^d x$ in the definition of R_2. The part of R_2 (Step (A)) involving integrations does not affect h since (2.44) contains only small wavevector components of σ.

The quantity h transforms like a spin vector under spin rotations. There are other scaling fields with other transformation properties under spin rotations or rotations in the coordinate space. If we consider these general scaling fields as small, i.e., as describing perturbations to the form (2.9), then they appear as additional terms to the $\mathscr{H}$ of (2.9) in a form similar to (2.44)

$$\sum_i \int d^d x\, g_i D_i \tag{2.46}$$

where D_i are functions of $\sigma(x)$ and its derivatives. They may also depend on the scaling fields g_m of (2.32). It is not trivial to determine the explicit form of D_i. For the case of $n \to \infty$, those few D_i with largest y_i have been worked out. We shall put off the interpretation of these y_i and D_i till the next section. Here we only note that the importance of a scaling field g_i depends crucially on the sign of y_i.

If $y_i > 0$, $g_i(R_s\mu) = s^{y_i}g_i(\mu)$ will go to ∞ as $s \to \infty$. If $y_i < 0$, $g_i(R_s\mu)$ will vanish as $s \to \infty$. If $y_i = 0$, $g_i(R_s\mu)$ becomes independent of s. In this last case, we say that there is a range of fixed points labelled by different values of g_i with $y_i = 0$. We shall fix our attention to a particular set of values of those g_i's with $y_i = 0$.

The critical surface is defined as the subspace of the parameter space where $g_i(\mu) = 0$ if $y_i > 0$. In other words, $\lim_{s\to\infty} R_s\mu = \mu^*$, the fixed point if μ is on the critical surface.

In view of Table I, the critical surface of our fixed point is defined by

$$g_\sigma \equiv h = 0,$$

$$g_\tau = 0,$$

$$g_1 \propto (r_0 - r_{0c}) = 0,$$

$$g_c = 0. \tag{2.47}$$

This means no magnetic field, no quadrupole field, $r_0 = r_{0c}$, and no cubic perturbation. The last restriction is often stated as "the fixed point is unstable against cubic perturbation". Cubic perturbation means a term which makes the n reference axes in spin space preferred axes for magnetization, and thereby reduces the isotropy in spin space to cubic symmetry. Further effects of cubic perturbation have been studied by Wallace (1973). See also Aharony's article, Chapter 6, in this volume.

The system is at its critical point when μ is on the critical surface. It is near the critical point if the scaling fields with $y_i > 0$, i.e., those listed in (2.47), are very small.

D. Transformation of the free energy

Define the free energy per unit volume associated with $\mathscr{H}[\sigma;\mu]$ by

$$\exp\left[-F(\mu)L^d\right] = \int \delta\sigma \exp\left(-\mathscr{H}[\sigma;\mu]\right) \tag{2.48}$$

and that associated with $R_2\mu$ by

$$\exp\left[-F(R_2\mu)(L/2)^d\right] = \int \delta\sigma' \exp\left(-\mathscr{H}[\sigma';R_2\mu]\right) \tag{2.49}$$

The relationship between $F(\mu)$ and $F(R_2\mu)$ is easily established through (2.3). We integrate both sides of (2.3) to obtain

$$\begin{aligned}
&\exp\left[-F(R_2\mu)(L/2)^d - F_2^>(\mu)L^d\right] \\
&= \int \delta(2^{-1+(d+\eta)/2}\zeta)\,\delta\tilde{\sigma} \exp\left(-\mathscr{H}[\zeta + \tilde{\sigma};\mu]\right) \\
&= \int \delta\sigma \exp\left(-\mathscr{H}[\sigma;\mu]\right) \prod_{k<\Lambda/2} 2^{1-(d+\eta)/2}
\end{aligned} \tag{2.50}$$

Therefore,

$$F(\mu) = 2^{-d}F(R_2\mu) + F_2^>(\mu) + B_2$$

where B_2 comes from the last product of 2's, and is independent of μ. More generally, we have the transformation law for the free energy:

$$F(\mu) = s^{-d}F(R_s\mu) + F_s^>(\mu) + B_s. \tag{2.51}$$

Since only derivatives of F are of interest, B_s is not important. The quantity $F_s^>(\mu)$ is the free energy contributed solely from fluctuations of large wavevectors (larger than Λ/s). Since we expect that only small wavevector fluctuations contribute to singular behaviour, $F_s^>(\mu)$ is expected to be non-singular for *a fixed s*.

If μ is slightly off the critical surface by satisfying (2.47) except for a non-zero but smaller $r_0 - r_{0c} \propto g_1$, we can set $s = \xi = |g_1|^{-\nu}$. The $g_1(R_\xi\mu) = \pm 1$ (i.e., sign of $r_0 - r_{0c}$) and all other $g_i(R_\xi\mu) \propto \xi^{y_i} \to 0$. Thus (2.51) becomes

$$F(\mu) \approx \xi^{-d} f_\pm + F_\xi^>(\mu) + B_\xi \tag{2.52}$$

where

$$f_\pm = F(g_1 = \pm 1, \text{other } g_i = 0). \tag{2.53}$$

Even though $F_s^>$, B_s are non-singular functions of $r_0 - r_{0c}$ at a fixed s, setting $s = \xi$ makes them singular. In fact, $B_\xi \propto \xi^{-d} \ln \xi$. One is usually interested in the derivative of F. We can differentiate (2.51) with respect to r_0 first to remove B_s. For example, the specific heat is

$$C(\mu) = -\frac{\partial^2 F}{\partial r_0^2} = s^{-d+2/\nu} C(R_s\mu) - \frac{\partial^2 F_s^>}{\partial r_0^2} \tag{2.54}$$

The last term in the $n \to \infty$ limit turns out to be proportional to s^{d-4} for large s. (See Ma, 1974a, Section VIII.) On setting $s = \xi$, we have

$$\left(\frac{\partial^2 F_s^>}{\partial r_0^2}\right)_{s=\xi} \propto \xi^{d-4} \propto |r_0 - r_{0c}|^{(4-d)/(d-2)}, \tag{2.55}$$

$$C(\mu) = |r_0 - r_{0c}|^{(2-d\nu)} C_\pm + \text{const.}\, |r_0 - r_{0c}|^{(4-d)/(d-2)}, \tag{2.56}$$

where $C_\pm$ are constants. Since $2 - d\nu = -\alpha$ is just $(4 - d)/(d - 2)$ here, we see that the two terms in (2.55) are equally singular. For small n this situation will remain. It shows that great care is needed in using RG to derive critical behaviour of physical quantities. Often certain "less singular" terms turn out to be just as important as singular terms.

III. Scaling Variables, Scaling Dimensions and Correlation Functions

A. Transformation of random variables, scaling variables

The probability distribution $P \propto \exp(-\mathscr{H}(g))$ is transformed under R_s into another probability distribution $P' \propto \exp(-\mathscr{H}(R_s g))$. Usually one defines the transformation of random variables by the criterion that any average involving transformed variables over P' is the same as the corresponding average involving the original variable over P. For example

$$\sigma_k \to (\sigma_k)' = s^{1-\eta/2} \sigma_{sk} \tag{3.1}$$

according to (2.6). Here we define the transformation of a variable A to A' under R_s by the following criterion:

$A \to A'$ if A' satisifes

$$\langle A\sigma_{k_1}\ldots\sigma_{k_m}\rangle_P - \langle A\rangle_P\langle\sigma_{k_1}\ldots\sigma_{k_m}\rangle_P$$
$$= [\langle A'\sigma_{sk_1}\ldots\sigma_{sk_m}\rangle_{P'} - \langle A'\rangle_{P'}\langle\sigma_{sk_1}\ldots\sigma_{sk_m}\rangle_{P'}]s^{m(1-\eta/2)} \tag{3.2}$$

for any $k_1, \ldots, k_m < \Lambda/s$.

Generally, variables of interest are functions of σ and also depend on the set of parameters μ. A variable $\mathscr{D}(\mu)$ which follows the transformation law

$$\mathscr{D}(\mu) \to (\mathscr{D}(\mu))' = s^y\mathscr{D}(\mu'), \tag{3.3}$$

where $\mu' = R_s\mu$, will be called a *scaling variable* with exponent y, or *scaling dimension* $-y$.

We can obtain a set of scaling variables by taking advantage of the simple transformation properties of the scaling fields g_i. Let us use g instead of μ as parameters and define

$$\mathscr{D}_i(g)\exp[-\mathscr{H}(g)] = -\frac{\partial}{\partial g_i}\exp[-\mathscr{H}(g)],$$

$$\mathscr{D}_i(g) = \partial\mathscr{H}(g)/\partial g_i. \tag{3.4}$$

Differentiating both sides of (2.6) with respect to g_i, we get, on the left hand side

$$-\frac{\partial}{\partial g_i}\left[\int\delta\sigma\, e^{-\mathscr{H}}\,\sigma_{k_1}\ldots\sigma_{k_m}\Big/\int\delta\sigma\, e^{-\mathscr{H}}\right]$$
$$= \langle(\mathscr{D}_i(g) - \langle\mathscr{D}_i(g)\rangle_P)\,\sigma_{k_1}\ldots\sigma_{k_m}\rangle_P \tag{3.5}$$

where the $-\langle\mathscr{D}_i(g)\rangle_P$ term comes from differentiating the denominator. On the right hand side we get

$$s^{y_i}\langle[\mathscr{D}_i(g') - \langle\mathscr{D}_i(g')\rangle_{P'}]\sigma_{sk_1}\ldots\sigma_{sk_m}\rangle_{P'}\, s^{\frac{1}{2}m(2-\eta)} \tag{3.6}$$

where $\mathscr{D}_i(g')$ is of course $\partial\mathscr{H}(g')/\partial g_i'$.

By the definitions (3.2) and (3.3), the equality of (3.5) and (3.6) implies that

$$\mathscr{D}_i(g) \to (\mathscr{D}_i(g))' = \mathscr{D}_i(g')s^{y_i} \tag{3.7}$$

and $\mathscr{D}_i$ is thus a scaling variable of scaling dimension $-y_i$.

Let us introduce a set of local variables $D_i(g, x)$ which satisfy

$$\mathscr{D}_i(g) = \int d^d x\, D_i(g, x). \tag{3.8}$$

If we write $\mathscr{H}(g)$ as

$$\mathscr{H}(g) = \int d^d x \, H(g, \sigma(x)) \tag{3.9}$$

then it is natural to identify

$$D_i(g, x) = \partial H / \partial g_i. \tag{3.10}$$

From the transformation rule (3.7) for $\mathscr{D}_i$, we expect that

$$D_i(g, x) \to s^{-d+y_i} D_i(g', x/s) \tag{3.11}$$

i.e., D_i is a scaling variable of scaling dimension $d - y_i$. However, (3.11) implies (3.7), but not vice versa. In general D_i needs to include further appropriate gradient terms in addition to $\partial H/\partial g_i$ in order to achieve (3.11).

The scaling variables associated with the scaling fields g_m of (2.32) are easily obtained We have

$$H = \tfrac{1}{2}((\nabla\sigma)^2 + U(\sigma^2)). \tag{3.12}$$

Therefore

$$D_m = \partial U / \partial g_m.$$

To calculate $D_m = \partial U/\partial g_m$, it is convenient to introduce the Legendre transform of U

$$\Omega(t) = U - \sigma^2 t \tag{3.13}$$

as a function of $t = \partial U/\partial \sigma^2$. We then have

$$\sigma^2 = -\partial\Omega/\partial t = \sum_{m=1}^{\infty} b_m t^{m-1}, \tag{3.14}$$

where $b_m = g_m + b_m^*$ as given by (2.25) and (2.32). It follows then

$$\Omega = \sum_{m=1}^{\infty} b_m t^m / m. \tag{3.15}$$

Since $(\partial U/\partial g_m)_{\sigma^2} = (\partial\Omega/\partial g_m)_t$, we have the very simple result

$$D_m = t^m/m, \quad m = 1, 2, \ldots. \tag{3.16}$$

The local scaling variables associated with some other scaling fields are those D_i listed in Table I. Note that we can obtain more scaling variables by applying the gradient operator ∇ to any D_i any number of times. The scaling dimension increases by one every time ∇ is applied. The set of all local scaling variables, including those obtained by applying ∇, and including

$D_0 = \delta(x - x_0)$ (which is a constant, i.e., a random variable without fluctuation) form a basis set. Any random variable can be expanded as a linear combination of scaling variables. This is entirely analogous to the expansion of any operator in quantum mechanics as a linear combination of tensor operators of a certain symmetry group. Note that only a scaling variable, not any variable, has a definite scaling dimension. Any variable is a linear combination of terms each of which has a definite scaling dimension and the critical behaviour of that variable depends mostly on the terms with the lowest scaling dimensions, as will be evident later.

B. Product of scaling variables

One might expect that $\mathscr{D}_i\mathscr{D}_j$ and $\partial^2\mathscr{H}/\partial g_i g_j$ are scaling variables of dimension $-y_i - y_j$, but in general neither of them is. The combination, called the "scaling product" of $\mathscr{D}_i$ and $\mathscr{D}_j$

$$\{\mathscr{D}_i\mathscr{D}_j\} \equiv \mathscr{D}_i\mathscr{D}_j - \partial^2\mathscr{H}/\partial g_i \partial g_j \tag{3.17}$$

is a scaling variable of dimension $-y_i - y_j$. This is easily verified by applying $\partial^2/\partial g_i \partial g_j$ to both sides of (2.6) and using the fact that (3.17) is simply the statement

$$\mathrm{e}^{-\mathscr{H}}\{\mathscr{D}_i\mathscr{D}_j\} = \frac{\partial^2}{\partial g_i \partial g_j}\mathrm{e}^{-\mathscr{H}}, \tag{3.18}$$

where $D_{ij} = \partial^2 H/\partial g_i \partial g_j$.

We can extend (3.17) to define the scaling product of the local scaling variables D_i and D_j

$$\{D_i(x)D_j(y)\} = D_i(x)D_j(y) - D_{ij}(x)\delta(x - y). \tag{3.19}$$

The dimension of $\{D_iD_j\}$ is $2d - y_i - y_j$. One can easily generalize (3.19) to define scaling products of more than two scaling variables.

C. Correlation functions

We introduce the notation

$$\begin{aligned} G_i(p_1 \ldots p_m; g) = \int & \mathrm{d}^d x_1 \ldots \mathrm{d}^d x_m \exp(-ip_1 \cdot x_1 \ldots - ip_m \cdot x) \\ & \times \langle D_i(0)\sigma(x_1)\sigma(x_2)\ldots\sigma(x_m)\rangle_P \\ = L^{md/2}&\langle D_i(0)\sigma_{p_1}\ldots\sigma_{p_m}\rangle_P \end{aligned} \tag{3.20}$$

$$\begin{aligned} \tilde{G}_{ij}(q, p_1 \ldots p_m; g) = & \int \mathrm{d}^d x_1 \ldots \mathrm{d}^d x_m \exp(-ip_1 \cdot x_1 - \ldots - ip_m \cdot x) \\ & \times \int \mathrm{d}^d y \langle\{D_i(0)D_j(y)\}\sigma(x_1)\ldots\sigma(x_m)\rangle \exp(-iq \cdot y) \end{aligned} \tag{3.21}$$

where q, $p_1 \ldots p_m$ are non-zero and small compared to Λ and no subset of them sums to zero. More complicated correlation functions can be defined in a similar way.

It follows from the transformation properties of the scaling variables and scaling products that

$$G_i(p_1 \ldots p_m; g) = s^{-d+y_i+m(d+2-\eta)/2}\, G_i(sp_1 \ldots sp_m; g'), \tag{3.22}$$

$$\tilde{G}_{ij}(q, p_1 \ldots p_m; g) = s^{-d+y_i+y_j+m(d+2-\eta)/2}\, \tilde{G}_{ij}(sq, sp_1 \ldots sp_m; g). \tag{3.23}$$

From these transformation properties, some critical behaviour of these correlation functions can be deduced. Suppose that μ is on the critical surface, i.e., all g_i with $y_i > 0$ vanish, then $g' \to 0$ as $s \to \infty$. Setting $s = k^{-1}$, we obtain, for example

$$G_i(k, -k; g) \propto k^{-y_i-2+\eta}, \tag{3.24}$$

$$\tilde{G}_{ij}(k, k, -k; g) \propto k^{-y_i-y_j-2+\eta} \tag{3.25}$$

for $k \to 0$.

If all g_i with $y_i > 0$ except one, g_1 with $y_1 = 1/\nu$, are zero, then, setting $s = \xi = g_1^{-\nu}$, we obtain for example

$$G_i(0, 0; g) \propto r^{-y_i/(2-\eta)-1}, \tag{3.26}$$

$$\tilde{G}_{ij}(0, 0, 0; g) \propto r^{-(y_i+y_j)/(2-\eta)-1}, \tag{3.27}$$

$$(\partial G_i(k, -k; g)/\partial k^2)_{k=0} \propto r^{-(y_i+\eta)/(2-\eta)-2}, \tag{3.28}$$

where $r = G^{-1}(0) \propto \xi^{-2+\eta}$ as defined before.

The correlation function of any given variable of physical interest can thus be expanded as a linear combination of correlation functions of scaling variables. In view of (3.22), (3.23) and the examples (3.24)–(3.28), we see that scaling variables with lower scaling dimensions are more important for small k, r.

In the above definition of transformation of random variables, the average value is always subtracted off. (See (3.2), (3.5) and (3.6).) We have been concerned with the fluctuations of various variables. The average values of $\mathscr{D}_i$ transform in a simple manner also. Since

$$\langle D_i(g) \rangle_P = -\partial F(g)/\partial g_i \tag{3.29}$$

where $F(g) = F(\mu)$ is the free energy defined by (2.48), and in view of (2.51), we have

$$\langle D_i(g) \rangle_P = s^{-d+y_i} \langle D_i(g') \rangle_{P'} - \frac{\partial F_s^>}{\partial g_i} \tag{3.30}$$

Apart from the last term the average of D_i transforms just like D_i. As remarked before, for a fixed s, $F_s^>$ is non-singular. However, if one sets $s = \xi \propto |r_0 - r_{0c}|^{-\nu}$, it becomes singular through ξ.

D. Discussion

Let us return to Table I and discuss the physical meaning of various D_i listed there. As we have mentioned earlier, D_τ is the quadrupole moment density and g_τ the quadrupole field. The variable $t = \partial H/\partial\sigma^2$ is the local potential seen by σ^2, and $\partial H/\partial\sigma^\alpha$ can be interpreted as the local magnetic field seen by σ. $D_c = \Sigma_\alpha(\sigma^\alpha)^4$ gives the cubic anisotropy. [Wallace (1973) has shown that when this term is non-zero, a first-order transition results.] $D_0 = \delta(x)$ is simply a constant. D_Δ and D_T give the stress tensor. $D_{\Delta\tau}$, $D_{\tau 1}$ can be viewed as stress tensors in spin space. $g_{\Delta\tau}D_{\Delta\tau}$ can be viewed as generated by a change of variable as we shall discuss shortly. D_Υ is simultaneously a stress in spin space and in coordinate space. The last, D_2, has no obvious physical interpretation. The associated scaling field g_2 is often called the first irrelevant variable. The exponent y_2 estimates how fast $R_s\mu$ approaches the fixed point when μ contains no anisotropic parameters. It should be emphasized that these explicit expressions are for $n \to \infty$. When n is not large, D_i may look very different and some of the physical interpretations given above will no longer be appropriate. For the physics of critical phenomena, it is more convenient to attach directly physical significance to the scaling variables and think of energy density, entropy density, stress and familiar quantities in terms of scaling variables.

$D_{\sigma 1}$ and $D_{\Delta\tau}$ in Table I and their associated scaling fields are examples of redundant variables and scaling fields (Wegner, 1974; see also Wilson and Kogut, 1974). They are redundant in the sense that the terms

$$\int g_{\sigma 1}^\alpha D_{\sigma 1\alpha}\, \mathrm{d}^d x, \int g_{\Delta\tau}^{\alpha\beta} D_{\Delta\tau\alpha\beta}\, \mathrm{d}^d x$$

in $\mathscr{H}$ can be generated by the "changes of variable":

$$\begin{aligned} \sigma^\alpha &\to \sigma^\alpha + g_{\sigma 1}^\alpha, \\ \sigma^\alpha &\to \sigma^\alpha + g_{\Delta\tau}^{\alpha\beta}\sigma_\beta, \qquad \beta \neq \alpha, \end{aligned} \tag{3.32}$$

respectively, because

$$D_{\sigma 1\alpha} = \partial H/\partial\sigma^\alpha = \delta\mathscr{H}/\delta\sigma^\alpha, \tag{3.33}$$

$$D_{\Delta\tau\alpha\beta} = (\delta\mathscr{H}/\delta\sigma^\alpha)\sigma^\beta. \tag{3.34}$$

Consequently, the free energy will not be a function of $g_{\sigma 1}$ or $g_{\Delta\tau}$. The correlation functions of these scaling variables will have trivial behaviours even though these behaviours are consistent with the general conclusions concerning correlation functions of scaling variables. The trivial behaviours are easily seen as follows.

$$\langle D_{\sigma 1 \alpha}(0)\sigma(x_1)\ldots\sigma(x_m)\rangle \propto \int \delta\sigma\left(-\frac{\delta}{\delta\sigma^\alpha(0)}\,\mathrm{e}^{-\mathscr{H}}\right)\sigma(x_1)\ldots\sigma(x_m) = 0 \qquad (3.35)$$

unless at least one of $x_1 \ldots x_m$ is 0, upon integrating $\sigma^\alpha(0)$ by parts. Note that

$$D_{\sigma 1 \alpha}(0)\,\mathrm{e}^{-\mathscr{H}} = -\frac{\delta}{\delta\sigma^\alpha(0)}\,\mathrm{e}^{-\mathscr{H}} \qquad (3.36)$$

in view of (3.33). Thus the correlation functions of $D_{\sigma 1}$ are short-ranged. Similar conclusions holds for $D_{\Delta\tau\alpha\beta}$.

This ends our discussion on RG and the related abstract concepts. We have shown how everything worked out explicitly. At the present stage of development of RG, such an explicit illustration is very valuable.

IV. Nonzero Magnetization and the Equation of State at $n \to \infty$

So far we have concentrated on RG and various transformation properties under RG. Transformation properties often cannot tell us some important details since they do not provide the explicit solution to the model. For example, there is no direct indication from RG that, below the critical point ($r_0 < r_{0c}$), there would be a non-zero magnetization $M = \langle\sigma\rangle$ even when $h \to 0$. In this section, we solve the Ginzburg–Landau model directly for the case $M \neq 0$, after a brief remark on the longitudinal and transverse susceptibilities.

A. Transverse and longitudinal susceptibilities

Suppose that M and h are along the 1 direction in the spin space. We define

$$\sigma^1(x) = M + \phi. \qquad (4.1)$$

Evidently $\langle\phi\rangle = 0$. Now there is a preferred direction, we need to define two correlation functions of spins:

$$G_1(k,\mu) = \langle|\phi_k|^2\rangle_P, \qquad (4.2)$$

$$\delta_{\alpha\beta}G_\perp(k,\mu) = \langle\sigma_k^\alpha\sigma_{-k}^\beta\rangle_P, \qquad \alpha, \beta \neq 1. \qquad (4.3)$$

If we add a small field δh to h, there will be a δM in addition to M as a result. If $\delta h \| h$, the longitudinal susceptibility $(\delta M/\delta h)_1$ is then given by

$$(\delta M/\delta h)_1 = G_1(0,\mu). \qquad (4.4)$$

If $\delta h \perp h$, the transverse susceptibility $(\delta M/\delta h)_\perp$ is given by

$$(\delta M/\delta h)_\perp = G_\perp(0, \mu). \tag{4.5}$$

We note that applying an infinitesimal δh perpendicular to h is the same as rotating h by an infinitesimal angle $\delta h/h$ in spin space. As a result, the original M must be rotated by the same angle, i.e.,

$$\delta h/h = \delta M/M, \tag{4.6}$$

which implies that $(\delta M/\delta h)_\perp = M/h$, i.e.,

$$G_\perp(0, \mu) = M/h. \tag{4.7}$$

This is a very important result. It fixes the small k limit of $G_\perp(k, \mu)$. In particular, if $h = 0$ and $M \neq 0$, we have

$$G_\perp^{-1}(0, \mu) = 0 \tag{4.8}$$

which applies for $r_0 < r_{0c}$, $h = 0$. The result (4.8) is a consequence of the invariance of $\mathscr{H}$ under rotations in spin space.

B. Solution below the critical point and the equation of state

Let us start with the Ginzburg–Landau model (1.11) with an additional term $-h_1\sigma^1$ and write $\sigma^1 = M + \phi$. We obtain

$$\begin{aligned}\mathscr{H} = \tfrac{1}{2}L^d(r_0 M^2 + \tfrac{1}{2}uM^4) - L^d hM \\ + \tfrac{1}{2}\int d^d x\, [(\nabla\sigma_\perp)^2 + (\nabla\phi)^2 + (r_0 + uM^2)\sigma_\perp^2 \\ + (r_0 + 3uM^2)\phi^2 \\ + 2uM\phi(\phi^2 + \sigma_\perp^2) \\ + \tfrac{1}{2}u(\sigma_\perp^2 + \phi^2)^2]\end{aligned} \tag{4.9}$$

where $\sigma_\perp^\alpha = \sigma^\alpha$ for $\alpha \neq 1$, and $\sigma_\perp' = 0$.

In the large-n limit, the effect of ϕ is overwhelmed by that of $\sigma_\perp$ (i.e., outnumbered $n - 1$ to 1). Thus, we can ignore ϕ altogether in (4.9), which then reduces to (1.11) with $r_0 + uM^2$ replacing r_0. Many results of Section I can be taken over without further calculation. In particular, from (1.18)–(1.20), we obtain, replacing r_0 by $r_0 + uM^2$,

$$G_\perp(k) = (r_\perp + k^2)^{-1} \tag{4.10}$$

$$r_\perp = r_0 + uM^2 + u\bar{n} = h/M \tag{4.11}$$

$$\bar{n} = nK_d \int_0^\Lambda dk\, k^{d-1}(r_\perp + k^2)^{-1}. \tag{4.12}$$

Here $r_\perp = h/M$ is required by (4.7). From (1.26), we obtain, for very small h/M, again replacing r_0 by $r_0 + uM^2$, r by h/M,

$$r_0 - r_{0c} + uM^2 = (h/M)^{d/2-1} unJ \bigg/ \left(\frac{d}{2} - 1\right). \tag{4.13}$$

This is the thermodynamic equation of state, i.e., an equation relating r_0, M and h, to the zeroth order in $1/n$.

When $h = 0$, (4.13) gives, for $r_0 < r_{0c}$,

$$M \propto (r_{0c} - r_0)^{\frac{1}{2}}$$

i.e.,

$$\beta = \tfrac{1}{2}, \tag{4.14}$$

for $r_0 = r_{0c}$, (4.13) gives

$$M \propto h^{(d-2)/(d+2)}$$

i.e.,

$$\delta = (d + 2)/(d - 2), \tag{4.15}$$

by the definitions of the exponents β and δ.

At $h = 0$, we have $r_\perp = 0$ for $r_0 < r_{0c}$. Thus, (4.10) tells us that $G_\perp = k^{-2}$, which is independent of r_0. This has the important consequence that $\langle\sigma_\perp^2\rangle$ is independent of r_0 and hence that the transverse spin fluctuations do not contribute to the specific heat for $r_0 < r_{0c}$, $h = 0$ and $n \to \infty$. More precisely,

$$\begin{aligned} C &= -\tfrac{1}{2}\frac{\partial}{\partial r_0}\langle\sigma^2\rangle \\ &= -\tfrac{1}{2}\frac{\partial}{\partial r_0}(M^2 + \langle\sigma_\perp^2\rangle + \langle\phi^2\rangle) \\ &= -\tfrac{1}{2}\frac{\partial}{\partial r_0}M^2(1 + O(1/n)) \\ &= \frac{1}{2u}(1 + O(1/n)), \quad r_0 < r_{0c}, \quad h = 0. \end{aligned} \tag{4.16}$$

Here $1/2u$ is the same as the constant in (1.33) for the case $r_0 > r_{0c}$. Does this mean $\alpha = O(1/n)$ (instead of $(d-4)/(d-2) + O(1/n)$), for $r_0 < r_{0c}$? Not really. It simply means that the $|r_0 - r_{0c}|^{-\alpha}$ term has a coefficient of $O(1/n)$ instead of $O(1)$ for $r_0 < r_{0c}$. To calculate α to $O(1)$ one has to calculate uC to $O(1/n)$.

The lesson we just learned is that studying critical behaviour by expanding a physical quantity in powers of $1/n$ requires a certain amount of luck. If the coefficient of the singular term of interest happens to be small for large n, one may easily miss it, and mistake some other term with a large coefficient.

For the $O(1/n)$ correction to the equation of state, see Brézin and Wallace (1973).

V. Calculation of Exponents to $O(1/n)$

In the preceding sections we have illustrated various concepts by explicit calculations to $O(1)$, i.e., for the $n \to \infty$ limit. The $O(1/n)$ corrections are desirable. Such corrections to the fixed point, scaling fields and scaling variables have not been available although there should be no difficulty besides hard labour in calculating them. The $O(1/n)$ corrections to exponents are much easier to calculate. For example, RG analysis tells us that

$$G(k, \mu) \propto k^{-2+\eta}(1 + O(k^{-y_2})) \tag{5.1}$$

for $r_0 = r_{0c}$. We also know that for $n \to \infty$, $\eta = 0$ and $y_2 = d - 4$. Suppose that $\eta = O(1/n)$, $y_2 = d - 4 + O(1/n)$, then, we expand k^η in powers of η to obtain

$$G^{-1}(k)k^{-2} \propto 1 - \eta \ln k + \tfrac{1}{2}\eta^2 \ln^2 k + \ldots + O(k^{4-d}, n^{-1}k^{4-d} \ln k, \ldots) \tag{5.2}$$

By calculating $G^{-1}k^{-2}$ to $O(1/n)$, using any method, we can identify η to $O(1/n)$ by inspecting the coefficient of the $\ln k$ term.

More generally, $O(1/n)$ corrections to any exponent can be obtained by looking at a logarithmic term in the $O(1/n)$ correction to the correlation function of the scaling variable with that exponent. Note that we only need to know the scaling variable to $O(1)$. This is analogous to perturbation calculation of energy levels in quantum mechanics. The first-order correction in energy can be obtained from zeroth order wavefunctions. When the zeroth order energy is degenerate, however, the calculation is slightly more complicated and one has to diagonalize the perturbing Hamiltonian in the space of degenerate zeroth order wavefunctions. The analogue here is that when two or more scaling variables have the same exponents to $O(1)$, one must construct appropriate linear combinations of the scaling variables in order to calculate exponents to $O(1/n)$.

We want to emphasize that extracting exponents by looking at logarithms is a very dangerous procedure, because a logarithmic term can appear for many reasons other than as a result of the change of an exponent. It is crucial that the sources of the logarithms are correctly identified. When the structure of RG is known, the identification becomes easier.

We shall illustrate the technique of extracting $O(1/n)$ corrections to exponents with a brief review of perturbation theory followed by a few examples.

The very interesting Abe–Hikami (1973a, 1974a) anomaly and another case of degeneracy will be included.

A. Perturbation theory

For the purpose of extracting exponents, one can choose any model, the simpler the better. We shall again use the Ginzburg–Landau model (1.11) with $u = O(1/n)$. Let us restrict discussion to $r_0 \geqslant r_{0c}$, $h = 0$.

Let the quadratic terms be $\mathscr{H}_0$ and the quartic terms be $\mathscr{H}_1$. We have

$$\mathscr{H}_0 = \tfrac{1}{2} \sum_{\alpha, k} |\sigma_k^\alpha|^2 (r_0 + k^2), \tag{5.3a}$$

$$\mathscr{H}_1 = \tfrac{1}{4} \sum_k \Big(\sum_{p,\alpha} \sigma_p^\alpha \sigma_{k-p}^\alpha\Big) u \Big(\sum_{q,\beta} \sigma_q^\beta \sigma_{k-q}^\beta\Big), \tag{5.3b}$$

$$\exp(-\mathscr{H}) = \exp(-\mathscr{H}_0 - \mathscr{H}_1),$$

and any average $\langle A \rangle$ over $P \propto \exp(-\mathscr{H})$ becomes

$$\langle A \rangle = \sum_{m=0}^{\infty} \frac{(-)^m}{m!} \langle \mathscr{H}_1^m A \rangle_0 \Big/ \sum_{m=0}^{\infty} \frac{(-)^m}{m!} \langle \mathscr{H}_1^m \rangle_0 \tag{5.4}$$

where $\langle \ldots \rangle_0$ means averaging over the Gaussian probability distribution $\exp(-\mathscr{H}_0)$. In general A, as well as $\mathscr{H}_1$, are products of σ_k's. The Gaussian average of a product of σ_k's is a product of pairwise averages. Each pairwise average is, according to (5.3a),

$$\langle \sigma_k^\alpha \sigma_{k'}^\beta \rangle_0 = \delta_{\alpha\beta} \delta_{-k,k'} G_0(k),$$
$$G_0(k) = (r_0 + k^2)^{-1}. \tag{5.5}$$

Terms in the expansion can then be represented by graphs. A pairwise average is represented by a solid line and factor u is represented by a dashed line joining two sets of σ's as (5.3b) shows. Note that the summations over components α and β are done separately. In our graphs a continuous string of solid lines have the same spin index. For a closed loop of solid lines, the summation over components gives a factor n. For large n, graphs with more loops are more important.

Let us define r and $\Sigma(k)$ by

$$r = G^{-1}(0),$$
$$G^{-1}(k) = r_0 + k^2 + \Sigma(k) = r + k^2 + \Sigma(k) - \Sigma(0) \tag{5.6}$$

It is more convenient to modify G_0 of (5.5) to include $\Sigma(0)$ and redefine

$$G_0^{-1} = r_0 + \Sigma(0) + k^2 = r + k^2 \tag{5.7}$$

and remember to subtract $\Sigma(0)$ from $\Sigma(k)$ in calculations. An advantage of doing this is that r is small for small $r_0 - r_{0c}$, while r_0 is not. For $r_0 - r_{0c} = 0$, we can set $r = 0$.

B. Properties of Π(k) and the dressed interaction

The expansion parameter is $u = O(1/n)$. However, as we mentioned earlier, a closed loop of solid lines gives rise to a factor n after summing over spin components. Thus, whenever we include a u, we must include all the graphs in Fig. 2 because each "bubble" is proportional to n and compensates the

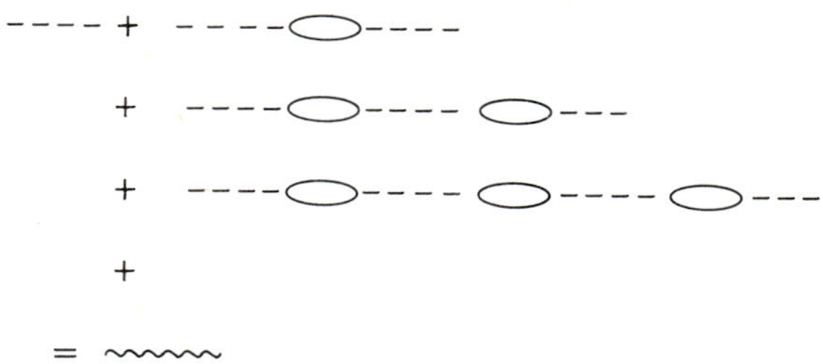

FIG. 2. The dressed interaction. See (5.10)–(5.15).

$1/n$ in the additional u needed to join the bubble to another bubble. The sum of these graphs will be denoted by a wavy line. Note that there are four powers of spins in $\mathscr{H}_1$ as shown in (5.3b). In averaging, one has to count all possibilities in pairing them up with spins in other $\mathscr{H}_1$'s or among themselves. The result of counting for a wavy line is

$$2[-u + (-u)^2 n\Pi(k) + (-u)^2(n\Pi(k))^2 + \ldots] = -2u(1 + un\Pi(k))^{-1}, \tag{5.8}$$

where Π, the bubble,

$$\Pi(k) \equiv \Pi(r, k^2) = (2\pi)^{-d} \int \mathrm{d}^d p \, G_0(p) \, G_0(p + k) \tag{5.9}$$

TABLE II. Properties of Π

$\Pi(r, k^2)$	$= \Pi(r/k^2, 1)k^{d-4}$
$\Pi(r, 1)$	$= \Pi(0, 1)(1 + 4r)^{(d-3)/2} + 2J(1 - \frac{1}{2}d)^{-1} r^{d/2-1} F(1, \frac{1}{2}, \frac{1}{2}, d; -4r)$
	$= \Pi(0, 1)[1 - 2(3 - d)r + O(r^2)] + 2J(1 - \frac{1}{2}d)^{-1} r^{d/2-1}[1 - (4/d)r + O(r^2)]$
$\Pi(0, 1)$	$= JB(\frac{1}{2}d - 1, \frac{1}{2}d - 1)$
$\Pi(r, 0)$	$= Jr^{d/2-2}$
J	$= \frac{1}{2}K_d \pi(\frac{1}{2}d - 1) \csc \pi(\frac{1}{2}d - 1)$
S_d	$= \frac{1}{2}K_d \Pi(0, 1)^{-1}$

is a function of k^2 and r. Table II gives some data concerning Π calculated with the cutoff Λ taken to be much larger than $\sqrt{r}$ and k. We shall derive them shortly. Clearly, $\Pi(k)$ is very large for small r and k. The quantity u is proportional to Λ^{4-d}. Thus $un\Pi \propto (\Lambda/k)^{4-d}$ which is large compared to 1. Therefore we can set (5.8) to be

$$2u/(1 + un\Pi(k)) = 2(n\Pi(k))^{-1}. \tag{5.10}$$

Thus a wavy line is independent of u.

We note that

$$\Pi(r, k^2) = (2\pi)^{-d} \int d^dp(r + p^2)^{-1} (r + (p + k)^2)^{-1}$$
$$= \Pi(r/k^2, 1)k^{d-4}. \tag{5.11}$$

It is sufficient to calculate $\Pi(r, 1)$. Use the identity

$$(AB)^{-1} = \int_0^1 d\alpha[(1 - \alpha)A + \alpha B]^{-2}. \tag{5.12}$$

Substituting $r + p^2$ for A and $r + (p + k)^2$ for B in (5.11), change variable from p to $p' = p + \alpha k$ and the p' integration is then available in tables. One finds

$$\Pi(r, 1) = J \int_0^1 d\alpha[r + \alpha(1 - \alpha)]^{d/2-2} \tag{5.13}$$

which reduces to what is in Table II. F and B are the hypergeometric function and the beta function, respectively. We remind the reader that

$$B(x, y) = \int_0^1 d\alpha\, \alpha^{x-1}(1 - \alpha)^{y-1} = \frac{\Gamma(x)\,\Gamma(y)}{\Gamma(x + y)} \tag{5.14}$$

Evidently the wavy line as given by (5.10) can be viewed as a "dressed" interaction between σ^2's. The set of graphs summed in Fig. 2 are just the kind of graphs summed in getting the screened Coulomb interaction in the theory of electron gas or plasma. This justifies the name "screening approximation" for this set of graphs used by many authors.

In view of (5.11), the Fourier transform of (5.10) is

$$V(x - x') = (2/n)\,(2\pi)^{-d} \int d^dk\, k^{4-d}\Pi^{-1}(r/k^2, 1) \exp[ik\cdot(x - x')]$$
$$= R^{-4}f(\sqrt{r}R), \tag{5.15}$$

where R is the magnitude of $x - x'$ and f is some function. $V(x - x')$ can be interpreted as the dressed interaction in coordinate representation. For

$r = 0$, i.e., at $r_0 = r_{0c}$, this interaction falls off like R^{-4}. For $r \neq 0$, the characteristic length is $\sqrt{r}$, which is just the correlation length to zeroth order in $1/n$. Note that the original or "bare" interaction u is of zero range, i.e., proportional to $\delta(x - x')$ in coordinate representation. The dressed interaction actually has a long range ξ due to the fluctuation of spins.

In the following, we go through a few representative calculations of exponents to $O(1/n)$ to illustrate the basic technique and the ambiguities one encounters in calculating. These calculations are taken from Ma (1973a, 1974b), which contain more detail.

C. Evaluation of η and γ

$G(k)$ is the simplest correlation function. At $r_0 = r_{0c}$, i.e., $r = 0$, we have, from (5.6), (5.2),

$$G^{-1}(k) = k^2 + \Sigma(k) - \Sigma(0)$$
$$= k^2(1 - \eta \ln k + \ldots). \tag{5.16}$$

To get η to $O(1/n)$, all we need is to get the $k^2 \ln k$ term in $\Sigma(k) - \Sigma(0)$.

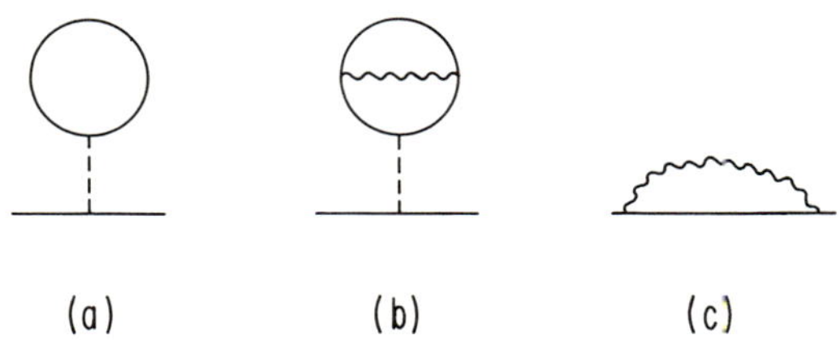

FIG. 3. (a) $\Sigma(k)$ to $O(1)$.
(b), (c) $O(1/n)$ terms of $\Sigma(k)$.

Figure 3 shows the $O(1)$ and $O(1/n)$ terms for $\Sigma(k)$. The terms (a) and (b) are independent of k and therefore do not contribute to $\Sigma(k) - \Sigma(0)$. The graph c gives

$$\Sigma(k) - \Sigma(0) = (2\pi)^{-d} \int d^d p \, 2/(n\Pi(k))((p + k)^{-2} - p^{-2})$$
$$= 2(n\Pi(1))^{-1} \int d^d p \, p^{4-d}((p + k)^{-2} - p^{-2}) \tag{5.17}$$

Write $(p + k)^{-2} = (p^2 + k^2 + 2\, pk \cos\theta)$ and

$$(2\pi)^{-d} \int d^d p = (K_{d-1}/2\pi) \int_0^\Lambda dp \, p^{d-1} \int_0^\pi d\theta \sin^{d-2\theta} \tag{5.18}$$

where K_{d-1} can be obtained from (1.21). These integrals are available in elementary tables. One indeed finds a $-\eta k^2 \ln k$ term with

$$\eta = 4(4/d - 1)S_d/n \tag{5.19}$$

where S_d is given by (1.39).

The exponent γ is defined by $1/r \propto (r_0 - r_{0c})^{-\gamma}$ for $(r_0 - r_{0c}) \to 0$. Since $r = r_0 + \Sigma(k = 0, r)$ and $0 = r_{0c} + \Sigma(0, 0)$, we have

$$r + \Sigma(0) - \Sigma(r) = r_0 - r_{0c} \propto r^{1/\gamma} \tag{5.20}$$

where $k = 0$ in Σ is understood. The zeroth order contribution to $\Sigma(r)$ is given by Fig. 3(a):

$$\begin{aligned}\Sigma_a(r) = \Sigma_a(0) &= nu(2\pi)^{-d} \int \mathrm{d}^d p((r + p^2)^{-1} - p^{-2}) \\ &= -nur^{d/2-1} J/(\tfrac{1}{2}d - 1)\end{aligned} \tag{5.21}$$

Therefore $\gamma = (d/2 - 1)^{-1}$ as we found in Section I. To find $O(1/n)$ correction to $1/\gamma$, we need to get the coefficient of $r^{d/2-1} \ln r$ in $\Sigma(r) - \Sigma(0)$. Figure 3(c) turns out not to contribute to such a term directly. Figure 3(b) can be viewed as a correction to Fig. 3(a):

$$\Sigma_b(r) = nu(2\pi)^{-d} \int \mathrm{d}^d q(r + q^2)^{-2}[\Sigma_c(r, q) - \Sigma_c(r, 0)], \tag{5.22}$$

$$\Sigma_c(r, q) = (2\pi)^{-d} \int \mathrm{d}^d p(2/n)\Pi(p)^{-1}[r + (q + p)^2]^{-1}. \tag{5.23}$$

Since $(r + q^2)^{-2} = -(\partial/\partial r)(r + q^2)^{-1}$, we rewrite (5.22) as

$$\begin{aligned}\Sigma_b(r) = nu(2\pi)^{-d} &\int \mathrm{d}^d p(2/n)\Pi(p)^{-1} \\ &\times \left(-\frac{1}{2}\frac{\partial \Pi(p)}{\partial r} - (r + p^2)^{-1}(2\pi)^{-d} \int \mathrm{d}^d q(r + q^2)^{-2}\right).\end{aligned} \tag{5.24}$$

The last q integral is easily done to give $Jr^{d/2-2}$. To get the logarithmic term, only the small-p region of the p integral of (5.24) is relevant. Thus, we arbitrarily cut off the p integral at $p = 1$ and write (5.24) as

$$\begin{aligned}\Sigma_b(r) &= uK_d \int_0^1 \mathrm{d}y\, y^{d/2-1}\Pi^{-1}(r, y)[-\tfrac{1}{2}\Pi'(r, y) - Jr^{d/2-2}(r + y)^{-1}] \\ &\equiv uK_d L r^{d/2-1},\end{aligned} \tag{5.25}$$

$$\Pi'(r, y) \equiv \partial\Pi/\partial r, \qquad y \equiv p^2,$$

and

$$L = \int_0^{1/r} \mathrm{d}z\, z^{d/2-1}\Pi^{-1}(1, z)\left[-\tfrac{1}{2}\Pi'(1, z) - J(1 + z)^{-1}\right], \tag{5.26}$$

$$\begin{aligned} -\partial L/\partial(\ln r) &= r^{-1}\Pi(r, 1)^{-1}\left[-\tfrac{1}{2}r^{-d/2+2}\Pi'(r, 1) - J/(1 + r)\right] \\ &= -3J\Pi(0, 1)^{-1} + (3 - d)r^{1-d/2} \\ &\quad + \text{higher order terms in } r. \end{aligned} \tag{5.27}$$

Thus, the logarithmic term in L is just $3J\Pi(0, 1)^{-1} \ln r$. Substituting L in (5.25) and combining the result with (5.21), we obtain

$$\begin{aligned} r + \Sigma(0) - \Sigma(r) &= nuJ(\tfrac{1}{2}d - 1)^{-1}r^{d/2-1}(1 - \lambda \ln r), \\ \lambda &= -(6/n)S_d(\tfrac{1}{2}d - 1) \end{aligned} \tag{5.28}$$

Note that $S_d = \frac{1}{2}K_d\Pi(0, 1)^{-1}$. It follows that

$$\gamma = (\tfrac{1}{2}d - 1)^{-1}(1 - (6/n)S_d) \tag{5.29}$$

D. Evaluation of α and the Abe–Hikami anomaly

We shall deduce α from the singular behaviour of

$$-\partial^2 F/\partial r_0^2 = C = \tfrac{1}{4}\int \mathrm{d}^d x(\langle\sigma^2(x)\sigma^2(0)\rangle - \langle\sigma^2(0)\rangle^2) \tag{5.30}$$

To zeroth order in $1/n$, the graphs for C are given in Fig. 4(a). We get

$$\begin{aligned} C_0 &= \frac{n}{2}\Pi(r, 0)/(1 + un\Pi(r, 0)) \\ &= \frac{1}{2u}(1 - r^{2-d/2}/(unJ) + \ldots) \end{aligned} \tag{5.31}$$

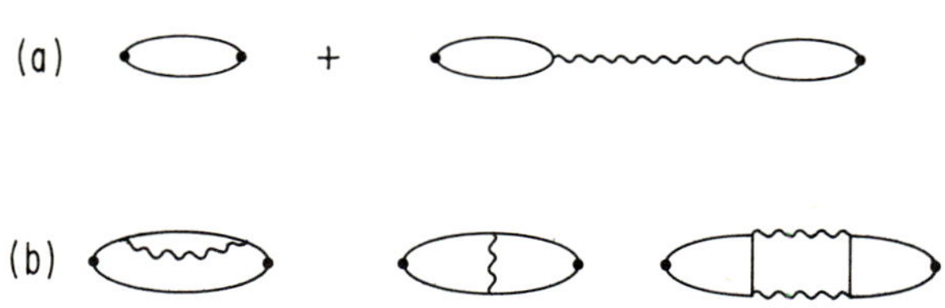

FIG. 4. (a) Leading order terms for C
(b) Correction.

and

$$\alpha = (d - 4)/(d - 2)$$

as was found in (1.32).

Note that, for the special values of d

$$d = 2 + 2/l, \qquad l = 2, 3, 4, \ldots \tag{5.32}$$

we have, according to (5.31),

$$-\alpha = l - 1. \tag{5.33}$$

Since $l - 1$ is a positive integer, we conclude that for d satisfying (5.32) and $n \to \infty$, the specific heat is non-singular.

The $O(1/n)$ term of α can be obtained by computing the next order correction C_1 to C. This correction can be written as

$$\begin{aligned} C_1 &= \Pi_1(1 + nu\Pi)^{-2} \\ &\approx (nu)^{-2}\Pi^{-2}\Pi_1, \end{aligned} \tag{5.34}$$

where Π_1 is shown in Fig. 4(b). Taking advantage of the fact that $(r + q^2)^{-2} = -(\partial/\partial r)(r + q^2)^{-1}$, we obtain

$$\begin{aligned} \Pi_1(r, 0) = \frac{n}{2}\frac{\partial}{\partial r}\Big\{(2\pi)^{-d}\int \mathrm{d}^d p(r + p^2)^{-2}\left[\Sigma_c(r, p^2) - \Sigma_c(r, 0)\right] \\ + \frac{n}{2}(2\pi)^{-d}\int \mathrm{d}^d p(r + p^2)^{-2}\frac{\partial}{\partial r}\Sigma_c(r, 0), \end{aligned} \tag{5.35}$$

where Σ_c is the graph given by Fig. 3(c) and by (5.23). The term in $\{\ldots\}$ is already calculated when we evaluated (5.22) and we can write down the answer after differentiating with respect to r:

$$3JS_d(\tfrac{1}{2}d - 1)(\ln r + \text{const.}) \tag{5.36}$$

The last term of (5.35) involves the evaluation of $\Sigma_c(r, 0)$, which is

$$\begin{aligned} \Sigma_c(r, 0) &= (2/n)(2\pi)^{-d}\int \mathrm{d}^d p(r + p^2)^{-1}\Pi(r, p^2)^{-1} \\ &= (2/n)\tfrac{1}{2}K_d rL, \end{aligned} \tag{5.37}$$

$$L \equiv \int_0^{1/r} \mathrm{d}z\, z^{d/2-1}(1 + z)^{-1}\Pi(1, z)^{-1}. \tag{5.38}$$

To extract $\ln r$ from L, we differentiate:

$$-\partial L/\partial(\ln r) = r^{-1}(1+r)^{-1}\Pi(r,1)^{-1}$$
$$= r^{-1}(1+r)^{-1}\Pi(0,1)^{-1}$$
$$\times\{1-2(3-d)r+O(r)-A_d r^{d/2-1}[1+O(r)]\}^{-1} \tag{5.39}$$

where we have used Table II for $\Pi(r, 1)$ of small r, and

$$A_d \equiv 2J(\tfrac{1}{2}d-1)^{-1}\Pi(0,1)^{-1} \tag{5.40}$$

Expanding (5.39), we obtain

$$-\partial L/\partial(\ln r) = \Pi(0,1)^{-1}r^{-1}\Big(1+(5-2d)r+O(r^2)$$
$$+A_d r^{d/2-1}[1+O(r)]+\sum_{j=2}^{\infty}A_d^j r^{(d/2-1)j}[1+O(r)]\Big) \tag{5.41}$$

Now we integrate (5.41) to get L, substitute it in (5.37) and obtain $\Sigma_c(r, 0)$. If we ignore the last sum over j in (5.41), we get

$$\Sigma_c(r,0) = \text{const.} + \text{const.}\ r^{d/2-1}$$
$$-(K_d/n)\Pi(0,1)^{-1}(5-2d)r\,(\ln r+\text{const.}). \tag{5.42}$$

Using this expression and (5.36) in (5.35), we get Π_1, and hence C_1 using (5.34). Combining with (5.31), we obtain

$$C_0+C_1 \propto \text{const.} - r^{2-d/2}(1+(2/n)S_d(8-\tfrac{7}{2}d)$$
$$\times(\ln r+\text{const.})) \propto \text{const.} - (r_0-r_{0c})^{-\alpha} \tag{5.43}$$

$$-\alpha = \gamma[2-d/2+(2/n)S_d(8-7d/2)]$$
$$= \left(\frac{4-d}{d-2}\right)\left(1+\frac{8(1-d)}{n(4-d)}S_d\right), \tag{5.44}$$

where (5.29) is used for γ. This value of α satisfies $\alpha = 2 - d\nu$.

Now let us look at the last sum over j in (5.41). If none of the terms in the sum is proportional to r, we should not get any contribution to $\ln r$ in L. We get a contribution to C_1 proportional to

$$r^{2-d/2}\sum_{j=2}^{\infty}\frac{A_d^j}{(\tfrac{1}{2}d-1)j-1}r^{(d/2-1)-1}[1+O(r)] \tag{5.45}$$

Since $r \propto (r_0 - r_{0c})^{\gamma}$, $\gamma = (\tfrac{1}{2}d-1)^{-1} + O(1/n)$, (5.45) is

$$\sum_{j=2}^{\infty}[(\tfrac{1}{2}d-1)j-1]^{-1}A_d^j(r_0-r_{0c})^{j-1}[1+O(r)] \tag{5.46}$$

which is a power series in $(r_0 - r_{0c})$ apart from the $O(r)$ term in the last factor. Thus the result (5.44) is not affected.

On the other hand, if a term $j = l \geqslant 2$ in the sum such that $(\frac{1}{2}d - 1)l - 1 = 0$, i.e., if d is one of $3, \frac{8}{3}, \frac{5}{2}, \frac{12}{5}, \ldots$ as given by (5.32), then this term will contribute a logarithm to L. As a result a term

$$C_A = -JS_d A_d^l(\ln r + \text{const.})r^{2-d/2}$$
$$\propto (r_0 - r_{0c})^{l-1}(\ln(r_0 - r_{0c}) + \text{const.}) \qquad (5.47)$$

This is the anomalous logarithmic term found by Abe and Hikami (1973a, 1974a). If this term is interpreted as the manifestation of an $O(1/n)$ correction to the exponent $-\alpha = l - 1$, then we have a contribution to $O(1/n)$ of α in addition to (5.44) for $d = 2 + 2/l$. This interpretation turns out to be incorrect. Abe and Hikami (1973) showed, by calculating $n^{-2}(\ln(r_0 - r_{0c}))^2$ and $n^{-3}(\ln(r_0 - r_{0c}))^3$ terms in C, that

$$C \propto [(1 + B_l)(r_0 - r_{0c})^{-\alpha} - B_l(r_0 - r_{0c})^{l-1}]$$
$$+ \ldots \qquad (5.48)$$

where the constant B_l is zero unless d is one of $2 + 2/l$, and α is given by (5.44). To the zeroth order in $1/n$, the B_l terms cancel. The anomalous logarithmic term thus reflects the discontinuity of the coefficients of the $(r_0 - r_{0c})^{-\alpha}$ term and the $(r_0 - r_{0c})^{l-1}$ term at $d = 2 + 2/l$. This discontinuity is not entirely surprising. At $d = 2 + 2/l$, the $(r_0 - r_{0c})^{-\alpha}$ and $(r_0 - r_{0c})^{l-1}$ have the same exponents for $n \to \infty$. This can be viewed as a case of degeneracy, which is absent for $d \neq 2 + 2/l$. When the $O(1/n)$ perturbation is turned on, the zeroth order amplitudes of the two powers must be readjusted for the degenerate case but not for the non-degenerate case.

E. Evaluation of $y_{\Delta\tau}$ and $y_{\tau 1}$

The exponents associated with the variables

$$\lambda_{\alpha\beta} = \nabla\sigma^\alpha \cdot \nabla\sigma^\beta - (1/n)(\nabla\sigma)^2\delta_{\alpha\beta} \qquad (5.49)$$

and

$$t\tau_{\alpha\beta} = t(\sigma^\alpha\sigma^\beta - (1/n)\sigma^2\delta_{\alpha\beta}) \qquad (5.50)$$

are both zero to the zeroth order in $1/n$. Thus, λ and $t\tau$ are scaling variables to that order. When $O(1/n)$ corrections to the exponents are considered, we have to consider linear combinations of λ and $t\tau$ owing to the degeneracy of their exponents to the zeroth order. Let $D = a\lambda + t\tau$, where a is a constant to be determined. Consider the correlation function

$$G_D \equiv aG_\lambda + G_{t\tau} \qquad (5.51)$$

where

$$G_\lambda(k,k') \equiv \int d^d x\, d^d x' \langle \lambda_{\alpha\beta}(0)\sigma^\alpha(x)\sigma^\beta(x')\rangle \exp(-ik\cdot x - ik'\cdot x')$$

$$G_{t\tau}(k,k') \equiv \int d^d x\, d^d x' \langle t(\sigma^2(0))\tau_{\alpha\beta}(0)\sigma^\alpha(x)\sigma^\beta(x')\rangle \times \exp(-ik\cdot x - ik'\cdot x'). \quad (5.52)$$

For the right choice of the coefficient a, we would get

$$r^2 G_D(0,0) \propto r^{1-y/(2-\eta)} \quad (5.53)$$

$$r^2\left[\frac{\partial G_D(k,-k)}{\partial k^2}\right]_{k=0} \propto r^{-(y+\eta)/(2-\eta)} \quad (5.54)$$

as (3.26) and (3.28) state for correlation functions of a scaling variable of exponent y.

There is no zeroth order contribution to $G_\lambda(0,0)$. The $O(1/n)$ contribution is shown in Fig. 5(a), which gives

$$r^2 G_\lambda(0,0) = -2(d-2)S_d(2/n)r\ln r \quad (5.55)$$

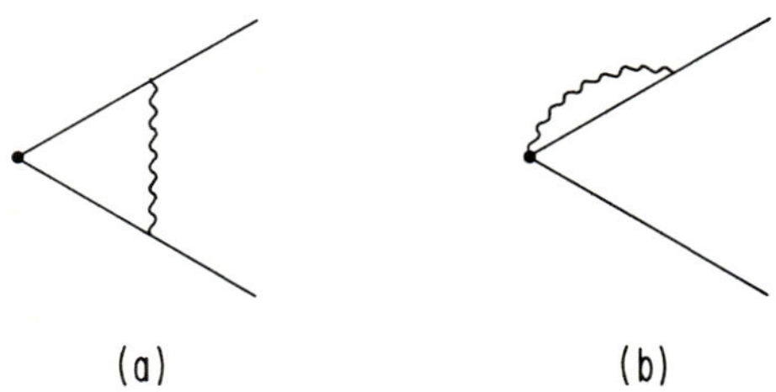

Fig. 5. (a) $O(1/n)$ contribution to G_λ, or G_τ.
(b) $O(1/n)$ term of $\langle \tau u(\sigma^2 - \langle\sigma^2\rangle)\sigma\sigma\rangle$.

The zeroth order term of $r^2\partial G_\lambda/\partial k^2$ is simply 1, and the first-order term is obtained from Fig. 5(a). We find

$$r^2[\partial G_\lambda(k,-k)/\partial k^2]_{k=0} = 1 + \tfrac{1}{2}\eta\ln r. \quad (5.56)$$

To calculate $G_{t\tau}$, we note that

$$t = r_0 + u\sigma^2$$
$$= r - \Sigma_c(0,r) + u(\sigma^2 - \langle\sigma^2\rangle) \quad (5.57)$$

since $r = r_0 + \Sigma(0, r) = r_0 + u\langle\sigma^2\rangle + \Sigma_c(0, r)$ to $O(1/n)$. The graphs in Fig. 3(a, b) give $u\langle\sigma^2\rangle$ and Σ_c is the contribution of Fig. 3(c). Thus, we have

$$G_{\tau\tau}(0, 0)r^2 = (r - \Sigma_c)\, G_\tau(0, 0)r^2 + 2\Sigma_c \tag{5.58}$$

where the last term $2\Sigma_c$ comes from the graph Fig. 5(b), which gives the $O(1/n)$ contribution of the $\tau u(\sigma^2 - \langle\sigma^2\rangle)$ term in $t\tau$ to $G_{\tau\tau}$. The zeroth order term in $G_\tau(k, -k)r^2$ is 1 and the $O(1/n)$ term is again given by Fig. 5(a). We find

$$G_{\tau\tau}(0, 0)r^2 = r[1 + 2(d - 2)(\ln r)S_d 2/n] \tag{5.59}$$

$$[r^2 \partial G_{\tau\tau}(k, -k)/\partial k^2]_{k=0} = -\eta \ln r \tag{5.60}$$

Now from (5.53) and (5.54) we have, to $O(1/n)$

$$G_D(0, 0)r^2 = r(1 - \tfrac{1}{2}y \ln r) = (G_{\tau\tau} + aG_\lambda)r^2 \tag{5.61}$$

$$\begin{aligned}[r^2 \partial G_D(k, -k)/\partial k^2]_{k=0} &= a(1 - \tfrac{1}{2}(y + \eta)\ln r)\\ &= ar^2 \partial G_\lambda/\partial k^2 + r^2 \partial G_{\tau\tau}/\partial k^2\end{aligned} \tag{5.62}$$

The factor a in (5.62) is a result of the factor a in (5.51) and the fact that $r^2 \partial G_\lambda/\partial k^2 = 1$ to the zeroth order. Substituting (5.55), (5.56), (5.59) and (5.60) in (5.61) and (5.62), we get the following coupled equations for y and a:

$$\begin{aligned}-\tfrac{1}{2}y &= (1 - a)2(d - 2)2S_d/n,\\ -\tfrac{1}{2}(y + \eta) &= \eta\left(\frac{1}{2} - \frac{1}{a}\right).\end{aligned} \tag{5.63}$$

There are two sets of solutions:

(i) $a = 1$, $y = y_{\Delta\tau} = 0$,

$$D_{\Delta\tau\alpha\beta} = \lambda_{\alpha\beta} + t\tau_{\alpha\beta}, \tag{5.64}$$

(ii) $a = (d - 4)d^{-1}(d - 2)^{-1}$,

$$\begin{aligned}y = y_{\tau 1} &= -8(d - 3 + 4/d)S_d/n\\ (D_{\tau 1})_{\alpha\beta} &= \frac{d - 4}{d(d - 2)}\lambda_{\alpha\beta} + t\tau_{\alpha\beta}\end{aligned} \tag{5.65}$$

F. Evaluation of y_c the cubic exponent

Consider the correlation function

$$\begin{aligned}G_c(k_1, \ldots, k_4) = \int d^d x_1 \ldots d^d x_4 \exp(ik_1 \cdot x - \ldots - ik_4 \cdot x_4)\\ \times \langle(\sigma^\alpha(0))^4 \sigma^\alpha(x_1)\sigma^\alpha(x_2)\sigma^\alpha(x_3)\sigma^\alpha(x_4)\rangle\end{aligned} \tag{5.66}$$

To $O(1)$, the exponent $y_c = 4 - d$. One easily derives the critical behaviour

$$r^4 G_c(0, 0, 0, 0) \propto r^{-(y_c' + 2\eta)/(2-\eta)} \tag{5.67}$$

whete y_c' is the $O(1/n)$ correction to $4 - d$. The $O(1/n)$ term of G_c is shown in Fig. 6. One obtains

$$-(y_c' + 2\eta)/(2 - \eta) = (12/n)S_d \tag{5.68}$$

which gives

$$y_c = 4 - d - (16/n)(1 + 2/d)Sd. \tag{5.69}$$

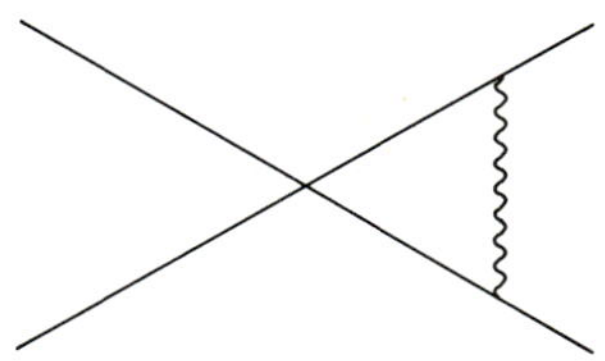

FIG. 6. $O(1/n)$ term of G_c.

G. Remarks

The above sampled calculations illustrate the basic technique and the fact that very often the extraction of exponents is not straightforward. The meaning of the logarithms is often easily misinterpreted. The example of $y_{\Delta\tau}$ and $y_{\tau 1}$ is particularly illustrative. If we had examined $\lambda_{\alpha\beta}$ of (5.49) alone, calculated G_λ and interpreted the coefficient of $\ln r$ as the $O(1/n)$ correction to an exponent, we should have made a bad mistake. Fortunately, the RG analysis tells us about $t\tau_{\alpha\beta}$. The extraction of exponents from logarithms must be guided by the RG analysis. The knowledge of scaling variables is often crucial, as this example shows.

The calculation of other exponents can be found in Ma (1973a, 1974b) and the work of Abe, Suzuki and coworkers cited in the list of references.

VI. Discussion

A. Other formulations

Much of the calculation in $1/n$ expansion has been done using a nearest-neighbour interaction model like (1.34) and (1.35), instead of a Ginzberg–Landau model. Most important is the work of Abe and coworkers, who solved their model systematically in powers of $1/n$. To our knowledge so far they have carried the calculation farther than other authors in $1/n$

expansion. Note in particular the work on the $O(n^{-2})$ term of η (Abe, 1973c), and on the discontinuity of specific heat amplitude (Abe and Hikami, 1973a, b). Similar techniques have been used by Suzuki and coworkers on long-range interactions, which will be mentioned later.

There are different ways to derive RG in the large n limit. Wegner and Houghton (1973) used a differential form of RG. One can also use a straightforward summation of leading graphs. [See Ma (1974a).]

We have not said much about the detailed form of correlation functions and the equation of state to $O(1/n)$. The reader is referred to the work of Aharony (1974), Abe and Hikami (1974), and Brézin and Wallace (1973).

Applications of $1/n$ expansion to quantum field theory have been made by Wilson (1973) and others.

B. Convergence and extrapolation

Information on higher-order terms in $1/n$ expansion has been lacking. There has been little systematic study on the convergence of the expansion. For d near 4, results of ε expansion give some hints. The appearance of powers of $(n + 8)^{-1}$ in the expression of exponents suggests that the radius convergence of $1/n$ expansion is small, i.e., $1/n \sim 1/8$. This of course implies that the $1/n$ expansion may not be useful for studying real systems, which usually have $n \lesssim 3$.

The convergence of the $1/n$ expansion for d near 2 is expected to be much worse. The theorem of Hohenberg (1967), and Mermin and Wagner (1966), states that for $n \geqslant 2$, there can be no long-range order at any finite temperature for short-range interacting systems of $d \leqslant 2$. On the other hand, for $d = 2$, $n = 1$ long-range order is allowed and the Ising model gives us an exactly solved example. This situation suggests that the $1/n$ expansion cannot be very useful for d near 2.

In spite of the bad convergence properties suggested above, various authors have made use of the $1/n$ expansion by making extrapolations. Suzuki (1973c) proposed an expansion in $1/(n + 8)$ and then to fit the coefficient by the known $1/n$ results. More recently, Bray (1974) studied the $1/n$ expansion including a $k^{2-\eta}$ term in addition to the k^2 term in $G_0^{-1}(k)$ and then calculated η self-consistently. Doniach (1973) was able to obtain various conclusions concerning phase transitions for $d = 2, n = 2$ based on plausible extrapolation from $1/n$ expansion. (The reader should be reminded that the Hohenberg–Mermin–Wagner theorems do not rule out phases which do not have a long-range order.)

For $d = 3$, the available $1/n$ expansion results for γ and η at $n = 3$ give

$$\gamma = 1{\cdot}19$$

$$\eta = 0{\cdot}090 - 0{\cdot}022 = 0{\cdot}068$$

where $-0{\cdot}022$ is the $O(n^{-2})$ term of η given by Abe (1973c). The values given by series expansion (as quoted by Suzuki, 1973c, Table 1) are $\gamma = 1{\cdot}375$ and $\eta = 0{\cdot}043$.

C. Long-range interaction

In this article we have studied only short-range interacting systems. There are various kinds of long-range interactions as consequences of dipolar interactions, non-rigid lattice, for example. They are discussed in Aharony's article, Chapter 6, in this volume. Here we mention some results on the isotropic long-range interaction between spins of the form

$$\sigma(x) \cdot \sigma(x')/|x - x'|^{d+\sigma}$$

with $\sigma < 2$. This kind of interaction gives rise to a term

$$\sum_k \sum_\alpha \sigma_k^\alpha \sigma_{-k}^\alpha k^\sigma$$

in $\mathscr{H}$. This term dominates over the $\sigma_k \sigma_{-k} k^2$ term for small k since $\sigma < 2$. The calculation of exponents can be done in the same way as described in Section V with $G_0^{-1} = r + k^\sigma$, and the $O(1/n)$ results have been reported by Fisher *et al.* (1972), by Ma (1973a) and by Suzuki (1972, 1973a, b). There are various ways of writing the results, which are complicated combinations of Γ functions. Let us summarize them as given by Suzuki (1973a):

$$\gamma = \frac{\sigma}{d - \sigma} + \gamma_1 \frac{1}{n},$$

$$\eta = 2 - \sigma,$$

$$\delta = (d + \sigma)/(d - \sigma),$$

$$\alpha = (d - 2\sigma)/(d - \sigma) + \alpha_1/n,$$

$$\beta = \tfrac{1}{2} + \beta_1/n,$$

$$\nu = (d - \sigma)^{-1} + \nu_1/n,$$

$$\gamma_1 = \frac{8\sigma\Gamma(\sigma/2)^2\Gamma(d - \sigma)}{(d - \sigma)^2\Gamma(d/2)\Gamma(\sigma - d/2)\Gamma((d - \sigma)/2)^2} \times \left[\frac{\Gamma(\sigma/2)\Gamma(d - \sigma)\Gamma(d/2 - \sigma)\Gamma((3\sigma - d)/2)}{\Gamma(\sigma)\Gamma(\sigma - d/2)\Gamma((d - \sigma)/2)\Gamma(d - 3\sigma/2)} - \frac{1}{2}\right]$$

$$\alpha_1 = -\gamma_1 d/\sigma$$

$$\beta_1 = \gamma_1(d - \sigma)/2\sigma$$

$$\nu_1 = \gamma_1/\sigma.$$

These results are for $d/2 < \sigma < 2$, and accurate to $O(1/n)$. For $\sigma < d/2$, i.e. $d > 2\sigma$, mean field theory applies. The exponents for $\sigma \to 2$ do not agree with those for short-range interactions. Sak (1973) studied the renormalization group for a system with this long-range interaction, and found a fixed point which is stable if $\sigma < 2 - \eta_s$ and unstable otherwise, where η_s is the η for short-range interacting systems. At $\sigma = 2 - \eta_s$, the exponents agree with the short-range exponents to $O(1/n)$.

D. Concluding remarks

So far there have been many indications that the $1/n$ expansions do not converge fast, as we have mentioned above. Furthermore, many interesting phenomena do not depend smoothly on n. In other words, their characteristics change abruptly at certain values of n. Sometimes, these special values of n are so large that $1/n$ expansion becomes useless for extrapolation to $n \lesssim 3$. An extreme example is the effect of a gauge field studied by Halperin, Lubensky and Ma (1974). There a new stable fixed point is found in $1/n$ expansion. But an ε-expansion calculation reveals that such a fixed point exists only for $n \sim 370$ or larger. Apparently, for certain phenomena, one must view the range $1 < n < \infty$ as a very large range in which many things can change in many ways as functions of n, instead of viewing $1 > 1/n > 0$ as a small range in which things change smoothly as a function of $1/n$.

There are of course advantages of $1/n$ expansion, in particular that there is no restriction on d. Some properties which are difficult to see in the ε expansion may be easily revealed by the $1/n$ expansion. An example is the Abe–Hikami anomaly discussed in Section V. How useful the $1/n$ expansion is in studying a certain problem depends ultimately on the special nature of that problem.

References

Abe, R. (1972). *Progr. Theor. Phys.* **48**, 1414.
Abe, R. (1973a). *Progr. Theor. Phys.* **49**, 113.
Abe, R. (1973b). *Progr. Theor. Phys.* **49**, 1074.
Abe, R. (1973c). *Progr. Theor. Phys.* **49**, 1877.
Abe, R. and Hikami, S. (1973a). *Phys. Lett.* **45A**, 11.
Abe, R. and Hikami, S. (1973b). *Progr. Theor. Phys.* **49**, 442.
Abe, R. and Hikami, S. (1974a). *Progr. Theor. Phys.* **51**, 1041.
Abe, R. and Hikami, S. (1974b). Preprint.
Aharony, A. (1973). *Phys. Lettrs* **A46**, 287.
Aharony, A. (1974). *Phys. Rev.* **B10**, 2834.
Berlin, T. H. and Kac, M. (1952). *Phys. Rev.* **86**, 821.
Bray, A. J. (1974). *Phys. Rev. Lett.* **32**, 1413.
Brézin, E. and Wallace, D. J. (1973). *Phys. Rev.* **B7**, 1967.
Doniach, S. (1973). *Phys. Rev. Lett.* **31**, 1450.
Ferrell, R. A. and Scalapino, D. J. (1972). *Phys. Rev. Lett.* **29**, 413.
Fisher, M. E., Ma, S. and Nickel, B. G. (1972). *Phys. Rev. Lett.* **29**, 917.

Halperin, B. I., Lubensky, T. C. and Ma, S. (1974). *Phys. Rev. Lett.* **32**, 292.
Hohenberg, P. C. (1967). *Phys. Rev.* **158**, 383.
Ma, S. (1972). *Phys. Rev. Lett.* **29**, 1311.
Ma, S. (1973a). *Phys. Rev.* **A7**, 2172.
Ma, S. (1973b). *Rev. Mod. Phys.* **45**, 589.
Ma, S. (1974a). *J. Math. Phys.* **15**, 1866.
Ma, S. (1974b). *Phys. Rev.* **A10**, 1818.
Mermin, N. D. and Wagner, H. (1966). *Phys. Rev. Lett.* **17**, 1133.
Sak, J. (1973). *Phys. Rev.* **B8**, 281.
Stanley, H. E. (1968). *Phys. Rev.* **176**, 718.
Suzuki, M. (1972). *Phys. Lettrs* **42A**, 5.
Suzuki, M. (1973a). *Progr. Theor. Phys.* **49**, 424.
Suzuki, M. (1973b). *Progr. Theor. Phys.* **49**, 1106.
Suzuki, M. (1973c). *In* "Proceedings of 1973 Conference on Renormalization Group in Critical Phenomena and Quantum Field Theory" (M. S. Green and J. D. Gunton, eds.)
Wallace, D. J. (1973). *J. Phys.* **C6**, 1390.
Wegner, F. J. (1972). *Phys. Rev.* **B5**, 4529,
Wegner, F. J. (1974). *J. Phys.* **C7**, 2098.
Wegner, F. J. and Houghton A. (1973). *Phys. Rev.* **A8**, 401.
Wilson, K. G. (1973). *Phys. Rev.* **D7**, 2911.
Wilson, K. G. and Kogut, J. (1974). *Phys. Report*, **12C**, 75.

5. The ε-Expansion for Exponents and the Equation of State in Isotropic Systems

D. J. Wallace

Department of Physics, The University, Southampton SO9 5NH, England

Introduction

This article describes one perturbative method, the ε-expansion, for attacking the problem of critical phenomena using field theory. The method was introduced by Wilson (1972) and provides a powerful tool for calculating phase transition behaviour since only standard Feynman graph expansion techniques for ϕ^4 interactions are required.

I have attempted to make the article self-contained and pedagogical particularly on questions of principle and method. However if it is read in isolation, the following remarks are important.

Firstly, I have given no general review of critical phenomenology. The reader who is completely unfamiliar with the concepts of critical exponents, scaling functions, universality etc. is referred to the established reviews on this subject, e.g. Stanley (1971), Kadanoff *et al.* (1967), Fisher (1967). Landau and Lifschitz (1958) also provides very useful background.

Secondly, I have assumed the reader has had some previous exposure to the standard Feynman graph expansion for a boson field. Suitable background references for this topic are, e.g. Bogoliubov and Shirkov (1959), Bjorken and Drell (1965). A general review of the path integral method in statistical mechanics is given by Feynman and Hibbs (1965).

Thirdly, I have attempted to avoid unnecessary duplication with other articles in this volume. As a result the role which the renormalization group itself plays in providing general understanding of phase transition behaviour is underweighted; see the articles by Brézin *et al.*, Wegner, and Jona-Lasinio and Di Castro. I have almost entirely omitted discussion of the behaviour of correlation functions at non-zero momentum; see Fisher and Jasnow in Vol 4 of this series. I have concentrated on short range isotropic interactions; for other interactions see the article by Aharony. In addition to these references, the earlier review article by Wilson and Kogut (1974) is very important.

The general outline of this article is as follows. In Section I the basic framework is set up for calculation of correlation functions and thermodynamic quantities above the critical temperature. The Hamiltonian which we consider involves an n-component boson field with short range ϕ^4 interactions which have $O(n)$ symmetry.

In Section II we see how the perturbation theory of Section I breaks down as the temperature is lowered to the critical temperature; if the dimension of space, d, is less than four, every order in perturbation theory becomes arbitrarily larger than a lower order because of infra-red divergences in integrals. If d is greater than four, there are no new infra-red divergences and the behaviour is that of mean field theory (see e.g. Landau and Lifschitz, 1958). Wilson's ε-expansion is explained; it gives systematic corrections to mean field theory results in powers of $\varepsilon = 4 - d$. The method is illustrated with the calculation of the critical exponents η and γ. To compare with physics in three dimensions one must set $\varepsilon = 1$. We review the convergence of the series in ε and the best results obtained so far.

In Section III we look at the effect of an external magnetic field (in magnetic language). The Feynman rules of Section I are appropriately modified and the equation of state of the model in its critical region (small external field and temperature close to the critical temperature) is calculated in the ε-expansion. Some interesting properties of the resulting equation of state are discussed.

In Section IV we look at the problem of classifying other interactions according to whether they do or do not change the phase transition behaviour of the model with only ϕ^4 interactions.

Finally, in a field which has grown so rapidly since 1971, I ask for the indulgence of authors to whose work I may have given insufficient credit and reference.

I. General Formalism

A. The n-component field model

The basic statistical-mechanical variable which we consider is a field $\phi_i(x)$ with n-components ($i = 1, 2 \ldots n$). For various values of n and the dimension of space d, the model we consider is intended to describe a wide variety of physical systems. For example one may hope to describe magnetic systems by interpreting $\phi_i(x)$ as a magnetisation density or structural phase transition in crystals by interpreting $\phi_i(x)$ as the mean (averaged over a small volume around x) of the displacement of the atom(s) responsible for the change in structure. For definiteness in this article we will always use the language of the ferromagnet.

A second basic ingredient of a model in statistical mechanics is the Hamiltonian. This is the functional of the field $\phi_i(x)$ which gives the energy of a system which is in a state of a given magnetization density $\phi_i(x)$. The starting point is a Hamiltonian of the form

$$\mathscr{H}(\phi)/kT = \int d^dx \left\{ \tfrac{1}{2}(\nabla\phi(x))^2 + \tfrac{1}{2}r_0\phi^2(x) + \frac{\bar{u}_0}{4!}(\phi^2(x))^2 + \frac{\Lambda^{-2}}{2}(\nabla^2\phi(x))^2 \right\} \tag{1.1}$$

where r_0 and $\bar{u}_0$ are assumed to be *analytic* functions of temperature in the phase transition region. The notation is

$$\phi^2 = \sum_{i=1}^{n} \phi_i\phi_i, \qquad (\nabla\phi)^2 = \sum_{a=1}^{d}\sum_{i=1}^{n} \nabla_a\phi_i\nabla_a\phi_i.$$

We have assumed the convention that the scale of the field ϕ is chosen so that the term $(\nabla\phi)^2$ appears with coefficient $\frac{1}{2}$. The factor $1/kT$ is included because the Hamiltonian $\mathscr{H}$ will always appear in the Boltzmann factor $\exp\text{-}\mathscr{H}(\phi)/kT$.

The physical principles which lead to this form of Hamiltonian may be listed as follows:

(i) We consider only *short range* interactions and hence restrict the form of the Hamiltonian to be an integral over space x of a *local* functional of the field $\phi_i(x)$, depending only on the field and its derivatives at the point x.

(ii) Initially we shall consider only systems in which the Hamiltonian is $O(d)$ symmetric in the co-ordinate variable and $O(n)$ symmetric in the field variable.

(iii) In order to describe a ferromagnetic system, the term $(\nabla\phi)^2$ must be given a *positive* coefficient; this ensures that at low temperatures where spontaneous magnetization exists, the lowest energy configuration is one of constant magnetization (x-independent) throughout the system.

(iv) In physical systems high wave-vector modes have energies which increase more rapidly with wave vector q than q^2, which arises from the term $(\nabla\phi)^2$ in $\mathscr{H}/\mathrm{kT}$. For example in lattice models the allowed wave-vectors are cut off at the edge of the Brillouin zone. To imitate this suppression of high wave-vectors the term $\frac{1}{2}\Lambda^{-2}(\nabla^2\phi)^2$ is included in (1.1). The quantity Λ has the dimensions of a wave-vector and plays the role of the cut-off in this model. (The reason why this produces sufficient suppression of high wave-vectors is discussed in Section II.)

(v) An explicit example of the analytic dependence of r_0 and $\bar{u}_0$ on temperature T is given in the formulation of a lattice model (e.g. Ising or classical Heisenberg) as a field model like (1.1). For a discussion of the relationship of lattice and field models see the article by Brézin *et al.* in this volume. More generally, the origin of the analyticity of r_0 and $\bar{u}_0$ lies in the fact that (1.1) is best considered as a semi-microscopic model for a thermodynamic system, in which the field $\phi_i(x)$ represents the mean magnetization density of a truly microscopic model averaged over a small volume—a few atomic spacings say—round the point x. Such a small volume average cannot produce non-analytic temperature dependence.

These general considerations produce a Hamiltonian of the form (1.1), when other terms of higher order in the fields and derivatives are neglected, e.g. $(\phi^2)^3$, $\phi^2(\nabla\phi)^2$ etc. We shall first study the phase transition behaviour of the system (1.1) and later (Section IV) see if the neglected terms can be expected to change the phase transition behaviour.

One may also study the extent to which *anisotropic* terms (i.e. terms which are not $O(n)$ or $O(d)$ invariant) can change the phase transition behaviour.

Principal among these is the effect on the system of a uniform external field term, given by

$$\mathscr{H}_H/kT = -\sum_{i=1}^{n} H_i \int \phi_i(x)\, \mathrm{d}^d x \tag{1.2}$$

with appropriate definition of the scale of the external field H_i. The resulting equation of state of the system is considered in Section III. The effect of more general short-range anisotropic perturbations is considered in Section IV.

In this article we shall restrict our attention to *perturbations* about the isotropic behaviour of the system described by (1.1). A more general review of anisotropic, long-range and other interactions is given in the article by Aharony in this volume.

We close this section with two remarks. Firstly the x integration in (1.1) and (1.2) runs over all space, so we are considering infinite volume systems. Secondly in order that (1.1) represents a thermodynamically stable system, the coefficient $\bar{u}_0$ must be positive, otherwise the energy of the system would become lower and lower as the magnetization $\phi_i(x)$ increased, for large ϕ. (For stability in a system with $\bar{u}_0$ negative one has to include in (1.1) terms of order $(\phi^2)^3$ or higher, resulting in a model for tricritical behaviour—see Riedel and Wegner (1972, 1973, 1974), Amit and de Dominicis (1973), Stephen and McCauley (1973)). We shall consider only the case $\bar{u}_0 > 0$.

Finally, we have already used the fact that all terms in (1.1) have the same length dimensions to state that Λ has dimension L^{-1} i.e. $[\Lambda] = [L^{-1}]$. Extending this analysis it is trivial to see that

$$[\phi] = [\Lambda]^{d/2-1}, \quad [r_0] = [\Lambda]^2, \quad [\bar{u}_0] = [\Lambda]^{4-d}. \tag{1.3}$$

It is particularly convenient to define a dimensionless coupling constant u_0 by

$$\bar{u}_0 = u_0 \Lambda^{4-d}. \tag{1.4}$$

Implicit in this statement is the physical assumption that the fundamental length scale in the problem Λ^{-1} (recall that Λ^{-1} is analogous to the lattice spacing) appears as the natural length scale for all dimensional parameters of the microscopic Hamiltonian.

B. Correlation functions

The third essential ingredient in a statistical mechanical model is a method for calculating observable quantities, which are the average, or expectation values, of functions of the basic variable. Generally, one is interested in the *correlation functions*; in our case, in which the basic variable is a field $\phi_i(x)$, the correlation functions are defined by

$$\langle \phi_{i_1}(x_1) \ldots \phi_{i_p}(x_p) \rangle = \frac{\int \mathrm{d}\phi\, \phi_{i_1}(x_1) \ldots \phi_{i_p}(x_p) \exp(-\mathscr{H}/kT)}{\int \mathrm{d}\phi \exp(-\mathscr{H}/kT)} \tag{1.5}$$

In this expression, $\exp(-\mathscr{H}/kT)$ is regarded as the conventional Boltzmann factor, weighting the function of the fields in whose mean value we are interested, and $\int \mathrm{d}\phi$ represents the sum over all possible configurations of the variable $\phi_i(x)$. This amounts to a sum over all the possible functional forms for $\phi_i(x)$—hence the term functional integral for $\int \mathrm{d}\phi$.

All the correlation functions (1.5) can be obtained from the generating functional

$$Z(r_0, u_0, \Lambda, \{H\}) = \int \mathrm{d}\phi \exp\left(-\mathscr{H}/kT + \int H_i(x)\phi_i(x)\right) \tag{1.6}$$

by differentiation with respect to $H_i(x)$. (Note that we shall generally use repeated index summation convention.) Comparison with (1.2) shows that $H_i(x)$ may also be thought of as a non-uniform external field. Thus (1.6) can also be used to obtain the correlation functions in the presence of a non-uniform external field. Suppressing indices by writing $\phi_{i_1}(x_1) = \phi(1)$, we have

$$\langle\phi(1)\ldots\phi(p)\rangle = Z^{-1}\frac{\delta}{\delta H(1)}\cdots\frac{\delta}{\delta H(p)}Z. \tag{1.7}$$

In general one cannot evaluate the integrals in (1.5) and (1.6) exactly. The principal tool at one's disposal is the evaluation by expansion in powers of the four field interaction u_0. This is the Feynman graph expansion.

Before we list the Feynman rules which summarize the expansion, it is perhaps worthwhile to sketch how the Feynman graph structure of the expansion comes about. Firstly one can extract the quartic interaction term from inside the integral in (1.6) by

$$Z = \exp\left[-\frac{\bar{u}_0}{4!}\int \mathrm{d}^d x\left(\frac{\delta}{\delta H_i(x)}\frac{\delta}{\delta H_i(x)}\right)^2\right].$$

$$\int \mathrm{d}\phi \exp -\int \mathrm{d}^d x[\tfrac{1}{2}(\nabla\phi)^2 + \tfrac{1}{2}r_0\phi^2 + \tfrac{1}{2}\Lambda^{-2}(\nabla^2\phi)^2 - H_i(x)\phi_i(x)] \tag{1.8}$$

The remaining integral is now the continuum limit of a Gaussian integral:

$$\int_{-\infty}^{\infty}\prod_i \mathrm{d}y_i \exp(-\tfrac{1}{2}y_i A_{ij} y_j + H_i y_i)$$

$$= \exp(\tfrac{1}{2}H_i(A^{-1})_{ij}H_j)\,.\int_{-\infty}^{\infty}\prod_i \mathrm{d}y_i' \exp(-\tfrac{1}{2}y_i' A_{ij} y_j') \tag{1.9}$$

where A_{ij} is a symmetric matrix; this identity is obtained by completing the square on the quadratic in y_i. The analogue of the inverse matrix $(A^{-1})_{ij}$ in (1.9) in the continuum case (1.8) is the propagator $\Delta(x, y)$, the inverse of the quadratic form in (1.8):

$$(r_0 - \nabla^2 + \Lambda^{-2}(\nabla^2)^2)\Delta(x, y) = \delta^d(x - y). \tag{1.10}$$

Thus in the continuum case, the generating function Z is given up to a multiplicative constant C by

$$Z = C \exp\left\{-\frac{\bar{u}_0}{4!}\int \mathrm{d}^d x \left(\frac{\delta}{\delta H_i(x)}\frac{\delta}{\delta H_i(x)}\right)^2\right\} \exp \tfrac{1}{2}\left\{\int \mathrm{d}^d x\, \mathrm{d}^d y\, H_j(x) \times \Delta(x, y)\, H_j(y)\right\}. \qquad (1.11)$$

The Feynman graph structure of this expression should be clear: the expansion in $\bar{u}_0$ gives rise to vertices which are connected to each other by the propagators $\Delta(x, y)$ through the action of $\delta/\delta H$ on $H\Delta H$. The reader who is unfamiliar with the functional formalism should convince himself that the result in (1.11) is identical to forming all possible contractions of pairs of fields ϕ in the expression

$$Z = \exp \int \mathrm{d}^d x \left\{H_i(x)\phi_i(x) - \frac{\bar{u}_0}{4!}(\phi^2(x))^2\right\} \qquad (1.12)$$

in which a contraction implies the replacement

$$\overbrace{\phi_i(x)\phi_j}(y) = \delta_{ij}\Delta(x, y) \qquad (1.13)$$

This last statement is of course the statement of the Wick expansion for the evaluation of Green functions (correlation functions) in the interaction picture.

It should be clear that the formalism in eqns (1.6)–(1.13) is also applicable for more general interactions than the quartic term $(\phi^2)^2$; one simply characterizes each new interaction by an appropriate vertex and extends the graphs considered to include vertices of this new type.

Finally in this section we give two examples of the relationship between correlation functions and thermodynamic quantities. The mean magnetization M_i is identified with the expectation value of the field ϕ_i. Thus we can write

$$M_i(x) = \langle \phi_i(x) \rangle = \frac{\delta}{\delta H_i(x)} \ln Z \qquad (1.14)$$

from (1.7). A second quantity of interest is the susceptibility tensor. It is simplest to consider the case of a uniform external field H_i and uniform magnetization M_i. The susceptibility tensor χ_{ij} is then defined by

$$\chi_{ij} = \frac{\partial M_i}{\partial H_j} \qquad (1.15)$$

Given that $M_i = \langle\phi_i(0)\rangle$, it is straightforward to use the definition of correlation functions in (1.5) to obtain

$$\chi_{ij} = \int d^d x \langle(\phi_i(x) - M_i)(\phi_j(0) - M_j)\rangle \tag{1.16}$$

Of course, it must be understood that this correlation function is to be calculated with the external field term $H \cdot \phi$ in the Hamiltonian (1.1). If we further specialize to consider the zero field susceptibility above the critical temperature, where there is no magnetization ($M_i = 0$), then by the $O(n)$ rotation invariance, χ_{ij} must be proportional to δ_{ij} and we can define a susceptibility χ by

$$\delta_{ij}\chi = \int d^d x \langle\phi_i(x)\phi_j(0)\rangle. \tag{1.17}$$

This equation will be of central importance later. We shall use it to characterize the critical temperature T_c as the point where $r_0(T_c)$ is such that the susceptibility χ becomes infinite.

C. Feynman Rules

In this section we shall develop an appropriate set of Feynman rules for calculating the symmetric correlation functions in zero external field above the critical temperature. Modifications of the rules to include an external field are given in Section IIIB.

For many practical purposes it is most convenient to work in momentum space, rather than co-ordinate space. This can be achieved from first principles by reformulating the previous section in terms of the Fourier transform field $\sigma_i(q)$

$$\sigma_i(q) = \int d^d x\, e^{iqx} \phi_i(x). \tag{1.18}$$

The correlation functions $G^{(p)}(q)$ of the spin $\sigma_i(q)$ are the Fourier transforms of the functions (1.5)

$$(2\pi)^d \delta^d(q_1 + q_2 + \ldots + q_p) G^{(p)}_{i_1 \ldots i_p}(q) = \prod_{j=1}^{p} \left(\int d^d x_j \exp iq_j x_j \right) \times \langle \phi_{i_1}(x_1) \ldots \phi_{i_p}(x_p) \rangle \tag{1.19}$$

The suffix p in $G^{(p)}_{i_1 \ldots i_p}(q)$ refers to the p fields $\phi_i(x_i)$ and the argument q means the set $(q_1, q_2 \ldots q_p)$. Because the initial Hamiltonian (1.1) is translation invariant (this holds also in the presence of the *uniform* external field as in (1.2)), there is always an overall momentum conserving factor in the Fourier transform (1.19) and this has been extracted explicitly in the definition of G.

As usual, the Fourier transform operation essentially diagonalizes the quadratic term in the Hamiltonian (1.1), and the propagator in momentum space,

$$\Delta(q) = (q^2 + r_0 + \Lambda^{-2}(q^2)^2)^{-1}. \tag{1.20}$$

The straightforward Feynman graph expansion summarized by expression (1.11) will then give the correlation functions G, as functions of u_0, r_0 and Λ. The result is not immediately useful for describing the phase transition behaviour because one does not know *a priori* the values $u_0(T_c)$ and $r_0(T_c)$ at the critical temperature. (Recall that u_0 and r_0 are taken as analytic functions of temperature). The qualitative input to determine the critical point is contained in (1.17); it can be summarized by stating that the two point correlation function at $q = 0$ becomes infinite at the critical temperature T_c.

For example, at zeroth order in perturbation theory (or if we set $u_0 = 0$) $G^{(2)}$ is given simply in terms of the free propagator $\Delta(q)$ (eqn (1.20)) and we have

$$\begin{aligned} \delta_{ij}\chi &= \delta_{ij} G^{(2)}(0) \\ &= \delta_{ij} \frac{1}{r_0} \end{aligned} \tag{1.21}$$

or

$$r_0 = \chi^{-1} + O(u_0) \tag{1.22}$$

Thus at zeroth order we obtain the simple Landau result that $r_0(T_c) = 0$, and since r_0 is assumed to be analytic in T, we can write

$$r_0 = t + O(u_0) + O(t^2) \tag{1.23}$$

where $t = a(T - T_c)$ is the reduced temperature in suitable units. Equation (1.22) states that at lowest order, one obtains the mean-field theory result, $\chi \propto (T - T_c)^{-1}$.

It is clear that the condition $\chi \to \infty$ could be used in principle at higher orders in perturbation theory in order to obtain correlation functions in terms of the reduced temperature t, u_0 and Λ.

However it is possible to obtain the correlation functions *explicitly* in terms of χ rather than r_0. This has the advantage that the critical point behaviour is immediately characterized by $\chi \to \infty$. The trick is to use the equivalent of mass renormalization in quantum field theory, by introducing a parameter r and splitting up the Hamiltonian (1.1) into the quadratic term

$$\mathcal{H}_0/kT = \int d^dx \{\tfrac{1}{2}(\nabla\phi)^2 + \tfrac{1}{2}r\phi^2 + \tfrac{1}{2}\Lambda^{-2}(\nabla^2\phi)^2\} \tag{1.24}$$

and the interaction term

$$\mathscr{H}_{\text{int}}/kT = \int \mathrm{d}^d x\{(4!)^{-1}\bar{u}_0(\phi^2)^2 + \tfrac{1}{2}(r_0 - r)\phi^2\}. \tag{1.25}$$

The free propagator arising from (1.24) is now

$$\Delta(q) = (q^2 + r + \Lambda^{-2}(q^2)^2)^{-1}. \tag{1.26}$$

We see that if we *choose* $r = \chi^{-1}$, then the correlation function $G^{(2)}$ is *exact* at $q = 0$, even in zeroth order in perturbation theory in $\mathscr{H}_{\text{int}}$ (eqn (1.25)). The higher order corrections to $G^{(2)}$ due to $\mathscr{H}_{\text{int}}$ must vanish identically at $q = 0$.

Now, the corrections to $G^{(2)}$ from $\mathscr{H}_{\text{int}}$ are expressed in terms of the self-energy graphs by

$$G^{(2)}_{i_1 i_2}(q) = \delta_{i_1 i_2}\frac{1}{q^2 + r + \Lambda^{-2}q^4 - (\Sigma(q) + r - r_0)}. \tag{1.27}$$

Graphically $G^{(2)}$ corresponds to all graphs with two external legs; the self energy $r - r_0$ comes from the "mass counterterm" in $\mathscr{H}_{\text{int}}$, and $\Sigma(q)$ from proper self energy graphs constructed from the ϕ^4 term in $\mathscr{H}_{\text{int}}$.

With the choice $r = \chi^{-1}$, setting $q = 0$ in (1.27) gives

$$r - r_0 + \Sigma(q = 0) = 0 \tag{1.28}$$

and hence $G^{(2)}$ can be written

$$G^{(2)}_{i_1 i_2}(q) = \delta_{i_1 i_2}\frac{1}{q^2 + r + \Lambda^{-2}q^4 - (\Sigma(q) - \Sigma(q = 0))}. \tag{1.29}$$

Thus the mass counterterm $\frac{1}{2}(r_0 - r)\phi^2$ in (1.25) can be entirely neglected provided that *all self energy graphs are taken to represent the graph itself subtracted by its value at* $q = 0$, as in (1.29). This prescription is to be observed when self-energy graphs occur within more complicated graphs.

The resulting correlation functions will clearly depend only on r, u_0 and Λ and not on r_0 and are denoted by $G^{(p)}_{i_1 \ldots i_p}(q, r, u_0, \Lambda)$. The Feynman rules for obtaining $G^{(p)}$ at a given order in perturbation theory can now be summarized:

(i) With the available interaction vertices, draw all possible graphs with p external legs, excluding disconnected vacuum-to-vacuum graphs. (These are always cancelled explicitly by the denominator in (1.5)—see e.g. Bjorken and Drell (1965) p. 187).

(ii) For each distinct graph, distribute momenta q_a and spin labels i_a ($a = 1, 2 \ldots p$) on the p external legs in all possible distinct ways and label the internal lines of the graph by a momentum so as to conserve momentum at vertices. If the graph has m closed loops, m independent internal momenta must be introduced.

(iii) For each line labelled by momentum k say, there is a propagator factor $\Delta(k)$ as in (1.26).

(iv) One integrates over each independent internal momentum k, according to

$$\int_k = \int d^dk(2\pi)^{-d}. \tag{1.30}$$

(v) There is an overall multiplicity factor for each graph given by the number of times the graph can be produced in the Wick expansion of

$$\frac{(-1)^l}{l!}\langle\phi_{i_1}\cdots\phi_{i_p}[\mathscr{H}_{\text{int}}/kT]^l\rangle. \tag{1.31}$$

In this factor is contained the coupling constant $\bar{u}_0$ for each interaction vertex appearing in the graph.

(vi) All self-energy insertions in a graph are to be subtracted by their value at $q = 0$.

We complete this introductory section with two comments.

Firstly, we are still interested in the relationship between $r(\chi^{-1})$ and the reduced temperature t. This relationship is given by (1.28).

Secondly it is generally much more convenient to work with the one particle irreducible (proper) vertex functions $\Gamma^{(p)}_{i_1\ldots i_p}(q, r, u_0, \Lambda)$ instead of the correlation functions $G^{(p)}$. The $\Gamma^{(p)}$ are defined as the set of all graphs with p external lines which cannot be cut into two parts by cutting a *single* line of the graph. With this restriction the rules for calculation are as before. For example the two-point vertex function $\Gamma^{(2)}$ is directly related to the self energy graphs Σ and is conventionally defined to be the inverse of $G^{(2)}$. Suppressing indices we have

$$\Gamma^{(2)}(q, r, u_0, \Lambda) = q^2 + r + \Lambda^{-2}q^4 - [\Sigma(q, r, u_0, \Lambda) - \Sigma(0, r, u_0, \Lambda)] \tag{1.32}$$

Similarly $\Gamma^{(4)}$ is given by the sum of all *connected* four-point graphs, with the external legs amputated.

There is a very elegant functional relationship between the G's and Γ's; they are related by a functional Legendre transform as follows. We have already discussed how $Z(H)$ in (1.6) is the generating functional for the correlation functions $G^{(p)}$. From $Z(H)$ one can define the functional $W(H)$ by

$$W(H) = \ln Z(H). \tag{1.33}$$

$W(H)$ is the generating functional for all *connected* correlation functions. From $W(H)$ one can define a Legendre transform $\Gamma(M)$ by

$$W(H) + \Gamma(M) = \int \mathrm{d}^d x \, M_i(x) H_i(x) \tag{1.34}$$

where $M_i(x)$ is defined by

$$M_i(x) = \frac{\delta W}{\delta H_i(x)}. \tag{1.35}$$

Differentiation of (1.34) then implies

$$H_i(x) = \frac{\delta \Gamma}{\delta M_i(x)}. \tag{1.36}$$

$\Gamma(M)$ defined in this way is actually the generating functional for the proper vertices $\Gamma^{(p)}$, or more correctly their Fourier transforms in co-ordinate space

$$\Gamma(M) = \sum_p \frac{1}{p!} \int \mathrm{d}^d x_1 \ldots \mathrm{d}^d x_p \Gamma^{(p)}_{i_1 \ldots i_p}(x, r, u_0, \Lambda) M_{i_1}(x_1) \ldots M_{i_p}(x_p). \tag{1.37}$$

We shall not need to use this functional formalism beyond noting its relationship to the usual Legendre transformation (see any standard text e.g. Stanley (1971)) between the Gibbs potential $G(H) = -\ln Z(H)$ (where $Z(H)$ is the usual partition function for a uniform magnetic field H) and the Helmholz free energy $F(M)$. This usual Legendre transform is simply a specialization of the functional case to $H_i(x) = H_i$, $M_i(x) = M_i$ (*independent of* x). *As is clear then from* (1.37), *the Helmholtz potential is the generating function for the proper vertices evaluated at zero external momenta.*

For further discussion and references on this aspect of the Feynman graph expansion, see the articles by Brézin *et al.* and Jona-Lasinio and Di Castro in this volume.

II. The ε-Expansion

A. Résumé

In the previous Section we set up a formalism for calculating the correlation functions (above the critical temperature and in zero external field—we shall call these the symmetric correlation functions) by perturbation theory in the interaction terms in a Euclidean field theory. The basic Hamiltonian is given in (1.1) and the Feynman rules for calculating the correlation functions in terms of the inverse susceptibility $r = \chi^{-1}$ are listed in Section I.C. A fundamental equation (1.28) gives r as a function of temperature through the temperature dependence of the bare mass parameter r_0. These results by no means exhaust the functions of interest in this theory. In particular one might wish to consider correlation functions involving higher powers of the field $\phi_i(x)$, e.g. $\phi^2(x)$, etc., and one is also interested in behaviour in a non-zero external field H_i both above and below the critical temperature. We shall consider the former in Section IV and the latter in Section III.

In this chapter we shall concentrate on the symmetric correlation functions only. In Section II.B we shall illustrate, using the relationship (1.28) between r and T, that direct expansion in u_0 can be trusted in the critical region $r/\Lambda^2 \to 0$, only if the dimension of space d is greater than 4. We shall see that if $d \leqslant 4$ then each order in perturbation in u_0 is arbitrarily larger than the previous order. In Section II.C we shall discuss Wilson's Feynman graph method (Wilson 1972) for determining the behaviour of (1.28) and more generally the correlation functions, as a power series expansion in $\varepsilon = 4 - d$. The results of higher order calculations are given in Section II.D. In Section II.E we shall discuss an attempt by Nickel (1974) to improve the convergence of the ε-expansion by resummation of the irreducible graphs. Some notes on the evaluation of graphs are contained in Section F.

B. Perturbation theory in the critical region

As an example of the way in which straightforward perturbation theory becomes inadequate in the critical region consider the relationship between inverse susceptibility r and temperature (1.28)

$$r_0(T) = r + \Sigma(0, r, u_0, \Lambda). \tag{2.1}$$

Here r_0 is assumed to be analytic in temperature T; u_0 could in principle depend on temperature, but let us consider it as constant for the time being (see Section IV for further comments). The critical temperature T_c occurs when r vanishes i.e.

$$r_0(T_c) = +\Sigma(0, 0, u_0, \Lambda). \tag{2.2}$$

Subtracting these equations and using the analyticity of r_0 to write

$$r_0(T) - r_0(T_c) = t, \tag{2.3}$$

where t is a measure (up to a suitable scale factor) of the reduced temperature $(T - T_c)$, we have

$$t = r + (\Sigma(0, r, u_0, \Lambda) - \Sigma(0, 0, u_0, \Lambda)). \tag{2.4}$$

Now the lowest order graph contributing to the self energy Σ is shown in Fig. 1. The corresponding analytic expression according to the Feynman rules of the previous chapter is

$$\Sigma(q, r, u_0, \Lambda) = -\frac{u_0}{4!}\Lambda^{4-d} \cdot 4(n + 2)\int_p [p^2 + r + \Lambda^{-2}(p^2)^2]^{-1} + O(u_0^2), \tag{2.5}$$

where $\int_p$ is defined in (1.30). Note that the value of this graph is independent of the momentum q of the external legs. The factor $4(n + 2)$ arises when evaluating the multiplicity of the graph.

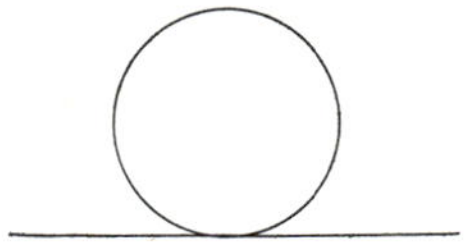

FIG. 1. One-loop graph contributing to the self-energy Σ (eqns 2.4 and 2.5).

Substituting (2.5) into (2.4) we have

$$t = r + \tfrac{1}{6}u_0(n + 2)\Lambda^{4-d}r \int_p [p^2 + r + \Lambda^{-2}(p^2)^2]^{-1}[p^2 + \Lambda^{-2}(p^2)^2]^{-1} + O(u_0^2) \qquad (2.6)$$

There is a question of convergence in two regions of integration of this integral. First, we see that in the ultraviolet region, $p \to \infty$, the integrand behaves like p^{-8} and the integral is hence convergent for any interesting dimension of space d. There is no ultraviolet problem in this model for phase transitions.

However the limit $r \to 0$ characterizing the critical behaviour of the system can become very singular because of contributions to the integral from the *small* p region, i.e. we are faced with an infrared problem. There are three cases to be considered.

(i) $d > 4$. When $r = 0$, the integral behaves like $\int d^d p[p^2]^{-2}$ for p small and is hence finite. The equation (2.6) becomes

$$t = r + cu_0 r + O(u_0^2) \qquad (2.7)$$

where c is some constant. Thus the relationship between temperature and susceptibility is still given by mean-field theory: $\chi \propto (T - T_c)^{-1}$. The perturbation in u_0 produces only a renormalization of the critical temperature through the term $\Sigma(0, 0, u_0, \Lambda)$ in (2.4) and a renormalization of the constant of proportionality between χ and $(T - T_c)^{-1}$.

This result remains true at higher orders in perturbation theory; at each order in u_0, one obtains *two* new internal lines and *one* new loop momentum to integrate over: when the dimension of the loop integral is greater than 4 there are no infrared divergences created in the limit $r \to 0$.

(ii) $d < 4$. The picture changes completely because the integral is now infrared divergent. In the limit $r \to 0$, we have

$$(2\pi)^{-d}\int d^d p[p^2 + r + \Lambda^{-2}(p^2)^2]^{-1}[p^2 + \Lambda^{-2}(p^2)^2]^{-1}$$

$$= S_d(2\pi)^{-d}\int_0^\infty dp \cdot p^{d-1}(p^2 + r + \Lambda^{-2}p^4)^{-1}(p^2 + \Lambda^{-2}p^4)^{-1}$$

$$= S_d(2\pi)^{-d} \cdot \frac{\pi\varepsilon/2}{\sin(\pi\varepsilon/2)} \cdot \frac{1}{\varepsilon}[r^{-\varepsilon/2} - (1 + \varepsilon/2)\Lambda^{-\varepsilon} + r^{-\varepsilon/2}O(r/\Lambda^2)]. \quad (2.8)$$

In evaluating this integral we have changed to radial and angular coordinates, pulling out the angular integral

$$S_d = \frac{2\pi^{d/2}}{\Gamma(d/2)}, \quad (2.9)$$

where Γ is the usual Euler gamma function (Gradshteyn and Ryzhik (1965) p. 933 *et seq.*). The remaining radial integrals can be performed using the standard result

$$\int^\infty dy\, y^{a-1}(1 + y)^{-a-b} = \frac{\Gamma(a)\Gamma(b)}{\Gamma(a+b)} \quad (2.10)$$

and properties of the Γ function. Here and henceforth, $\varepsilon = 4 - d$.

When (2.8) is substituted into (2.6), we see that the *effective* parameter in the Feynman graph expansion is the dimensionless coupling $u_0(\Lambda^2/r)^{(4-d)/2}$. No matter how small u_0 is, the effective expansion parameter becomes arbitrarily large in the critical region $r/\Lambda^2 \to 0$. Every term in perturbation theory becomes arbitrarily larger than any lower order term because the dimensional coupling constant $\bar{u}_0 = \Lambda^{4-d}u_0$ is rendered dimensionless by a power $r^{d/2-2}$ coming from the infrared region in the loop integration.

This is a general result of perturbation theory for the vertex functions: for a given number of external legs each new order in u_0 produces another loop integral involving two additional propagators and hence an additional infrared divergent factor. The statement of the strong coupling nature of the problem must be slightly generalized because vertex functions are momentum dependent and the dimensional coupling constant $\bar{u}_0$ can also be made dimensionless by powers of external momenta, i.e. $u_0(\Lambda/|q|)^{4-d}$ may become the effective expansion parameter ($|q|$ is a measure of the external momentum). Thus the problem of describing the region $r \ll \Lambda^2, q^2 \ll \Lambda^2$ is always a strong coupling problem for $d < 4$, in particular for the physical case $d = 3$.

(iii) $d = 4$. This special limiting case can be obtained simply by taking the limit $d \to 4$ in (2.8). The result is

$$(2\pi)^{-4} \int d^4p[p^2 + r + \Lambda^{-2}(p^2)^2]^{-1}[p^2 + \Lambda^{-2}(p^2)^2]^{-1}$$
$$= \tfrac{1}{2}S_4(2\pi)^{-4}\{-\ln(r/\Lambda^2) - 1 + 0(r/\Lambda^2)\} \qquad (2.11)$$

The strong coupling nature of this problem again arises from the logarithmic divergence of the integral in the infrared region.

The general picture which emerges from the above is the appearance of mean-field theory results for $d > 4$ and increasingly stronger corrections to mean field theory as d is decreased below four. We shall see in the next section how Wilson set up a perturbation expansion in $\varepsilon = 4 - d$ for obtaining the critical behaviour of vertex functions in less than four dimensions. Before that however we finish this section with some technical remarks.

First, one cannot expect to apply the above perturbation theory down to $d = 2$, because of new infrared divergences which appear. For example, infrared divergent integrals (e.g. $\int_q(q^2 + \Lambda^{-2}(q^2)^2)^{-1}$) appear on the R.H.S. of the equation defining $r_0(T_c)$ (see (2.2)), and the integral in (2.6) is divergent even for finite r when $d = 2$. Thus it is very difficult in the field models to make direct contact with the exactly soluble two-dimensional lattice models.

Second, it is clear that the loop integrals which appear can be defined for non-integer d by analytic continuation. For example all loop integrals can be made to involve only functions of the *magnitude* of momenta by using Feynman parameter or other techniques (we shall review the practical evaluation of integrals in more detail in Section II.F). The angular integrals then produce only factors of S_d as in (2.9), and the radial integrals are simply complicated but well-defined one-dimensional integrals, such as in (2.10).

Finally note that with the choice of $(\nabla^2\phi)^2$ as the form of cut off, the contribution of Fig. 1 to eqn (2.1) or (2.2) is in fact logarithmically divergent (ultraviolet) in $d = 4$. However the divergence cancels in the subtracted (2.4), as is clear from naive power-counting. Compare (2.5) and (2.6). The conventional power-counting arguments show that all higher graphs will also give ultraviolet convergent integrals in (2.4) and in vertex functions with more than two external legs, for $d \leqslant 4$.

C. The ε-Expansion

The prescription for calculating critical behaviour in the ε-expansion can be summarized as follows. List all graphs which contribute to the vertex function to some chosen order in perturbation in u_0—up to l loops say. Evaluate the analytic expressions for these graphs, using two approximations:

(a) Neglect terms which vanish with respect to the leading terms by integer powers of r/Λ^2 or $|q|/\Lambda$, since these are small in the critical region.

(b) Expand the integral as a power series in $\varepsilon = 4 - d$, either by expanding the analytic continuation in d dimensions about $d = 4$, or by using the explicit approximation for the radial integration

$$\int_0^\infty q^{d-1}\,dq f(q) = \int_0^\infty q^3\,dq\,q^{-\varepsilon} f(q)$$
$$= \int_0^\infty q^3\,dq\{1 - \varepsilon \ln q + \ldots\} f(q). \tag{2.12}$$

For example the result of doing this in expression (2.6) may be written

$$t = r\{1 + \tfrac{1}{6}[u_0 S_d (2\pi)^{-d}](n+2)(-\tfrac{1}{2}\ln(r/\Lambda^2) + \ldots) + O(u_0^2, u_0\varepsilon)\}. \tag{2.13}$$

The $\ln(r/\Lambda^2)$ arises because the integral has now become essentially a four dimensional one according to (2.12). One sees the appearance of higher powers of $\ln r/\Lambda^2$ not only as the order in ε is increased but also as the order in u_0 is increased because each new order in u_0 gives an extra loop which gives another infrared divergence factor of ln r.

The ε-expansion in this approach (Wilson 1972) is based upon the recognition that *there is a special value of* u_0*, of order* ε*, for which this power series in* ($u_0 \ln r$, $\varepsilon \ln r$) *exponentiates exactly to give the correct scaling behaviour in the critical region.*

For example we shall obtain shortly the result that this special value of u_0 is given by

$$u_0 S_d (2\pi)^{-d} = 6\varepsilon/(n+8) + O(\varepsilon^2). \tag{2.14}$$

Substituting this result into (2.13) gives

$$t = r\{1 - (n+2)\varepsilon/(n+8)\cdot(\tfrac{1}{2}\ln r/\Lambda^2 + \ldots)\} + O(\varepsilon^2). \tag{2.15}$$

The rule above tells us that the $\varepsilon \ln r/\Lambda^2$ in this expression is the ε-expansion of an exponential

$$t \propto r \cdot (r/\Lambda^2)^{-(n+2)\varepsilon/(2(n+8))}. \tag{2.16}$$

If we compare this with the usual definition of the exponent γ

$$\chi \propto (T - T_c)^\gamma \tag{2.17}$$

we can read off from (2.16) that (recall $\chi = r^{-1}$)

$$\gamma = 1 + \frac{(n+2)\varepsilon}{2(n+8)} + O(\varepsilon^2). \tag{2.18}$$

This simple example illustrates the basic idea behind the method: after approximations (a) and (b), the power series in $\ln r$ will exponentiate to the correct scaling behaviour if u_0 is set equal to a special value of order ε. The result is that critical exponents such as γ (and more generally scaling functions) are obtained as power series in ε, giving corrections to mean field theory. The predictions for the physical case $d = 3$ are obtained by putting $\varepsilon = 1$.

The theoretical basis for the method is as follows. Consider the p-point vertex function $\Gamma^{(p)}(q, r, u_0, \Lambda)$ (we suppress the indices $i_1 \dots i_p$ of the external legs) calculated in perturbation theory according to the Feynman rules of the previous chapter. They can in principle be written as functions of t, u_0 and Λ by solving (2.4) for r as a function of t, u_0 and Λ and substituting for r into the vertex functions $\Gamma^{(p)}$. This gives equivalent vertex functions $\tilde{\Gamma}^{(p)}(q, t, u_0, \Lambda)$. The advantage of working with these vertex functions was first pointed out in this context by Zinn–Justin; namely they obey a renormalization group equation of the form

$$\left\{\Lambda\frac{\partial}{\partial\Lambda} + \beta(u_0)\frac{\partial}{\partial u_0} - [\gamma_4(u_0) - \gamma_3(u_0)]t\frac{\partial}{\partial t} - \frac{p}{2}\gamma_3(u_0)\right\}\tilde{\Gamma}^{(p)}(q, t, u_0, \Lambda)$$
$$= \Delta\tilde{\Gamma}^{(p)}(q, t, u_0, \Lambda), \qquad (2.19)$$

where $\beta(u_0)$, $\gamma_3(u_0)$ and $\gamma_4(u_0)$ are functions of the dimensionless coupling constant u_0, and for any given graph in $\tilde{\Gamma}$ the corresponding term in the inhomogeneous part $\Delta\tilde{\Gamma}$ is smaller by terms of order r/Λ^2 or q^2/Λ^2; thus in the critical region one may hope to be able to set $\Delta\tilde{\Gamma} = 0$ and obtain a homogeneous differential equation obeyed by $\tilde{\Gamma}^{(p)}$. The neglect of $\Delta\tilde{\Gamma}$ corresponds to the approximation (a) at the beginning of this section.

The origin of (2.19) is seen most clearly by using renormalized perturbation theory. It is a consequence of the fact that *renormalized* vertex functions are *finite* as $\Lambda \to \infty$. (See Zinn–Justin 1973 and the article by Brézin *et al.* in this volume). A qualitative discussion of equivalent properties of the vertex functions is given by Wilson and Kogut (1974). A modification of the Wilson–Fisher (1972) recursion formula to a form involving only ϕ^4 coupling is given to second order in perturbation theory by Bruce *et al.* (1974). See also Aharony (1973).

We shall not reproduce here a full analysis of the differential equation (2.19); the discussion would parallel that in the article by Brézin *et al.* in this volume. Instead we shall simply use the homogeneous equation

$$\left\{\Lambda\frac{\partial}{\partial\Lambda} + \beta(u_0)\frac{\partial}{\partial u_0} - [\gamma_4(u_0) - \gamma_3(u_0)]t\frac{\partial}{\partial t} - \frac{p}{2}\gamma_3(u_0)\right\}\tilde{\Gamma}^{(p)}(q, t, u_0, \Lambda) = 0 \qquad (2.20)$$

obtained for the leading singularities contributing to $\tilde{\Gamma}^{(p)}$ from any given graph, to justify Wilson's method. The validity of this procedure is of course completely reliant on the assumption that the inhomogeneous part $\Delta\tilde{\Gamma}$, which is negligible by powers of Λ^{-2} order by order in perturbation theory, does not resum to a term which competes with the behaviour of the left-hand side of (2.19) in the critical region. Henceforth Γ and $\tilde{\Gamma}$ will denote the results of retaining only leading singularities of graphs as prescribed in approximation (a).

Suppose that u_0 has a special value u_0^* which is such that the function β in (2.20) vanishes:

$$\beta(u_0^*) = 0. \tag{2.21}$$

Then the form of (2.20) becomes even simpler; it is like an Euler equation expressing a homogeneity property of $\tilde{\Gamma}$. When combined with naive dimensional analysis, it yields the desired scaling behaviour of the vertex functions. For example, since $\tilde{\Gamma}^{(p)}$ has the same dimension as $G^{(p)}$ with p external propagators removed, then

$$\begin{aligned}[\tilde{\Gamma}^{(p)}] &= [G^{(p)}]/[G^{(2)}]^p \\ &= [\phi]^{-p}[x]^{-d} \\ &= [\Lambda]^{-p(d/2-1)+d}\end{aligned} \tag{2.22}$$

by eq. (1.3). Thus $\tilde{\Gamma}^{(p)}(q, t, u_0^*, \Lambda)$ must have the form

$$\tilde{\Gamma}^{(p)}(q, t, u_0^*, \Lambda) = \Lambda^{-p(d/2-1)+d} \cdot g^{(p)}(q/\Lambda, t/\Lambda^2, u_0^*) \tag{2.23}$$

In particular, if we consider the vertex functions at $q = 0$, eqn (2.20) implies that $g^{(p)}(0, t/\Lambda^2, u_0^*)$ must obey

$$\left[d - p\left(\frac{d}{2} - 1\right) - \frac{p}{2}\gamma_3(u_0^*)\right] g^{(p)}(0, x, u_0^*) = [2 + \gamma_4(u_0^*) - \gamma_3(u_0^*)] \times x\frac{\partial}{\partial x} g^{(p)}(0, x, u_0^*) \tag{2.24}$$

which has solution

$$g^{(p)}(0, x, u_0^*) \propto x^{\{d - p[d/2 - 1 + \gamma_3(u_0^*)/2]\}/[2 + \gamma_4(u_0^*) - \gamma_3(u_0^*)]}. \tag{2.25}$$

Thus at $q = 0$, the p-point vertex function scales like a power of the reduced temperature t. Similarly if we set $t = 0$,

$$\left(\Lambda\frac{\partial}{\partial\Lambda} - \gamma_3(u_0^*)\right)\tilde{\Gamma}^{(2)}(q, 0, u_0^*, \Lambda) = 0$$

and hence

$$\tilde{\Gamma}^{(2)} \propto q^2(q/\Lambda)^{-\gamma_3(u_0^*)} \tag{2.26}$$

Equations (2.23), (2.25) and (2.26) contain the scaling behaviour of immediate interest. They can be recast in terms of conventional exponents η and γ defined by

$$\tilde{\Gamma}^{(2)}(q = 0) = r \propto t^{\gamma} \tag{2.27}$$

and

$$\tilde{\Gamma}^{(2)}(q, t = 0) \propto q^{2-\eta}. \tag{2.28}$$

One therefore identifies

$$\gamma = [2 - \gamma_3(u_0^*)]/[2 + \gamma_4(u_0^*) - \gamma_3(u_0^*)] \tag{2.29}$$

$$\eta = \gamma_3(u_0^*) \tag{2.30}$$

It is straightforward to use eqns (2.27), (2.29) and (2.30) to re-express the vertex functions at $q = 0$ in powers of r. The result may be written

$$\Gamma^{(p)}(0, r, u_0^*, \Lambda) \propto \Lambda^{-p(d/2-1)+d}(r/\Lambda^2)^{[d-p(d/2-1+\eta/2)]/(2-\eta)} \tag{2.31}$$

In order to calculate η and γ explicitly, one needs the functions $\beta(u_0)$, $\gamma_4(u_0)$ and $\gamma_3(u_0)$. One obtains three simultaneous equations for them by using the fact that (2.20) is satisfied for any three distinct vertex functions. The conventional choices are (a) $\tilde{\Gamma}^{(2)}(0, t, u_0, \Lambda)$, (b) $\tilde{\Gamma}^{(4)}(0, t, u_0, \Lambda)$ and (c) $\tilde{\Gamma}^{(2)}(q, 0, u_0, \Lambda)$.

FIG. 2. One-loop graph contributing to $\Gamma^{(4)}$

In case (a), $\tilde{\Gamma}^{(2)}(0, t, u_0, \Lambda)$ is simply r itself and is hence given from (2.13) as

$$\tilde{\Gamma}^{(2)}(0, t, u_0, \Lambda) = t\{1 + \tfrac{1}{12}(u_0 S_d(2\pi)^{-d})(n + 2)(\ln t/\Lambda^2 + \ldots) + O(u_0^2, u_0\varepsilon)\}. \tag{2.32}$$

The lowest order contributions to $\Gamma^{(4)}$ come from Fig. 2; they give

$$\tilde{\Gamma}^{(4)}_{ijkl}(0, t, u_0, \Lambda) = \tfrac{1}{3}(\delta_{ij}\delta_{kl} + \delta_{il}\delta_{jk} + \delta_{ik}\delta_{jl})u_0\Lambda^{\varepsilon}$$
$$\times \left\{1 + \frac{n+8}{6}[u_0 S_d(2\pi)^{-d}](\tfrac{1}{2}\ln t/\Lambda^2 + \ldots) + O(u_0^2, u_0\varepsilon)\right\}. \tag{2.33}$$

The lowest-order self-energy graph Fig. 1 does not contribute to $\tilde{\Gamma}^{(2)}(q, 0, u_0, \Lambda)$ because as we have seen it is independent of the external momentum q.

The lowest-order contribution to $\tilde{\Gamma}^{(2)}(q, 0, u_0, \Lambda)$ is shown in Fig. 3. It is of order u_0^2 and when substituted into (2.20) implies $\gamma_3(u_0)$ is also of order u_0^2. The fact that expressions (2.32) and (2.33) obey (2.20) yields

$$\beta(u_0) = -\varepsilon u_0 + \tfrac{1}{6}(n+8)u_0(u_0 S_d(2\pi)^{-d}) + O(u_0^3, u_0^2\varepsilon) \tag{2.34}$$

$$\gamma_3(u_0) = O(u_0^2) \tag{2.35}$$

$$\gamma_4(u_0) = \tfrac{1}{6}(n+2)u_0 S_d(2\pi)^{-d} + O(u_0^2, u_0\varepsilon) \tag{2.36}$$

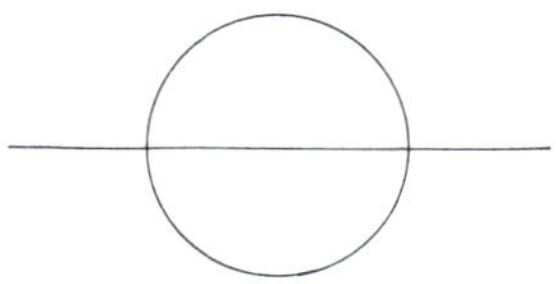

FIG. 3. Two-loop graph contributing to the self-energy Σ.

The reader can easily check that the special value u_0^* of (2.14) satisfies $\beta(u_0^*) = 0$ to order ε^2.

It should be clear in principle how to calculate η and γ to higher orders in ε; one requires $\tilde{\Gamma}^{(p)}$ to the appropriate order in (u_0, ε). The calculation would be rather involved because the Feynman rules of I.C give $\Gamma^{(p)}$ as a function of r and a substitution is required to give $\tilde{\Gamma}$ as a function of t. The prescription at the beginning of this section is just a streamlining of this procedure which eliminates the substitution procedure and requires only $\Gamma^{(p)}$ as functions of r.

Let us complete this section by summarizing the technique:

(a) To obtain the special value of the coupling constant u_0^* to order ε calculate the four-point function at zero external momentum up to one loop, discarding higher terms of order ε, as in (2.33). Noting that η comes at lowest order from the two-point graph in Fig. 3 and is hence of order ε^2, compare this expression for $\Gamma^{(4)}$ with the expansion of (2.31) for $p = 4$,

$$\Gamma^{(4)}(q = 0) \propto 1 + (\varepsilon/2)\ln r + O(\varepsilon^2)$$

to fix u_0^* to order ε as in (2.14).

(b) Knowing u_0^* to order ε, one can calculate $\Gamma^{(2)}(q, r = 0)$ to order ε^2 from Fig. 3. When higher orders in ε are discarded in evaluating the integrals in the graph, the result is proportional to $q^2 \ln(q/\Lambda)$, with a known coefficient. Specifically one finds

$$\Gamma^{(2)}(q, r = 0, u_0^*, \Lambda)$$

$$= q^2 - (n+2)\frac{u_0^{*2}}{18}\int_{p,k}\frac{1}{k^2}\frac{1}{(p-k)^2}\left[\frac{1}{(q+p)^2} - \frac{1}{p^2}\right] + O(u_0^{*3})$$

$$= q^2 - (n+2)\frac{u_0^{*2}}{18}\left[\frac{S_d}{2(2\pi)^d}\right]^2 q^2 \ln(q/\Lambda) + O(u_0^{*3}, u_0^{*2}\varepsilon)$$

Comparing this expression with the expansion of (2.28)

$$q^{2-\eta} = q^2[1 - \eta \ln q + O(\eta^2)]$$

and using the known value of u_0^* gives

$$\eta = \frac{(n+2)\varepsilon^2}{2(n+8)^2} + O(\varepsilon^3) \tag{2.37}$$

(c) Now the procedure can be iterated to higher order in ε: the correct behaviour of Γ^4 $(q = 0)$ is known from (2.31) to order ε^2 and enables u_0^* to be calculated to order ε^2. η can then be calculated to order ε^3 and so on.

(d) The exponent γ is calculated from (2.4) by similar exponentiation, as shown in eqns (2.13)–(2.17).

We shall discuss the known results of higher-order calculations in the next section.

The reader will appreciate that in order to avoid duplication with the article by Brézin *et al.* in this volume we have omitted a complete discussion of the solutions of the renormalization group equation (2.20). The reader is referred to that article for the general solution by the method of characteristics, in which u_0^* appears as a stable fixed point of the differential equation

$$\frac{du_0(\tau)}{d\tau} = \beta[u_0(\tau)].$$

It is also appropriate to mention that the $1/n$ expansion illustrates very clearly the existence and role of the special value u_0^* (see particularly Wilson, 1974).

D. Results for η and γ

The exponents γ and η were obtained to orders ε^2 and ε^3 respectively by Wilson (1972). The calculation was extended to order ε^3 and ε^4 respectively, by Nickel (1972, 1974) using the method discussed here. The same results were obtained using renormalized perturbation theory by Brézin *et al.* (1973b); see their article and further references in it. To calculate to this order, the special value of u_0^* must be calculated to order ε^3. The appropriate expressions must be known for all 4-point graphs with up to and including three loops. These graphs are listed completely in Fig. 4. The corresponding

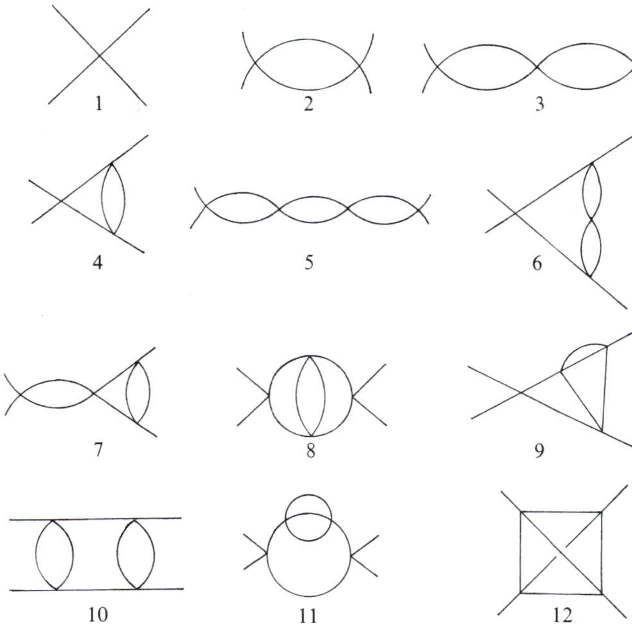

FIG. 4. All graphs with up to three loops contributing to $\Gamma^{(4)}$.

TABLE I. Multiplicities and analytic expressions for $\Gamma^{(4)}(0, r, u_0, \Lambda)$, from Nickel (1974). The graphs are numbered as in Fig. 4. The cut-off Λ has been set equal to 1. The multiplicities differ from those of Nickel because of our convention (1.1) for the normalization of u_0.

Graph	Multiplicity	Analytic expression
4.1	1	1
4.2	$6^{-1}(n+8)$	$\ln r + 2 - \varepsilon(\frac{1}{4}\ln^2 r + \frac{1}{2}\ln r)$ $+ \varepsilon^2(\frac{1}{24}\ln^3 r + \frac{1}{8}\ln^2 r + \frac{1}{24}\pi^2 \ln r)$
4.3	$6^{-2}(n^2 + 6n + 20)$	$\ln^2 r + 4\ln r + 4 - \varepsilon(\frac{1}{2}\ln^3 r + 2\ln^2 r$ $+ 2\ln r)$
4.4	$3^{-2}(5n + 22)$	$\frac{1}{2}\ln^2 r + \ln r - 2 - \frac{1}{2}\pi^2 + 4\lambda - \varepsilon(\frac{1}{4}\ln^3 r$ $+ \frac{1}{2}\ln^2 r + (\frac{1}{2} - \lambda)\ln r)$
4.5	$6^{-3}(n^3 + 8n^2 + 24n + 48)$	$\ln^3 r + 6\ln^2 r + 12\ln r$
4.6	$3^{-3}2^{-1}(3n^2 + 22n + 56)$	$\frac{1}{3}\ln^3 r + \ln^2 r - 2\lambda \ln r$
4.7	$3^{-3}2^{-1}(3n^2 + 22n + 56)$	$\frac{1}{2}\ln^3 r + 2\ln^2 r - (\frac{1}{2}\pi^2 - 4\lambda)\ln r$
4.8	$3^{-3}2^{-2}(3n^2 + 22n + 56)$	$\frac{1}{3}\ln^3 r + \ln^2 r - (4 + \pi^2 - 10\lambda)\ln r$
4.9	$2.3^{-3}(n^2 + 20n + 60)$	$\frac{1}{6}\ln^3 r - (2 + \frac{1}{2}\pi^2 - 4\lambda)\ln r$
4.10	$2^{-1}3^{-3}(n^2 + 20n + 60)$	$\frac{1}{3}\ln^3 r + \ln^2 r - 2\lambda \ln r$
4.11	$2^{-1}3^{-3}(n+2)(n+8)$	$\frac{1}{4}\ln^2 r + (1 - \frac{2}{3}\lambda)\ln r$
4.12	$3^{-3}(5n + 22)$	$6\zeta(3)\ln r$

multiplicity factors and analytic expressions are listed in Table I. At each order in u_0, there is an extra loop with an extra momentum variable and hence an extra loop integration; it is convenient to absorb the factor $S_d(2\pi)^{-d}$ into the coupling constant and define

$$u = \tfrac{1}{2}u_0 S_d(2\pi)^{-d}. \tag{2.38}$$

The remaining expression for the integral is what is shown in the table. The contribution of any graph to $\Gamma^{(4)}$ is obtained by multiplying the corresponding expressions in the columns of Table I and including a factor $u_0 u^{l-1}$ if there are l vertices (and hence $l - 1$ loops). Knowing η to order ε^2 as in (2.37) it should not be difficult for the reader to obtain u^* to order ε^2; he will probably require a little enthusiasm to obtain the third order result

$$u = \frac{3\varepsilon}{n+8}\left[1 + \left(\frac{9n+42}{(n+8)^2} - \tfrac{1}{2}\right)\varepsilon + \left(\tfrac{1}{4} + \frac{(\pi^2 - 10\Lambda)(5n-22) + 4(n+2)\Lambda/3}{2(n+8)^2} - \frac{\pi^2}{24} + \frac{3n^3 + 160n^2 + 1192n + 2632 - 12(5n+22)(n+8)\zeta(3)}{(n+8)^4}\right)\varepsilon^2\right] + O(\varepsilon^4). \tag{2.39}$$

Some comments are in order here.

(a) In Table I the cut-off Λ has been set to 1 by a suitable choice of length dimension.

(b) In high order calculations, the low order graphs must be calculated to the appropriate order in ε, using the expansion (2.12) or some other suitable means. This applies in all calculations with this method.

(c) At higher orders one has an explicit check on the calculation because the coefficients of $\varepsilon^p \ln^p r$ are fixed if the coefficient of $\varepsilon \ln r$ is known. For example when calculating $\Gamma^{(4)}(q = 0)$ to order ε^2, the coefficient of $\varepsilon^2 \ln^2 r$ must be 1/8 since it arises from the expansion of $r^{(\varepsilon/2)+O(\varepsilon^2)}$. Now, the coefficients of $\varepsilon^p \ln^p r$ are the leading logarithmic singularities coming from p loop graphs and from graphs with less than p loops evaluated to the compensating power in ε; they are fixed by the value of u_0^* at order ε and must have a coefficient which ensures exponentiation. There are similar exponentiation checks for the coefficients of $\varepsilon^p \ln^k r$ ($k \geqslant 2$); only the matching of $\varepsilon^p \ln r$ provides new information.

(d) At order ε^p, constant terms (non-logarithms) from Table I of order ε^{p-1} are also necessary. They must be carefully factored out into the constant of proportionality in (2.31).

(e) The evaluation of the integrals produces non-rational numbers at order ε^3. The symbol π has its usual value, $\zeta(3)$ is the Riemann ζ function

$$\zeta(3) = 1{\cdot}202\ldots \tag{2.40}$$

and λ is defined by the integral

$$\lambda = \int_0^\infty \frac{\theta d\theta}{2\cosh\theta - 1} = 1{\cdot}171\ldots \tag{2.41}$$

(f) The functions β, γ_4 and γ_3 may be different if one uses a different form of cut-off (compare the values for u_0^* given in the papers by Wilson (1972) and Brézin *et al.* (1973a) with the above result), but the exponents are the same in all cases. This is perhaps most clearly understood in the context of renormalized perturbation theory—two theories with a different cut-off can give the same renormalized vertex functions. See the article by Brézin *et al.* in this volume.

With the u_0^* of (2.39), one can calculate γ and η to order ε^3 and ε^4 respectively as described in the previous section. The two-point function graphs required are listed in Fig. 5. For γ, the integrals are evaluated at $q = 0$ with a subtraction of their value at $r = 0$ according to (2.4); the results (Nickel, 1972) are given in Table II. (Again units have been chosen so that $\Lambda = 1$). For

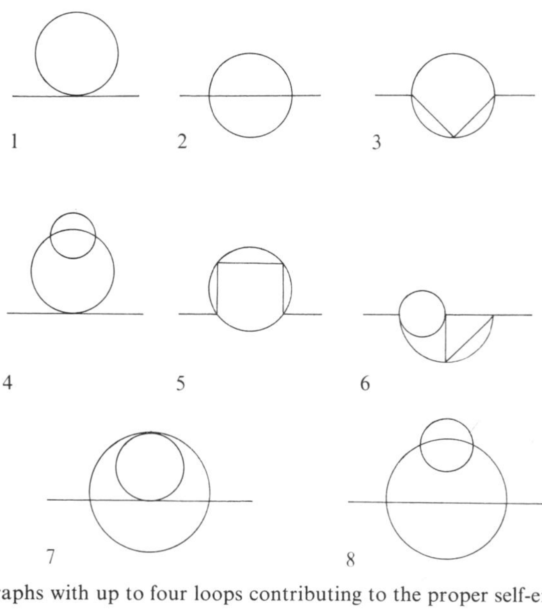

FIG. 5. All graphs with up to four loops contributing to the proper self-energy Σ.

TABLE II. Multiplicities and analytic expressions for $-r^{-1}(\Sigma(0, r) - \Sigma(0, 0))$, from Nickel (1972). The graphs are numbered as in Fig. 5. The cut-off Λ has been set equal to 1. The analytic expression for Fig. 5.1 differs from Nickel's because of the convention (2.38) for the normalisation of u.

Graph	Multiplicity	Analytic expression
5.1	$6^{-1}(n+2)$	$\ln r + 1 - \frac{\varepsilon}{4}\ln^2 r + \frac{\varepsilon^2}{24}(\ln^3 r + \pi^2 \ln r)$
5.2	$2^{-1}3^{-2}(n+2)$	$\frac{3}{2}\left[\ln^2 r - 4 - \pi^2 + 8\lambda - \frac{\varepsilon}{2}(\ln^3 r - \ln^2 r + 4(1-\lambda)\ln r)\right]$
5.3	$2^{-2}3^{-3}(n+2)(n+8)$	$\ln^3 r - 2\ln^2 r - (4 + 2\pi^2 - 14\lambda)\ln r$
5.4	$2^{-2}3^{-3}(n+2)^2$	$-\frac{1}{2}\ln^3 r - \ln^2 r - (8 + \frac{3}{2}\pi^2 - \frac{50}{3}\lambda)\ln r$

η, the integrals are evaluated at $r = 0$, with a subtraction of their value at $q = 0$ according to (1.29) or (1.32); the results (Nickel, 1974) are given in Table III. As remarked previously, the lowest order graph Fig. 5.1 is independent of q and does not contribute to η, so that η can be calculated to $O(\varepsilon^4)$ knowing u_0^* to order ε^3.

The exponents γ^{-1} and η are determined from the resulting coefficients of $\varepsilon^k \ln r$ or $\varepsilon^k \ln q$, higher powers of logarithms giving only checks on exponentiation.

TABLE III. Multiplicities and analytic expressions for $k^{-2}(\Sigma(k,0) - \Sigma(0, 0))$, from Nickel (1974). The graphs are numbered as in Fig. 5. The cut-off Λ has been set equal to 1.

Graph	Multiplicity	Analytic expression
5.2	$2^{-1}3^{-2}(n+2)$	$\ln k + \frac{1}{4} - \frac{2\lambda}{3} - \varepsilon(\ln^2 k - \frac{9}{4}\ln k) + \varepsilon^2(\frac{2}{3}\ln^3 k - \frac{9}{4}\ln^2 k + \frac{67}{16}\ln k)$
5.3	$2^{-2}3^{-3}(n+2)(n+8)$	$2\ln^2 k - 3\ln k - \varepsilon\left[\frac{10}{3}\ln^3 k - 10\ln^2 k + \left(\frac{63}{4} - \frac{\pi^2}{6}\right)\ln k\right]$
5.5	$2^{-3}3^{-4}(n+2)(n^2+6n+20)$	$4\ln^3 k - 9\ln^2 k + \frac{21}{2}\ln k$
5.6	$2^{-2}3^{-4}(n+2)(5n+22)$	$\frac{8}{3}\ln^3 k - 8\ln^2 k + (8 - \pi^2 + 10\lambda)\ln k$
5.7	$2^{-2}3^{-4}(n+2)(5n+22)$	$\frac{4}{3}\ln^3 k - 5\ln^2 k + (\frac{5}{2} - \pi^2 + 10\lambda)\ln k$
5.8	$2^{-2}3^{-3}(n+2)^2$	$\frac{1}{2}\ln^2 k - (1 + \frac{2}{3}\lambda)\ln k$

The results for the exponents are

$$\gamma = 1 + \frac{(n+2)}{2(n+8)}\varepsilon + \frac{(n+2)}{4(n+8)^3}(n^2+22n+52)\varepsilon^2$$
$$+ \frac{(n+2)}{8(n+8)^5}\{n^4+44n^3+664n^2+2496n+3104-48(5n+22)$$
$$\times (n+8)\zeta(3)\}\varepsilon^3 + O(\varepsilon^4) \qquad (2.42)$$

$$\eta = \frac{(n+2)}{2(n+8)^2}\cdot\varepsilon^2\left\{1 + \frac{(-n^2+56n+272)}{4(n+8)^2}\varepsilon + \frac{1}{16(n+8)^4}[-5n^4-230n^3\right.$$
$$\left.+1124n^2+17920n+46144-384(5n+22)(n+8)\zeta(3)]\varepsilon^2\right\}$$
$$+ O(\varepsilon^5). \qquad (2.43)$$

Numerical values of γ and η are listed for $n = 1$, 2 and 3 in Tables IV and V and compared with the results of high-temperature series (HTS) expansions.

Several comments are again worth making.

(a) The low-order terms in these ε-expansions clearly give desirable corrections to mean field theory when $\varepsilon = 1$ ($d = 3$). However the highest-order terms do not improve agreement with HTS expansions; this should not be puzzling since the ε-expansion involves a perturbation expansion in the coupling constant u_0, and this expansion is expected to be an asymptotic one with zero radius of convergence.

The asymptotic nature of the ε-expansion is also indicated numerically by the approximate recursion formula (Wilson and Kogut, 1974) but this evidence must be treated with caution since the approximate recursion formula is accurate at best to order η, i.e. to order ε^2 in the ε-expansion. An attempt by Nickel (1974) to study the convergence of the ε-expansion and to improve on these results is reviewed in the next section.

(b) Many of the other common exponents, e.g. α, β, δ, ν are obtained from η and γ by the well known scaling laws which are provable from the renormalization group equation (2.20). (See the article by Brézin *et al.* in this volume.) The results are tabulated in Wilson and Kogut (1974, p. 133).

(c) The exponent γ can also be obtained from the $\langle\phi^2\phi\phi\rangle$ vertex function at zero momentum (Brézin *et al.*, 1973c; Nickel, 1974). The graphs involved are the subset of Fig. 4 in which at least two external legs meet at a vertex (the ϕ^2 vertex). The same analytic expressions are required as for $\Gamma^{(4)}$ ($q = 0$), only the multiplicities of the graphs being different. Now the

$\langle \phi^2 \phi\phi \rangle$ Green function at zero external momentum is obtained by taking

$$\frac{\partial}{\partial T} \langle \phi\phi \rangle = \frac{\partial}{\partial T}\left(\frac{1}{r}\right)$$

and hence scales like $r^{-1}t^{-1}$. The two external propagator legs $(r)^{-2}$ are removed to obtain the vertex function which therefore behaves like rt^{-1} or $r^{(\gamma-1)/\gamma}$. Thus with the special value of u^* in eqn (2.39) the logarithms in the above diagrams will exponentiate to give $(\gamma - 1)/\gamma$ to order ε^3, and hence γ itself.

(d) The exponents η and γ have also been obtained for arbitrary d $(2 < d < 4$, i.e. $0 < \varepsilon < 2)$ by expansion in powers of n^{-1}, to order n^{-1}. (Abe 1972, 1973; Wilson 1974; Fisher *et al.*, 1972; Brézin and Wallace, 1973; Suzuki, 1972; Ferrell and Scalapino, 1972; Ma, 1973 and his article in this volume).

$$\gamma = \frac{1}{1 - \varepsilon/2} - \frac{3\varepsilon}{n} \frac{\sin(\pi\varepsilon/2)}{(\pi\varepsilon/2)} \frac{\Gamma(2-\varepsilon)}{\Gamma^2(2-\varepsilon/2)} + O\left(\frac{1}{n^2}\right) \tag{2.44}$$

$$\eta = \frac{\varepsilon^2}{n} \frac{\sin(\pi\varepsilon/2)}{(\pi\varepsilon/2)} \frac{\Gamma(2-\varepsilon)}{\Gamma(1-\varepsilon/2)\Gamma(3-\varepsilon/2)} + O\left(\frac{1}{n^2}\right) \tag{2.45}$$

Although this expansion is very poor numerically for small n, these results are interesting from many points of view. As far as the ε-expansion is concerned, the important fact is that expansions in powers of n^{-1} of the exponents (2.42) and (2.43) coincide with the ε-expansion of (2.44) and (2.45).

(e) The factors of $(n + 2)$ in (2.42) and (2.43) appear in all orders of perturbation theory (Balian and Toulouse, 1973; Fisher 1973) because the two point function, which is proportional to δ_{ij}, comes from contractions of the fields ϕ in vertices proportional to $\phi_i\phi_j\phi_k\phi_l$ $(\delta_{ij}\delta_{kl} + \delta_{il}\delta_{jk} + \delta_{ik}\delta_{jl})$. Thus any two point graph always involves at least one factor of a trace of the tensor coupling:

$$\delta_{kl}[\delta_{ij}\delta_{kl} + \delta_{ik}\delta_{jl} + \delta_{il}\delta_{jk}] = (n+2)\delta_{ij}$$

E. The irreducible diagram series

We have already remarked on the rather unsatisfactory convergence of the ε-expansion in Tables IV and V. Arbitrary re-summation methods such as Padé approximants are unlikely to give any more meaningful results with such short Taylor series, and in the absence of any definite justification for their use. Nickel (1974) has attempted with considerable success to improve on this situation along the following lines.

It is evident from Tables IV and V that the term which is dominant numerically and spoiling the convergence of the expansion is the $\zeta(3)$ term.

TABLE IV. Contributions to γ of orders 1 to ε^3, for $n = 1$, 2 and 3, from (2.42). $\zeta(3) \simeq 1{\cdot}202$ is the Riemann zeta function. The results for series expansions in three dimensions are from (a) Gaunt and Guttmann, Vol 3, p. 198 (b) Ferer *et al.* (1973) (c) Ferer *et al.* (1971).

n	1	ε	ε^2	ε^3	Sum ($\varepsilon = 1$)	Series
1	1	0·167	0·077	0·040–0·074$\zeta(3)$	1·195	1·250 ± 0·001[(a)]
2	1	0·2	0·1	0·056–0·077$\zeta(3)$	1·263	1·318 ± 0·01[(b)]
3	1	0·227	0·119	0·069–0·076$\zeta(3)$	1·324	1·405 ± 0·02[(c)]

TABLE V. Contributions to η of orders ε^2 to ε^4, for $n = 1$, 2 and 3, from (2.43). The results for series expansions in three dimensions are obtained from the scaling law $\eta = 2 - \gamma/\nu$ (see e.g. Stanley 1971, § 12), where γ is given in Table IV and ν comes from (a) M. A. Moore (private communication, $\nu = 0{\cdot}640 \pm 0{\cdot}003$), (b) and (c) as in Table IV.

n	ε^2	ε^3	ε^4	Sum ($\varepsilon = 1$)	Series
1	0·019	0·019	0·011–0·017$\zeta(3)$	0·029	0·047 ± 0·01[(a)]
2	0·020	0·019	0·011–0·016$\zeta(3)$	0·031	0·04 ± 0·01[(b)]
3	0·021	0·018	0·009–0·014$\zeta(3)$	0·031	0·040 ± 0·008[(c)]

This comes purely from the $\zeta(3)$ term in u^* which is required to balance the $\zeta(3)$ term from the tetrahedron graph Fig. 4.12 in order to obtain the correct exponent (2.31). This graph has the property that it diverges in four dimensions like a single power of $\ln r/\Lambda^2$ (See Table I). This property is associated with the fact that the graph is topologically *irreducible*, i.e. has no vertex or self-energy insertions. Thus if any line of such a graph is cut the resultant six point function has no ultraviolet divergent sub-integrations in four dimensions—only the final overall integration produces a single logarithmic divergence which appears as $\ln r/\Lambda^2$ since only r is available to give a dimensionless argument of the logarithm. Diagrams with vertex insertions, such as Figs. 4.3–4.10 diverge with correspondingly high powers of logarithms.

Nickel has suggested improving the ε-expansion by re-ordering perturbation theory according to the degree of reducibility (number of vertex and self-energy insertions) of a diagram. Thus if the tetrahedron diagram is included in calculating u_0^*, then so also should be the contribution of *all* irreducible diagrams in his resummation prescription.

Two main difficulties present themselves in this programme. Firstly, one cannot calculate all irreducible diagrams. However there are a controllable number of them up to six loops, whose multiplicities and logarithmic divergences Nickel has calculated. Secondly the results show that these first four terms of the sum of irreducible diagrams converge very badly for $\varepsilon = 1$.

However, they do respond very well to resummation by continued fractions.

This use of continued fractions for resumming the contributions of irreducible diagrams is supported numerically by the "single-field" model studied by Nickel in which (i) the total multiplicity of all irreducible graphs of a given number of loops is generated iteratively by generating function techniques and (ii) the coefficients of $\ln r/\Lambda^2$ of all irreducible graphs at order u_0^k are assumed to increase in a geometric progression a^k. This latter assumption is an approximate feature of the explicit calculations at "low" orders and is not nonsense from a general point of view—for example one expects the value of such integrals without self energy subtractions to be controllable by upper and lower bounds of the *same* sign. Anyway, the resultant Taylor series of coefficients $(\Sigma_n c_n \varepsilon^n) \ln r/\Lambda^2$ diverges dramatically (because of the increasing multiplicities), but the continued fraction sum converges numerically.

The modifications to u_0^* from the irreducible graphs produce modifications in γ and η. When these modifications are resummed by continued fractions the nett effect is a suppression of the coefficient of $\zeta(3)$ by a factor of about 0·4.

The final best estimates coming from this work are shown in Table VI, page 352. The error bars are Nickel's estimation of error from the convergence behaviour of the continued fraction and other considerations. Compare Tables IV and V.

Although the consistency of and justification for this resummation procedure require clarification, Nickel's paper surely establishes

(i) that the ε-expansion can with care and application be expected to yield very reliable results in three dimensions;

(ii) the existence of techniques which make feasible the evaluation of rather high-order diagrams.

F. Notes on evaluation of graphs

The most explicit and comprehensive review of techniques which enable one to evaluate the required logarithmic and constant terms of the quasi four-dimensional integrals is contained in the article by Nickel (1974). We refer the reader to that article for details; here we shall outline only the basic ideas behind the techniques. For brevity we choose units in which the cut-off $\Lambda = 1$.

The first point to note is that low-order diagrams are required to high order in ε and high-order diagrams to correspondingly lower orders in ε. It is useful that the first-order diagram in $\Gamma^{(4)}$ (Fig. 4.2) can be evaluated directly in d dimensions by contour integration techniques, or as discussed in (2.8) *et seq*:

$$\int_k (r + k^2 + k^4)^{-2} = \frac{S_d}{(2\pi)^d}\frac{1}{2}\int_0^\infty \frac{t^{1-\varepsilon/2}}{(r + t + t^2)^2}\,dt$$
$$= -\frac{1}{2}\frac{S_d}{(2\pi)^d}\left[\frac{(\pi\varepsilon/2)}{\sin(\pi\varepsilon/2)}\{-[(2/\varepsilon) - 1]r^{-\varepsilon/2} + (2/\varepsilon) + 1\}\right], \tag{2.46}$$

dropping all terms that vanish as $r \to 0$. The ε-expansion of the square brackets appears in Table I.

The evaluation of higher-order diagrams is much more tedious. In principle one can use Feynman parameter methods to perform angular integrals, and evaluate radial integrals to the required order in ε by the expansion (2.12), but this is not a practical method for complicated graphs. An improvement is to evaluate angular integrals directly with the measure

$$\int_0^\infty k^{3-\varepsilon}\,dk \int_0^\pi \sin^{2-\varepsilon}\psi\,d\psi \ldots. \tag{2.47}$$

If higher orders in ε are required one must expand $k^{-\varepsilon}$ and $(\sin\psi)^{-\varepsilon}$ as power series in ε, giving integrands with $\ln k$ and/or $\ln(\sin\psi)$.

The central point to bear in mind is that integrands should always be simplified by throwing away terms which are smaller than the leading terms by $O(r)$. For example, this allows the replacement in the four-point function $\Gamma^{(4)}$ of the free propagator by

$$\frac{1}{r + q^2 + q^4} \to \frac{1}{r + q^2} - \frac{1}{1 + q^2}. \tag{2.48}$$

Angular integrations such as (2.47) are much simpler with this form of propagator. (Note that one cannot use this approximation when calculating $\Gamma^{(2)}(q = 0, r)$, as the lowest-order graph illustrates.)

Generally speaking, the approach one takes depends on the degree of reducibility of the diagram.

(i) If a diagram is highly reducible with many e.g. bubble or "triple" subgraphs (see Fig. 6 for the labelling of momenta) then correspondingly there will be less complicated angular dependences. One can calculate exactly (in $d = 4$) the bubble and triple graphs as a function of their

FIG. 6. Bubble and triple subgraphs appearing frequently in reducible graphs: see e.g. (2.49).

external momentum. The integrals of high reducible graphs are then integrals over integrands formed mainly from these subintegrations. For example, if we denote the bubble and triple graphs by $P(k^2, r)$ and $T(k^2, r)$, then the graph (4.9) may be written

$$\int_0^\infty \left(\frac{1}{r + k^2} - \frac{1}{1 + k^2}\right) P(k^2, r)T(k^2, r)k^{3-\varepsilon}\, \mathrm{d}k \tag{2.49}$$

Although P and T are complicated functions (see Nickel 1974 eqns A1.2, A1.3) the logarithmic ($\ln r$) terms from the integral (2.49) are determined by asymptotic forms of P and T and can be evaluated explicitly with at most the introduction of constants π^2 and λ which appear in the integrals

$$\pi^2 = 6\int_0^\infty \frac{\mathrm{d}t}{t}\frac{1}{1+t}\ln(1+t)$$

$$\lambda = \int_0^\infty \frac{\theta\, \mathrm{d}\theta}{2\cosh\theta - 1}$$

$$\simeq 1{\cdot}171\ldots,$$

or equivalent variations.

(ii) At the other end of the spectrum are irreducible diagrams. These are characterized by complicated angular integrations (especially when expressed in momentum space), although this is balanced by the fact that the graphs diverge only like $\ln r$ in four dimensions so that one does not have to dig deeply beneath $\ln^p r$ divergences to get at the essential $\ln r$ terms. The simplification here is to write down the integral in co-ordinate space; if one is interested only in picking out the $\ln r$ coefficient of an irreducible diagram, one can approximate propagators by their $r = 0$ form, $G(x) \sim |x|^{-2}$, cutting off integrals where necessary with an upper limit $x^2 \sim r^{-1}$ and a lower limit $x^2 \sim 1$ (corresponding to $\Lambda = 1$ in momentum space). The coefficient of $\ln r$ can then be obtained explicitly or as infinite series which can be summed numerically.

In evaluating the two-point function $\Gamma^{(2)}$ $(q, r = 0)$ Nickel (1972) shows how a Mellin transform method can be used for the two-loop graph, Fig. 3. However, in higher orders angular integrations can become very complicated and are best handled by expanding propagators, e.g. $(k - K)^2$ in terms of polynomials which are orthogonal in $4 - \varepsilon$ dimensions. The same technique is also useful for the $\Gamma^{(4)}$ graphs to evaluate angular integrals directly in $4 - \varepsilon$ dimensions rather than using the ε-expansion of (2.47). The interested reader will find details of this technique in Nickel's paper.

Finally, it is probably wise to end this section by remarking that there are several checks which one can make on such calculations. For example:

(i) Exponentiation: as remarked before, the coefficients of final terms of the form $\varepsilon^p \ln^k r$ ($p \geqslant k \geqslant 2$) are all pre-determined by lower order terms and the fact that the logarithms exponentiate.

(ii) In evaluating reducible graphs there is sometimes more than one way of writing down a graph in terms of bubble and triple subintegrations. For example Fig. 4.4 may be written in the two ways

$$\int k^{3-\varepsilon}\, \mathrm{d}k \left(\frac{1}{r+k^2} - \frac{1}{1+k^2}\right) T(k^2, r)$$
$$= \int k^{3-\varepsilon}\, \mathrm{d}k \left(\frac{1}{r+k^2} - \frac{1}{1+k^2}\right)^2 P(k^2, r) \tag{2.50}$$

One can explicitly check that one obtains the same answer from the two expressions.

(iii) The logarithmic divergences of some irreducible graphs in four dimensions turn out to be the same; this was first pointed out by Nickel and shown to be the consequence of inversion symmetries between integrands. The general criterion for establishing logarithmically equivalent graphs was obtained by Wallace and Zia (1975a) and is as follows.

Consider forming vacuum–vacuum like graphs by pulling together the external legs of an irreducible four-point graph. Some examples are shown in Fig. 7. In some, all vertices of the vacuum–vacuum are completely equivalent—removal of any vertex gives the same four-point graph from which the "vacuum" graph was formed. In others, all vertices of the vacuum graph are not equivalent and removal of a four-point vertex can result in different four-point graphs—see Figs 7.4 and 7.5.

Topologically, each such graph can have no vertex or self-energy insertions and is hence irreducible. There exists an inversion symmetry between integrands such that the coefficient of the $\ln r$ term is the *same* for all irreducible graphs arising from a given "vacuum" graph in this way.

(iv) It is probably worth commenting that the multiplicities of graphs are also related by means of a "covering" vacuum–vacuum graph. This is true also for multiplicities of two-point graphs. The statement of the general result is: (a) if two four-point functions are related by coming from the same vacuum graph, then the ratio of their multiplicities is given by the ratio of the number of times each is formed when each vertex of the vacuum graph is removed in term: (b) all two-point functions obtained from a vacuum graph by cutting a line of the vacuum graph have multiplicities in the ratio of the number of times each two-point graph is obtained by cutting all lines of the vacuum graph in turn. These results are readily proved using generating functions for weights of graphs (Wallace and Zia, 1975a) and numerous examples of them can be found in Tables I and III.

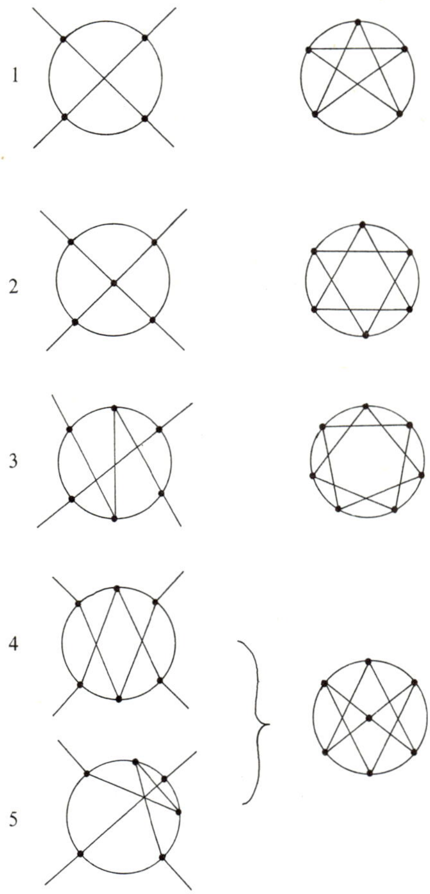

FIG. 7. Examples of irreducible four-point graphs and vacuum–vacuum-like graphs obtained by pulling together their four external legs at an extra vertex. Only the topological nature of the vacuum graphs is of interest; there is no implication of additional integrations by the contractions of the external legs.

III. Equation of State

A. Introduction

The equation of state describes the response of the system (measured by the mean magnetization M_i) to a constant external field (H_i) at a given temperature T. In general, if the system is isotropic, the equation relating the two vector quantities H_i and M_i may be written

$$H_i = M_i \tilde{f}(T, M^2) \tag{3.1}$$

The function $\tilde{f}$ is related to the free energy $F(T, M^2)$ (see eqns (1.35) *et seq.*) by

$$H_i = \frac{\partial F}{\partial M_i}(T, M^2). \tag{3.2}$$

Thus the function $\tilde{f}$ gives F up to an unknown function of T which we shall generally neglect in this article. (It is not unimportant—see D. R. Nelson, 1975, and references therein—but plays no role as far as the equation of state and other M-derivatives of F are concerned.)

The renormalization group transformation shows that in the critical region H and $T - T_c$ "small", the functions $\tilde{f}$ and F are homogeneous in $(T - T_c)$ and a power (critical exponent) of M. One may write the Widom scaling form (Widom, 1965; Domb and Hunter, 1965):

$$\tilde{f}(T, M^2) = M^{\delta - 1} f(t/M^{1/\beta}) \tag{3.3}$$

where t is the reduced temperature. There is strong phenomenological support for this homogeneity or scaling property both from experiment (see for example Vicentini Missoni *et al.* (1970)) and from numerical series expansion (Gaunt and Domb, 1970; Milošević and Stanley, 1972; Tarko and Fisher,1975).

In this Section we shall describe how the method of the previous Section may be used to obtain the scaling function f in (3.3) as a power series in ε, for the n-component Heisenberg system. The original papers on this calculation are by Avdeeva and Migdal (1972) and Brézin *et al.* (1972) on the $n = 1$ Ising system to ε^2, by Avdeeva (1973) and Brézin *et al.* (1973a) on the n-component system to ε^2, and by Wallace and Zia (1973, 1974) on the Ising system to ε^3.

The structure of this section is as follows. We shall first discuss the modification of the Feynman rules to handle the constant external field H and then show explicitly the calculation of the function f in low order. The function f has many interesting properties and we shall consider two of them: in Section D we shall discuss its analyticity properties (Griffiths 1967) and in Section E its behaviour near the coexistence curve $H \to 0$, $T < T_c$, where the $O(n)$ rotation invariance is spontaneously broken by the existence of the spontaneous magnetization M.

B. Setting up the calculation

Let us recall that the basic Hamiltonian we consider is given by

$$\mathscr{H}/kT = \int d^dx \left\{ \tfrac{1}{2}(\nabla\phi)^2 + \frac{\Lambda^{-2}}{2}(\nabla^2\phi)^2 + \tfrac{1}{2}r_0(T)\phi^2 + \frac{\bar{u}_0}{4!}(\phi^2)^2 - H\phi_1 \right\} \quad (3.4)$$

where we have chosen H in the direction of ϕ_1, by convention, and the same form of cut-off as previously. (In the papers by Brézin *et al.* (1972, 1973a) the cut-off is effected by putting an explicit upper limit Λ on the $|q|$ integration.) The starting point for all calculations of the equation of state is the identification of the mean magnetization density M with the expectation value of the field ϕ_1

$$M = \langle \phi_1(x) \rangle. \quad (3.5)$$

In a constant field H, the Hamiltonian (3.4) is translation invariant and $\langle \phi_1(x) \rangle$ is hence independent of x, so one is guaranteed a constant x-independent mean magnetization M as expected.

In principle, the right-hand side of (3.5) can be evaluated by Feynman graph expansion to give M as a function of H and T. The additional problem which arises now is that with the interaction $H\phi_1$, a ϕ_1 leg can "disappear into the vacuum" by contraction with $H\phi_1$. This gives rise to tadpole graphs such as in Fig. 8. It turns out that these tadpole graphs must be considered as the same order as the zeroth order diagram Fig. 9 if a valid description of the

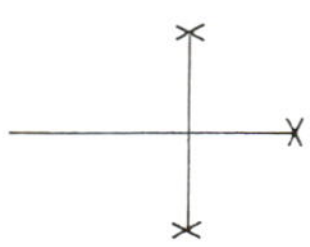

FIG. 8. Example of tadpole graph contributing to (3.5) when an external field is present.

FIG. 9. Zeroth order diagram contributing to the equation of state (3.5).

entire critical region H and $(T - T_c)$ small is to be obtained (see Lee, 1972, for references).

To see this, consider the zeroth order graph and the graph with one $(\phi^2)^2$ vertex and no loops (Fig. 8). These two graphs give

$$M = \frac{H}{r} + \frac{1}{6}\frac{\bar{u}_0 H^3}{r^4} + \ldots . \quad (3.6)$$

Now on the critical isotherm ($T = T_c$) the isotropic susceptibility parameter r, which appears in the propagators in the Feynman rules of Section I, becomes zero i.e. the second term (and correspondingly all higher terms) becomes increasingly important in this region.

There are at least two ways to handle this problem.

(a) One replaces the field $\phi_1(x)$ in the Hamiltonian by a new field $L(x)$ with zero expectation value i.e.

$$\phi_1(x) = L(x) + M \tag{3.7}$$

with

$$\langle L \rangle = 0. \tag{3.8}$$

By virtue of eqn (3.8) the field $L(x)$ never disappears into the vacuum to any order in perturbation theory and the problems involved in an expansion like (3.6) never appear. This is the method we shall use.

(b) The free energy F, whose derivative gives $f(t/M^{1/\beta})$ is given by the sum of all *zero-field* symmetric vertex functions with external legs at zero momentum as outlined in Section I. Thus in principle the free energy and hence the equation of state are calculable with the Feynman rules of the symmetric theory. It turns out that to obtain the free energy to a *given* order in ε, *all* symmetric vertex functions must be calculated to the same number of loops, i.e. terms in the ε-expansion are ordered by loops. One can do this at low orders by direct graphical calculation or more compactly by functional methods. This approach is discussed in the article by Brézin *et al.* in this volume (Chapter 3). See also Iliopoulos and Itzykson (1975) for references.

To return to the first method, when $\phi_1(x)$ is replaced by $L(x)$ according to (3.7), one obtains the Hamiltonian

$$\mathcal{H}/kT = \int \mathrm{d}^d x \Big\{ \tfrac{1}{2}(\nabla L)^2 + \tfrac{1}{2}(\nabla \phi_T)^2 + \tfrac{1}{2}\Lambda^{-2}(\nabla^2 L)^2 + \tfrac{1}{2}\Lambda^{-2}(\nabla^2 \phi_T)^2$$

$$+ \tfrac{1}{2}(r_0 + \tfrac{1}{2}\bar{u}_0 M^2)L^2 + \tfrac{1}{2}(r_0 + \tfrac{1}{6}\bar{u}_0 M^2)\phi_T^2 + \frac{\bar{u}_0}{4!}(L^2 + \phi_T^2)^2$$

$$+ \frac{\bar{u}_0}{3!} ML(L^2 + \phi_T^2) + [(r_0 + \tfrac{1}{6}\bar{u}_0 M^2)M - H]L \Big\}. \tag{3.9}$$

Here we have introduced the notation ϕ_T for the $n - 1$ fields transverse to the external field H, and

$$\phi_T^2 = \sum_{i=2}^{n} \phi_i^2.$$

Equation (3.8) is now the starting point for the calculation of the equation of state. In the "interaction picture" we must evaluate $\langle L \exp(-\mathcal{H}_{\text{int}}/kT)\rangle$

by Wick expansion. The simplest term which contributes when we expand in $\mathscr{H}_{\text{int}}$ is the term linear in L in the Hamiltonian (3.9). At the lowest order its coefficient must vanish according to (3.8) and we obtain the equation of state

$$H = r_0(T)M + \tfrac{1}{6}\bar{u}_0 M^3 + O(u_0) \tag{3.10}$$

where $O(u_0)$ represents higher-order terms in the perturbation expansion.

At this lowest order in perturbation theory we have already recognized that $r_0(T) \propto T - T_c$ (see eqn (1.23) or (2.3)) and with this identification we see that one has obtained the original Landau mean-field result for the equation of state. This is not surprising because the vanishing of the coefficient of $L(x)$ in expression (3.9) is precisely the statement that the value of M is the constant value of ϕ_1 which minimizes $\mathscr{H}/kT$ as a classical potential.

Equation (3.10) is very simply written, with the suitable choice of scales for the reduced temperature t and M

$$H/M^3 = t/M^2 + 1 \tag{3.11}$$

from which one reads off the mean field exponents $\delta = 3$, $\beta = \frac{1}{2}$, and

$$f(x) = x + 1 \qquad (x = t/M^{1/\beta}). \tag{3.12}$$

We shall use the ε-expansion to obtain systematic corrections to this function from the interaction terms in the Hamiltonian.

Before setting up the perturbation theory which enables us to do this we note that to obtain the mean-field result we must regard tM and $\bar{u}_0 M^3$ as the same order (zeroth) in perturbation theory. Thus if we regard $\bar{u}_0$ as $O(\varepsilon)$ then M^2 must be regarded as $O(\varepsilon^{-1})$. Now, the trilinear term in eqn (3.9) has a coefficient $u_0 M$ which must hence be regarded as $O(\varepsilon^{1/2})$. However a loop is generated in a graph with a fixed number of external legs by *one* additional quartic vertex or *two* additional trilinear vertices; this is one way of establishing the previous claim that the ε-expansion is an expansion by number of loops.

We now set up the systematic perturbation expansion for the calculation of higher order effects. The first point to notice is that the coefficients of L^2 and ϕ_T^2 in (3.9) are different. This is reflected in an anisotropic susceptibility tensor (1.16)

$$\chi_{ij} = \int \mathrm{d}^d x \, \langle (\phi_i(x) - M_i)(\phi_j(O) - M_j) \rangle, \qquad (i, j = 1 \ldots n). \tag{3.13}$$

There are two susceptibilities, longitudinal χ_L measuring response to a change of H_i parallel to H_i

$$\chi_L = \frac{\partial M}{\partial H_{\|}} \tag{3.14}$$

and transverse χ_T measuring response to a change in H_i perpendicular to H_i

$$\chi_T = \frac{\partial M}{\partial H_\perp} \tag{3.15}$$

In terms of the correlation functions, eqn (3.13) gives

$$\begin{aligned}\chi_L &= \int \mathrm{d}^d x \, \langle (\phi_1(x) - M)(\phi_1(O) - M) \rangle \\ &= \int \mathrm{d}^d x \, \langle L(x) \, L(O) \rangle \end{aligned} \tag{3.16}$$

and

$$\chi_T \delta_{ij} = \int \mathrm{d}^d x \, \langle \phi_i(x) \, \phi_j(O) \rangle, \qquad (i, j = 2, \ldots n). \tag{3.17}$$

We now parallel the treatment in the symmetric case by introducing two "renormalized masses" r_L and r_T for the longitudinal and transverse modes, and splitting the Hamiltonian (3.9) into a free part

$$\begin{aligned}\mathscr{H}_0/kT = \tfrac{1}{2} \int \mathrm{d}^d x \{ (\nabla L)^2 + (\nabla \phi_T)^2 + r_L L^2 + r_T \phi_T^2 \\ + \Lambda^{-2} (\nabla^2 L)^2 + \Lambda^{-2} (\nabla^2 \phi_T)^2 \} \end{aligned} \tag{3.18}$$

and an interaction

$$\begin{aligned}\mathscr{H}_1/kT = \int \mathrm{d}^d x \Big\{ \frac{\bar{u}_0}{4!} (L^2 + \phi_T^2)^2 + \frac{\bar{u}_0}{3!} M L (L^2 + \phi_T^2) \\ + \tfrac{1}{2}(r_0 - r_L + \tfrac{1}{2}\bar{u}_0 M^2) L^2 + \tfrac{1}{2}(r_0 - r_T + \tfrac{1}{6}\bar{u}_0 M^2) \phi_T^2 \\ + [(r_0 + \tfrac{1}{6}\bar{u}_0 M^2) M - H] L \Big\}. \end{aligned} \tag{3.19}$$

As in the symmetric case, one is free to choose r_L and r_T at will provided one takes the counter terms in (3.19) properly into account. With hindsight from the calculation of the symmetric correlation functions, we see that a particularly convenient calculational choice is to take

$$r_L \equiv \chi_L^{-1}, \qquad r_T \equiv \chi_T^{-1}. \tag{3.20}$$

The consequences of this are

(i) The critical region is immediately characterized by $r_L, r_T \to 0$ and again this fact will be used to make simplifying approximations when evaluating

Feynman graphs in which the free propagators $(q^2 + r_L + \Lambda^{-2}q^4)^{-1}$, $(q^2 + r_T + \Lambda^{-2}q^4)^{-1}$ appear.

(ii) The choice (3.20) in conjunction with eqns (3.16) and (3.17) implies that one takes account of the mass counter-terms in (3.19) in the same simple way as in the symmetric theory ((1.24) *et seq.*), namely the self-energy corrections to the propagators are subtracted at $q = 0$ and the mass counter-terms are simply forgotten. The equivalent equations to (1.28), (1.29) and (1.32) are now

$$[G_L^{(2)}]^{-1} = \Gamma_L^{(2)}(q, r_L, r_T, u_0, M) = q^2 + r_L + \Lambda^{-2}q^4 - \Sigma_L(q, r_L, r_T, u_0, M) + \Sigma_L(0, r_L, r_T, u_0, M), \tag{3.21}$$

$$[G_T^{(2)}]^{-1} = \Gamma_T^{(2)}(q, r_L, r_T, u_0, M) = q^2 + r_T + \Lambda^{-2}q^4 - \Sigma_T(q, r_L, r_T, u_0, M) + \Sigma_T(0, r_L, r_T, u_0, M) \tag{3.22}$$

$$-\Sigma_L(0, r_L, r_T, u_0, M) + r_0(T) - r_L + \tfrac{1}{2}\bar{u}_0 M^2 = 0 \tag{3.23}$$

$$-\Sigma_T(0, r_L, r_T, u_0, M) + r_0(T) - r_T + \tfrac{1}{6}\bar{u}_0 M^2 = 0. \tag{3.24}$$

Here Σ_L and Σ_T are the longitudinal and transverse self-energies. They are calculated purely from the trilinear and quartic interactions in (3.19) with the usual rule that all self-energy insertions on any internal lines of Σ are subtracted by their value at zero-momentum. One can also forget about the interaction linear in L; its role is precisely to cancel off all tadpole graphs in which an L field disappears by some interactions into the vacuum, since by construction $\langle L \rangle = 0$.

In this perturbation theory, this last statement gives the equation of state itself; it may be written in the general form

$$\frac{H}{M} = r_0(T) + \tfrac{1}{6}\bar{u}_0 M^2 + g(u_0, r_L, r_T, M). \tag{3.25}$$

The lowest-order term is from the graph Fig. 9. We have divided out a common factor M, and r_L^{-1} from the longitudinal propagator. The function g represents the sum of all graphs with one or more loops formed from the trilinear and quartic interactions in (3.19). In calculating g the mass counter-terms never appear explicitly if all self-energy graphs are subtracted at $q = 0$. Also the term linear in L in (3.19) never appears in g for the same reason that it does not appear in the self-energies Σ_L and Σ_T.

With these rules, the graphs for calculating g, Σ_L and Σ_T are much simpler than the form of (3.19) might suggest; there are only two types of vertex namely $(L^2 + \phi_T^2)^2$ and $L(L^2 + \phi_T^2)$.

Since we are interested in the critical behaviour, we can obtain an expression for the equation of state involving the reduced temperature by subtracting (3.25) by its form at the critical point $r_L = r_T = M = 0$. (Note that H/M also tends to zero at this point, independently of the "direction" in which one approaches the critical point). Thus

$$0 = r_0(T_c) + g(u_0, 0, 0, 0) \tag{3.26}$$

and in terms of reduced temperature t we have

$$\frac{H}{M} = t + \tfrac{1}{6}\bar{u}_0 M^2 + g(u_0, r_L, r_T, M) - g(u_0, 0, 0, 0). \tag{3.27}$$

This is an implicit form for the equation of state. To obtain an explicit form for $H(t, M)$, one must eliminate r_L and r_T in favour of t and M. This can be done by using the graphical expansion of (3.23) and (3.24), in exactly the same way that eqn (1.28) was used in the case $H = 0$ $(T > T_c)$ to obtain r as a function of t (see (2.1) *et seq.*). However it is simpler to obtain r_L and r_T iteratively from the one equation (3.27) by using

$$r_L = \frac{\partial H}{\partial M}, \tag{3.28}$$

$$r_T = \frac{H}{M}. \tag{3.29}$$

The first of these arises directly from identifying r_L with χ_L^{-1} (eqn 3.20). The second can be derived from $r_T = \chi_T^{-1}$, $\chi_T = \partial M/\partial H_\perp$ and the form of the equation of state for an isotropic system, $H_i = M_i \tilde{f}(t, M^2)$. (These are eqns (3.20), (3.15) and (3.1), respectively).

Equation (3.29) is a direct result of the fact that in zero external field the system is rotation invariant in the n-dimensional field space. In the region of the coexistence curve, $T < T_c$, $H \to 0$, $(M \neq 0)$ where there is spontaneous symmetry breaking, it shows explicitly that $r_T \to 0$, $\chi_T \to \infty$. This is an intuitively obvious result. The zero mass of the transverse modes is an illustration of Goldstone's Theorem (Goldstone *et al.*, 1962) which states that if a continuous symmetry of the Hamiltonian is spontaneously broken (i.e. a field which belongs to some non-trivial representation of the symmetry develops a non-vanishing vacuum expectation value) then there are massless modes in the theory. (There are circumstances in which the theorem may be broken, e.g. if vector fields with a gauge symmetry are present (see Abers and Lee, 1973, for a review) but we do not consider such interactions here).

C. The ε-expansion for the equation of state

In this section we shall illustrate explicitly the calculation of the exponents δ and β and the scaling function $f(x)$ in (3.3), to first order in $\varepsilon = 4 - d$.

The principle of the calculation is exactly as for the symmetric vertex functions of the previous chapter.

(i) The graphs which contribute to the function g in (3.27) are calculated to the appropriate order in $\varepsilon = 4 - d$, neglecting terms which vanish with respect to the leading terms by integer powers of r_L/Λ^2 or r_T/Λ^2.

(ii) The special value of the dimensionless coupling constant u_0 of order ε will ensure that the power series in ε obtained in this way will exponentiate to the correct scaling behaviour.

Before discussing the validity of this claim, let us show how to use it at lowest order in ε. Only two graphs Fig. 10 contribute to the function g at first order. They give

$$H/M = t + \tfrac{1}{6}u_0M^2 + \frac{u_0}{6}(n-1)\int_q [(q^2 + r_T + q^4)^{-1} - (q^2 + q^4)^{-1}]$$
$$+ \frac{u_0}{2}\int_q [(q^2 + r_L + q^4)^{-1} - (q^2 + q^4)^{-1}] + O(\varepsilon^2). \qquad (3.30)$$

For simplicity we have again chosen units in which $\Lambda = 1$.

FIG. 10. One-loop graphs contributing to the equation of state (3.27). The solid (dashed) lines denote free longitudinal (transverse) propagators.

The integrals to be evaluated have at this order exactly the same form as that in (2.6) arising from the lowest order (symmetric) self-energy graph. Expanding the result (2.8) to the lowest order in ε and substituting into expression (3.30) we obtain

$$\frac{H}{M} = t + \tfrac{1}{6}u_0M^2 + \tfrac{1}{2}\left(\frac{u_0S_d}{(2\pi)^d}\right)\left[\frac{(n-1)}{6}r_T \ln r_T + \frac{r_L}{2}\ln r_L\right] + O(\varepsilon^2). \qquad (3.31)$$

(Only the logarithms from the integrals contribute at this order; the constants produce only a change in normalization of fields as the calculation below will show). This equation is to be used with the special value of u_0 in (2.14):

$$\frac{u_0S_d}{(2\pi)^d} = \frac{6\varepsilon}{n+8} + O(\varepsilon^2)$$

and hence we obtain the expression

$$\frac{H}{M} = t + \tfrac{1}{6} \cdot u_0 \cdot M^2 + \frac{\varepsilon}{2(n+8)}[(n-1)r_T \ln r_T + 3r_L \ln r_L] + O(\varepsilon^2) \quad (3.32)$$

To obtain an explicit form for the equation of state we require r_L and r_T as functions of t and M^2 to zeroth order. Equations (3.28) and (3.29) give

$$r_L = \frac{\partial H}{\partial M} = t + \tfrac{1}{2}u_0 M^2 + O(\varepsilon), \quad (3.33)$$

$$r_T = \frac{H}{M} = t + \tfrac{1}{6}u_0 M^2 + O(\varepsilon) \quad (3.34)$$

and hence

$$\frac{H}{M} = t + \tfrac{1}{6}u_0 M^2 + \frac{\varepsilon}{2(n+8)}[(n-1)(t + \tfrac{1}{6}u_0 M^2)\ln(t + \tfrac{1}{6}u_0 M^2) + 3(t + \tfrac{1}{2}u_0 M^2)\ln(t + \tfrac{1}{2}u_0 M^2)] + O(\varepsilon^2) \quad (3.35)$$

This is the desired form of the equation of state from which one can calculate β, δ and the function f:

(i) Firstly as a check, we note that in the limit of zero field above the critical temperature, $H \to 0$, $M \to 0$ and $H/M \to r$ $(t > 0)$, we do retrieve the original expression $r \propto t^\gamma$ with $\gamma = 1 + \tfrac{1}{2}(n+2)\varepsilon/(n+8)$.

(ii) To obtain the exponent δ, relating H and M on the critical isotherm $t = 0$, set $t = 0$ in (3.25). Then

$$\frac{H}{M} = \tfrac{1}{6}u_0 M^2 \left[1 + \frac{\varepsilon}{2(n+8)}(n - 1 + 9)(\ln M^2 + \text{const.}) + O(\varepsilon^2)\right]$$

From the definition

$$H \propto M^\delta \qquad (t = 0)$$

we see that

$$\delta = 3 + \varepsilon + O(\varepsilon^2) \quad (3.36)$$

(iii) To obtain the exponent β, relating $(-t)$ and M on the coexistence curve $(H = 0, t < 0)$ let $H = 0$ in (3.35). The zeroth order result is

$$-t = \tfrac{1}{6}u_0 M^2 + O(\varepsilon)$$

and hence the first order is

$$-t = \tfrac{1}{6}u_0 M^2 \left[1 + \frac{3\varepsilon}{n+8}(\ln M^2 + \text{const.}) + O(\varepsilon^2)\right]$$

Thus if β is defined by

$$-t \propto M^{1/\beta} \qquad (H = 0,\ t < 0)$$

we have

$$\frac{1}{\beta} = 2 + \frac{6\varepsilon}{n+8} + O(\varepsilon^2)$$

or

$$\beta = \tfrac{1}{2} - \frac{3\varepsilon}{2(n+8)} + O(\varepsilon^2) \tag{3.37}$$

(iv) Finally one constructs the function $f(t/M^{1/\beta})$ by brute force. It is simplest first to redefine the normalization of M^2 to include the factor $\frac{1}{6}u_0$. Then defining the variable

$$x = t/M^{1/\beta}$$

one obtains

$$\frac{H}{M^\delta} - \frac{t}{M^{1/\beta}} = \frac{H}{M^3}(1 - \varepsilon \ln M) - \frac{t}{M^2}\left(1 - \frac{6}{n+8}\varepsilon \ln M\right) + O(\varepsilon^2)$$

$$= 1 + \frac{\varepsilon}{2(n+8)}[(n-1)(x+1)\ln(x+1) + 3(x+3)\ln(x+3)] + O(\varepsilon^2) \tag{3.38}$$

using (3.35). One verifies that all ln M terms disappear in this scaling form. We have neglected constants of order ε in the right hand side of (3.38) since they can be absorbed at this order by a change in scale of H, M and t. Thus at this order the Widom function f in (3.3) may be written

$$f(x) = x + 1 + \frac{\varepsilon}{2(n+8)}[(n-1)(x+1)\ln(x+1) + 3(x+3)\ln(x+3)] + O(\varepsilon^2) \tag{3.39}$$

This essentially completes the calculation of the order ε corrections to the mean-field equation of state, but several comments are worth making.

(i) The two basic principles of the calculation are (a) translate to fields (e.g. $L(x)$) with zero expectation value so that there are no tadpole graphs such as Fig. 8, and (b) work with fields in which the susceptibility tensor is diagonal, e.g. (χ_L, χ_T), and appears explicitly in free propagators so that the critical region is easily characterized. The same approach can be used to calculate the equation of state with more general interactions (Wallace,

1973a; Aharony, 1974; Aharony and Bruce, 1974). When anisotropic interactions are present the equation $r_T = H/M$ is no longer valid; one can obtain an expression for the transverse susceptibility either from the equivalent of (3.24) or by the use of Ward Identities (Brézin *et al.*, 1973; Wallace, 1973a).

(ii) It is easily checked that the results (3.37) and (3.38) for the exponents δ and β are in agreement with the traditional scaling laws (see e.g. Stanley, 1971)

$$\delta = 1 + 2(2 - \eta)/(d - 2 + \eta) \tag{3.41}$$

$$\beta = \gamma(d - 2 + \eta)/[2(2 - \eta)] \tag{3.42}$$

and the previous results (eqns 2.42, 2.43) for η and γ.

The proof of the scaling laws in the framework of the multiplicative renormalization group is contained in the article by Brézin *et al.* in this volume. The scaling property (3.3) of the equation of state when the coupling constant has the special value u_0^* is also discussed there. The proof is essentially as follows. The free energy is related to the symmetric vertex functions by (1.37). Now we have seen how the renormalization group (2.20) implies the power law behaviour (2.23), for *all* symmetric vertex functions if the coupling constant is such that $\beta(u_0) = 0$. This power law behaviour of the vertex functions implies scaling behaviour of the free energy and hence of the equation of state. See the article by Brézin *et al.* and eqns (4.1) *et seq.*

(iii) The function $f(x)$ has been calculated to order ε^2 (Brézin *et al.*, 1973a; Avdeeva, 1973). The graphs contributing at this order are the two-loop graphs in Fig. 11. Since some graphs contain both longitudinal and transverse propagators (for the non-Ising case, $n \neq 1$) the corresponding integrals may be complicated and are in general non-elementary functions of the ratio of r_T/r_L. Although each of these quantities is small in the

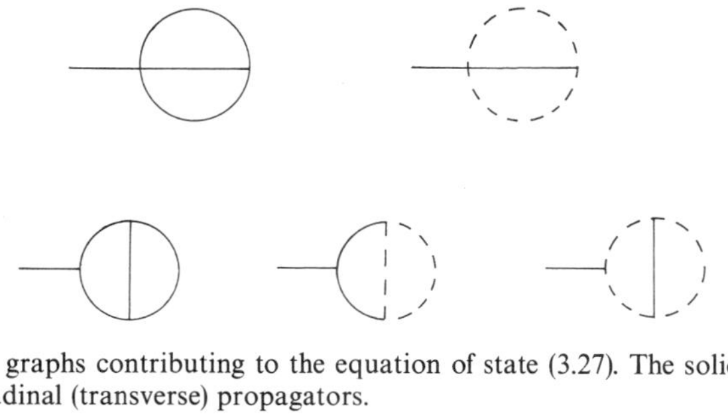

FIG. 11. Two-loop graphs contributing to the equation of state (3.27). The solid (dashed) lines denote free longitudinal (transverse) propagators.

critical region, this ratio varies from 0 on the coexistence curve ($H = 0$, $T < T_c$) through $1/\delta$ on the critical isotherm ($T = T_c$, $H \propto M^\delta$) to 1 in zero field above the critical temperature.

Before writing down the result to order ε^2 for the function $f(x)$ in the equation

$$H/M^\delta = f(x)$$

we remark that although the functional form of $f(x)$ is clearly uniquely obtained by the ε-expansion, the actual expression for $f(x)$ is arbitrary up to a change of scale of the variables H, M and t. To remove this arbitrariness, we choose units for H, M and t by the normalization conditions

$$\text{(i) } H = M^\delta \text{ at } t = 0 \quad \text{i.e. } f(0) = 1 \tag{3.43}$$

$$\text{(ii) } M^{1/\beta} = -t \ (t < 0, H = 0) \quad \text{i.e. } f(-1) = 0 \tag{3.44}$$

The result for $f(x)$ can then be written

$$f(x) = 1 + x + \varepsilon f_1(x) + \varepsilon^2 f_2(x) + O(\varepsilon^3) \tag{3.45}$$

where

$$f_1(x) = \frac{1}{2(n+8)}[(n-1)(x+1)\ln(x+1) + 3(x+3)\ln(x+3) - 9(x+1) \times \ln 3 + 6x\ln 2] \tag{3.46a}$$

$$\begin{aligned} f_2(x) = \left[\frac{1}{2(n+8)}\right]^2 \Big\{ & [n - 1 + 6\ln 2 - 9\ln 3 + (n-1)\ln(x+1)] \\ & \times [3(x+3)\ln(x+3) + (n-1)(x+1)\ln(x+1) + 6x\ln 2 - 9(x+1) \\ & \times \ln 3] + \tfrac{1}{2}(10-n)(x+1)[\ln^2(x+3) - \ln^2 3] + 36[\ln^2(x+3) \\ & - (x+1)\ln^2 3 + x\ln^2 2] - 54\ln 2[\ln(x+3) + x\ln 2 - (x+1)\ln 3] \\ & + 3(n-1)(\ln\tfrac{27}{4})(x+1)\ln(x+1) + \frac{212 + 17n - 4n^2}{n+8}[(x+3)\ln(x+3) \\ & + 2x\ln 2 - 3(x+1)\ln 3] + (n-1)(x+1)\ln(x+1)\ln(x+3) \\ & - \tfrac{1}{2}n(n-1)(x+1)\ln^2(x+1) + \frac{n-1}{n+8}(19n+92)(x+1)\ln(x+1) \\ & - 2(n-1)\{(x+6)I_1(\rho) - 6(x+1)I_1(\tfrac{3}{4})\} - 6(n-1)\{I_2(\rho) - (x+1) \\ & \times I_2(\tfrac{3}{4})\} + 4(n-1)\{I_3(\rho) - (x+1)I_3(\tfrac{3}{4})\} \Big\}. \end{aligned} \tag{3.46b}$$

The parameter ρ is given by

$$\rho = \frac{x + 3}{4(x + 1)} \tag{3.47}$$

which the reader will recognize from the previous section as the ratio $r_L/(4r_T)$, and $I_1(\rho)$, $I_2(\rho)$ and $I_3(\rho)$ may be written

$$I_1(\rho) = \int_0^\rho \frac{du \ln u}{u(1 - u)} \left[\left(1 - \frac{u}{\rho}\right)^{\frac{1}{2}} - 1 \right] - \int_\rho^\infty \frac{du \ln u}{u(1 - u)} \tag{3.48}$$

$$\begin{aligned} I_2(\rho) &= \rho \frac{\partial}{\partial \rho} I_1(\rho) \\ I_3(\rho) &= I_1(\rho) + 2I_2(\rho). \end{aligned} \tag{3.49}$$

Equation (3.46) exhibits the desired scaling property of the equation of state, and in addition since $f(x)$ is clearly analytic about $x = 0$, the equation of state is analytic about $t = 0$ for finite M and hence satisfies one of the analyticity requirements of Griffiths (1967). However it does not satisfy explicitly the requirement that H should be an analytic function of M for finite t; this condition translates into an asymptotic expansion for $f(x)$ for $x \to \infty$

$$f(x) = \sum_{k=0}^{\infty} a_k x^{\gamma - 2k\beta} \tag{3.50}$$

An asymptotic expansion of this form for the function f is reconstructed from (3.46) in the article by Brézin *et al.* in this volume; it gives results for the coefficients a_k which are in good agreement with the numerical results of Gaunt and Domb (1970). In the following section we shall discuss parametric models for the equation of state which give a representation in which the scaling property and the two analyticity properties are automatically satisfied exactly.

A further defect of (3.46) is its behaviour near the coexistence curve, $x \to -1$. The function $f(x)$ itself is well defined but derivatives of it (e.g. the longitudinal susceptibility) will diverge with increasing powers of $\ln(x + 1)$ which arise from increasing powers of $\ln r_T$. This behaviour is discussed in Section E of this chapter.

Despite these two defects of (3.46), it is compared directly with the numerical results of Gaunt and Domb (1970) for the Ising case ($n = 1$) and Milošević and Stanley (1972) for the Heisenberg case $n = 3$ by Brézin *et al.* (1973a). In both cases, agreement with the numerical results improves at each order of the ε-expansion. The convergence looks poor for the $n = 3$ case but this is largely due to the fact that the asymptotic property (3.50) is satisfied by eqns (3.45), (3.46) only up to the ε-expansion of the exponents γ and β in $x^{\gamma - 3k\beta}$.

D. Parametric models

It is clear that the basic form of the equation of state, (3.27) will satisfy order by order in the ε-expansion the Griffiths analyticity requirements; from the form of the zeroth order expressions (3.33) and (3.34) the propagators $(q^2 + r_L)^{-1}$ and $(q^2 + r_T)^{-1}$ can be expanded analytically in t (or M^2) without introducing infrared singularities in the integrals if M^2 (or t) is finite. Only the process of exponentiating logarithms to produce the *scaling* form (3.3) violates the analytical property in M^2.

The so-called parametric representation provides one way of writing the equation of state which automatically ensures both its analyticity and its scaling properties in the critical region. This property has been discussed at a fundamental level by Josephson (1969) and at a more explicit and phenomenological level by Schofield (1969), and Schofield *et al.* (1969). (An alternative resolution of the problem by writing the equation of state implicitly in terms of the susceptibility $\partial M/\partial H$ is given by Migdal (1972).)

The parametric model introduced by Schofield (1969) consists of expressing H, M and t in terms of two variables R and θ, according to

$$t = R\tau(\theta), \tag{3.51a}$$

$$M = R^{\beta}m(\theta), \tag{3.51b}$$

$$H = R^{\beta\delta}h(\theta). \tag{3.51c}$$

The conventional equation of state e.g. (3.1) is obtained by eliminating R and θ amongst t, M and H. The R dependence of the parametrization is chosen so that when R is eliminated one is guaranteed the scaling form (3.3). The θ dependence is in principle arbitrary, up to ensuring the symmetry $M \to -M$, $H \to -H$ of the equation of state. This can be done by taking $m(\theta)$ and $h(\theta)$ as *odd* functions of θ and $\tau(\theta)$ as an *even* function of θ. With this proviso, if τ, m and h are suitable analytic functions of θ, then the Griffiths analyticity requirements are automatically satisfied.

It is natural to exploit this property by taking lowest-order polynomials for these functions, to reproduce the qualitative features of a second-order phase transition (Schofield, 1969)

$$\tau(\theta) = 1 - b^2\theta^2 \tag{3.52a}$$

$$m(\theta) = c\theta \tag{3.52b}$$

$$h(\theta) = a\theta(1 - \theta^2) \tag{3.52c}$$

The parameter θ varies from 0 to 1 to give the physical region of the equation of state: $\theta = 0$ corresponds to $T > T_c$, $H = 0$; $\theta = \pm 1/b$ to $T = T_c$ and $\theta = \pm 1$ to the coexistence curve ($H = 0$, $T < T_c$). This is called the linear parametric model (m is linear in θ). Two of the parameters a, b, c are absorbed into the definition of the scales of H, M and t as in (3.43) and (3.44) for example. With the remaining free parameter (say b^2), the linear parametric model gives a rather good approximation phenomenologically (Schofield *et al.*, 1969; Gaunt and Domb, 1970).

It is clear that the general expression (3.46) for the equation of state to order ε^2 with its non-elementary functions $I_1(\rho)$ etc., cannot be represented by such simple polynomials as (3.52). One may proceed by taking the functions $\tau(\theta)$ and $h(\theta)$ as in (3.52a) and (3.52c) and determining the required function $m(\theta)$ from the equation of state (3.46). This is carried out explicitly to order ε by Brézin *et al.* (1973a). With a suitable choice of scale for the fields, H, M and t one finds

$$m(\theta) = \frac{1}{\sqrt{2}}\theta\left(1 - \frac{\varepsilon(n-1)}{2(n+8)}(1-\theta^2)\ln(1-\theta^2)\right) + O(\varepsilon^2) \tag{3.53}$$

An incomplete discussion of order ε^2 is given by Schofield (1974).

Despite the fact that in general the linear parametric model fails for the general case at order ε, the parametric representation (3.52a, c and 3.53) provides quite a compact representation of the equation of state. More interestingly, we see that for the *Ising* case $n = 1$, the linear parametric model *is* an exact representation of the equation of state to order ε. In fact this property holds (for the Ising case) to order ε^2 as well, and the linear parametric model provides a remarkably compact representation of the equation of state to this order in ε.

In order to understand this property, let us consider the Ising case in a little more detail. Since $n = 1$, there are no transverse fields, only one mass parameter r_L is required and at order ε^2 all graphs in (3.27) have the form, schematically,

$$\varepsilon(M^2, r_L) \times (\varepsilon \ln^2 r_L, \ln r_L, \text{const.}) \tag{3.54}$$

Now it is straightforward to work out the form of the inverse susceptibility r_L in the linear parametric model:

$$r_L = R^\gamma \frac{a}{c}[1 + \theta^2(b^2(2\beta\delta - 1) - 3) + \theta^4 b^2(3 - 2\beta\delta)] \times [1 + \theta^2 b^2(2\beta - 1)]^{-1} \tag{3.55}$$

This has the remarkable property that if we choose the (free) parameter

$$b^2 = \tfrac{3}{2} + b_1\varepsilon + O(\varepsilon^2) \tag{3.56}$$

then the expressions for β and δ in the ε-expansion (eqns (3.36) and (3.37)) imply r_L has an expansion in $(\varepsilon\theta^2)$

$$r_L = R^\gamma \frac{a}{c} \sum_l r_l (\varepsilon\theta^2)^l \tag{3.57}$$

where

$$r_0 = 1, r_1 = 2b_1 + \tfrac{1}{2} + O(\varepsilon), \qquad r_2 = b_1 + O(\varepsilon) \tag{3.58}$$

If the expansion (3.57) and the other parametric expressions (3.52a, b, c) are substituted into the calculated equation of state at order ε^2, two things happen. Firstly the only θ dependence which appears is the form $d_1 + d_2\theta^2$. Secondly all R dependence disappears as it should because we know that the calculated equation of state can be written in scaling form. Now if we ensure *a priori* that the *same* normalization conditions are applied to the calculated equation of state and the parametric form, then the polynomial $d_1 + d_2\theta^2$ must vanish identically at the two values of θ^2 corresponding to the two normalization points and hence it vanishes identically. Thus with b^2 chosen as in (3.56) (with b_1 still arbitrary) the linear parametric model is an exact representation of the Ising equation of state to order ε^2.

The calculation of the Ising equation of state is relatively simple, because there are no transverse modes, and can be extended to order ε^3 without much labour (Wallace and Zia, 1973, 1974). At the three loop level there are graphs with five ML^3 vertices, and terms of order M^4 appear in the equation (3.27). If the linear parametric model is substituted in as before, one can calculate the coefficient of the θ^4 term. It is

$$K = \tfrac{1}{3}\left(b_1^2 + \frac{\zeta(3)}{4} - \frac{\lambda}{8} - \tfrac{1}{16}\right) \tag{3.59}$$

$$\simeq \tfrac{1}{3}b_1^2 + 0{\cdot}03, \tag{3.60}$$

where $\zeta(3)$ and λ are given in (2.40) and (2.41). Although b_1 is a free parameter it cannot be chosen to make K vanish and hence at order ε^3 the linear parametric model must be modified by a suitable higher power of θ. This is foreseen by the numerical series results of Gaunt and Domb (1970). The required modifications are in surprisingly good agreement with the numerical results of Tarko and Fisher (1975) (see Wallace and Zia, 1974, for details).

Finally, we comment on the fact that even at order ε^2, the parameter b^2 of the parametric model is determined only to zeroth order from the calculated equation of state. The reasons for this are explained in Wallace and Zia (1974). Since this parameter is so poorly determined one might be led to suspect that the actual form of the equation of state itself is similarly poorly determined. This is not so however. The clue to this is contained in the remarks by Schofield *et al.* (1969) that in the linear parametric model certain quantities (e.g. ratio of susceptibility amplitudes for $T > T_c$, $T < T_c$) have an *extremum* as a function of the parameter b^2 for

$$b^2 = (\delta - 3)[(\delta - 1)(1 - 2\beta)]^{-1} \tag{3.61}$$

Indeed, the entire scaling function $f(t/M^{1/\beta})$ has an extremum in the linear parametric model for this value of b^2 (Wallace and Zia 1974)). Now in the ε-expansion, the special value (3.61) is given by

$$b^2 = \tfrac{3}{2} + O(\varepsilon^2) \tag{3.62}$$

which embraces the value of b^2 required to fit the linear parametric model to the calculated equation of state. The conclusion is that for b^2 close to $\frac{3}{2}$ the form of the scaling function $f(t/M^{1/\beta})$ is extremely insensitive to the precise value of b^2, i.e. the fact that b^2 is poorly determined in the ε-expansion is counteracted by the insensitivity of the equation of state to the precise value of b^2.

E. Behaviour near the coexistence curve

In the previous section we concentrated on a property of the Ising ($n = 1$) equation of state. In this section we consider a property of Heisenberg systems with a continuous symmetry ($n \neq 1$). As we discussed in (3.29) *et seq.* the most striking feature for Heisenberg systems is the existence of the $(n - 1)$ transverse modes which become massless ($r_T \to 0$) on the coexistence curve ($H \to 0$, for all $T < T_c$).

In principle this vanishing of r_T may induce new infrared singularities at all temperatures $T < T_c$. Now it is clear from (3.46) that the equation of state itself is well defined on the coexistence curve ($t/M^{1/\beta} \equiv x = -1$ with the normalization (3.44)); it has the form

$$\begin{aligned} H/M^{\delta} = (x + 1)\{1 + \varepsilon(A \ln(x + 1) + B + o(1)) + \varepsilon^2[C \ln^2(x + 1) \\ + D \ln(x + 1) + E + o(1)] + O(\varepsilon^3)\} \end{aligned} \tag{3.63}$$

where the terms $o(1)$ vanish like $(x + 1)$ near $x = -1$. The coefficients A, B, C, D and E can be obtained from (3.46) *et seq.* One finds

$$A = \frac{n - 1}{2(n + 8)}, \tag{3.64a}$$

$$B = \frac{3[1 + 3 \ln n(\frac{2}{3})]}{2(n + 8)}, \tag{3.64b}$$

$$C = \frac{(n - 10)(n - 1)}{8(n + 8)^2}, \tag{3.64c}$$

$$D = \frac{(n-1)}{4(n+8)^2}[n + 27 + 18 \ln 2 - 9 \ln 3 - 60/(n+8)]. \quad (3.64d)$$

(We shall see later that the coefficient E is not required at this order.)

Derivatives of (3.63) can become singular however. In particular the longitudinal mass parameter $r_L = \partial H/\partial M$ behaves like

$$\beta r_L/M^{\delta-1} = 1 + \varepsilon(A \ln(x+1) + A + B) + \varepsilon^2(C \ln^2(x+1) + (D + 2C) \times \ln(x+1) + E + D) + O(\varepsilon^3) + o(1) \quad (3.65)$$

as $x \to -1$.

The source of these singularities in r_L can also be seen directly in the self energy graphs such as Fig. 12 which contribute to r_L through (3.23).

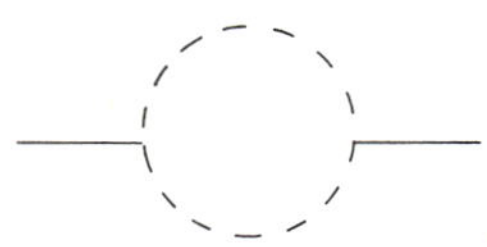

FIG. 12. Example of graph contributing directly to the singularity in r_L on the coexistence curve. See Bessis and Zinn-Justin (1972) for a discussion in the context of the linear σ-model.

The series (3.65) is typical of the expansion in ε of some critical exponent or exponents arising from this new critical behaviour as $r_T \to 0$. The question is to what form should these logarithms be exponentiated?

The possible singularity of the longitudinal susceptibility χ_L induced by the divergence of χ_T was realized long before the current interest in critical behaviour. It is a feature of the spherical model (Berlin and Kac, 1952; Joyce, Vol. 2, p. 417 *et seq.*) and spin-wave theories (Fisher, 1967, p. 645; Vaks *et al.*, 1967). A general argument, similar in spirit to the spin-wave approach, can be formulated (Patashinskii and Pokrovskii, 1973; K. G. Wilson (unpublished); Brézin and Wallace, 1973; Wallace 1973b). See also Symanzik (1973) and Applequist and Carazzone (1975). One predicts the behaviour

$$(\beta r_L/M^{\delta-1})^{-1} \sim c_1 + c_2(H/M^\delta)^{-\varepsilon/2} + o(1) \quad (3.66)$$

for $H \to 0$, $T < T_c$ fixed, to all orders in ε and up to neglect of terms which vanish like H for ε small. The quantities c_1 and c_2 are critical amplitudes depending only on n and ε; in particular c_2 must vanish for the Ising case $n = 1$. The divergence implies that one cannot define an exponent γ' associated with zero field susceptibility below the critical temperature except for the Ising system.

In order to establish (3.66) in the ε-expansion one must invert (3.65) and re-express the right-hand side in terms of H/M^δ from (3.63). The result is

$$(\beta r_L/M^{\delta-1})^{-1} = 1 - \varepsilon(A \ln y + A + B) - \varepsilon^2[\ln^2 y(C - A^2) + \ln y(D + 2C - 3A^2 - 2AB) + \text{const.}] + o(1), \qquad (3.67)$$

where $y \equiv H/M^\delta$. This equation provides only limited support for the form (3.66), however, since the logarithms must be exponentiated to *two* powers of y (0 and $-\varepsilon/2$) differing by order ε. Firstly from eqns (3.64) we find

$$C - A^2 = -\frac{(n-1)}{8(n+8)}. \qquad (3.68)$$

The coefficients of $\varepsilon \ln y$ and $\varepsilon^2 \ln^2 y$ in (3.67) are hence consistent with the expansion of ε in $y^{-\varepsilon/2}$. Secondly, comparison of (3.66) and (3.67) yields the coefficients c_1 and c_2, but only to order ε. One finds

$$c_1 = \frac{1}{n+8}\left[9 - \frac{\varepsilon}{2(n+8)}\left(9(n+8)\ln 2 - 81 \ln 3 + \frac{25n^2 + 142n + 76}{n+8}\right)\right] + O(\varepsilon^2) \qquad (3.69)$$

$$c_2 = \frac{n-1}{n+8}\left[1 + \frac{\varepsilon}{2(n+8)^2}(9(n+8)\ln 3 - n^2 + 6n + 52)\right] + O(\varepsilon^2) \qquad (3.70)$$

More convincing support for the general structure (3.66) is found in the $1/n$ expansion; one sees also explicitly that $c_1 = O(1/n)$ (Brézin and Wallace, 1973a), in agreement with (3.69).

With these values for c_1 and c_2, one sees that the diverging term in (3.66) will take over from the constant term when (note that the analysis of Brézin and Wallace 1973 is wrong in this respect—see Wallace and Zia, 1975c)

$$(H/M^\delta)^{\varepsilon/2} \lesssim \frac{c_2}{c_1} = \frac{n-1}{9}\left[1 + \frac{\varepsilon}{18}\left(16 - \frac{60}{n+8} + 9 \ln 2\right) + O(\varepsilon^2)\right]. \qquad (3.71)$$

Numerical values for this ratio are

$$n = 2 \qquad \frac{c_2}{c_1} = \tfrac{1}{9}[1 + 0{\cdot}90\,\varepsilon], \qquad (3.72a)$$

$$n = 3 \qquad \frac{c_2}{c_1} = \tfrac{2}{9}[1 + 0{\cdot}93\,\varepsilon]. \qquad (3.72b)$$

Although the expansion is clearly not very reliable for $\varepsilon = 1$, one deduces from (3.72b) that the divergent term should be noticeable for $H/M^\delta \lesssim 20\%$ of its value on the coexistence curve for $n = 3$ in three dimensions.

IV. Other Perturbations of the Hamiltonian

A. General considerations

Up until now in this article we have considered the behaviour of correlation functions in the phase transition region, principally at zero momentum. We have considered a Hamiltonian which is $O(n)$ symmetric apart from the external field term $-H \cdot \phi$. The $(\phi^2)^2$ coupling has a special value (2.39) to ensure that logarithms exponentiate to the appropriate scaling behaviour (2.31).

This treatment is clearly incomplete from at least two points of view. Firstly, for this special Hamiltonian we have not discussed the behaviour of correlation functions for general momentum $q \neq 0$. Secondly, one should ask how perburbations of this Hamiltonian by other interactions affect the phase transition behaviour. The moral one learns from both of these problems and in essence the critique of the entire method we describe is that one must know the form to which to exponentiate the extra logarithms which are obtained. The situation is very similar to that of the previous section; the control of the logarithms in, e.g. eqn (3.65) is achieved only by knowing the form (3.66) to which they exponentiate.

The problem of correlation functions at non-zero momentum is reviewed by Fisher and Jasnow in Volume 4 and by Brézin *et al.* in this volume. The typical question one faces is to find the corrections to the behaviour

$$\Gamma^{(2)}(q, 0, u_0^*, \Lambda) \propto q^{2-\eta} \tag{2.28}$$

as the system is moved away from its critical point $t = 0$, $H = 0$. The tool which accomplishes this is the operator product expansion (Wilson 1969; Kadanoff, 1969; Zimmermann, 1970; Wilson and Zimmermann, 1972), in conjunction with the renormalization group formalism. One finds that the correction terms take the form of specified powers of the variable $tq^{1/\nu}$, ($\nu = \gamma/(2 - \eta)$) with coefficients which are functions of the variable $t/M^{1/\beta}$. Knowing the form of the powers of q which appear, it is possible to obtain these functions of $t/M^{1/\beta}$ by the method described in this article; with the special value of u_0^* ((2.21) and (2.39)) any logarithms which appear can be exponentiated into the appropriate scaling form as is done for exponents and the equation of state itself. This programme has been pursued by Fisher and Aharony (1974a, b) to order ε^2 for $T > T_c$, $H = 0$, and by Combescot *et al.* (1974, 1975) to order ε^2 for $n = 1$ ($H \neq 0$). The reader is referred to Fisher and Jasnow, Volume 4, and Brézin *et al.* in this volume for comprehensive references on other methods and approaches.

Let us turn then to the question of the effect of other terms in the Hamiltonian. These terms can take many forms involving higher powers of derivatives, higher powers of fields, long-range interactions, terms which break the $O(d)$ or $O(n)$ rotation invariance etc. One can easily formulate in terms of the free energy a simple criterion for deciding if a given interaction will change the phase transition behaviour or not. For the Hamiltonian (1.1) with the special value u_0^* of the ϕ^4 coupling, all the symmetric vertex functions $\Gamma^{(p)}$ will scale like powers of r according to (2.31):

$$\Gamma^{(p)}(0, r, u_0^*, \Lambda) \propto r^{[d-p(d/2-1+\eta/2)]/(2-\eta)}. \tag{4.1}$$

The thermodynamic potential per unit volume $F(M, r)$ is given according to (1.37) *et seq.* by

$$F(M, r) = \sum_p \frac{1}{p!} M_{i_1} \ldots M_{i_p} \Gamma^{(p)}_{i_1 \ldots i_p}(0, r, u_0, \Lambda). \tag{4.2}$$

With vertex functions of the form (4.1), $F(M, r)$ is of the form $M^{\delta+1}\Phi(r/M^{\delta-1})$ where δ is given by the scaling law (3.41). This exhibits the scaling property of the free energy in the critical region.

If we add extra interactions into the Hamiltonian the form of the $\Gamma^{(p)}$ in (4.1) becomes modified and even new vertex functions may be obtained (if the perturbation is not $O(n)$ symmetric for example). Also new powers of r will appear in vertex functions.

Roughly speaking, if the extra terms in the free energy (4.2) are *less* singular in the limit r, $M^2 \to 0$ than the terms from (4.1), then the new interaction produces an infinitesimally small effect in the phase transition region and can be neglected. On the other hand, if the new terms are *more* singular then they become extremely important no matter how small their coupling strength. In order to obtain a true indication of their effect one must go beyond simple perturbation, just as one must for the ϕ^4 interaction itself. Except in the special case where there is only a shift in the transition temperature, the result will be a crossover to some new behaviour, e.g. to a first-order transition, or to a second-order transition with new exponents, or to no transition at all. (The external field term produces the last. The susceptibility never becomes infinite in finite external field; see (3.33) and (3.34).)

Detailed discussion of crossover to new behaviour is given in the article by Aharony in this volume. Here we shall restrict ourselves to the problem of classifying interactions according to whether they do or do not change the phase transition behaviour of the isotropic Heisenberg system (1.1).

The rather extensive evidence for universality (see e.g. Stanley 1971, Chapter 1), suggests that one does not expect to find many interactions which produce substantially different critical behaviour. One can intuitively expect this on theoretical grounds also from the discussion in Section II.B. For, just as the ϕ^4 interaction, with a coupling constant which has dimension

$$[u_0] = [\Lambda]^{4-d}, \tag{1.3}$$

produces infra-red singularities only if $d < 4$, so other interactions are expected to produce new infra-red divergences only if their coupling constant has positive dimension. Now it is trivial to extend the dimensional analysis of (1.3) to show that the coupling constant of a local interaction involving m_1 derivatives and m_2 powers of field has a dimension

$$[u(m_1, m_2)] = [\Lambda]^{d - m_1 - m_2(d/2 - 1)}. \tag{4.3}$$

For d close to 4 this dimension will be negative except for only a very few low values of m_1 and m_2. Hence the vast majority of interactions are not expected to produce new infra-red singularities provided d is greater than two.

However one sees from (4.3) that the ϕ^6 interaction is dimensionless in $d = 3$ so clearly a more refined treatment is essential.

B. Scaling variables of the Heisenberg fixed point

Considerable progress has been made in this problem although results are still incomplete.

A more precise formulation of the problem outlined above is presented by application of the renormalization group approach, namely one must look for special interactions called scaling variables which can be associated with a *single* new critical exponent (Kadanoff, 1966, 1969; Wilson and Kogut, 1974; Wegner, 1972a, b). A scaling variable is said to be *relevant* or *irrelevant* according to whether it does or does not change the phase transition behaviour in the sense of (4.2) *et seq*.

The problem then is one of enumerating the scaling variables and calculating their critical exponents. Leaving aside the work in lattice models, the $1/n$ expansion and long range interactions (Van Leeuwen and Niemeijer, Ma and Aharony respectively in this volume and Ma, 1974) the general direction of the development is as follows.

Firstly one remarks that the renormalization group transformation is concerned principally with changes of scale and commutes with $O(d)$ and $O(n)$ rotations in the co-ordinate or field spaces respectively. Hence a decomposition of all possible interactions into irreducible representations of $O(d)$ or $O(n)$ produces a partial factorization of the problem of finding scaling interactions for the Heisenberg fixed-point (in our case the Hamiltonian (1.1) with the special value u_0^* of (2.21) and (2.39)): only terms belonging to the *same* irreducible representation of $O(d) \times O(n)$ may be mixed.

This result is explicit in Wegner (1972b) where the approximate recursion formula (Wilson 1971) is used to derive exponents of scaling variables to order ε. Only powers of fields without derivatives can be handled in the approximate recursion formula. The scaling variables can be written as harmonic polynomials of degree k (Erdelyi, 1953) multiplied by polynomials in ϕ^2 specified by k, and the equivalent of a "principal quantum number" m.

It is also explicit in the treatment of arbitrary ϕ^4 perturbations by Brézin *et al.* (1974). See also Brézin *et al.* in this volume. The $O(\phi^4)$ scaling variables of the Heisenberg fixed point are the three classes of interactions $(\phi^2)^2$, $u_{ij}\phi^2\phi_i\phi_j$, $u_{ijkl}\phi_i\phi_j\phi_k\phi_l$ where u_{ij} and u_{ijkl} are traceless; $u_{ii} = 0$, $u_{iikl} = 0$. These correspond to $k = 0$, 2 and 4 respectively in Wegner's notation. One establishes that the first two are irrelevant interactions. We shall comment on the last shortly.

The calculation of exponents for higher powers of fields and derivatives is much more complicated, however; in the multiplicative renormalization group *all* interactions which belong to the *same* irreducible representation of the symmetry group of the fixed point Hamiltonian, in this case $O(d) \times O(n)$, and which have the *same* naive dimension in $d = 4$ can be expected to mix together (Zimmermann, 1970).

However it is possible to exploit the decomposition into irreducible representations of $O(d) \times O(n)$ in order to obtain critical exponents of special classes of interactions to rather high order in ε. Wilson and Kogut (1974, Chapter 9) have calculated critical exponents associated with interactions of the form $u_{a_1} \cdots {}_{a_m} \phi_i(\nabla_{a_1} \cdots \nabla_{a_m})\phi_i$ and $u_{ij}u_{a_1} \cdots {}_{a_m}\phi_i(\nabla_{a_1} \cdots \nabla_{a_m})\phi_j$ where u tensors are all traceless. These interactions are important in the theory of deep inelastic electron scattering (see Mueller *et al.*, 1971, 1972). Wallace and Zia (1975b) have calculated the exponents of interactions of the form

$$\mathcal{H}_{k,n} = u_{i_1} \cdots {}_{i_k} \int \mathrm{d}^d x\, \phi_{i_1}(x) \ldots \phi_{i_k}(x), \tag{4.3a}$$

where again u is traceless. Interactions of this type can appear in systems with only point group symmetry.

Because no other interactions exist which have the same tensorial properties and the same dimension in $d = 4$, as these pure harmonic polynomials, they are all scaling variables. In particular the interaction (4.3a) will give rise to new vertex functions $\Gamma^{(k)}_{i_1} \cdots {}_{i_k}$ proportional to $u_{i_1} \cdots {}_{i_k}$. If these vertex functions are evaluated at momentum $q = 0$ to first order in $u_{i_1} \cdots {}_{i_k}$ and to the appropriate order in u_0^*, they will scale with a *single* power of r:

$$\Gamma^{(k)}_{i_1} \cdots {}_{i_k}(0, r, u_0^*, \Lambda) \propto u_{i_1} \cdots {}_{i_k} r^{\alpha_k}. \tag{4.4}$$

The relevance or otherwise of the perturbation (4.3a) is obtained by writing the free energy (4.2) in the form

$$F(M, r) = M^{\delta+1}\{\Phi(r/M^{\delta-1}) + u_{i_1} \cdots {}_{i_k}(M_{i_1} \ldots M_{i_k})\,|M|^{-k}\,\Psi(r/M^{\delta-1}) \cdot r^{\psi_k}\} \tag{4.5}$$

where the exponent ψ_k is given in terms of α_k (using the scaling law (3.41)) by

$$\psi_k = \alpha_k - 1 + \frac{(k-2)(2-\varepsilon+\eta)}{2(2-\eta)}. \qquad (4.6)$$

The perturbation (4.3a) is then relevant if $\psi_k < 0$ and irrelevant if $\psi_k > 0$.

C. Calculations and conclusions

The exponent α_k has been calculated to order ε^3 by Wallace and Zia (1975b). The graphs required are those with k external legs and one vertex of the type (4.3a). The one loop graph is shown in Fig. 13. Other graphs are obtained from the four point graphs Fig. 4 by replacing each 4 point vertex in turn by the interaction (4.3a). At the three loop level there are also the graphs Fig. 14 which cannot be produced in this manner. The analytic expressions are those

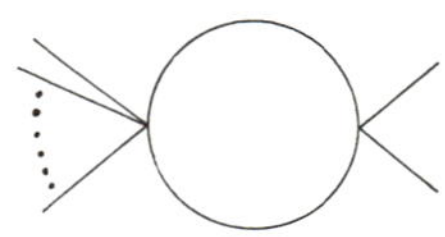

FIG. 13. The one-loop graph contributing to the exponent α_k (4.4). Other graphs are similarly obtained by insertion of the k-leg vertex in Fig. 4.

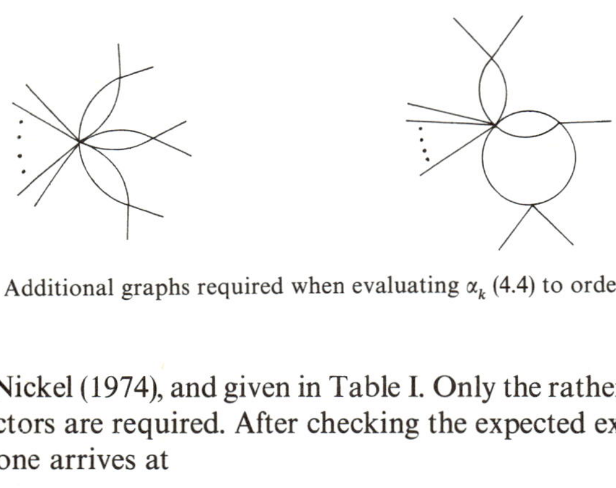

FIG. 14. Additional graphs required when evaluating α_k (4.4) to order ε^3.

calculated by Nickel (1974), and given in Table I. Only the rather complicated multiplicity factors are required. After checking the expected exponentiation of logarithms one arrives at

$$\alpha_k = \frac{\varepsilon k(k-1)}{2(n+8)}\left\{1 + \frac{\varepsilon}{2(n+8)}\left[2 - 4k - n + \frac{2(9n+42)}{n+8}\right]\right.$$

$$+ \frac{\varepsilon^2}{2(n+8)^4}\left[4(n+8)\left[(k-2)(n+14)(n+8) - 6(5n+22)\right]\zeta(3)\right.$$

$$-\frac{n^4}{2} - 18n^3 - 86n^2 - 168n + 592$$

$$+ 2(k - 4)(n + 8)[8k(n + 8) + n^2 - 24n - 136]\Big]\Big\}$$
$$+ O(\varepsilon^4) \tag{4.7}$$

where $\zeta(3)$ is again the Reimann ζ function. The exponent ψ_k may be obtained from this result using expression (4.6) and the known result for η, (2.43). This expression summarizes many interesting results.

For $k = 2$, ψ is negative and one obtains a crossover to new critical behaviour in which fewer of the fields are becoming critical at the phase transition temperature (Fisher and Pfeuty, 1972; Wilson, 1972; Yamazaki, 1974).

The case $k = 3$ has been considered mainly in the context of the Potts model (Potts, 1952; Golner, 1973; Amit and Shcherbakov, 1974; Zia and Wallace, 1975b). The result is

$$\psi_3 = -\tfrac{1}{2} + \frac{\varepsilon(4 - n)}{4(n + 8)} - \frac{5\varepsilon^2(n + 2)(n - 4)}{4(n + 8)^3} + \frac{6\varepsilon^3(n + 2)(n - 10)\zeta(3)}{(n + 8)^4}$$
$$- \frac{(n + 2)\varepsilon^3}{(n + 8)^5}(\tfrac{7}{8}n^3 + 27n^2 + 63n + 124) + O(\varepsilon^4). \tag{4.8}$$

This series indicates that ψ_3 will remain negative in three dimensions ($\varepsilon = 1$) and the trilinear coupling is expected to induce a first order phase transition (Golner, 1973). All higher orders in expression (4.8) also contain a factor of $n + 2$.

The case $k = 4$ is considered by Ketley and Wallace (1973) for the interaction $(\phi_1^4 + \ldots + \phi_n^4)$, invariant under hypercubic point-group symmetry. The analysis has been extended by Nickel (1974) to estimate the effects of higher irreducible diagrams as in Section II.E. Nickel concludes that in three dimensions the exponent ψ_4 is positive for $n \lesssim 3$ and negative for $n \gtrsim 3$. See Table VI. Further discussion of the implications for phase transition behaviour are given by Wallace (1973a, b), Aharony (1973), Ketley and Wallace (1973); see Aharony in this volume for a review.

It is obvious from the nature of the series (4.7) that it cannot be trusted for very large k. However one obtains a very convincing result for an exponent which *at a given* ε shows whether the interaction whose coupling becomes dimensionless *at that value of* ε is relevant or not for the Heisenberg fixed point. Now according to (4.3) an interaction of order ϕ^k is dimensionless when d is given by

$$k = \frac{2d}{d - 2}. \tag{4.9}$$

When k is replaced by this expression in (4.6) and (4.7), and d is expanded

TABLE VI. Best values for exponents in three dimensions using the "modified" tetrahedron contribution; η, γ, α and ψ_4 (see Sections IVB and IVC) are from Nickel (1974). The others are obtained from η and γ by scaling laws (see e.g. Stanley (1971) §11, §12). Error bars are Nickel's estimation from convergence behaviour of the continued fraction and other considerations. The series results are from (a) Gaunt and Guttmann, Vol. 3 (b) Ferer *et al.* (1973) (c) Ferer *et al.* (1971) (d) M. A. Moore (private communication) (e) Domb, Vol. 3 (f) From Ferer *et al.* (1973) using scaling laws.

Exponent	$n = 1$ Modified ε-expansion	$n = 1$ Series	$n = 2$ Modified ε-expansion	$n = 2$ Series	$n = 3$ Modified ε-expansion	$n = 3$ Series
γ	$1{\cdot}250 \pm 0{\cdot}005$	$1{\cdot}250 \pm 0{\cdot}001$ [a]	$1{\cdot}320 \pm 0{\cdot}005$	$1{\cdot}318 \pm 0{\cdot}010$ [b]	$1{\cdot}380 \pm 0{\cdot}005$	$1{\cdot}405 \pm 0{\cdot}02$ [c]
η	$0{\cdot}0413 \pm 0{\cdot}001$	$0{\cdot}047 \pm 0{\cdot}01$ [d]	$0{\cdot}0426 \pm 0{\cdot}001$	$0{\cdot}04 \pm 0{\cdot}01$ [b]	$0{\cdot}0419 \pm 0{\cdot}001$	$0{\cdot}040 \pm 0{\cdot}008$ [c]
α	$0{\cdot}086 \pm 0{\cdot}01$	$\simeq 1/8$ [e]	$-0{\cdot}027 \pm 0{\cdot}01$	$-0{\cdot}02 \pm 0{\cdot}03$ [b]	$-0{\cdot}125 \pm 0{\cdot}01$	$-0{\cdot}14 \pm 0{\cdot}06$ [c]
ψ_4	—	—	—	—	$0{\cdot}003 \pm 0{\cdot}005$	—
β	$0{\cdot}332 \pm 0{\cdot}002$	$0{\cdot}303 \rightarrow 0{\cdot}318$ [e]	$0{\cdot}351 \pm 0{\cdot}002$	$0{\cdot}348 \pm 0{\cdot}007$ [f]	$0{\cdot}367 \pm 0{\cdot}002$	$0{\cdot}373 \pm 0{\cdot}014$ [c]
δ	$4{\cdot}762 \pm 0{\cdot}006$	$5{\cdot}00 \pm 0{\cdot}05$ [e]	$4{\cdot}755 \pm 0{\cdot}006$	$4{\cdot}77 \pm 0{\cdot}06$ [f]	$4{\cdot}759 \pm 0{\cdot}006$	$4{\cdot}9 \pm 0{\cdot}4$ [c]
ν	$0{\cdot}638 \pm 0{\cdot}003$	$0{\cdot}640 \pm 0{\cdot}003$ [a]	$0{\cdot}674 \pm 0{\cdot}003$	$0{\cdot}670 \pm 0{\cdot}006$ [b]	$0{\cdot}705 \pm 0{\cdot}003$	$0{\cdot}717 \pm 0{\cdot}007$ [c]

to the appropriate order in ε one obtains

$$\psi = \frac{6\varepsilon}{n+8} + \frac{\varepsilon^2}{(n+8)^3}(n^2 + 49n + 148) + \frac{24\varepsilon^3}{(n+8)^4}(n^2 + 7n + 46)\zeta(3)$$

$$+ \frac{\varepsilon^3}{(n+8)^5}(-n^4 - 20n^3 + 117n^2 + 1132n + 2384) + O(\varepsilon^4). \qquad (4.10)$$

For example, when $\varepsilon = 1$, this exponent describes the effect of a harmonic polynomial of order 6, in three dimensions. The fact that ψ begins with a term of order ε shows that it does indeed describe a fluctuation effect.

Although the convergence of the series (4.10) is questionable, the fact that ψ is clearly positive provides a convincing confirmation of the extent of universality.

We end with one remark. One can obtain Wegner's result (1972b) for an interaction

$$(\phi^2)^{m-k/2} u_{i_1} \cdots {}_{i_k} \phi_{i_1} \cdots \phi_{i_k} \qquad (4.11)$$

by calculating the contribution of the graph Fig. 13 to the vertex function with $2m$ external legs, and naively exponentiating the logarithm. The result in our language is

$$\alpha_{m,k} = \frac{\varepsilon}{2(n+8)}\left[m(n+2) + 6m^2 - 6m - \tfrac{1}{2}k(k+n-2)\right] + O(\varepsilon^2) \qquad (4.12)$$

with again

$$\psi_{m,k} = \alpha_{m,k} - 1 + \frac{(m-1)(2-\varepsilon+\eta)}{(2-\eta)}$$

Of course this naive exponentiation of logarithms cannot be continued for the interactions (4.11) to higher order in ε, because they mix with other interactions of the same dimension (in $d = 4$) and tensor character. I am grateful for a discussion with P. H. Weisz which clarified in the context of the normal product algorithm (Zimmermann, 1970), the validity of the procedure at order ε.

The result (4.12) is again of limited use for m large. However we note that it has the interesting feature that *for a given m*, ψ is most likely to be negative (and the corresponding interaction relevant) when k has its maximum value 2 m corresponding to the harmonic perturbations considered above. This feature is true also in the known higher orders in the ε-expansion for the specific cases $m = 2$ and $m = 4$. Thus the fact that *harmonic* perturbations *are* irrelevant for $k \geqslant 5$ suggests that *all* high powers of field are irrelevant.

Acknowledgements

I am very grateful to B. G. Nickel, E. Brézin, J. C. le Guillou, J. Zinn-Justin, M. E. Fisher, A. Aharony, S. Ma, D. J. Amit and many others for sending me copies of their work prior to publication. I thank M. A. Moore for discussions on series results.

I thank all my colleagues at Southampton particularly R. K. P. Zia for stimulating comments and discussions.

It is a pleasure to thank Jan Hodgson for her conscientious and cooperative typing of the manuscript.

References

Abe, R. (1972). *Prog. Theor. Phys.* (Kyoto) **48**, 1414.
Abe, R. (1973). *Prog. Theor. Phys.* (Kyoto) **49**, 113.
Abers, S. and Lee, B. W. (1973). *Phys. Reports* **9**, 1.
Aharony, A. (1973). *Phys. Rev.* **B8**, 4270.
Aharony, A. (1974). *Phys. Rev.* **B10**, 3006.
Aharony, A. and Bruce, A. D. (1974). *Phys. Rev.* **B10**, 2973.
Amit, D. J. and de Dominicis, C. T. (1973). *Phys. Letters* **45A**, 193.
Amit, D. J. and Shcherbakov, A. (1974). *J. Phys.* **C7**, L96.
Avdeeva, G. M. and Migdal, A. A. (1972). *J.E.T.P. Letters* **16**, 178.
Avdeeva, G. M. (1973). *J.E.T.P.* **37**, 377.
Applequist, T. and Carazzone, J. (1975). *Phys. Rev.* **D11**, 2856.
Balian, R. and Toulouse, G. (1973). *Phys. Rev. Letts* **30**, 544.
Berlin, T. H. and Kac, M. (1952). *Phys. Rev.* **86**, 821.
Bessis, D. and Zinn-Justin, J. (1972). *Phys. Rev.* **D5**, 1313.
Bjorken, J. D. and Drell, S. D. (1965). "Relativistic Quantum Fields". McGraw-Hill, New York.
Bogoliubov, N. N. and Shirkov, D. V. (1959). "Introduction to the Theory of Quantized Fields", Interscience Publishers.
Brézin, E., Wallace, D. J. and Wilson, K. G. (1972). *Phys. Rev. Letts* **29**, 591.
Brézin, E., Wallace, D. J. and Wilson, K. G. (1973a). *Phys. Rev.* **B7**, 232.
Brézin, E., le Guillou, J. C., Zinn-Justin, J. and Nickel, B. G. (1973b). *Phys. Letts* **44A**, 227.
Brézin, E., le Guillou, J. C., Zinn-Justin, J. (1973c). *Phys. Rev.* **D8**, 434.
Brézin, E., le Guillou, J. C., Zinn-Justin, J. (1974). *Phys. Rev.* **B10**, 892.
Brézin, E. and Wallace, D. J. (1973). *Phys. Rev.* **B7**, 1967.
Bruce, A. D., Droz, M. and Aharony, A. (1974). *J. Phys.* **C7**, 3673.
Combescot, M., Droz, M. and Kosterlitz, J. M. (1974). *Phys. Rev. Letts* **33**, 705.
Combescot, M., Droz, M. and Kosterlitz, J. M. (1975). *Phys. Rev.* **B11**, 4661.
Domb, C. and Hunter, D. L. (1965). *Proc. Phys. Soc. (London)* **86**, 1147.
Erdelyi, A. (1953). "Higher Transcendental Functions", Vol. 2. McGraw-Hill, New York.
Ferer, M., Moore, M. A. and Wortis, M. (1971). *Phys. Rev.* **B4**, 3954.
Ferer, M., Moore, M. A. and Wortis, M. (1973). *Phys. Rev.* **B8**, 5205.
Ferrell, R. A. and Scalapino, D. J. (1972). *Phys. Rev. Letts* **29**, 413.
Feynman, R. P. and Hibbs, A. R. (1965). "Quantum Mechanics and Path Integrals", McGraw-Hill, New York.

Fisher, M. E. (1967). *Rept. Progr. Phys.* **30**, 615.
Fisher, M. E. (1973). *Phys. Rev. Letts* **30**, 679.
Fisher, M. E. and Aharony, A. (1974a). *Phys. Rev. Letts* **31**, 1238.
Fisher, M. E. and Aharony, A. (1974b). *Phys. Rev.* **B10**, 2818.
Fisher, M. E., Ma, S. and Nickel, B. G. (1972). *Phys. Rev. Letts* **29**, 917.
Fisher, M. E. and Pfeuty, P. (1972). *Phys. Rev.* **B6**, 1889.
Gaunt, D. S. and Domb, C. (1970). *J. Phys.* **C3**, 1442.
Goldstone, J., Salam, A. and Weinberg, S. (1962). *Phys. Rev.* **127**, 965.
Golner, G. R. (1973). *Phys. Rev.* **B8**, 3419.
Gradshteyn, I. S. and Ryzhik, I. W. (1965). "Tables of Integrals Series and Products", Academic Press, London and New York.
Griffiths, R. B. (1967). *Phys. Rev.* **158**, 176.
Iliopoulos, J. and Itzykson, C. (1975). *Rev. Mod. Phys.* **47**, 165.
Josephson, B. D. (1969). *J. Phys.* **C2**, 1113.
Kadanoff, L. P. (1966). *Physics* **2**, 263.
Kadanoff, L. P. (1969). *Phys. Rev. Letts* **23**, 1430.
Kadanoff, L. P. *et al.* (1967). *Rev. Mod. Phys.* **39**, 395.
Ketley, I. J. and Wallace, D. J. (1973), *J. Phys.* **A6**, 1667.
Landau, L. and Lifschitz, E. M. (1958). "Statistical Physics", Pergamon Press.
Lee, B. W. (1972). "Chiral Dynamics", Gordon and Breach.
Ma, S. (1973). *Phys. Rev.* **A7**, 2172.
Ma, S. (1974). *Phys. Rev.* **A10**, 1818.
Migdal, A. A. (1972). *J.E.T.P.* **35**, 816.
Milošević, S. and Stanley, H. E. (1972). *Phys. Rev.* **B5**, 2526.
Mueller, A. H. and Trueman, T. L. (1971). *Phys. Rev.* **D4**, 1635.
Mueller, A. H., Trueman, T. L. and Wilson, K. G. (1972). *Phys. Rev.* **D5**, 1026.
Nelson, D. R. (1975). *Phys. Rev.* **B11**, 3504.
Nickel, B. G. (1972). "Feynman Graph Expansions for Critical Exponents". Cornell preprint, unpublished.
Nickel, B. G. (1974). Unpublished.
Patashinskii, A. Z. and Pokrovskii, V. L. (1973). *J.E.T.P.* **37**, 733.
Potts, R. B. (1952). *Proc. Camb. Phil. Soc.* **48**, 106.
Riedel, E. K. and Wegner, F. J. (1972). *Phys. Rev. Letts* **29**, 349.
Riedel, E. K. and Wegner, F. J. (1973). *Phys. Rev.* **B7**, 248.
Riedel, E. K. and Wegner, F. J. (1974). *Phys. Rev.* **B9**, 294.
Schofield, P. (1969). *Phys. Rev. Letts* **22**, 606.
Schofield, P. (1974). *Phys. Letts* **46A**, 197.
Schofield, P., Litster, J. D. and Ho, J. T. (1969). *Phys. Rev. Letts* **23**, 1098.
Stanley, H. E. (1971). "Phase Transitions and Critical Phenomena", Clarendon Press, Oxford.
Stephen, M. J. and McCauley, J. L. (1973). *Phys. Letts* **44A**, 89.
Suzuki, M. (1972). *Phys. Letts* **42A**, 5.
Symanzik, K. (1973). D.E.S.Y. preprint T–733, unpublished.
Tarko, H. B. and Fisher, M. E. (1975). *Phys. Rev.* **B11**, 1217.
Vaks, V. G., Larkin, A. I. and Pikin, S. A. (1968). *J.E.T.P.* **26**, 647.
Vicentini-Missoni, M., Joseph, R. I., Green, M. S. and Levelt Sengers, J. M. H. (1970), *Phys. Rev.* **B1**, 2312.
Wallace, D. J. (1973a). *J. Phys.* **C6**, 1390, Corrigenda **C7**, 4551.
Wallace, D. J. (1973b). "Broken Symmetry and Phase Transitions". *In* "Cargèse Lectures 1973" (E. Brézin and J. M. Charap, eds.) Gordon and Breach.

Wallace, D. J. and Zia, R. K. P. (1973). *Phys. Letts* **46A**, 261.
Wallace, D. J. and Zia, R. K. P. (1974). *J. Phys.* **C7**, 3480.
Wallace, D. J. and Zia, R. K. P. (1975a). *Annals of Physics* (*N.Y.*) **92**, 142.
Wallace, D. J. and Zia, R. K. P. (1975b). *J. Phys.* **C8**, 839.
Wallace, D. J. and Zia, R. K. P. (1975c). *Phys. Rev.* **B12**, 5340.
Wegner, F. J. (1972a). *Phys. Rev.* **B5**, 4529.
Wegner, F. J. (1972b). *Phys. Rev.* **B6**, 1891.
Widom, B. (1965). *J. Chem. Phys.* **43**, 3898.
Wilson, K. G. (1969). *Phys. Rev.* **179**, 1449.
Wilson, K. G. (1971). *Phys. Rev.* **B4**, 3174, 3184.
Wilson, K. G. (1972). *Phys. Rev. Letts* **28**, 548.
Wilson, K. G. (1974). *Phys. Rev.* **D7**, 2911.
Wilson, K. G. and Fisher, M. E. (1972). *Phys. Rev. Letts* **28**, 240.
Wilson, K. G. and Kogut, J. (1974). *Phys. Reports* **12C**, 75.
Wilson, K. G. and Zimmermann, W. (1972). *Commun. Math. Phys.* **24**, 87.
Yamazaki, Y. (1974). *Phys. Letts* **49A**, 215.
Zia, R. K. P. and Wallace, D. J. (1975). *J. Phys.* **A8**, 1495.
Zimmermann, W. (1970). *In* "Lectures on Elementary Particles and Quantum Field Theory" (S. Deser, M. Grisaru and H. Pendleton, eds.) M.I.T. Press.
Zinn-Justin, J. (1973). "Wilson's Theory of Critical Phenomena and Renormalized Perturbation Theory". *In* "Cargèse Lectures, 1973" (E. Brézin and J. M. Charap, eds.) Gordon and Breach.

6. Dependence of Universal Critical Behaviour on Symmetry and Range of Interaction*

Amnon Aharony†

Department of Physics, Harvard University, Cambridge, Mass. 02138, U.S.A.

* Work supported in part by the National Science Foundation under Grant No. 32774.

† Permanent address: Department of Physics and Astronomy, Tel Aviv University, Ramat Aviv, Israel.

I. Introduction

The Hamiltonians describing systems which undergo phase transitions are usually quite complicated, and involve many parameters. The basic idea of universality (Kadanoff, 1966) is that the critical behaviour of these systems, when the correlation length becomes very large, depends only on a small number of general features of these Hamiltonians, and not on the actual values of the various parameters involved. Indeed, high-temperature series, performed on systems with short-range isotropic "exchange" interaction, lead to critical exponents and scaling functions which are universal, and depend only on the dimensionality of space, d, and on the number of components of the order parameter (or "spin"), n (see, e.g., volume 3 of this series). Further series expansions show, for example, crossovers from n-component behaviour to n'-component behaviour ($n' < n$), when the exchange interaction is not symmetric, and when the lower symmetry involves only n' "spin" components (see, e.g. Pfeuty *et al.*, 1974). However, series expansions become quite tedious when the symmetry of the system is complicated, or when long-range interactions are important.

The recent development of renormalization group approach to critical phenomena (Wilson and Kogut, 1974) gave a much easier tool to study such complicated interactions. Indeed, it enabled one for the first time to analyse the critical behaviour of complicated systems such as the dipolar systems (Fisher and Aharony, 1973a, Aharony and Fisher, 1973, Aharony, 1973a–e), or systems with complicated symmetry properties (see, e.g., Aharony, 1973f). It is thus the purpose of this article to review such applications, going into enough detail so that other applications can be easily followed. The main result is that, in addition to n and d, the *range* of the interaction and the *symmetry* of the Hamiltonian (either in spin-space or in the coordinate space) are also crucial in determining the ultimate critical behaviour of a system with a given Hamiltonian.

There are many approaches to the presentation of renormalization group. The approach we choose here is that of the exact recursion relations, as described in detail (for the short-range isotropic case) by Wilson and Kogut (1974). By choosing this approach, we are limited to a study of the critical behaviour in the vicinity of four dimensions, and to an expansion of all the resulting exponents in the small parameter $\varepsilon = 4 - d$. There does not yet exist a satisfactory practical way to do renormalization group studies exactly in three dimensions (see, however, Golner and Riedel, 1975; Kadanoff and Houghton, 1975). The Wilson approximate recursion relations (Wilson, 1971) have been used to study the stability of the isotropic short-range behaviour (Grover *et al.*, 1972), but do not allow complicated dependences of the interaction on the spatial coordinates, and involve approximations which are hard to estimate. Another alternative is the $1/n$-expansion (Abe, 1972, Ma, 1973), but most of the interesting problems involve $n = 1, 2, 3$, and the results from the $1/n$-expansion are not particularly useful in this range. (We shall refer to results from this expansion whenever they are available.)

A direct way to derive critical exponents is to use Feynman graph expansions (Wilson, 1972, Wallace, this volume). These are very useful for studying critical behaviour in the vicinity of one fixed point, but are not as easily applicable to the problem of searching for all possible types of critical behaviour. However, we shall cite such results from such expansions in the vicinity of given fixed points.

The structure of this Chapter is as follows: the recursion relation approach will be reviewed in Section II. This will include a study of the general form of Hamiltonian we shall be studying, a detailed definition of the renormalization group transformation, including a discussion of the relation between the notion of a fixed point and the derivation of critical exponents, a discussion of the Gaussian model, which is the basic model about which the perturbation expansion is performed, and a description of the diagrammatic expansion of the recursion relations, which finally leads to the ε-expansion.

Section III discusses the analysis of spin anisotropies, Section IV reviews applications to long-range and dipolar interactions, Section V describes various quartic symmetry breaking interactions and Section VI summarizes the study of a few lattice anisotropies. Naturally, I take advantage of the reviewer's freedom, and put more emphasis on problems in which I had personal involvement. However, it is hoped that most of the applications of the renormalization group to long-range or to asymmetric problems are at least mentioned. Note also that references are to the papers in which the renormalization group was applied, and not to all the papers relevant to a given problem. The fast growth of literature on the subject may have caused some omissions, and I apologize to authors whose work is not mentioned.

Finally, Section VII contains a critical discussion of the results, with emphasis on experiments and on remaining unsolved problems.

II. The Recursion Relation Approach

A. Reduced Hamiltonian: notations and normalizations

All the renormalization group calculations which we shall describe start with a classical Hamiltonian, which is a functional of the "spins", or local variables $\mathbf{S_R} \equiv \{S^\alpha_\mathbf{R}, \alpha = 1, 2, \ldots, n\}$, where $\mathbf{R}$ is a site on a d-dimensional lattice:

$$\mathscr{H} = \mathscr{H}\{\mathbf{S_R}\}. \tag{2A.1}$$

In what follows, we shall adopt the *continuous spin model* (Wilson and Kogut, 1974), in which the spin length

$$|\mathbf{S}| = \left[\sum_{\alpha=1}^{n} (S^\alpha)^2\right]^{1/2}$$

is allowed to vary continuously, and a spin weight function is introduced into the partitition function, to restrict fluctuations in the spin length:

$$Z[\mathscr{H}] = \int\left(\prod_\mathbf{R} d^n S_\mathbf{R}\right)\exp(\bar{\mathscr{H}}) \tag{2A.2}$$

with the reduced Hamiltonian

$$\bar{\mathscr{H}} = -\mathscr{H}/k_B T + W\{\mathbf{S_R}\} \tag{2A.3}$$

where $W\{\mathbf{S_R}\}$ grows very large and negative whenever any $|\mathbf{S_R}|$ becomes large. Typically,

$$W\{\mathbf{S_R}\} = -\sum_\mathbf{R} \omega(\mathbf{S_R}), \tag{2A.4}$$

with $\omega(\mathbf{x}) \to \infty$ as $|\mathbf{x}| \to \infty$. In what follows we shall use

$$\omega(\mathbf{x}) = \tfrac{1}{2}|\mathbf{x}|^2 + \tilde{u}|\mathbf{x}|^4 + \ldots. \tag{2A.5}$$

General abstract descriptions of the renormalization group procedure, as applied to (2A.2), have been given by many authors (Wilson and Kogut, 1974; Fisher, 1974a; Wegner, 1972a, 1973, and Chapter 2 of this volume; Aharony, 1974a; Ma, 1973, and Chapter 4 of this volume). We shall therefore not repeat them here. Instead, we shall consider here a form of $\bar{\mathscr{H}}$, which will be explicitly appropriate for application in the following sections.

Many terms which may appear in the Hamiltonian have been reviewed previously (Aharony, 1974a; Fisher, 1974). Most of these may be summarized by writing

$$\mathscr{H} = \mathscr{H}_{\mathrm{ex}} + \mathscr{H}_{\mathrm{s.i.}}, \tag{2A.6}$$

where $\mathscr{H}_{\mathrm{ex}}$ represents "exchange" terms, of the form

$$\mathscr{H}_{\mathrm{ex}} = -\tfrac{1}{2}\sum_{\mathbf{R},\mathbf{R}'}\sum_{\alpha,\beta} J^{\alpha\beta}(\mathbf{R}-\mathbf{R}')\, S^\alpha_\mathbf{R} S^\beta_{\mathbf{R}'}, \tag{2A.7}$$

and $\mathscr{H}_{\mathrm{s.i.}}$ represents "single ion" terms, of the form

$$\mathscr{H}_{\mathrm{s.i.}} = \sum_\mathbf{R}\left\{\tfrac{1}{2}\sum_{\alpha,\beta} A^{\alpha\beta} S^\alpha_\mathbf{R} S^\beta_\mathbf{R} + \sum_{\alpha,\beta,\gamma,\delta} B^{\alpha\beta\gamma\delta} S^\alpha_\mathbf{R} S^\beta_\mathbf{R} S^\gamma_\mathbf{R} S^\delta_\mathbf{R} + \ldots\right\}. \tag{2A.8}$$

We use here the "magnetic" language, since most of the examples to be discussed will be applicable to magnetic systems. There should be no difficulty in translating the names of the variables to other cases (e.g. the cases of fluids, $n = 1$, liquid helium four, $n = 2$, etc.). We shall discuss more complicated forms of $\mathscr{H}$ in Section VI. Explicit examples of physically interesting forms of $J^{\alpha\beta}(\mathbf{R})$ are summarized in Table I.

TABLE I. Examples of Exchange Interactions [Eqns (2A.7), (2A.13)]

Type	$J^{\alpha\beta}(\mathbf{R})$	$V_2^{\alpha\beta}(\mathbf{q})$ [to $O(q^2)$]	Reference
Isotropic	$J(R)\,\delta_{\alpha\beta}$	$(r + q^2)\,\delta_{\alpha\beta}$	Wilson and Fisher, 1972
Spin–anisotropic	$J^\alpha(R)\,\delta_{\alpha\beta}$	$(r_\alpha + q^2)\,\delta_{\alpha\beta}$	Fisher and Pfeuty, 1972
Cubic	$J(R^\alpha)\,\delta_{\alpha\beta}$	$[r + q^2 + f(q^\alpha)^2]\,\delta_{\alpha\beta}$	Aharony, 1973f, h
Isotropic "Dipolar"	$K(R)\,R^\alpha R^\beta/R^2$	$-hq^\alpha q^\beta + gq^\alpha q^\beta/q^2$	Aharony and Fisher, 1973
Dzyaloshinski–Moriya	$A^\gamma(R)\,\varepsilon_{\alpha\beta\gamma}$	$ia_\gamma q^\gamma \varepsilon_{\alpha\beta\gamma}$ + c.c.	Liu, 1973
Long-Range	$J_\infty R^{-d-\sigma}\,\delta_{\alpha\beta}$	$A\lvert\mathbf{q}\rvert^\sigma\,\delta_{\alpha\beta}$	Fisher, Ma and Nickel, 1972

The next step is to introduce the momentum space variables,

$$\sigma_{\mathbf{q}}^\alpha = \hat{c}\sum_{\mathbf{R}} \exp(i\mathbf{q}\cdot\mathbf{R})\,S_{\mathbf{R}}^\alpha, \tag{2A.9}$$

where $\hat{c}$ will be specified below. Equation (2A.7) then assumes the form

$$\mathscr{H}_{\text{ex}} = -\frac{1}{2\hat{c}^2\Omega}\int d^d q\,\hat{J}^{\alpha\beta}(\mathbf{q})\,\sigma_{\mathbf{q}}^\alpha\sigma_{-\mathbf{q}}^\beta, \tag{2A.10}$$

where

$$\hat{J}^{\alpha\beta}(\mathbf{q}) = \sum_{\mathbf{R}} \exp(i\mathbf{q}\cdot\mathbf{R})\,J^{\alpha\beta}(\mathbf{R}) \tag{2A.11}$$

and where (in the limit of an infinite system) the integral over $\mathbf{q}$ is over the appropriate Brillouin zone, which is of volume Ω. For a s.c. lattice of spacing a, Ω is equal to

$$\Omega_0 = (2\pi/a)^d. \tag{2A.12}$$

Fourier transforming (2A.4), (2A.7) and (2A.8) and substituting into (2A.3) leads to

$$\bar{\mathscr{H}} = -\tfrac{1}{2}\sum_{\alpha,\beta}\int_{\mathbf{q}} V_2^{\alpha\beta}(\mathbf{q})\,\sigma_{\mathbf{q}}^{\alpha}\sigma_{-\mathbf{q}}^{\beta} - \frac{1}{4!}\sum_{\alpha,\beta,\gamma,\delta} V_4^{\alpha\beta\gamma\delta}\int_{\mathbf{q}}\int_{\mathbf{q}'}\int_{\mathbf{q}''} \sigma_{\mathbf{q}}^{\alpha}\sigma_{\mathbf{q}'}^{\beta}\sigma_{\mathbf{q}''}^{\gamma}\sigma_{-\mathbf{q}-\mathbf{q}'-\mathbf{q}''}^{\delta} - \dots \tag{2A.13}$$

where [using the weight function (2A.5)]

$$V_2^{\alpha\beta}(\mathbf{q}) = \frac{1}{\hat{c}^2\hat{f}}\{\delta_{\alpha\beta} - [\hat{J}^{\alpha\beta}(\mathbf{q}) - A^{\alpha\beta}]/k_B T\}, \tag{2A.14}$$

while

$$\int_{\mathbf{q}} \equiv \frac{1}{\Omega_0}\int \mathrm{d}^d q, \tag{2A.15}$$

and $\hat{f} = \Omega/\Omega_0$. With our choice of $\mathscr{H}_{\text{s.i.}}$, (2A.8), and of the weighting function, (2A.5), $V_4^{\alpha\beta\gamma\delta}$ is a q-independent combination of $B^{\alpha\beta\gamma\delta}$ and of $\tilde{u}$. The q-dependent terms are usually irrelevant (see Aharony, 1973*g*, 1974b).

At this stage it is convenient to choose the normalization of the spins $\boldsymbol{\sigma}_{\mathbf{q}}$, so that one q-dependent term in $V_2^{\alpha\beta}(\mathbf{q})$ becomes independent of temperature. In general, $J^{\alpha\beta}(\mathbf{R})$ may include both short- and long-range terms. As we shall see, the Fourier transforms of the long-range terms usually include terms which are non-analytic in $\mathbf{q}$. Apart from those, $\hat{J}^{\alpha\beta}(\mathbf{q})$ may be expanded in powers of $\mathbf{q}$. Thus,

$$\hat{J}^{\alpha\beta}(\mathbf{q}) = \hat{J}^{\alpha\beta}(0) - jq^2\delta_{\alpha\beta} + \sum_{\gamma,\delta} j_1^{\alpha\beta\gamma\delta}q^{\gamma}q^{\delta} + \hat{J}^{\alpha\beta}_{\text{n.a.}}(\mathbf{q}) + O(q^4) \tag{2A.16}$$

where $\hat{J}^{\alpha\beta}_{\text{n.a.}}$ stands for the non-analytic terms, while the term $q^2\delta_{\alpha\beta}$ has been separated from the sum over γ and δ. The standard normalization at this stage is to choose $\hat{c}$ of (2A.9) as

$$\hat{c}^2 = j/\hat{f}\,k_B T, \tag{2A.17}$$

so that the coefficient of the term $q^2\delta_{\alpha\beta}$ in $V_2^{\alpha\beta}(\mathbf{q})$ becomes equal to unity.

Finally, it is helpful to approximate the geometry of the Brillouin zone by a sphere, and thereby restrict the range of all momenta by

$$|\mathbf{q}| \leqslant \Lambda, \qquad \Lambda = \lambda\pi/a, \tag{2A.18}$$

with λ chosen so as to match the volume Ω of the original Brillouin zone,

$$\lambda^d = [2/\sqrt{(\pi)}]^d\,\Gamma(\tfrac{1}{2}d + 1)\,\hat{f}. \tag{2A.19}$$

The cutoff (2A.18) is now to be understood to be a part of the definition (2A.15). In the following we shall measure distances in units of a, namely set $a \equiv 1$, or measure momenta in units of Λ, namely set $\Lambda = 1$.

Thus,

$$V_2^{\alpha\beta}(\mathbf{q}) = r_{\alpha\beta} + q^2\delta_{\alpha\beta} - \sum_{\gamma,\delta} \frac{1}{j} j_1^{\alpha\beta\gamma\delta} q^\gamma q^\delta - \frac{1}{j} \hat{J}^{\alpha\beta}_{\text{n.a.}}(\mathbf{q}) + O(q^4) \qquad \text{(2A.20)}$$

where

$$r_{\alpha\beta} = [k_B T\delta_{\alpha\beta} - \hat{J}^{\alpha\beta}(0) + A^{\alpha\beta}]/j. \qquad \text{(2A.21)}$$

Examples of the function $V_2^{\alpha\beta}(\mathbf{q})$ are given in Table I. Similarly, for this model,

$$\frac{1}{4!} V_4^{\alpha\beta\gamma\delta} = (k_B T/j)^2 \left[B^{\alpha\beta\gamma\delta}/k_B T + \frac{1}{3}\tilde{u}(\delta_{\alpha\beta}\delta_{\gamma\delta} + \delta_{\alpha\gamma}\delta_{\beta\delta} + \delta_{\alpha\delta}\delta_{\beta\gamma}) \right]. \qquad \text{(2A.22)}$$

We have worked out the explicit dependence of the final parameters on the temperature and on the parameters which appeared in the original Hamiltonian, first because these are the initial conditions for our renormalization procedure, second because they become important on considering non-universal normalization constants, and third in order to emphasize the way in which the temperature T enters into the problem. As T is lowered, one or more of the eigenvalues of the matrix $r_{\alpha\beta}$ may vanish. In mean field theory, this leads to criticality and may be used to define the critical temperature, T_0. If that happens simultaneously with a vanishing of an appropriate combination of $V_4^{\alpha\beta\gamma\delta}$'s, tricritical behaviour results. We shall return to these facts later, in Section V.

B. Renormalization group iterations

Summing up the previous section, the partition function has been written in the form

$$Z = \int \left(\prod_{\mathbf{q}} \mathrm{d}^n \sigma_{\mathbf{q}} \right) \exp(\bar{\mathscr{H}}[\boldsymbol{\sigma}_{\mathbf{q}}]) \qquad \text{(2B.1)}$$

with $\bar{\mathscr{H}}$ given by (2A.13).

The basic idea of the renormalization group iteration procedure is to perform the integral in (2B.1) in steps. We first integrate over the variables $\boldsymbol{\sigma}_{\mathbf{q}}$ which satisfy

$$\Lambda/b < |\mathbf{q}| < \Lambda, \qquad b > 1. \qquad \text{(2B.2)}$$

Denoting

$$\exp(\bar{\mathscr{H}}_1) = \int \left(\prod_{\Lambda/b < |\mathbf{q}| < \Lambda} \mathrm{d}^n \sigma_{\mathbf{q}} \right) \exp(\bar{\mathscr{H}}[\boldsymbol{\sigma}_{\mathbf{q}}]), \qquad \text{(2B.3)}$$

we can therefore write

$$Z = \int \left(\prod_{|\mathbf{q}| < \Lambda/b} d^n\sigma_{\mathbf{q}} \right) \exp(\bar{\mathscr{H}}_1), \tag{2B.4}$$

where now $\bar{\mathscr{H}}_1$ is a functional of all variables $\sigma_{\mathbf{q}}$ with $|\mathbf{q}| < \Lambda/b$. We have thus eliminated from the problem the fluctuations which had the shortest wavelengths. We now aim to bring (2B.4) to a form similar to (2B.1). This is done in two steps: we first preserve the spatial density of degrees of freedom, by rescaling all spatial vectors by a factor b,

$$\mathbf{q} \Rightarrow \mathbf{q}' = b\mathbf{q}. \tag{2B.5}$$

We then rescale all spin variables, in order to preserve the basic spin fluctuation magnitude,

$$\sigma^{\alpha}_{\mathbf{q}} \Rightarrow \sigma'^{\alpha}_{b\mathbf{q}} = \sigma'^{\alpha}_{\mathbf{q}'} = \zeta_{\alpha}^{-1}\sigma^{\alpha}_{\mathbf{q}}. \tag{2B.6}$$

Introducing these two rescalings into $\bar{\mathscr{H}}_1$ leads to a new effective Hamiltonian,

$$\bar{\mathscr{H}}'[\boldsymbol{\sigma}'_{\mathbf{q}'}] \equiv \bar{\mathscr{H}}_1[\boldsymbol{\sigma}_{\mathbf{q}}]. \tag{2B.7}$$

Expanding $\bar{\mathscr{H}}'$ in the form (2A.13) now enables us to identify the new functions $V_2^{\alpha\beta}$, $V_4^{\alpha\beta\gamma\delta}$, etc. In the applications to short range isotropic forces, the spin rescaling factors ζ_{α} are chosen so that the coefficient of $q^2\delta_{\alpha\beta}$ in $V_2^{\alpha\beta}$ [see eqn (2A.20)] remains equal to unity. However, we shall mention a few exceptions.

After the rescaling factors ζ_{α} are chosen, we can compare $\bar{\mathscr{H}}'$ with $\bar{\mathscr{H}}$, and write down a series of recursion relations for the coefficients $r_{\alpha\beta}$, $j_1^{\alpha\beta\gamma\delta}$, $V_4^{\alpha\beta\gamma\delta}$, etc. These form a representation of the renormalization group transformation

$$\bar{\mathscr{H}}' = \mathbb{R}\bar{\mathscr{H}}. \tag{2B.8}$$

The next step is to follow the "flow" of $\bar{\mathscr{H}}$, in the multi-dimensional Hamiltonian space, under the transformation $\mathbb{R}$. For certain values of the parameters in $\bar{\mathscr{H}}$, this flow will go to a *fixed point*, $\bar{\mathscr{H}}^*$, at which

$$\bar{\mathscr{H}}^* = \mathbb{R}\bar{\mathscr{H}}^*. \tag{2B.9}$$

In most cases of interest, we shall find more than one fixed point. One can draw an analogy between the flow under (2B.8) to the motion of a particle in a potential which may have local minima (stable fixed points), maxima (unstable ones) or saddle points (fixed points which are unstable with respect to some of the parameters).

By the structure of the transformation, we rescale our unit of length by a factor b under each iteration. Thus, if the true correlation length is ξ, then after l iterations we shall have an effective correlation length ξ_l, given by

$$\xi_l = b^{-l}\xi. \tag{2B.10}$$

Thus, it is clear that at the fixed point $\bar{\mathscr{H}}^*$ ξ must be infinite ($\xi_{l+1} = \xi_l$), and that is expected to happen at the critical point. One therefore relates to each fixed point a typical critical behaviour. Small deviations from $\bar{\mathscr{H}}^*$ enable us to study the critical behaviour in the vicinity of the critical point.

The whole process of iteration loses meaning when ξ_l becomes of order unity (the lattice unit). At this stage, there are no long-range correlations left, and mean field theory can be used for the effective Hamiltonian $\bar{\mathscr{H}}_l$. By studying the thermodynamic functions evaluated at $\bar{\mathscr{H}}_l$, which is a function of the iterated parameters $r_{\alpha\beta,l}$ etc., and by substituting l from $\xi_l = b^{-l}\xi \simeq 1$, we can thus find their form near criticality.

A direct way to obtain the critical exponents is to linearize (2B.8) about (2B.9),

$$\mathbb{R}(\bar{\mathscr{H}}^* + \mu Q) = \bar{\mathscr{H}}^* + \mu \mathbb{L} Q + O(\mu^2), \tag{2B.11}$$

and to find the eigenvalues Λ_i and the eigen-operators Q_i of the linear operator $\mathbb{L}$,

$$\mathbb{L} Q_i = \Lambda_i Q_i = b^{\lambda_i} Q_i. \tag{2B.12}$$

If $\lambda_i > 0$, then the coefficient of Q_i will grow larger under iterations. The operator Q_i is then called *relevant*. If $\lambda_i < 0$, this coefficient will decay, and Q_i is called *irrelevant*. If $\lambda_i = 0$, Q_i is called *marginal*.

At a fixed point representing a critical point of second order, only the parameters representing the temperature and the "magnetic" field are relevant. We shall set the magnetic field equal to zero in most of this paper. At fixed points representing critical points of higher order, more relevant operators occur (Riedel and Wegner, 1972). If the coefficients of these operators are not chosen carefully, then their relevance will lead to a "flow" away from these fixed points. This flow may be to another, more stable, fixed point, or to regions in parameter space in which the partition function has a thermodynamic instability, representing a first-order transition. In the first case one speaks of a *crossover* between one type of critical behaviour to another. The second case usually represents a first-order transition (Golner, 1973).

The free energy may be considered as a function of the "fields" μ_i, which appear in the expansion

$$\bar{\mathscr{H}} = \bar{\mathscr{H}}^* + \sum_i \mu_i Q_i. \tag{2B.13}$$

In the linear region (2B.11), the free energy per spin after l iterations is†

$$f(\mu_1, \mu_2, \ldots) = b^{-ld} f(b^{l\lambda_1}\mu_1, b^{l\lambda_2}\mu_2, \ldots). \tag{2B.14}$$

If $\xi_l = b^{-l}\xi = 1$, then

$$f(\mu_1, \mu_2, \ldots) = \xi^{-d} f(\xi^{\lambda_1}\mu_1, \xi^{\lambda_2}\mu_2, \ldots). \tag{2B.15}$$

† We omit here an additive term, which does not affect the derivation of the exponents but is important for calculations of the full free energy (Nauenberg and Nienhuis, 1974; Nelson and Rudnick, 1975).

Assuming

$$\xi \sim t^{-\nu}, \qquad t = (T - T_c)/T_c, \tag{2B.16}$$

we immediately identify the specific heat exponent α through

$$d\nu = 2 - \alpha, \tag{2B.17}$$

and the *crossover exponents*

$$\phi_i = \lambda_i \nu, \tag{2B.18}$$

so that

$$f(\mu_1, \mu_2, \ldots) \propto t^{2-\alpha} f(\mu_1 t^{-\phi_1}, \mu_2 t^{-\phi_2}, \ldots). \tag{2B.19}$$

Denoting by μ_1 the temperature field, $(T - T_c)$, we see that $\phi_1 = 1$, or

$$\nu = 1/\lambda_1. \tag{2B.20}$$

From (2A.21) one can see, that λ_1 is the eigenvalue associated with the variable r, which is the smallest eigenvalue of the matrix $r_{\alpha\beta}$.

For an irrelevant variable μ_i, $\lambda_i < 0$ and $\omega_i = -\phi_i > 0$. For small t, near the critical point, the variable $\mu_i t^{\omega_i}$ is small, and f can usually be expanded in it. Thus, we have

$$f(\mu_1, \mu_2, \ldots) \propto t^{2-\alpha} f_0(\mu_1 t^{-\phi_1}, \ldots) + \sum_i t^{2-\alpha+\omega_i} f_i'(\mu_1 t^{-\phi_1}, \ldots) + \ldots, \tag{2B.21}$$

where now f_0, f_i', etc. are functions only of the rescaled relevant variables $\mu_i t^{-\phi_i}$. The terms in the sum represent *corrections to the leading scaling form*, represented by f_0.

For a relevant variable μ_i, $\phi_i > 0$, and $\mu_i t^{-\phi_i}$ grows larger and larger as t approaches zero. The point at which

$$\mu_i |t|^{-\phi_i} \sim 1 \tag{2B.22}$$

is usually considered as the crossover point, below which one can no longer use the original fixed point to describe the critical behaviour of the system. Instead, one must solve the non-linear original recursion relations and try to find a new, stable, fixed point at which μ_i is irrelevant.

The rest of this paper will be devoted to the identification of the possible fixed points, and to the calculation of the critical exponents related to each of them, as represented in (2B.12).

C. The Gaussian model

The Gaussian model is obtained by keeping only the terms bilinear in the spins in $\bar{\mathscr{H}}$, (2A.13). Including a magnetic field in the Hamiltonian,

$$\bar{\mathscr{H}}_h = -\sum_{\mathbf{R}} \mathbf{H}_{\mathbf{R}} \cdot \mathbf{S}_{\mathbf{R}} = -\int_{\mathbf{q}} \mathbf{h}_{\mathbf{q}} \cdot \boldsymbol{\sigma}_{-\mathbf{q}},$$

the partition function (2A.2) becomes exactly

$$Z[\mathscr{H}] = (2\pi)^{N/2} \prod_{\mathbf{q}} [\det V_2^{\alpha\beta}(\mathbf{q})]^{-1/2} \exp\left[\tfrac{1}{2}\sum_{\alpha\beta}\int_{\mathbf{q}} h_{\mathbf{q}}^{\alpha} G^{\alpha\beta}(\mathbf{q}) h_{-\mathbf{q}}^{\beta}\right], \tag{2C.1}$$

with

$$G^{\alpha\beta}(\mathbf{q}) = [V_2(\mathbf{q})^{-1}]^{\alpha\beta}. \tag{2C.2}$$

$G^{\alpha\beta}(\mathbf{q})$ can be readily identified (by a simple differentiation) with the Fourier transform of the two-spin correlation function,

$$\langle \sigma_{\mathbf{q}}^{\alpha} \sigma_{\mathbf{q}'}^{\beta} \rangle = \delta(\mathbf{q} + \mathbf{q}') G^{\alpha\beta}(\mathbf{q}). \tag{2C.3}$$

[Here, $\delta(\mathbf{q})$ denotes $(2\pi)^d \delta^{(d)}(\mathbf{q})$]. Higher order spin correlations are similarly given, by the Wick theorem, as products of two-spin correlation functions (see Wilson and Kogut, 1974, for details).

The crucial step in the technical calculation of the partition function $Z[\mathscr{H}]$ for the non-Gaussian case will be based on a perturbation expansion in V_4 and higher order terms. As is well known, this will lead to Feynman-graph-like expansions, with $G^{\alpha\beta}(\mathbf{q})$ being the propagator. We shall go into details later.

Equation (2C.3) contains all the physical information we need: inverting $V_2^{\alpha\beta}$ of (2A.20), we can now readily identify the critical exponents ν and η, and also a few crossover exponents.

In the isotropic short-range case, $\hat{J}^{\alpha\beta}(\mathbf{q})$ of (2A.16) is of the form $\delta_{\alpha\beta} J(q^2)$. Hence,

$$V_2^{\alpha\beta}(\mathbf{q}) = (r + q^2)\delta_{\alpha\beta} + O(q^4), \tag{2C.4}$$

where (2A.21) reduces to

$$r = k_B(T - T_0)/j, \qquad k_B T_0 = \hat{J}(0) - A. \tag{2C.5}$$

Thus,

$$G^{\alpha\beta}(\mathbf{q}) = \delta_{\alpha\beta}/[r + q^2 + O(q^4)] = \frac{\delta_{\alpha\beta}}{r}\left\{1 + \frac{q^2}{r}[1 + O(q^2)]\right\}^{-1}. \tag{2C.6}$$

Therefore, the susceptibility is

$$\chi^{\alpha\beta} = G^{\alpha\beta}(0) = \delta_{\alpha\beta}/r \propto (T - T_0)^{-\gamma}, \tag{2C.7}$$

and $\gamma = 1$. For $q \ll \Lambda$, the term in parentheses in (2C.6) is a function of q^2/r, and thus we identify the correlation length as $\xi = r^{-1/2} \propto (T - T_0)^{-1/2}$, or $\nu = \frac{1}{2}$. At T_0, $G^{\alpha\beta} \sim q^{-2}$, and therefore $\eta = 0$. All the other exponents can similarly be derived, and identified as those of mean field theory.

In the anisotropic case (Riedel and Wegner, 1969; Fisher and Pfeuty, 1972; Pfeuty *et al.*, 1974),

$$\hat{J}^{\alpha\beta}(\mathbf{q}) = \delta_{\alpha\beta}\hat{J}^{\alpha}(\mathbf{q}) = \delta_{\alpha\beta}(\hat{J}(\mathbf{q}^2) - jg_\alpha),$$

and therefore

$$r_{\alpha\beta} = (r + g_\alpha)\delta_{\alpha\beta}, \tag{2C.8}$$

so that we have

$$G^{\alpha\beta}(\mathbf{q}) = \delta_{\alpha\beta}/[r + g_\alpha + q^2 + O(q^4)]. \tag{2C.9}$$

Suppose now

$$g_1 = \ldots = g_m = g,$$

$$g_{m+1} = \ldots = g_n = 0. \tag{2C.10}$$

As T is lowered towards T_0, with $g > 0$, only the susceptibilities related to the last $(n - m)$ spin components will diverge. However, if $g = 0$ all susceptibilities diverge. One can write for the susceptibilities

$$\chi^{\alpha\beta} \propto \delta_{\alpha\beta}(T - T_0)^{-1}[1 + g_\alpha/(T - T_0)]^{-1}, \tag{2C.11}$$

and therefore the crossover exponent related to g [compare with (2B.19)] is

$$\phi_g = 1. \tag{2C.12}$$

We shall see more examples later. We now want to demonstrate the iteration procedure, presented in the previous section, on the Gaussian model.

The first step, (2B.3), can simply be written as

$$\bar{\mathscr{H}}_1[\boldsymbol{\sigma}_\mathbf{q}] = \bar{\mathscr{H}}[\boldsymbol{\sigma}_\mathbf{q}] + \text{const.}, \tag{2C.13}$$

where the constant results from the integration over $\{\boldsymbol{\sigma}_\mathbf{q};\ 1/b < |\mathbf{q}| < 1\}$. The next two steps, (2B.5) and (2B.6), yield

$$[V_2^{\alpha\beta}(\mathbf{q})]' = \zeta_\alpha\zeta_\beta b^{-d} V_2^{\alpha\beta}(b^{-1}\mathbf{q}) \tag{2C.14}$$

The coefficient of $q^2\delta_{\alpha\beta}$ in $[V_2^{\alpha\beta}]'$ is thus $\zeta_\alpha\zeta_\beta b^{-d-2}$. Choosing it to be equal to unity we find

$$\zeta_\alpha^2 \equiv \zeta^2 = b^{d+2} \tag{2C.15}$$

With this value for ζ_α, the recursion relations for the other parameters in $V_2^{\alpha\beta}$ take the form

$$r'_{\alpha\beta} = b^2 r_{\alpha\beta}, \tag{2C.16}$$

$$(j_1^{\alpha\beta\gamma\delta})' = j_1^{\alpha\beta\gamma\delta}, \tag{2C.17}$$

$$[\hat{J}^{\alpha\beta}_{\text{n.a.}}(\mathbf{q})]' = b^2\hat{J}^{\alpha\beta}_{\text{n.a.}}(b^{-1}\mathbf{q}), \text{ etc.} \tag{2C.18}$$

For $\hat{J}^{\alpha\beta}_{\text{n.a.}} = 0$, these recursion relations yield the isotropic *Gaussian fixed point*, at which $r_{\alpha\beta}$ and all the higher-order coefficients vanish. The recursion relations (2C.16)–(2C.18) are already in diagonal linearized form near this fixed point, and we see that all the $r_{\alpha\beta}$'s are *relevant*, $j_1^{\alpha\beta\gamma\delta}$ is *marginal* and all the higher-order powers of q are *irrelevant* (the eigenvalue of $q^4\delta_{\alpha\beta}$ is b^{-2}, etc.)

In the isotropic short-range case, the single parameter r represents the temperature $(T - T_0)$, and we immediately identify [using (2B.20)]

$$\nu = \tfrac{1}{2}. \tag{2C.19}$$

In the anisotropic case, represented by (2C.8) and (2C.10), we have

$$r' = b^2 r, \qquad g' = b^2 g. \tag{2C.20}$$

The parameter g is thus relevant, and leads to a crossover from the isotropic Gaussian fixed point to the non-isotropic one, with a crossover exponent [see (2B.18)]

$$\phi_g = \lambda_g \nu = 2\nu = 1, \tag{2C.21}$$

in agreement with (2C.12).

As we set up our spin rescaling, (2C.15), any non-zero value of g will lead, under iterations, to a larger and larger value. The only non-zero fixed point solution of the recursion relation $g' = b^2 g$ is infinite. We are thus led to consider a new fixed point,

$$r^* = 0, \qquad g^* = \infty. \tag{2C.22}$$

At this new fixed point, the susceptibilities $\chi^{\alpha\beta}$ with $\alpha, \beta = 1, \ldots, m$ will be zero [see (2C.11), with $g_\alpha \to \infty$], so that we can say that there are no divergent fluctuations in the first m spin components. (There may still be non-zero non-divergent terms.) The susceptibilities related to the other spin components will be described by the isotropic $(n - m)$-components Gaussian model, with mean field exponents.

The critical temperature of this anisotropic critical behaviour is given, in this model, by $r = 0$. It is important to note that with our choice of r and g, (2C.8) and (2C.10), the operators related to them are not orthogonal. A more appropriate choice is

$$\bar{r} = r + \frac{m}{n} g, \qquad \bar{g} = \frac{n - m}{n} g, \tag{2C.23}$$

so that

$$r_\alpha = \bar{r} + \bar{g}, \qquad \alpha = 1, \ldots, m,$$

$$r_\alpha = \bar{r} - \frac{m}{n - m} \bar{g}, \qquad \alpha = m + 1, \ldots, n. \tag{2C.24}$$

Now, the parameter $\bar{r}$ is related to the energy operator and $\bar{g}$ is related to an operator perpendicular to the energy (Pfeuty *et al.*, 1974). The recursion relations for $\bar{r}$ and $\bar{g}$ are similar to (2C.20), and the true critical temperature (defined as the temperature at which iterations lead to the fixed point) is given by $\bar{r} = 0$, or by

$$T_c(g) - T_0 \propto \bar{g}^{1/\psi} \propto g^{1/\psi} \qquad (2C.25)$$

with the *shift exponent* (Fisher, 1971 Fisher and Jasnow, 1975)

$$\psi = 1. \qquad (2C.26)$$

To avoid infinite parameters at fixed points, we could change the choice of the spin rescaling factor (2C.15), so that

$$\zeta_\alpha^2 = b^d, \qquad \alpha = 1, \ldots, m$$
$$\zeta_\alpha^2 = b^{d+2}, \qquad \alpha = m + 1, \ldots, n. \qquad (2C.27)$$

With this choice, we preserve the magnitude of the fluctuations in the last, "critical", $(n - m)$ spin components, and also force r_α, $\alpha = 1, \ldots, m$, to remain constant. This makes the coefficient of $q^2\delta_{\alpha\beta}$ in $V_2^{\alpha\beta}(\mathbf{q})$, $\alpha, \beta = 1, \ldots, m$, an irrelevant variable, leading to no long-range fluctuations in these components. Mathematically, the results are the same (see also Nelson and Fisher, 1974).

Another interesting example relates to the case of isotropic long-range exchange forces, of the type

$$J^{\alpha\beta}(\mathbf{R}) \sim \delta_{\alpha\beta}/R^{d+\sigma} \qquad (2C.28)$$

These lead to a non-analytic Fourier transform (Fisher *et al.*, 1972, Sak, 1973),

$$J^{\alpha\beta}_{\text{n.a.}}(\mathbf{q}) = Aq^\sigma\delta_{\alpha\beta}. \qquad (2C.29)$$

Equation (2C.18) thus becomes

$$A' = b^{2-\sigma}A. \qquad (2C.30)$$

The parameter A is therefore relevant for $\sigma < 2$, marginal for $\sigma = 2$ and irrelevant for $\sigma > 2$. In the first case, we have a crossover, with a crossover exponent [see (2B.18)]

$$\phi_{\text{l.r.}} = \nu\lambda_A = \nu(2 - \sigma) = \tfrac{1}{2}(2 - \sigma). \qquad (2C.31)$$

This crossover leads to a new fixed point, at $A = \infty$. Changing the spin rescaling factor to

$$\zeta_\alpha^2 = \zeta_{\text{l.r.}}^2 = b^{\sigma+d} \qquad (2C.32)$$

now leads to a constant value of A, while the coefficient of $q^2\delta_{\alpha\beta}$ in $V_2^{\alpha\beta}(\mathbf{q})$ becomes irrelevant. This then yields

$$r' = b^\sigma r, \qquad (2C.33)$$

so that

$$\nu = 1/\sigma. \tag{2C.34}$$

Similarly, one finds $\gamma = 1$, $\eta = 2 - \sigma$, etc. These results agree with those obtained from the exact solution for the correlation function,

$$G^{\alpha\beta}(\mathbf{q}) = \delta_{\alpha\beta}/[r + q^2 + Aq^\sigma + O(q^4)]. \tag{2C.35}$$

An analysis of the Gaussian model is thus a very convenient way to study general features of the critical behaviour resulting from complicated Hamiltonians. Moreover, we shall see that this model does describe correctly the short-range critical behaviour at dimensions $d > 4$. This will follow from the fact that all $V_4^{\alpha\beta\gamma\delta}$ are irrelevant for $d > 4$, but lead to a crossover to new, non-trivial, fixed points for $d < 4$. For the long-range case, Gaussian behaviour "wins" for $d > 2\sigma$.

D. Diagrammatic expansions

We now consider $V_4^{\alpha\beta\gamma\delta}$, and all higher-order coefficients, as perturbations, and expand the partition function (2A.2) in powers of them. This expansion is discussed in other papers in this volume (e.g. by Wallace and by Brézin *et al.*), so we shall not go into much detail. However, the paper by Wallace reviews an alternative approach to the calculation of the critical exponents, based directly on the Feynman-graph expansions of correlation functions (Wilson, 1972). We feel that the approach based on recursion relations, which we present here, may give more intuitive feeling into the questions of crossover, competition between various types of critical behaviour, etc. After a certain fixed point is unambiguously identified, it is sometimes convenient to use the direct diagrammatic expansion around it to calculate various critical functions, and we shall mention examples.

We divide the reduced Hamiltonian into a Gaussian part,

$$\bar{\mathscr{H}}_0 = -\tfrac{1}{2}\sum_{\alpha\beta}\int_{\mathbf{q}} V_2^{\alpha\beta}(\mathbf{q})\sigma_{\mathbf{q}}^{\alpha}\sigma_{-\mathbf{q}}^{\beta} \tag{2D.1}$$

and a perturbative part,

$$\bar{\mathscr{H}}' = -\frac{1}{4!}\sum_{\alpha\beta\gamma\delta}\int_{\mathbf{q}}\int_{\mathbf{q}'}\int_{\mathbf{q}''} V_4^{\alpha\beta\gamma\delta}\sigma_{\mathbf{q}}^{\alpha}\sigma_{\mathbf{q}'}^{\beta}\sigma_{\mathbf{q}''}^{\gamma}\sigma_{-\mathbf{q}-\mathbf{q}'-\mathbf{q}''}^{\delta}\cdots. \tag{2D.2}$$

The partition function is now written as

$$Z = \int\Big(\prod_{\mathbf{q}} \mathrm{d}^n\sigma_{\mathbf{q}}\Big)\exp(\bar{\mathscr{H}}_0)\,[1 + \bar{\mathscr{H}}' + \tfrac{1}{2}\bar{\mathscr{H}}'^2 + \ldots]. \tag{2D.3}$$

We now follow the first step of iteration, namely the integration over $\boldsymbol{\sigma}_{\mathbf{q}}$

with $\Lambda/b < |\mathbf{q}| < \Delta$. To lowest order, the result for $\bar{\mathscr{H}}_1$, eqn (2B.3), is (apart from a spin-independent constant)

$$\bar{\mathscr{H}}_1[\boldsymbol{\sigma}_{\mathbf{q}}] = \mathscr{H}[\boldsymbol{\sigma}_{\mathbf{q}}] - \frac{1}{3!}\sum_{\alpha,\beta,\gamma,\delta}(V_4^{\gamma\delta\alpha\beta} + V_4^{\alpha\delta\gamma\beta} + V_4^{\gamma\beta\alpha\delta})\int_{\mathbf{q}}^{<} G^{\gamma\delta}(\mathbf{q})\int_{\mathbf{q}'}^{<}\sigma_{\mathbf{q}'}^{\alpha}\sigma_{-\mathbf{q}'}^{\beta} + O(V_4^2), \tag{2D.4}$$

where $\int_{\mathbf{q}}^{>}$ and $\int_{\mathbf{q}}^{<}$ denote integration over $\Lambda/b < |\mathbf{q}| < \Lambda$ and $|\mathbf{q}| < \Lambda/b$ respectively. The next two steps, (2B.5) and (2B.6), are easily performed, and the final results yield

$$[V_2^{\alpha\beta}(\mathbf{q})]' = \zeta_\alpha\zeta_\beta b^{-d}\left\{V_2^{\alpha\beta}(b^{-1}\mathbf{q}) - \frac{1}{3!}\sum_{\gamma\delta}(V_4^{\gamma\delta\alpha\beta} + V_4^{\alpha\delta\gamma\beta} + V_4^{\gamma\beta\alpha\delta}) \times \int_{\mathbf{q}'}^{>} G^{\gamma\delta}(\mathbf{q}') + O(V_4^2)\right\}, \tag{2D.5}$$

$$(V_4^{\alpha\beta\gamma\delta})' = \zeta_\alpha\zeta_\beta\zeta_\gamma\zeta_\delta b^{-3d}\{V_4^{\alpha\beta\gamma\delta} - O(V_4^2)\}, \tag{2D.6}$$

etc. The coefficient of $q^2\delta_{\alpha\beta}$ in $[V_2^{\alpha\beta}(\mathbf{q})]'$ is thus $\zeta_\alpha\zeta_\beta b^{-d-2} + O(V_4^2)$. We therefore write, for the short-range case,

$$\zeta_\alpha^2 \equiv \zeta^2 = b^{d+2-\eta}, \qquad \eta = O(V_4^2), \tag{2D.7}$$

with this choice, (2D.5) and (2D.6) yield

$$r'_{\alpha\beta} = b^{2-\eta}\left\{r_{\alpha\beta} - \frac{1}{3!}\sum_{\gamma\delta}(V_4^{\gamma\delta\alpha\beta} + V_4^{\alpha\delta\gamma\beta} + V_4^{\gamma\beta\alpha\delta})\int_{\mathbf{q}}^{>} G^{\gamma\delta}(\mathbf{q}) + \dots\right\}, \tag{2D.8}$$

$$(j_1^{\alpha\beta\gamma\delta})' = b^{-\eta}j_1^{\alpha\beta\gamma\delta} + O(V_4^2), \tag{2D.9}$$

$$[\hat{J}_{\text{n.a.}}^{\alpha\beta}(\mathbf{q})]' = b^{2-\eta}J_{\text{n.a.}}^{\alpha\beta}(b^{-1}\mathbf{q}), \tag{2D.10}$$

$$(V_4^{\alpha\beta\gamma\delta})' = b^{\varepsilon}[V_4^{\alpha\beta\gamma\delta} - O(V_4^2)], \qquad \varepsilon = 4 - d, \quad \text{etc.} \tag{2D.11}$$

Note that (2D.10) is exact, since no term in the diagrammatic expansion will contribute non-analytic terms (Sak, 1973, Aharony and Fisher, 1973).

From (2D.11) it is obvious that $V_4^{\alpha\beta\gamma\delta}$ is *irrelevant* near the Gaussian fixed point $V_4 \equiv 0$, for $\varepsilon < 0$ or $d > 4$. Thus, for these dimensions, iterations will lead us back to the Gaussian behaviour described in the previous section. It is only for $d < 4$ that we expect a new behaviour, since the relevance of V_4 must lead us away from the Gaussian behaviour, hopefully to new, non-trivial, fixed points. The crossover exponent for this "flow" is clearly $\phi = \nu\varepsilon = \frac{1}{2}\varepsilon$. It should be noted here that the dimensionality $d = 4$ is the borderline of mean field behaviour only for short-range interactions, when (2D.7) is appropriate.

Since our perturbation expansion is valid only for small V_4, one sees that (2D.11) may have a non-trivial fixed point only if ε is also small, and if V_4 is of order ε (Wilson and Fisher, 1972). Similar arguments may be given for large n, when V_4 is of order $1/n$ (see Ma, 1973, and Chapter 4 of this volume). In the main part of this Chapter we shall discuss the critical behaviour near four dimensions, due to the various possible fixed points, and derive ε-expansions for the various critical exponents and universal functions. When available, we shall also mention results from $1/n$ expansions.

It should be noted that, near $d = 4$, all the other parameters, e.g. the coefficients of $q^4\delta_{\alpha\beta}$ in $V_2^{\alpha\beta}(\mathbf{q})$, of q-dependent terms in $V_4^{\alpha\beta\gamma\delta}$, or of products of six or more spins in $\mathscr{H}$, are highly irrelevant [with eigenvalues of the form $b^{-2} + O(V_4)$]. We shall therefore ignore them in most of the following discussion. Note however that these parameters, although irrelevant, may effect the values of the relevant parameters during the first few iterations. It is therefore erroneous to give definite statements relating the initial values of relevant parameters leading to a given fixed point [e.g. the initial value of r, which determines the *critical temperature*, see (2C.5)].

Before going into examples, it is worth mentioning a few technicalities. Equation (2D.4) clearly has the form of a Feynman graph expansion (Wilson and Kogut, 1974): Each factor $V_4^{\alpha\beta\gamma\delta}$ is associated to a four-line vertex, each factor $G^{\gamma\delta}(\mathbf{q})$ is associated with an internal line, with the momentum to be integrated over $\Lambda/b < |\mathbf{q}| < \Lambda$, and each external line is associated with a factor $\zeta_\alpha\sigma_{\mathbf{q}}^\alpha$, integrated over $0 < |\mathbf{q}| < \Lambda/b$. Conservation of momenta at each vertex, and summation over the indices $\alpha, \beta, \ldots$ is understood. Since $\bar{\mathscr{H}}_1$ is a logarithm of an expansion of the form (2D.3), only connected graphs are to be included. In Figs 1 and 2 we show the diagrams which have to enter into (2D.5) and (2D.6), respectively, to orders V_4^2 and V_4^3. The explicit term in (2D.5) represents the first diagram in Fig. 1.

Bruce *et al.* (1974) gave a detailed study of the structure of the recursion relations described by the graphs shown in Figs 1 and 2. It turns out, that if all one wants is to calculate the critical exponents to order ε^2, one can simplify the above-mentioned rules, by taking the limit of large rescaling factor b. In this limit, one may ignore all the irrelevant variables (diagrams including six-line vertices, etc.) neglect "mass renormalization" graphs, like the last ones in Figs 1 and 2, and calculate the various integrals on momenta with propagators at zero "mass" [e.g. $r = 0$ in (2C.6)]. All the explicit recursion relations we shall give here will be based on these assumptions. The reader is referred to the paper by Bruce *et al.* (1974) for a detailed discussion of these simplifications (Note that these details affect only the order-ε^2 calculation, and not the order-ε one). An alternative approach, which takes the limit $b \to 1$ is also possible, leading to a set of differential equations in the iteration variable l (Wegner and Houghton, 1973; Wegner, this volume, Chapter 2).

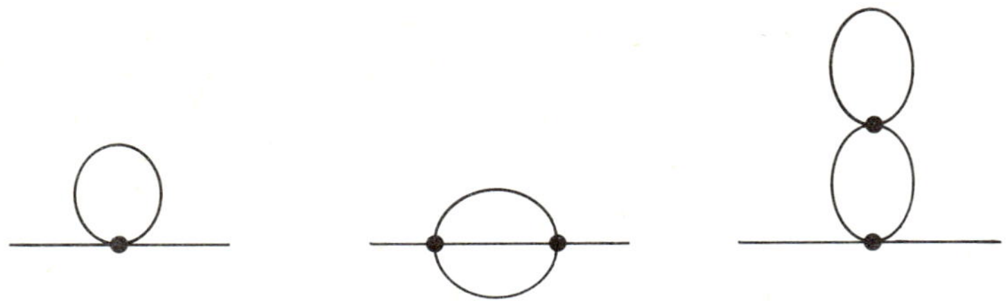

FIG. 1. Graphs contributing to the recursion relation for $V_2^{\alpha\beta}$ to order $(V_4)^2$, eqn (2D.5).

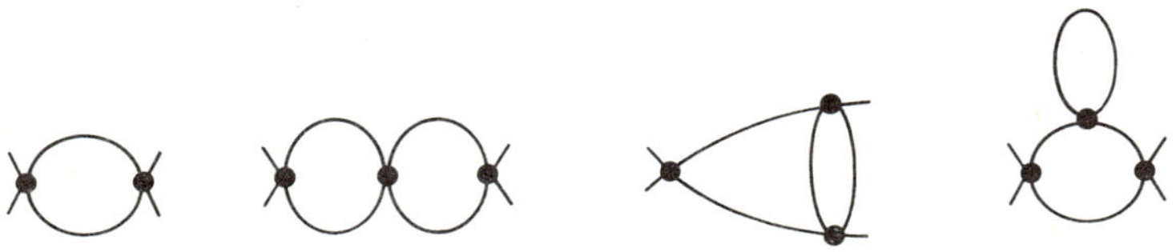

FIG. 2. Graphs contributing to the recursion relation for $V_4^{\alpha\beta\gamma\delta}$ to order $(V_4)^3$, eqn (2D.6).

Another remark concerns the d-dimensional integrals over propagators, for non-integral d. We shall follow Wilson (1972) and use

$$(2\pi)^{-d}\int d^d q f(|\mathbf{q}|) = K_d \int q^{d-1}\, dq f(|\mathbf{q}|), \tag{2D.12}$$

$$(2\pi)^{-d}\int d^d q f(|\mathbf{q}|, \mathbf{q}\cdot\mathbf{p}) = (2\pi)^{-1}K_{d-1}\int q^{d-1} dq \int_0^{\pi} d\theta \sin^{d-2}\theta f(|\mathbf{q}|, \mathbf{q}\cdot\mathbf{p}) \text{ etc.,} \tag{2D.13}$$

where $\mathbf{k}\cdot\mathbf{p} = |\mathbf{k}|\,|\mathbf{p}|\cos\theta$, and

$$K_d^{-1} = 2^{d-1}\pi^{d/2}\Gamma(\tfrac{1}{2}d). \tag{2D.14}$$

Many examples of angular integrals, like (2D.13), are given in Appendix B of Aharony and Fisher (1973). In practice, it is usually sufficient to calculate integrals as a short expansion in $\varepsilon = 4 - d$.

At the fixed points of interest, $V_4^{\alpha\beta\gamma\delta}$ will be of order ε. Linearizing the recursion relations (2D.8)–(2D.11) about a fixed point will lead to expressions of the form

$$\mu' - \mu^* = b^{\lambda_0}[1 + \lambda_1\varepsilon \ln b + \lambda_2\varepsilon^2 \ln b + \tfrac{1}{2}\lambda_1^2\varepsilon^2 \ln^2 b + O(\varepsilon^3)](\mu - \mu^*), \tag{2D.15}$$

and we shall always interpret this as

$$(\mu' - \mu^*) = b^{\lambda}(\mu - \mu^*), \qquad \lambda = \lambda_0 + \lambda_1\varepsilon + \lambda_2\varepsilon^2 + O(\varepsilon^3). \tag{2D.16}$$

In some cases we shall omit the terms of order $\varepsilon^2 \ln^2 b$, since they only serve for a consistency check of the exponentiation in (2D.16). Adopting the large b limit, we shall also ignore terms of order unity compared to those of order $\ln b$, etc.

As an example, and for future reference, we finish this section with the recursion relations for the short-range isotropic Heisenberg-like *n-vector* model, given by (2C.4) and by the isotropic part of (2A.22) ($B^{\alpha\beta\gamma\delta} \equiv 0$). In this case we have only the two parameters r and u, and the recursion relations are (with the assumptions of Bruce *et al.*, mentioned above)

$$r' = b^{2-\eta}[r + 4(n+2)uA(r) - 32(n+2)u^2B(r) + O(u^3)], \tag{2D.17}$$

$$u' = b^{\varepsilon-2\eta}\{u - 4(n+8)u^2C(r) + 16(n^2+6n+20)u^3[C(r)]^2 + 64(5n+22)u^3D(r) + O(u^4)\}, \tag{2D.18}$$

where

$$\eta = 8(n+2)(K_d u)^2 + O(u^3), \tag{2D.19}$$

$$A(r) = \int_{\mathbf{q}}^{>} (q^2 + r)^{-1} = K_d[(\tfrac{1}{2} + \varepsilon/4)\Lambda^2(1 - b^{-2}) - \tfrac{1}{2}\varepsilon\Lambda^2 \ln \Lambda + \varepsilon/2\,(\Lambda/b)^2 \ln(\Lambda/b) - rC(0) + O(r^2, \varepsilon^2), \tag{2D.20}$$

$$B(r) = \int_{\mathbf{q}_1}^{>}\int_{\mathbf{q}_2}^{>} \{(q_1^2 + r)(q_2^2 + r)[(\mathbf{q}_1 + \mathbf{q}_2)^2 + r]\}^{-1} = B(0) + O(r), \tag{2D.21}$$

$$C(r) = \int_{\mathbf{q}}^{>} (q^2 + r)^{-2} = K_d \ln b(1 + \tfrac{1}{2}\varepsilon \ln b - \varepsilon \ln \Lambda) + O(\varepsilon^2, r), \tag{2D.22}$$

$$D(r) = \int_{\mathbf{q}_1}^{>}\int_{\mathbf{q}_2}^{>} (q_1^2 + r)^{-2}(q_2^2 + r)^{-1}[(\mathbf{q}_1 + \mathbf{q}_2)^2 + r]^{-1} = \tfrac{1}{2}K_4^2 \ln b(1 + \ln b) + O(1). \tag{2D.23}$$

The fixed point values of u are zero (Gaussian fixed point) or

$$K_d u^* = \frac{\varepsilon\Lambda^\varepsilon}{4(n+8)}\left[1 + \frac{3(3n+14)}{(n+8)^2}\varepsilon + O(\varepsilon^2)\right]. \tag{2D.24}$$

Linearizing (2D.17) about this non-trivial fixed point (note that we do not need the explicit value of r^*), we find

$$\Delta r' = r' - r^* = b^{2-\eta}\{1 - 4K_d(n+2)u \ln b(1 + \tfrac{1}{2}\varepsilon \ln b - \varepsilon \ln \Lambda) + 48K_4^2(n+2)u^2 \ln b(\ln b + 1) + 16K_4^2(n+2)^2u^2 \ln^2 b + O(\varepsilon^3, u^3)\}\Delta r, \tag{2D.25}$$

and this is of the form (2D.16), with

$$\nu = \frac{1}{\lambda_1} = \tfrac{1}{2} + \frac{n+2}{4(n+8)}\varepsilon + \frac{(n+2)(n^2+23n+60)}{8(n+8)^3}\varepsilon^2 + O(\varepsilon^3). \tag{2D.26}$$

Equations (2D.19) and (2D.24) similarly yield

$$\eta = \frac{n+2}{2(n+8)^2}\varepsilon^2 + O(\varepsilon^3). \tag{2D.27}$$

Linearizing (2D.18) about (2D.24) yields $\Delta u' = b^{\lambda_u}\Delta u$, with (Aharony 1973c, f)

$$\lambda_u = -\varepsilon + \frac{3(3n+14)}{(n+8)^2}\varepsilon^2 + O(\varepsilon^3). \tag{2D.28}$$

Clearly, u is an irrelevant variable near this fixed point, and $\omega = -\nu\lambda_u$ usually represents the exponent of the *leading correction to scaling.*

In the following discussion we shall very often simplify expressions by choosing $\Lambda = 1$.

III. Spin Anisotropies

A. Crossover to a lower effective spin dimensionality

We now return to the spin-anisotropic case, described by the propagator (2C.9). Since the two-spin function $V_2^{\alpha\beta}(\mathbf{q})$ now has a lower symmetry, we must allow for a more general structure of the four-spin function, namely

$$\frac{1}{4!}V_4^{\alpha\beta\gamma\delta} = \tfrac{1}{3}(u_{\alpha\beta}\delta_{\alpha\gamma}\delta_{\beta\delta} + u_{\alpha\gamma}\delta_{\alpha\beta}\delta_{\gamma\delta} + u_{\alpha\beta}\delta_{\alpha\delta}\delta_{\beta\gamma}). \tag{3A.1}$$

We can now write recursion relations for r_α and for $u_{\alpha\beta}$. To order ε, these were first written, in the framework of the approximate recursion relations, by Fisher and Pfeuty (1972) (see also Wegner, 1972b). With the spin rescaling (2D.7), and with the assumptions of Bruce *et al.* (1974), they are

$$\begin{aligned} r'_\alpha = b^{2-\eta}\Big\{ r_\alpha &+ 4\sum_\gamma u_{\alpha\gamma}\Big[\Big[\int_{\mathbf{q}}^{>} G^{\gamma\gamma}(\mathbf{q}) + 2u_{\alpha\alpha}\int_{\mathbf{q}}^{>} G^{\alpha\alpha}(\mathbf{q})\Big] \\ &-32\sum_{\gamma\delta} u_{\alpha\gamma}u_{\alpha\delta}\int_{\mathbf{q}}^{>}\int_{\mathbf{p}}^{>} [G^{\alpha\alpha}(\mathbf{q})G^{\gamma\delta}(\mathbf{p})G^{\gamma\delta}(\mathbf{p}+\mathbf{q}) \\ &+ 2G^{\alpha\gamma}(\mathbf{q})G^{\gamma\delta}(\mathbf{p})G^{\delta\alpha}(\mathbf{p}+\mathbf{q})] + O(u^3) \end{aligned} \tag{3A.2}$$

and a similar general expression for $u'_{\alpha\beta}$. This expression becomes quite lengthy for a general propagator $G^{\alpha\beta}(\mathbf{q})$. However, for the diagonal propagator (2C.9), the recursion relation for $u_{\alpha\beta}$ becomes

$$
\begin{aligned}
u_{\alpha\beta'} = b^{\varepsilon-2\eta}\Bigg\{u_{\alpha\beta} - 4\Bigg[\sum_{\mu} u_{\alpha\mu}u_{\mu\beta}\int_{\mathbf{q}}^{>}[G^{\mu\mu}(\mathbf{q})]^2 \\
+ 2u_{\alpha\beta}\left(u_{\alpha\alpha}\int_{\mathbf{q}}^{>}[G^{\alpha\alpha}(\mathbf{q})]^2 + u_{\beta\beta}\int_{\mathbf{q}}^{>}[G^{\beta\beta}(\mathbf{q})]^2\right) \\
+ 4u_{\alpha\beta}^2\int_{\mathbf{q}}^{>} G^{\alpha\alpha}(\mathbf{q})G^{\beta\beta}(\mathbf{q})\Bigg]\Bigg\} + O(u^3).
\end{aligned}
\tag{3A.3}
$$

A list of the order u^3 terms is given by Bruce and Aharony (1974). The various integrals in (3A.2) and (3A.3) are of the same form as those given in (2D.20)–(2D.23).

If all r_α's are equal, then it is clear that (3A.2) and (3A.3) reduce to (2D.17) and (2D.18), and yield the usual isotropic n-component Heisenberg-like fixed point. For a small anisotropy [small g_α in (2C.8)], we thus linearize (3A.2) about the isotropic fixed point (2D.24). The linearized equation is

$$
\begin{aligned}
\Delta r'_\alpha = r_\alpha - r^* = b^{2-\eta^*}\Bigg\{\Delta r_\alpha - 4u^*\left[\sum_{\gamma}\int_{\mathbf{q}}^{>}\Delta G^{\gamma\gamma}(\mathbf{q}) + 2\int_{\mathbf{q}}^{>}\Delta G^{\alpha\alpha}\right] \\
-32K_d u^{*2}\int_{\mathbf{q}}^{>}\sum_{\gamma\delta}\Big[\Delta G^{\alpha\alpha}(\mathbf{q})J_2^{\gamma\delta\gamma\delta}(\mathbf{q}) + 2\Delta G^{\gamma\delta}(\mathbf{q})J_2^{\alpha\alpha\gamma\delta}(\mathbf{q}) \\
+ 2\Delta G^{\gamma\delta}(\mathbf{q})J_2^{\alpha\gamma\alpha\delta}(\mathbf{q}) + 4\Delta G^{\alpha\gamma}(\mathbf{q})J_2^{\gamma\delta\delta\alpha}(\mathbf{q})\Big] + O(u^{*3})\Bigg\},
\end{aligned}
\tag{3A.4}
$$

where

$$
J_2^{\alpha\beta\gamma\delta}(\mathbf{q}) = K_d^{-1}\int_{\mathbf{p}}^{>} G^{\alpha\beta}(\mathbf{p}+\mathbf{q})G^{\gamma\delta}(\mathbf{p}),
\tag{3A.5}
$$

and

$$
\Delta G^{\alpha\beta}(\mathbf{q}) = \sum_{\mu}\frac{\partial}{\partial r_\mu}[G^{\alpha\beta}(\mathbf{q})]\Delta r_\mu.
\tag{3A.6}
$$

For the diagonal propagator (2C.9), (3A.5) and (3A.6) reduce to

$$
\begin{aligned}
J_2^{\alpha\beta\gamma\delta}(\mathbf{q}) &= \delta_{\alpha\beta}\delta_{\gamma\delta}K_d^{-1}\int_{\mathbf{p}}^{>} p^{-2}(\mathbf{p}+\mathbf{q})^{-2} + O(r_\alpha, r_\gamma) \\
&= [-\ln q + \tfrac{1}{2} + O(q^2, r_\alpha, r_\gamma)]\delta_{\alpha\beta}\delta_{\gamma\delta}
\end{aligned}
\tag{3A.7}
$$

and

$$
\Delta G^{\alpha\beta}(\mathbf{q}) = -\delta_{\alpha\beta}[q^{-4} + O(r^*)]\Delta r_\alpha.
\tag{3A.8}
$$

As noted by Bruce *et al* (1974), we may ignore terms of order r^* in the various integrals, since these are exactly cancelled by "mass renormalization" terms, which we did not include to start with. Finally, (3A.4) becomes

$$\Delta r'_\alpha = b^{\lambda_2}\Delta r_\alpha + \frac{1}{n}(b^{1/\nu} - b^{\lambda_2}) \sum_\gamma \Delta r_\gamma, \tag{3A.9}$$

where ν is given by (2D.26) and where

$$\lambda_2 = 2\left[1 - \frac{1}{n+8}\varepsilon + \frac{n^2 - 18n - 88}{4(n+8)^3}\varepsilon^2 + O(\varepsilon^3)\right]. \tag{3A.10}$$

We now want to diagonalize the n recursion relations given by (3A.9). Having written these recursion relations in the form (3A.9), it is clear that one eigenvalue is $b^{1/\nu}$, with the eigenvector $\Delta r_\alpha \equiv \Delta r_\beta \equiv \ldots \equiv \Delta r$, corresponding to the temperature instability [represented by the exponent ν, see (2B.20)]. The second eigenvalue is b^{λ_2}, and it is $(n-1)$-fold degenerate, corresponding to the $(n-1)$ eigenvectors which satisfy

$$\sum_\gamma \Delta r_\gamma \equiv 0. \tag{3A.11}$$

These eigenvectors thus correspond to the instability due to the anisotropy, which is represented by an operator which is perpendicular [in the eigenvector space of (3A.9)] to the isotropic energy operator $\Sigma_{\ddot{\alpha}}(S^\alpha)^2$. An example of parameters which satisfy (3A.11) is given in (2C.24), where $\bar{r}$ represents the first eigenvector, and $\bar{g}$ represents a perpendicular eigenvector. We can therefore identify the crossover exponent ϕ_g, describing the instability due to the anisotropy, as

$$\phi_g = \nu\lambda_2 = 1 + \frac{n}{2(n+8)}\varepsilon\left[1 + \frac{n^2 + 24n + 68}{2(n+8)^2}\varepsilon\right] + O(\varepsilon^3). \tag{3A.12}$$

The order ε^3 term in ϕ_g has recently been calculated by Yamazaki (1974), using the Callan–Symanzik approach. Expansions in $1/n$ of ϕ are given by Hikami and Abe (1974) and Oppermann (1974a). Since $\phi_g > 0$, any small perturbation satisfying (3A.11) will lead to an instability, and will grow as b^{λ_2} at each iteration. Our present task thus becomes to identify the fixed point, if any, to which the iterative flow in Hamiltonian space leads. As noted in Section II.C, a recursion relation of the form $g' = b^{\lambda_2}g$ will lead to $g \to \infty$. Following Fisher and Pfeuty (1972), we thus consider the case of dominant $(n-m)$-isotropy, eqn (2C.10). If g is large, then the first m propagators $G^{\alpha\alpha}(\mathbf{q})$, $\alpha = 1, \ldots, m$, become negligibly small, and thus the recursion relations (3A.2) and (3A.3) include effectively only sums over $\alpha = m+1, \ldots, n$. By symmetry, $u_{\alpha\beta}$ is now independent of α and β (in this range). Hence, (3A.2) and (3A.3) reduce to (2D.17) and (2D.18), with $(n-m)$ replacing n everywhere.

Thus, the crossover will be to a fixed point which is isotropic in $(n - m)$ spin components, and otherwise has all the properties of the fixed point described at the end of Section II.D. In particular, all the exponents, scaling functions etc. have the same ε-expansions, with $(n - m)$ replacing n.

The critical temperature for the $(n - m)$-component-like critical behaviour will be given by the initial value of the temperature parameter $\bar{r}$ which gives a flow to the $(n - m)$-component fixed point. This will naturally depend on all the irrelevant parameters in the initial Hamiltonian, since these are finite for the first few iterations, and thus affect the value of $\bar{r}$. After enough iterations, say l, all the irrelevant variables become really negligible, and we are left with recursion relations for the values of $\bar{r}$ and $\bar{g}$ at the $(l + 1)$st iteration, given as complicated functions of the original $\bar{r}$, $\bar{g}$ and all the irrelevant variables. We can now demand that the last $(n - m)$ parameters r_α remain unchanged by the $(l + 1)$st iteration. This clearly gives an equation for the initial value of $\bar{r}$, and thus for the critical temperature. With this choice, only the first m r_α's will grow, and we shall approach the new, $(n - m)$-isotropic, fixed point.

Obviously, this is not an easy way to extract the shift exponent, defined in (2C.25). An alternative way, based on a direct diagrammatic expansion of the divergent susceptibilities related to the last $(n - m)$ spin components, yields (Bruce, 1974b)

$$\psi_g = \phi_g, \tag{3A.13}$$

to order ε^2.

As we mentioned previously, we could obtain the same results for the anisotropic critical behaviour by changing our spin rescaling convention (2D.7). Instead, we could choose

$$\begin{aligned} \zeta_\alpha^2 &= b^d[1 + O(\varepsilon)] \quad \text{for } \alpha = 1, \ldots, m \\ \zeta_\alpha^2 &= b^{d+2-\eta} \qquad\qquad \text{for } \alpha = m + 1, \ldots, n. \end{aligned} \tag{3A.14}$$

With this choice, the coefficients of q^2 in the terms $V_2^{\alpha\alpha}(\mathbf{q})$, $\alpha = 1, \ldots, m$ become irrelevant, as do the coefficients $u_{\alpha\beta}$, α, $\beta = 1, \ldots, m$. Thus, the spin variables $\sigma_{\mathbf{q}}^\alpha$, $\alpha = 1, \ldots, m$, can simply be integrated over, using Gaussian integrals, and we are left with an effective Hamiltonian involving only the last $(n - m)$ components (Nelson and Fisher, 1975).

An important application of the crossover from the n-component Heisenberg-like fixed point to the m- or $(n - m)$-component fixed points ($g < 0$ or $g > 0$) has to do with *bicritical points* (Fisher and Nelson, 1974, Aharony and Bruce, 1974a). For any non-zero value of g, the transition is m-component-like or $(n - m)$-component-like. Thus, in the g–T plane, there are two second-order transition lines leaving the $g = 0$ critical point. The line $g = 0\,(T < T_c)$ is a "spin-flop" line, at which the ordering flops from the first m components to the last $(n-m)$ components. The n-component critical point is thus called bicritical. The parameter g may be experimentally changed, by varying a uniform magnetic field in an anisotropic antiferromagnet (Fisher and Nelson, 1974) or an anisotropic stress (Aharony and Bruce, 1974a), and the g–T phase diagram may thus be investigated. By (3A.13), the two critical lines have the form

$$T_c(g) - T_c(0) \sim g^{1/\phi_g}. \tag{3A.15}$$

B. Crossover to mean field behaviour as $n \to \infty$ (extreme anisotropy)

In the last section, we considered a crossover from an n-vector behaviour to an m-vector behaviour (we change notation from n–m to m). The case $n \to \infty$ is of special interest, since in that case one can expand in $1/n$, instead of in ε (Ma, 1973, and Chapter 4 of this volume; Abe, 1972).

It turns out, that if both n and m are very large, so that the ratio m/n is of order unity, then the picture described in the last section still holds, and the ultimate critical behaviour is that of the m-component fixed point (Suzuki, 1973b; Hikami and Abe, 1974). However, if m/n approaches zero as $n \to \infty$, a new critical behaviour appears in that limit (Moore *et al.*, 1974). which is consistent with mean field theory. It has been shown by Aharony (1974c), that the crossover from the $n = \infty$ behaviour (mean field theory) to the finite n behaviour (m-component fixed point) is directly related to the crossover from the Gaussian fixed point to the non-trivial fixed point, discussed in Section II.D. To show this, it is sufficient to consider the extremely anisotropic model, in which only m spin components (out of n) interact,

$$\mathscr{H} = -\tfrac{1}{2}\sum_{i,j=1}^{N} J(|\mathbf{R}_i - \mathbf{R}_j|) \sum_{\alpha=1}^{m} S^\alpha(\mathbf{R}_i) S^\alpha(\mathbf{R}_j), \tag{3B.1}$$

with

$$\sum_{\alpha=1}^{n} [S^\alpha(\mathbf{R}_i)]^2 = n. \tag{3B.2}$$

The partition function for this model is

$$Z_N^{(n)} = \int \ldots \int \mathrm{d}^n S(\mathbf{R}_1) \ldots \mathrm{d}^n S(\mathbf{R}_N) \exp[-\beta\mathscr{H}] \sum_{j=1}^{N} \delta\left\{n - \sum_{\alpha=1}^{n} [S^\alpha(\mathbf{R}_j)]^2\right\}$$

$$= \int \ldots \int \mathrm{d}^m S(\mathbf{R}_1) \ldots \mathrm{d}^m(\mathbf{R}_N) \exp[-\beta\mathscr{H}] \prod_{j=1} I_{n-m}\left\{\sum_{\alpha=1}^{m} [S^\alpha(\mathbf{R}_j)]^2\right\}, \tag{3B.3}$$

where the integrations range over $-\infty < S^\alpha(\mathbf{R}_i) < \infty$, and where

$$I_{n-m}\left[\sum_{\alpha=1}^{m}(S^\alpha)^2\right] = \int dS^{m+1}\ldots dS^n\,\delta\left[n - \sum_{\alpha=1}^{n}(S^\alpha)^2\right]$$

$$= C_{n-m}\int_0^\infty x^{n-m-1}\,\delta\left[n - \sum_{\alpha=1}^{m}(S^\alpha)^2 - x^2\right]dx$$

$$= \begin{cases} \frac{1}{2}C_{n-m}\left[n - \sum_{\alpha=1}^{m}(S^\alpha)^2\right]^{(n-m-2)/2}, & \text{for } 0 < \sum_{\alpha=1}^{m}(S^\alpha)^2 < n, \\ 0, & \text{otherwise,} \end{cases} \qquad (3B.4)$$

with $C_n = 2\pi^{n/2}/\Gamma(n/2)$. We can now write $I_{n-m}[\Sigma_{\alpha=1}^m(S^\alpha)^2]$ for $0 < \Sigma_{\alpha=1}^m (S^\alpha)^2 < n$ in the form

$$I_{n-m}\left[\sum_{\alpha=1}^{m}(S^\alpha)^2\right] = \tfrac{1}{2}C_{n-m}\,n^{(n-m-2)/2}\exp\left[-w\left(\sum_{\alpha=1}^{m}(S^\alpha)^2\right)\right], \qquad (3B.5)$$

with

$$w\left[\sum_{\alpha=1}^{m}(S^\alpha)^2\right] = -\tfrac{1}{2}(n-m-2)\ln\left[1 - \sum_{\alpha=1}^{m}(S^\alpha)^2/n\right]$$

$$= \tfrac{1}{2}\left(1 - \frac{m+2}{n}\right)\sum_{\alpha=1}^{m}(S^\alpha)^2 + \frac{1}{4n}\left(1 - \frac{m+2}{n}\right)\left[\sum_{\alpha=1}^{m}(S^\alpha)^2\right]^2 + O(1/n^2).$$

$$(3B.6)$$

Although $I_{n-m}(x)$ vanishes for $x > n$, the errors made in (3B.3) by replacing it by (3B.5) and (3B.6) for this range become negligible for large n and finite m. Thus, the weight function $w(x)$ is exactly of the form (2A.5) for the m-component vector $(S^1, \ldots, S^m)$. For finite m and $n = \infty$, the four-spin term vanishes $(u \sim 1/n)$, and the partition function (3B.3) reduces to that of the Gaussian model. For any finite n, this four-spin term will lead to a crossover to the non-trivial m-component fixed point, with the usual crossover exponent $\phi = \frac{1}{2}\varepsilon$ (see Section II.D). From (2B.19) it thus follows that the free energy has the form

$$f(t, n) \propto |t|^2\,\mathfrak{F}(n^{-1}t^{-\varepsilon/2}). \qquad (3B.7)$$

For any finite n, $\mathfrak{F}$ has a singularity at $t = t_c$, and $f(t, n)$ then behaves as $|t - t_c|^{2-\alpha_m}$, with α_m being the specific heat exponent for the m-component non-trivial critical behaviour.

Since ϕ is smaller than unity (for $2 < d < 4$), it is clear that the leading term in the expression of t_c as function of $1/n$ will be the analytic one, namely (Aharony, 1974c)

$$T_c(n) - T_c(\infty) \sim n^{-1}. \qquad (3B.8)$$

Thus, the shift exponent is $\psi = 1$.

The case of extreme anisotropy, where only some of the spin components interact, is thus quite similar to the continuous spin model we are generally discussing in this Chapter. One can use identities like (3B.3) in many similar situations. For example, Burkhart and Gunton (1974) similarly considered an Ising model in a transverse field. One can integrate directly over the non-interacting spin components and over the component which interacts with the field, and again arrive at a form similar to (3B.3) with a weight function which depends on the field.

IV. Long-Range Interactions and Dipolar Systems

A. Isotropic long-range interactions

Before reviewing the dipolar problem, it is instructive to mention the case of *isotropic* long-range exchange interactions, given by (2C.28). The propagator for this case was given in (2C.35), and the recursion relation for the parameter A, the coefficient of q^σ, was given in (2C.30). This recursion relation is exact to all orders in perturbation theory, and therefore the crossover exponent from the isotropic short-range fixed point to the isotropic long-range one is exactly

$$\phi_{\text{l.r.}} = \nu(2 - \sigma), \tag{4A.1}$$

with ν given by (2D.26). In fact, this will hold for any "short-range" fixed point, with the appropriate value of ν.

To discuss the long-range fixed point it is convenient to rescale the spins by $\zeta_{\text{l.r.}}$, as given in (2C.32). With this spin rescaling, the exponent η [eqn (2D.7)] becomes *exactly*

$$\eta_{\text{l.r.}} = 2 - \sigma. \tag{4A.2}$$

The recursion relation for u (2D.18), (we assume an isotropic interaction), becomes

$$u' = b^{2\sigma - d}\{u - 4(n + 8)u^2 C_\sigma(r) + \ldots\}, \tag{4A.3}$$

where $C_\sigma(r)$ is given by (2D.22), with q^2 replaced by q^σ. Thus, u is irrelevant for $d > 2\sigma$. In this range, the Gaussian-classical critical behaviour, described in Section II.C, is maintained. For $d < 2\sigma$, it is obvious that expansions are possible in $\varepsilon' = (2\sigma - d)$. Indeed, such expansions yield (Fisher *et al.*, 1972) $u^* = O(\varepsilon')$, and thus

$$\frac{1}{\gamma_{\text{l.r.}}} = 1 - \frac{n+2}{n+8}\frac{\varepsilon'}{\sigma} - \frac{(n+2)(7n+20)}{(n+8)^3}\sigma[\psi(1) - 2\psi(\tfrac{1}{2}\sigma) + \psi(\sigma)]\left(\frac{\varepsilon'}{\sigma}\right)^2 + O(\varepsilon'^3), \tag{4A.4}$$

where $\psi(z)$ is the logarithmic derivative of the gamma function. Note that to lowest order, the expansion coefficients in (ε'/σ) are the same as those of the usual short range expansion in $(\varepsilon/2)$.

The above picture becomes slightly more complicated when σ approaches the value 2, since then the competition between the terms q^2 and q^σ is felt more strongly. It has been pointed out by Sak (1973), that to order ε^2 (now $\varepsilon = 4 - d$!), with the spin rescaling (2C.32), the recursion relation for the coefficient of the q^2 term in the function $V_2^{\alpha\beta}(q)$, $\bar{A}$, is

$$\bar{A}' = b^{-(2-\sigma-\eta_{\text{s.r.}})}\,\bar{A} + \eta_{\text{s.r.}}\,A\ln b, \tag{4A.5}$$

when $\eta_{\text{s.r.}}$ is given by (2D.19). Thus, at the long-range fixed point ($A \neq 0$) we have

$$\bar{A}^* = \eta_{\text{s.r.}}\,A/(2 - \sigma - \eta_{\text{s.r.}}). \tag{A4.6}$$

This fixed point is stable for

$$2 - \sigma - \eta_{\text{s.r.}} > 0, \tag{4A.7}$$

and unstable for $2 - \sigma - \eta_{\text{s.r.}} < 0$. In the latter case, $\bar{A}$ will grow to infinity, and one must return to the choice (2D.7) for ζ in order to obtain the stable short-range fixed point, with finite parameters.

At $\sigma = 2 - \eta_{\text{s.r.}}$, the two fixed points coincide, and their exponents become the same, except for additional logarithmic corrections (Wegner and Riedel, 1973; Wegner, 1972a). Indeed, $1/n$ expansions of both the short-range and the long-range exponents (Suzuki, 1973a) give the same continuity of exponents at $\sigma = 2 - \eta_{\text{s.r.}}$ for all $2 < d < 4$ to order $1/n$. It is thus reasonable to believe that this is a general rule. Extension of this continuity assumption to $d = 2$ leads to interesting results on the behaviour of the critical exponents at $d = 2$ for large n (Aharony and Imry, 1974).

One can now study various perturbations near the long-range fixed point, and thus determine additional exponents. All of these will have expansions similar to (4A.4). Alternatively, one can perform diagrammatic expansions in ε', similar to those performed in ε for the short-range case, and obtain the equation of state, the correlation function, etc.

B. Isotropic dipolar magnets; antiferromagnets

The dipole–dipole interaction in d-dimensions decays with distance as $1/R^d$, apart from angular-dependent terms. Had it been isotropic, then it should be described by the results of the previous Section, with $\sigma = 0$. For this value of σ, u is an irrelevant parameter, and one expects to find classical Gaussian behaviour. It turns out, however, that the angular dependence of the dipolar interaction is very important, and that it leads to an effective short-range-like behaviour.

The renormalization group approach to the problem of the critical behaviour of magnets with dipolar interactions has been described in detail in a series of papers (Fisher and Aharony, 1973a; Aharony and Fisher, 1973; Aharony, 1973a–e; Bruce and Aharony, 1974). We shall therefore only summarise the main results.

We first consider the isotropic Heisenberg case, with both an isotropic short-range exchange

$$\mathscr{H}_{\text{s.r.}} = -\tfrac{1}{2}\sum_{\mathbf{R}\boldsymbol{\delta}} J(\boldsymbol{\delta})\,(\mathbf{S}_{\mathbf{R}}\cdot\mathbf{S}_{\mathbf{R}+\boldsymbol{\delta}}), \tag{4B.1}$$

where $\boldsymbol{\delta}$ covers a few nearest-neighbour vectors, and a d-dimensional classical dipole–dipole interaction,

$$\mathscr{H}_{\text{d.d.}} = -\tfrac{1}{2}(g_S\mu_B)^2 \sum_{\mathbf{R}\neq\mathbf{R}'}\sum_{\alpha,\beta}\frac{\partial^2}{\partial R^\alpha \partial R^\beta}(|\mathbf{R}-\mathbf{R}'|^{2-d})\,S_{\mathbf{R}}^\alpha S_{\mathbf{R}'}^\beta. \tag{4B.2}$$

Due to the scalar product $\mathbf{S}_{\mathbf{R}}\cdot(\mathbf{R}-\mathbf{R}')$, which is implied in (4B.2), we must have $n = d$ (see, however, Section IV.D). The Fourier transform of the dipolar interaction in d dimensions is discussed at length elsewhere (Aharony and Fisher, 1973, Appendix A). For a cubic hyperlattice, $\mathscr{H}_{\text{d.d.}}$ can be written in the form (2A.10) with

$$\hat{J}_{\text{d.d.}}^{\alpha\beta}(\mathbf{q}) = (g_S\mu_B)^2\left\{a_1\frac{q^\alpha q^\beta}{q^2} - a_2 q^\alpha q^\beta - (a_3 + a_4 q^2 - a_5(q^\alpha)^2)\,\delta_{\alpha\beta}\right.$$
$$\left. + O\left((q^\alpha)^4, (q^\alpha)^2(q^\beta)^2, q^4, \ldots\right)\right\}, \tag{4B.3}$$

for $\mathbf{q} \neq 0$, where the a_i's are of order unity.† For $\mathbf{q} = 0$, $J_{\text{d.d.}}$ is shape-dependent, and may be written

$$\hat{J}_{\text{d.d.}}^{\alpha\beta}(0) = (g\mu_B)^2\, a_1 \mathscr{D}^{\alpha\beta}, \tag{4B.4}$$

where $\mathscr{D}^{\alpha\beta}$ is related to the demagnetization tensor.

Similarly, the short-range isotropic exchange energy has the form (2A.10) with

$$\hat{J}_{\text{s.r.}}^{\alpha\beta}(\mathbf{q}) = \hat{J}(\mathbf{q})\,\delta_{\alpha\beta}, \tag{4B.5}$$

where $\hat{J}(\mathbf{q})$ is analytic in q^2. Summing up (4B.3) and (4B.5), and going through the transformations described in Section II.1, we finally find the form (2A.13), with

$$V_2^{\alpha\beta}(\mathbf{q}) = (r + q^2 + f(q^\alpha)^2)\,\delta_{\alpha\beta} + g\frac{q^\alpha q^\beta}{q^2} - hq^\alpha q^\beta$$
$$+ O\left((q^\alpha)^4, (q^\alpha)^2(q^\beta)^2, \ldots\right). \tag{4B.6}$$

† Note that the values of a_2, a_4 and a_5 for the three-dimensional bcc and fcc cases, as given in Table I of Aharony and Fisher (1973), must be divided by v_a; see Cohen and Keffer (1974).

The parameter r is linear in the temperature T [see (2C.5)], while the parameters f, h and g are of order $(g\mu_B)^2/\hat{J}(0)$. The term involving f has cubic symmetry, and results from performing the Fourier transform over a cubic lattice. We shall delay discussion of the effects of cubic symmetry until Section V. In the meantime, we assume $f = 0$. Thus, $V_2^{\alpha\beta}(\mathbf{q})$ maintains rotational invariance, and it is sufficient to choose a weight function with the same symmetry, and ignore single ion terms with cubic symmetry. Thus, we write [see (2A.22)]

$$\frac{1}{4!} V_4^{\alpha\beta\gamma\delta} = \tfrac{1}{3} u(\delta_{\alpha\beta}\delta_{\gamma\delta} + \delta_{\alpha\gamma}\delta_{\beta\delta} + \delta_{\alpha\delta}\delta_{\beta\gamma}). \tag{4B.7}$$

Usually, if $V_2^{\alpha\beta}(\mathbf{q})$ contains terms of lower (e.g. cubic) symmetry, then we must allow for these in $V_4^{\alpha\beta\gamma\delta}$ as well. If we did not include them, they would be generated anyway, after one iteration.

We are now ready to consider the recursion relations for the parameters r, g, h and u.

We start by calculating the propagator, (2C.2):

$$G^{\alpha\beta}(\mathbf{q}) = \frac{1}{r + q^2}\left(\delta_{\alpha\beta} - \frac{q^\alpha q^\beta}{q^2}\right) + \frac{1}{r + g + (1 - h)\, q^2}\,\frac{q^\alpha q^\beta}{q^2}. \tag{4B.8}$$

This equation already includes all the information for the Gaussian model of this problem: for $r \gg g$ we have the isotropic correlation function (2C.6). For $r \ll g$, namely close to the critical temperature (at which $r = 0$), the second term in (4B.8) is negligible, and we have anisotropic correlations, preferring correlations between the transverse spin components ($\boldsymbol{\sigma}_\mathbf{q}$ perpendicular to $\mathbf{q}$). Both types of behaviour have classical exponents.

We can now follow Section II.D, and write down the recursion relations (2D.8)–(2D.11) explicitly. We start with the non-analytic piece, (2D.10). This immediately yields

$$g' = b^{2-\eta}\, g, \tag{4B.9}$$

exact to all orders in u. Thus, with our standard spin rescaling (2D.7) the fixed-point values of g are zero or infinity. The first one occurs only at *antiferromagnets*, when $\hat{J}_{\mathrm{d.d}}^{\alpha\beta}(\mathbf{q})$ can be analytically expanded about the ordering wave vector $\mathbf{k}_0$ (Aharony, 1973c). The second one will always ultimately be reached for ferromagnets, since the initial value of g is positive, and there is nothing to stop the growth of g to infinity under (4B.9). We thus separate our following discussion into two parts, and start with the *antiferromagnetic case*. In this case, the propagator (4B.8) reduces to

$$G^{\alpha\beta}(\mathbf{q}) = \frac{1}{r + q^2}\left(\delta_{\alpha\beta} + \frac{h q^\alpha q^\beta}{r + (1 - h)\, q^2}\right), \tag{4B.10}$$

and we can write all the other recursion relations. Especially, the one for h becomes, for small h,

$$h' = b^{\lambda_h} h, \tag{4B.11}$$

with (for $d = n = 4 - \varepsilon$)

$$\lambda_h = -\frac{64}{3}(K_4 u)^2 + O(u^3). \tag{4B.12}$$

Thus, for any non-zero u, h is an irrelevant variable, which decays to the fixed point value $h^* = 0$. It must be remembered, however, that λ_h is quite small (u is of order ε), so that h may still be non-negligible even after many iterations. If h is negligible, then the propagator (4B.10) reduces to the usual isotropic short-range propagator (2C.6), and the whole problem reduces to finding the isotropic short-range Heisenberg-like fixed point, as done e.g. by Wilson and Kogut (1974) and in Section II.D. In this case u^* is given by (2D.24), and therefore, at $n = d = 4 - \varepsilon$,

$$\lambda_h = -\frac{\varepsilon^2}{108} + O(\varepsilon^3). \tag{4B.13}$$

One thus expects corrections to the standard isotropic Heisenberg-like scaling behaviour, of order $h|t|^{\omega_h}$, with $\omega_h = -\nu\lambda_h$, [See eqn (2B.20)], and these can be ignored only if $|t| \ll |h|^{-1/\omega_h}$.

It is worth mentioning that these results hold only as long as $h < 1$. Clearly, when h crosses the value 1 integrations over the propagator (4B.10) become tricky, and in fact (4B.10) stops describing the correct propagator of the problem. (One must expand $V_2^{\alpha\beta}(\mathbf{q})$ about another point in $\mathbf{q}$-space).

Before going on, it is worth mentioning that the interaction (4B.6), with $g = 0$, may also occur in ferromagnets, resulting from "short-range" dipolar exchange interactions, for which $|\mathbf{R} - \mathbf{R}'|^{2-d}$ in (4D.2) is replaced by some short-range function of $|\mathbf{R} - \mathbf{R}'|$. Such interactions may result from averages over spin–orbit interactions. As a second remark, note that h is a special case of the general parameter $j_1^{\alpha\beta\gamma\delta}$ introduced in (2A.16). Since angular integrals over products of the type $q^\alpha q^\beta \ldots q^\omega$ usually decrease the final numerical result for integrals, compared to their isotropic analogs, it is felt safe to conjecture that in most cases $j_1^{\alpha\beta\gamma\delta}$ will turn out to be irrelevant, although almost marginal (Aharony and Fisher, 1973; Aharony, 1973c; Bruce, 1974a).

To investigate the actual critical behaviour of dipolar antiferromagnets, we now follow the rules of Section II.B, and iterate until $\xi_l = 1$ [see eqn (2B.10)]. At this point, the true two-spin correlation function can be obtained from the iterated one by

$$\Gamma^{\alpha\beta}(\mathbf{q}) = b^{l(2-\eta)}\,\Gamma_l^{\alpha\beta}(\mathbf{q}\,b^l)$$

$$\approx \frac{Ct^{-\gamma}}{1 + \xi^2 q^2}\left(\delta_{\alpha\beta} + \frac{h\,\xi^{2+\lambda_h} q^\alpha q^\beta}{1 + \xi^2 q^2} + O(\varepsilon^2, h^2)\right) \tag{4B.14}$$

with $\xi \sim t^{-\nu}$, and

$$\gamma = \nu(2 - \eta) = 1 + \tfrac{1}{4}\varepsilon + \tfrac{11}{96}\varepsilon^2 + O(\varepsilon^3). \tag{4B.15}$$

To obtain this result, we note that the order ε terms cancel, and that to this order $r_l = 1 + O(\varepsilon)$ (see Aharony and Fisher, 1973, for details.)

C. Isotropic ferromagnets

We now turn to the ferromagnetic case. As we mentioned before, (4B.9) now flows to $g^* = \infty$, and we expect to find a crossover from the isotropic Heisenberg-like behaviour, of $g = 0$, to a new, dipolar, behaviour. Near the Heisenberg fixed point, (4B.9) yields $\lambda_g = 2 - \eta$, and therefore the dipolar crossover exponent is exactly

$$\phi_{\text{d.d.}} = \nu\lambda_g = \nu(2 - \eta) = \gamma, \tag{4C.1}$$

where γ is the isotropic Heisenberg model susceptibility exponent. Dipolar behaviour is thus expected to be observed only if $|t| \ll g^{1/\gamma}$ [see (2B.22)].

Since g flows to infinity, we can now study the recursion relations in the vicinity of this fixed-point value. The propagator now becomes

$$G^{\alpha\beta}(\mathbf{q}) = \frac{1}{r + q^2}\left(\delta_{\alpha\beta} - \frac{q^\alpha q^\beta}{q^2}\right), \tag{4C.2}$$

and we can write explicit expressions for all the other parameters. Since the last term in (4B.8) is now eliminated, we can ignore the parameter h, as hq^2 will certainly become negligible compared to g after a few iterations. We thus consider only the equations for r and for u. The integrals over the propagator now involve angular integrations [resulting from substituting (4C.2) into (3A.2)], but otherwise the recursion relations are very similar to those of the isotropic case, discussed in Section II.D. To order u^3, with $\Lambda \equiv 1$, the recursion relation for u is (Bruce and Aharony, 1974)

$$u' = b^{\varepsilon - 2\eta}\left\{u - 4K_d u^2\left[7 + d - \frac{12}{d} + \frac{12}{d(d+2)}\right](b^\varepsilon - 1)/\varepsilon + K_d^2 u^3\left(533\frac{7}{9} + 1156 \ln b\right)\ln b + O(u^4, u^3\,\varepsilon)\right\}. \tag{4C.3}$$

with

$$\eta = \frac{80}{3}(K_d u)^2 + O(u^3), \qquad n = d = 4 - \varepsilon. \tag{4C.4}$$

This yields (ignoring the Gaussian dipolar fixed point $u^* = 0$, which is unstable for $\varepsilon > 0$)

$$K_d u^* = \frac{\varepsilon}{34}\left[1 + \frac{6211}{10404}\varepsilon + O(\varepsilon^2)\right]. \tag{4C.5}$$

The linearized recursion relation for r then becomes

$$\Delta r' = b^{2-\eta}\{1 - 4K_d u(d+2)(1-d^{-1})(b^{\varepsilon}-1)/\varepsilon + 126(K_d u)^2 \ln b + O(u^3)\}\,\Delta r \tag{4C.6}$$

yielding

$$\nu = \tfrac{1}{2}\left[1 + \frac{9}{34}\varepsilon + \frac{7013}{58956}\varepsilon^2 + O(\varepsilon^3)\right]. \tag{4C.7}$$

Equation (4C.4) gives

$$\eta = \frac{20}{867}\varepsilon^2 + O(\varepsilon^3), \tag{4C.8}$$

and all other exponents are easily calculable, using scaling relations. These are reproduced in Table II, in which they are also compared with their

TABLE II. Critical Exponents for Isotropic Dipolar and Short Range Exchange, to Order ε^2 ($n = d = 4 - \varepsilon$).

Exponent	Classical (Gaussian)	Isotropic short-range		Isotropic dipolar	
		ε-expansion	$\varepsilon = 1$	ε-expansion	$\varepsilon = 1$
η	0	$\frac{1}{48}\varepsilon^2$	0·0208	$\frac{20}{867}\varepsilon^2$	0·0231
2ν	1	$1 + \frac{1}{4}\varepsilon + \frac{1}{8}\varepsilon^2$	1·375	$1 + \frac{9}{34}\varepsilon + \frac{7013}{58956}\varepsilon^2$	1·384
γ	1	$1 + \frac{1}{4}\varepsilon + \frac{11}{96}\varepsilon^2$	1·365	$1 + \frac{9}{34}\varepsilon + \frac{2111}{19652}\varepsilon^2$	1·372
α_s	0	$-\frac{1}{8}\varepsilon^2$	−0·125	$-\frac{1}{34}\varepsilon - \frac{6223}{58956}\varepsilon^2$	−0·135
δ	3	$3 + \varepsilon + \frac{11}{24}\varepsilon^2$	4·458	$3 + \varepsilon + \frac{787}{1734}\varepsilon^2$	4·454
β	$\frac{1}{2}$	$\frac{1}{2} - \frac{1}{8}\varepsilon + \frac{1}{192}\varepsilon^2$	0·380	$\frac{1}{2} - \frac{2}{17}\varepsilon - \frac{55}{58956}\varepsilon^2$	0·381
ϕ	1	$1 + \frac{1}{6}\varepsilon + \frac{11}{144}\varepsilon^2$	1·243	$1 + \frac{10}{51}\varepsilon + \frac{10793}{132651}\varepsilon^2$	1·277
ϕ_v	$\frac{1}{2}\varepsilon$	$\frac{1}{24}\varepsilon^2$	0·0417	$\frac{3}{34}\varepsilon + \frac{15887}{176868}\varepsilon^2$	0·178
ω	$-\frac{1}{2}\varepsilon$	$\frac{1}{2}\varepsilon - \frac{7}{48}\varepsilon^2$	0·354	$\frac{1}{2}\varepsilon - \frac{785}{10404}\varepsilon^2$	0·425

isotropic short-range counterparts. Somewhat unexpectedly, the dipolar exponents are very close to their short-range counterparts, and are further away from their mean field values.

Linearization of (4C.3) about (4C.5) yields the exponent of the leading correction to scaling, ω, which is also given in Table II.

Iterating up to $\xi_l = 1$, we can now show that the true two-spin correlation function has the form

$$\Gamma^{\alpha\beta}(\mathbf{q}) = Ct^{-\gamma}\hat{D}(\xi^2 q^2)\left(\delta_{\alpha\beta} - \frac{q^\alpha q^\beta}{q^2}\right), \tag{4C.9}$$

where $\hat{D}(x^2)$ is a universal scaling function (see also Fisher and Aharony, 1973b, 1974; Bruce and Aharony, 1974), which has the form

$$\hat{D}^{-1}(x^2) = 1 + x^2 + O(\varepsilon^2). \tag{4C.10}$$

A similar recursion relation procedure of the dipolar recursion relations below T_c may lead to a check of the scaling relations involving the exponents β, δ, γ', etc. However, once we have identified the parameters characterizing the isotropic dipolar fixed point, it turns out to be more straightforward to derive these, as well as other quantities, by using a direct ε-expansion of the type described by Wallace in this volume. To order ε^2, it was shown by Bruce *et al.*, (1974) that the parameter u^*, as calculated above, is the same as the one needed in the ε-expansion to insure scaling. Indeed, an ε-expansion of the correlation function above T_c confirms (4C.9) (Aharony, 1973b). In another study (Aharony and Bruce, 1974b), an ε-expansion of the dipolar equation of state was performed, and the exponents β and δ derived directly, confirming scaling. It turns out that the internal magnetic field, h_i, which is obtained from the external one by subtracting a term proportional to $g\mathscr{D}m$, where m is the magnetization, obeys a scaled equation of state of the form

$$h_i/m^\delta = f(t/m^{1/\beta}), \tag{4C.11}$$

with $h_i/m^\delta = 1$ at $t = 0$, $t/m^{1/\beta} = -1$ at $h_i = 0$ and

$$\begin{aligned} f(x) = x + 1 + \frac{2\varepsilon}{17}\{&(x+1)\ln(x+1) + \tfrac{1}{2}(5x+13)\ln[\sqrt{(x+1)} + \sqrt{(x+3)}] \\ &+ \tfrac{1}{4}(x+1)^{1/2}(x+3)^{1/2}(\tfrac{1}{3}x^2 - \tfrac{5}{3}x - 10) - \tfrac{1}{12}x^3 + \tfrac{1}{4}x^2 \\ &+ \{\tfrac{5}{2}\sqrt{(3)} - \tfrac{13}{2}\ln[1+\sqrt{(3)}]\}(x+1) + (2\ln 2 + \tfrac{1}{3})x\} + O(\varepsilon^2). \end{aligned} \tag{4C.12}$$

Again, this is surprisingly close numerically to the isotropic equation of state, although it has a much more complicated form.

For practical purposes, to order ε^2, the critical exponents and the scaled equation of state at $d = 3$ near the isotropic dipolar fixed point are very close to their short-range Heisenberg-like counterparts. However, we do not know enough about the convergence properties of the ε-expansion to be able to extrapolate these to $\varepsilon = 1$. It may well be that although the ε-expansions at the short-range isotropic case gave best results for $\varepsilon = 1$ when truncated at order ε^2, another order must be used for the dipolar case. Since no high-temperature series exist for this problem, experiments remain the only available way to study the crossover from isotropic short-range to isotropic dipolar critical behaviour.

Beyond measuring exponents and scaling functions, a direct experimental check of the crossover in the two-spin correlation function, into the anisotropic form (4C.9), may be quite interesting.

D. Anisotropic ferromagnets; XY-like anisotropy

The Heisenberg-like fully isotropic exchange interaction (4B.1) is idealized. In many cases, the exchange interactions between different spin components are not equal, and one encounters a spin anisotropy of the type discussed in Section III.A. Thus, the single parameter r in (4B.6) must be replaced by r_α, as in (2C.8), and we have

$$V_2^{\alpha\beta}(\mathbf{q}) = [r_\alpha + q^2 + f(q^\alpha)^2]\,\delta_{\alpha\beta} + g\frac{q^\alpha q^\beta}{q^2} - hq^\alpha q^\beta + O[(q^\alpha)^4]. \tag{4D.1}$$

The propagator now becomes much more complicated, and has the form (Aharony 1973d)

$$G^{\alpha\beta}(\mathbf{q}) = \frac{1}{r_\alpha + q^2 + f(q^\alpha)^2} \times \left(\delta_{\alpha\beta} - \frac{g - hq^2}{q^2 + (g - hq^2)\,Q^2}\,\frac{q^\alpha q^\beta}{r_\beta + q^2 + f(q^\beta)^2}\right). \tag{4D.2}$$

with

$$Q^2 = \sum_\gamma (q^\gamma)^2/[r^\gamma + q^2 + f(q^\gamma)^2]. \tag{4D.3}$$

Since the two-spin function $V_2^{\alpha\beta}$ has now a lower symmetry, we must allow for a more general structure of the four-spin function, namely (3A.1).

We are now ready to study the recursion relations. As before, the diagrams do not contribute to the non-analytic term, and thus the recursion relation for g remains unchanged, as in (4B.9). Thus, we have to discuss the cases $g^* = 0$ and $g^* = \infty$.

For $g = 0$, and for small f, h and $g_\alpha = r_\alpha - r$, we are in the vicinity of the isotropic Heisenberg-like fixed point, at which $g_\alpha = f = h = 0$. We shall leave the cubic-symmetry term f for later consideration. As discussed in Section IV.B, the parameter h is irrelevant, and therefore can also be ignored. We are thus left with the anisotropic short-range propagator, as given in (2C.9). This case was discussed at length in Section III.A.

If $g > 0$, then g will flow to infinity, independently of the explicit form of the propagator. Thus, we can simply follow the discussion of Section III.A, using as propagator the $g \to \infty$ limit of (4D.2). In this limit, ignoring the cubic parameter f, one has

$$G^{\alpha\beta}(\mathbf{q}) \simeq \frac{1}{r_\alpha + q^2}\left(\delta_{\alpha\beta} - \frac{1}{Q^2}\,\frac{q^\alpha q^\beta}{r_\beta + q^2}\right). \tag{4D.4}$$

We can now substitute this into the recursion relation (3A.2), and linearize it about the isotropic dipolar fixed point (4C.5). In the linearization process, one must be careful to take into account the full dependence of $G^{\alpha\beta}$ on r_γ. An explicit, but lengthy calculation finally leads to the form (3A.9), with ν given by (4C.7) and with (Bruce and Aharony, 1974)

$$\lambda_2 = 1 - \frac{7}{102}\varepsilon - \frac{5153}{265302}\varepsilon^2 + O(\varepsilon^3). \tag{4D.5}$$

Thus, the dipolar crossover exponent due to spin anisotropy, is

$$\phi = \nu\lambda_2 = 1 + \frac{10}{51}\varepsilon + \frac{10793}{132651}\varepsilon^2 + O(\varepsilon^3). \tag{4D.6}$$

In the above discussion, we assumed that the initial dipolar parameter g was much larger than the spin anisotropy parameters $g_\alpha = r_\alpha - r$, so that $g \simeq \infty$ while the g_α's are still small. One could, of course, also consider the opposite case, in which $g \ll g_\alpha$. In this case, the system will first cross over to the $(d - m)$-component short-range fixed point (considering the example of Section III.A), and only then feel the dipolar instability. Near the $(d$–$m)$-component short-range fixed point, (4B.9) will yield

$$\phi_{\text{d.d.}} = \nu(2 - \eta) = \gamma, \tag{4D.7}$$

with γ now being the $(d$–$m)$-component susceptibility exponent. We are thus facing a whole hierarchy of crossovers, depending on the relative magnitude of the initial symmetry-breaking parameters.

Having identified the crossover exponent from the isotropic dipolar fixed point, we next have to discuss the ultimate asymptotic critical behaviour of the system. In that limit, both the dipolar parameter g and the spin anisotropy parameter are infinite. Going back to the propagator (4D.4), this means that effectively (for the spin anisotropy discussed in Section III.A) all propagators involving the first m spin components are negligible. Moreover, (4D.3) then reduces to

$$Q^2 = \sum_{\gamma=m+1}^{d} (q^\gamma)^2, \tag{4D.8}$$

so that the final propagator is

$$\begin{aligned} G^{\alpha\beta}(\mathbf{q}) &\approx \frac{1}{r + q^2}\left(\delta_{\alpha\beta} - \frac{q^\alpha q^\beta}{Q^2}\right), \quad m + 1 \leqslant \alpha, \beta \leqslant d \\ &\approx 0, \qquad\qquad \text{otherwise.} \end{aligned} \tag{4D.9}$$

This represents a projection operator into the transverse $(d$–$m)$-component spin space. All angular integrals thus become $(d$–$m)$-dimensional, whereas the original q-integrals are still d-dimensional. Substituting into the recursion relations, this finally yields, for $n = d - m > 1$

$$4K_d u^* = \frac{n\varepsilon}{(n-1)[n+8-4/(n+2)]} + O(\varepsilon^2), \qquad (4D.10)$$

and thus
$$2\nu = 1 + \frac{n+2}{2[n+8-4/(n+2)]}\varepsilon + O(\varepsilon^2), \qquad (4D.11)$$

to be compared with the isotropic n-component result (2D.26). The change, of order $2\varepsilon/(n+8)[n+8-4/(n+2)]$, is again quite small.

We finish with a few comments. The XY-like case can be described both by $n = 2$ and by $n = d - 1 = 3 - \varepsilon$. Substitution into (4D.11) gives a measure of the uncertainties imbedded in the ε-expansion. One could also argue that the usual, fully isotropic, case is better described by $n = 3$, and not by $n = d$. This question cannot be resolved without a better understanding of the convergence properties of the ε-expansion.

E. Anisotropic Ising-like ferromagnets

As n approaches 1, the fixed point value (4D.10) diverges to infinity. It is thus clear that (4D.11) cannot be used for $n = 1$. The basic reason for this is that for a single spin component, (4D.9) is identically zero, and no longer represents the true propagator of the system. Going back to (4D.1), it is clear that if only a single spin component, e.g. S^z, becomes critical then the appropriate propagator is (Aharony, 1973a, e)

$$G(\mathbf{q}) = (r + q^2 + g(q^z/q)^2 - h(q^z)^2)^{-1}. \qquad (4E.1)$$

The recursion relation for g is still (4B.9), and therefore the crossover exponent from the short-range Ising-like behaviour to the dipolar one is given by the value of γ appropriate for the short-range behaviour. For large values of g, h is negligible, and it is clear that the propagator (4E.1) will contribute to integrals in q-space only for very small values of $(q^z/q) = \cos\theta$. Thus, using (2D.13), we can write

$$\int_{\mathbf{q}} G(\mathbf{q}) = (2\pi)^{-1} K_{d-1} \int q^{d-1}\,\mathrm{d}q \int \mathrm{d}\theta \sin^{d-2}\theta\,(r + q^2 + \cos^2\theta)^{-1}$$
$$\simeq (2\pi)^{-1} K_{d-1}\, g^{-1/2} \int q^{d-1}\,\mathrm{d}q \int \mathrm{d}\mu\,(r + q^2 + \mu^2)^{-1}. \qquad (4E.2)$$

This looks exactly like a $(d+1)$-dimensional integral, where μ is used as a $(d+1)$-st component of the q-vector. Using this argument, Larkin and Khmel'nitskii (1969) concluded that the leading singularity in the dipolar Ising behaviour in d dimensions is the same as that of the short-range Ising behaviour in $(d + 1)$ dimensions. This leads to classical behaviour plus logarithmic corrections at $d = 3$.†

A more direct recursion relations study (Aharony, 1973a, e) indeed confirms these results. The important detail in this derivation is the fact, that the four spin coefficient u is no longer the small expansion parameter. Instead, one must use $u\,g^{-1/2}$. The recursion relation for u, in the limit of large g, becomes

$$u' = b^{\varepsilon - 2\eta}(u - 36K_d b_d u^2 g^{-1/2} + \dots), \tag{4E.3}$$

where b_d is a constant of order unity, and η is of order $u^2 g^{-1}$. Using (4B.9) for g', we thus find

$$(ug^{-1/2})' = b^{\varepsilon - 1}(ug^{-1/2} - 36K_d b_d (ug^{-1/2})^2 + \dots) \tag{4E.4}$$

Hence, $ug^{-1/2}$ is relevant for $\varepsilon > 1$, or for $d < 3$, and irrelevant for $d > 3$. Thus, classical behaviour is found for $d > 3$, and expansions in $3 - d$ represent the non-classical behaviour for $d < 3$. Note that except for b_d, which appears in all the integrals, and therefore cancels at the end, the coefficients of the recursion relation (4E.4) are the same as for the short-range Ising problem in $(d + 1)$ dimensions. This also applies to the recursion relation for r. Thus, for $d < 3$ we have†

$$2\nu = 1 + \tfrac{1}{6}(3 - d) + O((3 - d)^2), \tag{4E.5}$$

$$\eta = O((3 - d)^2).$$

At $d = 3$, $ug^{-1/2}$ is a marginal parameter. After l iterations, the solution of (4E.4) is then of the form

$$u_l g_l^{-1/2} \propto (l + \bar{l})^{-1}, \tag{4E.6}$$

where $\bar{l}$ is a constant. Substituting into the recursion relation for r, and choosing the initial value of r carefully, to avoid an exponential growth of the form

$$r_l \propto b^{2l}(l + \bar{l})^{-1/3}\, r_0, \tag{4E.7}$$

one finds that r_l will also decay to zero as $(l + \bar{l})^{-1}$. The solution (4E.7), together with (2B.10), then yields

$$\xi^2 \propto b^{2l} \propto r_0^{-1}|\ln r_0|^{1/3} \propto |T - T_c|^{-1}\,|\ln|T - T_c||^{1/3}. \tag{4E.8}$$

The correlation function is then obtained as in (4B.14),

$$\Gamma(\mathbf{q}) \propto \xi^2\{1 + \xi^2(q^2 - h(q^z)^2 + g(q^z/q)^2)\}^{-1}. \tag{4E.9}$$

† A recent analysis of *higher*-order terms (Brézin and Zinn-Justin, 1975) shows that there are small differences between the two cases.

V. Generalized "Cubic" Systems: Various Effective Quartic Spin Interactions

A. The competition between Heisenberg, Ising and Cubic fixed points

As we mentioned in Section IV.B, the parameter f in (4B.6) represents terms of cubic symmetry, and appears whenever the spins are on a real cubic lattice. If f is included, the propagator $G^{\alpha\beta}(\mathbf{q})$ must be given by (4D.2) (with all r_α's having the same value). Obviously, analytic calculations with this propagator are quite difficult, even in the antiferromagnetic, or non-dipolar, case $g = 0$. For small values of f, one can expand $G^{\alpha\beta}(\mathbf{q})$ in powers of f, and thus construct recursion relations. It can then easily be seen that the terms involving f will generate terms of fourth and higher order in the spins, which will have cubic symmetry. In fourth order, these have the form

$$\mathscr{H}_c = v_1 \sum_\alpha \sum_{\mathbf{R}} (S^\alpha_{\mathbf{R}})^4. \tag{5A.1}$$

Such terms may, of course, also appear in the original Hamiltonian of the system, as single ion interactions. Similar terms are very important at structural phase transitions (Cowley and Bruce, 1973). It turns out that most of the important details of the cubic critical behaviour can be obtained by ignoring the parameter f. Moreover, f finally turns out to be irrelevant (Aharony, 1973c, Bruce, 1974a), similarly to the parameter h discussed in Section IV.B. We shall therefore ignore f in the following discussion (note, however, that f represents slowly decaying correction terms!).

We thus start with the hyper-cubic short-range n-component system, with $V_2^{\alpha\beta}(\mathbf{q})$ given simply by (2C.4), and $V_4^{\alpha\beta\gamma\delta}$ given by (3A.1) where,

$$u_{\alpha\beta} = u + v\delta_{\alpha\beta}. \tag{5A.2}$$

The recursion relations for u and for v follow from (3A.3), and the results, to order u^2, v^2, ε^2, are (Aharony, 1973f)

$$\begin{aligned} u' = b^{\varepsilon-2\eta}\{&u - 4K_d \ln b(1 + \tfrac{1}{2}\varepsilon \ln b)[(n+8)u^2 + 6uv] \\ &+ 16K_4^2 \ln^2 b[(n^2+6n+20)u^3 + 9(n+4)u^2v + 27uv^2] \\ &+ 32K_4^2 \ln b(1+\ln b)[(5n+22)u^3 + 36u^2v + 9uv^2]\} \end{aligned} \tag{5A.3}$$

and

$$\begin{aligned} v' = b^{\varepsilon-2\eta}\{&v - 4K_d \ln b(1 + \tfrac{1}{2}\varepsilon \ln b)(12uv + 9v^2) \\ &+ 16K_4^2 \ln^2 b(36u^2v + 54uv^2 + 27v^3) \\ &+ 32K_4^2 \ln b(1+\ln b)[3(n+14)u^2v + 72uv^2 + 27v^3]\} \end{aligned} \tag{5A.4}$$

where

$$\eta = 8K_4^2[(n+2)u^2 + 6uv + 3v^2]. \tag{5A.5}$$

These recursion relations have four fixed points:

Gaussian, $$u^{\mathrm{G}} = v^{\mathrm{G}} = 0, \tag{5A.6}$$

Ising, $$u^{\mathrm{I}} = 0,\ v^{\mathrm{I}} = \frac{\varepsilon}{36\,K_d}\left(1 + \frac{17}{27}\varepsilon\right) + O(\varepsilon^3), \tag{5A.7}$$

Heisenberg, $$v^{\mathrm{H}} = 0,\quad u^{\mathrm{H}} = \frac{\varepsilon}{4K_d(n+8)}\left(1 + \frac{3(3n+14)}{(n+8)^2}\varepsilon\right) + O(\varepsilon^3) \tag{5A.8}$$

and Cubic,

$$u^{\mathrm{C}} = \frac{\varepsilon}{12\,K_d n}\left(1 + \frac{(n-1)(106-19n)}{27n^2}\varepsilon\right) + O(\varepsilon^3)$$

$$v^{\mathrm{C}} = \frac{\varepsilon}{12\,K_d n}\left(\frac{n-4}{3} + \frac{(n-1)(17\,n^2 + 110n - 424)}{81\,n^2}\varepsilon\right) + O(\varepsilon^3). \tag{5A.9}$$

The "Heisenberg" fixed point is the isotropic n-component fixed point discussed in Section II.D. The "Ising" fixed point has $u^{\mathrm{I}} = 0$, so that its Hamiltonian simply divides into a sum of independent Ising-like one-component Hamiltonians, each having the Ising-like exponents. The cubic fixed point is new, and has a fixed-point Hamiltonian which preserves the cubic symmetry.

To study stability of these fixed points, we linearize (5A.3) and (5A.4) about each of them. In doing so, it is important to remember also to linearize η, eqn (5A.5). The resulting exponents [as defined in (2B.12)] are

$$\lambda_u^{\mathrm{G}} = \lambda_v^{\mathrm{G}} = \varepsilon, \tag{5A.10}$$

$$\lambda_u^{\mathrm{I}} = \tfrac{1}{3}\varepsilon - \frac{19}{81}\varepsilon^2 + O(\varepsilon^3), \qquad \lambda_v^{\mathrm{I}} = -\varepsilon + \frac{17}{27}\varepsilon^2 + O(\varepsilon^3), \tag{5A.11}$$

$$\lambda_u^{\mathrm{H}} = -\varepsilon + \frac{9n+42}{(n+8)^2}\varepsilon^2 + O(\varepsilon^3), \qquad \lambda_v^{\mathrm{H}} = \frac{n-4}{n+8}\varepsilon + \frac{5n^2+14n+152}{(n+8)^3}\varepsilon^2 + O(\varepsilon^3) \tag{5A.12}$$

and

$$\lambda_1^{\mathrm{C}} = -\varepsilon + \frac{(n-1)(17n^2-4n+212)}{27n^2(n+2)}\varepsilon^2 + O(\varepsilon^3),$$

$$\lambda_2^{\mathrm{C}} = \frac{4-n}{3n}\varepsilon + \frac{(n-1)(19n^3-72n^2-660n+848)}{81n^3(n+2)}\varepsilon^2 + O(\varepsilon^3). \tag{5A.13}$$

Thus, the Gaussian fixed point is never stable. Considering the Ising fixed point, it is interesting to note that to the order calculated,

$$\lambda_u^{\mathrm{I}} = \alpha_{\mathrm{I}}/\nu_{\mathrm{I}}, \tag{5A.14}$$

where α_{I} and ν_{I} are the Ising-model specific heat and correlation length exponents. In fact, this seems to be a general identity, independent of the ε-expression (Sak, 1974): The operator $u\Sigma_{\alpha\neq\beta}(S_{\mathbf{R}}^{\alpha})^2(S_{\mathbf{R}}^{\beta})^2$, which is related to the exponent λ_u^{I}, is the product of two independent Ising-like energy operators. Thus, it has the dimension $x = 2(1 - \alpha_{\mathrm{I}})/\nu_{\mathrm{I}} = d - \alpha_{\mathrm{I}}/\nu_{\mathrm{I}}$. Following Wegner (1972b) we thus identify $\alpha_{\mathrm{I}}/\nu_{\mathrm{I}}$ as λ_u^{I}. Since α_{I} is always positive, we conclude that the Ising fixed point is unstable .The two-dimensional case is an interesting exception, which probably leads to Baxter-like models, in which the exponents depend on the marginal operator u (Baxter, 1971). The crossover exponent leading away from the Ising fixed point is thus

$$\phi = \nu_{\mathrm{I}}\lambda_u^{\mathrm{I}} = \alpha_{\mathrm{I}}. \tag{5A.15}$$

We now turn to a discussion of the stability of the Heisenberg and Cubic fixed points. For isotropic systems, the Heisenberg fixed point is stable, and λ_u^{H} represents corrections to scaling. However, for large enough n it is clear that λ_v^{H} becomes positive, and the Heisenberg fixed point becomes unstable. For large values of n, both λ_1^{C} and λ_2^{C} are negative (at $\varepsilon = 1$), and the Cubic fixed point becomes stable. To find exactly at which value of n the Heisenberg fixed point becomes unstable, and the Cubic one becomes stable, we must solve the equation $\lambda_v^{\mathrm{H}}(n_c) = 0$, or $\lambda_2^{\mathrm{C}}(n_c) = 0$. In fact, it is easy to convince oneself that at n_c one also has $v^{\mathrm{C}}(n_c) = 0$, and the two fixed points must coincide. This leads to logarithmic corrections to the isotropic Heisenberg exponents. From (5A.9), (5A.12) or (5A.13), one finds

$$n_c = 4 - 2\varepsilon + O(\varepsilon^2). \tag{5A.16}$$

Ketley and Wallace (1973) calculated λ_v^{H} to order ε^3, using direct diagrammatic expansions (see also Wallace's paper in this volume). Using their results, we find

$$n_c = 4 - 2\varepsilon + \frac{5}{12}(6\zeta(3) - 1)\,\varepsilon^2 + O(\varepsilon^3). \tag{5A.17}$$

Clearly, it is difficult to estimate the value of n_c at $\varepsilon = 1$. A Padé approximant to (5A.17) yields

$$n_c \simeq \frac{4 + 3{\cdot}176\,\varepsilon}{1 + 1{\cdot}294\,\varepsilon} \simeq 3{\cdot}128 \text{ at } \varepsilon = 1. \tag{5A.18}$$

Thus, the interesting case $n = 3$ is very close to n_c, and it is practically impossible to determine which fixed point is stable, although it is more probable that $3 \lesssim n_c$†. However, it is clear that for $n = 2$ the isotropic Heisenberg fixed point is the stable one. It is therefore important to study both types of critical behaviour, and let other techniques determine which one wins at $n = 3$. The fact that n_c is so close to 3 means that both λ_v^{H} and λ_2^{C} are very small. Thus, even if one of the fixed points is stable, the irrelevant parameter leading to it will represent very slowly decaying corrections to the leading scaling behaviour.

In Fig. 3 we give the flow diagrams of the iterative recursion relations in the

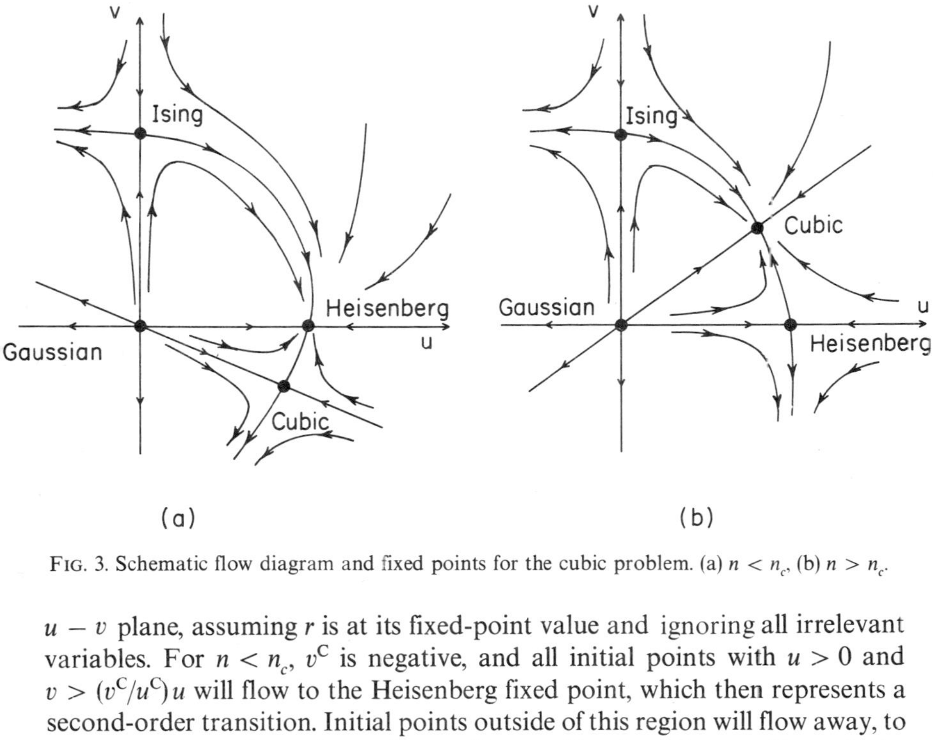

FIG. 3. Schematic flow diagram and fixed points for the cubic problem. (a) $n < n_c$, (b) $n > n_c$.

$u - v$ plane, assuming r is at its fixed-point value and ignoring all irrelevant variables. For $n < n_c$, v^{C} is negative, and all initial points with $u > 0$ and $v > (v^{\mathrm{C}}/u^{\mathrm{C}})u$ will flow to the Heisenberg fixed point, which then represents a second-order transition. Initial points outside of this region will flow away, to more negative values of v and/or u. Eventually, these flows will reach the region in which the four spin terms in the reduced Hamiltonian no longer satisfy the conditions for positivity,

$$u + v > 0, \qquad nu + v > 0. \tag{5A.19}$$

It is then believed that the transition becomes of first order. In mean field theory, the violation of (5A.19) indeed leads to first order transitions. After

† In a numerical study of Wilson's (1971) approximate recursion relations at $d = 3$, Grover *et al.* (1972) find that $\lambda_v^H \simeq -0{\cdot}058$ at $n = 3$, so that $n_c \gtrsim 3$ at $\varepsilon = 1$.

many iterations, it is reasonable to assume that the effective correlation length (2B.10) is small enough, so that mean field arguments may be used. Indeed, a direct calculation of the equation of state for $v < 0$ and $n \to \infty$ yields a first order transition (Wallace, 1973). These studies also yield λ_v^{H} (and hence n_c) to order $1/n$ (see also Oppermann, 1974b; Ketley and Wallace, 1973).

If the initial point is on the v-axis, the flow will lead to the Ising fixed point. Thus, this axis divides between first-order and second-order behaviour. Some thermodynamic field has to be adjusted so that $u = 0$, and one starts on the v-axis (Note that we ignore irrelevant variables, e.g. sixth order terms. These may affect the initial value of u, and the following arguments must be modified, although the general picture remains the same). Thus, the Ising fixed point is a *tricritical* point, with Ising-like exponents.

Similarly, if the initial values of u and v satisfy $v = (v^{\mathrm{C}}/u^{\mathrm{C}})\,u$ (ignoring irrelevant variables), the Cubic fixed point will be reached, leading to *Cubic-like tricritical behaviour.*

For $n > n_c$, the Heisenberg and Cubic fixed points interchange: the Cubic fixed point then represents a second-order transition, and the Heisenberg fixed point becomes tricritical.

We shall not go here into the details of the exponents describing the non-trivial tricritical points found above, since this demands a large introduction on tricritical exponents relating to the "non-ordering" densities (Aharony, 1974f). However, it is important to note that the exponents describing the ordering density (the magnetization, susceptibility, etc.) will have different values for the special initial parameters mentioned above. In many cases, u and v can be varied by applying an external stress (Aharony, 1973g; Aharony and Bruce, 1974a). This may lead to a possible experimental verification of these results.

To remain at the Gaussian fixed point, the initial values of both u and v must be set equal to zero (ignoring irrelevant variables). Together with the temperature field, and the ordering (magnetic) field, one thus has to adjust four thermodynamic fields, and the Gaussian fixed point now represents a critical point of order four.

B. Cubic critical behaviour

For $n < n_c$, the Heisenberg fixed point, discussed in Section II.D, is stable. The cubic terms in the Hamiltonian represent corrections to scaling, of the form $v|t|^{\omega_c}$, with $\omega_c = -\nu_{\mathrm{H}}\lambda_v^{\mathrm{H}}$. Since λ_v^{H} is small at $\varepsilon = 1$, these corrections are important. They have various effects, the most important of which has to do with easy axes, and with transverse susceptibilities (Bruce and Aharony, 1975). As long as $v|t|^{\omega_c}$ is finite, the magnetization will tend to align along an axis ($v < 0$) or along a diagonal ($v > 0$). The transverse susceptibility in zero field, which is divergent for the isotropic case, is of order $[v|t|^{\gamma+\omega_v}]^{-1}$, and diverges only at T_c (see also Wallace, 1973; Ketley and Wallace, 1973). The two effects are, of course, related, since the transverse susceptibility describes the fluctuations in the direction of the magnetization vector. For $v > 0$, the corrections to scaling due to v are also instrumental in changing the bicritical point, discussed in Section III.A. into a tetracritical point (Bruce and Aharony, 1975).

For $n > n_c$, we expect a crossover from the Heisenberg fixed point to the Cubic fixed point, with the crossover exponent

$$\phi_v^{\mathrm{H}} = \nu_{\mathrm{H}}\lambda_v^{\mathrm{H}} = \frac{n-4}{2(n+8)}\varepsilon + \frac{n^3 + 16n^2 + 4n + 240}{2(n+8)^3}\varepsilon^2 + O(\varepsilon^3) \quad (5B.1)$$

(see Ketley and Wallace, 1973, for the next term). Again, this is small at $\varepsilon = 1$, meaning that the crossover is very slow and that the real asymptotic cubic critical behaviour may be felt only very close to T_c.

We now turn to study this cubic critical behaviour. To obtain critical exponents near the new cubic fixed point we must linearize the relevant recursion relations about it. From (5A.5) it immediately follows that

$$\eta_{\mathrm{C}} = \frac{(n+2)(n-1)}{54n^2}\varepsilon^2 + O(\varepsilon^3). \quad (5B.2)$$

Linearization of (3A.2) around (5A.9) yields

$$\Delta r' = b^{2-\eta}\{1 - 4[(n+2)u + 3v]\,C(r) + 96[(n+2)u^2 + 6uv + 3v^2]\,D(r) + \ldots\}\,\Delta r \quad (5B.3)$$

[see (2D.21), (2D.23)], so that

$$2\nu_{\mathrm{C}} = 1 + \frac{n-1}{3n}\varepsilon + \frac{n-1}{324n^3}(17n^2 + 290n - 424)\,\varepsilon^2 + O(\varepsilon^3). \quad (5B.4)$$

The introduction of a magnetic field, and linearization of the appropriate recursion relations, yields the exponents β, δ, etc. As in the dipolar case, it is more convenient at this stage to use the values u^{C} and v^{C} of (5A.9) in an ε-expansion. For $n > n_c$, $v^{\mathrm{C}} > 0$ and the magnetization $\mathbf{M}$ tends to align along a diagonal, e.g. $(1, 1, \ldots, 1)$. Introducing a magnetic field $\mathbf{H}$ in the same direction, and following the standard diagrammatic procedure, yield (Aharony, 1974d)

$$H/M^\delta = f(t/M^{1/\beta}), \quad (5B.5)$$

with

$$f(x) = x + 1 + \frac{\varepsilon}{3n^2}\Big\{2(n-1)(x+3)\ln(x+3) + (n-1)(n-2)\left(x + \frac{3(n-2)}{2(n-1)}\right)\ln\left(x + \frac{3(n-2)}{2(n-1)}\right)$$

O

$$-\left[6(n-1)\ln 3 + \frac{3}{2}(n-2)^2 \ln\frac{3(n-2)}{2(n-1)}\right](x+1)$$
$$+\left[4(n-1)\ln 2 + \frac{1}{2}(n-2)(n-4)\ln\frac{n-4}{2(n-1)}\right]x\Bigg\} + O(\varepsilon^2), \tag{5B.6}$$

and with δ and β related to ν_C and η_C by standard scaling.

There are a few important distinctions between cubic and isotropic critical behaviour. First, the fact that $v^C \neq 0$ means that cubic symmetry is retained even asymptotically close to the critical point. This means that the notion of *easy axes* remains valid, and that near the cubic fixed point ($n > n_c$) the magnetization will align along diagonals. In comparison, the cubic term represents only a *correction to scaling* near the Heisenberg fixed point for $n < n_c$. Second, both the longitudinal and the transverse susceptibilities are not divergent, in the cubic case, in zero field below T_c.

For large n, the limits of (5B.2) and (5B.4) are

$$\eta_C = \eta_I + O(1/n) \tag{5B.7}$$

and

$$\nu_C = \nu_I/(1-\alpha_I) + O(1/n). \tag{5B.8}$$

(This was also noted by Rudnick, 1973.) A direct generalization of the $1/n$-expansion (Aharony, 1973h) indeed proves (5B.7) and (5B.8) for all values of ε, by replacing the Gaussian propagators by Ising-like propagators and treating u as a perturbation. Note that for $u = 0$, we have n uncoupled Ising models. The term involving u introduces a coupling between the spin components, which originates from the constraint on the length of the total spin, $\Sigma_{\alpha=1}^{n}(S^\alpha)^2$. Indeed, one can give general proofs (Fisher, 1974b; Emery, 1975), independent of perturbation expansions, that in the limit $n \to \infty$ the cubic free energy per spin component is the same as that of a constrained Ising model, leading to a renormalization of the critical exponents (Fisher, 1968), as is indeed implied by (5B.7) and (5B.8).

For $n < n_c$, the cubic fixed point represents a tricritical point. If the initial values of u and v are near the line $v = u\, v^C/u^C$ (ignoring irrelevant variables), then the exponents given above will describe the critical behaviour for a wide range of temperatures. For $v > uv^C/u^C$, there will be a slow crossover to the Heisenberg behaviour, given by the exponent $\nu_C\lambda_2^C$.

We can now complicate the cubic problem, by adding various terms of lower symmetry to the initial Hamiltonian. In the spin-anisotropic case, with the propagator (2C.9), the recursion relation (5B.3) is replaced by n equations of the form (3A.9). This yields (Aharony, 1974e)

$$\phi_g^C = 1 + \tfrac{1}{6}\varepsilon - \frac{5n^2 - 49n + 53}{81n^2}\varepsilon^2 + O(\varepsilon^3). \tag{5B.9}$$

In the limit $n \to \infty$, a generalization of the above-mentioned $1/n$-expansion yields

$$\phi_g^C = 1/(1 - \alpha_I) + O(1/n). \tag{5B.10}$$

Following the arguments of Section III.A, we conclude that the critical behaviour of the cubic system with a leading m-component exchange isotropy crosses over to that of an m-component cubic fixed point. For $m = 2$, this is identical to two decoupled Ising models, rotated by 45° from the original spin axes (Wilson and Fisher, 1972). For $m = 1$, the cubic fixed point becomes a Gaussian fixed point. In neither case is the "cubic" point stable, and the ultimate critical behaviour is thus m-isotropic Heisenberg-like.

Near the Cubic fixed point, for $n > n_c$, $v > 0$. Thus, the easy axes are along the cube diagonals. Adding a spin anisotropy, of the type (2C.9), creates a competition between these easy axes and those along the cube edges. This results with the appearance of a new ordered phase, in which the magnetization rotates continuously from the m-component direction ($g < 0$) to the $(n - m)$-component direction ($g > 0$). In this case, the $g = 0$ flop line splits into two second order lines, and the bicritical point becomes tetracritical (Bruce and Aharony, 1975).

The problem becomes more difficult when the dipolar interaction is introduced, with the propagator (4C.2). As in Section IV.C, this propagator changes the integrals which appear in the recursion relations, and finally (5A.3) and (5A.4) are replaced by ($n = d = 4 - \varepsilon$)

$$u' = b^{\varepsilon}\{u - K_d(34u^2 + 20\,uv + \tfrac{3}{2}v^2) + \ldots\}, \tag{5B.11}$$

$$v' = b^{\varepsilon}\{v - K_d(28\,uv + 21\,v^2) + \ldots\}. \tag{5B.12}$$

It is easy to check that these have no real fixed-point solution with v of order ε. The isotropic dipolar fixed point, (4C.5), is unstable with respect to the cubic parameter v, and the crossover exponent, analogous to (5B.1), is (Bruce and Aharony, 1974)

$$\phi_v^{\text{d.d.}} = \frac{3}{34}\varepsilon + \frac{15887}{176868}\varepsilon^2 + O(\varepsilon^3). \tag{5B.13}$$

It is not yet clear where this crossover leads to. It is possible that there exists a cubic dipolar fixed point, with the cubic parameter f of order unity and with v of order ε, but the complicated form of the propagator (4D.2) makes the discovery of such a fixed point very difficult. Other possibilities are that v

is not small, so that our perturbative procedure breaks down, or that no dipolar cubic fixed point exists. In the latter case, v may drive the transition to become of first order. It should be noted that the dipolar fixed point is unstable with respect to the cubic perturbation even in the anisotropic XY-like case, discussed in Section V.D. For the n-component dipolar fixed point,

$$\phi_v^{\text{d.d.}} = \frac{n^3 - 3n^2 + 2n + 12}{2(n^3 + 9n^2 + 2n - 12)} \varepsilon + O(\varepsilon^2) \tag{5B.14}$$

(Aharony, 1974a).

C. n Coupled m-vector models

In general, the fourth-order terms in the spins may be more complicated than presented by the cubic case (5A.2). Brézin *et al.* (1974) considered the most general $V_4^{\alpha\beta\gamma\delta}$ consistent with one quadratic invariant, which must obey the condition

$$\sum_{\alpha} V_4^{\alpha\alpha\beta\gamma} = \text{const.}\ \delta_{\beta\gamma}. \tag{5C.1}$$

They concluded that even in this general case, the general fixed point structure is similar to that presented in Section V.A: There are usually four fixed points, which are the Gaussian, the isotropic Heisenberg, and two others. The Heisenberg fixed point is stable against any anisotropic quartic perturbation for $n < n_c$, and one of the other non-trivial fixed points is stable for $n > n_c$.

In particular, Brézin *et al.* consider a system in which each spin is composed of n m-component vectors, $\mathbf{S} \equiv (\mathbf{S}_1, \ldots, \mathbf{S}_n)$, $\mathbf{S}_i \equiv \{S_i^\alpha, \alpha = 1, \ldots, m\}$, with a quartic "single ion" coupling of the form

$$u|\mathbf{S}|^4 + v \sum_{i=1}^{n} |\mathbf{S}_i|^4 = \sum_{i\alpha j\beta} (u + v\delta_{ij})(S_i^\alpha)^2 (S_j^\beta)^2. \tag{5C.2}$$

This reduces to the cubic case for $m = 1$, and has special applications to the random m-vector problem for $n = 0$ (Grinstein and Luther, 1976), and to the problem of a $2m$-component antiferromagnet in a magnetic field along m spin axes for $n = 2$ (Nelson and Fisher, 1975; Nelson *et al.*, 1974). We shall return to these applications later.

The recursion relations for u and for v now follow from (3A.3). The only modification is that now each summation becomes a summation over both $i(= 1, \ldots, n)$ and $\alpha(= 1, \ldots, m)$. The results are

$$u' = b^{\varepsilon}\{u - 4K_d \ln b((nm + 8)u^2 + 2(m + 2)uv) + \ldots\} \tag{5C.3}$$

and

$$v' = b^{\varepsilon}\{v - 4K_d \ln b((m + 8)v^2 + 12uv) + \ldots\}. \tag{5C.4}$$

We can now easily generalize equations (5A.6)–(5A.9), with the four fixed points (to order ε)

Gaussian, $$u^{\mathrm{G}} = v^{\mathrm{G}} = 0, \tag{5C.5}$$

m-component, $$u^m = 0, \qquad v^m = \frac{\varepsilon}{4K_d(m+8)} \tag{5C.6}$$

Heisenberg, $$v^{\mathrm{H}} = 0, \qquad u^{\mathrm{H}} = \frac{\varepsilon}{4K_d(nm+8)} \tag{5C.7}$$

and "mixed",

$$u^X = \frac{4-m}{(m+8)\,nm - 16(m-1)} \frac{\varepsilon}{4K_d},$$

$$v^X = \frac{nm-4}{(m+8)\,nm - 16(m-1)} \frac{\varepsilon}{4K_d}. \tag{5C.8}$$

Again, the Gaussian fixed point is never stable, and the m-component decoupled fixed point has [see (5A.14)]

$$\lambda_u^m = \alpha_m/\nu_m, \tag{5C.9}$$

with α_m and ν_m the exponents of the m-vector model. At $d = 3$, α_m becomes negative slightly below $m = 2$ (Ahlers, 1973). Thus, this decoupled fixed point seems to be stable for all $m > 1$. To order ε, one finds that $\lambda_u^m < 0$ for $m > m_\alpha = 4 - 4\varepsilon$, and a Padé approximant at $\varepsilon = 1$ gives $m_\alpha = 2$. To order zero in ε, $m_\alpha = 4$. It is very important to have a reliable estimate for m_α. Otherwise, one may be led to wrong conclusions for $m = 3$, $d = 3$ (Grinstein, 1974; Harris and Lubensky, 1974).

Near the Heisenberg fixed point, λ_v^{H} is now given by (5A.12), with nm replacing n. Thus, this fixed point is stable for

$$nm < n_c, \tag{5C.10}$$

with n_c given by (5A.17) or (5A.18).

Near the mixed fixed point, we have

$$\lambda_1^X = -\varepsilon, \qquad \lambda_2^X = \frac{(4-m)(4-nm)}{(m+8)\,nm - 16(m-1)}\varepsilon. \tag{5C.11}$$

Thus, to this order, this fixed point is stable for

$$4 < nm < 4n \qquad \text{if } n > 1,$$

$$4n < mn < 4 \qquad \text{if } 1 > n > \frac{16(m-1)}{m(m+8)},$$

$$m < 4 \text{ or } mn > 4 \quad \text{if } n < \frac{16(m-1)}{m(m+8)}, \qquad m > 1. \tag{5C.12}$$

Note that the numbers "4" appearing here are of two origins, (n_c and m_α), and each has to be replaced by an appropriate ε-expansion. For $n > 1$, it is reasonable to conjecture that (5C.12) must be replaced by

$$n_c < nm < m_\alpha n. \tag{5C.13}$$

Since $m_\alpha < 2$, we conclude that the mixed fixed point is stable (for $n > 1$) only if $m = 1$. Thus, the cubic fixed point is the only stable fixed point for $nm > n_c$. Similarly, for $n = 0$, it is reasonable to conjecture that (5C.12) reduces to

$$1 < m < m_\alpha. \tag{5C.14}$$

Since $m_\alpha < 2$, this means that the mixed fixed point is never stable for $n = 0$. For $n = 0$ and $m = 1$, the "mixed" fixed point [identical to the cubic fixed point of (5A.9)] seems to diverge to infinity. In reality, it is replaced by a fixed point for which u and v are of order $\varepsilon^{\frac{1}{2}}$ (Khmel'nitskii, 1974).

In each case we can now follow the discussion of Section V.A., and draw flow diagrams such as those presented in Fig. 3. For $n > 1$ and $nm < n_c$, the Heisenberg fixed point describes a second-order transition, while the other two non-trivial fixed points represent tricritical behaviour. In the region (5C.13), the mixed (or "cubic") fixed point gives the second-order behaviour and in the region $m > m_\alpha$ ($n > 1$) the decoupled m-component fixed point gives the second-order behaviour. The case $n = 0$ is more complicated, and will be discussed separately.

The exponents for the "mixed" fixed point are easily obtained from the appropriate linearization. In particular,

$$2\nu_X = 1 + \frac{3m(n-1)}{(m+8)\,nm - 16(m-1)}\varepsilon + O(\varepsilon^2). \tag{5C.15}$$

Note that in the limit $n \to \infty$, this reduces to the renormalized m-component $2\nu_m/(1 - \alpha_m)$ [See (5B.8) and Aharony, 1973h].

It is now easy to generalize this model to include some of the previously discussed interactions. For example, if each of the m-vector spins also has cubic symmetric single ion interactions, then (5C.2) will be replaced by

$$\sum_{i\alpha j\beta} (u + v\delta_{ij} + w\delta_{ij}\,\delta_{\alpha\beta})\,(S_i^\alpha)^2\,(S_j^\beta)^2. \tag{5C.16}$$

One can easily construct recursion relations for u, v and w. In addition to the four fixed points (5C.5)–(5C.8), with $w = 0$, one also finds an m-cubic fixed point ($u = 0$), an Ising fixed point ($u = v = 0$), an nm-cubic fixed point ($v = 0$)

and a new mixed fixed point

$$u^{XX} = \frac{4 - m}{(m - 4)^2 + m^2(n - 1)} \frac{\varepsilon}{12 K_d}$$

$$v^{XX} = \frac{mn - 4}{(m - 4)^2 + m^2(n - 1)} \frac{\varepsilon}{12 K_d}$$

$$w^{XX} = \frac{(m - 4)(mn - 4)}{(m - 4)^2 + m^2(n - 1)} \frac{\varepsilon}{36 K_d}. \tag{5C.17}$$

For $n = 0$, this fixed point may be applicable to random cubic systems.

A simple generalization of the model (5C.2) is obtained when the coupled spin vectors are allowed to have different numbers of components. As an example, consider a model of coupled m_1- and m_2-vector models, with the effective Hamiltonian

$$\bar{\mathscr{H}} = -\int_{\mathbf{x}} \{\tfrac{1}{2}(r_1 \mathbf{S}_1^2 + r_2 \mathbf{S}_2^2 + (\nabla \mathbf{S}_1)^2 + (\nabla \mathbf{S}_2)^2)$$

$$+ u|\mathbf{S}_1|^4 + v|\mathbf{S}_2|^4 + 2w|\mathbf{S}_1|^2 |\mathbf{S}_2|^2\}. \tag{5C.18}$$

Such an effective Hamiltonian is of interest in the study of the bicritical point ($r_1 \simeq r_2$) of an anisotropic antiferromagnet in a uniform field, when both the anisotropy and the field single out m_1 spin components (Nelson *et al.*, 1974). It is easy to write down the recursion relations, which have the form

$$\begin{aligned} r_1' &= b^2 \{r_1 + 4((m_1 + 2) uA(r_1) + m_2 wA(r_2)) + \ldots\} \\ r_2' &= b^2 \{r_2 + 4((m_2 + 2) vA(r_1) + m_1 wA(r_2)) + \ldots\} \\ u' &= b^{\varepsilon} \{u - 4K_d \ln b((m_1 + 8) u^2 + m_2 w^2) + \ldots\} \\ v' &= b^{\varepsilon} \{v - 4K_d \ln b((m_2 + 8) v^2 + m_1 w^2) + \ldots\} \\ w' &= b^{\varepsilon} \{1 - 4K_d \ln b((m_1 + 2) u + (m_2 + 2) v + 4w) + \ldots\} w, \end{aligned} \tag{5C.19}$$

where $A(r)$ is given in (2D.20). For $w = 0$, these equations give two decoupled sets of equations, yielding four combinations of Gaussian and m-component critical behaviour. One can easily extend the argument which led to (5C.9) and (5A.14), to find that the exponent related to the coupling w at each of these is

$$\lambda_w = \tfrac{1}{2}\left(\frac{\alpha_1}{\nu_1} + \frac{\alpha_2}{\nu_2}\right). \tag{5C.20}$$

Since all the Gaussian fixed points are unstable, with respect to u or v, it is sufficient to study the decoupled (m_1, m_2) non-trivial fixed point. To lowest order in ε, this is stable against the w-coupling for

$$m_1 m_2 + 2(m_1 + m_2) > 32. \tag{5C.21}$$

However, higher-order terms are very important. For example, in the case discussed by Nelson *et al.* (1974), $m_1 = 1$, $m_2 = n - 1$, λ_w is negative for $n > n_w$, with

$$n_w = 11 - 16\varepsilon + O(\varepsilon^2) \simeq \frac{121}{11 + 16\varepsilon}, \tag{5C.22}$$

which gives $n_w \simeq 4{\cdot}5$ at $\varepsilon = 1$. The great advantage of expressions like (5C.20) is, that we can calculate them directly from known results (e.g. from high-temperature series, or experiments), independently of the ε-expansion. Indeed, these give $n_w \gtrsim 4$.

In addition to the decoupled fixed points, one finds the $(m_1 + m_2)$-isotropic fixed point $(u = v = w)$, stable for $(m_1 + m_2) < n_c$, and a new, *biconical* fixed point, at which u, v and w are not equal and are usually irrational functions of m_1 and m_2. Thus, the critical exponents resulting from the linearization of the first two recursion relations are also irrational functions. (See Nelson *et al.*, 1974, for details.)

In the next few Sections we shall review the applications of the models presented here to physical problems of interest. The important step in any such application is the transformation of the real Hamiltonian of the system into the form (2A.13), with $V_2^{\alpha\beta}$ having the simple isotropic short range form (2C.4) and with the quartic terms being of the form (5C.2). This usually involves averaging over some non ordering parameters, or a redefinition of the appropriate spin-like variables.

D. The random m-component magnet

There are many ways to represent the microscopic Hamiltonian of a quenched random system. Randomness may be introduced through the random occupation of lattice sites by magnetic ions (Grinstein and Luther, 1976), through the random distribution of the exchange parameters (Harris and Lubensky, 1974), through a random direction of a uniaxial anisotropy (Aharony, 1975), etc. In the context of a continuous-spin model, all of these are equivalent, although the last one gives a richer structure. We start by reviewing the problem of random occupation of lattice sites (Grinstein and Luther, 1974). The Hamiltonian is written in the form

$$\beta\mathscr{H} = -\tfrac{1}{2}\sum_{\langle ij\rangle} K_{ij} P_i P_j \sum_{\alpha=1}^{m} S_i^\alpha S_j^\alpha, \tag{5D.1}$$

where P_i is zero if the site $\mathbf{R}_i$ is vacant and one if the site is occupied. The average of P_i is denoted by P, and the value of P_i at any site is independent of its value at any other site. To obtain the free energy, one wants to calculate the partition function for a given distribution $\{P_i\}$, and then average the free energy over all distributions. Grinstein and Luther (1976) show, that the partition function $Z\{P_i\}$ can be written in the form

$$Z\{P_i\} = \int\left(\prod_i \mathrm{d}^m S_i\, \delta\left[m - \sum_\alpha (S_i^\alpha)^2\right]\right) \exp(-\mathscr{H}/k_B T)$$
$$\propto \int \prod_i d^m \sigma_i \exp(\overline{\mathscr{H}}\{\boldsymbol{\sigma}_i, P_i\}), \tag{5D.2}$$

with

$$\overline{\mathscr{H}}\{\boldsymbol{\sigma}_i, P_i\} = \tfrac{1}{2}\sum_{ij} L_{ij} \sum_\alpha \sigma_i^\alpha \sigma_j^\alpha - P \sum_{r=2}^{\infty} C_{2r} \sum_j \left[\sum_\alpha (\sigma_j^\alpha)^2\right]^r$$
$$- \sum_{r=1}^{\infty} C_{2r} \sum_j Q_j \left[\sum_\alpha (\sigma_j^\alpha)^2\right]^r, \tag{5D.3}$$

where the C_{2r} are numbers, of order unity, defined by

$$\exp\left[-\sum_r C_{2r} |\boldsymbol{\sigma}|^{2r}\right] = \int \mathrm{d}^m S\, \delta(m - |\mathbf{S}|^2) \exp(\boldsymbol{\sigma}\cdot\mathbf{S}), \tag{5D.4}$$

and where

$$Q_j = P_j - P, \tag{5D.5}$$
$$L_{ij} = (K^{-1})_{ij} - P\delta_{ij}. \tag{5D.6}$$

Note that for $P = 1$, (5D.2) is identical with (2A.2), with the weight function (2A.5). This is indeed a direct way to derive that form.

Grinstein and Luther then expand $Z\{P_i\}$ in a perturbation series in the two last sums in (5D.3), and average over $\{\mathrm{P}_i\}$ term by term. Since $\langle Q_j \rangle = 0$, the first term in the last sum of (5D.3) contributes to lowest order only in the combination

$$\tfrac{1}{2} C_2^2 \sum_{ij} \langle Q_i Q_j \rangle \left[\sum_\alpha (\sigma_i^\alpha)^2\right]\left[\sum_\beta (\sigma_j^\beta)^2\right] = \tfrac{1}{2} C_2^2\, P(1-P) \sum_j \sum_{\alpha\beta} (\sigma_j^\alpha)^2 (\sigma_j^\beta)^2, \tag{5D.7}$$

which is translationally invariant. However, this is to be exponentiated, and we must not include in the diagrammatic sums any disconnected graphs, which would result from closed loops, coming from contracting the two σ_j^α's or the two σ_j^β's. To eliminate these, Grinstein and Luther suggested replacing each m-component vector $\boldsymbol{\sigma}$ by an nm-component vector, $\{\sigma^{\alpha i}\}$, with indices $\alpha = 1, \ldots, m$ and $i = 1, \ldots, n$, and finally setting $n = 0$ (each closed loop will give a factor n and will be eliminated). Thus, the effective four-spin terms become of the form (5C.2), with $v = C_4 P$ and with $u = -\frac{1}{2}C_2^2 P(1 - P)$. Similar arguments can be used to derive all higher-order terms in the spin, but we need not consider these near $d = 4$, except for a few initial iterations. Emery (1975) gave a direct proof, independent of perturbation theory, that the $n = 0$ limit of the nm-vector model indeed represents a random m-vector model.

At this stage we can simply turn back to the last section, and apply the results for $n = 0$. For $m > m_\alpha$ we found that the m-component fixed point is stable. The "mixed" fixed point now has $u^X > 0$, and the flow diagram is sketched in Fig. 4. This is more complicated than Fig. 3, since we now have two stable fixed points. The line connecting the Gaussian and mixed fixed points now serves as a saddle line, dividing between the regions of attraction of the m-component and the $n = 0$ Heisenberg fixed points.

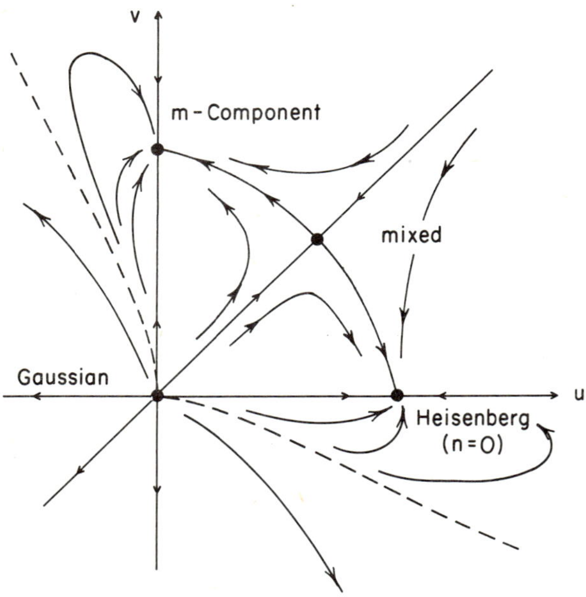

FIG. 4. Schematic flow diagram and fixed points for the random magnet, $n = 0$ and $m > m_\alpha$.

We now have an interesting situation for $u < 0$ or $v < 0$. Note that the initial value of u in this problem is *negative*. *A priori*, one might expect that a few iterations which include various irrelevant variables may change the sign of u, so that the $n = 0$ Heisenberg-like fixed point or the mixed fixed point will eventually be approached. It turns out, however, that $(-u)$ is the second cumulant of the distribution function of the random variables, and as such it must remain negative (Harris and Lubensky, 1974; Lubensky, 1975). Indeed, inclusion of all higher-order coefficients in the first iterations confirms this conclusion (Aharony *et al.*, 1976). Thus, the effective value of u after many iterations will remain negative. In this case, the flow will either be to the m-component fixed point, yielding corrections to scaling of the form $(1 - P)|t|^{\omega}$, where $\omega = -v_m \lambda_u^m = -\alpha_m$, or away to large negative values of u. In the latter case, Aharony *et al.* (1976) show that the transition is *not* of first order. It is probably *smeared* (Lubensky, 1975), but its exact nature remains to be studied. The broken line in Fig. 4 is a schematic sketch of the dividing line between a flow to the m-component fixed point and away to "no man's land" of "smeared" transition. Unlike the case discussed in Section V.A., we cannot find a simple fixed point describing the behaviour on this line. A similar situation occurs for $v < 0$.

As m crosses the value m_α, u^X becomes negative and the mixed fixed point becomes stable. As m approaches 1, the mixed fixed point becomes of order $\varepsilon^{\frac{1}{2}}$ (Khmel,nitskii, 1974). The Ising fixed point is now unstable, and the flow for negative values of u leads to this "mixed", or "random", fixed point. If the effective value of u, after a few iterations, could be positive, then the behaviour would be described by the $n = 0$ exponents, e.g.

$$2\nu = 1 + \tfrac{1}{8}\varepsilon + \frac{15}{256}\varepsilon^2 + O(\varepsilon^3). \tag{5D.8}$$

It should be clear at this point that the extrapolation of these results to $d = 3$ is quite difficult: the flow diagrams must then include all six-spin terms, and higher-order terms also change the parameters during the first few iterations. However, the general picture of competing fixed points should still be true, and one should search experimentally for effects of the type described here.

The situation becomes more complex for a model of amorphous magnets, in which a single ion uniaxial anisotropy is assumed, which has a random direction (Harris *et al.*, 1973). In this case, we start with the Hamiltonian

$$\mathscr{H} = -\sum_{\langle ij\rangle} J_{ij}\mathbf{S}_i \cdot \mathbf{S}_j - D\sum_i (\hat{X}_i \cdot \mathbf{S}_i)^2, \tag{5D.9}$$

where $\hat{X}_i$ is an m-component unit vector, with a random direction. We can separate $\mathscr{H}$ in the form $\mathscr{H} = \mathscr{H}_0 + \mathscr{H}_1$, with

$$\mathscr{H}_0 = -\sum_{\langle ij\rangle} J_{ij}\mathbf{S}_i \cdot \mathbf{S}_j - \frac{1}{m} D \sum_i \mathbf{S}_i^2 \tag{5D.10}$$

and

$$\mathcal{H}_1 = -D \sum_i \sum_{\alpha\beta} (X_i^\alpha X_i^\beta - \delta_{\alpha\beta} m^{-1}) S_i^\alpha S_i^\beta. \tag{5D.11}$$

Just as in (5D.7), we can expand $\exp(-\beta\mathcal{H})$ perturbatively in powers of $\mathcal{H}_1$, and calculate the average over the random directions of $\hat{X}_i$ term by term. The resulting effective four-spin Hamiltonian now becomes of the form discussed above, with an additional term, i.e.,

$$w \sum_{ij\alpha\beta} \sigma^{\alpha i} \sigma^{\beta i} \sigma^{\alpha j} \sigma^{\beta j}, \tag{5D.12}$$

($i, j = 1, \ldots, n$; $\alpha, \beta = 1, \ldots, m$; and $n \to 0$ at the end). Again, one finds all the previous fixed points, and a few additional ones. A detailed analysis of this problem is given elsewhere (Aharony, 1975).

E. Metamagnetic, bicritical and tetracritical behaviour in antiferromagnets

Nelson *et al.* (1974) have recently applied some of the results described in the previous sections to various problems arising when ferromagnetic and antiferromagnetic ordering compete. Again, the basic idea is first to transform the original Hamiltonian of the system into an effective one, which is of the form discussed above, and then to apply to it renormalization group iterations.

The basic problem with antiferromagnets which also have ferromagnetic interactions is, that the two types of order are related to two distinct wave vectors, say $\mathbf{k}_0$ and zero. If we proceed with the standard technique of integrating over large wave vectors, then eventually $\mathbf{k}_0$ will be integrated out, and we shall lose the antiferromagnetic coupling. To avoid this, Nelson and Fisher (1975) invented a transformation of the spin variables which is equivalent to a mapping of the outer half of the Brillouin zone onto the inner one, so that $\mathbf{k}_0$ is mapped onto zero. Now, the $\mathbf{q}$-integrals run only over the inner half of the Brillouin zone, but each $\mathbf{q}$-vector has two spin vectors attached to it.

Technically, when the lattice is ordered antiferromagnetically then it is decomposed into two interpenetrating sublattices, A and B. The Nelson–Fisher new spin variables are then defined as

$$\boldsymbol{\sigma}_\pm(\mathbf{q}) = \tfrac{1}{2} \sum_{\mathbf{R} \subset A} \exp(i\mathbf{q} \cdot \mathbf{R}) \mathbf{S}_\mathbf{R} \pm \tfrac{1}{2} \sum_{\mathbf{R} \subset B} \exp(i\mathbf{q} \cdot \mathbf{R}) \mathbf{S}_\mathbf{R}. \tag{5E.1}$$

In real space, this is equivalent to replacing each pair of lattice points, one in A and one in B, by a single new lattice point, at which two new spin variables are defined (the sum and difference of the two original spins). These new variables are chosen so as to have a diagonal quadratic interaction in the $2n$ variables $\boldsymbol{\sigma}_+$ and $\boldsymbol{\sigma}_-$.

We shall now summarize the applications of the Nelson–Fisher transformation. Consider an antiferromagnet, in which the interaction inside a sublattice is ferromagnetic, while the interaction between sites on two different sublattices is antiferromagnetic. Changing the signs of all spins on one sublattice, all interactions become effectively ferromagnetic, and the Hamiltonian is of the form

$$\mathscr{H} = -\tfrac{1}{2}\sum_{\mathbf{R},\mathbf{R}'} J(\mathbf{R}-\mathbf{R}')\mathbf{S}_{\mathbf{R}}\cdot\mathbf{S}_{\mathbf{R}'} - H\sum_{\mathbf{R}} \exp(i\mathbf{k}_0\cdot\mathbf{R})S^1_{\mathbf{R}} \qquad (5E.2)$$

(we take the uniform magnetic field to be in the 1-direction). Adding the standard weight function (2A.5), substituting (5E.1) and remembering that $J(\mathbf{R}-\mathbf{R}')$ has different values for sites in the same or in different sublattices, Nelson and Fisher (1975) finally find an effective reduced Hamiltonian of the form

$$\begin{aligned}\bar{\mathscr{H}} = {} & -\tfrac{1}{2}N_a^{-1}\sum_{\mathbf{q}}\{[1-\hat{K}_+(\mathbf{q})]\boldsymbol{\sigma}_+(\mathbf{q})\cdot\boldsymbol{\sigma}_+(-\mathbf{q}) + [1-\hat{K}_-(\mathbf{q})]\boldsymbol{\sigma}_-(\mathbf{q})\cdot\boldsymbol{\sigma}_-(-\mathbf{q})\}\\ & + 2L\sigma^1_-(0) - 2uN_a^{-3}\sum_{\mathbf{q}}\sum_{\mathbf{q}'}\sum_{\mathbf{q}''}\sum_{\alpha,\beta}\{\sigma^\alpha_+(\mathbf{q})\sigma^\alpha_+(\mathbf{q}')\sigma^\beta_+(\mathbf{q}'')\sigma^\beta_+(-\mathbf{q}-\mathbf{q}'-\mathbf{q}'')\\ & + \sigma^\alpha_-(\mathbf{q})\sigma^\alpha_-(\mathbf{q}')\sigma^\beta_-(\mathbf{q}'')\sigma^\beta_-(-\mathbf{q}-\mathbf{q}'-\mathbf{q}'') + 2\sigma^\alpha_+(\mathbf{q})\sigma^\alpha_+(\mathbf{q}')\sigma^\beta_-(\mathbf{q}'')\\ & \times\sigma^\beta_-(-\mathbf{q}-\mathbf{q}'-\mathbf{q}'') + 4\sigma^\alpha_+(\mathbf{q})\sigma^\beta_+(\mathbf{q}')\sigma^\alpha_-(\mathbf{q}'')\sigma^\beta_-(-\mathbf{q}-\mathbf{q}'-\mathbf{q}''),\end{aligned} \qquad (5E.3)$$

where $L = H/k_BT$, $\hat{K}_+(\mathbf{q}) = \hat{J}(\mathbf{q})/k_BT$ and $\hat{K}_-(\mathbf{q}) = \hat{K}_+(\mathbf{q}) - 2\hat{J}_{ab}(\mathbf{q})/k_BT$. [$\hat{J}(\mathbf{q})$ and $\hat{J}_{ab}(\mathbf{q})$ are the Fourier transforms of $J(\mathbf{R})$ and of $J_{ab}(\mathbf{R})$, the latter being non-zero only if $\mathbf{R}$ separates two points on different sublattices].

We can now proceed as in Section II.A, expanding $[1-\hat{K}_\pm(\mathbf{q})]$ in powers of q^2, etc. Note that $\hat{K}_\pm$ need not be isotropic functions of $\mathbf{q}$ (indeed, if the sublattices are $(d-1)$-dimensional layers then $\hat{K}_\pm$ will be functions of $q^{\parallel 2}$ and of $q^{\perp 2}$, $\parallel$ and $\perp$ meaning parallel or perpendicular to the layers). It is clear already from (5E.3) that $\bar{\mathscr{H}}$ resembles the reduced Hamiltonian of a $2n$-component vector model. However, note that the last term is of a new form, which we did not consider in Section V.C.

At zero field, the coefficient of $\mathbf{S}^2_+$, r_+, is smaller than that of $\mathbf{S}^2_-$, r_-, and the anisotropic crossover, described in Section III, occurs. The final critical components are then those of $\mathbf{S}_+$, exhibiting n-component-like exponents.

At non-zero field, we must define new, shifted, spin variables, e.g.

$$\sigma^1_-(\mathbf{q}) \rightarrow \sigma^1_-(\mathbf{q}) + M\delta(\mathbf{q}), \qquad (5E.4)$$

and choose M so that the linear term in $\sigma^1_-(0)$ is eliminated. Including an appropriate rescaling of the spins, so that the coefficient of $\boldsymbol{\sigma}_+(\mathbf{q})\cdot\boldsymbol{\sigma}_+(-\mathbf{q})$ has a term q^2, with a coefficient equal to one, the Hamiltonian finally becomes of the form

$$\bar{\mathscr{H}} = -\tfrac{1}{2}\int_{\mathbf{q}}\Big\{(r^\parallel_+ + q^2)\sigma^1_+(\mathbf{q})\sigma^1_+(-\mathbf{q}) + (r^\perp_+ + q^2)\sum_{\alpha\neq 1}\sigma^\alpha_+(\mathbf{q})\sigma^\alpha_+(-\mathbf{q})$$

$$+ [r_-^{\parallel} + O(q^2)]\sigma_-^1(\mathbf{q})\sigma_-^1(-\mathbf{q}) + [r_-^{\perp} + O(q^2)] \sum_{\alpha \neq 1} \sigma_-^{\alpha}(\mathbf{q})\sigma_-^{\alpha}(-\mathbf{q})\Big\}$$

$$- \int_{\mathbf{q}}\int_{\mathbf{q}'} [w_1\sigma_-^1 \sum_{\alpha} \sigma_-^{\alpha}\sigma_-^{\alpha} + w_2\sigma_+^1 \sum_{\alpha} \sigma_+^{\alpha}\sigma_-^{\alpha} + w_3\sigma_-^1 \sum_{\alpha} \sigma_+^{\alpha}\sigma_+^{\alpha}]$$

$$- \int_{\mathbf{q}}\int_{\mathbf{q}'}\int_{\mathbf{q}''} \sum_{\alpha\beta} [u^+\sigma_+^{\alpha}\sigma_+^{\alpha}\sigma_+^{\beta}\sigma_+^{\beta} + u^-\sigma_-^{\alpha}\sigma_-^{\alpha}\sigma_-^{\beta}\sigma_-^{\beta} + v\sigma_+^{\alpha}\sigma_+^{\alpha}\sigma_-^{\beta}\sigma_-^{\beta}$$

$$+ w\sigma_+^{\alpha}\sigma_-^{\alpha}\sigma_+^{\beta}\sigma_-^{\beta}]. \tag{5E.5}$$

The non-explicit terms of order q^2 may depend on the direction of $\mathbf{q}$. The coefficients $r_+^{\parallel}$, $r_+^{\perp}$, $r_-^{\parallel}$, $r_-^{\perp}$, w_1, w_2 and w_3 are all functions of M, and through it functions of H. The explicit expressions show, that $r_-^{\parallel}$ and $r_-^{\perp}$ will always remain larger than $r_+^{\parallel}$ and $r_+^{\perp}$. Therefore, we shall always have a crossover to a transition in which only $\boldsymbol{\sigma}_+$ may become critical. Choosing the spin rescaling factors as in (3A.14), so that $r_-^{\parallel}$, $r_-^{\perp}$ and the coefficients of $q^2\boldsymbol{\sigma}_+\cdot\boldsymbol{\sigma}_+$ remain unchanged, a few iterations lead to the effective Hamiltonian

$$\bar{\mathscr{H}} = -\tfrac{1}{2}\int_{\mathbf{q}} [r_+^{\parallel} + q^2)\,\sigma_+^1(\mathbf{q})\,\sigma_+^1(-\mathbf{q}) + (r_+^{\perp} + q^2) \sum_{\alpha \neq 1} \sigma_+^{\alpha}(\mathbf{q})\,\sigma_+^{\alpha}(-\mathbf{q})$$

$$+ r_-^{\parallel}\sigma_-^1(\mathbf{q})\,\sigma_-^1(-\mathbf{q}) + r_-^{\perp} \sum_{\alpha \neq 1} \sigma_-^{\alpha}(\mathbf{q})\,\sigma_-^{\alpha}(-\mathbf{q})]$$

$$- \int_{\mathbf{q}}\int_{\mathbf{q}'} [w_2\sigma_+^1 \sum_{\alpha} \sigma_+^{\alpha}\sigma_-^{\alpha} + w_3\sigma_-^1 \sum_{\alpha} \sigma_+^{\alpha}\sigma_+^{\alpha}]$$

$$- \int_{\mathbf{q}}\int_{\mathbf{q}'}\int_{\mathbf{q}''} \sum_{\alpha\beta} u^+\,\sigma_+^{\alpha}\sigma_+^{\alpha}\sigma_+^{\beta}\sigma_+^{\beta}. \tag{5E.6}$$

We can now explicitly integrate over the variables $\boldsymbol{\sigma}_-$. This will lead to a similar expression, without the $\boldsymbol{\sigma}_-$'s and with new coefficients of $\sigma_+^{\alpha}\sigma_+^{\alpha}\sigma_+^{\beta}\sigma_+^{\beta}$, u_{11} ($\alpha = \beta = 1$), u_{12} ($\alpha = 1, \beta \neq 1$) and u_{22} ($\alpha \neq 1, \beta \neq 1$). Explicitly,

$$u_{11} = u^+ - w_2^2/r_-^{\parallel}, \qquad u_{12} = u^+ - \frac{w_2^2}{r_-^{\perp}} - \frac{2w_2w_3}{r_-^{\parallel}}.$$

$$u_{22} = u^+ - w_3^2/r_-^{\parallel}. \tag{5E.7}$$

We thus finally end with an effective n-component Hamiltonian, with anisotropic exchange, and the problem reduces to that discussed in Section III. If the initial exchange Hamiltonian of (5E.2) also includes a uniaxial spin anisotropy of the right sign, then at zero field we shall have an Ising-like transition. Increasing the field, we may reach a stage at which $r_+^{\parallel} = r_+^{\perp}$, and this leads to the coupled $[1 + (n - 1)]$-components model discussed at the end of Section V.C., with a bicritical or tetracritical point. For large enough fields, relations like (5E.7) may drive the transition to become first-order, by moving the initial point in the $u_{\alpha\beta}$ space to a region from which there is a "run-away." This leads to the metamagnetic tricritical behaviour. Since at this stage we have only one effective four-spin parameter, u_{11} or u_{12}, the tricritical exponents are Gaussian.

F. Compressible magnets

There have been many recent renormalization group studies of the problem of compressible magnets (Wegner, 1974; Sak, 1974; Rudnick *et al.*, 1974; Imry, 1974; Aharony, 1973g). In this review, we shall follow mainly the approach of Sak, as modified by Imry.

We start with an isotropic system. The continuous-spin Hamiltonian, including elastic degrees of freedom, may be written in the form

$$\bar{\mathscr{H}}_0 = -\int d^d x \Big\{ \tfrac{1}{2}[r_0 \mathbf{S}^2 + (\nabla \mathbf{S})^2] + u_0 \mathbf{S}^4 + (\tfrac{1}{2}K - \frac{1}{d}\mu)\Big(\sum_{\alpha=1}^{d} u_{\alpha\alpha}\Big)^2 + \mu \sum_{\alpha,\beta} u_{\alpha\beta}{}^2 + g\mathbf{S}^2 \sum_{\alpha} u_{\alpha\alpha} \Big\}, \tag{5F.1}$$

where $\mathbf{u}(\mathbf{x})$ is the displacement vector, and

$$u_{\alpha\beta} = \tfrac{1}{2}\left(\frac{\partial u_\alpha}{\partial x_\beta} + \frac{\partial u_\beta}{\partial x_\alpha} + \sum_\gamma \frac{\partial u_\gamma}{\partial x_\alpha}\frac{\partial u_\gamma}{\partial x_\beta}\right). \tag{5F.2}$$

The last term in (5F.1) represents the spin–lattice coupling. We now divide $u_{\alpha\beta}$ into a static uniform deformation, $u^0_{\alpha\beta}$, and a phonon part:

$$u_{\alpha\beta}(\mathbf{x}) = u^0_{\alpha\beta} + \frac{i}{V}\sum_{k\neq 0} k_\beta u_\alpha(\mathbf{k}) \exp(i\mathbf{k}\cdot\mathbf{x}) + O(u_\alpha^2). \tag{5F.3}$$

We first consider the case of constant homogeneous static strain (or constant volume, $V = V_0$). In this case, the $u^0_{\alpha\alpha}$'s are constant, and we integrate only over the variables $u_\alpha(\mathbf{k})$ and the shear part of the tensor $u^0_{\alpha\beta}$ in the partition function. The resulting magnetic partition function has the form $\int \exp(\bar{\mathscr{H}})$, with

$$\bar{\mathscr{H}} = -\tfrac{1}{2}\int_{\mathbf{q}} (r + q^2)\,\boldsymbol{\sigma}_{\mathbf{q}}\cdot\boldsymbol{\sigma}_{-\mathbf{q}} - u\int_{\mathbf{q}}\int_{\mathbf{q}'}\int_{\mathbf{q}''} (\boldsymbol{\sigma}_{\mathbf{q}}\cdot\boldsymbol{\sigma}_{\mathbf{q}'})(\boldsymbol{\sigma}_{\mathbf{q}''}\cdot\boldsymbol{\sigma}_{-\mathbf{q}-\mathbf{q}'-\mathbf{q}''}) - \frac{v}{V}\left[\int_{\mathbf{q}} \boldsymbol{\sigma}_{\mathbf{q}}\cdot\boldsymbol{\sigma}_{-\mathbf{q}}\right]^2, \tag{5F.4}$$

where

$$u = u_0 - \tfrac{1}{4}g^2\Big/\left(\tfrac{1}{2}K + \frac{d-1}{d}\mu\right), \qquad v = \tfrac{1}{4}g^2\Big/\left(\tfrac{1}{2}K + \frac{d-1}{d}\mu\right). \tag{5F.5}$$

For constant pressure, P, we have to add to $\mathscr{H}_0$ a term (Imry, 1974)

$$-PV = -PV_0\left[1 + \sum_{\alpha=1}^{d} u^0_{\alpha\alpha} + \sum_{\alpha\neq\beta} u^0_{\alpha\alpha}u^0_{\beta\beta} + O(u^3_{\alpha\alpha})\right], \tag{5F.6}$$

and then integrate over the variables $u^0_{\alpha\beta}$. The effect of this is only to change the coefficient v in (5F.4), yielding

$$v = \tfrac{1}{4}g^2\left[\left(\tfrac{1}{2}K + \frac{d-1}{d}\mu\right)^{-1} - \left(\tfrac{1}{2}K + \frac{d-1}{d}P\right)^{-1}\right] \tag{5F.7}$$

In both cases, the critical behaviour can be analysed by considering the effective spin Hamiltonian (5F.4). The renormalization group recursion relations are straightforward, yielding four fixed points: Gaussian ($u^{\mathrm{G}} = v^{\mathrm{G}} = 0$), Heisenberg, $[v^{\mathrm{H}} = 0,\ 4K_d u^{\mathrm{H}} = \varepsilon/(n+8) + O(\varepsilon^2)]$,

"spherical," $$u^{\mathrm{S}} = 0,\ 4K_d v^{\mathrm{S}} = \varepsilon/n + O(\varepsilon^2) \tag{5F.8}$$

and

"renormalized," $$4K_d u^{\mathrm{R}} = \varepsilon/(n+8) + O(\varepsilon^2)$$
$$4K_d v^{\mathrm{R}} = (4-n)\,\varepsilon/n\,(n+8) + O(\varepsilon^2). \tag{5F.9}$$

The fixed point (5F.8) was named "spherical" (Rudnick *et al.*, 1974), since it is directly related to the spherical model: the Hamiltonian (5F.4), with $u = 0$, is equivalent to a constrained Gaussian model, and thus to the spherical model. Indeed, the ε-expansion of the exponents at this fixed point are identical with those of the spherical model.

The fixed point (5F.9) was named "renormalized", since the critical exponents associated with it can be shown to be the renormalized exponents of the n-component Heisenberg fixed point (Fisher, 1968). Again, there exist general arguments, independent of the ε-expansion, which yield this result (Rudnick *et al.* 1974; Sak, 1974).

The Hamiltonian (5F.4) is very similar to the cubic Hamiltonian we considered in Section V.A. In particular, the last term has the structure of energy–energy coupling. Thus, it follows that the crossover exponent away from the Heisenberg fixed point is exactly α_{H} [see (5A.15)]. For the Ising case, $\alpha_{\mathrm{I}} > 0$, and this fixed point is unstable. The flow is either to the "renormalized" fixed point, if $v > 0$, or to a first-order transition, if $v < 0$. The initial sign of v depends, of course, on many irrelevant variables. In particular, (5F.5) indicates that a second-order transition, with renormalized exponents, is possible for constant volume, and—by (5F.7)—for high pressures ($p > \mu$). Ignoring irrelevant variables, the point $P = P_t = \mu$ seems to be a tricritical point.†

For the XY or Heisenberg cases, $\alpha < 0$, and the unrenormalized fixed point ($v^* = 0$) becomes stable. The "renormalized" fixed point now has $v^* < 0$, and represents a tricritical point. For a large range of values of v, the original unrenormalized critical behaviour is observed, with corrections to scaling involving $|t|^{-\alpha}$. For large negative values of v, a first-order transition may occur.

Up to this point, we assumed that $u > 0$. From (5F.5) it is clear that u may become negative for strong spin lattice coupling (large g). Again, this may lead to a first-order transition (Aharony, 1973g). For $u = 0$, the spherical fixed point will describe the tricritical behaviour. The Gaussian fixed point is thus a critical point of order four, to be reached only if both u and v are zero, and appropriate sixth-order terms are introduced.

The picture given above was idealized, as we included in (5F.1) only fully isotropic interactions, both in the spins and in the elastic degrees of freedom. For cubic lattices, (5F.1) should be replaced by (Aharony and Bruce, 1974a)

$$\begin{aligned}\bar{\mathscr{H}} = -\int_{\mathbf{x}}\Big\{&\tfrac{1}{2}[r_0\mathbf{S} + (\nabla\mathbf{S})^2] + u_0\mathbf{S}^4 + v_0\sum_\alpha S_\alpha^4\\ &+\tfrac{1}{2}C_{11}\sum_\alpha u_{\alpha\alpha}^2 + C_{12}\sum_{\alpha<\beta} u_{\alpha\alpha}u_{\beta\beta} + \tfrac{1}{2}C_{44}\sum_{\alpha<\beta} u_{\alpha\beta}^2\\ &- g_1\sum_\alpha u_{\alpha\alpha}S_\alpha^2 - g_2\sum_\alpha u_{\alpha\alpha}(\mathbf{S}^2 - S_\alpha^2) - g_3\sum_{\alpha<\beta} u_{\alpha\beta}S_\alpha S_\beta\Big\}.\end{aligned} \tag{5F.10}$$

This reduces to (5F.1) only if $C_{11} = C_{12} + 2C_{44}$, and $g_3 = 0$, $g_1 = g_2$. We can now substitute (5F.3), and integrate over $u_\alpha(\mathbf{k})$. This generates many additional quartic spin terms, e.g.

$$\sum_\alpha\left[\int_{\mathbf{q}}\sigma_{\mathbf{q}}^\alpha\sigma_{-\mathbf{q}}^\alpha\right]^2, \quad \sum_{\alpha\beta}\int_{\mathbf{q}}\sigma_{\mathbf{q}}^\alpha\sigma_{-\mathbf{q}}^\beta\int_{\mathbf{p}}\sigma_{\mathbf{p}}^\alpha\sigma_{-\mathbf{p}}^\beta, \ldots. \tag{5F.11}$$

These were ignored in previous discussions of (5F.10). Clearly, an analysis of all possible fixed points involving these quartic terms is quite lengthy, and is left for the future. A partial discussion is given by de Moura *et al.* (1976).

Superimposed on the symmetry of the lattice, one can then add an external anisotropic stress, $\sigma_{\alpha\beta}$, leading to an additional energy of the form

$$\bar{\mathscr{H}}' = -\sum_{\alpha\beta}\sigma_{\alpha\beta}u_{\alpha\beta}. \tag{5F.12}$$

† However, the harmonic approximation (5F.1) breaks down at this pressure, and the lattice becomes mechanically unstable (Halperin, 1975, private communication). Thus, a more precise treatment for high pressure is needed.

Integration over $u_{\alpha\beta}$ will now lead to spin anisotropies, and thus to bicritical or tetracritical behaviour (Aharony and Bruce, 1974a; Bruce and Aharony, 1975). Since we find that the XY and Heisenberg fixed points are stable against the interaction generated by the elastic degrees of freedom, it is reasonable to expect that the existing analysis of bicritical and tetracritical points (e.g. Bruce and Aharony, 1975), in which these are described as Heisenberg-like fixed points, is still valid.

G. Systems with larger "spin" dimensionality

In most magnetic cases, the number of spin components, n, is not larger than 3. It is clear from the above discussion that larger values of n will lead to interesting effects, since the isotropic Heisenberg fixed point becomes unstable for $n > n_c$ ($\simeq 3{\cdot}13$ at $d = 3$). The search for systems with large n can go in several directions. One possibility is to try to find antiferromagnets for which several Fourier transforms $\{\boldsymbol{\sigma}_{\mathbf{q}_i}, i = 1, \ldots, n\}$ become critical simultaneously, thus yielding an example of the nm-vector model considered in Section V.C (Alben, 1974). This may typically involve $n = 4$, $m = 3$, so that the effective number of spin components may be as high as 12. Similar situations may in principle occur in structural phase transitions, when several modes become soft simultaneously (Mukamel, 1975).

A much wider scope of possibilities involves tensor order parameters. For example, the p-wave superconducting transition of ^{3}He has an order parameter which is a 3×3 complex matrix, and thus $n = 18$! However, these are well described by mean field theory.

Another example involves the transition from the isotropic phase to the nematic phase in liquid crystals, recently considered by Lubensky and Priest (1974). In this case, the order parameter is an $m \times m$ symmetric traceless tensor Q_{ij}, and the reduced Hamiltonian is of the form

$$\overline{\mathscr{H}} = -\int \mathrm{d}^d x \, \{\tfrac{1}{4} r\, Q_{ij} Q_{ij} + \tfrac{1}{4} \nabla_i Q_{jk} \nabla_i Q_{jk} + w Q_{ij} Q_{jk} Q_{ki} + \tfrac{1}{32} u (Q_{ij} Q_{ij})^2 + \tfrac{1}{32} v Q_{ij} Q_{jk} Q_{kl} Q_{li}\}. \tag{5G.1}$$

For $w \neq 0$, this leads to a first-order transition. However, if w is zero, Lubensky and Priest analyse the second-order transition by writing down recursion relations in r, u and v. As before, these yield four fixed points: Gaussian, n-Isotropic, with $n = \frac{1}{2}(m^2 + m) - 1$, and two additional ones, with u and v given by complicated irrational functions of n (both of order ε). The "isotropic" fixed point is stable for $n < n_c$. One of the other fixed points is stable for $n_c < n \lesssim n_d$ with $n_d \simeq 7{\cdot}38 + O(\varepsilon)$, and no stable fixed point is found for $n > n_d$. In this region, one probably has a first order transition. Note that for $m = 3$, the two quartic terms are proportional, and no new behaviour is found.

However, for $m = d = 4 - \varepsilon$ we have $n = 9 + O(\varepsilon)$, and a first-order transition is probably to be expected.

For small values of w, one can still use the above formalism, and study the variation of w under iterations. This has the form

$$w' = b^{\lambda_w} w, \tag{5G.2}$$

with $\lambda_w = 1 + O(\varepsilon)$. Thus, w is a relevant variable, with a crossover exponent $\phi_w = \nu\lambda_w = \frac{1}{2} + O(\varepsilon)$. Unlike the usual case, this crossover is to a first-order transition, and not to a new fixed point. A similar situation occurs in many of the problems considered in this chapter, when the initial Hamiltonian is outside the region of attraction of the stable fixed point (e.g. $u < 0$ in the cubic problem, Fig. 3). For small enough w, when $w|t|^{-\phi_w}$ is small, we can still see second-order effects, e.g. a large correlation length, etc. Only as $t \to 0$, when this variable diverges, will the first-order effects emerge. A similar effect occurs in the Potts model (Golner, 1973).

The Hamiltonian (5G.1) can now be generalized, by the introduction of electric and magnetic fields (Priest, 1974). These introduce terms like $Q_{ij}H_iH_j$ and $Q_{ij}E_iE_j$, and may lead to crossover to a tricritical behaviour, probably of the Riedel–Wegner type (Riedel and Wegner, 1972), or to lower effective values of n.

Another example which involves many order parameters is the spectic-A phase of liquid crystals. Here we have both the fluctuations in density, represented by a complex order parameter ψ, and the fluctuations in the director vector $\mathbf{n}$. The Hamiltonian for this system was recently analysed by Halperin *et al.* (1974), and by Halperin and Lubensky (1974). In the nematic phase, $\mathbf{n}$ is "ordered," and one can integrate over the $\mathbf{n}$-degrees of freedom. This generates a cubic term in $|\psi|$, and thus leads to a first-order nematic–smectic A transition, similar to the one we mentioned above. Alternatively, Halperin *et al.* (1974) generalized the problem to that of an $n/2$-dimensional complex order parameter, and performed the renormalization group iterations over both ψ and $\mathbf{n}$. The result yielded a non-trivial fixed point only for $n > n_X$, with $n_X = 365{\cdot}9 + O(\varepsilon)$. Again, for $n < n_X$ there is no fixed point, and the transition is presumably of first order. The same theory applies to the phase transition in superconductors, where the director vector $\mathbf{n}$ is replaced by the vector potential.

VI. Spatial Anisotropies

A. Lattice anisotropies

A lattice anisotropy leads to different exchange interactions in different directions in space. In particular, consider a system of d-dimensional hyperplanes of ferromagnetically coupled spins, which form a $\bar{d}$-dimensional lattice by interacting with adjacent hyperplanes (Chang and Stanley, 1973; Grover, 1973). We can write the Hamiltonian in the form

$$\overline{\mathscr{H}} = -\sum_m \int d^d x \left\{ \tfrac{1}{2}[r\mathbf{S}_m^2 + (\nabla \mathbf{S}_m)^2] + u\mathbf{S}_m^4 - R\sum_i \mathbf{S}_m \cdot \mathbf{S}_{m+i} \right\}, \tag{6A.1}$$

where m labels the hyperplanes, i labels "nearest neighbour" hyperplanes and the integration is over the d-dimensional hyperplanes. At $R = 0$, $\overline{\mathscr{H}}$ separates into noninteracting d-dimensional Hamiltonians, yielding d-dimensional critical exponents. The recursion relation for R, near $R = 0$, is exactly

$$R' = b^{2-\eta} R, \tag{6A.2}$$

resulting directly from the rescaling of the spins, (2B.6), (2D.7). Thus, the crossover exponent is exactly

$$\phi_R = \nu(2 - \eta) = \gamma, \tag{6A.3}$$

where γ is the d-dimensional exponent.†

For any non-zero R, (6A.2) will lead to $R \to \infty$, and a different spin rescaling is necessary. It is easy to show that (6A.1) is equivalent to

$$\mathscr{H} = -\tfrac{1}{2}\int_{\mathbf{q}} \left[r + q^2 - f \sum_{\alpha = d+1}^{d+\bar{d}} (q^\alpha)^2 \right] \boldsymbol{\sigma}_{\mathbf{q}} \cdot \boldsymbol{\sigma}_{-\mathbf{q}} + u \int_{\mathbf{q}_1}\int_{\mathbf{q}_2}\int_{\mathbf{q}_3} \left(\boldsymbol{\sigma}_{\mathbf{q}_1} \cdot \boldsymbol{\sigma}_{\mathbf{q}_2} \right)\left(\boldsymbol{\sigma}_{\mathbf{q}_3} \cdot \boldsymbol{\sigma}_{-\mathbf{q}_1 - \mathbf{q}_2 - \mathbf{q}_3} \right), \tag{6A.4}$$

where the q-integrals are now $(d + \bar{d})$-dimensional, $(1 - f)$ is proportional to R and the last $\bar{d}$ directions are perpendicular to the interacting hyperplanes (Bruce, 1974a). A simple coordinate rescaling $[(1 - f)(q^\alpha)^2 \to (q^\alpha)^2$ for $\alpha = d + 1, \ldots, d + \bar{d}]$ yields the standard $(d + \bar{d})$ dimensional Hamiltonian, which gives the usual $(d + \bar{d})$-dimensional exponents.‡ Thus, d-dimensional behaviour crosses over to $(d + \bar{d})$ dimensional behaviour.

B. Semi-infinite systems

Up to this point, we have mentioned only infinite systems, so that we have had full translational invariance. Lubensky and Rubin (1973, 1975) recently considered the problem of semi-infinite systems. The main difference here is, that one has to introduce a boundary condition on the plane which bounds the system, e.g.

$$\mathbf{S}(x_\perp = 0, \mathbf{x}_\parallel) = 0, \tag{6B.1}$$

† This could also be derived using general dimensional arguments, similar to those we used in proving (5A.14).

‡ The possiblity of this rescaling means that f is a marginal operator, as long as $f \neq 1$.

where $x_\perp$ measures the coordinate perpendicular to this plane. Equation (2A.9) thus has to be replaced by

$$\mathbf{S}(\mathbf{x}) = \sqrt{2}\int \mathrm{d}^d q\, \boldsymbol{\sigma}_{\mathbf{q}} \exp(i\mathbf{q}_{\|}\cdot\mathbf{x}_{\|}) \sin q_\perp x_\perp, \tag{6B.2}$$

where $\mathbf{q} = (q_\perp, \mathbf{q}_{\|})$, and the Hamiltonian assumes the form

$$\mathscr{H} = -\tfrac{1}{2}\int_{\mathbf{q}} (r + q^2)(\boldsymbol{\sigma}_{\mathbf{q}}\cdot\boldsymbol{\sigma}_{-\nu\mathbf{q}}) - u\int_{\mathbf{q}_1}\int_{\mathbf{q}_2}\int_{\mathbf{q}_3}\int_{\mathbf{q}_4} (\boldsymbol{\sigma}_{\mathbf{q}_1}\cdot\boldsymbol{\sigma}_{\mathbf{q}_2})(\boldsymbol{\sigma}_{\mathbf{q}_3}\cdot\boldsymbol{\sigma}_{\mathbf{q}_4})\,\delta^d(\mathbf{q}_1, \mathbf{q}_2, \mathbf{q}_3, \mathbf{q}_4), \tag{6B.3}$$

where $\nu\mathbf{q} = (-q_\perp, \mathbf{q}_{\|})$,

$$\delta^d(\mathbf{q}_1, \mathbf{q}_2, \mathbf{q}_3, \mathbf{q}_4) = \delta^{d-1}(\mathbf{q}_{1_{\|}} + \mathbf{q}_{2_{\|}} + \mathbf{q}_{3_{\|}} + \mathbf{q}_{4_{\|}})\,\Delta(q_{1_\perp}, q_{2_\perp}, q_{3_\perp}, q_{4_\perp}) \tag{6B.4}$$

and

$$\Delta(q_{1_\perp}, q_{2_\perp}, q_{3_\perp}, q_{4_\perp}) = \frac{2}{\pi}\sum_{x_\perp > 0} \sin q_{1_\perp} x_\perp \sin q_{2_\perp} x_\perp \sin q_{3_\perp} x_\perp \sin q_{4_\perp} x_\perp. \tag{6B.5}$$

The renormalization group iteration procedure is now exactly the same, with the q-integrals being over the cylinders $\Lambda/b < |\mathbf{q}| < \Lambda$, $\Lambda/b < |q_\perp| < \Lambda$. These generate a non-local spin Hamiltonian, which then makes it necessary to replace the constant spin rescaling (2B.6) by two steps, the first eliminating the non-local two-spin interaction and the second rescaling the spin length. This leads to a more complicated behaviour of the correlation function as function of $q_\perp$ and $q_{\|}$. Otherwise, the bulk exponents, determined from the recursion relation for r, remain unchanged. A study of the peculiar dependence of the correlation function on $\mathbf{q}$ leads to a determination of various exponents related to the surface critical behaviour.

VII. Discussion

The critical behaviour of a system with a complicated Hamiltonian is described by a series of crossovers, determined by an arrangement of the symmetry-breaking interactions in a decreasing order. For example, an anisotropic Ising-like dipolar system may have various routes to its ultimate critical behaviour: Mean field (Gaussian) → Isotropic Heisenberg (short-range) → Isotropic dipolar → XY dipolar → Ising dipolar, Mean field → Ising short-range → Ising dipolar, etc. (see Fisher, 1974a, for a pictorial representation). In order to be able to perform a quantitative analysis of these crossover processes, it is essential to know the crossover function, e.g. f_0 in (2B.21). A full analysis of this function, for all the possible parameters, remains a task for future research. However, preliminary ε-expansions of this function already exist (Nelson, 1975; Bruce *et al.*, 1975), and it is reasonable to expect better quantitative tools for a study of these crossovers in the near future.

In the meantime, the best way to obtain the true asymptotic critical exponents is to perform the measurements as close to T_c as possible, and to try to correct for crossover effects by introducing into the analysis corrections to scaling, such as those given in (2B.21). Near the true asymptotic fixed point, all parameters except the temperature and the magnetic field are irrelevant, and should lead to such corrections.

We next address ourselves to the numerical accuracy of the ε-expansion, when used at $\varepsilon = 1$. For the isotropic short-range critical exponents, truncation of the ε-expansion after ε^2 gives the best fit to the high temperature results at $\varepsilon = 1$ (Brézin *et al.*, 1973). This may be explained by the asymptotic nature of the ε-expansion (Wilson, 1972). It is not at all clear that the same truncation should work for other problems, e.g. the dipolar problem. In fact, experiments on the isotropic dipolar ferromagnet EuO (Kornblit and Ahlers, 1975) give $\alpha \simeq -0{\cdot}04$, in agreement with the ε-expansion truncated after the first order (see Table II)!†

A related problem arises when one tries to identify the most stable fixed point [e.g. in the cubic problem, on the basis of (5A.17)], or the actual Hamiltonian flows to that fixed point. At $d = 3$, one must include many irrelevant variables in the initial stages of iteration, and only then can one decide if a given Hamiltonian is in the range of attraction of a given stable fixed point, Thus, Figs 3 and 4 are quite oversimplified.

One must thus be cautioned against using simplified truncated ε-expansions at $\varepsilon = 1$. An ultimate answer to the above questions (what are the exact dipolar exponents at $d = 3$? Is the cubic fixed point stable at $n = 3$, $d = 3$? Will an Ising random magnet have a "smeared" transition or "random" exponents? etc.) will be reached only by a direct renormalization group procedure, appropriate for $d = 3$ (probably similar to the one described by Van Leewen and Niemeijer in this volume, by Kadanoff and Houghton, 1975, or by Golner and Riedel, 1975).

Should one thus conclude that the renormalization group analysis near four dimensions is useless? Obviously, my answer is negative. Using the ε-expansion, one is now able to identify systems which belong to different universality classes, by looking at their microscopic Hamiltonians. It is reasonable to

† However, these experiments are probably also consistent with $\alpha \simeq -0{\cdot}10$ (Ahlers and Kornblit, 1975), so that it is premature to make final conclusions.

expect that this identification, or the number of possible fixed points, will not change drastically as ε grows towards one. One can also discuss trends in the change of critical exponents going from one fixed point to another (e.g., the dipolar exponent γ is slightly larger than the non-dipolar one, at the same order in ε; the "random" ("mixed") value of α is negative, while the "pure" Ising value is positive, etc.). It will now be very helpful if many (real or numerical) experiments are performed, on systems which might be expected to have new exponents, so that a better feeling of the usefulness of the ε-expansion will be obtained.

Some conclusions from the renormalization group are not limited to the vicinity of four dimensions. In particular, we mention the dipolar Ising-like system (Section IV.E), which at $d = 3$ should exhibit classical exponents with logarithmic corrections. Indeed, measurements on $LiTbF_4$ (Als-Nielsen *et al.*, 1974, Ahlers *et al.*, 1975) seem to confirm these predictions. Similarly, relations like (4C.9), (5A.14), (5C.9). (6A.3) etc. are also independent of the ε-expansion.

In conclusion, although the accuracy of the numerical results from the ε-expansion at $\varepsilon = 1$ is not always clear, the recursion relations study near $d = 4$ still gives a deep insight into the various possible types of critical behaviour, and into the crossovers between them.

Acknowledgments

The author is very grateful to Professor P. C. Martin and Professor H. Ehrenreich for their hospitality at Harvard University, to Professor M. E. Fisher, Professor K. G. Wilson, Dr. A. D. Bruce, Dr. D. R. Nelson and Professor A. H. Luther for many fruitful discussions and to many colleagues (especially Drs. G. Ahlers, V. J. Emery, D. R. Nelson, A. H. Luther, T. C. Lubensky, S. Hikami, Y. Imry, J. Sak, D. J. Wallace and others) for preprints of their work prior to publication. Comments from Drs. B. I. Halperin, M.E. Fisher, T. C. Lubensky, and E. K. Riedel are also gratefully acknowledged.

References

Abe, R. (1972). *Prog. Theor. Phys.* **48**, 1414.
Aharony, A. (1973a). *Phys. Letters* A**44**, 313.
Aharony, A. (1973b). *Phys. Rev.* **B8**, 3342.
Aharony, A. (1973c). *Phys. Rev.* **B8**, 3349.
Aharony, A. (1973d). *Phys. Rev.* **B8**, 3358.
Aharony, A. (1973e). *Phys. Rev.* **B8**, 3363; **B9**, 3946(E).
Aharony, A. (1973f). *Phys. Rev.* **B8**, 4270.
Aharony, A. (1973g). *Phys. Rev.* **B8**, 4314.
Aharony, A. (1973h). *Phys. Rev. Letts* **31**, 1494.

Aharony, A. (1974a), *Proc. 19th Conf. on Magnetism and Magnetic Materials* [C. D. Graham, Jr., and J. J. Rhyne, eds.], AIP Conf. Proc. Series **18**, 863.
Aharony, A. (1974b). *Phys. Rev.* B**9**, 2416.
Aharony, A. (1974c). *J. Phys.* C**7**, L63.
Aharony, A. (1974d). *Phys. Rev.* B**10**, 3006.
Aharony, A. (1974e). *Phys. Letters* A**49**, 221.
Aharony, A. (1974f). Paper presented at the Conference on Critical Phenomena in multicomponent systems, Athens, Ga.
Aharony, A. (1975). *Phys. Rev.* B**12** 1038.
Aharony, A. and Bruce, A. D. (1974a). *Phys. Rev. Letts* **33**, 427.
Aharony, A. and Bruce, A. D. (1974b). *Phys. Rev.* B**10**, 2973.
Aharony, A. and Fisher, M. E. (1973). *Phys. Rev.* B**8**, 3323.
Aharony, A. and Imry, Y. (1974). *Bull. Am. Phys. Soc.* **19**, 306.
Aharony, A., Imry, Y. and Ma, S. K. (1976). *Phys. Rev.* B**13**, 466.
Ahlers, G. (1973). *In* Proceedings of the Conference on Renormalization Group in Critical Phenomena, Temple University, May 1973 (J. D. Gunton and M. S. Green, eds.). p. 137; *Phys. Rev.* A**8**, 530.
Ahlers, G. and Kornblit, A. (1975). *Phys. Rev.* B**12**, 1938.
Ahlers, G., Kornblit, A. and Guggenheim, H. J. (1975) *Phys. Rev. Letts* **34**, 1227.
Alben, R. (1974). *C.R. Acad. Sc. Paris* **279**B, 111.
Als-Nielsen, J., Holmes, L. M. and Guggenheim, H. J. (1974). *Phys. Rev. Letts* **32**, 610.
Baxter, R. J. (1971). *Phys. Rev. Letts* **26**, 832.
Brézin, E., Le Guillou, J. C., Zinn-Justin, J. and Nickel, B. G. (1973). *Phys. Letters* **44**A, 227.
Brézin, E., Le Guillou, J. C. and Zinn-Justin, J. (1974). *Phys. Rev.* B**10**, 892.
Brézin, E. and Zinn-Justin, J. (1975). Preprint.
Bruce, A. D. (1974a). *J. Phys.* C**7**, 2089.
Bruce, A. D. (1974b). *Phys. Letters.* A**48**, 317.
Bruce, A. D. and Aharony, A. (1974). *Phys. Rev.* B**10**, 2078.
Bruce, A. D. and Aharony, A. (1975). *Phys. Rev.* B**11**, 478.
Bruce, A. D., Droz, M. and Aharony, A. (1974). *J. Phys.* C**7**, 3673.
Bruce, A. D., Kosterlitz, J. M. and Nelson, D. R. (1975). *Proc. 20th Conference on Magnetism and Magnetic Materials*, AIP Conf. Proc., **24**, 321.
Burkhart, T. W. and Gunton, J. D. (1974). *Phys. Rev.* A**9**, 2802.
Chang, T. S. and Stanley, H. E. (1973). *Phys. Rev.* B**8**, 4435.
Cohen, M. H. and Keffer, F. (1974). *Phys. Rev.* B**10**, 787 (E).
Cowley, R. A. and Bruce, A. D. (1973). *J. Phys.* C**6**, L191.
de Moura, M., Lubensky, T. C., Imry, Y. and Aharony, A. (1976). *Phys. Rev.* B**13**, 2176.
Emery, V. J. (1975). *Phys. Rev.* B**11**, 239.
Fisher, M. E. (1968). *Phys. Rev.* **176**, 257.
Fisher, M. E. (1971). *In* "Critical Phenomena", *Proc. 51st Enrico Fermi Summer School* (M. S. Green, ed.) Academic Press, New York, p. 1.
Fisher, M. E. (1974a). *Rev. Mod. Phys.* **46**, 597.
Fisher, M. E. (1974b). Reported at the 31st Yeshiva University Statistical Mechanics Meeting, May 1974, and unpublished.
Fisher, M. E. and Aharony, A. (1973a). *Phys. Rev. Letts* **30**, 559.
Fisher, M. E. and Aharony, A. (1973b). *Phys. Rev. Letts* **31**, 1238, 1537(E).
Fisher, M. E. and Aharony, A. (1974). *Phys. Rev.* B**10**, 2812.
Fisher, M. E. and Jasnow, D. (1975). Vol. 4 of this series.
Fisher, M. E. and Nelson, D. R. (1974). *Phys. Rev. Letts* **32**, 1350.

Fisher, M. E. and Pfeuty, P. (1972). *Phys. Rev.* **B6**, 1889.
Fisher, M. E., Ma, S.-K. and Nickel, B. G. (1972). *Phys. Rev. Letts* **29**, 917.
Golner, G. R. (1973). *Phys. Rev. B***8**, 3419.
Golner, G. R. and Riedel, E. K. (1975). *Phys. Rev. Letts* **34**, 856.
Grinstein, G. (1974). Ph.D. thesis, Harvard University.
Grinstein, G. and Luther, A. (1976). *Phys. Rev.* **B13**, 1329.
Grover, M. K. (1973). *Phys. Letters* **A44**, 253.
Grover, M. K., Kadanoff, L. P. and Wegner, F. J. (1972). *Phys. Rev.* **B6**, 311.
Halperin, B. I. and Lubensky, T. C. (1974). *Sol. St. Comm.* **14**, 997.
Halperin, B. I., Lubensky, T. C. and Ma, S.-K. (1974). *Phys. Rev. Letts* **32**, 292.
Harris, A. B. and Lubensky, T. C. (1974). *Phys. Rev. Letts* **33**, 1540.
Harris, R., Plischke, M. and Zimmerman, M. J. (1973). *Phys. Rev. Letts* **31**, 160.
Hikami, S. and Abe, R. (1974). *Prog. Theor. Phys.* **52**, 369.
Imry, Y. (1974). *Phys. Rev. Letts.* **33**, 1304.
Kadanoff, L. P. (1966). *Physics* **2**, 263.
Kadanoff, L. P. and Houghton, A. (1975). *Phys. Rev.* **B11**, 377.
Ketley, I. J. and Wallace, D. J. (1973). *J. Phys.* **A6**, 1667.
Khmel'nitskii, D. E. (1974). Preprint.
Kornblit, A. and Ahlers, G. (1975). *Phys. Rev.* **B11**, 2678.
Larkin, A. I. and Khmel'nitskii, D. E. (1969). *Zh. Eksp. Teor. Fiz.* **56**, 2087 [*Sov. Phys. JETP* **29**, 1123].
Liu, L. L. (1973). *Phys. Rev. Lett.* **31**, 459.
Lubensky, T. C. (1975). *Phys. Rev.* **B11**, 3573.
Lubensky, T. C. and Priest, R. G. (1974). *Phys. Letters* **A48**, 103.
Lubensky, T. C. and Rubin, M. H. (1973). *Phys. Rev. Letts* **31**, 1469.
Lubensky, T. C. and Rubin, M. H. (1975). *Phys. Rev.* **B11**, 4533.
Ma, S.-K. (1973). *Rev. Mod. Phys.* **45**, 589.
Moore, M. A., Saul, D. M. and Wortis, M. (1974). *J. Phys.* **C7**, 162.
Mukamel, D. (1975). *Phys. Rev. Letts* **34**, 481.
Nauenberg, M. and Nienhuis, B. (1974). *Phys. Rev. Letts* **33**, 1598.
Nelson, D. R. (1975). *Proc. 20th Conference on Magnetism and Magnetic Materials*, AIP Conf. Proc., **24**, 319, and to be published.
Nelson, D. R. and Fisher, M. E. (1975). *Phys. Rev.* **B11**, 1030; **B12**, 263.
Nelson, D. R., Kosterlitz, J. M. and Fisher, M. E. (1974). *Phys. Rev. Letts* **33**, 813.
Nelson, D. R. and Rudnick, J. (1975). *Phys. Rev. Letts* **35**, 178.
Oppermann, R. (1974a). *Phys. Letters* **A47**, 383.
Oppermann, R. (1974b). *J. Phys.* **A7**, L137.
Pfeuty, P., Jasnow, D. and Fisher, M. E. (1974). *Phys. Rev.* **B10**, 2088.
Priest, R. G. (1974). *Phys. Letters* **A47**, 475.
Riedel, E. K. and Wegner, F. J. (1969). *Z. Phys.* **225**, 195.
Riedel, E. K. and Wegner, F. J. (1972). *Phys. Rev. Letts* **29**, 349.
Rudnick, J. (1973). Unpublished.
Rudnick, J., Bergman, D. J. and Imry, Y. (1974). *Phys. Letters* **A46**, 449.
Sak, J. (1973). *Phys. Rev.* **B8**, 281.
Sak, J. (1974). *Phys. Rev.* **B10**, 3957.
Suzuki, M. (1973a). *Prog. Theor. Phys.* **49**, 424, 1106, 1440.
Suzuki, M. (1973b). *Prog. Theor. Phys.* **49**, 1451.
Wallace, D. J. (1973). *J. Phys.* **C6**, 1390.
Wegner, F. J. (1972a). *Phys. Rev.* **B5**, 4529.
Wegner, F. J. (1972b). *Phys. Rev.* **B6**, 1891.

Wegner, F. J. (1973). *Proc. 8th Finnish Summer School in Theoretical Solid State Physics*, August 1973.
Wegner, F. J. (1974). *J. Phys.* C**7**, 2109.
Wegner, F. J. and Houghton, A. (1973). *Phys. Rev.* A**8**, 401.
Wegner, F. J. and Riedel, E. K. (1973). *Phys. Rev.* B**7**, 248.
Wilson, K. G. (1971). *Phys. Rev.* B**4**, 3174, 3184.
Wilson, K. G. (1972). *Phys. Rev. Letts* **28**, 548.
Wilson, K. G. and Fisher, M. E. (1972). *Phys. Rev. Letts* **28**, 240.
Wilson, K. G. and Kogut, J. (1974). *Phys. Reports* C**12**, 77.
Yamazaki, Y. (1974). *Phys. Letters.* A**49**, 215.

7. Renormalization Theory for Ising-like Spin Systems

Th. Niemeijer and J. M. J. van Leeuwen

Laboratorium voor Technische Natuurkunde, Technische Hogeschool, Delft, the Netherlands

Introduction

Most of the applications of the renormalization theory, as proposed by Wilson (Wilson, 1971a, b; Wilson and Fisher, 1972; Wilson and Kogut,

1974) for many-body systems, concern lattices with classical (i.e. continuous), n-component spin variables, where $\varepsilon = 4 - d$ (d is the dimensionality of the lattice) has to be used as an expansion parameter for physical quantities. Here we shall be concerned with spins which can assume only the values ± 1. Such a limitation, though certainly not essential (e.g. Nihat Berker, 1975) for the method outlined here, greatly simplifies the introduction of renormalization transformations and the actual calculation of the free energy, including its critical singularities, from these transformations.

The essence of the renormalization transformation approach to systems with many degrees of freedom is a stepwise evaluation of the free energy. Starting with the partition function of N spins s_i on a lattice, as the sum over all configurations $\{s\}$, this summation is carried out under the restriction that the configurations $\{s\}$ match with a given cell spin configuration $\{s'\}$. Here the N' cell spins $s'_{i'}$ assume the values ± 1 and are located on a lattice which is isomorphic with the original site lattice. The cell spin lattice is smaller by a scale factor $N'/N = l^{-d}$, where l is the linear dimension of the cell in units of the site lattice. The result can be written as the free energy of the cell spin configurations, which in turn defines uniquely a Hamiltonian for the cell spins.

Instead of summing over all the cell spins to obtain the free energy of the original site system, one considers the transition of the Hamiltonian describing the site system to the Hamiltonian describing the cell system. This transition obviously is a transformation in the space of Hamiltonian parameters for the site lattice: the renormalization transformation. Since the cell lattice is isomorphic with the original lattice, this transformation in the space of Hamiltonian parameters can be iterated and one is most interested in the cases where the sequence of Hamiltonians approaches a non-trivial limit (fixed point).

For the usefulness of a renormalization transformation it is essential that it is regular in the domain of interest in the space of Hamiltonian parameters. In that case the free energy of the cell system and the site system must exhibit the same singularities and one might call the two systems, which are connected by the transformation, thermodynamically equivalent. In general it is quite easy to write down a renormalization transformation, but hard to assess whether it is regular, particularly near its non-trivial fixed points. A special criterion that has to be satisfied is that critical systems are carried over into critical systems. For a certain class of renormalization transformations, those where the cell spins and the site spins are linearly related, this criterion is satisfied automatically.

We shall, however, pay little attention to these "linear" transformations but concentrate on transformations where the relation between the cell and site spins is non-linear. One of the main points, which now becomes much more cumbersome, is to discuss the behaviour of the correlation functions, which will be done in Section III. But apart from considerations about the correlation functions one can indicate the usefulness of the renormalization theory for critical phenomena by considering the fluctuations.

One knows from the work of Landau that the critical singularities are caused by long wavelength fluctuations. A long wavelength fluctuation will manifest itself more in the configuration of cell spins than in the site spin configurations compatible therewith. One might say that by leaving the cell spin variables in the problem the important fluctuations are still kept in the system. Therefore the transformation, which involves only a sum over site spins for a given cell spin configuration, does not involve the "dangerous" fluctuations and will thus be regular.

The cell spin Hamiltonian may even be seen as an expression for the energy of the fluctuations of the site spins. The fluctuation energy is expressed with respect to a mean value, which is going to play a fundamental rôle in our treatment. This mean value is a kind of background energy with respect to which fluctuations are measured energetically. On repeating the renormalization transformation the free energy is gradually expressed as a sum over this background energy. Near the fixed point this sum builds up the typical critical singularities.

The theory for the correlation functions is set up in a similar way. In terms of fluctuations the argument runs parallel; when a spin pair correlation function is long-ranged, this is a result of the long wavelength fluctuations. It turns out that by applying the renormalization transformation repeatedly, the full correlation function is expressed in terms of a regular short-ranged part, in some analogy with the Ornstein–Zernike theory.

In Section I we start with definitions and introduce the basic concepts such as cell spins, fixed points, eigenvectors, critical exponents and scaling fields. In Section II the thermodynamic implications of the renormalization theory, including the spontaneous magnetization, are discussed, and similarly in Section III the implications of the renormalization transformations for the correlation functions. In Section IV a number of computational methods are introduced and in Section V a survey of the results obtained so far is given. The paper closes with a discussion, Section VI, of several questions that have been raised in the preceding sections.

I. Definitions

We consider a lattice with a spin $s_i = \pm 1$ on every site. The subscript i labels the lattice sites; for the case of d-dimensional lattices i usually stands for d labels $i_1, \ldots, i_d$ to mark the coordinate in every direction respectively.

Let a be a subset of sites and define s_a as

$$s_a = \prod_{i \in a} s_i \tag{1.1}$$

The most general Hamiltonian describing an Ising-like system can then be written as

$$\mathscr{H}(s) = \sum_a K_a s_a \tag{1.2}$$

where the usual Boltzmann factor $-\beta = -1/k_B T$ has already been absorbed in the Hamiltonian, i.e. the "actual" interaction parameters are $J_a = K_a/k_B T$. For future convenience the total energy is shifted in such a way that

$$\sum_{\{s\}} \mathscr{H}(s) = 0 \tag{1.3}$$

where the symbol $\Sigma_{\{s\}}$ will always mean a summation over all the spin configurations. Also a in Σ_a of (1.2) runs through all non-empty subsets of sites.

The interaction parameters K_a can be formally obtained from $\mathscr{H}(s)$ via the relation

$$K_a = 2^{-N} \sum_{\{s\}} s_a \mathscr{H}(s) \tag{1.4}$$

where N stands for the number of lattice sites. Equation (1.4), recovering the interaction parameters K_a from the Hamiltonian $\mathscr{H}(s)$, is of importance for actual numerical calculations.

In practice we shall mostly deal with homogeneous Hamiltonians (1.2), i.e. Hamiltonians with the following property: let α be the class of all subsets a of sites which can be identified by a symmetry operation of the lattice; α may e.g. stand for "single site" (K_α then being a magnetic field), "nearest neighbour pair", etc. We shall restrict ourselves to systems where all K_a of $a \in \alpha$ have the same value K_α. For these homogeneous Hamiltonians, (1.2) can be rewritten as

$$\mathscr{H}(s) = \sum_a K_a s_a = \sum_\alpha K_\alpha \sum_{a \in \alpha} s_a \tag{1.5}$$

Obviously (1.5) can be written in a less formal way as

$$\mathscr{H}(s) = H \sum_i s_i + K \sum_{\langle i,j \rangle} s_i s_j + \ldots \tag{1.6}$$

where the symbol $\langle i,j \rangle$ as usual means that the sites i and j are nearest neighbours. The restriction to homogeneous systems, however, is by no means necessary. In fact this approach to renormalization theory for Ising-like spin systems can even be applied to systems with random interactions (Harris and Lubensky, 1974).

Specific examples of homogeneous systems (1.5) for which exact solutions are known are e.g.:

(a) "Pure" Ising systems defined by the Hamiltonian

$$\mathscr{H}(s) = K \sum_{\langle i,j\rangle} s_i s_j \tag{1.7}$$

for the linear and 2-dimensional lattices (Onsager, 1944).

(b) The triple-spin interaction model, which is defined on the 2-dimensional triangular lattice by the Hamiltonian (Baxter and Wu, 1973)

$$\mathscr{H}(s) = K \sum_{\langle i,j,k\rangle} s_i s_j s_k \tag{1.8}$$

where the symbol $\langle i, j, k\rangle$ stands for the sites of the vertices of any triangle.

(c) The symmetric 8-vertex model (or Baxter (1971, 1972) model), which can also be expressed as an Ising-like system on a quadratic lattice by the following Hamiltonian ($i = (i_1, i_2)$ is the site position):

$$\mathscr{H}(s) = K \sum_{i_1, i_2} [s_{i_1, i_2} s_{i_1+1, i_2+1} + s_{i_1+1, i_2} s_{i_1, i_2+1}] + \lambda \sum s_{i_1, i_2} s_{i_1+1, i_2+1} s_{i_1+1, i_2} s_{i_1, i_2+1} \tag{1.9}$$

Note that, by the absence of a magnetic field, the Hamiltonians (1.7) and (1.9) both only contain terms which are even in the spin variables, whereas (1.8) only contains a term which is odd in the spin variables.

Now, returning to the general theory, consider a set of N' so-called cell spins $s'_{i'} = \pm 1$ (the reason for this nomenclature will soon become clear). The index i' refers to the i'-th cell of a lattice isomorphic to the original lattice. All quantities pertaining to the cell spin system will be labelled with primes. Furthermore, let $P(s', s)$ be a weight factor, which depends on the cell spin and site spin configurations $\{s'\}$ and $\{s\}$, with the properties

$$P(s', s) \geqslant 0, \qquad \forall s', s \tag{1.10a}$$

and

$$\sum_{\{s'\}} P(s', s) = 1 \tag{1.10b}$$

We now define a Hamiltonian $\mathscr{H}'(s')$ for the cell spin system by

$$\exp[G + \mathscr{H}'(s')] = \sum_{\{s\}} P(s', s) \exp \mathscr{H}(s) \tag{1.11}$$

and

$$\sum_{\{s'\}} \mathscr{H}'(s') = 0 \tag{1.12}$$

G is independent of the cell spin configurations $\{s'\}$ and is defined by imposing the condition (1.12) on the cell spin Hamiltonian $\mathscr{H}'(s')$.

The restriction (1.10a) has been imposed on $P(s', s)$ in order to guarantee that the right-hand side of (1.11) is positive and that thus G and $\mathscr{H}'(s')$ are real quantities. As (1.10a) is sufficient but by no means necessary, this condition is sometimes relaxed in some calculations.

The second condition, (1.10b), leads to the following important relationship between the free energy F' of the cell system and the free energy F of the site system:

$$G + F' = \ln \sum_{\{s'\}} \exp\left[G + \mathscr{H}'(s')\right] = \ln \sum_{\{s'\}} \sum_{\{s\}} P(s', s) \exp \mathscr{H}(s)$$
$$= \ln \sum_{\{s\}} \exp \mathscr{H}(s) = F \tag{1.13}$$

A third condition will be imposed on $P(s', s)$ as well, in order to guarantee that the cell spin Hamiltonian $\mathscr{H}'(s')$ has the same symmetries as the site spin Hamiltonian $\mathscr{H}(s)$. To elucidate this we decompose $\mathscr{H}'(s')$ by means of (1.4) as

$$\mathscr{H}'(s') = \sum_{a'} K'_{a'} s'_{a'} \tag{1.14}$$

where a' runs through all non-empty subsets of cells. Then we require $P(s', s)$ such that for subsets a' of a certain class α, the cell interaction parameters $K'_{a'}$ are all equal to a certain value K'_{α}, i.e. $\mathscr{H}'(s')$ can again be written as

$$\mathscr{H}'(s') = \sum_{\alpha} K'_{\alpha} \sum_{a' \in \alpha} s'_{a'} \tag{1.15}$$

Note that we do not use primes on the classes of sets of cells; they are superfluous since e.g. the class of nearest neighbours is a definition which is independent of whether we are working in the cell or in the site system.

Now consider F' and F as functions of the interaction parameters K'_{α} and K_{α} (abbreviated to K' and K). In the thermodynamic limit (no long-range forces present) F' and F assume the form $F' = N'f(K')$ and $F = Nf(K)$, with in both cases the *same* function f. G also becomes an extensive function $G = Ng(K)$ in the thermodynamic limit with $N/N' = l^d$ constant (d = dimensionality of the system). Inserting these relations into (1.13) leads to the following basic renormalization relation for the free energy per site:

$$f(K) = g(K) + l^{-d} f(K') \tag{1.16}$$

Obviously l is also the linear distance between the cells measured in units of the site lattice.

The relation (1.11) is called a renormalization transformation. The three above-mentioned restrictions on $P(s', s)$ still allow an enormous variety in choice for $P(s', s)$, each leading to a different renormalization transformation.

A trivial choice would be to take $P(s', s)$ independent of s', which in view of (1.10b) leads to

$$P(s', s) = 2^{-N'} \tag{1.17}$$

This makes the right-hand side of (1.11) independent of s' and therefore $\mathscr{H}'(s') = 0$ and $G = F$.

As an example, to which we shall return repeatedly in Sections IV and V and for which extensive calculations have been carried out, we give here a more interesting choice for a 2-dimensional triangular lattice (Niemeijer and van Leeuwen, 1973, 1974). Consider a triangular lattice where cell spins have been placed in the centre of three sites as indicated in Fig. 1a.

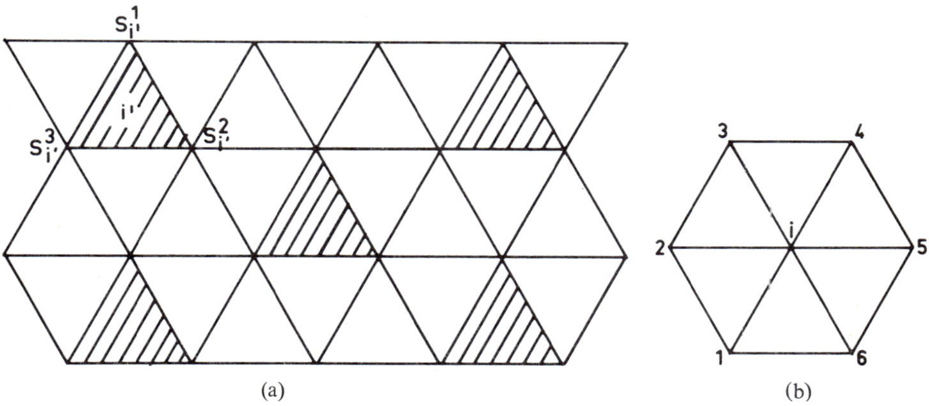

FIG. 1

Note that the cell spins are again located on a triangular lattice. Now define the weight-factor

$$P(s', s) = \prod_{i'} \tfrac{1}{2}[1 + s'_{i'}(s^1_{i'} + s^2_{i'} + s^3_{i'} - s^1_{i'}s^2_{i'}s^3_{i'})/2] \tag{1.18}$$

with $s^1_{i'}$, $s^2_{i'}$ and $s^3_{i'}$ denoting the sites in cell i'. One easily verifies that (1.18) fulfils the conditions (1.10a and b). The symmetry condition on this particular $P(s', s)$ is not quite fulfilled, but this can be remedied by averaging over all six triangle positions $i' = (i, 1, 2)$ etc. indicated in Fig. 1b. The choice (1.18) is special, since $P(s', s)$ is either 0 or 1. $P(s', s)$ equals 1 when

$$s'_{i'} = \text{sign}\,(s^1_{i'} + s^2_{i'} + s^3_{i'}) \tag{1.19}$$

and is zero otherwise. With the $P(s', s)$ given by (1.18) the transformation

(1.11) can be interpreted differently. For a given value of $s'_{i'}$ the three neighbouring site spins can still be in four configurations, labelled by $\sigma_{i'}$:

$$\begin{aligned} \sigma_{i'} &= 0: \quad s_{i'}^1 = s_{i'}^2 = s_{i'}^3 = s'_{i'}, \\ \sigma_{i'} &= 1: \quad -s_{i'}^1 = s_{i'}^2 = s_{i'}^3 = s'_{i'}, \\ \sigma_{i'} &= 2: \quad s_{i'}^1 = -s_{i'}^2 = s_{i'}^3 = s'_{i'}, \\ \sigma_{i'} &= 3: \quad s_{i'}^1 = s_{i'}^2 = -s_{i'}^3 = s'_{i'}. \end{aligned} \tag{1.20}$$

So effectively the summation in (1.11) over $\{s\}$ runs over the possible configurations $\{\sigma\}$ and one may write (1.11) as

$$\exp[G + \mathscr{H}'(s')] = \sum_{\{\sigma\}} \exp \mathscr{H}(s). \tag{1.21}$$

A choice as (1.18) has computational advantages in limiting the number of configurations to be summed over. The relation between cell and site spins as implied by (1.18) may be compared to the election of delegates in a district system with only two parties present. The site spins (voters) of a cell (district) determine the cell spin (delegate) by a majority rule. (We note here in passing that the square of the transformation (1.18) is not equal to the transformation based on taking a "super cell" of three cells as the basic unit, just as the minority may elect the president by an advantageous distribution of the votes over the districts. In this respect (1.18) is an example of a non-linear weight-factor which we shall discuss in Section III.)

Returning to (1.11), we view this as a transformation from the interaction parameters K_α to the renormalized interaction parameters K'_α, in short

$$K' = K'(K). \tag{1.22}$$

One can consider $G(K)$ as a special K'_ϕ with ϕ corresponding to the empty set. Equations (1.22) will be taken as the renormalization equations, which together with (1.16) determine the free energy $f(K)$. The strategy of the renormalization theory will be to obtain the singularities of the free energy $f(K)$ from the regular functions $g(K)$ and $K'(K)$. There is, however, no *a priori* guarantee that $g(K)$ and $K'(K)$ will be regular in the region of interest, since the choice (1.17) shows that $g(K) = f(K)$ and $g(K)$ then has all the singularities of $f(K)$. It is part of the renormalization problem to choose $P(s', s)$ in such a way that $g(K)$ and $K'(K)$ are regular in the relevant domain.

We now introduce the important concept of a fixed point of the equations (1.22). A fixed point K^* is a set K^*_α such that

$$K'_\alpha(K^*) = K^*_\alpha. \tag{1.23}$$

Of course the transformation (1.22) can in general have several fixed points, but not every one of them will have physical significance for critical phenomena; e.g. one trivial fixed point is the set K_α^* with

$$K_\alpha^* = 0 \tag{1.24}$$

corresponding to the infinite temperature limit of any spin system (since the parameter $-\beta = -1/k_B T$ has been absorbed in the interaction parameters).

It is one of the basic requirements of the renormalization approach that, for the determination of the critical exponents of a class of Ising systems, which have the same critical behaviour, the K'_α are regular functions of the K_α near the non-trivial fixed points. For the determination of the free energy of a specific spin system we shall need $g(K)$ and the transformation (1.22) to be regular in a much larger region of the parameter space.

For later results we shall need the linearized form of the transformation (1.22), i.e. the matrix

$$\partial K'_\alpha / \partial K_\beta = T_{\alpha\beta}. \tag{1.25}$$

A very special rôle is played by this linearized transformation in the fixed point (the fact that it is linearized at the fixed point will be denoted by an asterisk)

$$(\partial K'_\alpha / \partial K_\beta)_{K^*} = T^*_{\alpha\beta}. \tag{1.26}$$

So in the neighbourhood of K^* the transformation can be written as:

$$K'_\alpha - K^*_\alpha = \sum_\beta T^*_{\alpha\beta}(K_\beta - K^*_\alpha). \tag{1.27}$$

The matrix $T^*_{\alpha\beta}$ is not symmetric, so *a priori* one cannot be sure that its eigenvalues, which will turn out to be directly related to the critical exponents, are real. A general feature of $T^*_{\alpha\beta}$ in many applications is, that its elements below the diagonal are usually much smaller than those above the diagonal.

It is convenient to introduce a type of normal coordinates in which the transformation (1.27) takes a simple form. Let $T^*_{\alpha\beta}$ have eigenvalues λ_i with associated left eigenvectors φ^i_α:

$$\sum_\alpha \varphi^i_\alpha T^*_{\alpha\beta} = \lambda_i \varphi^i_\beta. \tag{1.28}$$

Then construct "normal" coordinates u_i as

$$u_i = \sum_\alpha \varphi^i_\alpha (K_\alpha - K^*_\alpha) \tag{1.29}$$

such that the u_i according to (1.27) and (1.28) transform as

$$u'_i = \sum_\alpha \varphi^i_\alpha (K'_\alpha - K^*_\alpha) = \lambda_i \sum_\beta \varphi^i_\beta (K_\beta - K^*_\beta) = \lambda_i u_i \tag{1.30}$$

One may extend the definition of u_i beyond the linear regime in such a way that (1.30) also holds for finite u_i. For that purpose (1.27) has to be expanded further as

$$K'_\alpha - K^*_\alpha = \sum_\alpha T^*_{\alpha\beta}(K_\beta - K^*_\beta) + \frac{1}{2!}\sum_{\beta,\gamma} T^*_{\alpha,\beta\gamma}(K_\beta - K^*_\beta)(K_\gamma - K^*_\gamma)\ldots \tag{1.31}$$

as well as (1.29)

$$u_i = \sum_\alpha \varphi^i_\alpha(K_\alpha - K^*_\alpha) + \frac{1}{2!}\sum_{\alpha\beta} \varphi^i_{\alpha\beta}(K_\alpha - K^*_\alpha)(K_\beta - K^*_\beta)\ldots \tag{1.32}$$

where $T^*_{\alpha,\beta\gamma} = (\partial^2 K_\alpha/\partial K_\beta \partial K_\gamma)_{K^*}$. In order to ensure that

$$u'_i = \lambda_i u_i \tag{1.33}$$

also holds to second order, one should relate $\varphi^i_{\alpha\beta}$ to $T^*_{\alpha,\beta\gamma}$ as the solutions of the linear set of equations

$$\sum_{\mu,\nu} \varphi^i_{\mu\nu}[\lambda_i \delta_{\mu\alpha}\delta_{\nu\beta} - T^*_{\mu\alpha}T^*_{\nu\beta}] = \sum_\mu \varphi^i_\mu T^*_{\mu,\alpha\beta}. \tag{1.34}$$

The solutions of (1.34) are

$$\varphi^i_{\mu\nu} = \sum_{j,k} \frac{1}{\lambda_i - \lambda_j\lambda_k} \sum_{\gamma\alpha\beta} \varphi^i_\gamma T^*_{\gamma,\alpha\beta} \psi^j_\alpha \psi^k_\beta \varphi^j_\mu \varphi^k_\nu. \tag{1.35}$$

(where ψ^i_α is the αth component of the right eigenvector of $T^*_{\alpha\beta}$ with eigenvalue λ_i).

To find the u_i as functions of K, it is for practical purposes more convenient to use the relation

$$\lambda_i u_i(K) = u_i(K') \text{ or } \lambda_i^n u_i(K) = u_i(K^{(n)}) \tag{1.36}$$

in which $K^{(n)}$ is the image of K after n transformations. By choosing K in the neighbourhood of the fixed point K^* one finds (except under special circumstances) values of $u_i(K^{(n)})$ further away from K^*. The set $\{u_i(K)\}$ may be considered as the curvilinear coordinates adapted to the transformation (Wegner's scaling fields, Wegner, 1972). Note that the u_i are defined up to a constant factor, since the eigenvectors are only defined up to a constant factor. It should also be pointed out that both procedures fail in case of an eigenvalue $\lambda = 1$ (a so-called marginal eigenvalue).

II. General Theory

This Section will be devoted to the consequences of the basic formula (1.16), viz.

$$f(K) = g(K) + l^{-d} f(K')$$

which in terms of the scaling fields u_i can be written as:

$$f(u_1, u_2, \ldots) = g(u_1, u_2, \ldots) + l^{-d} f(\lambda_1 u_1, \lambda_2 u_2, \ldots). \tag{2.1}$$

Our basic assumption is that the scaling fields u_i are regular functions of the interaction parameters K_α and that $g(u_1, u_2, \ldots)$ is a regular function of the scaling fields u_i.

If a field u_i scales under the renormalization transformation with a factor $|\lambda_i| > 1$ (<1) it is called relevant (irrelevant). In the special case that $\lambda_i = 1$ the field is called marginal. In order to appreciate further the notion of scaling fields we point out that, obviously, the values of the relevant fields increase upon performing a transformation, whereas the irrelevant fields decrease. The fixed point itself, as defined in (1.23) corresponds to all $u_i = 0$. The set of points which ultimately end up in the fixed point is called the domain of attraction; this set of points forms a hypersurface (Fig. 2) in the parameter space of the interactions. If there are n interactions and m relevant variables, the dimension of this hyper surface is $n - m$. For reasons which will become clear later on, the domain of attraction is also called the *surface of criticality*. If e.g. u_1 and u_2 are relevant fields, the surface of criticality is defined by $u_1 = u_2 = 0$.

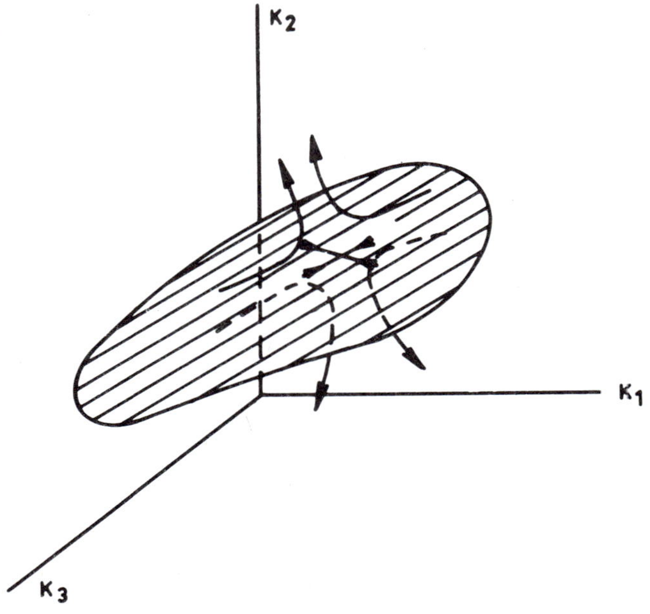

FIG. 2

Another general distinction between the scaling fields can be made on the basis of symmetry. Each $\mathscr{H}(s)$ is made up of an even part, which is invariant under a flip of all spins, and an odd part, which changes sign under a spin flip. Now, if all $P(s', s)$ are chosen such that they are invariant under a simultaneous flip of s' and s, then it follows that the subspace of the even interactions of $\mathscr{H}(s)$ is invariant under the transformation. Thence it follows that the subspace of the even part of $\mathscr{H}(s)$ is invariant under the transformation, and as a consequence, in most cases, such as (1.9) and (1.7), fixed points are found in the subspace of even interactions. For an "even fixed point", the matrix $T^*_{\alpha\beta}$ breaks up into an even–even and an odd–odd part as can be seen by differentiating (1.11) with respect to a K_α:

$$[\partial G/\partial K_\alpha + \partial \mathscr{H}'(s')/\partial K_\alpha]\, e^{\mathscr{H}'(s')} = \sum_{\{s\}} P(s', s)\, e^{\mathscr{H}(s)}\, [\partial \mathscr{H}(s)/\partial K_\alpha] \tag{2.2}$$

Then, using that

$$\partial \mathscr{H}(s)/\partial K_\alpha = \sum_{a \in \alpha} s_a$$

and

$$\partial \mathscr{H}'(s')/\partial K_\alpha = \sum_{\alpha'} [\partial K'_{\alpha'}/\partial K_\alpha] \left[\sum_{a' \in \alpha'} s'_{a'} \right] \tag{2.3}$$

one may write (2.2) in the form

$$\partial G/\partial K_\alpha + \sum_{\alpha'} \left[\sum_{a' \in \alpha'} s'_{a'} \right] [\partial K'_{\alpha'}/\partial K_\alpha] = \left\langle \sum_{a \in \alpha} s_a \right\rangle_{s'} \tag{2.4}$$

where the average with the subscripts $\langle\ \rangle_{s'}$, to remind one of the dependence on $\{s'\}$, is defined as

$$\langle A \rangle_{s'} = \sum_{\{s\}} P(s', s)\, e^{\mathscr{H}(s)}\, A(s) \Big/ \sum_{\{s\}} P(s', s)\, e^{\mathscr{H}(s)} \tag{2.5}$$

From (2.5) it follows that for a Hamiltonian $\mathscr{H}(s)$ with only even interactions, $\langle A \rangle_{s'}$ and A will have the same symmetry. So in (2.4) for α even only even α' appear, and for odd α only odd α' will appear. Thus $T^*_{\alpha\beta}$ has even and odd eigenvalues and related even and odd scaling fields.

The distinction between scaling fields in odd and even symmetry character can be extended to more complicated symmetries; this will be discussed in connection with the symmetric eight-vertex model in Section VI.

In this Section we discuss:

A. the connection between the eigenvalues of $T^*_{\alpha\beta}$ and critical exponents;
B. the determination of the critical temperature;
C. the calculation of the free energy from (2.1);
D. the spontaneous magnetization.

A. Eigenvalues and critical exponents

We shall start off with a short reminder of the nomenclature of a few critical exponents for magnetic systems. For the sake of convenience we shall restrict ourselves to ferromagnetic systems; extensions to antiferromagnetic systems can be found e.g. in Stanley (1971).

We define the dimensionless distance τ from the critical temperature T_c by

$$\tau = (T - T_c)/T_c \tag{2.6}$$

The singular behaviour of the specific heat, spontaneous magnetization, magnetic susceptibility and the response to an external magnetic field at T_c are respectively characterized by the exponents α, β, γ and δ: $C_H \simeq \tau^{-\alpha}$, $m \simeq (-\tau)^\beta$, $\chi_H \simeq \tau^{-\gamma}$ and $H = |m|^\delta$. If $\tau < 0$ the critical exponents usually are primed. In Table II.1 we give the numerical values of the critical exponents for the Ising-like systems defined by the Hamiltonian (1.7) on a quadratic and a triangular lattice (1.8) and that of the symmetric eight-vertex model defined by (1.9).

TABLE II.1. Critical exponents for several Ising-like systems. The models have been defined in the text. Those marked by * and † are conjectures by Suzuki (1974) and Barber and Baxter (1973) respectively. $s(\lambda) = \pi/[\pi - \arccos(\tanh 2\lambda)]$; λ has been defined in (1.9).

Model	$\alpha = \alpha'$	β	$\gamma = \gamma'$	δ
2-*d* Ising	0(log)	1/8	7/4	15
2-*d* triplet interation	2/3	1/12*	7/6*	15*
symmetric 8-vertex	$2 - s(\lambda)$	$s(\lambda)/16$†	$7s(\lambda)/8$†	15†

The connection between the eigenvalues λ_i of the matrix $T^*_{\alpha\beta}$ and the critical exponents is most easily established by asking whether a singular behaviour in powers of the u_i for the free energy is compatible with (2.1). To begin with let us put $u_2 = u_3 = \ldots = 0$ and assume a behaviour like $A|u_1|^{a_1}$ for $f(u_1, 0, 0, \ldots)$. Inserting this into (2.1) and equating the powers of $u_1^{a_1}$ (which by the regularity assumption of $g(u_1, 0, 0, \ldots)$ do not appear in g) leads to the equality

$$A|u_1|^{a_1} = l^{-d}A[\lambda_1 u_1]^{a_1} \tag{2.7}$$

implying for the exponent a_1

$$a_1 = \mathrm{d} \log l/\log|\lambda_1|. \tag{2.8}$$

The conclusion on the one hand is that if a singular power in u_1 would occur, the exponent a_1 is given by λ_1 through (2.8). On the other hand we see from (2.8) that only the relevant scaling fields ($\lambda_i > 1$) qualify on physical grounds for a singular behaviour as the exponent should be positive to keep the free energy finite at $u_1 = 0$. The exponent a_1 is a physical quantity and thus λ_1 should vary as a power of l for different renormalization transformations with different l. Therefore, following Wegner (1972), we write for the (relevant) eigenvalues (which are assumed to be real and positive)

$$\lambda_i = l^{y_i} \tag{2.9}$$

such that a_1 and y_1 are related as

$$a_1 = d/y_1. \tag{2.10}$$

We note that the amplitude A drops out of (2.7). Since A could be zero, we may not invert the argument. A relevant eigenvalue does not necessarily lead to a singular behaviour. This point is investigated further under C.

Next we look into the dependence on two scaling fields u_1 and u_2 and determine the singular powers $|u_1|^{a_1}|u_2|^{a_2}$ which are allowed by (2.1). Comparing exponents left and right yields the relation

$$a_1 y_1 + a_2 y_2 = d. \tag{2.11}$$

For positive a_i at least one of the y_i has to be positive. A combination of a relevant scaling field ($y_1 > 0$) and an irrelevant one ($y_2 < 0$) leads to a value of a_1 larger than d/y_1 and therefore is a less singular contribution in u_1 than the previously found power d/y_1. Such corrections to the most singular part will be discussed in C. For the remaining case of two relevant scaling fields a combination of the possible powers $u_1^{a_1}u_2^{a_2}$ forms with (2.11) the functions

$$|u_1|^{d/y_1} f_1(u_2/|u_1|^{y_2/y_1}) \tag{2.12}$$

or

$$|u_2|^{d/y_2} f_2(u_1/|u_2|^{y_1/y_2}). \tag{2.13}$$

These are solutions of the scaling relation

$$f_{\text{sing}}(l^{y_1}u_1, l^{y_2}u_2) = l^d f_{\text{sing}}(u_1, u_2). \tag{2.14}$$

Except for the fact that here l is a discrete variable, (2.14) is similar to Widom's static scaling hypothesis (Widom, 1965a, b), which is usually written in the form

$$f_{\text{sing}}(x^{a_\varepsilon}\tau, x^{a_h}h) = x f_{\text{sing}}(\tau, h) \tag{2.15}$$

where h is a dimensionless magnetic field.

In order to make the connection with the usual scaling theory of critical phenomena we have to show that in the appropriate cases a fixed point has two relevant variables u_1 and u_2, one of which, e.g. u_1, is temperature-like and the other, u_2, field-like.

As has already been mentioned (if the Hamiltonian contains only even interactions), a "normal" fixed point will lie in the subspace of even interactions. A change of temperature is a change of magnitude of all interaction constants K_α. Thus a variation of the temperature in the fixed point keeps the Hamiltonian in the even subspace and the temperature will couple to the even scaling fields. Similarly a change of the magnetic field couples to the odd scaling fields. We leave the confirmation of this picture to Section V, where several specific cases are treated, and give here only the relation of the thermodynamic critical exponents and the two exponents y_T (even) and y_H (odd) of the relevant variables

$$\begin{aligned} \alpha &= 2 - d/y_T \\ \beta &= (d - y_H)/y_T, \\ \gamma &= (2y_H - d)/y_T, \\ \delta &= y_H/(d - y_H), \end{aligned} \tag{2.16}$$

from which immediately follow the well-known scaling laws

$$\alpha + 2\beta + \gamma = 2 \quad \text{and} \quad \alpha + \beta(\delta + 1) = 2. \tag{2.17}$$

We note here again that the relation (2.14) does not determine the functions f_1 or f_2. A determination of f_1 and f_2 follows in C.

As noted before, a difference between (2.14) and (2.15) is that x is a continuous parameter whereas l has a fixed value for a renormalization transformation.

For more than two relevant variables (which happens in special fixed points) one has similarly to (2.14) for the singular part the scaling relation:

$$f_{\text{sing}}(l^{y_1}u_1, l^{y_2}u_2, \ldots, l^{y_n}u_n) = l^d f_{\text{sing}}(u_1, u_2, \ldots, u_n). \tag{2.18}$$

Up till now we have only been discussing the behaviour in the fixed point and we must extend the conclusion to the domain of attraction, which is by (2.1) thermodynamically equivalent to the fixed point. However, having observed that the (possible) singularities occur for small values of the relevant scaling fields, we may already conclude that the critical systems must be found in the domain of attraction as this is the collection of systems with all relevant scaling fields $u_1 = u_2 = \ldots = u_n = 0$.

B. Determination of the critical temperature

The domain of attraction or the surface of criticality is determined by setting the relevant scaling fields equal to zero. Restricting ourselves to the even subspace and assuming that there is only one relevant eigenvalue λ_T with corresponding scaling field u_T, the equation for the surface of criticality becomes with (1.32)

$$0 = u_T = \sum_\alpha \varphi_\alpha^T (K_\alpha - K_\alpha^*) + \frac{1}{2!} \sum_{\alpha\beta} \varphi_{\alpha\beta}^T (K_\alpha - K_\alpha^*)(K_\beta - K_\beta^*) + \ldots \tag{2.19}$$

For a given set of interaction parameters $J_\alpha = k_B T K_\alpha$ equation (2.19) then determines the critical temperature $T_c(J)$. Writing $K_\alpha = J_\alpha / k_B T_c(J)$ and substituting in (2.19) the function $T_c(J)$ has to be such that (2.19) is fulfilled.

If one approximates the surface of criticality by a plane which is tangent to it in the fixed point, (2.19) reduces to the following equation for $T_c(J)$

$$k_B T_c(J) = \sum_\alpha \varphi_\alpha^T J_\alpha \Big/ \sum_\alpha \varphi_\alpha^T K_\alpha^*. \tag{2.20a}$$

This yields the critical temperature with the fixed point as reference point. Dalton and Wood (1969) have studied the variation of the critical temperature $T_c(J)$ due to the admixture of nearest-neighbour interactions taking the "pure" Ising system (1.7) as a reference point. In order to compare (2.20a) with their data we first notice that (2.20a) yields for the pure Ising system critical temperature T_c

$$k_B T_c = \varphi_{\text{n.n.}}^T J_{\text{n.n.}} \Big/ \sum_\alpha \varphi_\alpha^T K_\alpha^*, \tag{2.20b}$$

where the subscript n.n. stands for nearest neighbour. Then using (2.20b) in (2.20a) one obtains the equation for $T_c(J)$ in the form of Dalton and Wood (1969):

$$T_c(J) = T_c \left\{ 1 + \sum_{\alpha \neq \text{n.n}} (\varphi_\alpha^T / \varphi_{\text{n.n}})(J_\alpha / J_{\text{n.n}}) \right\}. \tag{2.20c}$$

C. Calculation of the free energy†

From the previous section it follows that from (2.1) alone, the free energy cannot be determined in terms of $g(u_1, u_2, \ldots)$ since the singular part fulfils the homogeneous equation with $g = 0$. As in the theory of differential equations one needs a boundary condition to specify the amplitude of the solution of the inhomogeneous equation which has to be added to a particular solution of the inhomogeneous equation.

A point in the neighbourhood of the domain of attraction (see Fig. 2)

† Nauenberg and Nienhuis (1974b), Nightingale and 't Hooft (1975), Nelson and Fisher (1975), and van Leeuwen (1975a).

of a non-trivial fixed point first moves, upon repeated application of the renormalization transformation, along the domain of attraction towards the fixed point. If it originally was not exactly in the domain of attraction, it bends away from the fixed point, usually towards another fixed point with a higher dimensional domain of attraction. For the ferromagnetic fixed point in the even subspace a point outside the domain of attraction moves either to a region with very large interaction parameters (i.e. $T = 0$) or to a region with very small interaction parameters (i.e. $T = \infty$). Both limiting situations can be considered as fixed points. The behaviour of the free energy in the neighbourhood of these (trivial) fixed points determines the singular behaviour in the neighbourhood of the domain of attraction of the original fixed point. It is of course possible that there are several fixed points, each having its own domain of attraction, which leads to a more complicated picture.

Let us illustrate the general procedure by taking the following one-dimensional example. Consider the subspace of one relevant (even) scaling field u_1 setting all other $u_2 = u_3 = \ldots = 0$. Then (2.1) simplifies to

$$f(u) = g(u) + l^{-d} f(\lambda u) \tag{2.21}$$

(dropping the index 1 of u_1). Iterating (2.21) n times yields

$$f(u) = \sum_{j=0}^{n-1} l^{-jd} g(\lambda^j u) + l^{-nd} f(\lambda^n u) \tag{2.22}$$

Now we take as boundary condition that we may let $n \to \infty$ in (2.22) and drop the term $l^{-nd} f(\lambda^n u)$ in that limit:

$$\lim_{n \to \infty} l^{-nd} f(\lambda^n u) = 0, \tag{2.23}$$

Whether this is permitted or not must be verified in specific cases. The idea behind this boundary condition is that the free energy behaves sufficiently well for large values of the scaling fields ($T = 0$ or $T = \infty$), in the sense that it does not compensate the power l^{-nd}. Thus we take for f the expression

$$f(u) = \sum_{j=0}^{\infty} l^{-jd} g(\lambda^j u). \tag{2.24}$$

Without detailed calculation this expression already suggests that $u = 0$ is a possible singular point. For $u > 0$, $f(u)$ is determined by the values of g for $u > 0$, and f for $u < 0$ is determined by g for $u < 0$. If there is no symmetry relation between g for $u > 0$ and $u < 0$, two different functions f may result for $u > 0$ and $u < 0$ respectively, which join continuously at the value

$$f(0) = \sum_{j=0}^{\infty} l^{-jd} g(0) = g(0)/(1 - l^{-d}). \tag{2.25}$$

We shall define the scaling field u in such a way that $u > 0$ corresponds to temperatures below the critical temperature.

We shall now first construct the regular part $f_{\text{reg}}(u)$ of $f(u)$. Since $g(u)$ is a regular function it will, by definition, have a power series expansion

$$g(u) = \sum_{n=0}^{\infty} g_n u^n. \tag{2.26}$$

Inserting this into (2.24) and substituting for $f(u)$ yields

$$f_{\text{reg}}(u) = \sum_{n=0}^{\infty} f_n u^n \tag{2.27}$$

One finds the coefficients f_n as

$$f_n = g_n/(1 - l^{-d}\lambda^n). \tag{2.28}$$

We see that the case $\lambda^n = l^d$ must be discussed separately as will be done at the end of this section. f_{reg} is a particular solution of (2.21) characterized by the fact that it is regular at the origin. From (2.28) we see that the radius of convergence of f is larger by a factor λ than that of g.

Returning to the expression (2.24) we shall denote by n_0 the first power for which $\lambda^n > l^d$ and write

$$g(u) = \sum_{n=0}^{n_0-1} g_n u^n + g_{\text{rem}}(u) \tag{2.29}$$

(where the subscript "rem" stands for remainder). We note that for an irrelevant scaling field ($\lambda < 1$), λ^n will never exceed l^d, such that $n_0 = \infty$ and $g_{\text{rem}}(u)$ will disappear. Inserting (2.29) into (2.24) one observes that for the first n_0 powers the summation over j converges and using (2.28) one finds

$$f(u) = \sum_{n=0}^{n_0-1} f_n u^n + \sum_{j=0}^{\infty} l^{-jd} g_{\text{rem}}(\lambda^j u) \tag{2.30}$$

For the remainder we write

$$\sum_{j=0}^{\infty} l^{-jd} g_{\text{rem}}(\lambda^j u) = \sum_{j=-\infty}^{\infty} l^{-jd} g_{\text{rem}}(\lambda^j u) - \sum_{j=-\infty}^{-1} l^{-jd} g_{\text{rem}}(\lambda^j u) \tag{2.31}$$

The part that is added and subtracted can be evaluated by a power series expansion of $g_{\text{rem}}(u)$

$$\sum_{j=-\infty}^{-1} l^{-jd} g_{\text{rem}}(\lambda^j u) = \sum_{m=n_0}^{\infty} f_m u^m \tag{2.32}$$

Collecting (2.30), (2.31) and (2.32) together one arrives at

$$f(u) = f_{\text{reg}}(u) + f_{\text{sing}}(u) \tag{2.34}$$

with $f_{\text{sing}}(u)$ given as

$$f_{\text{sing}}(u) = \sum_{j=-\infty}^{\infty} l^{-jd} g_{\text{rem}}(\lambda^j u) \tag{2.34}$$

Of course $f_{\text{sing}}(u)$ fulfils the condition of homogeneity

$$f_{\text{sing}}(\lambda u) = l^d f_{\text{sing}}(u) \tag{2.35}$$

since $f_{\text{reg}}(u)$ is a solution of (2.21). Equation (2.34) is an explicit solution of (2.35). It should be noted that only the remainder of g appears in (2.34), otherwise the summation could not be extended to $j = -\infty$ and that in turn is necessary to fulfil (2.35). On the upper side $j = \infty$ the convergence is determined by the behaviour of $g(u)$ for large u, since the subtracted terms do not influence the convergence. This again has to be checked in every specific case. If the convergence at the upperside $j = \infty$ would become questionable it should be taken as a signal that the boundary condition in (2.23) has to be reconsidered.

Having obtained an explicit expression for f_{sing} we are in a position to discuss the question whether f_{sing} indeed behaves as $A|u|^{d/y}$ and to determine the value of the amplitude A. We define an amplitude $A(u)$ as

$$A(u) = |u|^{-d/y} \sum_{j=-\infty}^{\infty} l^{-jd} g_{\text{rem}}(\lambda^j u). \tag{2.36}$$

$A(u)$ then automatically satisfies the relation $A(\lambda u) = A(u)$. Thus $A(u)$ takes on equal values in all arguments $\ldots \lambda^{-2}u, \lambda^{-1}u, u, \lambda u, \lambda^2 u, \ldots$ of which sequence $u = 0$ (the fixed point) is a limit point. This, however, is not sufficient to guarantee that $A(u)$ is a constant (depending only on the sign of u), since (2.36) allows for a periodic function in $\log|u|$ with period $\log\lambda$. Thus we write

$$\begin{aligned} A(u) &= \sum_{n=-\infty}^{\infty} A_n^{\pm} \exp[2\pi i n \log|u|/\log\lambda] \\ &= \sum_{n=-\infty}^{\infty} A_n^{\pm} |u|^{2\pi i n/\log\lambda}. \end{aligned} \tag{2.37}$$

The signs $\pm$ are introduced to distinguish the functions $A^{\pm}(u)$ for $u \gtrless 0$. The fourier coefficients are obtained from (2.36) as

$$A_n^{\pm} = \frac{1}{\log\lambda} \int_0^{\log\lambda} \mathrm{d}\log u \left\{ |u|^{-d/y} \sum_{j=-\infty}^{\infty} l^{-jd} g_{\text{rem}}(\lambda^j u) \right\} |u|^{-2\pi i n/\log\lambda}. \tag{2.38}$$

Changing to the variable $t = \lambda^j|u|$ in the jth term, the summation over j leads to a sum over intervals in t which spans the range $[0, \infty]$

$$A_n^{\pm} = \frac{1}{\log \lambda} \int_0^{\infty} \frac{dt}{t^{(1+d/y)}} g_{\text{rem}}(\pm t) t^{-2\pi i n/\log \lambda}. \tag{2.39}$$

The result (2.37) with $A_n^{\pm}$ given by (2.39) is somewhat disconcerting since it does not follow from other renormalization procedures. However, it is unlikely that the oscillatory terms ($n \neq 0$) will actually exist, because they imply that the free energy has an oscillation in $\log|u|$ of period $\log \lambda$, superimposed on the power $|u|^{d/y}$. But λ contains the scaling length l which can be changed by choosing a different renormalization transformation. In fact, if two renormalization transformations exist with asymptotic, i.e. for $u \to 0$, equivalent scaling fields and incommensurate $\log l_1$ and $\log l_2$, such oscillating terms cannot exist.† But if we stick to one particular renormalization transformation, as we shall do in this paper, we can only hope that the $A_n^{\pm}$ for $n \neq 0$ are found to be zero from (2.39). Actual calculations so far have always led to no noticeable $A_n^{\pm}$ for $n \neq 0$.

So we assume that only $A_0^{\pm}$ exists with the result that

$$A(u) = A_0^{\pm} = \frac{1}{\log \lambda} \int_0^{\infty} \frac{dt}{t(1 + d/y)} g_{\text{rem}}(\pm t) = A^{\pm}. \tag{2.40}$$

Thus the resulting $f_{\text{sing}}(u)$ reads, with (2.34), (2.36), and (2.40),

$$f_{\text{sing}}(u) = A^{\pm} |u|^{d/y}, \qquad u \gtrless 0. \tag{2.41}$$

The general situation, with more than one scaling field present, can be treated quite similarly. With the same boundary condition

$$\lim_{n \to \infty} l^{-nd} f(\lambda_1^n u_1, \lambda_2^n u_2, \ldots) = 0 \tag{2.42}$$

the solution of (2.1) is given by

$$f(u_1, u_2, \ldots) = \sum_{j=0}^{\infty} l^{-jd} g(\lambda_1^j u_1, \lambda_2^j u_2, \ldots). \tag{2.43}$$

Again g is split into two parts: one part containing the power combinations for which (2.43) converges and a remainder. The first part leads to a piece of the regular part of the free energy $f_{\text{reg}}(u_1, u_2, \ldots)$ which is constructed from the power series of $g(u_1, u_2, \ldots)$

$$g(u_1, u_2, \ldots) = \sum_{n_1, n_2, \ldots = 0}^{\infty} g_{n_1, n_2} u_1^{n_1} u_2^{n_2} \ldots \tag{2.44}$$

as

$$f_{\text{reg}}(u_1, u_2, \ldots) = \sum_{n_1, n_2, \ldots = 0}^{\infty} f_{n_1, n_2, \ldots} u_1^{n_1} u_2^{n_2} \ldots \tag{2.45}$$

† This argument is due to D. R. Nelson (private communication). The possible existence of oscillatory terms was noted by Nauenberg (1975b).

with $f_{n_1, n_2, \ldots}$ given by

$$f_{n_1, n_2, \ldots} = g_{n_1, n_2, \ldots}/(1 - l^{-d}\lambda_1^{n_1}\lambda_2^{n_2}\ldots) \tag{2.46}$$

In the second part of the summation over $j = -\infty$ to -1 is added and subtracted and evaluating this sum through power series expansion one recovers the second piece of f_{reg}. So again one finds f as

$$f(u_1, u_2, \ldots) = f_{\text{reg}}(u_1, u_2, \ldots) + f_{\text{sing}}(u_1, u_2, \ldots) \tag{2.47}$$

with f_{sing} given by

$$f_{\text{sing}}(u_1, u_2, \ldots) = \sum_{j=-\infty}^{\infty} l^{-jd} g_{\text{rem}}(\lambda_1^j u_1, \lambda_2^j u_2, \ldots) \tag{2.48}$$

f_{sing} fulfils the homogeneity property.

An interesting aspect of more than one scaling field is the interplay of the relevant variables and the rôle of the irrelevant variables. If we take u_1 as a relevant variable and write $u_2 = v_2|u_1|^{y_2/y_1}$, $u_3 = v_3|u_1|^{y_3/y_1}$ etc., then

$$A_1(v_2, v_3, \ldots; u_1) = |u_1|^{-d/y_1} \sum_{j=-\infty}^{\infty} l^{-jd} g_{\text{rem}}(\lambda_1^j u_1, \lambda_2^j v_2 |u_1|^{y_2/y_1}, \ldots) \tag{2.49}$$

has the property that $A_1(v_2, v_3, \ldots; \lambda_1 u_1) = A_1(v_2, v_3, \ldots; u_1)$, as before implying that A_1 is periodic in $\log|u_1|$ with period $\log \lambda_1$. We shall again assume that the oscillatory terms vanish and that A_1 is completely determined by its zeroth fourier component, which reads in analogy with (2.40)

$$A_1^{\pm}(v_2, v_3, \ldots) = \frac{1}{\log \lambda_1} \int_0^{\infty} \frac{dt}{t^{(1+d/y)}} g_{\text{rem}}(\pm t, v_2 t^{y_2/y_1}, v_3 t^{y_3/y_1}, \ldots). \tag{2.50}$$

Thus we obtain for f_{sing} the expression

$$f_{\text{sing}}(u_1, u_2, u_3, \ldots) = |u_1|^{d/y_1} A_1^{\pm}(u_2/|u_1|^{y_2/y_1}, u_3/|u_1|^{y_3/y_1}, \ldots). \tag{2.51}$$

This is a generalization of (2.12). We now point out a difference between the relevant and irrelevant variables. Suppose that u_2 is another relevant variable and u_3, etc. irrelevant variables. Then $y_3 < 0, \ldots$ and for small u_1 the second, third etc. argument become small. To leading order we have

$$f_{\text{sing}}^{\text{lead}}(u_1, u_2, u_3, \ldots) = |u_1|^{d/y_1} A_1^{\pm}(u_2/|u_1|^{y_2/y_1}, 0, 0 \ldots) \tag{2.52}$$

showing that the most singular part is independent of the irrelevant variables. So around the domain of attraction we have to leading order a universal singular behaviour in the relevant variables. Eqn (2.52) shows that (2.12) is not only restricted for the fixed point but holds for the whole domain of attraction, which is therefore correctly identified with the *surface of criticality*.

For the next order we obtain (due to the u_3-dependence)

$$f_{\text{sing}}^{\text{next}}(u_1, u_2, u_3, \ldots) = u_3 |u_1|^{(d-y_3)/y_1} \left[\frac{\partial A_1^{\pm}}{\partial v_3} (u_2/|u_1|^{y_2/y_1}, v_3, \ldots) \right]_{v_3=0} \tag{2.53}$$

and similar expressions for the corrections to scaling (Wegner, 1972) due to the other variables. The largest y_i (<0) yields the smallest difference in power of $|u_1|$ with respect to the leading power.

For the relevant variable u_2 we have to keep the whole dependence of A_1 on v_2, since $u_2/|u_1|^{y_2/y_1}$ may be small or large depending on the values of u_2 and u_1. A has been given a subscript 1 simply because u_1 has been singled out as a relevant variable. Choosing u_2 as a relevant variable would have led to different expressions (of the second type of (2.13)) where the rôles of u_1 and u_2 would have been interchanged.

Finally we have to consider the case, as mentioned before, that a power λ^n is equal to l^d, or more generally that a combination $\lambda_1^{n_1}\lambda_2^{n_2} \ldots = l^d$. Now one can follow two different procdures: either treat the "dangerous" power combination separately from the start or take the limit in the final formulae as obtained by treating (one of) the eigenvalue(s) as a continuous parameter. In either case the result is the appearance of a logarithmic term. As the derivation closely parallels the one already given, we shall present here only the result.

Define $f_{\text{reg}}(u_1, u_2, \ldots)$ as before with the exception that for the coefficient $f_{n_1, n_2, \ldots}$ of the "dangerous" power for which $\lambda_1^{n_1}\lambda_2^{n_2} \ldots = l^d$, not the relation (2.46) is used but instead

$$f_{n_1, n_2, \ldots} = \tfrac{1}{2} g_{n_1, n_2, \ldots}. \tag{2.54}$$

Now $g_{\text{rem}}(u_1, u_2, \ldots)$ is defined as $g(u_1, u_2, \ldots)$ minus the "dangerous" power and the summable powers. The amplitude $A_1^{\pm}(v_1, v_2, \ldots)$ is defined as

$$A_1^{\pm}(v_2, v_3, \ldots) = \frac{1}{\log \lambda_1} \left[\int_0^1 \mathrm{d}t \, \frac{g_{\text{rem}}(\pm t, v_2 t^{y_2/y_1}, \ldots)}{t^{(1+d/y_1)}} \right.$$
$$\left. + \int_1^{\infty} \mathrm{d}t \, \frac{g_{\text{rem}}(\pm t, v_2 t^{y_2/y_1}, \ldots) + g_{n_1, n_2 \ldots}(\pm t)^{n_1}(v_2 t^{y_2/y_1})^{n_2} \ldots}{t^{(1+d/y_1)}} \right]. \tag{2.55}$$

Then the singular part of the free energy reads:

$$f_{\text{sing}}(u_1, u_2, \ldots) = |u_1|^{d/y_1} A_1^{\pm}(u_2/|u_1|^{y_2/y_1}, \ldots)$$
$$- \frac{g_{n_1, n_2, \ldots}}{\log \lambda_1} u_1^{n_1} u_2^{n_2} \ldots \log |u_1| \tag{2.56}$$

Here of course $n_1, n_2 \ldots$ is the combination for which $\lambda_1^{n_1}\lambda_2^{n_2} \ldots = l^d$. For the case of a single relevant variable, $u_2 = u_3 = \ldots = 0$, the first term in (2.56) is a regular power of u_1 since then d/y is an integer and the singular part of the free energy is a pure logarithmic term. As with most logarithmic terms, its coefficient is simply given by the coefficient of the "dangerous" power in $g(u_1, u_2, \ldots)$. Whereas in general the coefficient of the singular power $|u_1|^{d/y_1}$ is different for $u_1 > 0$ and $u_1 < 0$ respectively (i.e. below and above T_c), the coefficient of the logarithmic term is equal, as was already noted by Widom (1965a, b).

D. The spontaneous magnetization

So far only critical singularities have been discussed. However, in a ferromagnetic phase transition the critical point is also the onset of the appearance of a spontaneous magnetization for lower temperatures. In terms of the free energy the problem is to understand from (1.16) why, for zero magnetic field H, the derivative $\partial f/\partial H$ (magnetization) vanishes for $T > T_c$ and has a non-zero value for $T < T_c$. In order to discuss this phenomenon we have to restrict ourselves to a ferromagnetic phase transition. How a ferromagnetic phase transition manifests itself in the context of the renormalization theory has yet to be explained. To simplify the discussion we restrict ourselves to a renormalization transformation where the weights $P(s', s)$ are either 1 or 0, such that a well-defined class of allowed site spin configurations belongs to every cell spin configuration.

We will sketch here, without proof, the picture that emerges from the example to be treated in Section V. Consider a fixed point in the even subspace and a neighbouring point in the direction of the relevant (even) eigenvector. We take the direction of the eigenvector such that it points to systems with larger K_α (i.e. in the direction of lower temperatures). If the neighbouring point is repeatedly transformed the K_α will grow indefinitely to arbitrarily large values. On the other hand, a point close to the fixed point but in the direction opposite to the relevant eigenvector will decrease to vanishing values of K_α. In the latter case the point approaches the trivial fixed point $K = 0$ (∞ temperature). We shall argue that in the low-temperature limit also a new fixed point is approached, which has, however, the somewhat awkward position $K = \infty$ or at least a position with some of the $K_\alpha = \infty$.

The ∞-temperature properties of the matrix $\partial K'_{\alpha'}/\partial K_\alpha$ are easily obtained by the cumulant expansion given in the next Section. It suffices here to state that all the eigenvalues stay below l^d. The $T = 0$ limit is more difficult to discuss. The key idea is that for the transformation (1.11) in the limit $K_\alpha \to \infty$ *one* (or a finite number) of the configurations $\{s\}$ starts to contribute dominantly. In appendix B it will be shown that excitations with respect to this dominant configuration contribute only to order $\exp(-\Delta)$ where Δ is the energy of the excitation. The dominant state $\{\bar{s}\} = \{\bar{s}(s')\}$ does depend on the cell state $\{s'\}$. In fact it is the state which maximizes $P(s', s) \exp \mathscr{H}(s)$ as function of $\{s\}$.

For the triangular lattice transformation (1.18) it is possible to substantiate this idea by examples. We denote by $\{s_+\}$ the site spin configuration with all spins $s_i = 1$ and by $\{s_-\}$ the configuration with all $s_i = -1$. When all the $K_\alpha \geqslant 0$ (for even and odd α) the state $\{s_+\}$ maximizes $\mathscr{H}(s)$. For the cell spin configuration $\{s'_+\}$ the site spin configuration $\{s_+\}$ is allowed and thus

$$\{\bar{s}(s'_+)\} = \{s_+\}, \qquad K_\alpha \geqslant 0. \tag{2.57}$$

When $\{s'\}$ differs from $\{s'_+\}$ in one cell i' with $s'_{i'} = -1$, we must find $\{\bar{s}(s')\}$ between the site spin configurations with at least two site spins down in the cell i'. Without the relative strengths of the K_α known, one does not know whether $\mathscr{H}(s)$ possibly increases by turning some more spins, but when the magnetic field is sufficiently strong one should keep the number of down spins to a minimum and the 3 states with only two spins down maximize $\mathscr{H}(s)$ for that $\{s'\}$. For the case where all even $K_\alpha \geqslant 0$ and odd $K_\alpha \leqslant 0$ one has

$$\{\bar{s}(s'_-)\} = \{s_-\}\begin{cases} K_\alpha \geqslant 0 & \alpha \text{ even} \\ K_\alpha \leqslant 0 & \alpha \text{ odd} \end{cases} \tag{2.58}$$

and similar considerations for $\{s'\}$ differing from $\{s'_-\}$ in a few cells.

Neglecting the excitations and using that $P(s', \bar{s}(s')) = 1$, the transformation reduces for low temperatures to

$$G + \mathscr{H}'(s') = \mathscr{H}(\bar{s}(s')) + \log n(s') \tag{2.59}$$

where $n(s')$ is the degeneracy of the maximizing state $\bar{s}(s')$. For sufficiently low temperature the $\log n(s')$ term becomes negligible with respect to the large energy term $\mathscr{H}$. Under the circumstances given in (2.57), viz. (2.58), $\mathscr{H}'(s')$ will also be maximal for $\{s'_+\}$ viz. $\{s'_-\}$ showing that these absolute maximum states $\{s_\pm\}$ (ground states) are invariant under the transformation. The subspace with the odd $K_\alpha = 0$ separates the two domains where $\{s_+\}$ and $\{s_-\}$ are the groundstates. In the even subspace $\mathscr{H}(s_+)$ and $\mathscr{H}(s_-)$ are equal.

Now we apply the low-temperature limit also to (2.4) in the same sense as (2.59), by only taking the dominant state(s) into account in the average $\langle\,\rangle'_{s'}$. Then (2.4) becomes for $T \to 0$

$$\partial G/\partial K_\alpha + \sum_{\alpha'}\Big(\sum_{a' \in \alpha} s'_{a'}\Big)(\partial K'_{\alpha'}/\partial K_\alpha) = [n(s')]^{-1} \sum_{\text{deg}} \sum_{a \in \alpha} \bar{s}_a(s') \tag{2.60}$$

where Σ_{deg} is the sum over the degenerate maximizing states $\bar{s}(s')$. We consider (2.60) for $\{s'_\pm\}$ and write

$$\sum_{a \in \alpha} (s_\pm)_a = \chi_\alpha^\pm N. \tag{2.61}$$

Here $\chi_\alpha^\pm$ is the number of interactions of type α per site in state $\{s_\pm\}$. One has e.g.

$$\chi_\alpha^\pm = \pm 1 \quad \alpha = \text{single site}$$

$$\chi_\alpha^\pm = z/2 \quad \alpha = \text{nearest neighbours, } z = \text{ coordination number.}$$

Then (2.60) becomes

$$\partial G/\partial K_\alpha + l^{-d} \sum_{\alpha'} \chi_{\alpha'}^\pm (\partial K_{\alpha'}/\partial K_\alpha) = \chi_\alpha^\pm \tag{2.62}$$

provided that (2.57) and (2.58) apply, which is still the case in the even subspace $K_\alpha \geqslant 0$ ($\bar{s}(s'_\pm)$ are non degenerate). Subtracting the $-$ from the $+$ equation yields

$$\sum_{\alpha'} (\chi_{\alpha'}^+ - \chi_{\alpha'}^-)(\partial K_{\alpha'}/\partial K_\alpha) = l^d(\chi_\alpha^+ - \chi_\alpha^-) \tag{2.63}$$

showing that $\chi_\alpha^+ - \chi_\alpha^-$ is an eigenvector with eigenvalue l^d. The eigenvector has components only in the odd directions, since $\chi_\alpha^+ = \chi_\alpha^-$ for the even α.

Equation (2.63) constitutes an important result. Going from a fixed point K^* along an eigendirection in the even subspace to a low-temperature regime where $K_\alpha \geqslant 0$, (2.63) applies and one moves towards a $T = 0$ fixed point having an eigenvalue l^d. The eigenvalue clearly arises out of competition of the two ferromagnetic states $\{s_+\}$ and $\{s_-\}$. A fixed point with these properties will be associated with a ferromagnetic phase transition. As we shall argue in the discussion, the transformation (1.18) is implicitly designed to focus attention on the ferromagnetic phase transition and so are most other proposed renormalization transformations.

With this preparation we now return to the problem of the spontaneous magnetization. Let K be a point in the even subspace and K_α an odd interaction constant. Differentiation of (1.16) with respect to K_α yields

$$\frac{\partial f(K)}{\delta K_a} = l^{-d} \sum_{\alpha'} \frac{\partial f(K')}{\partial K'_{\alpha'}} \frac{\partial K'_{\alpha'}}{\partial K_\alpha} \qquad (\alpha \text{ odd}) \tag{2.64}$$

where we use again the fact that $\partial g/\partial K_\alpha$ vanishes for α odd. Iterating (2.64) n times leads to

$$\frac{\partial f(K)}{\partial K_\alpha} = l^{-nd} \sum_{\alpha_n} \cdots \sum_{\alpha_n} \frac{\partial f(K^{(n)})}{\partial K^{(n)}_{\alpha_n}} \cdot \frac{\partial K^{(n)}_{\alpha_n}}{\partial K^{(n-1)}_{\alpha_{n-1}}} \cdots \frac{\partial K'_{\alpha'}}{\partial K_\alpha} \qquad (\alpha \text{ odd}). \tag{2.65}$$

The derivative $\partial K^{(j+1)}_{\alpha_{j+1}}/\partial K^{(j)}_{\alpha_j}$ must be evaluated at the point $K^{(j)}$, the jth times transform of the initial point K.

Now first take K in the neighbourhood of the fixed point in the direction of (or opposite to) the relevant (even) eigenvector. As has been argued the matrix $\partial K_{\alpha_n}^{(n)}/\partial K_{\alpha_{n-1}}^{(n-1)}$ will in both cases approach a limiting value. For $K^{(n)} \to 0$ the eigenvalues are smaller than l^d and the factor l^{-nd} in front of the expression (2.65) will give a zero limit for $n \to \infty$ as the derivatives $\partial f/\partial K_\alpha$ stay finite. For $K^{(n)} \to \infty$ the expression (2.65) yields a finite limit, since the matrices start to develop an eigenvalue l^d as seen from (2.63) for α odd. The limiting derivative $\partial f/\partial K_\alpha$ can also be found by observing that $f(K)$ is determined by the ground states $\{s_\pm\}$ as

$$f(K) = \sum_\alpha K_\alpha \chi_\alpha^\pm + N^{-1} \log 2. \tag{2.66}$$

An infinitesimal perturbation δK_α (α odd) removes the degeneracy (=2 for a ferromagnet: all up or down) and adds an amount

$$\delta f = \chi_\alpha^+ |\delta K_\alpha|.$$

So we conclude that for $K \to \infty$

$$\partial f(K)/\partial K_\alpha = \pm \chi_\alpha^+ = \chi_\alpha^\pm \qquad (\alpha \text{ odd}) \tag{2.67}$$

and we obtain for (2.59)

$$\frac{\partial f(K)}{\partial K_\alpha} = \pm \sum_{\alpha'} \chi_{\alpha'}^+ \lim_{n\to\infty} \left[\prod_{j=1}^{n} l^{-d} \partial K^{(j)}/\partial K^{(j-1)} \right]_{\alpha'\alpha} \tag{2.68}$$

Since l^d becomes an eigenvalue for $j \to \infty$ the infinite product will have a limiting value.

The result (2.68) is not limited to a specific path from the fixed point to $K = \infty$ but does equally well apply to points on the same temperature side of the critical surface. When only one even relevant eigenvalue is present points on the low-temperature side of the critical surface move towards the path along the relevant eigenvector ("gully bottom" in Wilson's (1971a, b) terminology). Similarly we find no spontaneous magnetization on the high-temperature side of the critical surface.

We point out that the expression (2.68) relates the derivatives to those of (2.67) at a more attractive fixed point. In this case the derivatives at $K = \infty$ could be determined directly, but in general one should also use the renormalization theory to determine the derivatives in terms of the derivatives of even more attractive fixed points, etc.

III. Correlation Functions

So far only the thermodynamic implications of the renormalization transformation (1.11) have been discussed. In this section we shall analyse the consequences for the correlation functions. The treatment given here differs from the usual discussion of correlation functions in renormalization theory, because the arbitrariness of the weight function $P(s', s)$ does not permit us to follow the usual arguments, as we shall show below.

The standard correlation function $g(\mathbf{r})$ is defined as

$$g(\mathbf{r}_i - \mathbf{r}_j) = \langle s_i s_j \rangle \tag{3.1}$$

where $\mathbf{r}_i - \mathbf{r}_j$ is the relative position vector between sites i and j, which is measured in units of the site lattice spacing. For simplicity we restrict ourselves to even hamiltonians such that $g(\mathbf{r})$ decays to zero for large separations. We intend to establish a connection between $g(\mathbf{r})$ and the corresponding quantity $g'(\mathbf{r}')$ for the cell system. $g'(\mathbf{r}')$ is defined as

$$g'(\mathbf{r}'_{i'} - \mathbf{r}'_{j'}) = \langle s'_{i'} s'_{j'} \rangle' \tag{3.2}$$

where $\mathbf{r}'_{i'} - \mathbf{r}'_{j'}$ gives the relative position between the cells i' and j' measured in units of the cell lattice spacing. The primes on $g'(\mathbf{r}')$ and $\langle\ \rangle'$ are to remind us that the quantities refer to the renormalized hamiltonian $\mathscr{H}'(s')$.

It is not possible to derive directly a connection between the g and g' as was done for the free energies, unless $P(s', s)$ has special additional properties. To see this we form g' according to the prescription (3.2) and use the renormalization transformation (1.11)

$$g'(\mathbf{r}'_{i'} - \mathbf{r}'_{j'}) = \sum_{\{s'\}} s'_{i'} s'_{j'} \exp\left[G + \mathscr{H}'(s')\right] \Big/ \sum_{\{s'\}} \exp\left[G + \mathscr{H}'(s')\right] =$$

$$= \sum_{\{s'\}} \sum_{\{s\}} s'_{i'} s'_{j'} P(s', s) \exp \mathscr{H}(s) \Big/ \sum_{\{s\}} \exp \mathscr{H}(s) \tag{3.3}$$

such that

$$g'(\mathbf{r}'_{i'} - \mathbf{r}'_{j'}) = \sum_{\{s'\}} s'_{i'} s'_{j'} \langle P(s', s) \rangle . \tag{3.4}$$

Without further specification of $P(s', s)$ the relation (3.4) is of little use. For a special class of $P(s', s)$, sometimes misleadingly called linear transformations, (3.4) is an interesting relation. We shall discuss this in part A. In B we return to the more general problem and give a simplified treatment of $g(\mathbf{r})$, which brings out the most essential features. The complete discussion is rather involved and is outlined in C. In the last part, D, we return to relation (3.4) and present an interesting connection between eigenvectors and critical exponents which is due to Kadanoff and Houghton (1975).

A. Linear weight-factors P(s′, s)†

In practically all applications the weight-factor $P(s', s)$ is a product of weight-factors associated with the cells. Then $P(s', s)$ can be written as

$$P(s', s) = \prod_{i'} \tfrac{1}{2}(1 + s'_{i'} p(s^1_{i'}, s^2_{i'}, \ldots)). \tag{3.5}$$

† The relation of linear versus non-linear weight-factors is discussed by Bell and Wilson (1974).

Note that $P(s', s)$ fulfils condition (1.10b). In order that (1.10a) is fulfilled the function $|p|$ should not exceed 1 for any combination of its arguments. If $p(s_{i'}^1, \ldots)$ is a linear function of the spins of cell i', we call $P(s', s)$ a linear weight-factor (rather than calling the transformation based on $P(s', s)$ a linear transformation). Let us consider the simplest linear weight-factor

$$p(s_{i'}^1, s_{i'}^2, \ldots) = (s_{i'}^1 + s_{i'}^2 + \ldots)p \tag{3.6}$$

where p is as yet an arbitrary constant. Inserting (3.6) and (3.5) in (3.4) and first performing the summation over $\{s'\}$, one selects out of the product over the cells precisely the factors corresponding to cell i' and j' (all the other cell factors yield 1). So one finds

$$g'(\mathbf{r}'_{i'} - \mathbf{r}'_{j'}) = p^2 \langle (s_{i'}^1 + \ldots)(s_{j'}^2 + \ldots) \rangle. \tag{3.7}$$

The interesting property of (3.6) is that $g'(\mathbf{r}')$ is related to $g(\mathbf{r})$ for different distances $\mathbf{r}$ formed by taking a site in cell i' and one in cell j'. Although in general complicated, (3.7) becomes simple when the distance between the cells i' and j' is large with respect to the difference in position inside the cells. Then

$$|\mathbf{r}_{i'}^s - \mathbf{r}_{j'}^t| \simeq r = lr' = l|\mathbf{r}'_{i'} - \mathbf{r}'_{j'}| \tag{3.8}$$

for all choices s and t out of cells i' and j'. Thus (3.7) becomes asymptotically

$$g'(r') \simeq p^2 l^{2d} g(lr') \tag{3.9}$$

(ignoring the possible angular dependence of $g(\mathbf{r})$).

This equation is very similar in structure to the general homogeneous equation for the free energy (2.35). Let us make the notation more explicit by considering the dependence on thermodynamic variables through a (relevant) scaling field u_T. Then (3.9) may be written as

$$g(u_T, r) = l^{-d+2-\eta} g(u'_T, r') = l^{-d+2-\eta} g(l^{y_T} u_T, l^{-1} r) \tag{3.10}$$

where we have put tentatively:

$$p^2 = l^{-d-2+\eta}. \tag{3.11}$$

Using the property (2.13) for generalized homogeneous functions we find for $g(u_T, r)$

$$g(u_T, r) = r^{-(d-2+\eta)} A(u_T r^{y_T}) \tag{3.12}$$

(one should realize that $1/r$ has the status of a scaling field for $r \to \infty$). Usually one writes the correlation function as a function of r/ξ, where ξ is the correlation length. (3.11) can be cast in that form by defining

$$\xi = u_T^{-1/y_T} \sim \varepsilon^{-\nu}. \tag{3.13}$$

This relation implies that the correlation length ξ depends asymptotically on the temperature difference τ with the exponent $\nu = 1/y_T$. The relation (3.11) is, however, curious, because it relates the physical exponent η to a seemingly arbitrary mathematical parameter p of the transformation (3.6). It would be better to reverse the argument and see relation (3.11) as a restriction on the choice of a meaningful p. Linear weights, more general than (3.6), lead to the same problem, to which we shall return in the discussion†.

For non-linear weight-factors e.g. (1.18) for the triangular lattice, the spin pair correlation function is, by (3.4), coupled to higher-order correlation functions (triplet–singlet and triplet–triplet correlation functions in the case of (1.18), as we shall show in *D*).

B. Simplified treatment of g(r)

The discussion of the correlation functions will be undertaken by considering them as derivatives of the free energy. Suppose we add a perturbation to the Hamiltonian $\mathscr{H}(s)$ of the type

$$\delta\mathscr{H}(s) = \sum_i \delta H_i s_i \tag{3.14}$$

which is a spatially dependent magnetic field H_i at site i. $\delta\mathscr{H}$ will lead to a change in the free energy δF

$$e^{F+\delta F} = \sum_{\{s\}} \exp\left[\mathscr{H}(s) + \delta\mathscr{H}(s)\right]. \tag{3.15}$$

Differentiating (3.15) with respect to δH_i and δH_j and setting $\delta H(s) = 0$ afterwards, yields the correlation function

$$G(\mathbf{r}_i - \mathbf{r}_j) = \delta^2 F/\delta H_i \delta H_j = \langle s_i s_j \rangle - \langle s_i \rangle \langle s_j \rangle. \tag{3.16}$$

Now we apply the renormalization transformation (1.11) to $\mathscr{H}(s) + \delta\mathscr{H}(s)$

$$\exp\left[G + \mathscr{H}'(s') + \delta C + \delta\mathscr{H}'(s')\right] = \sum_{\{s\}} P(s', s) \exp\left[\mathscr{H}(s) + \delta\mathscr{H}(s)\right] \tag{3.17}$$

and make the simplifying assumption that $\delta\mathscr{H}'(s')$ is given by

$$\delta\mathscr{H}'(s') = \sum_{i'} \delta H'_{i'} s'_{i'} \tag{3.18}$$

i.e. that only new (spatially dependent) magnetic fields $\delta H'_{i'}$ are generated for the cells by $\delta\mathscr{H}(s)$.

† The condition (1.10a) is violated by (3.11) since (1.10a) requires $|p| < l^{-d}$, while for $\eta > 2 - d$ (which holds for $d > 1$) (3.11) implies $|p| > l^{-d}$.

For the excess free energy δF we have corresponding to (1.13)†

$$\delta F = \delta C + \delta F'. \tag{3.19}$$

By (3.18) $\delta F'$ is a function of the $\delta H'_{i'}$ such that we have for (3.16)

$$G(\mathbf{r}_i - \mathbf{r}_j) = C(\mathbf{r}_i - \mathbf{r}_j) + \sum_{i'} \sum_{j'} G'(\mathbf{r}'_{i'} - \mathbf{r}'_{j'})(\delta H'_{i'}/\delta H_i)(\delta H'_j/\delta H_j)$$
$$+ \sum_{i'} (\delta F'/\delta H'_{i'})(\delta^2 H'_{i'}/\delta H_i \delta H_j) \tag{3.20}$$

with $C(\mathbf{r}_i - \mathbf{r}_j)$ defined by

$$C(\mathbf{r}_i - \mathbf{r}_j) = \delta^2 C/\delta H_i \delta H_j. \tag{3.21}$$

The derivative matrix

$$T_{i'i} = \delta H'_{i'}/\delta H_i \tag{3.22}$$

follows by differentiating (3.17) with respect to δH_i and use of (3.18)

$$\delta C/\delta H_i + \sum_{i'} s'_{i'} T_{i'i} = \langle s_i \rangle_{s'}, \tag{3.23}$$

where $\langle\ \rangle_{s'}$ is defined in (2.5).

Differentiating once more yields

$$C(\mathbf{r}_i - \mathbf{r}_j) + \sum_{i'} s'_{i'} \delta^2 H'_{i'}/\delta H_i \delta H_j = \langle s_i s_j \rangle_{s'} - \langle s_i \rangle_{s'} \langle s_j \rangle_{s'}. \tag{3.24}$$

For further simplification we may restrict ourselves to an even $\mathscr{H}(s)$. Then the averages $\langle\ \rangle_{s'}$ on the right-hand side of (3.23) and (3.24) are odd and even in s' respectively and thus $\delta C/\delta H_i$ and $\delta^2 H'_{i'}/\delta H_i \delta H_j$ vanish in that case. So (3.20) may be written as

$$G(\mathbf{r}_i - \mathbf{r}_j) = C(\mathbf{r}_i - \mathbf{r}_j) + \sum_{i'} \sum_{j'} G'(\mathbf{r}'_{i'} - \mathbf{r}'_{j'}) T_{i'i} T_{j'j}. \tag{3.25}$$

This is the desired transformation law for the correlation function which is the analogue of the free energy transformation (1.16). By choosing the symbols G and C we want to stress also the analogy with the Ornstein–Zernike relation, where the correlation function G is expressed in terms of the supposedly short-range direct correlation C. Here we shall also assume that $C(\mathbf{r})$ is short-ranged, which is a natural extension of the regularity assumption for $g(K)$ entering in the free energy equation. By summing (3.24) over j and using the derivative of (2.4) with respect to the magnetic field one finds the following relation between C and g.

† We use the same symbol G for the correlation function and the "empty set" contribution G to the free energy, which we hope will not lead to confusion. However, we have denoted the counterpart of the free energy G by C in (3.17) and (3.19) to avoid misunderstanding.

$$\sum_j C(\mathbf{r}_i - \mathbf{r}_j) = \frac{1}{N}\left[\left\langle \sum_i s_i \sum_j s_i \right\rangle_{s'} - \left\langle \sum_i s_i \right\rangle_{s'} \left\langle \sum_j s_j \right\rangle_{s'}\right] = \partial g^2/\partial H^2 \qquad (3.26)$$

(ignoring in both (3.24) and (2.4) higher-order terms in s').

In fact one should run the argument backwards and base the regularity of g on the short-rangedness of $C(\mathbf{r})$. The right-hand side of (3.24) may be considered as the correlation function for a given cell configuration s' as constraint. Essentially one supposes that the *constrained* correlation function is short-range. The degrees of freedom left over in the partial average do not include the long-wavelength fluctuations which are responsible for the long-range correlations (and the thermodynamical singularities). These come about by the summation over the cell configurations, which amongst others describe the long-wavelength fluctuations. So the restricted averages on the right-hand side of (3.24) do not lead to long-range behaviour and associated thermodynamic singularities.

The analysis of (3.25) is, as for the Ornstein–Zernike equation, relatively simple, due to the translational invariance of $T_{i'i}$. We define the fourier transform $\tilde{T}(\mathbf{k})$ of $T_{i'i}$ as

$$\tilde{T}(\mathbf{k}) = \sum_i T_{i'i} \exp i\mathbf{k}\cdot(\mathbf{r}_{i'} - \mathbf{r}_i) \qquad (3.27)$$

where both $\mathbf{r}_{i'}$ and $\mathbf{r}_i$, being the positions of cell i' and site i, are measured in units of the site lattice spacing. We note that $\tilde{T}(\mathbf{0})$ in the approximation (3.18) (or (3.23)) has the meaning of the magnetic eigenvalue λ_H, when worked out at the fixed point

$$\tilde{T}^*(\mathbf{0}) = \lambda_H = l^{y_H}. \qquad (3.28)$$

Now multiplying (3.25) with $\exp i\mathbf{k}\cdot(\mathbf{r}_i - \mathbf{r}_j)$ and summing over i and j yields

$$N\tilde{G}(\mathbf{k}) = N\tilde{C}(\mathbf{k}) + N'\tilde{T}(-\mathbf{k})\tilde{T}(\mathbf{k})\tilde{G}'(\mathbf{k}'). \qquad (3.29)$$

The argument $\mathbf{k}'$ of $\tilde{G}'$ follows by rewriting the position vectors $\mathbf{r}_{i'}$ and $\mathbf{r}_{j'}$ in terms of the positions $\mathbf{r}'_{i'}$ and $\mathbf{r}'_{j'}$ which are measured in the cell lattice spacing. For some renormalization transformations we have a simple scale factor

$$\mathbf{r}_{i'} - \mathbf{r}_{j'} = l(\mathbf{r}'_{i'} - \mathbf{r}'_{j'}) \qquad (3.30)$$

but there are cases (as in the triangular example) where the cell system is also rotated over an angle with respect to the site system. This complication is of little importance because in all examples we know of, one can do away with this rotation by considering higher powers of the transformation. So we restrict the discussion to case (3.30) which implies $\mathbf{k}' = \mathbf{k}l$. Thus (3.29) becomes

$$\tilde{G}(\mathbf{k}) = \tilde{C}(\mathbf{k}) + l^{-d}\tilde{T}(-\mathbf{k})\tilde{T}(\mathbf{k})\tilde{G}'(\mathbf{k}l). \qquad (3.31)$$

The form (3.31) is almost the same in mathematical nature as the transformation law (1.16) for the free energy. The wave vector **k** plays the rôle of an additional scaling field with exponent $y = 1$. The difference from (1.16) is that the **k**-dependent combination $l^{-d}\tilde{T}(-\mathbf{k})\tilde{T}(\mathbf{k})$ replaces the factor l^{-d}. This complication will be removed in a way that prepares for the general discussion of the next section. Define the regular function $\psi(\mathbf{k})$ through the recurrence relation

$$l^{y_H}\psi'(\mathbf{k}l) = \tilde{T}(\mathbf{k})\psi(\mathbf{k}) \tag{3.32}$$

with boundary condition $\psi^*(\mathbf{0}) = 1$. In order to explain (3.32) consider first this relation at the fixed point

$$l^{y_H}\psi^*(\mathbf{k}l) = \tilde{T}^*(\mathbf{k})\psi^*(\mathbf{k}). \tag{3.32a}$$

By power series expansion in **k** the $\psi^*(\mathbf{k})$ may be found for small **k** and then recursively for larger **k** by direct application of (3.32a) as in the case of scaling fields. Near the fixed point and for small **k** one may obtain $\psi(k)$ from (3.32) by simultaneous power series expansion in **k** and the scaling fields. Through application of (3.32) the functions $\psi(\mathbf{k})$ may be obtained further out.

Now multiply (3.31) by $\psi(-\mathbf{k})\psi(\mathbf{k})$ and with (3.32) one obtains

$$\tilde{\chi}(\mathbf{k}) = \tilde{\gamma}(\mathbf{k}) + l^{-d+2y_H}\tilde{\chi}'(\mathbf{k}l) \tag{3.33}$$

with the definitions

$$\begin{aligned} \tilde{\chi}(\mathbf{k}) &= \tilde{G}(\mathbf{k})\psi(\mathbf{k})\psi(-\mathbf{k}), \\ \tilde{\gamma}(\mathbf{k}) &= \tilde{C}(\mathbf{k})\psi(\mathbf{k})\psi(-\mathbf{k}). \end{aligned} \tag{3.34}$$

The analogy with (1.16) is now complete and we may use the analysis of Section IIB for the discussion of the small **k** behaviour. We shall do this in the next section for the more general case. It suffices here to point out that the singular part of $\tilde{G}(\mathbf{k})$ behaves at the fixed point as

$$\tilde{G}(\mathbf{k}) \sim \tilde{\chi}(\mathbf{k}) \sim |\mathbf{k}|^{d-2y_H}, \tag{3.35}$$

which implies for η (see (3.12)) the scaling relation

$$\eta = d + 2 - 2y_H. \tag{3.36}$$

This is a hyperscaling law equivalent to the relation

$$\delta = (d + 2 - \eta)/(d - 2 + \eta) \tag{3.37}$$

due to Fisher (1969).

C. General discussion of the correlation functions

For a generalization of the previous discussion, which hinged on the simplifications (3.18) and (3.25), we have to generalize the idea of scaling to spatially dependent interactions. Once this has been achieved the treatment follows rather closely the lines of the simplified version IIB.

Instead of considering the special magnetic perturbation (3.14) we employ the general expression

$$\delta\mathcal{H}(s) = \sum_a \delta K_a s_a. \tag{3.38}$$

By differentiation of the excess free energy δF with respect to δK_a and δK_b (setting $\delta\mathcal{H} = 0$ afterwards) we obtain the general correlation function

$$G_{ab} = \delta^2 F/\delta K_a \delta K_b = \langle s_a s_b \rangle - \langle s_a \rangle \langle s_b \rangle. \tag{3.39}$$

By taking a and b equal to the sites i and j one recovers the correlation function $G(\mathbf{r}_i - \mathbf{r}_j)$.

Applying the renormalization transformation (1.11) to $\mathcal{H} + \delta\mathcal{H}$ yields an excess $\delta C + \delta\mathcal{H}'$ where $\delta\mathcal{H}'$ will now (instead of (3.18)) be represented by the general expression

$$\delta\mathcal{H}'(s) = \sum_{a'} \delta K'_{a'} s'_{a'}. \tag{3.40}$$

Again the excess free energy fulfils the relation (cf. (3.19))

$$\delta F = \delta C + \delta F' \tag{3.41}$$

from which the following relation between the correlation functions can be derived by differentiating with respect to δK_a and δK_b

$$G_{ab} = C_{ab} + \sum_{a'}\sum_{b'} G'_{a'b'}(\delta K'_{a'}/\delta K_a)(\delta K'_{b'}/\delta K_b) + \sum_{a'} (\delta F'/\delta K'_{a'})(\delta^2 K'_{a'}/\delta K_a \delta K_b) \tag{3.42}$$

with the definition

$$C_{ab} = \delta^2 C/\delta K_a \delta K_b. \tag{3.43}$$

The other derivatives in (3.42) will be denoted by

$$\begin{aligned} T_{a'a} &= \delta K'_{a'}/\delta K_a, \\ T_{a',ab} &= \delta^2 K'_{a'}/\delta K_a \delta K_b \end{aligned} \tag{3.44}$$

and follow from the constraint averages

$$\begin{aligned} \langle s_a \rangle_{s'} &= \delta C/\delta K_a + \sum_{a'} s'_{a'} T_{a'a} \\ \langle s_a s_b \rangle_{s'} - \langle s_a \rangle_{s'} \langle s_b \rangle_{s'} &= C_{ab} + \sum_{a'} s'_{a'} T_{a',ab}. \end{aligned} \tag{3.45}$$

$T_{a'a}$ is a generalization of $T_{\alpha'\alpha}$ and yields $T_{\alpha'\alpha}$ upon summation over a, as follows from (2.4)

$$\sum_{a\in\alpha} T_{a'a} = T_{\alpha'\alpha}, \tag{3.46}$$

where α' is the class to which a' belongs. In contrast to the previous discussion we may not drop $T_{a',ab}$. We shall incorporate the additional term in (3.42) involving $T_{a',ab}$, by taking suitable combinations of the δK_a as new coordinates (scaling fields) for which the second derivatives vanish. The new coordinates $u_i(\mathbf{k})$ depend on a wavenumber $\mathbf{k}$ and an index i giving the type of combination similarly to the index of the scaling fields. We will require that $u_i(\mathbf{k})$ transform as:

$$\delta u_i'(\mathbf{k}l) = l^{y_i}\delta u_i(\mathbf{k}) \tag{3.47}$$

where y_i is the same as the exponent of the scaling field u_i defined earlier. For the formation of the $\delta u_i(\mathbf{k})$ we consider first the fourier transforms†

$$\begin{aligned}\delta\tilde{K}_\alpha(\mathbf{k}) &= N^{-1}\sum_{a\in\alpha}\delta K_a \exp i\mathbf{k}\cdot\mathbf{r}_a,\\ \delta K_a &= \sum_{\mathbf{k}}\delta\tilde{K}_\alpha(\mathbf{k})\exp -i\mathbf{k}\cdot\mathbf{r}_a\end{aligned} \tag{3.48}$$

where $\mathbf{r}_a$ is the position of set a on the site lattice and where the summation over $\mathbf{k}$ runs through all wave vectors fitting the periodic boundary conditions of the site lattice. Then we construct in analogy of (3.27) a $\mathbf{k}$-dependent matrix

$$\tilde{T}_{\alpha\beta}(\mathbf{k}) = \sum_{b\in\beta} T_{a'b}\exp i\mathbf{k}\cdot(\mathbf{r}_{a'} - \mathbf{r}_b) \tag{3.49}$$

In view of (3.46) we have $T_{\alpha\beta}(\mathbf{0}) = T_{\alpha\beta}$. With (3.49) we derive the (first order) transformation law for the fourier coefficients δK_α as (see also appendix A)

$$\begin{aligned}\delta\tilde{K}_\alpha'(\mathbf{k}') &= (N')^{-1}\sum_{a'\in\alpha}\delta K_{a'}'\exp i\mathbf{k}'\cdot\mathbf{r}_{a'}'\\ &= (N')^{-1}\sum_{\mathbf{k}}\sum_{a'\in\alpha}\left\{\sum_{\beta}\tilde{T}_{\alpha\beta}(\mathbf{k})\delta\tilde{K}_\beta(\mathbf{k})\right\}\exp i[\mathbf{k}'\cdot\mathbf{r}_{a'}' - \mathbf{k}\cdot\mathbf{r}_{a'}].\end{aligned} \tag{3.50}$$

In the sum over a', it should be remembered that $\mathbf{r}_{a'}'$ measures the position of set a' (of cells) in the cell system while $\mathbf{r}_{a'}$ gives the position of a' in the site system. So $\mathbf{r}_{a'} = l\mathbf{r}_{a'}'$ and thus

$$(N')^{-1}\sum_{a'\in\alpha}\exp i[\mathbf{k}'\cdot\mathbf{r}_{a'}' - \mathbf{k}\cdot\mathbf{r}_{a'}] = \delta_{\mathbf{k}',\mathbf{k}l}. \tag{3.51}$$

† In order not to complicate the notation we change in the following the interpretation of α from being the class of all sets a which can be identified with each other by a symmetry operation on the lattice to the class of sets a which can be identified by a translation.

Therefore (3.50) reads

$$\delta\tilde{K}'_\alpha(\mathbf{k}l) = \sum_\beta \tilde{T}_{\alpha\beta}(\mathbf{k})\delta\tilde{K}_\beta(\mathbf{k}). \tag{3.52}$$

Now if we define a set of regular functions $\varphi^i_\alpha(\mathbf{k})$ through the relation

$$\sum_\alpha \varphi^{i'}_\alpha(\mathbf{k}l)\tilde{T}_{\alpha\beta}(\mathbf{k}) = l^{y_i}\varphi^i_\beta(\mathbf{k}) \tag{3.53}$$

then the combinations

$$\delta u_i(\mathbf{k}) = \sum_\alpha \varphi^i_\alpha(\mathbf{k})\delta\tilde{K}_\alpha(\mathbf{k}) \tag{3.54}$$

fulfil the transformation law (3.47). Equation (3.53) is the generalization of (3.32) but now referring to left eigenvectors. At the fixed point and for $\mathbf{k} = 0$, (3.53) reduces to

$$\sum_\alpha \varphi^{i*}_\alpha(\mathbf{0})T^*_{\alpha\beta} = l^{y_i}\varphi^{i*}_\beta(\mathbf{0}) \tag{3.55}$$

such that the values $\varphi^{i*}_\alpha(\mathbf{0})$ can be chosen to be equal to the left eigenvectors φ^i_α defined in (1.28). With this boundary condition the functions $\varphi^i_\alpha(\mathbf{k})$ may be obtained for non-zero $\mathbf{k}$ and in the neighbourhood of the fixed point by making a power series expansion of (3.53) in $\mathbf{k}$ and the scaling fields u_i. By repeated use of (3.53) one may solve for $\varphi^i_\alpha(\mathbf{k})$ for larger values of $\mathbf{k}$ and further away from the fixed point.

In order to change back and forth from the $\delta u_i(\mathbf{k})$ to the $\delta\tilde{K}_\alpha(\mathbf{k})$ we need also the right eigenvector version of (3.53)

$$\sum_\beta \tilde{T}_{\alpha\beta}(\mathbf{k})\psi^i_\beta(\mathbf{k}) = l^{y_i}\psi^{i'}_\alpha(\mathbf{k}l) \tag{3.56}$$

which defines the functions $\psi^i_\alpha(\mathbf{k})$ starting from the $\psi^{i*}_\alpha(\mathbf{0}) = \psi^i_\alpha$ defined as the right eigenvectors of $T^*_{\alpha\beta}$. By multiplying (3.56) on the left by $\varphi^{j'}_\alpha(\mathbf{k}l)$ and summing over α we find with (3.53)

$$l^{y_j}\sum_\alpha \varphi^j_\alpha(\mathbf{k})\psi^i_\alpha(\mathbf{k}) = l^{y_i}\sum_\alpha \varphi^{j'}_\alpha(\mathbf{k}l)\psi^{i'}_\alpha(\mathbf{k}l). \tag{3.57}$$

Since the φ^i_α and the ψ^i_α can be chosen orthogonal at the fixed point for $\mathbf{k} = 0$, this orthogonality propagates by (3.57) to $\mathbf{k} \neq 0$ and away from the fixed point:

$$\sum_\alpha \varphi^j_\alpha(\mathbf{k})\psi^i_\alpha(\mathbf{k}) = \delta_{ij}. \tag{3.58}$$

The completeness of the φ^i_α and ψ^i_α then also carries over from the fixed point and $\mathbf{k} = 0$ to

$$\sum_i \psi^i_\alpha(\mathbf{k})\varphi^i_\beta(\mathbf{k}) = \delta_{\alpha\beta}. \tag{3.59}$$

With (3.59) one can invert (3.54) to

$$\delta \tilde{K}_\alpha(\mathbf{k}) = \sum_i \psi^i_\alpha(\mathbf{k})\delta u_i(\mathbf{k}). \tag{3.60}$$

The next step would be to extend (3.54) to second order in δK_α in order to take the second-order derivatives in the last term in (3.42) into account. The extension should be such that (3.47) remains true to second order. We leave this point for the moment and investigate first the merits of having constructed scaling coordinates obeying (3.47). We define new correlation functions $\tilde{\chi}_{ij}(\mathbf{k})$ as

$$\tilde{\chi}_{ij}(\mathbf{k}) = \delta^2 f/\delta u_i(-\mathbf{k})\delta u_j(\mathbf{k}) \tag{3.61}$$

and a short-range correlation function $\tilde{\gamma}_{ij}(\mathbf{k})$ ($c = C/N$)

$$\tilde{\gamma}_{ij}(\mathbf{k}) = \delta^2 c/\delta u_i(-\mathbf{k})\delta u_j(\mathbf{k}) \tag{3.62}$$

From (3.41) and (3.47) it follows that they are related as

$$\tilde{\chi}_{ij}(\mathbf{k}) = \tilde{\gamma}_{ij}(\mathbf{k}) + l^{-d+y_i+y_j}\tilde{\chi}'_{ij}(\mathbf{k}l) \tag{3.63}$$

This is obviously a transformation law for correlation functions of the desired form. Note that in deriving (3.63) we have used the fact that new and old scaling coordinates are related linearly.

Now two points remain to be dealt with: the solution of $\tilde{\chi}_{ij}(\mathbf{k})$ in terms of $\tilde{\gamma}_{ij}(\mathbf{k})$ (for which the machinery developed in section IIC can be used), and the connection between $\tilde{\chi}_{ij}$ and G_{ab} on the one hand and $\tilde{\gamma}_{ij}$ and C_{ab} on the other hand. For the latter point we do need the second-order extension of (3.54) and (3.60). In Appendix A the straightforward but rather involved formulae are given. We present here the result.

Consider the fourier transforms of G_{ab} and C_{ab}

$$\tilde{G}_{\alpha\beta}(\mathbf{k}) = \sum_{b\in\beta} G_{ab} \exp i\mathbf{k}\cdot(\mathbf{r}_a - \mathbf{r}_b)$$
$$\tilde{C}_{\alpha\beta}(\mathbf{k}) = \sum_{b\in\beta} C_{ab} \exp i\mathbf{k}\cdot(\mathbf{r}_a - \mathbf{r}_b). \tag{3.64}$$

Then $\tilde{G}_{\alpha\beta}(\mathbf{k})$ and $\tilde{\chi}_{ij}(\mathbf{k})$ are related through the formula

$$\tilde{G}_{\alpha\beta}(\mathbf{k}) = \sum_{ij} \varphi^i_\alpha(-\mathbf{k})\varphi^j_\beta(\mathbf{k})\tilde{\chi}_{ij}(\mathbf{k}) + \sum_\gamma (\partial f/\partial K_\gamma) \sum_i \psi^i_\gamma(\mathbf{0})\varphi^i_{\alpha\beta}(-\mathbf{k},\mathbf{k}). \tag{3.65}$$

Here $\varphi^i_{\alpha\beta}(\mathbf{k}_1, \mathbf{k}_2)$ is a regular function defined in appendix (A.7). Thus the singularities of $G_{\alpha\beta}(\mathbf{k})$ in the $\mathbf{k}$ variable are completely contained in $\tilde{\chi}_{ij}(\mathbf{k})$. The short-range part of the correlation function contains also some thermodynamic singularities through the appearance of $\partial f/\partial K_\alpha$ in the last term.

The connection between $\tilde{\gamma}_{ij}(\mathbf{k})$ and $\tilde{C}_{\alpha\beta}(\mathbf{k})$ is given by

$$\tilde{\gamma}_{ij}(\mathbf{k}) = \sum_{\alpha\beta} \tilde{C}_{\alpha\beta}(\mathbf{k})\psi^i_\alpha(-\mathbf{k})\psi^j_\beta(\mathbf{k}) - \sum_\alpha (\partial g/\partial K_\alpha) \sum_l \varphi^l_\alpha(\mathbf{0})\varphi^{lij}(-\mathbf{k}, \mathbf{k}). \tag{3.66}$$

The function $\varphi^{lij}(\mathbf{k}_1, \mathbf{k}_2)$ is regular and defined in appendix (A.10). So $\tilde{\gamma}_{ij}(\mathbf{k})$ is indeed short-ranged and regular in the thermodynamic variables.

The line of calculation is now as follows. First compute for a certain transformation the regular functions $\varphi^i_\alpha(\mathbf{k})$, $\psi^i_\alpha(\mathbf{k})$, $\varphi^i_{\alpha\beta}(\mathbf{k}, -\mathbf{k})$ and $\varphi^{lij}(-\mathbf{k}, \mathbf{k})$ as well as from (3.45) and (3.64) the $\tilde{C}_{\alpha\beta}(\mathbf{k})$ such that $\tilde{\gamma}_{ij}(\mathbf{k})$ can be calculated. Then solve (3.63) by the methods of Section IIC yielding $\tilde{\chi}_{ij}(\mathbf{k})$, which gives with (3.65) the desired $\tilde{G}_{\alpha\beta}(\mathbf{k})$. Of this process we outline the determination of the long-range behaviour of the correlation functions.

At criticality the singular part in $\mathbf{k}$ of $\tilde{\chi}_{ij}(\mathbf{k})$ is given by

$$[\tilde{\chi}_{ij}(\mathbf{k})]_{\text{sing}} = \tilde{A}_{ij}(\hat{\mathbf{k}})|\mathbf{k}|^{d-y_i-y_j}, \tag{3.67}$$

and the amplitude $\tilde{A}_{ij}(\hat{\mathbf{k}})$ given by

$$\tilde{A}_{ij}(\hat{\mathbf{k}}) = [\log l]^{-1} \int_0^\infty \mathrm{d}t\, t^{-1-d+y_i+y_j}[\tilde{\gamma}_{ij}(\hat{\mathbf{k}}t)]_{\text{rem}} \tag{3.68}$$

where the subscript "rem" is defined as in Section IIC: subtract the powers k^p from $\tilde{\gamma}_{ij}(\mathbf{k})$ for which $l^p < l^{d-y_i-y_j}$. (For the irrelevant eigenvalues $y_i < 0$, $y_j < 0$, now also a remainder and thus singular parts and amplitudes exist.) The behaviour (3.67) implies that the exponent η_{ij} for the correlation function $\tilde{\chi}_{ij}$ is given by (cf. (3.36))

$$\eta_{ij} = d + 2 - y_i - y_j. \tag{3.69}$$

The behaviour near criticality will be described by taking one relevant scaling field u_T into account, which couples to the temperature difference with the critical point. Using expression (2.41) we have

$$[\tilde{\chi}_{ij}(u_T, \mathbf{k})]_{\text{sing}} = |\mathbf{k}|^{\eta_{ij}-2}\tilde{A}_{ij}(\hat{\mathbf{k}}; u_T|\mathbf{k}|^{-y_T}) \tag{3.70}$$

with the amplitude given by the integral

$$\tilde{A}_{ij}(\hat{\mathbf{k}}; v_T) = [\log l]^{-1} \int_0^\infty \mathrm{d}t\, t^{-1-d+y_i+y_j}[\tilde{\gamma}_{ij}(\hat{\mathbf{k}}t; v_T t^{y_T})]_{\text{rem}}. \tag{3.71}$$

For the spatial behaviour (3.70) implies

$$\tilde{\chi}_{ij}(\mathbf{r}) \simeq \frac{A_{ij}(\hat{\mathbf{r}}; r u_T^{1/y_T})}{r^{d-2+\eta_{ij}}}. \tag{3.72}$$

Thus a correlation length ξ

$$\xi = u_T^{-1/y_T} \tag{3.73}$$

appears, with exponent $\nu = 1/y_T$ (cf. (3.13)), which is the same for all correlation functions.

For the correlation function $G_{\alpha\beta}(\mathbf{r})$ the result (3.67) yields at criticality

$$G_{\alpha\beta}(\mathbf{r}) \simeq \sum_{ij} \phi_\alpha^i \phi_\beta^j A_{ij}(\hat{\mathbf{r}})/r^{(d-2+\eta_{ij})}. \tag{3.74}$$

As we are interested in the dominant decay we could drop the **k**-dependence in $\varphi_\alpha^i(\mathbf{k})$ in (3.65). For the same reason in the summation in (3.74) over i and j one should take only the term with the *lowest* η_{ij} for which the coefficient in front does not vanish; e.g. for α and β both referring to single sites $G_{\alpha\beta}(\mathbf{r})$ couples to the combination i, j both equal to the relevant (odd) magnetic mode H.

The case where α and β both refer to nearest-neighbour pair (the energy–energy correlation function for the Ising model) couples to i, j both equal to the relevant (even) temperature mode T.

D. Connection between exponents and eigenvectors

After having analysed the general behaviour of the correlation function we come back to the relation (3.4) for the correlation function which we generalize to:

$$g'_{\alpha\beta}(\mathbf{r}'_{i'} - \mathbf{r}'_{j'}) = \sum_{\{s'\}} s'_{\alpha,i'} s'_{\beta,j'} \langle P(s', s) \rangle. \tag{3.75}$$

Rather than trying to present general formulae for arbitrary $P(s', s)$ we illustrate the content of (3.75) for the triangular (non-linear) weight-factor (1.18)). First we study the case of the pair correlation function where α and β both stand for "single site". Then the sum over (3.75) selects out of the product for $P(s', s)$ the factors referring to cells i' and j'. Instead of (3.7) we obtain

$$g'(\mathbf{r}'_{i'} - \mathbf{r}'_{j'}) = \tfrac{1}{4}\langle (s_{i'}^1 + s_{i'}^2 + s_{i'}^3 - s_{i'}^1 s_{i'}^2 s_{i'}^3)(s_{j'}^1 + s_{j'}^2 + s_{j'}^3 - s_{j'}^1 s_{j'}^2 s_{j'}^3)\rangle. \tag{3.76}$$

Now there appear on the right-hand side singlet–singlet, singlet–triplet and triplet–triplet correlation functions. Writing $\alpha = s$ for a single site and $\alpha = t$ for a triplet and using (3.74) for the asymptotic behaviour we obtain at the fixed point

$$\sum_{i,j} \frac{\phi_s^i \phi_s^j A_{ij}}{r'^{d-2+\eta_{ij}}} = \tfrac{1}{4} \sum_{i,j} \frac{(3\phi_s^i - \phi_t^i)(3\phi_s^j - \phi_t^j) A_{ij}}{(lr')^{d-2+\eta_{ij}}}. \tag{3.77}$$

The leading terms on both sides have to be equal. For the odd $\alpha = s, t$ the slowest decay comes from the magnetic (odd) eigenvalue y_H. So the terms with $i = j = H$ must be equal on both sides in (3.77), which is the case of

$$l^{d-y_H} \phi_s^H = \tfrac{3}{2}\phi_s^H - \tfrac{1}{2}\phi_t^H \tag{3.78}$$

where we used (3.69) for $\eta_{HH} = d + 2 - 2y_H$.

Equation (3.78) is a direct connection between the exponent y_H and the components ϕ_α^H of the (left) eigenvector. It replaces the relation (3.11) which was derived for the linear weight-factor (3.6). The numbers $\frac{3}{2}$ and $\frac{1}{2}$ occurring at the right-hand side of (3.78) are typical for the triangular weight-factor (1.18). Since (3.77) includes only the *dominant* decay for each mode, we may not draw further conclusions for the other modes in (3.77). Kadanoff and Houghton (1975) have used a relation equivalent to (3.78) to optimize a free parameter in their renormalization transformation. (See Section V.B.3).

Relations like (3.78) can be derived for other choices of α and β in (3.75), but the equation soon becomes of little practical value. By taking α and β to be a pair of neighbouring cells ($\alpha = \beta = p$) one can derive a relation between the even relevant exponent y_T and the corresponding eigenvector φ_α^T of the type

$$l^{d-y_T}\phi_p^T = \sum_\alpha w_\alpha \phi_\alpha^T. \tag{3.79}$$

where the w_α are numbers involving nearest-neighbour pairs ($\alpha = p$), other more distant pairs, quartets and a sextet of sites (e.g. $w_p = \frac{1}{2}$).

IV. Computational Methods

The crux of the renormalization transformation (1.22) is to evaluate the functions $K'(K)$ to a sufficient accuracy, such that reliable conclusions can be drawn from them. To that purpose the summation over the site spin configurations $\{s\}$ in (1.11) has to be carried out. $\mathscr{H}(s)$ is in general more complicated than e.g. a "pure" Ising Hamiltonian. Moreover the summation over $\{s\}$ is constrained by the weight-factor $P(s', s)$, so its exact evaluation is, except for a few cases, hardly feasible. Also, it is not the purpose of the renormalization theory to add to the number of exactly solvable models, but rather to provide for more realistic cases a computational scheme for the critical singularities and the free energy.

In this Section we treat approximation schemes of general character, which are partly complementary. These schemes are more or less copies of comparable approximation schemes for the Ising model. Here we shall review briefly the main methods that have been used so far:

A. simple renormalization transformations on finite lattices;
B. the cumulant approximation;
C. the cluster approximation.

In all the applications we are going to deal with, the site lattice will be subdivided into cells, the cell lattice having the same symmetry (or approximately the same symmetry) as the underlying site lattice. For the triangular lattice a possible subdivision in cells has already been given in Fig. 1a. For

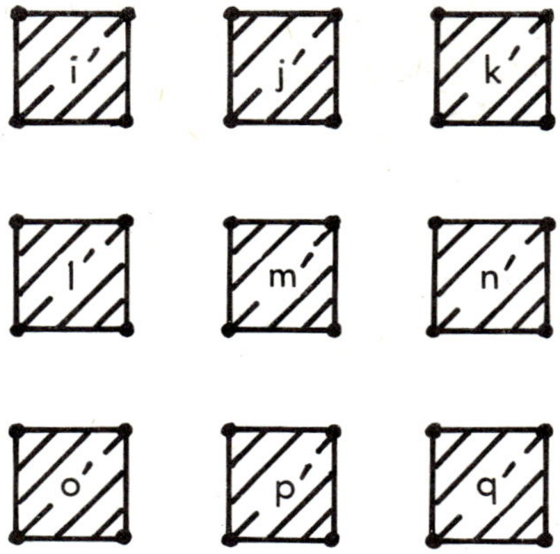

FIG. 3

the square lattice one might for example make a subdivision in cells as in Fig. 3 or Fig. 4.

The subdivision of Fig. 3 will be worked out in some detail in Section V.B.1. As can be seen from Fig. 4 it is not at all necessary that all site spins need be part of cell spins. Clearly the cell sub divisions of Fig. 3 and Fig. 4 preserve

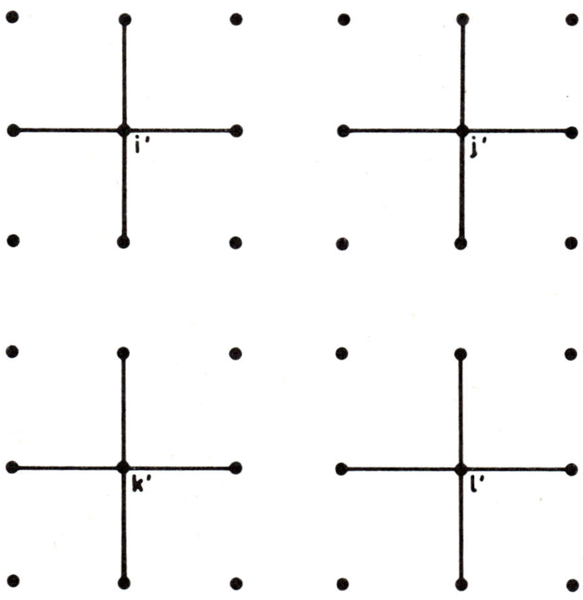

FIG. 4

completely the symmetry of the underlying site lattice, whereas the cell subdivision in Fig. 1a of the triangular lattice slightly distorts it.

A particular property of all the weight-functions of the applications given in Section V is that for a given cell spin $s'_{i'}$ the weight-factor $P(s', s)$ depends only on the site spins $s^j_{i'}$ "inside" the cell spin, i.e. the weight-factor $P(s', s)$ factorizes:

$$P(s', s) = \prod_{i'=1}^{N'} P(s'_{i'}; s^1_{i'}, s^2_{i'}, \ldots, s^n_{i'}). \tag{4.1}$$

The property (4.1) turns out to be extremely useful in actual calculations.

A. Simple renormalization transformations on finite lattices

In this approach (Tjon, 1974; Subbarao, 1975) one takes a symmetric arrangement of cells, gives a prescription for $P(s', s)$ and, normally on a computer, performs the summation over the configurations $\{s\}$. Starting with a certain (numerical) input K_α, this then directly yields (numerical) values for the coupling parameters $K'_{a'}$ between the cells, depending on the values of K_α on the original site lattice.

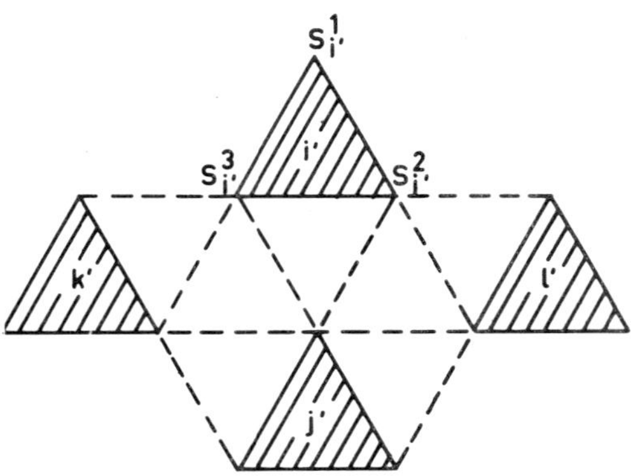

FIG. 5

Notice that we have used the notation $K'_{a'}$, instead of K'_α for the interaction parameters between the cells. This is because a simple transformation like this on a finite lattice usually immediately introduces certain asymmetries into the cell lattice. If one applies this procedure for example on the block of cells drawn in Fig. 5, the nearest-neighbour coupling between the cells i' and l' will in general be different from the nearest-neighbour coupling between the cells i' and j'. This, however, can be remedied as follows: modify the transformation by simply taking the linear average of the new couplings belonging to the same class, such that one can again speak about K'_α depending on the type α only.

One may refine this simple procedure of finding renormalization transformations on finite lattices slightly (Nauenberg and Nienhuis, 1974b) by applying *periodic boundary conditions*, e.g. as follows. Take the four-cell cluster $\{i', j', l', m'\}$ on a square lattice as drawn in Fig. 3 and surround this block by *one* additional layer of spins of which the directions are completely determined by periodic boundary conditions in the sense that $s^1_{k'} = s^1_{i'}, \ldots s^4_{k'} = s^4_{i'}$, etc.

Setting aside this particular example and assuming that the system under consideration contains sufficiently many cells that one may impose periodic boundary conditions, one may view the periodic system also as an infinite system in which the possible (site and cell) spin configurations are restricted to periodic configurations. Now for low temperatures the periodic configurations are the dominant ones, so this method may be expected to yield a correct low-temperature description of the renormalization transformation. Un-

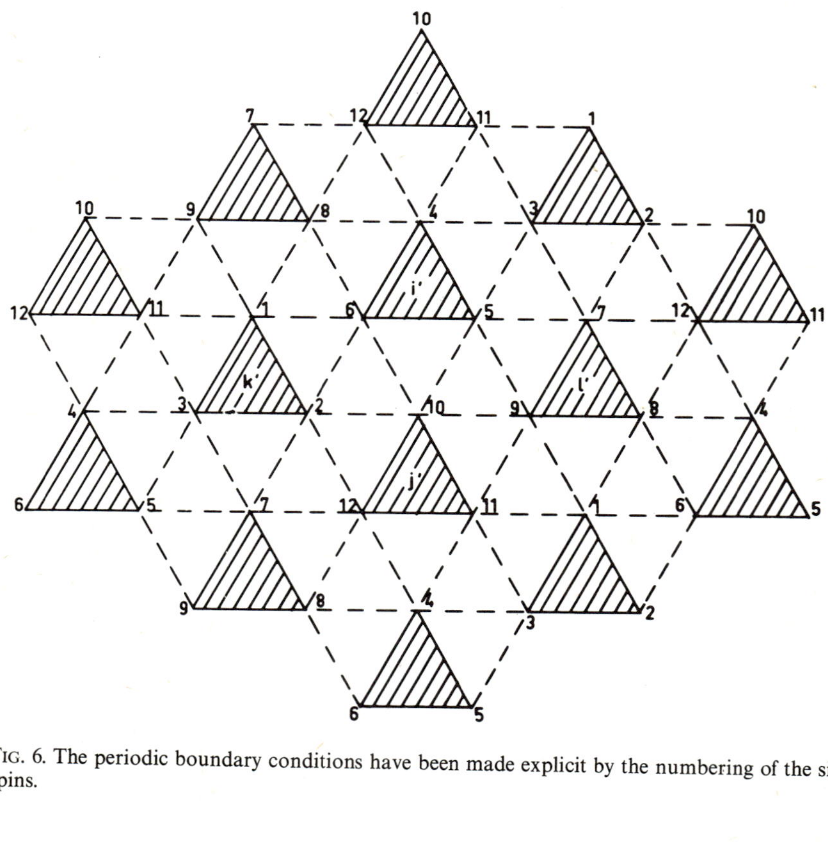

FIG. 6. The periodic boundary conditions have been made explicit by the numbering of the site spins.

fortunately there exists no justification for the restriction to periodic configurations at critical temperatures.

Another drawback of applying periodic boundary conditions appears e.g. when one uses the method for the block in Fig. 5 for which periodic boundary conditions have been drawn in Fig. 6. Upon setting up the renormalization transformation for this system it turns out that there is not even a fixed point (Tjon, private communication). This may be due to the fact that these periodic boundary conditions are not compatible with the lattice symmetry, but the main trouble probably is that some interactions have to be identified too soon, e.g. the cells l' and k', although they are next-nearest neighbours, become nearest neighbours through the periodic boundary conditions

We conclude this section with remarking that simple renormalization transformations obtained from finite lattices without periodic boundary conditions may be influenced in an *ad hoc* manner by boundary effects, whereas the use of periodic boundary is only justified at low temperatures.

B. The cumulant approximation

This method (Niemeijer and van Leeuwen, 1974) is based on a splitting of the spin Hamiltonian into a zeroth part $\mathscr{H}_0$ and a perturbation $\mathscr{V}$. The separation should be such that $\mathscr{H}_0$ is manageable, in a certain way as we shall see, and that $\mathscr{V}$ may be considered small in comparison:

$$\mathscr{H}(s) = \mathscr{H}_0(s) + \mathscr{V}(s). \tag{4.2}$$

Now define the zeroth order transformation as

$$\exp[G_0 + \mathscr{H}'_0(s')] = \sum_{\{s\}} P(s', s) \exp \mathscr{H}_0(s) \tag{4.3}$$

and the zeroth order average $\langle\ \rangle_0$ as

$$\langle A \rangle_0 = \sum_{\{s\}} P(s', s) A(s) \exp[\mathscr{H}_0(s) - G_0 - \mathscr{H}'_0(s')]. \tag{4.4}$$

The transformation (1.11) can then be rewritten as

$$G + \mathscr{H}'(s') = G_0 + \mathscr{H}'_0(s') + \ln \langle \exp \mathscr{V} \rangle_0. \tag{4.5}$$

Since we have assumed that the evaluation of G_0 and $\mathscr{H}'_0(s')$ should pose no problems and that $\mathscr{V}$ could be considered as small compared to $\mathscr{H}_0(s)$, we are entitled to make the cumulant expansion for $\langle \exp \mathscr{V} \rangle_0$

$$\ln \langle \exp \mathscr{V} \rangle_0 = \langle \mathscr{V} \rangle_0 + \frac{1}{2!} \langle [\mathscr{V} - \langle \mathscr{V} \rangle_0]^2 \rangle_0 + \frac{1}{3!} \langle [\mathscr{V} - \langle \mathscr{V} \rangle_0]^3 \rangle_0 + \ldots \tag{4.6}$$

The applicability of the cumulant method is based on the possibility of finding a suitable separation of $\mathscr{H}$ into $\mathscr{H}_0$ and $\mathscr{V}$. It can be applied quite generally for weakly interacting systems (high temperatures), and we present here as an example a cumulant expansion for the renormalization transformation of Section I on the triangular lattice.

$\mathscr{H}_0$ is taken to be the set of interactions inside the cells whereas $\mathscr{V}$ consists of the interactions between the sites of different cells. Inside the cells we can only have a magnetic field H on the sites, pair interactions and three-spin coupling between the three sites of one cell. The essential point is that $\mathscr{H}_0$ now is a sum of uncoupled cell hamiltonians $h_{0i'}(s_{i'}^1, s_{i'}^2, s_{i'}^3)$ (where $s_{i'}^l$ is the lth site of the i'th cell). Since the weight-factor (1.18) is also a product over cells, the zeroth order transformation reduces to a product of transformations over cells. For the i'th cell on the triangular lattice we have for only magnetic field and pair interaction present:

$$\exp[3g_0 + h'_{0i'}(s'_{i'})] = \sum_{\{s_{i'}^1, s_{i'}^2, s_{i'}^3\}} \tfrac{1}{2}[1 + s'_{i'}(s_{i'}^1 + s_{i'}^2 + s_{i'}^3 - s_{i'}^1 s_{i'}^2 s_{i'}^3)\tfrac{1}{2}]$$
$$\exp[H(s_{i'}^1 + s_{i'}^2 + s_{i'}^3) + K(s_{i'}^1 s_{i'}^2 + s_{i'}^2 s_{i'}^3 + s_{i'}^3 s_{i'}^1)]. \tag{4.7}$$

Carrying out the summation over $s_{i'}^1$, $s_{i'}^2$ and $s_{i'}^3$ yields

$$\begin{cases} g_0 = (\tfrac{1}{6}) \log(z_+ z_-) \\ h'_{0i'} = H'_0 s'_{i'} = s'_{i'}(\tfrac{1}{2}) \log(z_+/z_-) \\ z_\pm = \exp(3K \pm 3H) + 3\exp(-K \pm H). \end{cases} \tag{4.8}$$

Here $H'_0(K, H)$ is the zeroth order contribution to the renormalized magnetic field H' and similarly $g_0(K, H)$ to g.

The evaluation of $\langle \mathscr{V} \rangle_0$ is relatively easy. All one has to know are the following basic averages in the i'th cell

$$\begin{cases} \langle s_{i'}^1 \rangle_0 = \langle s_{i'}^2 \rangle_0 = \langle s_{i'}^3 \rangle_0 = a_1 s'_{i'} + b_1, \\ \langle s_{i'}^1 s_{i'}^2 \rangle_0 = \langle s_{i'}^2 s_{i'}^3 \rangle_0 = \langle s_{i'}^3 s_{i'}^1 \rangle_0 = a_2 + b_2 s'_{i'}, \\ \langle s_{i'}^1 s_{i'}^2 s_{i'}^3 \rangle_0 = a_3 s'_{i'} + b_3 \end{cases} \tag{4.9}$$

where the a_i and b_i are given by

$$a_i = \tfrac{1}{2}\left[\frac{n_{i+}}{z_+} + \frac{n_{i-}}{z_-}\right], \quad b_i = \tfrac{1}{2}\left[\frac{n_{i+}}{z_+} - \frac{n_{i-}}{z_-}\right],$$

$$n_{1\pm} = \exp(3K \pm 3H) + \exp(-K \pm H),$$

$$n_{2\pm} = \exp(3K \pm 3H) - \exp(-K \pm H),$$

$$n_{3\pm} = \exp(3K \pm 3H) - 3\exp(-K \pm H). \tag{4.10}$$

Note that for $H = 0$ the b_i vanish.

Any interaction $K_a s_a$ having sites of set a in various cells can now be computed as the product over the averages of the cells according to the formulae (4.9). E.g. we find for an interaction $K\, s^l_{i'} s^m_{j'}$ (with $l, m = 1, 2$ or 3) the value

$$\langle K s^l_{i'} s^m_{j'} \rangle_0 = K(a_1 s'_{i'} + b_1)(a_1 s'_{j'} + b_1) = K[a_1^2 s'_{i'} s'_{j'} + a_1 b_1 (s'_{i'} + s'_{j'}) + b_1^2] \tag{4.11}$$

producing a renormalized interaction between the cell spins $s'_{i'}$ and $s'_{j'}$ of strength Ka_1^2, a magnetic field Ka_1b_1 on both cell spins and a contribution Kb_1^2 to g. For nearest-neighbour cells the total first-order contribution K' due to nearest-neighbour interaction is

$$K' = 2a_1^2 K. \tag{4.12}$$

It is interesting to note that, restricting ourselves to the subspace of nearest-neighbour interactions only, relation (4.12) implies a fixed point for $2a_1^2 = 1$. Taking a_1 as given by (4.10) for $H = 0$ the equation for K^* reads

$$2(\exp(3K^*) + \exp(-K^*))^2 = (\exp(3K^*) + 3\exp(-K^*))^2 \tag{4.13}$$

with solution

$$K^* = (\tfrac{1}{4}) \log[1 + 2\sqrt{2}] \simeq 0{\cdot}3356. \tag{4.14}$$

The computation of higher orders proceeds straightforwardly along the same lines, although it soon becomes quite involved. Here we shall only point out an important property of preserving the short-rangedness of an interaction for every cumulant. Consider the nearest-neighbour interaction $Ks^l_{i'}s^m_{j'}$, (with $i' \neq j'$). Now

$$\langle (K \sum_{l,m} \sum_{i',j'} s^l_{i'} s^m_{j'})^2 \rangle_0 - \langle K \sum_{l,m} \sum_{i',j'} s^l_{i'} s^m_{j'} \rangle_0^2 =$$
$$K^2 \sum_{l,m,p,q} \sum_{i',j',k',l'} [\langle s^l_{i'} s^m_{j'} s^p_{k'} s^q_{l'} \rangle_0 - \langle s^l_{i'} s^m_{j'} \rangle_0 \langle s^p_{k'} s^q_{l'} \rangle_0]. \tag{4.15}$$

If the cells i', j', k' and l' are all different, then

$$\langle s^l_{i'} s^m_{j'} s^p_{k'} s^q_{l'} \rangle_0 = \langle s^l_{i'} \rangle_0 \langle s^m_{j'} \rangle_0 \langle s^p_{k'} \rangle_0 \langle s^q_{l'} \rangle_0 \tag{4.16}$$

and the right hand side of (4.15) vanishes. There must be at least one pair of equal indices, so $i' = k'$ or $j' = l'$. Since $s^l_{i'}$ and $s^m_{j'}$ are neighbouring spins, i' and j' must be neighbouring cells, and similarly k' and l'. Thus, in order to obtain a non-vanishing contribution in (4.15) i', j', k' and l' must form a connected chain of two neighbouring pairs having at least one cell in common. On the other hand the product (4.16) contributes to interactions involving the cell spins $s'_{i'}$, $s'_{j'}$, $s'_{k'}$ and $s'_{l'}$, and thus produces only short-range interactions.

As one observes in (4.11) and (4.15) the orders in the cumulants are proportional to the powers of the interaction constants that are involved. Accordingly this cumulant expansion will converge better for higher temperatures. It is also essentially an approximation for large magnetic fields, since in a large magnetic field the energy levels are mainly determined by $\mathscr{H}_0(s)$. From (4.8) it follows that the subspace consisting of the axis $H \neq 0$ and all other $K_\alpha = 0$ is invariant under the transformation, and thus the approximation is most effective in the neighbourhood of this axis.

Around a fixed point, however, as indicated by (4.14), serious questions about the convergence of the approximation arise. We shall come back to this point in the example treated in Section V.B.2 and Section VI.

The method can also be used to determine the high-temperature limit of the matrix $T_{\alpha\beta}$. Around $K_\alpha = 0$ the matrix follows from $\mathscr{H}'_0(s)$ and $\langle \mathscr{V} \rangle_0$ and is given in Table II for the triangular lattice. Here K stands for the nearest-neighbour interaction, L for the next-nearest neighbours, N for four-spin interaction (diamond of neighbouring sites), H for the single site (magnetic field interaction), and M for the triplet-spin interaction (triangle of neighbouring sites). We note that the matrix of Table II, as belonging to the fixed point $K^* = 0$, has three eigenvalues $\neq 0$: the (odd) relevant eigenvalue $\frac{3}{2}$ and two (even) irrelevant eigenvalues $\frac{1}{2}$ and $\frac{1}{8}$.

TABLE II. $T_{\alpha\beta}$ for $K^* = 0$, for the triangular lattice.

α' \ α	K	L	N	H	M
K'	$\frac{1}{2}$	$\frac{3}{4}$	$-\frac{1}{4}$	0	0
L'	0	0	0	0	0
N'	0	0	0	0	0
H'	0	0	0	$\frac{3}{2}$	$-\frac{1}{2}$
M'	0	0	0	0	$\frac{1}{8}$

C. The cluster approximation

Both the finite-system and the cumulant method have no claim to accuracy for typical critical conditions of a system where neither energy nor entropy dominate in the free energy. By nature the finite-system approach restricts the interaction parameters involved to those fitting in the finite system, i.e. to short-range interactions. The method discussed in this section is also based on taking a limited number of short-range interaction parameters, but tries to combine it with properly accounting for the non-periodic configurations as happens in the cumulant expansion (Niemeijer and van Leeuwen, 1973, 1974).

To see the connection let us compare the results of the calculation for a finite system consisting of two cells with the first-order cumulant. Take the triangular lattice with weight factor (1.18) as an example. For an isolated set of two cells (Fig. 7) we find with (4.5)

$$G + \mathscr{H}'(s'_{i'}, s'_{j'}) = G_0 + \mathscr{H}'_0(s'_{i'}, s'_{j'}) + \ln \langle \exp K\, s^1_{j'}(s^2_{i'} + s^3_{i'}) \rangle_0. \quad (4.17)$$

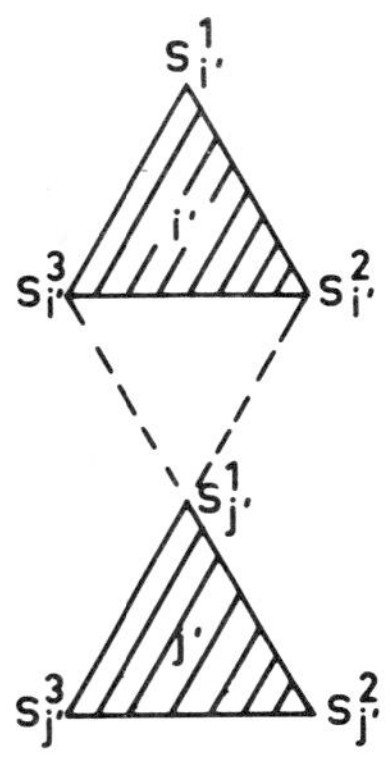

FIG. 7. Two isolated cells of the triangular lattice.

So for zero field the renormalized nearest-neighbour interaction is

$$K' = \operatorname{arctanh}[2a_1^2 \tanh K/(1 + a_2 \tanh^2 K)] \quad (4.18)$$

with a_1 and a_2 given by (4.10) ($H = 0$). One observes that for $K \to 0$ (4.18) reduces to the cumulant expansion (4.12). Clearly the full evaluation of the transformation for the isolated set of two cells incorporates the lowest-order cumulant, but contains in addition higher-order contributions. Making a diagrammatic expansion for the cumulants one may say that (4.18) contains all contributions of the cumulants for which the diagrams are completely contained in the two cells.

It is evident that a larger set will include more of the cumulant contributions. The problem is to make an optimal use of the information contained in a finite-system calculation. Besides that we encounter, for larger systems, the problem of keeping track of which cluster a certain coupling belongs to.

Let a be an arbitrary cluster of cells†. By using the word cluster we want to indicate that we have a connected set of cells in mind. We assume that $P(s', s)$ is a product over cell weight-factors, as in (4.1), such that we may

† In this part we shall omit primes on cell sets, since they are the only ones that will occur in the remainder of this section.

speak about $P_a(s', s)$ as the weight factor of a cluster a. For a cluster a we have formally

$$\sum_{\{s\}} P_a(s', s) \exp \mathscr{H}_a(s) = \exp(G_a + \sum_{b \subseteq a} K_b'^a s_b') \tag{4.19}$$

where the renormalized energy for cluster a has been decomposed into its constituent contributions $K_b'^a s_b'$. The upper index a is a reminder that the $K_b'^a$ do depend on the choice of the cluster a. If a were the whole lattice, the index a could be dropped and the resulting K_b' would depend only on the type β to which b belongs. Thus for increasing a, $K_b'^a$ approaches K_β'. In order to see K_β' as a sum of increasing cluster contributions we make the Ursell expansion

$$K_b'^a = \sum_{b \subseteq c \subseteq a} U_b^c \tag{4.20}$$

in the hope that the U_b^c decrease rapidly with increasing c. The U_b^c can be expressed in terms of the $K_b'^a$ which can be contained in c, by the inversion of (4.20). As we are dealing with a partially ordered set, Moebius's inversion theorem (Rota, 1964) applies, yielding

$$U_b^c = \sum_{b \subseteq a \subseteq c} (-1)^{c-a} K_b'^a \tag{4.21}$$

where c and a in $(-1)^{c-a}$ stand for the numbers of cells in c and a respectively. The proof of (4.21) is based on the identity

$$\sum_{a \subseteq b \subseteq c} (-1)^{b-a} = \delta_{a,c} \tag{4.22}$$

where $\delta_{a,c} = 1$ if and only if the sets a and c are identical.†

If we were to let a in (4.20) be the whole lattice of cells we would still have the full problem

$$K_b' = \sum_{b \subseteq c} U_b^c \tag{4.23}$$

and recover K_β'.

The problem is that, for practical reasons, we can only compute U_b^c through (4.21) for c of sufficiently small size. By selecting c out of a set which has the full symmetry of the lattice we shall automatically find a K_b' which

† The proof of (4.22) runs as follows. Let there be n cells in c and not in a. Then every b with $a \subseteq b \subseteq c$ can be characterized by a row $\{\mu_i\}$, $i = 1, \ldots, n$ and $\mu_i = 1$ or 0. Then $(-1)^{b-a} = (-1)^{\sum_i \mu_i}$ thus

$$\sum_{a \subseteq b \subseteq c} (-1)^{b-a} = \sum_{\{\mu_i\}} (-1)^{\sum_i \mu_i} = \prod_{i=1}^{n} \left(\sum_{\mu_i} (-1)^{\mu_i} \right) = (1 - 1)^n = \delta_{a,c}$$

depends only on the type β to which b belongs. So the most efficient choice is to take c out of subsets of a basic cluster type δ. In formulae:

$$K'_\beta = \sum_{\beta \subseteq \gamma \subseteq \delta} \sum_{b \subseteq c \in \gamma} U_b^c \tag{4.24}$$

where we have transferred the inclusion relation to types (by inspecting whether members can include each other). Inserting (4.21) into (4.24) we may write it in the form

$$K'_\beta = \sum_{\beta \subseteq \alpha \subseteq \delta} C_\delta(\alpha) K'^{\alpha}_\beta \tag{4.25}$$

with the following definitions

$$\begin{aligned} C_\delta(\alpha) &= \sum_{\alpha \subseteq \gamma \subseteq \delta} \sum_{a \subseteq c \in \gamma} (-1)^{c-a}, \\ K'^{\alpha}_\beta &= \sum_{b \subseteq a \in \alpha} K'^{a}_b. \end{aligned} \tag{4.26}$$

K'^{α}_β is the sum over all possible ways that a K'^{a}_b can be found with an a of type α covering b of type β. K'^{α}_β may be seen as the interaction coefficient of type β in cluster α (multiplied by the number of times β can be covered by α). The combinatorial factors $C_\delta(\alpha)$ then determine how the subclusters α of the basic cluster δ have to be combined. The interesting aspect of (4.25) is that these combinatorial factors do not depend on the type of interaction K_β that one is computing. In the next Section, V.A.3, a table of $C_\delta(\alpha)$ is included for the triangular lattice. Many $C_\delta(\alpha)$ turn out to be zero for a symmetric δ, which is a pleasant feature of the cluster approximation.

We finally remark that (4.25) can also be used for the computation of the derivative matrix, or linearized transformation, in the form

$$T_{\alpha\beta} = \sum_{\alpha \subseteq \gamma \subseteq \delta} C_\delta(\gamma)(\partial K'^{\gamma}_\alpha / \partial K_\beta). \tag{4.27}$$

V. Applications

In this section we shall deal with several applications of renormalization procedures for the triangular and quadratic lattices. It is subdivided as follows:

A. *The triangular lattice*
1. Finite lattice approximation (without periodic boundary conditions) (Tjon, 1974; Subbarao, 1975).
2. Cumulant approximation (Niemeijer and van Leeuwen, 1974).
3. Cluster approximation and spontaneous magnetization (Niemeijer and van Leeuwen, 1973, 1974; van Leeuwen, 1975a).

B. *The quadratic lattice*
1. Finite lattice approximation (using periodic boundary conditions) (Nauenberg and Nienhuis, 1974a, b).
2. Cumulant expansion (straightforward) (Hsu *et al.*, 1975).
3. Cumulant expansion (sophisticated) (Kadanoff and Houghton, 1975).

The applications discussed here all employ a renormalization transformation of the type (1.11).

A. The triangular lattice

1. *Finite lattice approximation* (*without periodic boundary conditions*)
Tjon has made a numerical study of the renormalization transformation in the four-cell approximation with the weight-factor (1.18) (Fig. 5). As

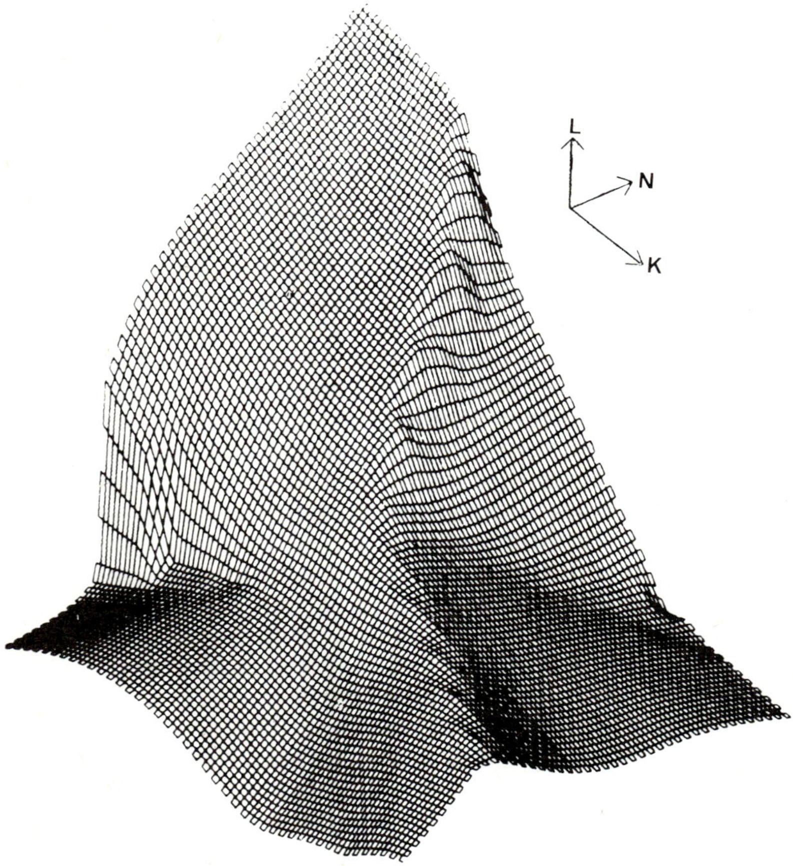

FIG. 8. The critical surface of the triangular lattice in a finite lattice approximation.

parameter space he used the nearest-neighbour (K), next-nearest-neighbour (L) and the four-spin interaction (N). The renormalization transformation was studied numerically; since an asymmetry was introduced into the cell lattice, a linear averaging was applied to the resulting interaction strengths. Only one non-trivial fixed point was found at $K = 0{\cdot}2933$, $L = 0{\cdot}0007$, $N = 0{\cdot}0029$.

Also the surface of criticality connected with this fixed point was determined and found to consist of one single surface. It is known that in the absence of the four-spin interaction ($N = 0$) only a ferromagnetic transition is possible and correspondingly, as expected, on the critical line in the K–L plane, L decreases with increasing K. For $K > 0$ and $1 > \alpha = L/K > 0$ the line is almost straight, whereas for $\alpha < 0$ there is considerably more curvature. The value of the slope for $\alpha < 1$ is given by $m = 1{\cdot}20$ and should

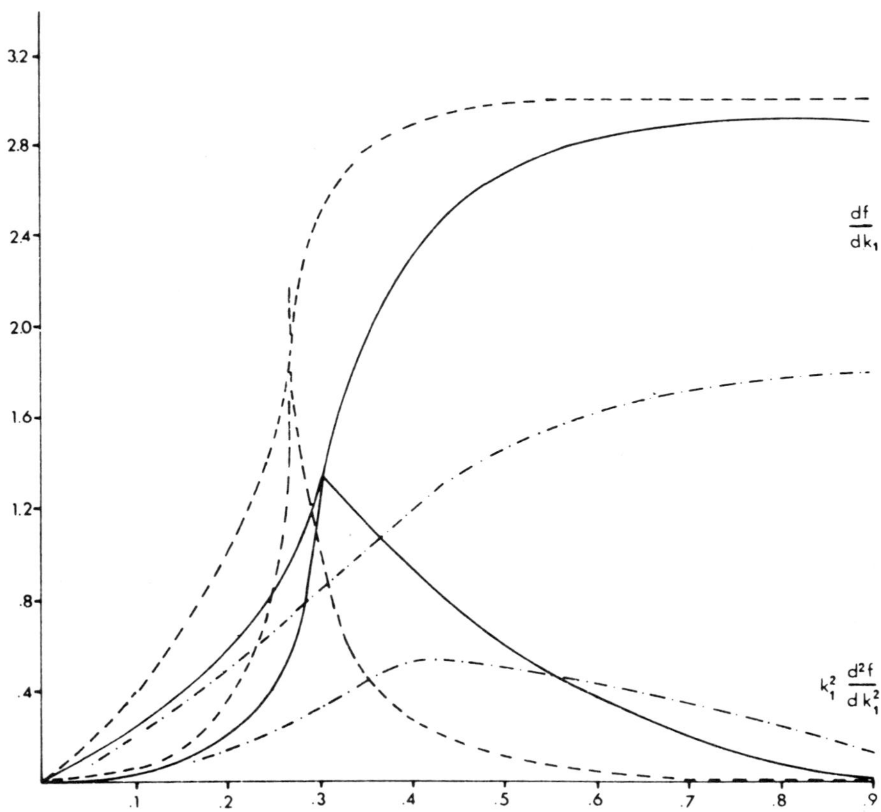

FIG. 9. Internal energy and specific heat for the pure Ising system. Dashed curve: exact solution; full curve: renormalization results; dashed–dotted curve: results for 12 spins.

be compared with the high-temperature series result $m = 1{\cdot}35$ of Dalton and Wood (1969).

With the aid of the technique developed in Section II.C, Tjon also obtained several thermodynamic quantities using his renormalization transformation. The results for the internal energy df/dK and the specific heat $K^2 d^2f/dK^2$ for the Ising system with $L = N = 0$ have been plotted in Fig. 9. For comparison the exact results for twelve spins have also been plotted. It is to be noticed that the thermodynamic quantities as derived from the renormalization transformation are markedly better than the finite lattice results.

We refer for a discussion of Tjon's conclusion about the shape of the critical surface for $N \neq 0$ to the original paper (Tjon, 1974).

2. *Cumulant approximation*

In Section IV.B the first-order cumulant approximation to the renormalization transformation was derived with the result (4.12)

$$K' = 2a_1^2 K, \tag{5.1}$$

where a_1 is defined in (4.10). Equation (5.1) not only determines the location of the fixed point

$$K^* = (1/4)\log[1 + 2\sqrt{2}] \simeq 0{\cdot}3356 \tag{5.2}$$

but also the thermal eigenvalue and exponent

$$\begin{cases} \lambda_T = [\partial K'/\partial K]_{K^*} = 1{\cdot}634, \\ y_T = \log \lambda_T / \log l = 0{\cdot}894. \end{cases} \tag{5.3}$$

One should compare these two numbers with the Ising values $K_c = 0{\cdot}27465 = \ln 3^{1/4}$ and $y_T = 1{\cdot}000$ for the triangular lattice.

It can be seen from (5.1) that for subcritical systems, $K > K^*$, the new value K' is larger than the old K, whereas for supercritical systems $0 < K < K^*$ one has $K' < K$.

The next-order cumulant has been evaluated by Niemeijer and van Leeuwen (1974). The ordering in the contribution is obtained by considering the nearest-neighbour coupling K between the cells as a small parameter. Then in second order in K next-nearest-neighbour cells and third-neighbour cells start to interact. We denote the newly arising interactions by L' and M' respectively. Besides a second-order contribution to L' and M', a second-order contribution to the nearest-neighbour cell coupling K' is formed. The second-order contributions read:

$$\begin{aligned} (K')_2 &= 4(p + q)a_1^2 K^2, \\ (L')_2 &= (p + 7q)a_1^2 K^2, \\ (M')_2 &= 4qa_1^2 K^2. \end{aligned} \tag{5.4}$$

with p and q given by

$$p = 1 - a_1^2, \qquad q = a_2 - a_1^2 \tag{5.5}$$

and the a_i given by (4.10).

The expressions (5.4) show that L and M may be considered as second-order quantities and all the other coupling constants of higher order. To be complete in second order, one must introduce L and M in the first-order cumulant $\langle \mathscr{V} \rangle_0$. This leads to the additional contribution

$$\begin{aligned}(K')_1 &= 3a_1^2 L + 2a_1^2 M,\\ (L')_1 &= a_1^2 M.\end{aligned} \tag{5.6}$$

Adding the expressions (5.1), (5.4) and (5.6) yields the full second-order transformation equations:

$$\begin{aligned}K' &= 2a_1^2 K + 4(p + q)a_1^2 K^2 + 3a_1^2 L + 2a_1^2 M,\\ L' &= (p + 7q)a_1^2 K^2 + a_1^2 M,\\ M' &= 4qa_1^2 K^2.\end{aligned} \tag{5.7}$$

This transformation has a non-trivial fixed point at

$$K^* = 0{\cdot}2789, \qquad L^* = -0{\cdot}0143, \qquad M^* = -0{\cdot}0152. \tag{5.8}$$

The derivative matrix $T^*_{\alpha\beta}$ at the fixed point follows for the even-even block by differentiation of (5.7) with respect to K, L and M, and after insertion of the fixed-point parameters one finds:

$$T^*_{\alpha\beta} = \begin{pmatrix} 1{\cdot}8966 & 1{\cdot}3446 & 0{\cdot}8964 \\ -0{\cdot}0403 & 0{\cdot}0 & 0{\cdot}4482 \\ -0{\cdot}0782 & 0{\cdot}0 & 0{\cdot}0 \end{pmatrix} \tag{5.9}$$

with the eigenvalues

$$\lambda_T = \lambda_1 = 1{\cdot}7835, \qquad \lambda_2 = 0{\cdot}2286, \qquad \lambda_3 = -0{\cdot}1156. \tag{5.10}$$

The even relevant eigenvalue λ_T has the exponent $y_T = 1{\cdot}053$. The left eigenvector ϕ^T_α corresponding to λ_T is (with $\phi^T_{\text{n.n.}} = 1$)

$$\phi^T_\alpha = (1{\cdot}0,\ 0{\cdot}7539,\ 1{\cdot}0961). \tag{5.11}$$

Using ϕ^T_α in (2.20c) yields the value $K_c = 0{\cdot}2514$ (versus the exact value $K_c = 0{\cdot}27465$).

The magnetic (odd) eigenvalues are found with the aid of expression (2.4). One finds in the cumulant expansion

$$\langle \sum_{a \in \alpha} s_a \rangle_{s'} = \langle \sum_{a \in \alpha} s_a \rangle_0 + \langle \sum_{a \in \alpha} (s_a - \langle s_a \rangle_0)\mathscr{V} \rangle_0 + \ldots. \tag{5.12}$$

For α equal to the single site one has in zeroth and first order

$$\begin{aligned} &\langle s_{i'}^1 \rangle = \langle s_{i'}^2 \rangle = \langle s_{i'}^3 \rangle = a_1 s'_{i'}, \\ &\sum_{i'} \langle (s_{i'}^1 - a_1 s'_{i'}) \mathscr{V} \rangle_0 = 4Ka_1(p + 2q) \sum_{i'} s'_{i'}. \end{aligned} \tag{5.13}$$

Thus in these orders a single site perturbation produces only a magnetic field on the cells. So in $T^*_{\alpha\beta}$ one has

$$T^*_{\alpha\alpha} = 3a_1[1 + 4K(p + 2q)], \quad \alpha \text{ single site} \tag{5.14}$$

and $T^*_{\alpha\beta} = 0$ for $\alpha \neq \beta =$ single site. The results of the first- and second-order calculations can be collected in the following table:

Order	y_T	y_H
0th	—	1·369
1st	0·894	2·022
2nd	1·053	1·847*
Exact	1·0	1·8750

(5.15)

They were derived by Niemeijer and van Leeuwen (1974) except the one marked by *, which was derived by Braathen and Hemmer (1975).

For the triangular lattice also the discussion given in Section II.D for the spontaneous magnetization can be illustrated with the lowest-order cumulant expansion. For the function $g(K)$ and the transformation one has in lowest order with (4.10), (5.1) and (5.13)

$$\begin{aligned} g(k) &= (1/3)\log(\exp 3K + 3\exp - K) \\ K' &= 2a_1^2(K)K \\ H' &= 3a_1(K)H \end{aligned} \tag{5.16}$$

with $a_1(K)$ given in (4.10). The function $a_1(K)$ varies from the value 1/2 for $K = 0$, via $2^{-1/2}$ at the fixed point K^* to 1 for $K = \infty$.

As mentioned, the value of K grows upon application of the transformation when starting with $K > K^*$, ultimately by a factor 2 at every step, whereas for $K < K^*$ the value decreases ultimately by a factor 1/2 at every step. This shows that all points (with $H = 0$) end up either at $K = \infty$ or at $K = 0$. For the matrix $\partial K'_\alpha/\partial K_\beta$ we obtain with (5.16) the limiting behaviour

$$\begin{aligned} &\partial K'/\partial K = 2, \qquad \partial H'/\partial H = 3 \qquad K \to \infty \\ &\partial K'/\partial K = \tfrac{1}{2}, \qquad \partial H'/\partial H = \tfrac{3}{2} \qquad K \to 0 \end{aligned} \tag{5.17}$$

which indeed is independent of the point K in the limit.

For the triangular lattice we have for the number χ_α (defined in (2.61)):

$$\chi_\alpha = 1 \qquad \alpha = \text{single site}$$

$$\chi_\alpha = 3 \qquad \alpha = \text{nearest-neighbour pair}. \tag{5.18}$$

Since one finds for $\partial g(K)/\partial K$

$$\partial g(K)/\partial K = [\exp 3K - \exp - K]/[\exp 3K + 3 \exp - K] \simeq 1 \tag{5.19}$$

for $K \to \infty$, it is seen with (5.17) and (5.18) that indeed (2.62) is satisfied both for α the single site and for α the nearest-neighbour pair. Thus $\partial K'_{\alpha'}/\partial K_\beta$ has an odd eigenvalue $l^d = 3$ for $K = \infty$ as was generally stated in II.D.

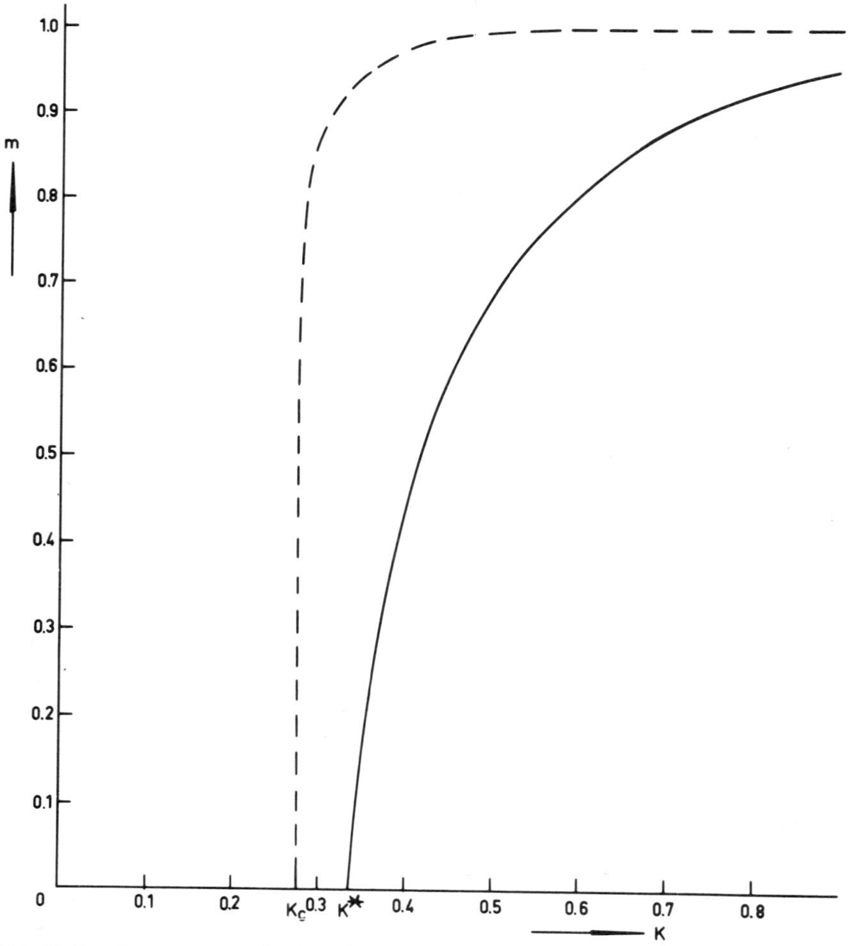

FIG. 10. Spontaneous magnetization; dashed curve: exact solution; full curve: renormalization transformation result.

For $K \to \infty$ the magnetization $m = \partial f/\partial h$ becomes complete: ± 1 (see (2.67)). For the odd derivative we find with (5.16)

$$\partial H^{(j)}/\partial H^{(j-1)} = 3a_1(K^{(j-1)}). \tag{5.20}$$

Thus we find according to (2.68) for the spontaneous magnetization

$$m(K) = \pm \prod_{j=0}^{\infty} a_1(K^{(j)}) \tag{5.21}$$

where $K^{(j)}$ is the j-times-transformed image of K. (5.21) has been evaluated for $a_1(K)$ given by (4.10) and the transformation (5.16) for K. The result is shown in Fig. 10 together with the exact result for the Ising model. The poor agreement shows how delicately the magnetization depends on $\beta = (d - y_H)/y_T$ which here equals 0·715 instead of the exact value of 0·125.

3. *The cluster approximation*

The cluster approximation, which was outlined in Section IV.C, has been evaluated for the triangular lattice for 6 different cell configurations. The two-cell approximation was already discussed as an example of the cluster approximation. It was seen that the two-cell approximation involves only the nearest-neighbour interaction, since no other interactions can be formed by the transformation.

As far as magnetic perturbations around the fixed point are concerned, this figure contains only single cells as odd subsets. Thus the odd-odd part of the matrix $T_{\alpha\beta}$ is 1-dimensional. In this case it is not difficult to work out the formula for $T_{\alpha\beta}$, and thus λ_H, analytically with result

$$\lambda_H = 3a_1 + \frac{6u(2a_1 + u)(p + 2q)}{1 + 2a_1^2u^2 + a_2u^2} \tag{5.22}$$

with $u = \text{tgh } K^*$ and a_i defined in (4.10) and p and q in (5.5).

For approximation with clusters consisting of more than two cells we have to make use of the combinatorial factors $C_\delta(\alpha)$, defined in (4.26), for the different clusters chosen on the triangular lattice. Calculations have been performed for the 2, 3, 4, 5, 6 and 7 clusters; the necessary non-zero combinatorial factors have been listed in Table III.

The results for the eigenvalues and the critical temperature are listed in Table IV. We give an explicit description of the formulae relating to the three-cell cluster, and quote the results for larger clusters. The three cells are located in a triangle, such that all pairs of cells are nearest neighbours. Then the transform (in zero field) concerns again only the nearest-neighbour interaction parameter. With Table III we have explicitly:

$$K' = K_2'^3 - K_2'. \tag{5.23}$$

TABLE III. $C_\delta(\alpha)$ as defined in (4.26). Only the non-zero values are listed.

δ \ α								
	−5	1	–	–	–	–	–	–
	1	−1	1	–	–	–	–	–
	1	0	−2	–	1	–	–	–
	−2	2	1	−1	−3	1	–	–
	−2	2	0	−1	−1	0	1	0
	0	0	1	0	−1	0	–	1

Without too much difficulty the formulae can be worked out analytically and the fixed point and the eigenvalue can be determined numerically. The result is:

$$K^* = 0{\cdot}255, \quad \lambda_T = 1{\cdot}501, \quad y_T = 0{\cdot}739. \tag{5.24}$$

Magnetic perturbations give rise to a single- and a triple-cell potential. Thus the odd–odd block of $T_{\alpha\beta}$ is two-dimensional and follows from differentiation of (cf. Table III):

$$\begin{aligned} K_1' &= K_1'^3 - K_1'^2 + K_1'^1 \\ K_3' &= K_3'^3. \end{aligned} \tag{5.25}$$

The results yields at the fixed point $T^*_{\alpha\beta}$:

$$T^*_{\alpha\beta} = \begin{pmatrix} 2{\cdot}707 & 7{\cdot}031 \\ -0{\cdot}063 & 0{\cdot}364 \end{pmatrix}$$

which yields eigenvalues

$$\begin{cases} \lambda_H = \lambda_1 = 2{\cdot}501, & \lambda_2 = 0{\cdot}571. \\ y_H = 1{\cdot}669 \end{cases}$$

TABLE IV. The eigenvalues and critical temperatures in the successive approximations as obtained by Niemeijer and van Leeuwen (1973). The results of the six-cell cluster have been obtained by Subbarao (1975).

CLUSTER	y_T	y_H	K_c
	.7908	2.0217	0.365
	.7394	1.6688	0.255
	.8177	1.7562	0.253
	1.0518	1.9232	0.281
	1.0518	1.9219	0.281
	1.0281	1.87375	0.27416
EXACT	1.000	1.87500	0.27465

In the four-, five-, six- and seven-cell approximations there came into play three, six, eight and eleven even interactions respectively. These calculations have been performed for all but the six-cell approximation by Niemeijer and van Leeuwen (1973), and for the six cell approximation by Subbarao (1975). In each case the configurations of the site spins have been generated systematically for every constraint given by the configuration of the cell spins and the resulting free energies have been decomposed into their contribution for the several possible cell interactions. A slight complication arises because the cell interactions, which should receive equal weights on account of the symmetry of the lattice, actually receive different weights even if the corresponding interactions on the site lattice have been given equal values.

TABLE V. Fixed-point values for the interaction parameters K_α^* and components $r_\alpha = \varphi_\alpha^T/T_{nn}$ of the left eigenvector belonging to λ_T. α again corresponds to the particular geometrical arrangement of interacting spins or cells, which are schematically indicated.

Approximation		α			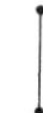						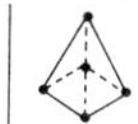		
1st order pert		K_α^*	0·3356	—	—	—	—	—	—	—	—	—	—
Cluster		K_α^*	0·365	—	—	—	—	—	—	—	—	—	—
Cluster		K_α^*	0·255	—	—	—	—	—	—	—	—	—	—
Cluster		K_α^*	0·257	-0·0022	—	-0·00085	—	—	—	—	—	—	—
Cluster		K_α^*	0·331	-0·0275	-0·0267	0·0086	0·0080	—	-0·0037	—	—	—	—
Cluster		K_α^*	0·3069	-0·0183	-0·0214	0·0034	0·0066	0·0036	-0·0022	-0·0016	-0·0009	0·0003	0·00004
Cluster		r_α	1·000	1·205	—	2·990	—	—	—	—	—	—	—
Cluster		r_α	1·000	1·708	1·917	1·237	5·978	—	2·742	—	—	—	—
Cluster		r_α	1·000	1·607	1·811	1·248	5·782	1·083	2·808	1·372	3·081	0·452	2·777

This, as already noted after (1.18), is due to a slight asymmetry in the weight-factor $P(s', s)$ and can be remedied by averaging over the orientationally inequivalent cell distributions of Fig. 1(b).

The numerical results for the fixed-point interaction parameters K_α^* and the components $r_\alpha = \varphi_\alpha^T/\varphi_{nn}^T$ of the left eigenvector belonging to λ_T have been listed in Table V.

B. The quadratic lattice

1. *Finite-size lattice calculation with periodic boundary conditions*

Nauenberg and Nienhuis (1974a, b) have studied a renormalization transformation of the quadratic lattice in a four-cell, each of 4 spins, approximation, with periodic boundary conditions (Fig. 11).

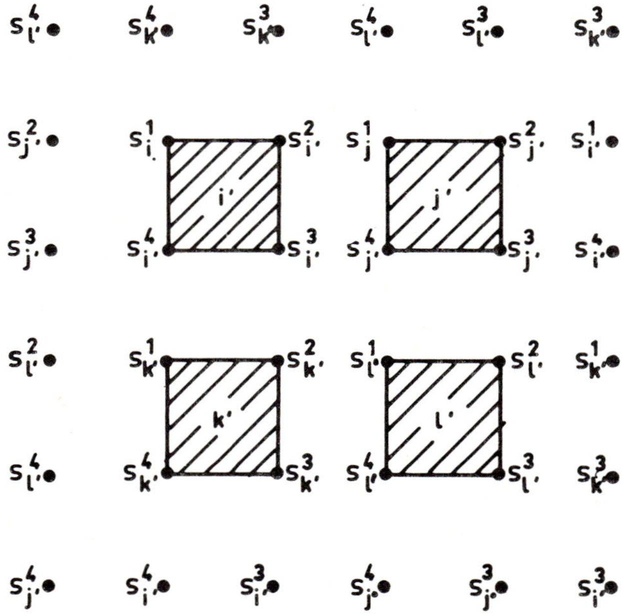

FIG. 11. Four cells with periodic boundary conditions.

The cell spin definition $s'_{i'} = \text{sign}(s_{i'}^1 + s_{i'}^2 + s_{i'}^3 + s_{i'}^4)$ cannot be used here, since the cell contains an even number of spins and thus the argument of the sign function may be zero and therefore undefined. There are several ways out of this problem; the one taken by Nauenberg and Nienhuis is as follows: every cell has sixteen configurations, which can be labelled by the set $(s_{i'}^1 s_{i'}^2 s_{i'}^3 s_{i'}^4)$. A spin variable $s'_{i'} = \pm 1$ is then assigned to each cell according to the following rules: The eight configurations $(++++)$, $(-+++)$, $(+-++)$, $(++-+)$, $(+++-)$, $(++--)$, $(+-+-)$ and $(+--+)$ correspond to cell spin up, $s'_{i'} = +1$, whereas a cell spin $s'_{i'} = -1$ is obtained by flipping all the spins of one of the configurations corresponding with one of the $s'_{i'} = 1$ configurations. These four choices are all equivalent for clusters of a finite size arranged on a square lattice. In the absence of odd (magnetic) interactions, only three types of interaction are possible in this approximation, namely, K: nearest neighbours, L: next-nearest neighbours, and N: the four-spin interaction of the spins lying on the vertices of an elementary square of the lattice, i.e. the last term in the Hamiltonian (1.9).

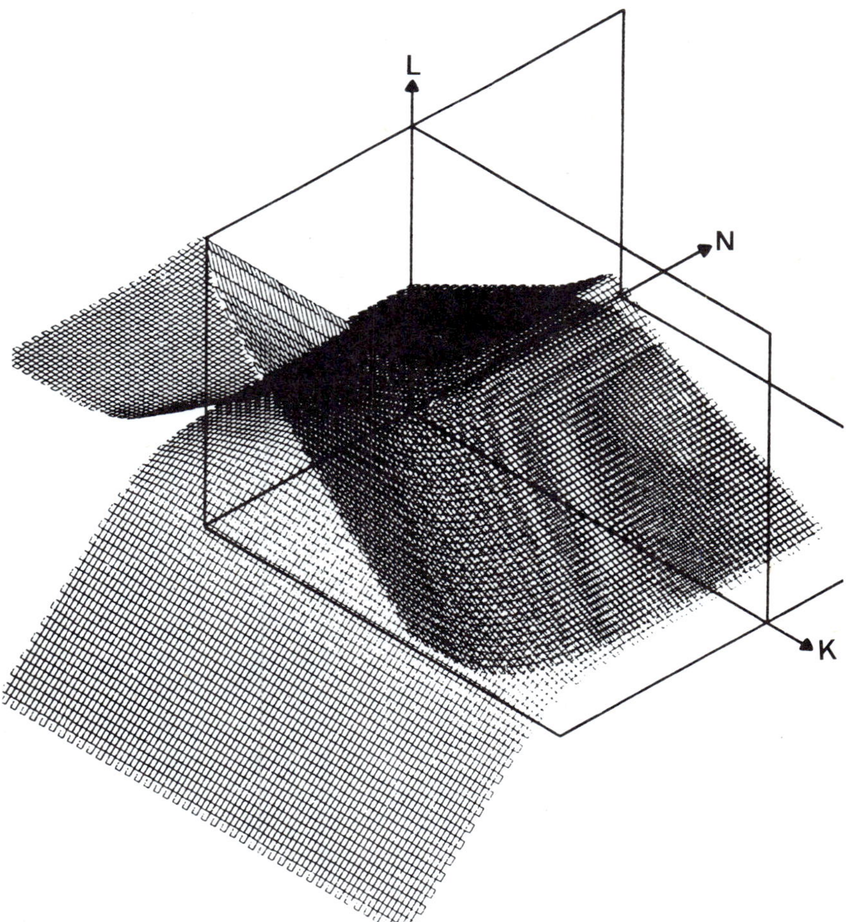

FIG. 12. The critical surface of the square lattice in the four-spin cell approximation with periodic boundary conditions (after Nauenberg and Nienhuis, 1974a, b).

It is easily verified that the free energy of this system has the following symmetries

$$f(K, L, N) = f(-K, L, N) \quad \text{and} \quad f(0, L, N) = f(0, -L, N) \tag{5.26}$$

These symmetries are only approximately preserved under the present renormalization transformation. We shall return to this problem in Section VI.

Nauenberg and Nienhuis (1974a) have evaluated numerically the renormalization transformations

$$K'_\alpha = K'_\alpha(K, L, N), \tag{5.27}$$

and have found one single fixed point at $K^* = 0{\cdot}300$, $L^* = 0{\cdot}0871$ and $N^* = -0{\cdot}00126$ and the two trivial fixed points $K_\alpha = 0$ and $K_\alpha = \infty$. The critical surface, which is determined by the subset of points which are mapped into the fixed point, is shown in Fig. 12 for the domain $-2 \leqslant K$, $N \leqslant L$ and is seen along the direction $K = 1$, $L = 1$ and $N = 1$.

Actually the surface consists of two sheets, which are joined in the limit $K = \pm\infty$, $L = -\frac{1}{2}|K|$ with N fixed. The fixed point is situated on the upper surface, whereas the lower surface, which contains no fixed point, after the first transformation is directly mapped completely on the upper surface. Special points on the surface $\bar{K}_\alpha \neq K^*_\alpha$ even map directly into the fixed point. These last peculiarities, viz that the lower surface is mapped directly onto the upper surface and has no fixed point itself, are due to the facts, as we shall in Section VI, that this particular choice of the renormalization transformation does not leave the ground state of the system invariant for several ranges of the K_α's—in this special case the ranges of K_α for which the ground state is antiferromagnetic (see Nelson and Fisher, 1975).

Apart from determining the critical surface with this renormalization transformation, Nauenberg and Nienhuis also calculated $f(K)$, $\partial f/\partial K$ and $K^2 \partial f(K)/\partial K^2$ corresponding to the free energy, energy and specific heat respectively for the square lattice along the nearest-neighbour axis K, by the method outlined in Section II. Their results are shown in Fig. 13, where for comparison the exact Onsager results have also been drawn. In the limit $K \to 0$ the free energy is correct up to second order in K, and for $K \to \infty$ at least up to the first order.

2. *Cumulant expansion* (*straightforward*)

The convergence of the cumulant expansion applied on the triangular lattice in Section V.A.2 looked rather slow, and of course convergence is not even

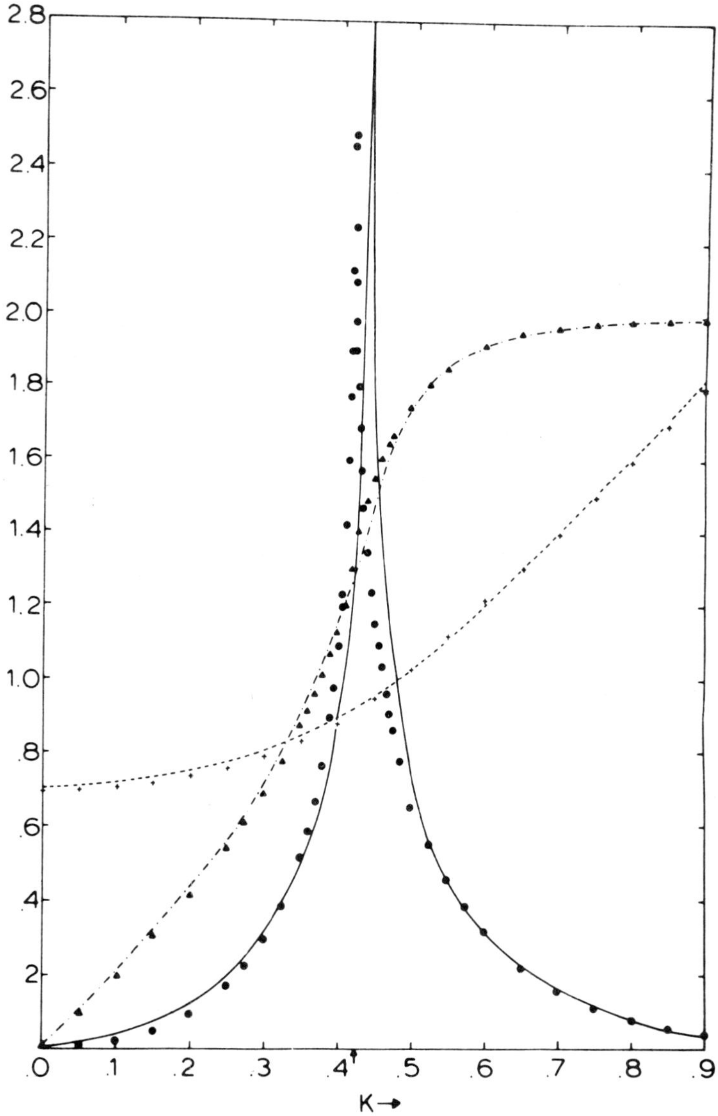

FIG. 13. Dashed curve, dashed–dot curve and solid curve: Onsager's free energy, internal energy and specific heat respectively. The crosses +, triangles ▲ and dots ● are the results for the same quantities respectively obtained by Nauenberg and Nienhuis by their renormalization transformation.

guaranteed. Observing the poor convergence of the cumulant expansion for the triangular lattice, Hsu *et al.* (1975) suggested that the rapidity of the convergence might be improved by taking a cell that contained a larger number of spins, so that $\mathscr{V}$ would decrease with respect to $\mathscr{H}_0$. In order to test this idea they considered the square 2-dimensional lattice and divided it up into square cells each containing 9 spins (Fig. 14a) (although in hind-sight, for comparison with the previous calculation on the triangular lattice, it might have been better to perform the same calculation again on the triangular lattice with cells containing 7 spins, as drawn in Fig. 14b).†

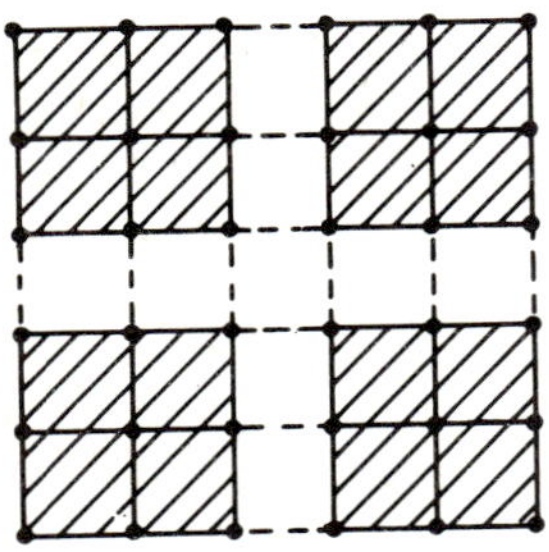

FIG. 14a

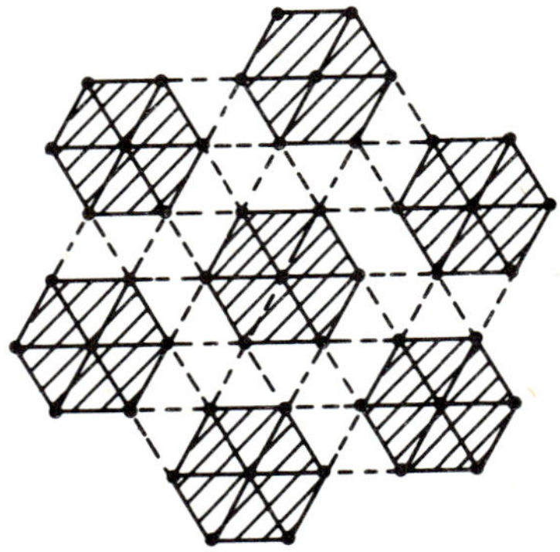

FIG. 14b

The same definition for the cell spin was used as for the triangular lattice, viz

$$s'_{i'} = \mathrm{sign}\left(\sum_{j=1}^{9} s^j_{i'}\right). \tag{5.28}$$

Since the calculation is completely straightforward and identical to the one in Section V.A.2, and is given in full by Hsu *et al.* we shall here quote only the results. The calculations have been performed up to third order and are listed in the Table VI, in which for comparison the exact Onsager results are also shown.

As can be seen the thermal and magnetic eigenvalues are determined to within 0·2% and 1·1% of the exact values in the second-order approximation, and the critical temperature is $K_c = 0{\cdot}4302$ in this approximation (compared to 0·4407 . . . exact). The expansion appears to be rather erratic since, as

† This calculation has in the meantime been performed by Sudbø and Hemmer (1975). See Note added in proof, p. 505.

TABLE VI. Thermal and magnetic eigenvalues for the square Ising lattice resulting from successive orders in the cumulant expansion. The third-order results were derived by Shih Chieh Hsu (private communication).

	y_T	y_H	K_c
Exact	1·0	1·8750	0·4407
1st order	0·9270	1·9433	0·4697
2nd order	1·0021	1·8845	0·4302
3rd order	1·0795	1·8994	0·4312

can be seen from the table the results from the third order are worse than the second-order approximation. A tentative explanation of this phenomenon will be given in Section VI.

3. *Cumulant expansion* (*sophisticated*)

Kadanoff and Houghton (1975) have applied the cumulant expansion, using a definition of the cell spin with an adjustable parameter $\bar{K}$. Via an alternative way to determine λ_H, the optimal choice is made by adjusting $\bar{K}$ in such a way that λ_H gives the same value for both procedures. We shall only sketch their method; for a detailed discussion the reader is referred to the original paper.

The lattice is divided in cells as shown in Fig. 15. Every cell contains four spins. Obviously the increase in lattice constant is $l = 2$. With every cell of four spins we associate a cell spin $s'_{i'}$ and as weight-factor we choose

$$P(s', s) = \prod_{i'} \tfrac{1}{2}[1 + s'_{i'} \tanh \bar{K}(s^1_{i'} + s^2_{i'} + s^3_{i'} + s^4_{i'})]. \tag{5.29}$$

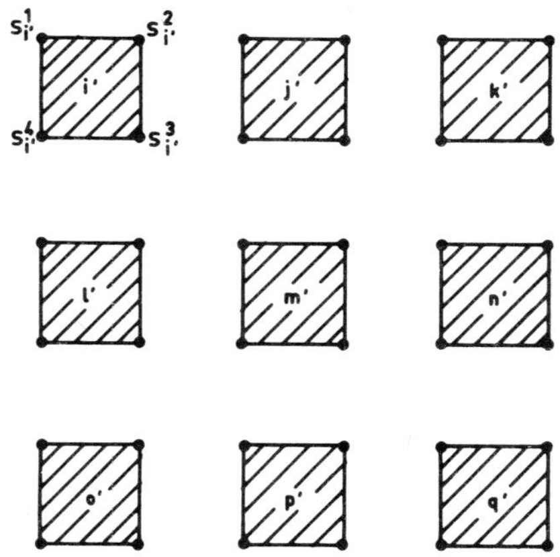

FIG. 15. Kadanoff and Houghton's second block transformation.

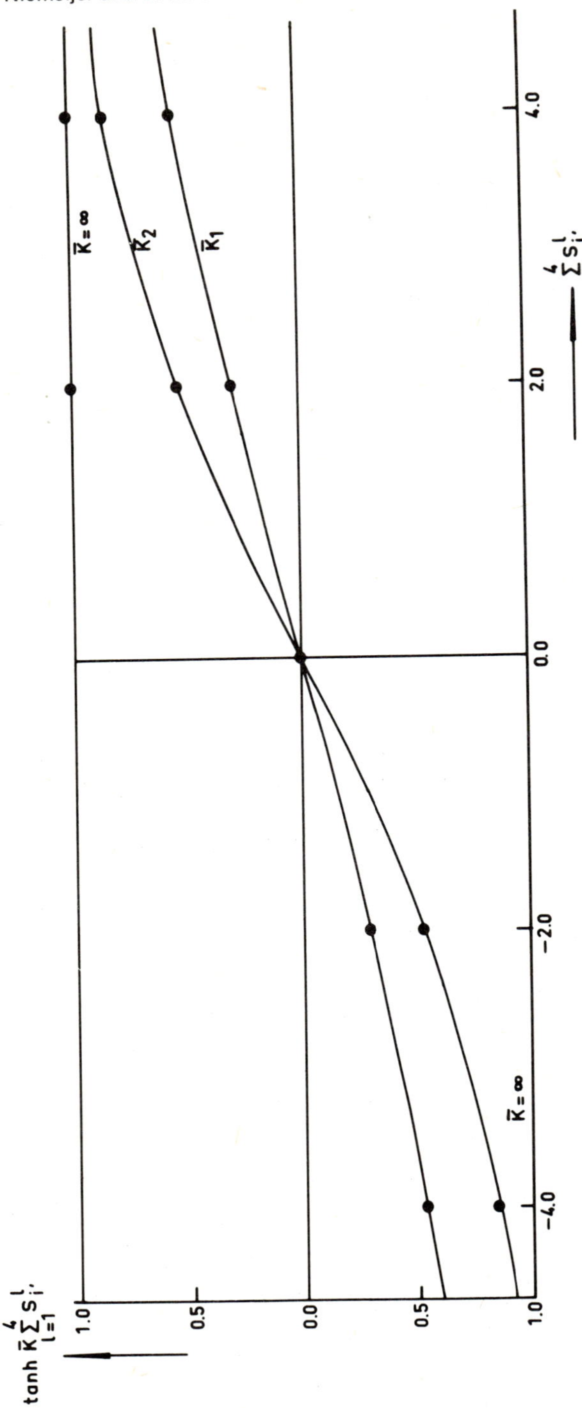

FIG. 16. The function tanh $\bar{K}(s_{i'}^1 + s_{i'}^2 + s_{i'}^3 + s_{i'}^4)$ for several values of $\bar{K}$: $\bar{K}_1 = 0{\cdot}137$, $\bar{K}_2 = 0{\cdot}275$.

As can easily be seen from Fig. 15, the weight-factor (5.29) becomes identical to (1.18) of the triangular lattice if there were only three spins inside a cell, and one would take the limit $\bar{K} \to \infty$.

The renormalization transformation, defined by the weight-factor (5.29) for the second block transformation, has been calculated approximately using the second-order cumulant method sketched in Section III.B, and the values of y_T and y_H have been determined. They should of course not depend on $\bar{K}$, and to check this y_T and y_H have been plotted in Fig. 17 and Fig. 18

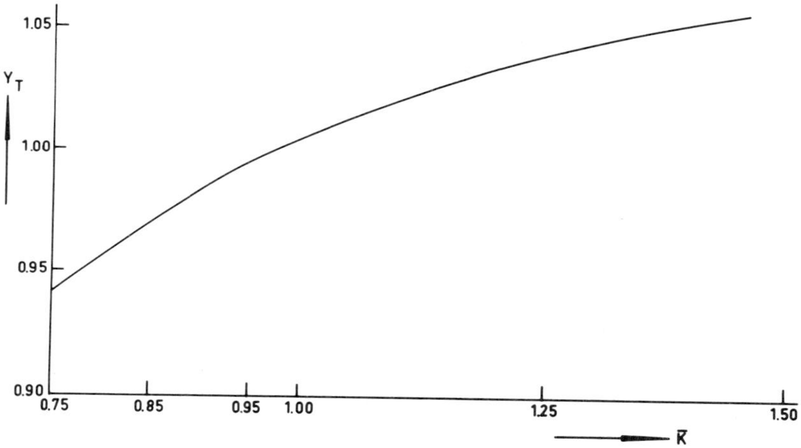

FIG. 17. y_T as a function of $\bar{K}$.

(full curve) to see whether this independence is indeed realized. It is noticed that the calculated eigenvalues are roughly constant over a whole range of $\bar{K}$ but that some $\bar{K}$-dependence is indeed present. It would be helpful to be able to choose some particular value of $\bar{K}$ as better than others, and as a matter of fact, since the results of Section III enable us to determine y_H in an independent way, there is a way to choose a $\bar{K}$ which is better than others. The weight-factor (5.29) can be rewritten as

$$P(s', s) = \prod_{i'} \tfrac{1}{2}[1 + s'_{i'}(s^1_{i'} + s^2_{i'} + s^3_{i'} + s^4_{i'})(w_s + w_t s^1_{i'} s^2_{i'} s^3_{i'} s^4_{i'})]. \tag{5.30}$$

Using the representation

$$\tanh \bar{K}(s^1 + s^2 + s^3 + s^4) = (s^1 + s^2 + s^3 + s^4)(w_s + w_t s^1 s^2 s^3 s^4) \tag{5.31}$$

with

$$\begin{aligned} w_s &= (\tanh 4\bar{K} + 2 \tanh 2\bar{K})/8 \\ w_t &= (\tanh 4\bar{K} - 2 \tanh 2\bar{K})/8. \end{aligned} \tag{5.32}$$

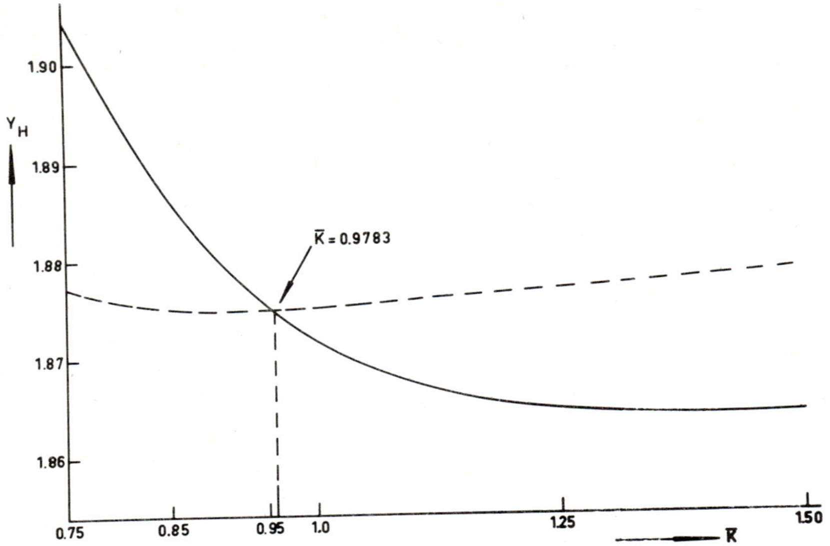

FIG. 18. y_H as obtained from the second block transformation (solid curve) and as obtained from (5.33) (dashed curve).

Now (5.30) has obtained the same form as (1.18) but with more complicated w_s and w_t instead of the factors $\frac{1}{2}$ and $-\frac{1}{2}$ in (1.18). So we may directly apply the arguments given in Section III.D, and eqn (3.78) is replaced by

$$l^{d-y_H}\phi_s^H = 4(w_s\phi_s^H + w_t\phi_t^H). \tag{5.33}$$

The two graphs intersect at $\bar{K} = 0{\cdot}9783$, which will be taken as the best $\bar{K}$. With this value of $\bar{K}$ one finds

$$y_T = 0{\cdot}999964 \tag{5.34}$$

and

$$y_H = 1{\cdot}87553 \tag{5.35}$$

These should be compared with the exact Onsager values 1·0 and 1·8750 respectively. One may conclude that the trick of adjusting $\bar{K}$, through two different determinations of y_H, greatly improves the method of calculation.

VI. Discussion

"One cannot write a renormalization cookbook"
(K. G. Wilson, 1976).

By the time this article was closed (April 1975), the developments in the applications (even for $d = 3$) and the improvements in calculational schemes were in rapid progress. This renders a discussion and evaluation of the results obtained and the problems to be dealt with a risky, at worst ephemeral, enterprise. Yet a number of conclusions seem to emerge from this approach.

1. The renormalization transformation (1.11) has been studied for some weight-factors $P(s', s)$. In approximation these renormalization transformations always exhibit at least one non-trivial fixed point, which can be associated with the critical point of a ferromagnetic phase transition. No actual proof of the existence of this fixed point has been given (but the reader should be reminded that no proof exists of the convergence, for whatever temperature region, of high-temperature expansions, to which the cumulant expansion IVB bears some resemblance); nor can the transformation be studied systematically around the fixed point in terms of an expansion parameter (like the ε-expansion). Yet the results of the approximations do converge so well towards the exact results for ferromagnetic transitions, known from other sources, that it seems fair to conclude that the true transformation has at least a ferromagnetic fixed point with eigenvalues $\lambda_T = l$ and $\lambda_H = l^{15/8}$ for 2-dimensional lattices. This fixed point appears to be isolated (not on a line of fixed points), which means that the critical behaviour in the neighbourhood of the fixed point is universally the same as that of the 2-dimensional Ising model. Also the hyperscaling relations $\nu d = 2 - \alpha$ and $\delta = (d + 2 - \eta)/(d - 2 + \eta)$ are found to be valid. It should be stressed that other fixed points with their respective domains of attraction may be present in the complete phase diagram for different lattices, e.g.

(i) antiferromagnetic fixed points and Baxter-like (super) fixed points which may be part of a line of fixed points. We shall return to this question at the end of this Section.

(ii) the "Baxter–Wu" fixed point for the triangular lattice, corresponding to the triplet interaction system (1.8).

It is doubtful whether one particular choice of $P(s', s)$ will lead to the completely acceptable phase diagram and flow pattern for a certain lattice with given interactions. (E.g. it has been shown (Braathen and Hemmer, 1975) that the weight-function defined in (1.18) for the triangular lattice does not lead to a Baxter–Wu fixed point but only to the ferromagnetic (Ising) fixed point.)

The approximations all involve only a finite number of interactions. The longer-range interactions seem to contribute in a minor way (except perhaps for linear transformations, where η is so to speak "loose", as can be brought about by long-range interactions).

2. Of the proposed approximation schemes, the cluster approximation,

judging by its results, looks the most reliable. The convergence of the cumulant expansion becomes questionable for interaction strengths of the order of the fixed-point values. One might try to improve this convergence by enlarging $\mathscr{H}_0$ with respect to $\mathscr{V}$ through increase of the size of the cell (Hsu *et al.*, 1975), but the results obtained so far along these lines are not conclusive yet. The tentative idea is that the cumulant expansion is a conditional convergent series. The "optimal" cumulant order depends on the cell size chosen and seems to be related to the type of correlation function of site spins in a cell needed for a certain cumulant order. Since the cell size is finite correlation functions exist only up to a certain range in a cell; beyond that maximum order no further information can be extracted from cell correlation functions, and the cumulant expansion starts to diverge rather than to improve the result.

The cluster approximation converges more steadily with the increase of the basic cluster. The method, however, becomes very laborious for large clusters, and unless one learns more about selecting the important configurations, the method is practically restricted to 2-dimensional lattices.

A new, very promising feature is to build into the renormalization transformation adjustable parameters and find means to optimalize the free parameter(s) as was done by Kadanoff and Houghton (1975).

3. The renormalization approach is more than a theory of critical phenomena, as Nauenberg and Nienhuis (1974b), Tjon (1974), Nauenberg (1975), Nelson and Fisher (1975), and Niemeijer and Ruijgrok (1975) have shown by calculating the thermodynamic properties in the whole temperature range. Also the first-order properties of the ferromagnetic phase transition, i.e. the spontaneous magnetization, can be deduced from the zero-temperature behaviour of the renormalization transformation, as shown in Section IID. Little has been calculated along these lines so far (Nienhuis and Nauenberg, 1975). The same is true for the determination of the correlation functions by renormalization theory methods both at criticality and in non-critical regions.

So far for the achievements. The list of problems to be dealt with is yet longer. The obvious conclusion—that more has to be calculated and by better methods—has already been mentioned in passing. We concentrate here on two aspects given little attention in the previous sections.

(a) The examples treated show that it is meaningful to treat the renormalization transformation in a subspace of a limited number of interaction parameters. In such a subspace the attention has been focussed on the region of parameters, where the ferromagnetic phase transition occurs. As already mentioned, however, one should study the transformation in the whole parameter subspace and locate all the fixed points in order to get an *overall* view of the flow pattern as induced by the renormalization transformation.

Fixed points at zero and infinite temperatures have been mentioned somewhat implicitly, especially in the discussion of the spontaneous magnetization in Section IID, but this was all associated with the ferromagnetic fixed point.

Nauenberg and Nienhuis (1974a and Section V.B.1) have treated a renormalization transformation for the square lattice in a fairly large region of the nearest (K), next-nearest (L) and quadruple (four spin) (N) interactions. They find that, in the region where one expects an antiferromagnetic transition ($K < 0$, $L > 0$), the renormalization transformation maps the system into a ferromagnetic system and so no fixed point is found which can be associated with an antiferromagnetic critical point. This should be regarded as a shortcoming of their specific choice for the renormalization transformation, since one cannot expect the ferro- and antiferromagnetic phase transitions to be of the same nature, because for a ferromagnetic transition the magnetic field is a relevant direction, whereas it is not for the antiferromagnetic transition. The conclusion must be that this renormalization transformation is of value only in a limited region. The transformation $K'(K)$ must become singular when K is an antiferromagnetic critical point.

It is, however, easy to remedy this shortcoming by changing the definition of the cell as shown in Fig. 19 and by defining a cell spin again as

$$s'_i = \operatorname{sign} \sum_{j=1}^{5} s_i^j \tag{6.1}$$

(van Leeuwen, 1975b). Then not only does a ferromagnetic fixed point appear, but at the same time two antiferromagnetic fixed points, corresponding to phase transitions to Néel-type and layered ordering respectively. More details will be given further on. (See also Nelson and Fisher, 1975.)

As a first step to a full analysis of the behaviour of the free energy in the parameter space, one should (in general) analyse the low-temperature properties of the renormalization transformation, and see to it that the renormalization transformation projects every possible ground state onto itself. At a surface in the parameter space where two ground states are competing, generally a phase transition occurs. Such considerations, amongst others, lead for the square lattice to the choice of the cells in Fig. 19.

This brings us to a second question:

(b) How to choose the weight-function $P(s', s)$? Some comments have already been made in relation to the question of linear versus non-linear weight-functions (III.A; see also Bell and Wilson (1974)). The linear weights have the drawback that one has to determine in advance a certain seemingly arbitrary constant in order to get meaningful results, or show by trial and error that for only one choice of the arbitrary parameter it is possible to obtain a fixed point. Moreover the linear weight-function $P(s', s)$ is such, that for the right choice of the arbitrary parameter it may become negative, which is "officially" forbidden by (1.10a). For the description of critical phenomena, however, this seems a minor point, as it is hard to imagine that the few configurations with negative weight would win over the positive majority. For the low-temperature properties this is a more serious point, since for $T = 0$ only a few configurations do count and one has to make sure that the ones with negative weight do not become dominant. Linear weight-functions, on the other hand, have the important advantage that they clearly map critical systems onto each other, because they scale the correlation length. Since the theory of the correlation functions is possible also for non-linear weight-functions (III), generalizing the linear theory, we feel that there is no longer any compelling reason to have a preference for linear weight-functions.

As has already been mentioned, one should look for weight-functions that improve the results for the same computational effort. But we would like to address ourselves here to a more fundamental question: which phase transitions can be described by a certain weight-function, or, the other way

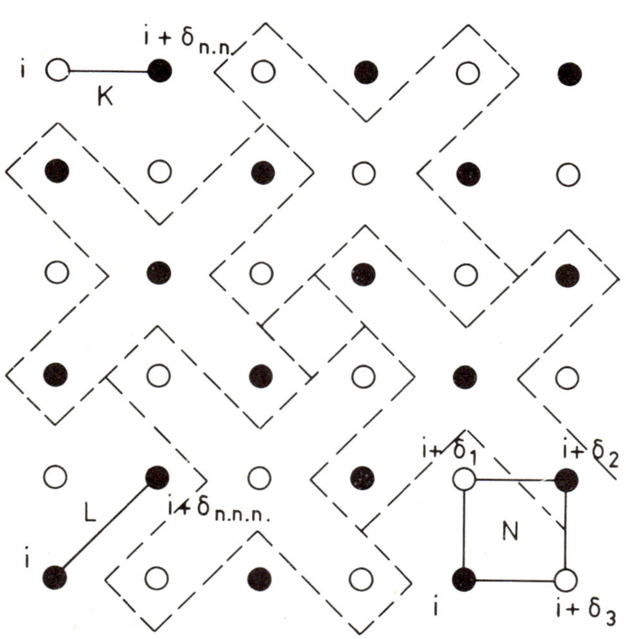

FIG. 19. Quadratic lattice with black and white sublattices. K = nearest-neighbour, L = next-nearest-neighbour, and N = four-spin coupling. Cells involve either black or white spins.

round: how does one have to design a weight-function $P(s', s)$ in order to describe a certain phase transition? In our opinion the key to the answer is in the low-temperature behaviour. Suppose that we restrict ourselves to a limited set of interaction parameters. For the various relative strengths of these parameters one has different configurations of spins which maximize $\mathscr{H}(s)$, i.e. different ground states. A phase transition occurs at the dividing surface of two competing ground states. If one wants to describe this phase transition one should choose the weight-function $P(s', s)$ such that each of the possible ground states is invariant under the renormalization transformation. Also, by extending these ideas to temperatures $T \neq 0$, one should choose $P(s', s)$ such that as many symmetries as possible in the free energy of the system are explicitly conserved in the transformation.

We close this discussion with an example (van Leeuwen, 1975b) where the choice of the weight-function $P(s', s)$ is deliberately adapted to the type of phase transition that has to be described. Also this example touches upon the intriguing question how it is possible that the symmetric 8-vertex model has interaction-dependent exponents. The weights chosen are based on the cell division of Fig. 19 and $P(s', s)$ is 1 when (6.1) is fulfilled for all cells and 0 otherwise.

The renormalization transformation is then studied in the subspace of a nearest-neighbour coupling K, a next-nearest-neighbour coupling L, and a 4-spin coupling N. When $L = N = 0$ the Hamiltonian reduces to the Ising model, for $K = 0$ to the symmetric 8-vertex model, and for $K = N = 0$ to an Ising Hamiltonian for two uncoupled sublattices (the black and white sites in Fig. 19). The free energy exhibits in the K, L, N subspace the following symmetries

$$
\begin{aligned}
f(K, L, N) &= f(-K, L, N) \\
f(0, L, N) &= f(0, -L, N).
\end{aligned}
\tag{6.2}
$$

The choice of $P(s', s)$ is such that the first symmetry is obeyed by any approximation of the renormalization transformation. The reason is that $P(s', s)$ does not couple the black and white sublattices and K represents an interaction which is *odd* in both sublattices while L and N are even. So, seen from the sublattices, the first symmetry (6.2) is analogous to the up–down symmetry in the magnetic field in a usual Ising system. The second symmetry (6.2) is not automatically fulfilled in approximations. By drawing more complicated cells it could be made so, but as it is not so important and moreover nearly fulfilled in approximations, we do not bother to guarantee it strictly.

The transformation is such that it leaves the ferromagnetic state, the Néel antiferromagnetic state and the layered antiferromagnetic state invariant. Approximate calculations show that there is:

(i) A ferromagnetic fixed point F (near $K = 0{\cdot}3$, $L = 0{\cdot}07$, $N = -0{\cdot}01$) with two relevant eigenvalues with exponents $y_T(=1)$ and $y_H(=15/8)$.

(ii) The first symmetry (6.2) implies the existence of an antiferromagnetic fixed point AF (near $K = -0{\cdot}3$, $L = 0{\cdot}07$ and $N = -0{\cdot}01$). The relevant exponent y_T remains but the magnetic exponent y_H becomes irrelevant.†

In the subspace with the interaction between the sublattices absent (i.e. $K = N = 0$), one has two uncoupled Ising systems, which have

(iii) A (sublattice) ferromagnetic fixed point SF (near $K = 0$, $L = 0{\cdot}3$, $N = 0$).

(iv) A (sublattice or) layered antiferromagnetic fixed point SAF (near $K = 0$, $L = -0{\cdot}3$, $N = 0$), which is a consequence of the second symmetry relation (6.2).

Figure 20 gives a schematic picture of the fixed points, flow pattern and critical curves projected in the K–L plane.

The eigenvalue structure of the sublattice fixed points SF and SAF is partially determined by combinations of eigenvalues of the corresponding fixed points F and AF. So one finds additional relevant exponents for SF:

$$\begin{aligned} y_s &= 2y_H - d = \gamma/\nu \\ y_H &= y_H + y_T - d = (\eta - \beta)/\nu \\ y_B &= 2y_T - d = d\alpha/(2 - \alpha) \end{aligned} \tag{6.3}$$

where α, β, γ, ν are the Ising exponents. For SAF only y_B exists. The interesting point is that for 2-dimensional lattices $y_T = 1$ (or $\alpha = 0$) such that y_B becomes marginal ($y_B = 0$), indicating that SF and SAF are points of a fixed line which runs in the $K = 0$ subspace. Such a line of fixed points is necessary to explain the interaction-dependence of the symmetric 8-vertex model critical exponents.

Another consequence of this analysis is that the existence of y_s in SF implies a cusp in the critical surface (in contradistinction to Nauenberg and Nienhuis's (1974b) picture, Fig. 12). Near SF the critical surface behaves as

$$K = A(L - L_c)^\gamma$$

where A is some amplitude and L_c the location of SF on the L axis.

The typical exponents of the triangular lattice with triplet interaction (1.8) can be explained similarly by constructing a renormalization transformation acknowledging the symmetries associated with a subdivision of the lattice into 3 interpenetrating triangular sublattices.

† See Nienhuis and Nauenberg (1975) for a different picture.

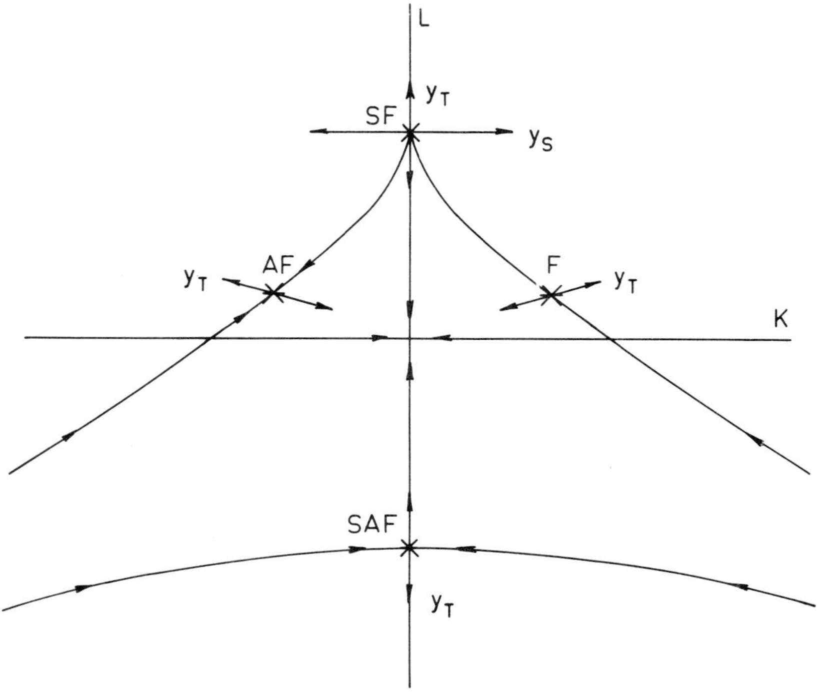

FIG. 20. Schematic location of fixed points, even eigenvalues and critical curves in $K_1 - K_2$ projection.

References

Barber, M. N. and Baxter, R. J. (1973). *J. Phys.* **C6**, 2193.
Baxter, R. J. (1971). *Phys. Rev. Letts* **26**, 832.
Baxter, R. J. (1972). *Ann. Phys. (N.Y.)* **70**, 193.
Baxter, R. J. and Wu, F. Y. (1973). *Phys. Rev. Letts* **31**, 1294.
Bell, T. L. and Wilson, K. G. (1974). *Phys. Rev.* **B10**, 3935.
Braathen, H. J. and Hemmer, P. C. (1975). *Arkiv for Det Fysiske Seminar i Trondheim*, No. 2.
Dalton, N. W. and Wood, D. W. (1969). *J. Math. Phys.* **10**, 1271.
Fisher, M. E. (1969). *Phys. Rev.* **180**, 594.
Harris, A. B. and Lubensky, T. C. (1974). *Phys. Rev. Letts* **33**, 1540.
Hsu, S. C., Niemeijer, Th. and Gunton, J. D. (1975). *Phys. Rev.* **B11**, 2699.
Kadanoff, L. P. and Houghton, A. (1975). *Phys. Rev.* **B11**, 377.
Van Leeuwen, J. M. J. (1975a). *In* "Fundamental Problems in Statistical Mechanics III", (E. G. D. Cohen, ed.). North Holland/American Elsevier.
Van Leeuwen, J. M. J. (1975b). *Phys. Rev. Letts* **34**, 1056.
Nauenberg, M. and Nienhuis, B. (1974a, b). *Phys. Rev. Letts* **33**, 344, 1598.
Nauenberg, M. (1975a). *J. Math. Phys.* **16**, 703.
Nauenberg, M. (1975b). *J. Phys.* **A8**, 925.

Nelson, D. R. and Fisher, M. E. (1975). *Ann. Phys.* **91**, 226.
Nihat Berker, A. (1975). Preprint.
Niemeijer, Th. and van Leeuwen, J. M. J. (1973). *Phys. Rev. Letts* **31**, 1411.
Niemeijer, Th. and van Leeuwen, J. M. J. (1974). *Physica* **71**, 17.
Niemeijer, Th. and Ruijgrok, Th. W. (1975). *Physica* **81A**, 427.
Nienhuis, B. and Nauenberg, M. (1975). *Phys. Rev. Lett.* **35**, 477.
Nightingale, M. P., and 't Hooft, A. H. (1975). To be published.
Onsager, L. (1944). *Phys. Rev.* **65**, 117.
Rota, G. C. (1964). *Zeitschrift fur Wahrscheinlichkeitstheorie* **2**, 340.
Stanley, H. E. (1971). "Introduction to Phase Transitions and Critical Phenomena", Clarendon Press, Oxford, England.
Subbarao, K. (1975). *Phys. Rev.* **B11**, 1165.
Sudbø, A. S. and Hemmer, P. C. (1975). Arkiv for Det Fysiske Seminar i Trondheim.
Suzuki, M. (1974). *Progr. Theor. Phys.* **51**, 1992.
Tjon, J. A. (1974). *Physics Letts* **49A**, 289.
Wegner, F. J. (1972). *Phys. Rev.* **B5**, 4529.
Widom, B. (1965a, b). *J. Chem. Phys.* **43**, 3892, 3898.
Wilson, K. G. (1971a, b). *Phys. Rev.* **B4**, 3174, 3184.
Wilson, K. G. and Fisher, M. E. (1972). *Phys. Rev. Letts* **28**, 240.
Wilson, K. G. and Kogut, J. (1974). *Physics Reports* **12C**, 76.
Wilson, K. G. (1976). *Advances in Mathematics*, to appear.

Appendix A

This appendix is concerned with the definition of the scaling coordinates $u_i(\mathbf{k})$ of (3.47) to second order in the deviation $\delta K_\alpha(\mathbf{k})$. The basis is the relation

$$\delta K'_{a'} = \sum_b T_{a'b}\delta K_b + \frac{1}{2!}\sum_{bc} T_{a',bc}\delta K_b \delta K_c \tag{A.1}$$

connecting to second order the induced cell interaction with the site perturbations. In the text the first-order term was treated; here we shall concentrate on the second-order term. Corresponding to (3.49) we define

$$\tilde{T}_{\alpha,\beta\gamma}(\mathbf{k}_1,\mathbf{k}_2) = \sum_{\substack{b\in\beta\\ c\in\gamma}} T_{a',bc} \exp i[\mathbf{k}_1\cdot(\mathbf{r}_{a'} - \mathbf{r}_b) + \mathbf{k}_2\cdot(\mathbf{r}_{a'} - \mathbf{r}_c)]. \tag{A.2}$$

For the fourier component $\delta K'_\alpha(\mathbf{k}')$ we find in second order

$$\begin{aligned}[\delta K'_\alpha(\mathbf{k}')]_2 &= (2N')^{-1} \sum_{a'\in\alpha}\sum_{bc} T_{a',bc}\delta K_b \delta K_c \exp i\mathbf{k}'\cdot\mathbf{r}'_{a'} \\ &= (2N')^{-1} \sum_{a'\in\alpha}\sum_{bc} T_{a',bc} \sum_{\mathbf{k}_1\mathbf{k}_2} \delta\tilde{K}_\beta(\mathbf{k}_1)\delta\tilde{K}_\gamma(\mathbf{k}_2) \exp i[\mathbf{k}'\cdot\mathbf{r}'_{a'} \\ &\qquad - \mathbf{k}_1\cdot\mathbf{r}_b - \mathbf{k}_2\cdot\mathbf{r}_c] \\ &= (2N')^{-1} \sum_{a'\in\alpha}\sum_{\mathbf{k}_1\mathbf{k}_2}\sum_{\beta\gamma} T_{\alpha,\beta\gamma}(\mathbf{k}_1,\mathbf{k}_2)\delta\tilde{K}_\beta(\mathbf{k}_1)\delta\tilde{K}_\gamma(\mathbf{k}_2) \\ &\qquad \times \exp i[\mathbf{k}'\cdot\mathbf{r}'_{a'} - (\mathbf{k}_1 + \mathbf{k}_2)\cdot\mathbf{r}_{a'}]. \end{aligned} \tag{A.3}$$

As in (3.51), one has

$$(N')^{-1} \sum_{a' \in \alpha} \exp i[\mathbf{k}' \cdot \mathbf{r}'_{a'} - (\mathbf{k}_1 + \mathbf{k}_2) \cdot \mathbf{r}_{a'}] = \delta_{\mathbf{k}', (\mathbf{k}_1 + \mathbf{k}_2)l}. \tag{A.4}$$

Inserting (A.4) into (A.3) and combining this second-order result with (3.52):

$$\delta K'_\alpha(\mathbf{k}l) = \sum_\beta \tilde{T}_{\alpha\beta}(\mathbf{k}) \delta \tilde{K}_\beta(\mathbf{k}) + \frac{1}{2!} \sum_{\mathbf{k}_1} \sum_{\beta\gamma} T_{\alpha,\beta\gamma}(\mathbf{k}_1, \mathbf{k} - \mathbf{k}_1) \delta \tilde{K}_\beta(\mathbf{k}_1) \delta \tilde{K}_\gamma(\mathbf{k} - \mathbf{k}_1). \tag{A.5}$$

Thus we put tentatively

$$\delta u_i(\mathbf{k}) = \sum_\alpha \varphi^i_\alpha(\mathbf{k}) \delta \tilde{K}_\alpha(\mathbf{k}) + \frac{1}{2!} \sum_{\mathbf{k}_1} \sum_{\alpha\beta} \varphi^i_{\alpha\beta}(\mathbf{k}_1, \mathbf{k} - \mathbf{k}_1) \delta \tilde{K}_\alpha(\mathbf{k}_1) \delta \tilde{K}_\beta(\mathbf{k} - \mathbf{k}_1) \tag{A.6}$$

as the extension of (3.54). In order that $\delta u_i(\mathbf{k})$ fulfils (3.47), also the second order, the $\varphi^i_{\alpha\beta}(\mathbf{k}_1, \mathbf{k}_2)$ have to be chosen to satisfy

$$l^{y_i} \varphi^i_{\alpha\beta}(\mathbf{k}_1, \mathbf{k}_2) = \sum_\gamma \varphi^i_\gamma((\mathbf{k}_1 + \mathbf{k}_2)l) \tilde{T}_{\gamma,\alpha\beta}(\mathbf{k}_1, \mathbf{k}_2) + \sum_{\gamma\delta} \varphi^{i'}_{\gamma\delta}(\mathbf{k}_1 l, \mathbf{k}_2 l) \times \tilde{T}_{\gamma\alpha}(\mathbf{k}_1) \tilde{T}_{\delta\beta}(\mathbf{k}_2). \tag{A.7}$$

The solution of (A.7) is similar to that of (3.53), determined by the condition that $\varphi^i_{\alpha\beta}(\mathbf{k}_1, \mathbf{k}_2)$ is regular in $\mathbf{k}_1, \mathbf{k}_2$ and the scaling field u_i and merges for the fixed point and $\mathbf{k}_1 = \mathbf{k}_2 = 0$ into the $\varphi^i_{\alpha\beta}$ defined in (1.35)

$$\varphi^{i*}_{\alpha\beta}(\mathbf{0}, \mathbf{0}) = \varphi^i_{\alpha\beta}. \tag{A.8}$$

The solution of (A.7) can be simplified by introducing

$$\phi^{ijl}(\mathbf{k}_1, \mathbf{k}_2) = \sum_{\alpha\beta} \varphi^i_{\alpha\beta}(\mathbf{k}_1, \mathbf{k}_2) \psi^j_\alpha(\mathbf{k}_1) \psi^l_\beta(\mathbf{k}_2) \tag{A.9}$$

for which the following simpler equation holds

$$l^{y_i} \varphi^{ijl}(\mathbf{k}_1, \mathbf{k}_2) = \sum_{\alpha\beta\gamma} \varphi^i_\gamma((\mathbf{k}_1 + \mathbf{k}_2)l) \tilde{T}_{\gamma,\alpha\beta}(\mathbf{k}_1, \mathbf{k}_2) \psi^j_\alpha(\mathbf{k}_1) \psi^l_\beta(\mathbf{k}_2) + l^{y_j + y_l} \varphi^{ijl}(\mathbf{k}_1 l, \mathbf{k}_2 l). \tag{A.10}$$

Using the closure relation (3.59), $\varphi^i_{\alpha\beta}$ can be recovered as

$$\varphi^i_{\alpha\beta}(\mathbf{k}_1, \mathbf{k}_2) = \sum_{jl} \varphi^{ijl}(\mathbf{k}_1, \mathbf{k}_2) \varphi^j_a(\mathbf{k}_1) \varphi^l_\beta(\mathbf{k}_2). \tag{A.11}$$

To extend (3.60) to second order we must invert (A.6). Multiplying (A.6) by $\psi^{i'}_\alpha(\mathbf{k})$ and summing over i, using (3.60) for the second-order term, gives

$$\delta \tilde{\mathbf{K}}_\alpha(\mathbf{k}) = \sum_i \psi^i_\alpha(\mathbf{k}) \delta u_i(\mathbf{k}) - \frac{1}{2!} \sum_i \psi^i_\alpha(\mathbf{k}) \sum_{\mathbf{k}_1} \sum_{\beta\gamma} \varphi^i_{\beta\gamma}(\mathbf{k}_1, \mathbf{k}_1 - \mathbf{k}_1) \sum_j \psi^j_\beta(\mathbf{k}_1) \times \delta u_j(\mathbf{k}_1) \sum_l \psi^l_\gamma(\mathbf{k} - \mathbf{k}_1) \delta u_l(\mathbf{k} - \mathbf{k}_1). \tag{A.12}$$

Using (A.9) this reduces to

$$\delta\tilde{K}_\alpha(\mathbf{k}) = \sum_i \psi^i_\alpha(\mathbf{k})\delta u_i(\mathbf{k}) - \frac{1}{2!}\sum_{ijl}\sum_{k} \psi^i_\alpha(\mathbf{k})\varphi^{ijl}(\mathbf{k}_1, \mathbf{k} - \mathbf{k}_1)\delta u_j(\mathbf{k}_1)\delta u_l(\mathbf{k} - \mathbf{k}_1). \tag{A.13}$$

With (A.6) and (A.13) we can compute the derivatives needed for the conversion (3.65) from χ_{ij} to $\tilde{G}_{\alpha\beta}$ and (3.66) from $\tilde{C}_{\alpha\beta}$ to γ_{ij}. For the first order we have

$$\begin{aligned} \delta u_i(\mathbf{k})/\delta\tilde{K}_\alpha(\mathbf{k}_1) &= \varphi^i_\alpha(\mathbf{k})\delta_{\mathbf{k},\mathbf{k}_1} \\ \delta\tilde{K}_\alpha(\mathbf{k})/\delta u_i(\mathbf{k}_1) &= \psi^i_\alpha(\mathbf{k})\delta_{\mathbf{k},\mathbf{k}_1} \end{aligned} \tag{A.14}$$

and the second-order derivatives are

$$\begin{aligned} \delta u_i(\mathbf{k})/\delta\tilde{K}_\alpha(\mathbf{k}_1)\delta K_\beta(\mathbf{k}_2) &= \varphi^i_{\alpha\beta}(\mathbf{k}_1, \mathbf{k}_2)\delta_{\mathbf{k},\mathbf{k}_1+\mathbf{k}_2} \\ \delta\tilde{K}_\alpha(\mathbf{k})/\delta u_i(\mathbf{k}_1)\delta u_j(\mathbf{k}_2) &= -\sum_l \psi^l_\alpha(\mathbf{k})\varphi^{lij}(\mathbf{k}_1, \mathbf{k}_2)\delta_{\mathbf{k},\mathbf{k}_1+\mathbf{k}_2}. \end{aligned} \tag{A.15}$$

The conversion formulae (3.65) and (3.66) then result by observing that

$$\begin{aligned} \tilde{G}_{\alpha\beta}(\mathbf{k}) &= \delta^2 f/\delta\tilde{K}_\alpha(-\mathbf{k})\delta\tilde{K}_\beta(\mathbf{k}) \\ \tilde{C}_{\alpha\beta}(\mathbf{k}) &= \delta^2 c/\delta\tilde{K}_\alpha(-\mathbf{k})\delta\tilde{K}_\beta(\mathbf{k}) \end{aligned} \tag{A.16}$$

and using (A.14) and (A.15) to convert the derivatives. The following have also been used:

$$\begin{aligned} \delta f/\delta K_\alpha(0) &= \partial f/\partial K_\alpha \\ \delta c/\delta K_\alpha(0) &= \partial g/\partial K_\alpha \end{aligned} \tag{A.17}$$

as follows from the definition of $\delta K_\alpha(0)$.

Appendix B

In this appendix we consider the system for very low temperatures, where one configuration (or a finite number of degenerate states) starts to dominate the sum over the configurations in the renormalization transformation (1.11). In Section IID this dominant state, maximizing $P(s', s)\exp\mathscr{H}(s)$, is denoted by $\{\bar{s}(s')\}$. Any other configuration $\{s\}$, obtained by flipping a number of spins in $\{\bar{s}\}$, will lead to a contribution, which is smaller by a factor $\exp -\Delta$, where Δ is the excitation energy:

$$\Delta = -\mathscr{H}(s) + \mathscr{H}(\bar{s}). \tag{B.1}$$

For $T \to 0$, Δ grows as $1/T$ for fixed interaction constants and exp-Δ becomes arbitrarily small. The problem is that many states $\{s\}$ exist, all differing by the same amount Δ from the dominant state, and that their number may compensate the smallness of exp-Δ. Consider e.g. the ferromagnetic states $\{s_{\pm}\}$ (all $s_i = +1$ or all $s_i = -1$). Let the cell system be in state $\{s'_{\pm}\}$ and let the dominant state again be ferromagnetic: $\{\bar{s}(s'_{\pm})\} = \{s_{\pm}\}$ (see (2.57) and (2.58)). Flipping one site spin with respect to $\{s_{\pm}\}$ leads to an excitation energy Δ, but one can make this excited state in N possible ways. Flipping two site spins will lead generally to a higher Δ but it can be achieved in $N(N-1)/2$ ways, etc. The point is to show that the multitude of excited states (the entropy) still leads to exponentially small corrections for low temperatures.

The problem of the degeneracy of the excited states occurs mainly in states $\{s'\}$ of high symmetry such as the ferromagnetic states. Thus we restrict the discussion to $\{s'_{\pm}\}$ and configurations differing from $\{s'_{\pm}\}$ in a finite number of cells.

We write the Hamiltonian $\mathscr{H}(s)$ as

$$\mathscr{H}(s) = \mathscr{H}\{\bar{s}(s')\} - \mathscr{V}(s', s) \tag{B.2}$$

and for the new Hamiltonian $\mathscr{H}'(s')$:

$$\mathscr{H}'(s') = \bar{\mathscr{H}}(s') - \mathscr{V}(s') \tag{B.3}$$

where $\bar{\mathscr{H}}(s')$ is the image of $\mathscr{H}\{\bar{s}(s')\}$ according to (2.59). Here we discuss the effect of $\mathscr{V}(s', s)$ on the renormalization transformation.

$$\exp - [\mathscr{U} + \mathscr{V}'(s')] = \sum_{\{\sigma\}} \exp - \mathscr{V}(s', s) \tag{B.4}$$

We follow the notation of (1.21) as we assume that $P(s', s) = 0$ or 1, such that $\{\sigma\}$ runs through the configurations which are allowed in state $\{s'\}$. In order to describe the deviations from the dominant state we introduce

$$\begin{aligned} \mu_i &= [s_i - \bar{s}_i(s')]/2 \\ s_i &= \bar{s}_i(s') + 2\mu_i \end{aligned} \tag{B.5}$$

and rewrite

$$\mathscr{V}(s', s) = \sum_{a} \mathscr{V}_{\alpha}(s')\mu_a, \qquad \mu_a = \prod_{i \epsilon a} \mu_i \tag{B.6}$$

by expressing all s_i with (B.5) in terms of the μ_i. (B.2) shows that $\mathscr{V}(s', s)$ is zero in the state $\{s\} = \{\bar{s}(s')\}$ or all $\mu_i = 0$. Thus in (B.6) the sum runs over all non-empty sets of sites. $\mathscr{V}(s', s)$ is further decomposed as

$$\mathscr{V}(s', s) = \mathscr{V}_0(s', s) + \Delta\mathscr{V}(s', s) \tag{B.7}$$

where $\mathscr{V}_0$ contains all terms in (B.6) where the sets a are contained in one and the same cell. So $\mathscr{V}_0(s', s)$ is a sum over terms referring to the cells i':

$$\mathscr{V}_0\,(s', s) = \sum_{i'} \mathscr{V}_{0i'}\,(s'; \mu^1_{i'} \ldots \mu^n_{i'}). \tag{B.8}$$

The reason for separating off the term $\mathscr{V}_0$ is that the sum

$$\sum_{\{\sigma\}} \exp - \mathscr{V}_0\,(s', s) = \prod_{i'} \left[\sum_{\mu^1_{i'} \ldots}' \exp - \mathscr{V}_{0i'}\,(s'; \mu^1_{i'} \ldots \mu^n_{i'}) \right] \tag{B.9}$$

can be performed. E.g. for the triangular lattice with weight-factor (1.18) the sum over the allowed configurations inside a cell runs over the four possibilities (see (1.25))

$$\begin{aligned} \sigma_{i'} &= 0 \quad \text{or} \quad \mu^1_{i'} = \mu^2_{i'} = \mu^3_{i'} = 0 \\ \sigma_{i'} &= 1 \quad \text{or} \quad \mu^1_{i'} = 1, \mu^2_{i'} = \mu^3_{i'} = 0 \\ \sigma_{i'} &= 2 \quad \text{or} \quad \mu^2_{i'} = 1, \mu^1_{i'} = \mu^3_{i'} = 0 \\ \sigma_{i'} &= 3 \quad \text{or} \quad \mu^3_{i'} = 1, \mu^1_{i'} = \mu^2_{i'} = 0. \end{aligned} \tag{B.10}$$

The first term yields 1 in the bracket of (B.9), and with the three other terms (B.9) becomes:

$$\sum_{\{\sigma\}} \exp - \mathscr{V}_0\,(s', s) = \prod_{i'} [1 + \exp -\Delta^1_{i'}(s') + \exp -\Delta^2_{i'}(s') + \exp -\Delta^3_{i'}(s')] \tag{B.11}$$

where $\Delta^1_{i'}(s')$ is the energy associated with flipping spin $s^1_{i'}$ in the cell configuration $\{s'\}$. For the ferromagnetic states $\{s_\pm\}$ the three Δ's become equal and independent of i'

$$\Delta^1_{i'}(s'_\pm) = \Delta^2_{i'}(s'_\pm) = \Delta^3_{i'}(s'_\pm) = \Delta_0. \tag{B.12}$$

For states $\{s'\}$ with only a few deviations from $\{s'_\pm\}$, (B.12) holds for most cells and only in the neighbourhood of the cells where the cell spin $s'_{i'}$ is opposite to the majority, deviations from Δ_0 occur.

Defining $\mathscr{V}'_0(s')$ and $\mathscr{U}_0$ as

$$\exp -[\mathscr{U}_0 + \mathscr{V}'_0(s')] = \sum_{\{\sigma\}} \exp - \mathscr{V}_0(s', s) \tag{B.13}$$

it follows from (B.9) that these quantities are given by

$$\mathscr{U}_0 + \mathscr{V}'_0(s') = - \sum_{i'} \ln \left[\sum_{\mu^1_{i'} \ldots}' \exp - \mathscr{V}_{0i'}(s'; \mu^1_{i'} \ldots \mu^n_{i'}) \right]. \tag{B.14}$$

For the special case (B.11) and $\Delta_{i'}^{\alpha}(s')$ large, (B.14) reduces to

$$\mathscr{U}_0 + \mathscr{V}'_0(s') \simeq -\sum_{i'} \sum_{\alpha=1}^{3} \exp -\Delta_{i'}^{\alpha}(s'), \tag{B.15}$$

This shows that at least the contribution of $\mathscr{V}_0(s', s)$ leads to exponentially small corrections of the renormalized Hamiltonian despite the large number of the excitations.

The inclusion of $\Delta\mathscr{V}(s', s)$ can be approximated by the cumulant method based on $\mathscr{V}_0(s', s)$ as unperturbed Hamiltonian

$$\mathscr{U} + \mathscr{V}'(s') = \mathscr{U}_0 + \mathscr{V}'_0(s') - \ln\langle \exp -\Delta\mathscr{V}\rangle_0 \tag{B.16}$$

where

$$\langle A\rangle_0 = \left[\sum_{\{\sigma\}} A(s) \exp -\mathscr{V}_0(s', s)\right] \exp\left[(\mathscr{U}_0 + \mathscr{V}_0(s')\right]. \tag{B.17}$$

The evaluation of the first cumulant

$$\langle \exp -\Delta\mathscr{V}\rangle_0 \simeq \exp\left[-\langle\Delta\mathscr{V}\rangle_0 + \ldots\right] \tag{B.17}$$

involves the computation of $\langle\mu_a\rangle_0$. As an example we give the $\langle\mu_i\rangle$ for the ferromagnetic state in the triangular lattice:

$$\langle\mu_i\rangle_0 = [\exp -\Delta_0]/[1 + 3\exp -\Delta_0]. \tag{B.18}$$

The other $\langle\mu_a\rangle$ (involving 2 or more sites in one cell) are zero. The point is that $\langle\mu_i\rangle_0$ is exponentially small for $T \to 0$ which makes $\langle\mathscr{V}\rangle_0$ vanish exponentially. For the same reasons as given by the cumulant method in the high-temperature limit, the higher-order cumulants involve only connected structures, keeping the dependence on the number of sites of the right order. We conclude therefore that the excitations from the dominant configuration will lead only to exponentially small corrections for $T \to 0$.

Note added in proof

The cumulant expansion for the triangular lattice with 7-spin cells, Fig. 14b, referred to on page 488, has in the meantime been performed by Sudbø and Hemmer (1975) for several orders. Their results are listed in the following table.

	y_T	y_H	K_c
Exact	1·000	1·875	0·2744
1st order	0·918	1·287	0·3003
2nd order	1·124	1·879	0·2647
3rd order	1·072	1·875	0·2752

8. The Renormalization Group Approach to Critical Phenomena

C. Di Castro, G. Jona-Lasinio

Istituto di Fisica dell'Università, Roma
Gruppo G.N.S.M., Roma

1. Introduction

Renormalization group ideas originated in field theory where they had some interesting although not very conspicuous applications. The existence of a special invariance property of renormalizable quantum field theories had been recognized in 1953 by Stueckelberg and Petermann (1953). However, the first paper dealing with problems typical of the present domain of application of the renormalization group was the 1954 article by Gell-Mann and Low. Their aim was the investigation of the small-distance behaviour of quantum electrodynamics and in the attempt to control the sum of the entire perturbation series they introduced certain global functional relationships for the Green's functions of the theory. These relationships did not provide by themselves a new calculational tool but they provided a language within which it was possible to analyse different asymptotic behaviour of the theory in terms of very simple mathematical hypotheses. It was a consequence of the G-M–L analysis that the theory could exhibit a simple scaling behaviour at small distances.

From the above description it is clear that the renormalization group appeared at first as a purely technical aid without any immediate intuitive physical basis.

This circumstance to a certain extent may explain why R.G. ideas for several years played only a minor role in the developments of field theory. R.G. was discussed in detail in the book of Bogoliubov and Shirkov (1959) but its interest was still mainly formal.

The situation was radically changed in the late sixties by new developments in a subject apparently unrelated to field theory: the scaling theory of critical phenomena. In those years Kadanoff (1966) proposed and developed to a certain extent an idea which in a short while became the basis for an intuitive and very rich physical interpretation of the renormalization group. Kadanoff showed that the scaling properties of thermodynamic systems near a critical point could be explained on the assumption that the effective interaction governing a system would not qualitatively change if several degrees of freedom were "grouped" together and considered as a single entity in the mathematical description. In other words, nothing would qualitatively change if the system were substituted by another one with fewer degrees of freedom. In particular Kadanoff's picture led to certain scaling equations for the correlation functions used in the statistical description (Kadanoff *et al.*, 1967).

In view of the formal analogy between field theory and statistical mechanics all this suggested quite naturally the possibility of establishing a connection with the scaling properties of Green's functions in quantum electrodynamics provided by the G-M–L approach. This step was made by the present authors in 1969 (Di Castro, Jona-Lasinio, 1969).

At the end of the 60's therefore the scaling properties in critical phenomena had been given a qualitative physical interpretation and a tool for their mathematical investigation had been introduced. It had become possible to discuss the scaling laws within a field theoretical approach using either the renormalization group method (De Pasquale *et al.*, 1972; Jona-Lasinio 1971; Di Castro 1971) or the Ward identities as proposed by Migdal (1969). Still one did not see how to perform explicit calculations.

Substantial progress was achieved with the direct mathematical implementation by Wilson (1971) of Kadanoff's description. Wilson's approach, besides furnishing a much more precise and articulate version of Kadanoff's idea (Wilson and Kogut, 1974) provided also the language within which the first non-trivial approximation methods were formulated (Wilson and Fisher, 1972; Wilson 1972a, b). This might appear as an historical accident because everything that can be done within the Wilson scheme, can be obtained also from the field theoretical renormalization group (Di Castro, 1972; Brézin *et al.*, 1973a, 1973b, 1974; Di Castro *et al.*, 1974). However, it is our impression that the greater appeal to intuition of the Wilson method was really instrumental in determining the approximation schemes.

As we have just said the field theoretic renormalization group and the Wilson approach, which are often referred to as the old and the modern version of the R.G., seem to be equivalent in so far as the derivation of approximate results is concerned. This situation poses a theoretical problem. Is this coincidence of results accidental, as some authors have implicitly suggested, or is there a deeper connection between the two methods? Furthermore a whole variety of transformations is possible within the Kadanoff–Wilson line, as indicated originally by Wilson himself, while more than one type of renormalization group equations has been used in implementing the field theoretic point of view. As far as the latter point is concerned we mention in particular the Callan–Symanzik equation (Callan, 1970; Symanzik, 1970) and the modified G-M–L equations introduced by the authors of the present article (Di Castro *et al.* 1974).

The necessity of a unification of the different ways to scaling seems rather urgent at this point.

The present article has been written with the aim of providing the reader with a thread connecting different realizations of the R.G. idea. In particular we discuss explicitly the connection between the Kadanoff–Wilson program and the field theoretic program taking into account the recent articulations of both lines especially in the direction of non-linear transformations.

The guiding principle for understanding their differences is the following. In the Kadanoff–Wilson program the operation of "grouping" together variables referring to different degrees of freedom induces a transformation of the statistical ensemble describing the thermodynamic system. Equivalently one can argue in terms of a transformation of the Hamiltonian. Whatever notion one uses, one is interested in the result of many successive applications of the transformation which is a way of probing the system over large distances.

In the field theoretic approach on the contrary the statistical ensemble is not changed while the stochastic variables undergo a local transformation. By means of the local transformation one probes the statistical ensemble in the region of large values of the fluctuating variables. This is equivalent via the renormalization group equations to an analysis of the system over large distances.

We now outline the content of the various sections of the paper and their logical connection.

In Section II we briefly review the phenomenological theory of the critical point in order to establish what a more fundamental theory should explain.

In Section III we describe the block variable transformation which is a direct mathematical translation of Kadanoff's picture. This approach is then developed until all the important concepts of Wilson theory are introduced and a comparison with the field theoretic approach becomes possible.

Section IV contains a minimum of field theoretic formalism and a description of its connection with the statistical mechanics of lattice systems.

Section V is a self-contained and systematic introduction to the field theoretic renormalization group. It is sufficiently detailed to understand the basic ideas of the ε-expansion given in Section VI.

Section VII is very important because it provides a bridge between the point of view of Section III and that of Section V.

Section VIII deals with possible extensions of the renormalization group idea. Non-linear transformations are introduced both from the standpoint of block variables and by the field theoretic approach. There emerges a possibility of formal unification of different methods introduced by various authors in these years. However, a systematic treatment is not yet in sight.

Appendix A describes a connection of the renormalization group with probability theory. Although not essential for understanding the present paper, this connection opens new possibilities for the renormalization group idea.

Appendix B recalls the diagrammatic expansions needed in the field theoretic approach.

Appendix C shows how to switch from the homogeneous renormalization group introduced in Section V to the inhomogeneous Callan–Symanzik equation on which much of the work within the field theoretic approach relies.

II. Scaling

Near the critical point thermodynamic systems are characterized by long-range effective interaction among fluctuations of local quantities such as the order parameter and the energy density. Their asymptotic scaling behaviour is a peculiar feature which allows for the construction of a phenomenological theory (Kadanoff *et al.*, 1967; Pokrovskii, 1968).

At the critical point (i.e. $t \sim T - T_c = 0$ and the external field $h = 0$) the two-point correlation functions of the order parameter φ and the energy E for large separation of the positions assume the scaling form:

$$f_\varphi(r) = \langle \hat{\varphi}(r)\hat{\varphi}(0)\rangle - \langle \hat{\varphi}\rangle^2 \simeq \frac{\text{const}}{r^{2x_\varphi}} \tag{2.1}$$

$$f_E(r) = \langle \hat{E}(r)\hat{E}(0)\rangle - \langle \hat{E}\rangle^2 \simeq \frac{\text{const}}{r^{2x_E}} \tag{2.2}$$

The parameters x_φ and x_E determine what is called the dimension of φ and E in terms of an inverse of a length. Both theory and experiments show that they are different from those obtained either by perturbation theory or by a naïve dimensional analysis.

x_φ is usually expressed with the critical index η as

$$x_\varphi = \frac{d}{2} - 1 + \frac{\eta}{2} = 1 - \frac{\varepsilon}{2} + \frac{\eta}{2}, \qquad \eta \neq 0 \tag{2.3}$$

and x_E will be identified with

$$x_E = d - \frac{1}{\nu} = 2 - \varepsilon - \left(\frac{1}{\nu} - 2\right), \qquad \nu \neq \tfrac{1}{2} \tag{2.4}$$

where $\varepsilon = 4 - d$. The index ν specifies how the coherence distance ξ goes to infinity at the critical point

$$\xi \sim |t|^{-\nu,\nu'}, \qquad t \to 0_\pm \tag{2.5}$$

i.e. how the correlation among fluctuations becomes long-ranged. $1/\nu$ (or $1/\nu'$) gives the scaling form of t in terms of an inverse of a length; it therefore co-incides with the dimension x_t of t.

When the indices are classical ($\nu = \frac{1}{2}$ and $\eta = 0$), as in the standard Ornstein–Zernike results, x_E and x_φ assume the values $x_E^0 = d - 2$ and $x_\varphi^0 = (d/2) - 1$ which are the "bare" dimensions given by pure dimensional analysis. The Ginzburg criterion on the validity of classical theories suggests that these values should only hold for $d \geqslant 4$. The border line $\varepsilon = 0$ will play a special role in the following.

Near criticality, i.e. for t small but different from zero and in the absence of the external field, the correlation functions become homogeneous functions of r and the coherence distance ξ

$$f_{\varphi}(r, t) \sim \frac{1}{r^{2x_{\varphi}}} f\left(\frac{r}{\xi}\right), \qquad \xi \to \infty, \quad r \to \infty, \quad \frac{r}{\xi} \ll 1 \tag{2.6}$$

or

$$f_{\varphi}(r, t) \sim |t|^{2\nu x_{\varphi}} \bar{f}\left(\frac{r}{\xi}\right), \qquad \xi \to \infty, \quad r \to \infty, \quad \frac{r}{\xi} \gg 1 \tag{2.6'}$$

with the assumption that $f(0)$ and $\bar{f}(\infty)$ are finite, while the order parameter can be expressed by a simple power law

$$\varphi \sim |t|^{\beta}, \qquad t \to 0_{-} \tag{2.7}$$

If a small external field h coupled to the order parameter is present, the correlation function is still homogeneous in terms of the variables r, $|t|^{-1/x_t}$, $|h|^{-1/x_h}$, where x_h is the dimension of h.

The homogeneity property is also shared by the singular part of the thermodynamic potential density, e.g. the logarithm of the partition function per unit volume

$$F(t, h) = |t|^{d/x_t} \bar{F}\left(\frac{t}{h^{x_t/x_h}}\right), \qquad t, h \to 0 \tag{2.8}$$

where d is the dimension of the space, while the critical isotherm is

$$|\varphi| \sim |h|^{1/\delta}, \qquad t = 0, \quad h \to 0 \tag{2.9}$$

The correlations extending over large distances are responsible for the divergence of the thermodynamic quantities.

These singularities being simple power laws (with the possibility of logarithms) can be parametrized in terms of critical indices as we have already done in the case of the order parameter, the coherence distance and the critical isotherm, i.e. the susceptibility is

$$\chi_{\varphi} = \frac{1}{kT} \int f_{\varphi}(r, t)\, \mathrm{d}r \sim |t|^{-\gamma, \gamma'}, t \to 0_{\pm}, \quad h = 0 \tag{2.10}$$

and the specific heat is

$$\chi_E = \frac{1}{kT} \int f_E(r, t)\, \mathrm{d}r \sim \frac{\partial^2 F}{\partial t^2} \sim |t|^{-\alpha, \alpha'}, \quad t \to 0_{\pm}, \quad h = 0 \tag{2.11}$$

x_{φ} and x_E are defined at the critical point and specify the long-range behaviour of the correlation among fluctuations, all the other indices, particularly x_t and x_h, specify the approach of the system to the critical point.

It is now natural that all of them can be expressed in terms of x_φ and x_E only.

First of all consider any pair of conjugate thermodynamic variables defined by the condition that one is obtained as the derivative of the thermodynamic potential with respect to the other as for instance

$$\varphi = \frac{\partial F}{\partial h}, \qquad E = -\frac{\partial F}{\partial t}. \tag{2.12}$$

It follows from (2.8) that for a pure scaling behaviour, the sum of the dimensions of a pair of conjugate variables is always equal to the dimensionality of the space d:

$$x_\varphi + x_h = d, \qquad x_E + x_t = d \tag{2.13}$$

so that we can use alternatively the dimension of either pair of conjugate variables. All the previous scaling and homogeneity properties also imply

$$\gamma = \frac{d - 2x_\varphi}{x_t}, \quad \alpha = 2 - \frac{d}{x_t}, \quad \beta = \frac{x_\varphi}{x_t}, \quad \frac{1}{\delta} = \frac{x_\varphi}{x_h} \tag{2.14}$$

which together with

$$\alpha = \alpha', \qquad \nu = \nu', \qquad \gamma = \gamma' \tag{2.15}$$

substantiate the statement that the previous scaling invariant picture is governed by the knowledge of two independent critical indices.

Systems with different symmetry properties like the Ising model, the X–Y model, or the isotropic Heisenberg model show different critical behaviour characterized by different sets of critical indices.

For instance if we start with an isotropic system and introduce a small anisotropy field h_A then the scaling hypothesis, for example for the susceptibility, has to be modified (Riedel and Wegner, 1969) as follows

$$\chi(t, h_A) \simeq |t|^{-\gamma}\bar{\chi}(h_A^{1/\phi}/t), \qquad h_A^{1/\phi} \ll t \tag{2.16}$$

where the "cross-over" index ϕ is defined as the ratio of the dimension of h_A to x_t

$$\phi = x_{h_A}/x_t = x_{h_A}\nu \tag{2.17}$$

As long as $h_A^{1/\phi}$ is negligible compared to t, the behaviour of the system is characterized by the isotropic index γ.

When $h_A^{1/\phi}/t \sim O(1)$ (cross-over region), the system starts to deviate from the isotropic behaviour and finally switches to the anisotropic form characterized by another index $\dot{\gamma}$

$$\chi \sim |\dot{t}|^{-\dot{\gamma}}, \quad \dot{t} \sim T - T_c(h_A) \tag{2.18}$$

The extension of the cross-over region obviously depends on the value of ϕ.

To see how this scheme follows by simple general considerations we discuss the Kadanoff approach in terms of block variables, which makes also clear why we introduce a "renormalization" group transformation.

III. Block Variables and the Kadanoff–Wilson Approach

A. Block variable transformation

Consider the simple Hamiltonian with nearest-neighbours spin interaction, in the presence of an external field:

$$-H = J \sum_{(\mathbf{n},\mathbf{m})} \sigma_{\mathbf{n}}\sigma_{\mathbf{m}} + h \sum_{\mathbf{n}} \sigma_{\mathbf{n}} \tag{3.1}$$

The factor $1/kT$ is included in its definition. H is defined on a d-dimensional cubic lattice, with lattice constant a_0, which specifies the cut-off $\Lambda = a_0^{-1}$. The vectors **n** and **m** label the lattice sites.

It is clear from the discussion of the previous Section that near T_c we are interested in the long-range properties of the system; the detailed nature of the system over short distances should then be immaterial.

We therefore construct a new lattice by means of block variables, i.e. by grouping the variables contained in cells of size $|A^L| = (La_0)^d = (\Lambda_L^{-1})^d$:

$$M_{A_{\mathbf{l}}^L} = \sum_{\mathbf{n}\in A_{\mathbf{l}}^L} \sigma_{\mathbf{n}}. \quad \mathcal{E}_{A_{\mathbf{l}}^L} = \sum_{(\mathbf{m},\mathbf{n})\in A_{\mathbf{l}}^L} \sigma_{\mathbf{n}}\sigma_{\mathbf{m}} \tag{3.2}$$

where $\mathbf{l} \equiv (l_1, \ldots, l_d)$ labels the position of the new cells by means of the condition

$$A_{\mathbf{l}}^L : \{Ll_i \leqslant n_i < L(l_i + 1)\} \tag{3.2'}$$

In Fig. 1 the example for $L = 2$ and $d = 1$ is shown

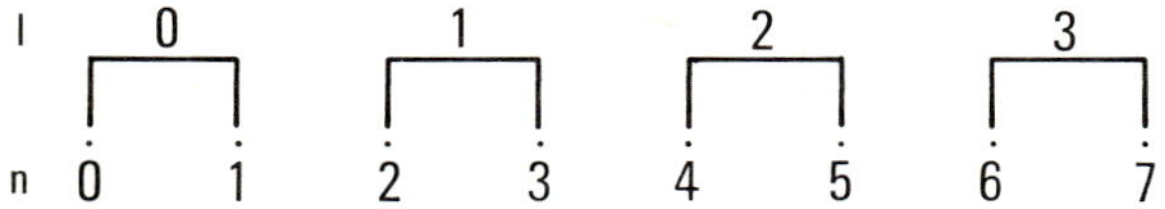

FIG. 1.

In order that the new variables may have non-trivial expectations in the limit of large L, we have to normalize them. The appropriate normalization, as it will be clear later, is

$$\varphi_{\mathbf{l}} = \frac{M_{A_{\mathbf{l}}^L} - \langle M_{A_{\mathbf{l}}^L}\rangle}{|A_{\mathbf{l}}^L|^{(1-x_\varphi/d)}}, \qquad E_{\mathbf{l}} = \frac{\mathcal{E}_{A_{\mathbf{l}}^L} - \langle \mathcal{E}_{A_{\mathbf{l}}^L}\rangle}{|A_{\mathbf{l}}^L|^{(1-x_E/d)}} \tag{3.3}$$

where x_φ and x_E turn out to be the dimensions introduced in (2.1) and (2.2).

The correlation functions among block variables depend on L

$$f_{\varphi}^{(L)} = \langle \varphi_{\mathbf{l}} \varphi_{\mathbf{l}+\mathbf{l}'} \rangle_L, \qquad f_E^{(L)} = \langle E_{\mathbf{l}} E_{\mathbf{l}+\mathbf{l}'} \rangle_L \tag{3.4}$$

where the average has to be taken on their new statistical ensemble.

If $\rho\{\sigma\}$ is the statistical ensemble for the original Hamiltonian (3.1)

$$\rho\{\sigma\} = \frac{1}{Z} e^{-H\{\sigma\}}, \tag{3.5}$$

the distribution $\rho_L\{\varphi\}$ for the block variables changes according to the following law when the size of the blocks is varied from $|A^L|$ to $|A^{L'}|$ with $L' = sL$:

$$\rho_{sl}\{\varphi_{\mathbf{l}}\} = \int \prod_{\mathbf{k}} d\varphi'_{\mathbf{k}} \rho_L\{\varphi'_{\mathbf{k}}\} \delta(\varphi_{\mathbf{l}} - s^{-(d-x_\varphi)}(\sum_{\mathbf{k} \in A_{\mathbf{l}}^{sL}} \varphi'_{\mathbf{k}})) \tag{3.6}$$

The scaling factor $s^{-(d-x_\varphi)}$ takes into account the change in normalization as specified in (3.3). Of course as long as the variables are discrete the integral must be interpreted as a summation.

Equation (3.6) can be formally written as

$$\rho_{sL}\{\varphi_{\mathbf{l}}\} = K_s(\rho_L\{\varphi'_{\mathbf{k}}\}) \tag{3.6'}$$

This block variable transformation is closer to Kadanoff's original picture than the Wilson transformation and it has been considered by various authors. See in particular Gallavotti and Martin Löf (1975), Cassandro and Gallavotti (1975), Jona-Lasinio (1975)†. We shall pursue it here as far as it is necessary to establish the comparison with the field theoretic approach.

The transformation preserves the partition function of the system apart from a multiplicative factor

$$Z_{sL} = c_{s,L} Z_L \tag{3.7}$$

(3.6) can also be interpreted as a mapping in an appropriate space of Hamiltonians where each point is characterized by the strengths of the different types of couplings:

$$H_{sL} = R_s(H_L). \tag{3.8}$$

This is what is usually called a renormalization group transformation.

Since the parameters appearing in the Hamiltonian are also the arguments of the thermodynamic potentials, the invariance property (3.7) suggests that

† There is an important connection between the block spin transformation as described in this Section and probability theory. This aspect, developed in the last reference, is briefly recalled in Appendix A.

in general we can define as a renormalization transformation any substitution on the arguments of a thermodynamic potential which leaves it invariant in value.

We now consider the relation among correlation functions induced by (3.6). The correlation function for the sL blocks reads:

$$\langle \varphi_{\mathbf{l}} \varphi_{\mathbf{l}+\mathbf{l}'} \rangle_{sL} = \int \Pi \, \mathrm{d}\varphi \rho_{sL}\{\varphi\} \varphi_{\mathbf{l}} \varphi_{\mathbf{l}+\mathbf{l}'}$$
$$= \int\int \Pi \, \mathrm{d}\varphi \Pi \, \mathrm{d}\varphi' \rho_L\{\varphi'\} \varphi_{\mathbf{l}} \varphi_{\mathbf{l}+\mathbf{l}'} \Pi \delta(\varphi_{\mathbf{l}} - s^{-(d-x_\varphi)}(\sum_{\mathbf{k} \in A^{sL}_{\mathbf{l}}} \varphi'_{\mathbf{k}})) \tag{3.9}$$

where we used (3.6).

For $|\mathbf{l}|$ and $|\mathbf{l} + \mathbf{l}'|$ much larger than one, each variable within the new cell can be approximately labelled by the same index, i.e. either $s\mathbf{l}$ or $s(\mathbf{l} + \mathbf{l}')$. Since the number of variables within each cell is varied by a factor s^d, from (3.9) we obtain

$$\langle \varphi_{\mathbf{l}} \varphi_{\mathbf{l}+\mathbf{l}'} \rangle_{sL} = s^{2d} s^{-2(d-x_\varphi)} \langle \varphi_{s\mathbf{l}} \varphi_{s(\mathbf{l}+\mathbf{l}')} \rangle_L. \tag{3.10}$$

With $\mathbf{r} = s\mathbf{l}' a_0 L > s\Lambda_L^{-1}$ equation (3.10) can also be written as

$$s^{-2x_\varphi} f_\varphi\left(\frac{r}{s}; R_s(H_L)\right) = f_\varphi(r; H_L). \tag{3.10'}$$

In the next subsection we shall indicate why an identical argument cannot be repeated for the energy correlation function.

B. Scaling properties at the critical point

When the size of the blocks is sufficiently large, so that the small-scale details have been averaged out, and $T \to T_c$, $h \to 0$ so that the correlations among fluctuations extend over large distances ($\xi \gg L a_0 \gg a_0$), Kadanoff argued that the correlation functions should become independent of L.

This means that at the critical point and for L sufficiently large, if we iterate the block transformation further, the ensemble distribution transforms into itself

$$\rho^* = K_s(\rho^*), \qquad t = h = 0, \quad L \gg 1. \tag{3.11}$$

ρ^* is called a "fixed point" distribution and according to the Kadanoff hypothesis it should characterize the system at the critical point. To ρ^* should be associated a Hamiltonian H^* satisfying the condition

$$H^* = R_s(H^*) \tag{3.12}$$

Critical Hamiltonians are then those H_c's for which

$$\lim_{s\to\infty} R_s(H_c) = H^* \tag{3.13}$$

Under the previous hypothesis from (3.10′) for $L \to \infty$ at the critical point we obtain the pure scaling behaviour (2.1) for the correlation function

$$f_\varphi \sim \frac{1}{r^{2x_\varphi}} \qquad t = 0,\ h = 0,\ r \to \infty \tag{3.14}$$

The exponent x_φ of the normalization of the block variables (3.3) is thus identified with the dimension (2.3) of φ.

We now turn our attention to the energy correlation function $f_E^{(L)}$. The argument which led to (3.10′) for f_φ cannot be repeated in this case. The reason is the following. The block variable transformation does not preserve the functional form of the energy density in the sense that $E_{\mathbf{l}}(\varphi)$ will be a different functional of φ at each iteration. To take into account this circumstance it is convenient to introduce an explicit L dependence by writing $E_{\mathbf{l}}^L(\varphi)$. When we consider the expectation value

$$f_E^{(L)} = \langle E_{\mathbf{l}}^L(\cdot) E_{\mathbf{l}+\mathbf{l}'}^L(\cdot)\rangle_L, \tag{3.15}$$

we cannot therefore expect to find any simple relationship connecting $f_E^{(sL)}$ with $f_E^{(L)}$. Furthermore the non-trivial functional form of $E_{\mathbf{l}}^L(\varphi)$ for $L \to \infty$ explains why the naïve dimensional analysis which would assign to $E_{\mathbf{l}}$ the dimension $2x_\varphi$, fails.

However, we will now show that if $\lim_{L\to\infty} f_E^{(L)}$ exists, this implies a scaling form for the asymptotic energy correlation function f_E calculated in the original ensemble as well as for $f_E^{(\infty)}$ itself.

Going back to the definition of the energy block variables (3.3) we have

$$f_E^{(\infty)} = \lim_{L\to\infty} \left\langle \frac{\sum_{(\mathbf{n},\mathbf{m})\in A_{\mathbf{l}}^L} \sigma_{\mathbf{n}}\sigma_{\mathbf{m}} - \langle \sum_{(\mathbf{m},\mathbf{n})\in A_{\mathbf{l}}^L} \sigma_{\mathbf{n}}\sigma_{\mathbf{m}}\rangle}{|A_{\mathbf{l}}^L|^{1-x_E/d}} \right.$$
$$\left. \frac{\sum_{(\mathbf{n}',\mathbf{m}')\in A_{\mathbf{l}+\mathbf{l}'}^L} \sigma_{\mathbf{n}'}\sigma_{\mathbf{m}'} - \langle \sum_{(\mathbf{n}',\mathbf{m}')\in A_{\mathbf{l}+\mathbf{l}'}^L} \sigma_{\mathbf{n}'}\sigma_{\mathbf{m}'}\rangle}{|A_{\mathbf{l}+\mathbf{l}'}^L|^{1-x_E/d}} \right\rangle \tag{3.16}$$

The averages are taken in the original ensemble. Simple manipulations now give

$$f_E^{(\infty)} = \lim_{L\to\infty} \frac{1}{L^{2(d-x_E)}} \sum_{(\mathbf{m},\mathbf{n})\in A_{\mathbf{l}}^L} \sum_{(\mathbf{m}',\mathbf{n}')\in A_{\mathbf{l}+\mathbf{l}'}^L} f_E(r_{(\mathbf{m},\mathbf{n})(\mathbf{m}',\mathbf{n}')})$$
$$\simeq \lim_{L\to\infty} \frac{1}{L^{d-2x_E}} \sum_{(|\mathbf{l}'|-1)<|r|<(|\mathbf{l}'|+1)L} f_E(r) \tag{3.17}$$
$$\sim \lim_{L\to\infty} \frac{1}{L^{-2x_E+y}} \frac{1}{|\mathbf{l}'|^{y}}, \qquad |\mathbf{l}'| \gg 1$$

if

$$f_E(r) \underset{r\to\infty}{\sim} \frac{1}{r^y} \tag{3.18}$$

From this equation it follows that the limit exists only if f_E scales asymptotically and $y = 2x_E$. Both normalization exponents of the block variables (3.3) therefore coincide with the dimensions introduced in (2.1) and (2.2) respectively.

The pure scaling behaviour of the system at the critical point has been recovered. The analysis of the transformation near criticality will complete our discussion in terms of block variables.

C. Linearized group transformation

As we have seen in Section II, the indices which specify the long-range behaviour of the fluctuations are defined at the critical point. All the other critical indices, such as ν, ν', γ, γ', α, α', β and δ, specify the approach to the critical point of the various physical quantities. The scaling laws (2.14) relate all these indices to x_φ and x_E. Furthermore corrections to the pure asymptotic power law behaviour of the physical quantities are necessary to define the limit of validity of the asymptotic theory. In order to reach these results we have therefore to study how the group transformation behaves when the Hamiltonian deviates slightly from H_c.

If the system deviates slightly from the critical point, the Hamiltonian can be written as

$$H = H_c + \tau H_1 \tag{3.19}$$

where τ is a small parameter.

For example according to our previous discussion in the case of the Hamiltonian (3.1), whose system becomes critical at $J = J_c$ and $h = 0$, we should have

$$H_c = -J_c \sum_{(\mathbf{n},\,\mathbf{m})} \sigma_{\mathbf{n}}\sigma_{\mathbf{m}} \tag{3.20}$$

and

$$\tau H_1 = t \sum_{(\mathbf{n},\,\mathbf{m})} \sigma_{\mathbf{n}}\sigma_{\mathbf{m}} - h \sum_{\mathbf{n}} \sigma_{\mathbf{n}}; \qquad t \sim J_c - J \to 0, \quad h \to 0 \tag{3.21}$$

In general the deviating part of the Hamiltonian is a sum of deviating fields $\hat{\mu}_i$ such as t and h times their conjugate operators 0_i

$$\tau H_1 = \sum_i \int \hat{\mu}_i 0_i. \tag{3.22}$$

We now linearize the transformation by assuming the following "smoothness" postulate (Cassandro and Gallavotti, 1975), since in general there is no reason why the transformation should introduce singularities.

$$R_s(H_c + \tau H_1) = R_s(H_c) + \tau L_s(H_c) \cdot H_1 \tag{3.23}$$

where L_s is a linear operator which depends on H_c.

As we shall see, much of the physical interpretation of the theory is related to the properties of $L_s(H_c)$. So we discuss it in some detail. By construction, the transformation R_s obeys the composition rule:

$$R_{s'}(R_s(H)) = R_{ss'}(H) \tag{3.24}$$

Applying the smoothness postulate (3.23), we obtain

$$L_{s'}(R_s(H)) \cdot L_s(H) = L_{ss'}(H) \tag{3.25}$$

Because of (3.12) the linear operator satisfies the group property at the fixed point

$$L_{s'}(H^*)L_s(H^*) = L_{ss'}(H^*). \tag{3.26}$$

Suppose that $L_s(H^*)$ can be diagonalized so that we can determine the eigenvalues λ_i and the corresponding eigenvectors e^i, then Eq. (3.26) implies

$$\lambda_i(s)\lambda_i(s') = \lambda_i(ss') \tag{3.27}$$

or

$$\lambda_i(s) = s^{x_i} \tag{3.28}$$

The assumption that $L_s(H^*)$ can be diagonalized is not trivial, since in general it will not be symmetric.

To see the meaning of these eigenvalues we expand the deviating part of the Hamiltonian τH_1 in terms of the eigenvectors e^i of $L_s(H^*)$

$$\tau H_1 = \sum_i \mu_i e^i \tag{3.29}$$

For large s, $R_s(H_c) \to H^*$ and the transformed Hamiltonian is then obtained by specifying (3.23) in H^*

$$R_s(H^* + \tau H_1) = H^* + \sum_i \mu_i s^{x_i} e^i \tag{3.30}$$

This transformation must be read as acting on the parameters μ_i's. A comparison of (3.30) with (3.22) shows that this decomposition of τH_1 in terms of eigenvectors of $L_s(H^*)$ corresponds to an asymptotic parametrization of the system in terms of the fields μ_i (in general linear combination of the original physical fields $\hat{\mu}_i$), which scale under the action of the group transformation. The x_i's coincide with the corresponding dimensions or critical indices.

Perturbations to H^* in the direction e^i with $x_i > 0$ will drive the system away from criticality corresponding to an effective increase of the value of μ_i as s increases, e.g. the temperature and the external field must have this property.

According to Kadanoff's picture for example, all the details of the system over short distances should modify the Hamiltonian only with terms along those directions e^i with $x_i < 0$, which drive the system towards the scaling invariant form.

We can therefore say that the relevant fields (or their conjugate variables) are those with positive dimension ($0 < x_i < d$). The variables with negative dimension disappear as a consequence of the repeated application of the group transformation and therefore do not influence the critical behaviour, i.e. they are irrelevant variables. The marginal fields are those with zero dimension ($x_i = 0$). They have a special rôle, which will not be discussed here.

Extension of the scaling property (3.30) to the case when non-linear terms of the group transformation are considered leads to the concept of scaling fields introduced by Wegner (1972), which under the action of R_s on the Hamiltonian transform exactly as

$$g_i' = s^{x_i} g_i$$

Asymptotically near to the fixed point when the group transformation can be linearized, they must reduce to the μ_i.

The knowledge of the scaling variables would solve completely the problem of critical phenomena. The exact solution of such a problem is almost impossible. A classification of scaling variables has been given by Ma (1974b) in the limit of large n (number of components of the order parameter).

General references for this subsection are F. Wegner (1972), Ma (1973), Wilson and Kogut (1974).

D. Properties of approach to the critical point

We are now in a position to discuss the behaviour of the system near criticality to obtain the homogeneous properties (2.6), (2.6′) and (2.8) which then imply the scaling laws (2.14).

Assume that our Hamiltonian (3.1) is critical except for the presence of a small deviating field say t, as in (3.21), with $h = 0$. The distance of the Hamiltonian from the critical surface specified by μ_1 with $x_1 > 0$, is proportional to t, i.e. $\mu_1 \sim t$. For t sufficiently small and s sufficiently large from (3.30) we then expect

$$R_s(H_c + \tau H_1) \underset{\text{"s large"}}{\sim} H^* + ts^{x_t}e^1 \tag{3.31}$$

What do we mean by "s large"?

Roughly speaking the system will deviate from the scaling behaviour associated with the distribution ρ^* when

$$ts^{x_t} \sim O(1). \tag{3.32}$$

If s becomes larger the system is driven away from criticality. Equation (3.10′) for L large and "s large" can now be written as

$$f_\varphi(r, t) \sim s^{-2x_\varphi} f_\varphi\left(\frac{r}{s}, ts^{x_t}\right), \qquad t \to 0, \quad r \to \infty \tag{3.33}$$

The homogeneous forms (2.6), (2.6′) follow from (3.33), with the identification

$$\xi \sim |t|^{-1/x_t} \tag{3.34}$$

Similar expressions can be built when the small deviating term in the Hamiltonian is proportional to the external field h and the "magnetic exponent" x_h appears.

When we consider the thermodynamic potential F we can refer to the invariance property (3.7). Some considerations are necessary on the terms related to the coefficient $c_{s,L}$. It is made up of two parts. One is simply related to the rescaling of the variables φ due to the normalization factor s^{x_φ}. If the transformed Hamiltonian H_{sL} is defined to be zero when all the φ's are zero, the other term of $c_{s,L}$ is related to the additional constant $A(T)$ introduced by the transformation

$$\exp - A = \int \prod_{\mathbf{k}} \mathrm{d}\varphi'_{\mathbf{k}} \exp(-H\{\varphi'\}) \prod_{\mathbf{l}} \delta(-s^{-(d-x_\varphi)} \sum_{\mathbf{k}\in A_{\mathbf{l}}^L} \varphi'_{\mathbf{k}}) \tag{3.35}$$

The first term is obviously inessential to discuss the critical behaviour. This term is usually neglected with the assumption that it leads to less singular terms for F. This assumption is not always fully justified (Ma, 1974a; Nelson, 1975). For simplicity we shall assume it without further comment.

We are then left with the transformation property

$$F(H) = s^{-d}F(R_s(H)) \tag{3.36}$$

where the factor s^{-d} comes out from the simple consideration that the trace of H_s is taken over $N' = N/S^d$ degrees of freedom.

Asymptotically near to the fixed point we have the simple homogeneity property

$$F(\{\mu_i\}) = s^{-d}F(\{\mu_i s^{x_i}\}) \tag{3.37}$$

All the critical indices can now be expressed in terms of $x_E = d - x_t$ and $x_\varphi = d - x_h$ as in (2.14).

If the next largest eigenvalue x_u, besides the temperature and the magnetic exponents, is considered, correction terms to scaling of the order of us^{x_u} are present. With the definition (3.32) of the critical region they read as

$$ut^{-\nu x_u}. \tag{3.38}$$

The study of the singularities of thermodynamic quantities has been reduced to the study of a set of critical exponents $\{x_i\}$, which have been connected to derivatives of the transformation evaluated at the fixed point. The initial values of the parameters specifying the Hamiltonian determine only to which fixed point of the transformation the system approaches, i.e. to which "class of universality" it belongs. Different classes of universality correspond to different fixed points with different symmetry properties.

IV. Functional Form of Many-body Theory

A. *Introductory remarks*

The mathematical picture we have given turns out to be useful if a physical system can be described by a distribution which, being on the critical surface, is driven by the transformation toward a fixed point.

The actual verification of this scheme resides in the possibility of performing explicit calculations. In general, exact renormalization group equations are too difficult to deal with and one is confined to few tractable examples or approximation procedures.

These are either the ε-expansion, the $(1/n)$-expansion, by which most of the hypotheses of the renormalization group approach to critical phenomena have been qualitatively and semi-quantitatively verified, or an approximate recursion relation (Wilson 1971) for the interaction parameters of the transformed Hamiltonian in terms of analytic function of the original interaction parameters, which allows for numerical evaluation of critical indices but does not give much insight into the problem. The recent finite lattice approximation by Niemeijer and Van Leeuwen (1973, 1974) gives very good numerical values for the critical indices, but requires an "ad hoc" renormalization group for each system under consideration. The general criteria to be satisfied in order to obtain good results are still far from clear.

Field theory offers an alternative approach which, like the previous picture, gives both a complete theoretical scheme for critical phenomena and analytic methods of computation such as renormalized perturbation theory. It takes advantage of over thirty years of research in both relativistic and non-relativistic many-body theory. It is a simple technical point (Parisi, 1973) to show that the divergences of fluctuations at the critical point as given by (2.1) and (2.2) lead to infrared divergences in bare perturbation theory, just as ultraviolet divergences in quantum field theory forbid the use of an expansion in the bare coupling constant. Renormalization has overcome these difficulties in this last case. In the same way the renormalization procedure combined with asymptotic properties of the solution of the group equations turns out to be useful in critical phenomena.

The coincidence of the results of the Gell-Mann and Low approach we are going to describe with the Kadanoff–Wilson method, seems at first sight accidental. We shall make an effort to underline the deep conceptual unity which is implicit in the various aspects of the renormalization group idea by making explicit in Section VII the connections between the Kadanoff–Wilson and the Gell-Mann–Low approach.

B. The model

We consider the standard "effective model Hamiltonian" of an n-component classical field ($\hat{\varphi} \equiv \{\hat{\varphi}_j(r)\}; j = 1, 2, \ldots, n$), specified *via* a set of i-body interaction functions $(1/i!)\ \hat{\mu}_i\ (r_1, \ldots, r_i)$ which include the temperature factor $1/kT$.

Since the relevant features of critical phenomena are connected to fluctuations of local variables of systems with short-range interactions, we can use local interaction functions

$$\hat{\mu}_i(r, r_2, \ldots, r_i) \to \delta(r - r_2) \ldots \delta(r - r_i)\hat{\mu}_i(r) \tag{4.1}$$

In field-theoretic language this means that we are dealing with local composite operators, which involve powers or gradients of $\hat{\varphi}$:

$$H = \int \mathrm{d}^d r \left\{ \tfrac{1}{2} (\nabla\hat{\varphi})^2 + \sum_i \frac{\hat{\mu}_i(r)}{i!}\, \hat{\varphi}^i \right\} \tag{4.2}$$

where d is the space dimensionality and

$$(\nabla\hat{\varphi})^2 = \sum_{j=1}^{n} (\nabla\hat{\varphi}_j(r))^2, \qquad \hat{\varphi}^2 = \sum_{j=1}^{n} \varphi_j^2, \qquad \hat{\varphi}^4 = (\hat{\varphi}^2)^2, \ldots$$

The set of $\{\hat{\mu}_i\}$ thus includes the external field $(\hat{\mu}_1\hat{\varphi}_j = \Sigma_{j=1}^{n} h_i\, \hat{\varphi}_j)$, the "temperature" $t = \hat{\mu}_2$ chosen in such a way that $t = 0$ always corresponds to the true critical temperature, and the coupling constant $\hat{\mu}_4 = u$ of the quartic interaction.

For a detailed correspondence between statistical mechanical models and field theory one should refer to the original paper by Polyakov (1969). As stressed in the Varenna lecture by the authors, any classical statistical system can be written in a form close to the S-matrix of an euclidean field-theory by using a method first formulated by Hori (1952). This was the starting point of our application of the renormalization group to classical systems.

A more recent and specific formulation is given in the Wilson and Kogut (1974) review paper.

Let us simply recall (Mühlschlegel and Zittartz, 1963) that the partition function of a spin-$\frac{1}{2}$ Ising Hamiltonian on a lattice can be written in terms of an auxiliary continuous variable ψ_i defined on the lattice as

$$Z = \mathrm{tr}\,\exp(-H_{\mathrm{Ising}}[\sigma]) \sim \int \delta[\psi]\exp(-H[\psi]) \tag{4.3}$$

where $\int\delta[\psi]$ stands for a functional integral and

$$H[\psi] = \tfrac{1}{2}\sum_{ij}\psi_i J_{ij}^{-1}\psi_j - \sum_i \ln 2\cosh[\psi_i + h_i] \tag{4.4}$$

The summations run over the lattice sites and J_{ij}^{-1} is the matrix inverse of the spin interaction J_{ij}. Equation (4.3) can also be derived in a simple way from the Hori formula.

The mean field approximation is obtained by minimizing $H[\psi]$, which implies the standard equation

$$\sum_j J_{ij}^{-1}\psi_j^0 = \tanh(\psi_i^0 + h_i) \tag{4.5}$$

The Landau thermodynamic functional comes out by approximating $\mathscr{F} = \ln Z$ with the maximum value of the integrand

$$\mathscr{F}^{(0)} = \ln Z^{(0)} = -H[\psi^0] + \mathrm{const} \tag{4.6}$$

and retaining only up to the fourth order term in ψ^0 and to the first in the external field h.

The Ornstein–Zernike form for the correlation function is recovered within the Gaussian approximation of (4.3) and (4.4) (second-order terms in ψ in (4.4)).

We have already stated that the results of all these "classical" approximations do not reproduce the correct asymptotic behaviour of critical systems for $d < 4$.

In general we shall have to retain the infinite terms of the expansion of (4.4). This leads to a Hamiltonian of the form (4.2) provided the three following operations are performed: (1) for small external field h, its first-order term only is retained. (2) The lattice spacing is sent to zero so that a continuous model is recovered. (3) The field ψ is rescaled into a field $\hat{\varphi}$ in such a way that the gradient term originated by the second operation has the coefficient equal to $\frac{1}{2}$.

Since the lattice spacing is usually assumed to be irrelevant for the description of critical behaviour, the original Ising model should be equivalent to a model of type (4.2) with $n = 1$.

$H[\hat{\varphi}]$ specifies our statistical ensemble and being the argument of an exponential must be a pure number. This implies that all $\hat{\mu}_i$'s have bare dimensions x_i^0 expressed in terms of an inverse of a length

$$\left[\frac{\mathrm{d}}{\mathrm{d}x}\right] = 1, \qquad x_\varphi^0 = 1 - \frac{\varepsilon}{2}, \qquad x_h^0 = 1 + d/2,$$

$$x_2^0 = x_t^0 = 2, \qquad x_4^0 = x_u^0 = \varepsilon, \ldots, x_n^0 = d - n\left(1 - \frac{\varepsilon}{2}\right) \tag{4.7}$$

where $\varepsilon = 4 - d$.

A simple inspection shows that the case $d = 4$ and $d = 2$ are exceptional.

An argument based on power counting to determine the superficial divergence of Feynman diagrams and renormalizability of the theory (Zinn-Justin, 1973), shows that we can use the $u\hat{\varphi}^4$ field-theory model instead of (4.2), which leads to the simplest non-trivial ensemble beyond the Gaussian form,

$$H = \int \left[\tfrac{1}{2}(\nabla\hat{\varphi})^2 + \tfrac{1}{2}t\hat{\varphi}^2 + \frac{u}{4!}\hat{\varphi}^4 - h\hat{\varphi}\right] d^d r \tag{4.8}$$

An intuitive plausibility argument based on the relevance of interactions gives the same result. For $d > 4$ all the x_n^0 except x_2^0 are negative. The Gaussian approximation should be valid and in fact, according to the Ginzburg criterion, the classical indices should be correct.

For $3 < d < 4$ the quartic term, which is marginal at $d = 4$, is the most relevant interaction. At $d = 3$ the coupling of $\hat{\varphi}^6$ becomes marginal and for $d \leqslant 3$ should be added to the field-theoretic Hamiltonian (4.8). At $d = 2$ all x_n^0 become equal to 2 and we should go back to the Hamiltonian (4.2) in its general form.

The above argument is confirmed by the simple consideration that the true expansion parameters of unrenormalized perturbation theory are $u_n/t^{x_n^0/2}$, which goes to infinity or to zero as $t \to 0$ according to the previous distinction of relevant and irrelevant interactions.

From now on we shall consider only the simple model Hamiltonian (4.8), which described either a relativistic field-theory model in $d - 1$ space and one time dimension or a classical ferromagnet in d-space dimension of Ising type for $n = 1$, where the correlation functions take the place of the Schwinger functions of the relativistic theory.†

† This statement is not obvious. However we shall not discuss it further here, as it has been widely treated in the literature (Symanzik, 1966).

By varying the number n of components of the field $\hat{\varphi}$, the symmetry of the order parameter changes.

Accordingly our model should belong to the same class of universality as the XY model or the complex two-component field model for liquid helium when quantum effects associated with the discrete frequencies in the Matsubara technique are neglected for $n = 2$.

The case $n = 3$ should correspond to the Heisenberg model and in the limit of n going to infinity one should recover the spherical model.

The n-dependence of the model shows up in the multiplicity factors of the Feynman diagrams in the perturbative calculation of the correlation functions. Since the basic features of the theory are the same we shall for simplicity mainly deal with the case $n = 1$ except where explicitly stated.

It should be noted that suitable modifications of the terms appearing in (4.8) allow for the discussion of a much larger variety of critical phenomena. Long-range forces of the $1/r^{d+\sigma}$ type can be included by modifying the gradient term, since in the Fourier transformed space this potential introduces a term proportional to k^σ $(0 < \sigma < 2)$.

The exchange anisotropy is included by adding a term $t_j\hat{\varphi}_j^2$ to the quadratic interaction.

Cubic anisotropy leads instead to an additional term $(v/4!)\ \Sigma_{j=1}^n\ (\hat{\varphi}_j^4)$ in the quartic interaction.

The $\hat{\varphi}^6$ term allows for the discussion of the tricritical behaviour at $d \leqslant 3$.

C. Functional equations

To overcome the difficulty of the asymptotic theory as $T \to T_c$ we shall introduce an improved version of the Gell-Mann and Low transformation and the corresponding group equations.

Before doing so we have to develop further our formalism. The logarithm of the partition function $\mathscr{F}$ is a functional of t, h and u

$$\mathscr{F} = \ln Z = \mathscr{F}(t, h, u) \tag{4.9}$$

Its functional derivatives define the corresponding conjugate variables:

$$\frac{\delta\mathscr{F}}{\delta t} \sim -E, \qquad \frac{\delta\mathscr{F}}{\delta h} = \varphi \tag{4.10}$$

where φ is the order parameter and E is the energy fluctuation density.

The nth functional derivatives give the connected part of the nth correlation functions

$$f_\varphi^n(r, \ldots, r_n) = \frac{\delta^n\mathscr{F}}{\delta h(r_1)\ldots\delta h(r_n)}, \quad f_E^n(r_1, \ldots, r_n) = \frac{\delta^n\mathscr{F}}{\delta t(r_1)\ldots\delta t(r_n)} \tag{4.11}$$

Obviously mixed correlation functions are also generated.

Starting from $\mathscr{F}$ by successive Legendre transformations we can eliminate the "field" variables in favour of the "density" variables and generate a set of thermodynamic functionals.

For instance if we use φ as an independent variable instead of h, the corresponding functional Γ is (Jona-Lasinio, 1964)

$$\Gamma\{t, \varphi, u\} = \mathscr{F}\{t, h(\varphi, t, u), u\} - \int h(\varphi, t, u)\varphi \, \mathrm{d}^d r \tag{4.12}$$

where $h = h(\varphi, t, u)$ is solution of the second equation (4.10). It is easy to verify that

$$\frac{\delta\Gamma}{\delta\varphi(r)} = -h(r), \qquad \frac{\delta\Gamma}{\delta t} = -E \tag{4.13}$$

and

$$\hat{\Gamma}^{(2)}(r_1, r_2) = -\frac{\delta^2\Gamma}{\delta\varphi(r_1)\delta\varphi(r_2)} = f_\varphi^{-1}(r_1, r_2) \tag{4.14}$$

The nth functional derivative of Γ with respect to φ generates the one-particle irreducible n-point vertex function $\hat{\Gamma}^{(n)}(r_1, \ldots, r_n)$:

$$\hat{\Gamma}^{(n)}(r_1, \ldots, r_n) = -\frac{\delta^n\Gamma}{\delta\varphi(r_1)\ldots\delta\varphi(r_n)} \tag{4.15}$$

whose meaning is specified in the Appendix B.

In the Fourier transformed space the $\Gamma^{(n)}(k_1, \ldots, k_{n-1})$ are defined as

$$\Gamma^{(n)}(k_1, \ldots, k_{n-1}) = \int \mathrm{d}^d r_1 \ldots \int \mathrm{d}^d r_{n-1} \exp[i(r_1 k_1 + \ldots r_{n-1}k_{n-1})] \hat{\Gamma}^{(n)}(r_1, \ldots, r_{n-1}) \tag{4.16}$$

The bare dimensionality of $\Gamma^{(n)}$ is

$$x_n^0 = d - n\left(1 - \frac{\varepsilon}{2}\right) \tag{4.17}$$

which once again shows that the cases $d = 4$ and $d = 2$ are exceptional.

The functional Γ can be formally expanded in terms of $\hat{\Gamma}^{(n)}$

$$\Gamma = \sum_{n=0}^{\infty} \frac{1}{n!} \int \ldots \int \hat{\Gamma}^{(n)}(r_1, \ldots, r_n)\varphi(r_1)\ldots\varphi(r_n) \tag{4.18}$$

where for simplicity we deal only with the case $T > T_c$.

V. The Field Theoretic Renormalization Group

A. Group equations

According to the discussion of the previous section our simplest model Hamiltonian is

$$H = \int \left[\tfrac{1}{2}(\nabla\hat{\varphi})^2 + \tfrac{1}{2}t\hat{\varphi}^2 + \frac{u}{4!}\hat{\varphi}^4 - h\hat{\varphi} \right] \mathrm{d}^d r \tag{5.1}$$

The zeroth order approximation of the theory corresponds to the Gaussian approximation.†

$$f^0_\varphi(k^2, t) \sim \frac{1}{k^2 + t} \tag{5.2}$$

which in the ordinary space is of the form (2.6), (2.6′) with

$$x_\varphi = 1 - \frac{\varepsilon}{2}, \qquad \xi^{-1} \sim t. \tag{5.3}$$

We have then the classical indices

$$\eta = 0, \quad \gamma = 1, \qquad \nu = \tfrac{1}{2}. \tag{5.3′}$$

We shall now discuss a general approach to critical behaviour which exploits certain invariance properties of the theory under local multiplicative transformations of the field variables. As we indicated in the introduction, in the present approach one works all the time with the original statistical ensemble. The lesson we learnt from the block-spin transformation was that the interesting scaling properties were always connected with the long-range behaviour of appropriate correlation functions in the original ensemble. In the present method we shall obtain, starting from the local transformations of the fields, certain non-linear functional equations for the correlation functions which are especially suited for studying their asymptotic behaviour. These equations will constitute a generalization of those originally studied by Gell-Mann and Low in quantum electrodynamics.

According to the analysis of the previous Sections, in order to have a complete control of the critical point we need two correlation functions expressing the order parameter and energy fluctuations together with a correlation function describing the fluctuations associated with the first irrelevant field. In terms of the latter quantity we may discuss the approach to the critical point and therefore the first corrections to scaling. In our model problem, we shall need an additional quantity related to the fluctuations of

† In the case of a quantum system of interacting bosons the rôle of t is assumed by the chemical potential subtracted at its critical value. Discrete frequencies $i\omega_n = i2\pi nT$ would also be present. On the assumption that the only relevant contribution to the critical behaviour is given by $\omega_n = 0$, it reduces to our model with $n = 2$.

φ^4. For computational purposes it is convenient to use the following three quantities

$$\gamma_2 = \frac{\partial\Gamma^{(2)}}{\partial k^2}, \qquad \gamma_t = \frac{\partial\Gamma^{(2)}}{\partial t}, \qquad \gamma_4 = \frac{\Gamma^{(4)}}{u}. \tag{5.4}$$

We notice the following trivial invariance properties.

First of all the functional Γ is invariant if we let

$$\varphi \to \varphi z^{-\frac{1}{2}}, \qquad \hat{\Gamma}^{(n)} \to \hat{\Gamma}^{(n)} z^{n/2} \tag{5.5}$$

On the other hand the diagrammatic expressions of γ_2, γ_t and γ_4 given in the Appendix B imply the obvious invariance of the theory

(1) wave function

$$\gamma_2 \to z\gamma_2, \qquad \varphi \to \varphi z^{-\frac{1}{2}}, \qquad u \to uz^2 \tag{5.6}$$

(2) mass

$$\gamma_t \to z_t\gamma_t, \qquad t \to tz_t^{-1} \tag{5.7}$$

(3) vertex renormalization

$$\gamma_4 \to z_v\gamma_4, \qquad u \to uz_v^{-1} \tag{5.8}$$

It is clear that (1), (2) and (3) imply (5.5).

These relations all together imply that our system is invariant if we make the substitution:

$$\gamma_2 \to z\gamma_2 = \gamma_2', \qquad \gamma_t \to z_t\gamma_t = \gamma_t', \qquad \gamma_4 \to z_v\gamma_4 = \gamma_4' \tag{5.9}$$

and simultaneously rescale the variables

$$\varphi \to \varphi z^{-\frac{1}{2}} = \varphi', \qquad t \to tzz_t^{-1} = t', \qquad u \to uz^2 z_v^{-1} = u' \tag{5.10}$$

The importance of this invariance is that now we have at our disposal the multiplicative factors z, z_t, z_v to choose the normalization of the "correlation functions" γ''s at a given normalization point (n.p.)

$$\gamma_2'|_{\text{N.P.}} = 1, \qquad \gamma_t'|_{\text{N.P.}} = 1, \qquad \gamma_4'|_{\text{N.P.}} = 1 \tag{5.11}$$

Equations (5.11) determine the z's in terms of the γ''s.

Different choices of the normalization point give various parametrizations of the transformation, as will be clarified later.

The standard Gell–Mann and Low normalization point is†

$$\text{N.P.} \equiv k^2 = \lambda^2, \varphi' = 0 \tag{5.12}$$

independently of the value of t' and u' for γ_2' and γ_4'.

† For γ_4, $k^2 = \lambda^2 \to \text{S.P.}\,(k) = \lambda^2$, where S.P. (k) means that $k_1, \ldots, k_4$ are chosen in such a way that $k_i k_j = \frac{1}{3}k^2(1 - 4\delta_{ij})$. The situation for γ_t' is more involved and will be discussed in Section VIII.

In this case the z's will depend on λ, t' and u'. Since we expect both t' and φ' to be relevant variables, which determine the critical surface $t' = \varphi' = 0$, whereas u' should go to its fixed point value by repeated applications of the group transformation, it is very convenient in critical phenomena to parametrize the transformation in terms of u' only.†

This is achieved by the choice

$$\text{N.P.} \equiv k^2 = \varphi' = 0, \qquad t' = \lambda^2 \tag{5.13}$$

It is easier to work with dimensionless variables

$$t_\lambda = \frac{t'}{\lambda^2}, \qquad \varphi_\lambda = \frac{\varphi'}{\lambda^{1-\varepsilon/2}}, \qquad u_\lambda = \frac{u'}{\lambda^\varepsilon} \tag{5.14}$$

where each variable is divided by λ to a power which is the corresponding bare dimensionality.

For a given λ we set

$$\begin{aligned} z_\lambda \gamma_2 &= \bar{\gamma}_2\left(\frac{k^2}{\lambda^2}, t_\lambda, \varphi_\lambda, u_\lambda\right) \\ z_{t,\lambda}\gamma_t &= \bar{\gamma}_t\left(\frac{k^2}{\lambda^2}, t_\lambda, \varphi_\lambda, u_\lambda\right) \\ z_{v,\lambda}\gamma_4 &= \bar{\gamma}_4\left(\frac{k_i}{\lambda}, t_\lambda, \varphi_\lambda, u_\lambda\right) \end{aligned} \tag{5.15}$$

If we switch from one value of λ to another value λ' and define

$$z_{\lambda'/\lambda} = \frac{z_{\lambda'}}{z_\lambda}, \qquad z_{t,\lambda'/\lambda} = \frac{z_{t,\lambda'}}{z_{t,\lambda}}, \qquad z_{v,\lambda'/\lambda} = \frac{z_{v,\lambda'}}{z_{v,\lambda}}, \tag{5.16}$$

The invariance properties are now expressed as:

$$\begin{aligned} \bar{\gamma}_2\left[\frac{k^2}{\lambda^2}, t_\lambda, \varphi_\lambda, u_\lambda\right] &= z_{\lambda'/\lambda}^{-1}\bar{\gamma}_2\left[\frac{k^2}{\lambda'^2}, t_{\lambda'}, \varphi_{\lambda'}, u_{\lambda'}\right] \\ \bar{\gamma}_t\left[\frac{k^2}{\lambda^2}, t_\lambda, \varphi_\lambda, u_\lambda\right] &= z_{t,\lambda'/\lambda}^{-1}\,\bar{\gamma}_t\left[\frac{k^2}{\lambda'^2}, t_{\lambda'}, \varphi_{\lambda'}, u_{\lambda'}\right] \\ \bar{\gamma}_4\left[\frac{k_i}{\lambda}, t_\lambda, \varphi_\lambda, u_\lambda\right] &= z_{v,\lambda'/\lambda}^{-1}\bar{\gamma}_4\left[\frac{k_i}{\lambda'}, t_{\lambda'}, \varphi_{\lambda'}, u_{\lambda'}\right] \end{aligned} \tag{5.17}$$

† This is the case also for the Callan and Symanzik equation. This method is reviewed in the paper by Zinn-Justin (1973). Its relation with our approach will be discussed in Appendix C.

with

$$\bar{\Gamma}(t_\lambda, \varphi_\lambda, u_\lambda | \lambda) = \bar{\Gamma}(t_{\lambda'}, \varphi_{\lambda'}, u_{\lambda'} | \lambda') \tag{5.18}$$

where $\bar{\Gamma}$ is the thermodynamic potential Γ expressed in terms of the dimensionless variables and

$$t_{\lambda'} = \frac{t_\lambda z_{\lambda'/\lambda} z_{t,\lambda'/\lambda}^{-1}}{(\lambda'/\lambda)^2}, \qquad \varphi_{\lambda'} = \frac{\varphi_\lambda z_{\lambda'/\lambda}^{-1/2}}{(\lambda'/\lambda)^{1-\varepsilon/2}}, \qquad u_{\lambda'} = \frac{u_\lambda z_{\lambda'/\lambda}^2 z_{v,\lambda'/\lambda}^{-1}}{(\lambda'/\lambda)^\varepsilon} \tag{5.19}$$

or in general

$$v_{\lambda'} = T_{\lambda'/\lambda}[v_\lambda] \tag{5.20}$$

where v_λ is any one of the dimensionless variables.

In the following we can omit the subscript λ'/λ in the z's.

So far the relationships between z, z_t, z_v and λ'/λ have not been specified. The normalization condition now reads

$$\bar{\gamma}_2(0, 1, 0, u_{\lambda'}) = 1, \bar{\gamma}_t(0, 1, 0, u_{\lambda'}) = 1, \bar{\gamma}_4(0, 1, 0, u_{\lambda'}) = 1 \tag{5.21}$$

from which it follows

$$z^{-1} = \bar{\gamma}_2\left[0, \left(\frac{\lambda'}{\lambda}\right)^2 z_t z^{-1}, 0, u_\lambda\right]$$

$$z_t^{-1} = \bar{\gamma}_t\left[0, \left(\frac{\lambda'}{\lambda}\right)^2 z_t z^{-1}, 0, u_\lambda\right] \tag{5.22}$$

$$z_v^{-1} = \bar{\gamma}_4\left[0, \left(\frac{\lambda'}{\lambda}\right)^2 z_t z^{-1}, 0, u_\lambda\right]$$

From (5.22) we obtain

$$z = z\left(\frac{\lambda'}{\lambda}, u_\lambda\right), z_t = z_t\left(\frac{\lambda'}{\lambda}, u_\lambda\right), z_v = z_v\left(\frac{\lambda'}{\lambda}, u_\lambda\right) \tag{5.23}$$

The renormalization transformation depends only on the scaling factor λ'/λ and the "renormalized dimensionless coupling constant" u_λ as we had required.

The compatibility of our previous positions is of course based on the experience with perturbation theory and strongly relies on it. We have used only general properties, which we expect to maintain their validity in the exact theory.

By means of (5.21) the transformation has been expressed in terms of the very correlation functions we want to determine. If we susbstitute (5.22) in (5.17) and (5.18) we obtain a set of equations that we hope to solve self-consistently. If after the substitution we take the derivatives with respect to λ'/λ and then set $\lambda' = \lambda$, we obtain a set of homogeneous differential group equations for $\bar{\gamma}_2$, $\bar{\gamma}_t$, $\bar{\gamma}_4$ and $\bar{\Gamma}$:

$$\left[-\frac{k}{\lambda}\frac{\partial}{\partial k/\lambda} - \sigma_t(u_\lambda)t_\lambda\frac{\partial}{\partial t_\lambda} - \sigma_\varphi(u_\lambda)\int \varphi_\lambda\frac{\delta}{\delta\varphi_\lambda} + \psi_u(u_\lambda)\frac{\partial}{\partial u_\lambda} + \left.\frac{\partial z^{-1}}{\partial\lambda'/\lambda}\right|_{\lambda'=\lambda}\right]\bar{\gamma}_2 = 0$$

$$\left[-\frac{k}{\lambda}\frac{\partial}{\partial k/\lambda} - \sigma_t(u_\lambda)t_\lambda\frac{\partial}{\partial t_\lambda} - \sigma_\varphi(u_\lambda)\int \varphi_\lambda\frac{\delta}{\delta\varphi_\lambda} + \psi_u(u_\lambda)\frac{\partial}{\partial u_\lambda} + \left.\frac{\partial z_t^{-1}}{\partial\lambda'/\lambda}\right|_{\lambda'=\lambda}\right]\bar{\gamma}_t = 0 \tag{5.24}$$

$$\left[-\frac{k_i}{\lambda}\frac{\partial}{\partial k_i/\lambda} - \sigma_t(u_\lambda)t_\lambda\frac{\partial}{\partial t_\lambda} - \sigma_\varphi(u_\lambda)\int \varphi_\lambda\frac{\delta}{\delta\varphi_\lambda} + \psi_u(u_\lambda)\frac{\partial}{\partial u_\lambda} + \left.\frac{\partial z_v^{-1}}{\partial\lambda'/\lambda}\right|_{\lambda'=\lambda}\right]\bar{\gamma}_4 = 0$$

and

$$\left[\lambda\frac{\partial}{\partial\lambda} - \sigma_t(u_\lambda)t_\lambda\frac{\partial}{\partial t_\lambda} - \sigma_\varphi(u_\lambda)\int \varphi_\lambda\frac{\delta}{\delta\varphi_\lambda} + \psi_u(u_\lambda)\frac{\partial}{\partial u_\lambda}\right]\bar{\Gamma} = 0 \tag{5.25}$$

If the system is homogeneous $\lambda(\partial/\partial\lambda)$ can be replaced by d in the last equation.

The coefficients σ_t, σ_φ and ψ_φ are obviously determined by the derivatives of (5.19)

$$\frac{\lambda \mathrm{d}v_\lambda}{\mathrm{d}\lambda} = \psi_v[u_\lambda] = \frac{\partial}{\partial\lambda'/\lambda}[T_{\lambda'/\lambda}v_\lambda]_{\lambda'=\lambda}, \qquad v_\lambda = [\varphi_\lambda, t_\lambda, u_\lambda] \tag{5.26}$$

We made explicit the linear dependence of ψ_t and ψ_φ on the two variables t_λ and φ_λ which are zero at the critical point

$$\psi_t = -t_\lambda\sigma_t(u_\lambda), \qquad \psi_\varphi = -\varphi_\lambda\sigma_\varphi(u_\lambda) \tag{5.27}$$

We have finally

$$\sigma_\varphi(u_\lambda) = -\frac{\partial}{\partial\lambda'/\lambda}[z^{-\frac{1}{2}}/(\lambda'/\lambda)^{1-(\varepsilon/2)}]_{\lambda'=\lambda} = 1 - \frac{\varepsilon}{2} - \left.\frac{\partial z^{-\frac{1}{2}}}{\partial\lambda'/\lambda}\right|_{\lambda'=\lambda} \tag{5.28}$$

$$\sigma_t(u_\lambda) = -\frac{\partial}{\partial\lambda'/\lambda}[zz_t^{-1}/(\lambda'/\lambda)^2]_{\lambda'=\lambda} = 2 - \left.\frac{\partial z}{\partial\lambda'/\lambda}\right|_{\lambda'=\lambda} - \left.\frac{\partial z_t^{-1}}{\partial\lambda'/\lambda}\right|_{\lambda'=\lambda} \tag{5.29}$$

$$\psi_u(u_\lambda) = \frac{\partial}{\partial\lambda'/\lambda}\left[\frac{u_\lambda z^2 z_v^{-1}}{(\lambda'/\lambda)^\varepsilon}\right]_{\lambda'=\lambda} = u_\lambda\left[-\varepsilon + 2\frac{\partial z}{\partial\lambda'/\lambda}\bigg|_{\lambda'=\lambda} + \frac{\partial z_v^{-1}}{\partial\lambda'/\lambda}\bigg|_{\lambda'=\lambda}\right] \tag{5.30}$$

Equations (5.24) can be also easily obtained by taking functional derivatives of (5.25) as shown by Di Castro *et al.* (1974). Renormalization groups similar to the one discussed here have been considered also by Gomes and Schroer (1974) and Jegerlehner (1974). In a different context this group has been introduced also by Weinberg (1973).

A different version of the group equations with ψ_u of the form (5.30) containing explicitly the ε dependence which allows for a perturbative expansion, was introduced by Di Castro (1972) with the normalization (5.12). The case $t = 0$ was treated with this version.

B. Scaling

We want to discuss the group equations of the last sub-section to derive the asymptotic scaling picture, i.e. the homogeneity properties, the scaling laws and the distinction between relevant and irrelevant variables.

It is clear from (5.26) that fixed points of our transformation are solutions of the equations

$$\psi_\nu(u^*) = 0, \qquad \nu \equiv [\varphi_\lambda, t_\lambda, u_\lambda] \tag{5.31}$$

since in this case the variables do not change for further application of the transformation.

On the critical surface $\psi_t = \psi_\varphi = 0$ are trivially satisfied and the fixed point is determined by the only condition

$$\psi_u(u^*) = 0 \tag{5.31'}$$

This implies that σ_t and σ_φ assume a value independent of the initial coupling:

$$\sigma_t(u^*) = \sigma_t^*, \qquad \sigma_\varphi(u^*) = \sigma_\varphi^* \tag{5.32}$$

Equation (5.31′) has two solutions. One is the trivial case $u^* = 0$. A second possibility is the vanishing of the expression in brackets of (5.30)

$$\varepsilon = \frac{\partial z_v^{-1}}{\partial\lambda'/\lambda}\bigg|^*_{\lambda'=\lambda} + 2\frac{\partial z}{\partial\lambda'/\lambda}\bigg|^*_{\lambda'=\lambda} \tag{5.33}$$

Equations (5.28) and (5.29) lead to the following relations:

$$\left.\frac{\partial z^{-\frac{1}{2}}}{\partial \lambda'/\lambda}\right|^{*}_{\lambda'=\lambda} = d - 2\sigma_{\varphi}^{*} - 2, \qquad \left.\frac{\partial z_t^{-1}}{\partial \lambda'/\lambda}\right|^{*}_{\lambda'=\lambda} = d - 2\sigma_{\varphi}^{*} - \sigma_t^{*} \tag{5.34}$$

The fixed point condition (5.33) then implies

$$\left.\frac{\partial z_v^{-1}}{\partial \lambda'/\lambda}\right|^{*}_{\lambda'=\lambda} = d - 4\sigma_{\varphi}^{*} \tag{5.35}$$

Simple inspection of the differential equations (5.24) and (5.25) near the fixed point shows that $\bar{\gamma}_2$, $\bar{\gamma}_t$, $\bar{\gamma}_4$ and the singular part of $\bar{\Gamma}$ are homogeneous functions of the variables

$$k, \qquad t_{\lambda}^{1/\sigma_t^*}, \qquad \varphi_{\lambda}^{1/\sigma_{\varphi}^*}$$

with degrees of homogeneity:

$$\begin{aligned} &d - 2\sigma_{\varphi}^{*} - 2 \text{ for } \bar{\gamma}_2, && d - 2\sigma_{\varphi}^{*} - \sigma_t^{*} \text{ for } \bar{\gamma}_t, \\ &d - 4\sigma_{\varphi}^{*} \text{ for } \bar{\gamma}_4, && d \text{ for } \bar{\Gamma} \end{aligned} \tag{5.36}$$

This implies the form (2.6) and (2.6′) for the correlation function and the analogue of (2.8) for the thermodynamic potential with

$$x_{\varphi} = \sigma_{\varphi}^{*} = 1 - \frac{\varepsilon}{2} + \frac{\eta}{2}, \qquad x_t = \sigma_t^{*} = \frac{1}{\nu} \tag{5.37}$$

At the critical point $t = \varphi = 0$, $\bar{\gamma}_2$ and $\bar{\gamma}_t$ scale therefore as

$$\bar{\gamma}_2 \sim \left(\frac{k}{\lambda}\right)^{d - 2\sigma_{\varphi}^{*} - 2}, \qquad \bar{\gamma}_t \sim \left(\frac{k}{\lambda}\right)^{d - 2\sigma_{\varphi}^{*} - \sigma_t^{*}} \tag{5.38}$$

or at $k = 0$, $\varphi = 0$

$$\bar{\gamma}_2 \sim \left(\frac{t}{\lambda^2}\right)^{(d - 2\sigma_{\varphi}^{*} - 2)/\sigma_t^{*}}, \qquad \bar{\gamma}_t \sim \left(\frac{t}{\lambda^2}\right)^{(d - 2\sigma_{\varphi}^{*} - \sigma_t^{*})/\sigma_t^{*}} \tag{5.38$'$}$$

Equation (5.35), which corresponds to the fixed point condition (5.34), gives now the scaling index x_4 for $\bar{\gamma}_4$ i.e.

$$x_4 = \varepsilon - 2\eta \tag{5.39}$$

Equation (5.39) corresponds to one of the Patashinskii and Pokrovskii scaling laws for the $\Gamma^{(n)}$. It amounts to a matching condition between the asymptotic form of $\Gamma^{(4)}$ and $\Gamma^{(2)}$ to give a finite value for the dimensionless renormalized coupling constant at the fixed point.

It is obvious now that, since the theory is asymptotically controlled by the two parameters (5.37) x_φ and x_t, as in Section II all the critical indices can be expressed in terms of them, giving rise to the scaling laws (2.14).

For a few technical points especially related to the specific heat and the equation of state the reader is referred to the original paper (Di Castro *et al.*, 1974).

To give a global set-up to what we have discussed so far, by way of example we solve explicitly the equation (5.25) for $\bar{\Gamma}$

$$\left[-\sigma_t t_\lambda \frac{\partial}{\partial t_\lambda} - \sigma_\varphi \int \varphi_\lambda \frac{\delta}{\delta\varphi_\lambda} + \psi_u \frac{\partial}{\partial u_\lambda} + d\right]\bar{\Gamma} = 0 \tag{5.40}$$

The results for $\bar{\gamma}_2$ and $\bar{\gamma}_t$ are similarly obtained with the simple substitution of d with $d - 2\sigma_\varphi^* - 2$ and $d - 2\sigma_\varphi^* - \sigma_t^*$ respectively and adding the further dependence on k.

The standard characteristic curves method will be used. These are defined as

$$\frac{\mathrm{d}\bar{\Gamma}}{\mathrm{d}l} = -\,\mathrm{d}\bar{\Gamma}, \qquad \frac{\mathrm{d}t(l)}{\mathrm{d}l} = -\,\sigma_t(u(l)), \qquad \frac{\mathrm{d}\varphi(l)}{\mathrm{d}l} = -\,\sigma_\varphi(u(l)),$$
$$\frac{\mathrm{d}u(l)}{\mathrm{d}l} = \psi(u(l)) \tag{5.41}$$

With the boundary condition

$$t(0) = t_\lambda, \qquad \varphi(0) = \varphi_\lambda, \qquad u(0) = u_\lambda \tag{5.42}$$

It is easy to check that the equation is satisfied along these curves.

Solutions of eqs. (5.41) are

$$\bar{\Gamma}(t_\lambda, \varphi_\lambda, u_\lambda) = \mathrm{e}^{dl}\,\bar{\Gamma}(t(l), \varphi(l), u(l)) \tag{5.43}$$

with

$$t(l) = t_\lambda \exp\left[-\int_0^l \sigma_t(u(l'))\,\mathrm{d}l'\right], \quad \varphi(l) = \varphi_\lambda \exp\left[-\int_0^l \sigma_\varphi(u(l'))\,\mathrm{d}l'\right] \tag{5.44}$$

and

$$u(l) = \rho^{-1}\left[\rho(u_\lambda) + l\right], \qquad \rho[u_\lambda] = \int^{u_\lambda} \frac{\mathrm{d}u'}{\psi_u(u')} \tag{5.45}$$

As $|l| \to \infty$, $u(l)$ tends to the fixed point u^*, solution of (5.31′).

If

$$\sigma_t^* > 0 \quad \text{and} \quad \sigma_\varphi^* > 0 \tag{5.46}$$

$t(l)$ and $\varphi(l)$ explode when $l \to -\infty$. In this case the two relevant variables must be zero

$$t_\lambda = \varphi_\lambda = 0 \quad \text{with} \quad u(l) \to u^* \neq 0 \tag{5.47}$$

If such a fixed point exists we can make σ_t^* and σ_φ^* appear explicitly into the solutions by writing

$$t(l) = t_\lambda \exp(-\sigma_t^* l)\frac{\xi_t(u(l))}{\xi_t(u_\lambda)},\ \varphi(l) = \varphi_\lambda \exp(-\sigma_\varphi^* l)\frac{\xi_\varphi(u(l))}{\xi_\varphi(u_\lambda)} \tag{5.48}$$

where

$$\xi(u(l)) = \exp\left\{-\int_{l^*}^{l}[\sigma(u(l')) - \sigma^*]\,\mathrm{d}l'\right\}$$
$$= \exp\left\{-\int_{u^*}^{u(l)}[\sigma(u') - \sigma^*)/\psi_n(u')]\,\mathrm{d}u'\right\} \tag{5.49}$$

with

$$\xi(u^*) = 1, \quad l^* = -\infty$$

For simplicity we take $t(l) = \xi_t(u(l))$ so that

$$l = \frac{1}{\sigma_t^*}\ln\frac{t_\lambda}{\xi_t(u)},\quad u(l) \to \tilde{u}_t = \rho^{-1}\left[\rho(u_\lambda) + \frac{1}{\sigma_t^*}\ln\frac{t_\lambda}{\xi_t(u_\lambda)}\right] \tag{5.50}$$

Equation (5.43) now reads

$$\overline{\Gamma}(t_\lambda, \varphi_\lambda, u_\lambda) = \left[\frac{t_\lambda}{\xi_t(u_\lambda)}\right]^{d/\sigma_t^*}\overline{\Gamma}\left\{\xi_\varphi(\tilde{u}_t)\frac{\varphi_\lambda/\xi_\varphi(u_\lambda)}{[t_\lambda/\xi_t(u_\lambda)]^{\sigma_\varphi^*/\sigma_t^*}}, \xi_t(\tilde{u}_t), \tilde{u}_t\right\} \tag{5.51}$$

which is a substitution law of the type hypothesized by Kadanoff.

When t_λ and φ_λ tend to zero, we reach a fixed point (5.47) as

$$\tilde{u}_t - u^* = u(l) - u^* \underset{l\to-\infty}{\sim} (u_\lambda - u^*)\exp(\psi' l) \underset{t_\lambda\to 0}{\sim} (u_\lambda - u^*)(t_\lambda)^{\psi'/\sigma_t^*} \tag{5.52}$$

only if

$$\psi' = \left(\frac{\partial\psi_u}{\partial u}\right)_{u=u^*} > 0 \tag{5.53}$$

In fact the fixed point is stable against changes of u_λ if $u_\lambda - u^*$ is an irrelevant variable, i.e. if its dimension is negative:

$$x_u = \sigma_u^* = -\psi' < 0 \tag{5.54}$$

In this case $\bar{\Gamma}$ assumes the required homogeneous form

$$\bar{\Gamma}(t_\lambda, \varphi_\lambda, u_\lambda) \underset{t_\lambda,\, \varphi_\lambda \to 0}{\sim} \left[\frac{t_\lambda}{\xi_t(u_\lambda)}\right]^{d/\sigma_t^*} Y\left\{\frac{[\varphi_\lambda/\xi_\varphi(u_\lambda)]^{1/\sigma_\varphi^*}}{[t_\lambda/\xi_t(u_\lambda)]^{1/\sigma_t^*}}, u^*\right\} \tag{5.55}$$

with the scaling properties (2.14)

$$\frac{d}{\sigma_t^*} = 2 - \alpha, \quad \frac{\sigma_\varphi^*}{\sigma_t^*} = \beta$$

The index ψ' controls the approach to the fixed point and determines the corrections to the asymptotic scaling behaviour through the formula (5.52).

VI. ε-Expansion and Approximate Treatment of Scaling Variables and Cross-over Effects

A. *ε-expansion*

We now introduce the ε-expansion to check explicitly the existence of a non-trivial fixed point with the required properties (5.46) and (5.54). In this way our mathematical apparatus will find at least a model verification and the expressions (5.37) of the critical indices in terms of correlation functions can be evaluated in terms of Feynman diagrams. Therefore the ε-expansion will be discussed only for its conceptual relevance, since its actual applications are treated extensively by Brézin *et al.* in Chapter 3 and by Wallace in Chapter 5.

The non-perturbative interpretation of the fixed point condition (5.31′) leads to the scaling law (5.39). We now analyse it in perturbation theory assuming ε small.

$u^* = 0$ is always a solution of (5.31′). The other possibility leads to the condition (5.33). From their definitions in terms of $\bar{\gamma}_2$ and $\bar{\gamma}_4$, z and z_v can be computed as power series in u_λ if u_λ is small. Equation (5.33) gives

$$u^* = u^*(\varepsilon) \tag{6.1}$$

The first diagram contributing to z_v is

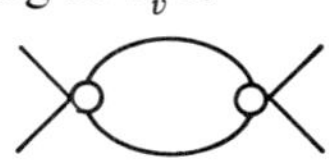

(6.2)

which is a first-order contribution to $\bar{\gamma}_4$.

We thus obtain

$$u^* \sim O(\varepsilon) \tag{6.3}$$

which is small if d is near to 4. It remains small for higher-order terms also, so that an expansion of the indices in powers of ε is possible.

We can calculate the critical indices substituting this value for u^* in the diagrammatic expression for σ_φ^* and σ_t^* given by (5.28) and (5.29) evaluated at the fixed point.

The first diagrams contributing to z^{-1} and z_t^{-1} are a) 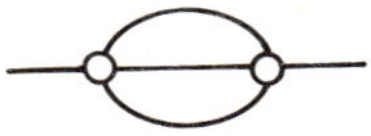

and b)  respectively, which lead to

$$\sigma_\varphi^* = x_\varphi = 1 - \frac{\varepsilon}{2} + \frac{\eta}{2} = 1 - \frac{\varepsilon}{2} + \frac{n+2}{4(n+8)^2}\varepsilon^2,$$

$$\sigma_t^* = x_t = \frac{1}{\nu} = 2 - \frac{n+2}{n+8}\varepsilon \tag{6.4}$$

The n-dependence of the critical indices goes in the right direction. This procedure can be iterated to higher orders. The actual calculations proceed just as in the Callan and Symanzik equation. We therefore refer the reader to Chapter 3 by Brézin *et al.*

The index ψ' of approach to the fixed point is easily evaluated from (5.53) and (5.30) to be

$$\psi' = \begin{cases} -\varepsilon & \text{for } u^* = 0 \\ \varepsilon & \text{for } u^* = O(\varepsilon) \end{cases} \tag{6.5}$$

which means that the trivial fixed point $u^* = 0$ is stable for $d > 4$ and unstable for $d < 4$. The opposite is true for the non-trivial fixed point $u^* \sim 0(\varepsilon)$.

B. *Scaling variables and cross-over effects*

We begin by discussing some properties of the differential equations (5.26) describing the transformation acting on the variables:

$$\lambda' \frac{dv_\lambda^i}{d\lambda} = \psi_{v^i}\{v_\lambda^j\} \quad v_\lambda^i \equiv [\varphi_\lambda, t_\lambda, u_\lambda] \tag{6.6}$$

with the ψ's given by Eqs (5.27), (5.28), (5.29) and (5.30). For simplicity we assume $\varphi = 0$ and omit the subscript λ.

If we linearize these equations around a fixed point u^*, solution of (5.31′), we have

$$\lambda \frac{dt}{d\lambda} = - x_t t \qquad \lambda \frac{d(u-u^*)}{d\lambda} = - x_u(u - u^*) \tag{6.7}$$

with

$$x_t = \sigma_t(u^*), \quad x_u = -\psi'(u^*) \tag{6.8}$$

giving rise to the asymptotic scaling properties.

For each set of critical indices $\{x_i\}$, the scaling variables g_i, are functions of the original variables $\{v_i\}$, and under the action of the group scale exactly as

$$g_i' = \left(\frac{\lambda'}{\lambda}\right)^{-x_i} g_i \tag{6.9}$$

and in terms of them the group equations become

$$\lambda \frac{dg_i}{d\lambda} = -x_i g_i \tag{6.10}$$

in substitution of (6.6).

First of all the scaling variables depend on the transformation used to define them. Two different transformations defined by different ψ_v may lead to different sets of scaling variables with the same critical exponents $\{x_i\}$. In this case we say that the two groups are equivalent. The conditions under which this obtains will be specified in Section VIII, where these arguments will be considered in a more general form.

Various sets of scaling variables may differ in a more substantial way. For instance two sets of critical exponents $\{x_i^\circ\}$ and $\{x_i\}$, the first being connected for example to the trivial fixed point $u^* = 0$ and the second to the non-trivial fixed point $u^* \sim O(\varepsilon)$, involve two sets of scaling variables $\{g_i^\circ\}$ and $\{g_i\}$.

A transformation connecting the two sets when specified in terms of the original physical variables, allows the studying of the cross-over from one critical behaviour to the other.

Such a transformation is non-equivalent in the sense that it does not conserve the value of the indices.

The equations may show simultaneously features related to two different fixed points only if non-linear terms are taken into account. A systematic procedure for discussing the scaling variables when non-linear terms are present was introduced by Wegner (1972) and used by Riedel and Wegner (1973, 1974) to discuss the tricritical point and general cross-over effects. It was recently extended by Fisher and Nelson (1975) to a discrete group.

We shall give a simple example only, related to (6.6).

If we expand these group equations for t and u at second order we obtain

$$\lambda \frac{dt}{d\lambda} = \psi_t(u) \simeq -x_t^\circ t - (x_t - x_t^\circ)\frac{u}{u^*} t, \tag{6.11}$$

$$\lambda \frac{du}{d\lambda} = \psi_u(u) \simeq -x_u^0 u\left(1 - \frac{u}{u^*}\right), \tag{6.12}$$

where t, u and ε are all assumed to be small and of the same order.

The transformation becomes completely parametrized by the critical exponents

$$\{x_i^\circ\} = \{x_u^\circ = \varepsilon, x_t^\circ = 2\} \quad \text{and} \quad \{x_i\} = \left\{x_u = -\varepsilon, x_t = 2 - \frac{\varepsilon}{3}\right\}$$

evaluated at the lowest order as in (6.4) and (6.5) and related to the trivial ($u^* = 0$) and non-trivial $u^* \sim O(\varepsilon)$ fixed point respectively for $n = 1$.

If we now assume that $\{x_i^\circ\}$ and $\{x_i\}$ are two sets of indices obtained from other sources and consider (6.11) and (6.12) as exact, this model coincides with the one hypothesized by Riedel and Wegner to discuss cross-over effects in various cases such as tricritical systems, the Ashkin and Teller model, and Fisher exponent renormalization.

The two sets of scaling variables are now easily determined

$$g_t = t\left(\frac{u}{u^*}\right)^x, \quad g_t^0 = t\left(1 - \frac{u}{u^*}\right)^x$$

$$g_u = \frac{1 - u/u^*}{u/u^*}, \quad g_u^0 = \frac{u'/u^*}{1 - u/u^*} \tag{6.13}$$

with $x = (x_t - x_t^\circ)/x_u^\circ$.

They, asymptotically near to each one of the two fixed points, reduce to t, $u - u^*$ or t and u respectively. They are related by the non-equivalent transformation

$$g_t(\lambda) = g_t^\circ(\lambda)[g_u^\circ(\lambda)]^x, \qquad g_u(\lambda) = [g_u^\circ(\lambda)]^{-1} \tag{6.14}$$

which allows for the discussion of the whole cross-over region.

As previously indicated in the discussion of our model Hamiltonian (4.8), when a more involved symmetry property is taken into account, a new cross-over field is present, which leads to a further renormalization condition of type (5.11). The related critical exponent gives the new cross-over index. The explicit evaluation of critical indices by means of the ε-expansion in all the cases previously indicated is described in various papers, see for example Yamazaki (1975) for the field theoretic approach and the review paper by Fisher (1974) for the Wilson method.

VII. Formal Equivalence of Kadanoff–Wilson and Field Theoretic Renormalization Groups

Both methods previously discussed lead to an asymptotically scaling invariant theory, with equal critical exponents at least in the first few orders in the ε-expansion. Even so the coincidence of the results seems almost accidental rather than related to a profound unity of the renormalization group approach in its various forms.

In both cases the thermodynamic potential is invariant. However, the Kadanoff–Wilson method gives the transformation (3.6) of the distribution for larger block variables in terms of the distribution for smaller block variables. A linear relation is then induced among correlation functions for variables of different sizes as in (3.10′).

The Gell–Mann and Low transformation does not make any reference to the block variables and does not change the form of the initial distribution (i.e. if we start with a $u\hat{\varphi}^4$ model we remain within this model under successive application of the transformation).

It reflects the invariance of the system under a simultaneous change of the correlation functions (5.9) and a rescaling of the variables (5.10). The rescaling factors z, z_t and z_v are expressed by means of the normalization conditions in terms of the same correlation functions as in (5.22). This gives rise to a set of differential equations (5.24) for the correlation functions and (5.26) for the variables, which have to be solved self-consistently.

If we consider the asymptotic solutions (5.38) and (5.38)′ of the equations for the correlation functions $\bar{\gamma}_2$ and $\bar{\gamma}_t$ and the expressions of the z and z_t in terms of them, it is easy to check that their asymptotic self-consistent expressions are:

$$z^{-1} \xrightarrow[\lambda'/\lambda \to 0]{} (\lambda'/\lambda)^{-\eta}, \qquad z_t^{-1} \xrightarrow[\lambda'/\lambda \to 0]{} (\lambda'/\lambda)^{2-\eta-(1/\nu)} \tag{7.1}$$

By means of Eq. (7.1) the Gell-Mann and Low equations for $\bar{\gamma}_2$ and $\bar{\gamma}_t$ at $\varphi = 0$ asymptotically become:

$$\bar{\gamma}_2\left(\frac{k^2}{\lambda^2}, t_\lambda, 0, u_\lambda\right) = (\lambda'/\lambda)^{-\eta}\bar{\gamma}_2\left(\frac{k^2}{\lambda'^2}, \frac{t_\lambda}{(\lambda'/\lambda)^{1/\nu}}, 0, u^*\right) \tag{7.2}$$

and

$$\bar{\gamma}_t\left(\frac{k^2}{\lambda^2}, t_\lambda, 0, u_\lambda\right) = (\lambda'/\lambda)^{2-\eta-(1/\nu)}\bar{\gamma}_t\left(\frac{k^2}{\lambda'^2}, \frac{t_\lambda}{(\lambda'/\lambda)^{1/\nu}}, 0, u^*\right) \tag{7.3}$$

with the next largest correction term of the order of

$$u_{\lambda'} - u^* \sim (u_\lambda - u^*)(\lambda'/\lambda)^{\psi'} \tag{7.4}$$

as given by (5.52).

We will show that the same equations come out from the Kadanoff–Wilson approach as conditions for the existence of the asymptotic theory, where however the scaling factors (7.1) are expressed in terms of Wilson critical indices.

The transformation property (3.10′) of the order parameter correlation function $f_\varphi(r)$ under the Kadanoff–Wilson transformation in Fourier space reads

$$f_\varphi(k\,; H) = s^{2-\eta} f_\varphi(k s; R_s(H)) \qquad k < \Lambda/s \tag{7.5}$$

Since $\Gamma^{(2)}$ is the inverse of f_φ we simply obtain

$$\Gamma^{(2)}(k\,; H) = s^{\eta-2}\Gamma^{(2)}(k\,s; R_s(H)) \qquad k < \Lambda/s \tag{7.6}$$

First recall that for a critical Hamiltonian H_c and for large s

$$\Gamma^{(2)}(k\,; H_c) = s^{\eta-2}\Gamma^{(2)}(k\,s; H^*)$$

which means that $\Gamma^{(2)}(k; H_c)$ becomes independent of H_c when $k \to 0$ and assumes a pure scaling form.

In the neighbourhood of criticality, i.e. in the presence of a small perturbing Hamiltonian τH_1, using the smoothness postulate (3.23), we have

$$\Gamma^{(2)}(k\,; H_c + \tau H_1) = s^{\eta-2}\Gamma^{(2)}(k\,s; R_s(H_c) + \tau L_s(H_c)\cdot H_1) \qquad k < \Lambda/s \tag{7.7}$$

The structure of $L_s(H^*)$ has been studied in subsection III.C. We must now investigate the form of $L_s(H_c)$ for any critical H_c. From (3.25) for large s we obtain

$$L_{s'}(H^*)L_s(H_c) = L_{ss'}(H_c) \tag{7.8}$$

By interchanging s and s', we have for s and s' large enough

$$L_{s'}(H^*)L_s(H_c) = L_s(H^*)L_{s'}(H_c) \tag{7.9}$$

which implies that $L_s(H_c)$ for large s must have the form

$$L_s(H_c)_{ij} = s^{x_i} c_{ij}(H_c) \tag{7.10}$$

where $c_{ij}(H)$ is supposed to be diagonal for $H = H^*$.

Recall the decomposition (3.30) of τH_1. Then if $L_s(H_c)\cdot H_1$ has a component $\bar{\mu}_1$ in the direction e^1 but not along other relevant directions, using (7.10), the relation (7.7) for $\Gamma^{(2)}$ asymptotically becomes:

$$\Gamma^{(2)}(k\,; H^* + s^{x_1}\bar{\mu}_1 e^1 + O(s^{x_2}) + \ldots) = s^{2-\eta}\Gamma^{(2)}\left(\frac{k}{s}\,; H_c + \sum_i \mu_i\, e^i\right) \qquad k < \Lambda \tag{7.11}$$

where $\bar{\mu}_i = \Sigma_j c_{ij}(H_c)\mu_j$, $x_2 < 0$ is the next largest eigenvalue of $L_s(H^*)$.

Equation (7.11) clarifies the meaning of the matrix $c_{ij}(H_c)$: it provides the transformation between the scaling directions at H_c and the scaling directions at H^*.

We now try to eliminate the s-dependence from the dominating term in the left-hand side by introducing the "renormalized" field

$$\mu_{R1} = s^{x_1}\bar{\mu}_1 \tag{7.12}$$

With this substitution (7.11) goes into

$$\Gamma^{(2)}(k\,;H^* + \mu_{R1}e^1 + O(s^{x_2}) + \ldots) =$$

$$s^{2-\eta}\Gamma^{(2)}\left(\frac{k}{s}\,;H_c + \sum_i (c^{-1})_{i1}s^{-x_1}\mu_{R1}e^i + \sum_{j\neq 1,i} (c^{-1})_{ij}\bar{\mu}_j\,e^i\right) \tag{7.13}$$

having used the relationship

$$\mu_i = \sum_j (c^{-1})_{ij}\bar{\mu}_j$$

From Eq. (7.13) we can draw several interesting conclusions. We see that the relevant direction at H_c is given by the expression

$$\sum_i (c^{-1})_{i1}\,e^i = \tilde{e}^1 \tag{7.14}$$

and the tangent plane to the critical surface at H_c is defined by the condition

$$\bar{\mu}_1 = \Sigma_j c_{1j}\,\mu_j = 0$$

Once the transformation is specified, since the matrix elements are functions of H_c, (7.14) can be interpreted as an equation for H_c in the following sense. Suppose we fix the direction $\tilde{e}^1$. Then (7.14) has to be satisfied by all the Hamiltonians which admit $\tilde{e}^1$ as the relevant direction and whose critical behaviour is described by the fixed point H^* of R_s. By letting $\tilde{e}^1$ vary we can determine in principle a class of Hamiltonians which contains all the critical systems attracted by H^*. The matrix elements therefore contain non-trivial information on the structure of the critical surface.

We now analyse the meaning of the change of variable (7.12). If μ_{R1} is constant this implies that $\bar{\mu}_1$ has acquired a non-trivial s-dependence which compensates for the factor s^{x_1}. Since the only length present in the theory is the inverse cut-off Λ^{-1} this means that $\bar{\mu}_1$ has become cut-off dependent. In this sense the theory has been "renormalized" (Wilson 1972b, 1973).

With an obvious notation we can now write down the relationships

$$\lim_{s\to\infty} s^{2-\eta}\Gamma^{(2)}\left(\frac{k}{s}; s^{-x_1}\mu_{R_1}\right) = \text{finite} \tag{7.15}$$

$$\lim_{s\to\infty} s^{2-\eta}\frac{\partial\Gamma^{(2)}}{\partial\mu_{R_1}}\left(\frac{k}{s}, s^{-x_1}\mu_{R_1}\right)\bigg|_{\mu_{R_1}=0} = \text{finite} \tag{7.16}$$

These should be compared with (7.2) and (7.3). They are identical once one identifies the rôle of the cut-off in (7.15) and (7.16) with that of the normalization point of the field theoretic renormalization group equations. But the interesting fact is that the exponents appearing in (7.15) and (7.16) are those of the Kadanoff–Wilson approach. This implies that there should be a solution of the field theoretic equations which leads to the same exponents.

It is clear that a similar argument can be repeated for the next largest eigenvalue and this leads to the identification $x_2 = -\psi'$.

We conclude with a comment on the possibility of interchanging the rôles of the cut-off and the normalization point. We can always start with a model Hamiltonian with a built-in cut-off and then renormalize the theory *à la* Gell-Mann and Low. If the theory is renormalizable we can at this point have $\Lambda \to \infty$ and the only surviving length is the normalization point. This is why we never mentioned the cut-off when discussing approximation methods like the ε-expansion.

A more extended discussion of the content of this section and of related problems will appear in a joint paper with G. Benettin, L. Peliti and A. L. Stella.

VIII. Generalized and Non-linear Transformations

The transformations we have introduced both in the field theoretic and in the Kadanoff–Wilson approach can be called linear if we focus our attention on the transformation properties of the dynamical variables and of their correlation functions. In fact, the block spins are linearly related to the original spins by Eqs (3.3). Our Gell-Mann and Low multiplicative transformation is obviously linear if viewed as acting on the variables φ and t as in (5.19), since z and z_t are parametrized in terms of u only.

The induced linear relations among correlation functions are specified in Eqs (3.10′) for the first case and in (5.17) for the second case. The adjustable normalization parameters of the block transformation are completely determined by the behaviour of the correlation functions at the critical point as in (3.14). In the Gell-Mann and Low case z and z_t are self-consistently determined by the scaling behaviour of the correlation functions as specified in (7.1).

General aspects of non-linear transformations were considered by Jona-Lasinio (1973) and Wegner (1974) and more recently by Di Castro (1975). These groups have been successfully used to discuss lattice systems in one and two dimensions where the results of the ε-expansion cannot be extended. (Niemeijer and Van Leeuwen, 1973, 1974; Fisher and Nelson, 1975; Nauenberg and Nienhuis, 1974). The general properties of a particular non-linear group given as a generalization of the block variables transformation have also been studied (Bell and Wilson, 1975).

We shall now introduce various generalizations of the transformations introduced so far (to include also non-linear groups) and will show that there is a considerable freedom in their choice in the sense that they lead to the same critical behaviour. Developments in this direction are still very preliminary. Still we think it is interesting to confront the reader with these new perspectives.

Let us start with an explicit example related to the block variable transformation.

Consider the relation (3.6) among distributions of block variables of different sizes. The most obvious generalization consists of introducing a non-linear relationship between the block spins at one iteration and block spins at the next iteration. We can set for example

$$\varphi_{\mathbf{l}} = b\Big(\sum_{\mathbf{K}\in A_l^{sL}} \varphi'_{\mathbf{K}}\Big) + c\Big(\sum_{\mathbf{K}\in A_l^{sL}} \varphi'_{\mathbf{K}}\Big)^3 + \dots \tag{8.1}$$

and the corresponding recursion relation for the distribution ρ reads

$$\rho_{sL}\{\varphi\} = \int \pi\, \mathrm{d}\varphi'\, \rho_L\{\varphi'\}\pi\delta\left(\varphi_l - b\Big(\sum_{K\in A^s{}_l^L} \varphi'_K\Big) - c\Big(\sum_{K\in A^s{}_l^L} \varphi'_K\Big)^3 \dots\right) \tag{8.2}$$

It is evident that the correlation functions in the ensemble ρ_{sL} are no longer related to those evaluated in the ensemble ρ_L by a simple multiplicative relationship.

If for example we retain only the cubic term in (8.1), we find that the two-point correlation function f_φ in ρ_{sL} is related to the two-block, four-block and six-block correlation functions in ρ_L. This is a particular case of the model investigated by Bell and Wilson (1975) who emphasize that because of the last property for the correlation functions, a non-trivial fixed point can be reached without having to adjust the parameter b of the transformation.

We shall now generalize the ideas introduced in the field theoretic approach to give a unified version of all the groups previously mentioned. Let us recall that a general property shared by both linear transformations previously discussed is that a group transformation can be viewed as a transformation on the arguments of the thermodynamic potential and leaving it invariant.

In the field theoretic approach this property is reflected by the set of differential equations (5.26)

$$\lambda \frac{\mathrm{d}v^i_\lambda}{\mathrm{d}\lambda} = \psi_{v^i}\{v^j_\lambda\} \tag{8.3}$$

which defined the transformation on the variables and the substitution law (5.18) for the thermodynamic potential $\overline{\Gamma}$ which in the differential form (5.25) reads

$$\frac{\mathrm{d}\overline{\Gamma}}{\mathrm{d}\lambda} = \left[\lambda \frac{\partial}{\partial\lambda} + \sum_i \int \psi_{v^i} \frac{\delta}{\delta v^i}\right]\overline{\Gamma} = 0 \tag{8.4}$$

We have again omitted the subscript λ on the variables. This last equation is the necessary and sufficient condition for the thermodynamic potential to be an integral or invariant of the system of differential equations (8.3), which are the Lie equations of the group transformation.

Our special parametrization in terms of u by means of the choice (5.13) of the normalization point has provided us with the simple linear form (5.27) for ψ_t and ψ_φ, i.e.

$$\psi_t = -t\sigma_t(u), \qquad \psi_\varphi = -\varphi\sigma_\varphi(u) \tag{8.5}$$

We can proceed to generalize our transformation in various directions. We can for instance allow that the $\bar{z}$ and therefore σ_t, ψ_φ and ψ_u depend also on t. The transformation is non-linear in t and correlation functions involving its conjugate variable, i.e. the energy E, are mixed like the order parameter correlation functions in the previous example of a transformation non-linear in φ.

We can then increase the number of variables to include all the $\hat{\mu}_i$ of the Hamiltonian (4.2) instead of confining ourselves to the simple $u\hat{\varphi}^4$ case. By functional Legendre transformations we can switch from the local field variables $\hat{\mu}_i$ to their conjugate local density variables ω_i defined by the functional derivative of the logarithm of the partition function $\mathscr{F}$ as in (4.10)

$$\frac{\delta\mathscr{F}}{\delta\hat{\mu}_i} = -\frac{1}{i!}\omega_i \tag{8.6}$$

We can therefore equally well use the "fields" $\hat{\mu}_i$ or the "densities" ω_i or a mixture of them

$$\{v^i\} = \{\hat{\mu}_l, \omega_j\} \qquad l \neq j \tag{8.7}$$

as for the Γ functional.

We shall still for simplicity call $\bar{\Gamma}$ the thermodynamic potential in terms of the transformed dimensionless variables $\{v'_\lambda\}$. The corresponding group equations are still given by (8.3) and (8.4) where the index i is now labelling the infinite set of parameters v^i. The Wilson transformation is then easily included by letting the $\bar{z}_i$ be non-local.

In general we write

$$\{v^i_{\lambda'}\} = T_{\lambda'/\lambda}\{v^i_\lambda\} \tag{8.8}$$

The corresponding ψ_i may depend on the position.

Non-linearity in φ now means that ψ_φ depends non-linearly on φ.

The fixed point conditions are the generalization of (5.31)

$$\psi_{v^i}\{v^{*j}\} = 0 \quad \text{for all } i \tag{8.9}$$

We can linearize our generalized equations (8.3) and (8.4) around a fixed point, solution of (8.9).

The matrix

$$\sigma^*_{ij} = -\left.\frac{\partial \psi_{v^i}}{\partial v^j}\right|_{\text{fixed point}} \tag{8.10}$$

specifies the asymptotic behaviour of the system.

If the matrix σ^*_{ij} can be diagonalized, its eigenvectors determine the asymptotic scaling variables g_n and its eigenvalues their dimensions x_n.†

Equations (8.3) and (8.4) when linearized and expressed in terms of the scaling variables g_n assume the scaling form

$$\lambda \frac{dg_n}{d\lambda} = -x_n g_n, \qquad \left[\lambda \frac{\partial}{\partial \lambda} + \sum_n x_n \int g_n \frac{\delta}{\delta g_n}\right]\bar{\Gamma} = 0. \tag{8.11}$$

The scaling variables are defined in general by (8.11) considered as exact, to include non-linear terms in the original variables.

The critical behaviour is determined by the matrix σ^*_{ij}; two transformations T and T' acting on the v_i as in (8.8) are therefore equivalent if their cor-

† Notice that the transformation has been already differentiated with respect to the scaling parameter $s = \lambda/\lambda'$ at $s = 1$. The eigenvalues are therefore the critical exponents x_n instead of s^{x_n}, which we would obtain linearizing directly the transformation in the form (8.8).

responding σ_{ij}^* have the same eigenvalues.† This is the analogue of $L_s(H^*)$ discussed in Section II.

We want to show that this is the case if T and T' are related by a differentiable transformation.

Assume that a third transformation h exists such that

$$hT = T'h. \tag{8.12}$$

Equation (8.12) means that the application of h followed by the application of T' gives the same results as T followed by h. Clearly if $\{v^{*i}\}$ is a fixed point of T, $h\{v^{*i}\}$ is a fixed point of T'.

We differentiate both sides of (8.12), and obtain

$$Dh\{T\{v^i\}\} \cdot DT\{v^i\} = DT'\{h\{v^i\}\} \cdot Dh\{v^i\} \tag{8.13}$$

which is a relationship among linear operators. When (8.13) is evaluated at the fixed point of T,

$$DT'\{h\{v^{*i}\} = Dh\{v^{*i}\}^{-1} DT\{v^{*i}\} Dh\{v^{*i}\} \tag{8.14}$$

DT' and DT are related by a similarity transformation and have therefore the same eigenvalues but not necessarily the same eigenvectors. We may conclude that the fixed points of T and T' describe the same critical behaviour with the same critical exponents but not necessarily with the same asymptotic scaling variables.

It is now clear what we meant in Section VI by saying that equivalent transformations may lead to different sets of scaling variables with the same critical exponents.

The transformation connecting two sets of scaling variables related to two different fixed points must be non-differentiable as one can simply verify in the example (6.14).

We now discuss some simple generalizations of our previous field theoretic renormalization group. These will be obtained by letting the functions ψ_φ, ψ_u and ψ_t in (8.4) depend on t and u. The equations (5.24) for $\bar{\gamma}_2$ and $\bar{\gamma}_4$ are still valid, though for $\bar{\gamma}_t$ the situation is more complicated.

For reasons that will become clear in the course of the discussion, it is convenient to work with the Fourier transform of the function $\delta\bar{\Gamma}^{(2)}/\delta t(x)$ which depends on two momenta k, q. This function, which we shall denote

† Actually it is sufficient that their positive and the largest negative (or premarginal as for example $x_n = -\psi'$ in (5.23)) eigenvalues be equal. In this case their asymptotic critical behaviour and the approach to it are the same.

$\bar{\Gamma}_t^{(2)}(k, q)$, is simply related to $\bar{\gamma}_t$ as $\bar{\gamma}_t(k) = \lim_{q\to 0} \bar{\Gamma}_t^{(2)}(k, q)$ and satisfies the equation

$$\left[-k\frac{\partial}{\partial k} - q\frac{\partial}{\partial q} - \sigma_t t\frac{\partial}{\partial t} - \sigma\varphi\int \varphi\frac{\delta}{\delta\varphi} + \psi_u\frac{\partial}{\partial u} + d - 2\sigma_\varphi - \sigma_t\right]\bar{\Gamma}_t^{(2)}$$

$$-\sigma_{tt}(q)\, t\bar{\Gamma}_t^{(2)}(k, q) - \sigma_{\varphi t}(q)\,\varphi\bar{\Gamma}^{(3)}(k, q) + \psi_{ut}(q)\,\bar{\Gamma}_u^{(2)} - 2\sigma_{\varphi t}(q)\bar{\Gamma}^{(2)}(k) = 0 \tag{8.15}$$

where we maintain also in this case the notation $\psi_t = -\sigma_t t$ and

$$\sigma_{tt}(q) = F.T.\left(\frac{\delta\sigma_t}{\delta t(x)}\right), \sigma_{\varphi t}(q) = F.T.\left(\frac{\delta\sigma_\varphi}{\delta t(x)}\right), \psi_{ut}(q) = F.T.\left(\frac{\delta\psi_u}{\delta t(x)}\right),$$

$$\bar{\Gamma}_u^{(2)}(k, u) = F.T.\left(\frac{\delta\bar{\Gamma}^{(2)}}{\delta u}\right).$$

Furthermore at q = 0

$$\sigma_{tt}(0) = \frac{\partial\sigma_t}{\partial t}, \sigma_{\varphi t}(0) = \frac{\partial\sigma_\varphi}{\partial t}, \psi_{ut}(0) = \frac{\partial\psi_u}{\partial t}, \bar{\Gamma}_u^{(2)}(k, 0) = \frac{\partial\bar{\Gamma}^{(2)}(k)}{\partial t}$$

(8.15) has been derived by taking functional derivatives of (8.4). The more complicated structure of (8.15) compared to (5.24) reflects the circumstance that $\bar{\Gamma}_t^{(2)}$ does not transform multiplicatively.

We now indicate two possibilities for the choice of the coefficients $\sigma_t, \sigma_\varphi, \psi_u$. A very natural choice is to impose the condition that the transformation for $\bar{\Gamma}_t^{(2)}$ becomes multiplicative for some fixed t e.g. $t = 0$. This is achieved if $\sigma_{\varphi t}(q)|_{t=0} = \psi_{ut}(q)|_{t=0} = 0$.

We then impose the normalization conditions at the $N.P. \equiv t = \varphi = 0$, $k = q = \lambda$

$$\bar{\gamma}_2|_{N.P.} = 1, \qquad \bar{\Gamma}_t^{(2)}|_{N.P.} = 1, \qquad \bar{\gamma}_4|_{N.P.} = 1$$

In this way σ_φ, σ_t, ψ_u at $t = 0$ are determined as usual. Any choice of $\sigma_t(u, t)$, $\sigma_\varphi(u, t)$, $\psi_u(u, t)$ compatible with the above five conditions then leads, as is easily checked, to a group which is asymptotically multiplicative and equivalent to the one we have described in detail in Section V. For $t \neq 0$ these functions can be determined for example by setting up an iteration scheme in which one gives conditions for the corrections to the asymptotic behaviour. In this way one constructs a "non-linear" group essentially by perturbing with the second line of (8.15) the usual multiplicative transformation.

A different possibility is the following. Suppose we normalize $\bar{\gamma}_2$ and $\bar{\gamma}_4$ with the usual Gell–Mann and Low conditions

$$\bar{\gamma}_2|_{k^2=\lambda^2,\,\varphi=0} = 1, \qquad \bar{\gamma}_4|_{k=\lambda,\,\varphi=0} = 1, \qquad \forall t$$

In this case σ_φ and ψ_u are completely determined, at least for $t(x) = t$ independent of x. We then choose $\sigma_{tt}(q)$ by requiring that the second line in (8.15) vanishes for $\forall t$, $q = k = 0$ and $q = k = \lambda$. The first condition is necessary to eliminate infrared divergences in σ_{tt}, ψ_{ut} and $\sigma_{\varphi t}$, while the second makes possible the determination of $\sigma_t(u, t = 0)$ by means of the normalization $\bar{\Gamma}_t^{(2)}|_{k=q=\lambda,\,\varphi=t=0} = 1$.

$\sigma_t(u, t)$ is known because our previous conditions fix $\partial\sigma_t/\partial t = \sigma_{tt}(q = 0)$. In this example the non-locality of the transformation has been used explicitly. The case we have just considered has some interest because it shows that the extension of the Gell–Mann and Low group to include the temperature behaviour is a non-trivial problem.

Some comments are in order. It is clear that non-linearity for the transformation of a given variable means that the correlation functions related to its conjugate variable do not transform multiplicatively. If we insist on multiplicativity and we work with t, φ and u alone, the group we described in Section V is practically unique. In the examples given in this section we required multiplicativity in some limit situation. Of course this is not the only possibility. For example, by imposing cancellations between the first and second lines in Equation (8.15), one can introduce changes in the structure of the fixed point, a possibility which may be of considerable interest when additional variables besides t and u are involved. The presence of adjustable functions of several variables gives the freedom to impose requirements like the minimization of some interactions in the fixed point Hamiltonian. However, it is clear that a general theory for the non-linear case is very difficult and each particular problem should suggest, on the basis of a specific analysis, the most appropriate transformation.

Acknowledgements

The authors would like to express their gratitude to many colleagues for stimulating discussions and collaboration over several years. In particular to G. Benettin, D. Capocaccia, M. Cassandro, G. Ciccotti, S. D'Agata, F. De Pasquale, G. Gallavotti, G. Parisi, L. Peliti, A. L. Stella, P. Tombesi, W. Young, E. Tabet.

A preliminary version of some parts of this paper was prepared for graduate courses given by the authors on various occasions. In particular one of us (G. J-L) would like to thank the Laboratoire de Physique Théorique of the Ecole Polytechnique Féderale—Lausanne and the Department of Theoretical

Physics of the University of Geneva for their warm hospitality, while lecturing at the 3-ème Cycle de la Suisse Romande.

Finally we would like to thank A. Di Silvestro for her excellent work in editing the manuscript.

References

Bell T. L. and Wilson K. G. (1975). *Phys. Rev.* **B10**, 3935.

Bogoliubov N. N. and Shirkov D. V. (1959). "Introduction to the Theory of Quantized Fields", pp. 511–553. Interscience Publishers, Inc., New York.

Brezin E., Le Guillou J. C. and Zinn-Justin J. (1973a). *Phys. Rev.* **D8**, 434.

Brezin E., Le Guillou J. C. and Zinn-Justin J. (1973b). *Phys. Rev.* **D8**, 2418.

Brezin E., Le Guillou J. C., Zinn-Justin J. and Nickel B. G. (1974). *Phys. Letters* **44A**, 227.

Callan C. G. (1970). *Phys. Rev.* **D2**, 1541.

Cassandro M. and Gallavotti G. (1975). *Nuovo Cimento* **25B**, 691.

De Pasquale F., Di Castro C. and Jona-Lasinio G. (1972). *In* "Proceedings of the International School of Physics" E. Fermi LI corso, pp. 123, 156. Academic Press, New York.

Di Castro C. and Jona-Lasinio G. (1969). *Phys. Letters* **29A**, 322.

Di Castro C. (1971). *Riv. Nuovo Cimento* **1**, 199.

Di Castro C. (1972). *Lett. Nuovo Cimento* **5**, 69.

Di Castro C., Jona-Lasinio G. and Peliti L. (1974). *Annals of Phys.* **87**, 327.

Di Castro C. (1975). *In* "International Symposium on Mathematical Problems in Theoretical Physics" (H. Araki, Ed.), pp. 342–348. Springer Verlag, Berlin.

Fisher M. E. (1974). *Rev. Mod. Phys.* **46**, 597.

Fisher M. E. and Nelson D. (1975). Preprint.

Gallavotti G. and Martin-Löf A. (1975). *Nuovo Cimento* **25B**, 425.

Gell-Mann M. and Low F. E. (1954). *Phys. Rev.* **95**, 1300.

Gomes M. and Schroer B. (1974). Preprint.

Hori S. (1952). *Progr. Theor. Phys.* **7**, 578.

Jegerlehner F. (1974). Lectures given at the International Institute for Theoretical Physics, Bonn.

Jona-Lasinio G. (1964). *Nuovo Cimento* **34**, 1790.

Jona-Lasinio G. (1971). *In* "Critical Phenomena in Alloys, Magnets and Superconductors" (R. E. Mills, E. Ascher and R. I. Jaffe, Eds), p. 189. McGraw Hill, New York.

Jona-Lasinio G. (1973). *In* "Proceedings of the Nobel Symposium" vol. 24, p. 38. Academic Press, New York.

Jona-Lasinio G. (1975). *Nuovo Cimento* **26**, 99.

Kadanoff L. P. (1966). *Physics* **3**, 255.

Kadanoff L. P., Götze W., Hamblen D., Hecht R., Lewis E. A. S., Palciauskas V. V., Rayl M., Swift J., Aspnes D. and Kane J. (1967). *Rev. Mod. Phys.* **39**, 395.

Ma S. K. (1973). *Rev. Mod. Phys.* **45**, 589.

Ma S. K. (1974a). *Journal Math. Phys.* **15**, 1866.

Ma S. K. (1974b). *Phys. Rev.* **A10**, 1818.

Migdal A. A. (1969). *Sov. Phys. JETP*, **28**, 1036.

Mühlschlegel B. and Zittartz H. (1963). *Zeitschrift für Physik*, **175**, 553.

Nauenberg M. and Nienhuis B. (1974). *Phys. Rev. Letters*, **33**, 944.

Nelson D. (1975). Preprint.
Niemeijer Th. and van Leeuwen J. M. J. (1973). *Phys. Rev. Letters*, **31**, 1411.
Niemeijer Th. and van Leeuwen J. M. J. (1974). *Physica*, **71**, 17.
Parisi G. (1973). Lectures given at the Cargèse Summer School.
Patashinskii A. Z. and Pokrowskii V. L. (1964). *Sov. Phys. JETP*, **19**, 677.
Pokrovskii V. L. (1968). *Sov. Phys. Uspeki*, **11**, 66.
Polyakov A. M. (1969). *Sov. Phys. JETP*, **28**, 533.
Riedel E. and Wegner F. (1969). *Zeitschrift für Physik*, **225**, 195.
Riedel E. and Wegner F. (1973). *Phys. Rev.* **B7**, 248.
Riedel E. and Wegner F. (1974). *Phys. Rev.* **B9**, 294.
Stueckelberg E. C. G. and Petermann (1953). *Helv. Phys. Acta*, **26**, 499.
Symanzik K. (1966). *Journ. Math. Phys.* **7**, 510.
Symanzik K. (1970). *Comm. Math. Phys.* **18**, 227.
Wegner F. (1972). *Phys. Rev.* **D5**, 4529.
Wegner F. (1974). *J. Phys. C.* **7**, 2098.
Weinberg S. (1973). *Phys. Rev.* **D10**, 3497.
Wilson K. G. (1971). *Phys. Rev.* **B4**, 3174 and 3184.
Wilson K. G. and Fisher M. E. (1972). *Phys. Rev. Letts*, **28**, 240.
Wilson K. G. (1972a). *Phys. Rev. Letts*, **28**, 548.
Wilson K. G. (1972b). *Phys. Rev.* **D6**, 419.
Wilson K. G. (1973). *Phys. Rev.* **D7**, 2911.
Wilson K. G. and Kogut J. (1974). *Phys. Rep.* **12C**, 76.
Yamazaki Y. (1975). *In* "International Symposium on Mathematical Problems in Theoretical Physics" (H. Araki, Ed.), pp. 349–352. Springer Verlag, Berlin.
Zinn-Justin J. (1973). Lectures given at the Cargèse Summer School.

Appendix A

A connection with probability theory

There is a very interesting connection between the renormalization group idea in its various formulations and a classical problem of probability theory. This is the problem of limit distributions for sums of random variables. Actually, it turns out that the renormalization group can be viewed to a large extent as a generalization of that problem. The connection is straightforward from the point of view of the Kadanoff–Wilson approach.

Let us apply the recursion formula (3.6) to the particular case in which the φ are non-interacting identical variables. In this case the distribution ρ_L will be of the form

$$\rho_L\{\varphi\} = \Pi u_L(\varphi) \tag{A.1}$$

where u_L is the distribution for a single independent block variable.

Also ρ_{2L} will have a similar structure and (3.6) reduces for $s = 2$, $d = 1$ to the equation

$$u_{2L}(\varphi) = \int \mathrm{d}\varphi' \, \mathrm{d}\varphi'' u_L(\varphi') u_L(\varphi'') \delta(\varphi - a(\varphi' + \varphi'')) \tag{A.2}$$

with $a = 2^{-(1-x_\varphi)}$ which can be further simplified to

$$u_{2L}(\varphi) = \frac{1}{a^2}\int d\varphi' u_L\left(\frac{1}{a}\varphi'\right) u_L\left(\frac{1}{a}(\varphi' - \varphi)\right) \tag{A.3}$$

For $L \to \infty$ we obtain

$$u(\varphi) = \frac{1}{a^2}\int d\varphi' u\left(\frac{1}{a}\varphi'\right) u\left(\frac{1}{a}(\varphi' - \varphi)\right) \tag{A.4}$$

This equation (or more precisely a slight generalization of it) defines in probability theory the so-called stable distributions. The stable distributions are the only possible limit distributions for normalized sums of identically distributed random variables.

The solutions of (A.4) can be calculated explicitly by Fourier transformation. If we limit ourselves to the case in which $u(\varphi)$ is symmetrical around the origin, its Fourier transform is

$$f(t) = \exp(-c|t|^\alpha) \tag{A.5}$$

where c is an arbitrary positive constant and

$$1 < \alpha \leqslant 2 \tag{A.6}$$

t and α are used in this appendix to conform with the standard notations of probability theory, and should not be confused with those in the text. Actually (A.4) has more general solutions of the form

$$f(t) = \exp[-c|t|^\alpha \theta(\ln t)] \tag{A.7}$$

where θ is a periodic function of period $T = |\ln a|$. Distributions like (A.7) are usually considered a generalization of stable distributions.

The parameter α is immediately connected with the normalization exponent χ_φ. In fact:

$$\alpha = 2, \qquad x_\varphi = \tfrac{1}{2} \qquad \text{Gaussian}$$

$$\alpha \neq 2, \qquad x_\varphi = 1 - \frac{1}{\alpha}$$

In the present case, the transformation K_2 (see (3.6′)) has Gaussian and non-Gaussian fixed points.

Given a particular stable distribution, it is interesting to determine all the distributions which tend to it upon iteration of (A.3). In the language of probability, this is called the Domain of Attraction of the given stable distribution.

This problem can also be solved explicitly.

Given an initial u_0 one can immediately tell whether it will tend to a

Gaussian or to a distribution with exponent α. For example if:

$$\lim_{X\to\infty} \frac{X^2 \int_{|x|>X} u_0(x)\,dx}{\int_{|x|<X} x^2 u_0(x)\,dx} \to 0$$

u_0 will tend to a Gaussian. If on the other hand it has the asymptotic scaling property:

$$\lim_{x\to\infty} \frac{u_0(x)}{u_0(kx)} \sim k^{\alpha+1}$$

it will tend to the stable distribution with exponent α.

It should be sufficiently clear by now that the study of the renormalization group eqn (3.6) is the generalization to the case of dependent variables in a multidimensional space of the problem we have just discussed. In particular the notion of universality class corresponds to the domain of attraction of probability theory. For further details and references see Jona-Lasinio (1975).

Appendix B

Summary of diagrammatic expansions

The two-particle correlation function satisfies the Dyson equation

$$f_\varphi = f_\varphi^\circ + f_\varphi^\circ \Sigma f_\varphi \tag{B.1}$$

where f_φ° is the expression of f_φ for the non-interacting system ($u = 0$) and Σ is the self-energy part. The self-energy contains only diagrams "One-particle-irreducible", which cannot be split into two separate parts by cutting just one line. Diagrammatically

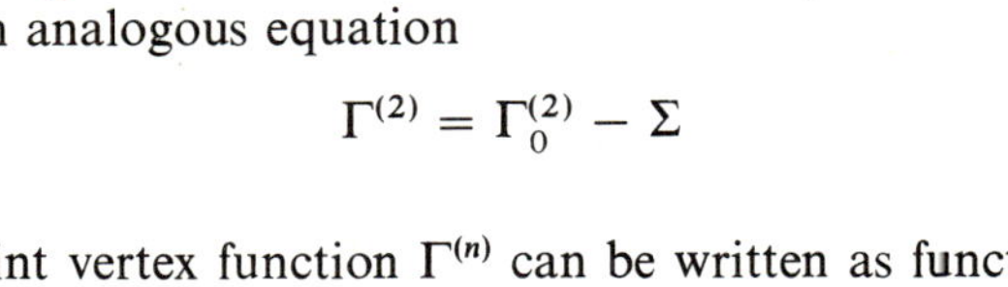

(B.2)

where the heavy lines are full f_φ, the light lines are the bare f_φ°, the circles are the coupling u and the dark circle is the vertex part $\Gamma^{(4)}$. Since $\Gamma^{(2)} = f_\varphi^{-1}$ it satisfies an analogous equation

$$\Gamma^{(2)} = \Gamma_0^{(2)} - \Sigma \tag{B.3}$$

The n-th point vertex function $\Gamma^{(n)}$ can be written as functional derivatives of $\Gamma^{(2)}$ and therefore contain only one-particle-irreducible diagrams

$$\hat{\Gamma}^{(n)}(x_1, \ldots, x_n) = \frac{\delta^{n-2}\hat{\Gamma}^{(2)}(x_1, x_2)}{\delta\varphi(x_3)\ldots\delta\varphi(x_n)} \tag{B.4}$$

For example the diagrammatic expression of $\Gamma^{(4)}$ is

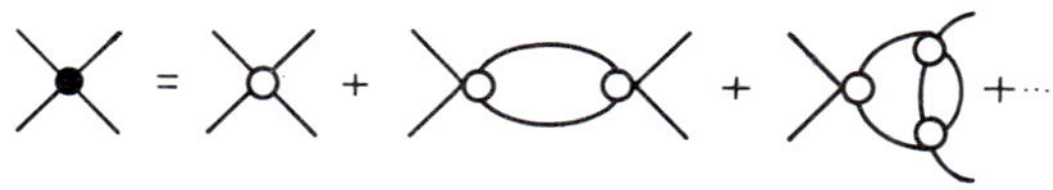

(B.5)

with no external lines.

The diagrammatic expression for the mixed vertex function $\gamma_t = \partial\Gamma^{(2)}/\partial t$ can be obtained from (B.3) for $\Gamma^{(2)}$. Since the derivative with respect to t splits each line into two and introduces a vertex γ_t we have

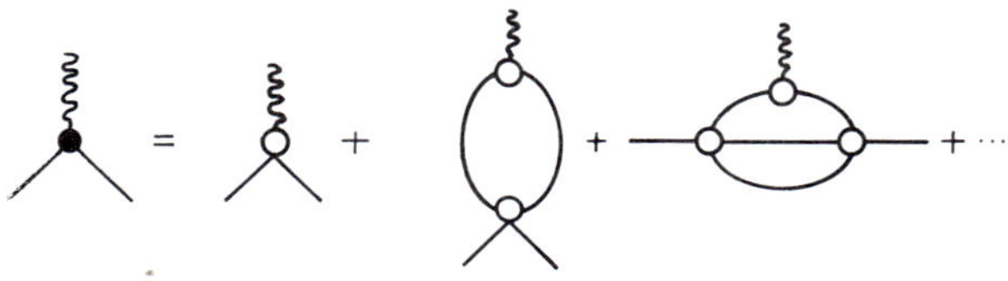

(B.6)

Appendix C

Relation to the Callan–Symanzik equation

The connection between our group equation (5.25) for the Γ-functional and the corresponding Callan–Symanzik equation is easier established if we use the "true mass" $m \sim \xi^{-1}$ as a variable instead of t.

It can be defined as

$$\xi^{-2} \sim m^2 = \frac{\Gamma^{(2)}\big|_{k^2=0,\ \varphi=0}}{\partial\Gamma^{(2)}/\partial k^2\big|_{k^2=0,\ \varphi=0}} \tag{C.1}$$

which being the ratio of two $\Gamma^{(2)}$ is obviously invariant under the transformation (5.6), (5.7) and (5.8).

It is simpler now to switch from a normalization parameter λ to $\bar{\lambda}$ defined in such a way that our normalization is for a value $\bar{t}$ of t_λ for which

$$t_{\bar{\lambda}} = \bar{t} \rightarrow m(\bar{t}, u_{\bar{\lambda}}) = \bar{\lambda} \tag{C.2}$$

instead of $t_\lambda = 1$.

All our group equations remain the same with the minor substitutions:

$$\begin{aligned} z^{-1}\left(\frac{\bar{\lambda}'}{\bar{\lambda}}, u_{\bar{\lambda}}\right) &= \bar{\gamma}_2\left(0, \left(\frac{\bar{\lambda}'}{\bar{\lambda}}\right)^2, 0, u_{\bar{\lambda}}\right), \\ z_t^{-1}\left(\frac{\bar{\lambda}'}{\bar{\lambda}}, u_{\bar{\lambda}}\right) &= \bar{\gamma}_t\left(0, \left(\frac{\bar{\lambda}'}{\bar{\lambda}}\right)^2, 0, u_{\bar{\lambda}}\right), \\ z_v^{-1}\left(\frac{\bar{\lambda}'}{\bar{\lambda}}, u_{\bar{\lambda}}\right) &= \bar{\gamma}_4\left(0, \left(\frac{\bar{\lambda}'}{\bar{\lambda}}\right)^2, 0, u_{\bar{\lambda}}\right). \end{aligned} \tag{C.3}$$

where the $\bar{\gamma}$'s functions are given in terms of the variable m instead of t.† We now proceed to establish the connection with the Callan–Symanzik equation.

The Γ-functional Γ_c leading to the Callan–Symanzik equation is specified by the following normalization conditions

$$\Gamma^{(2)}\big|_{k^2=0,\,\varphi=0} = m^2, \qquad \left.\frac{\partial\Gamma^{(2)}}{\partial k^2}\right|_{k^2=0,\,\varphi=0} = 1,$$

$$\Gamma^{(4)}\big|_{k_i=0,\,\varphi=0} = u, \qquad \left.\frac{\partial\Gamma^{(2)}}{\partial t}\right|_{k^2=0,\,\varphi=0} = 1 \tag{C.4}$$

at all temperatures and therefore for all values of m^2.

Our theory of section V, however, is normalized for a given value of the temperature.

It is customary in the C.S. equation to work with dimensional variables. The invariance property (5.18) reads

$$\Gamma(\varphi, m', u|\bar{\lambda}) = \Gamma(\varphi z^{-\frac{1}{2}}, m', u z^2 z_v^{-1}|\bar{\lambda}'). \tag{C.5}$$

Due to the normalization condition (C.4) the relation between Γ and Γ_c is obtained by specializing $\bar{\lambda}' = m'$ in Γ

$$\Gamma_c(\varphi', m', u') = \Gamma(\varphi', m', u'|m'). \tag{C.6}$$

By means of the group transformation (C.5) $\Gamma(\varphi', m', u'|m')$ can be connected with $\Gamma(\varphi, m', u|m)$ to give the basic equations for the identification

$$\Gamma_c(\varphi', m', u') = \Gamma(\varphi, m', u|m) \tag{C.7}$$

with

$$\varphi' = \varphi z^{-\frac{1}{2}}\left(\frac{m'}{m}, u\right), \qquad u' = u z^2\left(\frac{m'}{m}, u\right) z_v^{-1}\left(\frac{m'}{m}, u\right) \tag{C.8}$$

Let us take the derivative of (C.7) with respect to m' and then set $m' = m$. We have

$$\left[m\frac{\partial}{\partial m} - \sigma_\varphi^0 \int \varphi \frac{\delta}{\delta\varphi} + \psi_0 \frac{\partial}{\partial u}\right]\Gamma_c(\varphi, m, u)$$

$$= m' \frac{\partial}{\partial m'}\Gamma(\varphi, m', u|m)\big|_{m'=m} \tag{C.9}$$

† We have not introduced a different notation for the γ's to avoid further complications.

where

$$m' \frac{\partial \varphi'}{\partial m'}\bigg|_{m'=m} = \varphi m' \frac{\partial z^{-\frac{1}{2}}}{\partial m'}\bigg|_{m'=m} = -\sigma_\varphi^0 \varphi \tag{C.10}$$

$$m' \frac{\partial u'}{\partial m'}\bigg|_{m'=m} = m' \frac{\partial}{\partial m'}(u z^2 z_v^{-1})\big|_{m'=m} = \psi_0 \tag{C.11}$$

Equation (C.9) is already the C.S. equation with the inhomogeneous term in an unusual form.

To obtain it in its standard form we have to make the right-hand side of (C.9) explicit:

$$m' \frac{\partial \Gamma}{\partial m'}\bigg|_{m'=m} = 2m^2 \left(\frac{\partial m'^2}{\partial t}\right)^{-1}\bigg|_{m'=m} \frac{\partial \Gamma(\varphi, m', u|m)}{\partial t}\bigg|_{m'=m} \tag{C.12}$$

which because of (C.6) leads to

$$m' \frac{\partial \Gamma}{\partial m'}\bigg|_{m'=m} = 2m^2 \frac{\partial \Gamma_c}{\partial t}\left(\frac{\partial m'^2}{\partial t}\right)^{-1}\bigg|_{m'=m} \tag{C.13}$$

where $m' = m$ means that the derivative of m'^2 with respect to t is to be evaluated at the normalization point of $\Gamma(\varphi, m', u|m)$. Define

$$r(m', u|m) = \Gamma^{(2)}(k^2 = 0, \varphi = 0, m', u|m),\ r(m, u|m) = m^2 \tag{C.14}$$

then its derivative with respect to t at the normalization point is equal to one, i.e. $\partial r/\partial t|_{m'=m} = 1$. We therefore have

$$\frac{\partial t}{\partial m'^2}\bigg|_{m'=m} = \left(\frac{\partial r}{\partial t}\right)^{-1}\bigg|_{m'=m} \left(\frac{\partial r}{\partial m'^2}\right)_{m'=m} = \left(\frac{\partial r}{\partial m'^2}\right)_{m'=m} \tag{C.15}$$

From Eq. (C.1) and (C.3) it follows that at $\varphi = 0$

$$r = z^{-1}\left(\frac{m'}{m}, u\right) m'^2 \tag{C.16}$$

We then have

$$\frac{\partial r}{\partial m'^2}\bigg|_{m'=m} = 1 + m'^2 \frac{\partial z^{-1}}{\partial m'^2}\bigg|_{m'=m} = 1 - \sigma_\varphi^0 \tag{C.17}$$

i.e.

$$\left(\frac{\partial t}{\partial m'^2}\right)_{m'=m} = 1 - \sigma_\varphi^0 \tag{C.18}$$

Equation (C.9) together with (C.13) and (C.18) lead to the Callan–Symanzik equation

$$\left[m\frac{\partial}{\partial m} + \psi_0\frac{\partial}{\partial u} - \sigma_\varphi^0 \int \varphi \frac{\delta}{\delta\varphi}\right]\Gamma_c(\varphi, m, u) = 2(1 - \sigma_\varphi^0)m^2\frac{\partial \Gamma_c}{\partial t} \qquad \text{(C.19)}$$

Group equations without inhomogeneous terms are more suitable for studying asymptotic theory, since the inhomogeneous terms appear only when iterative processes to calculate corrections to scaling are set up.

Author Index

Z

Subject Index

D

G

H

I

P

Q

R